Proteases in Physiology and Pathology

Sajal Chakraborti • Naranjan S. Dhalla
Editors

Proteases in Physiology and Pathology

Springer

Editors
Sajal Chakraborti
Department of Biochemistry and Biophysics
University of Kalyani
Kalyani, West Bengal, India

Naranjan S. Dhalla
St. Boniface Hospital Research Centre,
University of Manitoba, Faculty of
Health Sciences, College of Medicine
Institute of Cardiovascular Sciences
Manitoba, Winnipeg, Canada

ISBN 978-981-10-2512-9 ISBN 978-981-10-2513-6 (eBook)
DOI 10.1007/978-981-10-2513-6

Library of Congress Control Number: 2017951624

Printed on acid-free paper

This Springer imprint is published by Springer Nature
The registered company is Springer Nature Singapore Pte Ltd.
The registered company address is: 152 Beach Road, #21-01/04 Gateway East, Singapore 189721, Singapore

This book on ***Proteases in Physiology and Pathology*** *is dedicated to Professor Kambadur Muralidhar, a distinguished biologist and an outstanding teacher/ educationist of our country. Professor Muralidhar was born in 1948 at Coimbatore, Tamil Nadu, India. He received his PhD from the Indian Institute of Science, Bangalore, in 1976. In the PhD work under Prof. N.R. Moudgal, he demonstrated that the beta subunit of luteinizing hormone can bind to ovarian receptors, which fetched him the Prof KV Giri Memorial Gold medal for the best PhD thesis of the year. He was selected for the first Lectureship in the School of Life*

Sciences at the newly founded Central University of Hyderabad in August 1976. During 1979–1981, he has been a Research Associate in the Department of Cell and Molecular Biology at SUNY-Buffalo, New York, USA, and worked with Prof. O.P. Bahl. His work led to the development of the most sensitive RIA for HCG, a pregnancy hormone. He also demonstrated the immuno-contraceptive vaccine potential of DS5-hCG beta subunit. He joined the University of Delhi in 1983 as Reader in Biochemistry, Department of Zoology, and became a Professor in Endocrinology and Biochemistry in 1988. He was Chairman of the Department during 2001–2004.

His laboratory discovered the presence of tyrosine-O-sulphate in sheep and buffalo PRL. Using biophysical techniques, his group has studied the protein unfolding and folding of buffalo GH. Naturally occurring size isoforms of buffalo prolactin, cathepsin derived peptides from buffalo prolactin, and a synthetic peptide based on the internal sequence of buffalo prolactin were discovered to be powerful inhibitors of angiogenesis stimulated by VEGF or bradykinin. His group has published over 140 articles in peer-reviewed journals and several books and chapters.

Professor Muralidhar was Chairman of the University Grants Commission's (Govt. of India) Curriculum Development Committee for Zoology (2001). He assisted University of Delhi in restructuring Undergraduate Science Education and introduced a new Honors course in Integrated Biology, which is the first of its kind in the country. He was the Chairman of the PAC-Animal Science of the Department of Science and Technology (DST), Govt. of India, for three terms (2001–2004, 2004–2007, and 2010–2013) and nurtured research in this area. He assisted different funding agencies of the Govt. of India in different capacities. He also served as a member of Research Advisory Committees of several nationally important research institutions in India.

Professor Muralidhar is an elected Fellow of all the Science Academies of India, i.e., NASI, Allahabad; INSA, New Delhi; and IASc, Bangalore. He has delivered the Sadaksharaswami Endowment Lecture for SBC (I) (1996), US Srivastava Memorial Lecture for NASI (2007), V. Gopalakrishna Rao Endowment Lecture for Osmania University (2004), Y Subba Row Memorial Lecture for GGS Indraprastha University (2007), Prof MRN Prasad Memorial Lecture for INSA (2010), Hargobind Khorana Memorial Lecture for GGS IP University (2012), Distinguished Lecture for Kalyani University (2016), and GP Sharma Memorial Lecture for Punjab University (2011). He served as member of the Executive Council of NASI (2003–2005) and INSA (1999–2001). He was the Chief Editor of the Proceedings of Indian National Science Academy for over 6 years (2000–2005) when he transformed the journal and brought international recognition. He was Chairman of the INSA National Committee for Cooperation with IUBS for a term. He has recently become a JC Bose National Fellow (DST). Following superannuation in

December 2013, Kambadur Muralidhar has joined the South Asian University as an Honorary Professor in January, 2014.

Professor Kambadur Muralidhar undoubtedly is a legendary figure in Indian science. He has excellent ability to motivate young researchers. We feel honored to dedicate this book to Professor Kambadur Muralidhar and wish him good health and success in his long, fruitful activities.

Preface

I thought that my voyage was at its end at the last
limit of my power that the path before me was closed,
provisions were all exhausted and the time had come for me
to take shelter in a silent obscurity.
But, I find that thy will knows no end in me.
And, when old words die out on the tongue
new melodies break forth from the heart and where
the old tracks are all lost new country is unveiled
with its wonders.

(Rabindranath Tagore (Gitanjali: Song of offerings)

The history of research on proteases is relatively old, which has been initiated in the late eighteenth century, although in recent times it has gained a tremendous momentum because of their widespread applications, especially in biotechnology and medicine. There are many ways in which proteases elicit both the beneficial and detrimental effects on the functioning of living beings, and this has prompted researchers to study their roles in health and disease.

Recent research revealed that about 2% of all gene products are proteases, indicating that it is one of the important functional groups of proteins. Notably, it seems difficult to know how a protease can be distinguished from another related one. On one side, scientists are engaged in understanding the basic mechanisms of the potentiality of different types of proteases in a variety of disease progression and evaluation of relevant therapeutics; on the other side, researchers are trying to answer two fundamental questions: How does knowledge of one protease help in the understanding of related proteases? How can a novel protease's role be truly ascertained?

Research on renin inhibitors as potential anti-hypertensive drugs started in the early 1970s. Some early peptide-like inhibitors showed significant inhibitory activity towards renin but lacked adequate bioavailability. After decades of research, the first bioavailable renin inhibitor aliskiren was approved and marketed in 2007; however it was discontinued in 2011 due to its side effects. Notably, the success of inhibitors of angiotensin-converting enzyme (ACE) and β-blockers for the treatment of hypertension have supported the concept that protease inhibitors can prove useful as successful drugs. Peptidomimetic inhibitors, which bind at the active site of matrix metalloproteases (MMPs), have been tested, and most MMP inhibitors in clinical development are hydroxamate derivatives such as batimastat and marimastat, although nonpeptide MMP inhibitors such as AG3340 and Bay-12-9566 have also shown their efficacy in preventing different types of cancer. Despite the success achieved in understanding fundamental scientific information and in designing some highly valuable drugs by exploring active site targeted inhibitors, the limited number of protease inhibitors introduced during the past decade as well as several failures indicates a need for basic research on disease-causing proteases for more details. In this book, we believe that different authors in their respective chapters provided some novel information, which will eventually help to unravel many barriers that pharmacologists and drug designers are experiencing currently.

This book is intended to provide comprehensive treaties of physiological and pathological implications of some proteases. We would like to express our appreciation to all the contributors for their enthusiasm and perseverance in bringing this book to fruition. We wish to thank Dr. Madhurima Kahali (Biomedicine, Springer, New Delhi), Sowndarya Kumaravel and F. Avilapriya for all the very important initiating effort towards achieving this goal. Finally, we like to thank Prof. Sankar Kumar Ghosh (Vice Chancellor, University of Kalyani) for his encouragement.

Kalyani, West Bengal, India Sajal Chakraborti
Manitoba, Winnipeg, Canada Naranjan S. Dhalla

Contents

About the Editors

Sajal Chakraborti is a Senior Professor of Biochemistry at the University of Kalyani, West Bengal, India. His research covers the role of proteases in regulating pulmonary vascular tone under oxidant and calcium signalling phenomena. He has been engaged in teaching and research in biochemistry for the past 40 years.

Naranjan S. Dhalla is a distinguished Professor at the University of Manitoba, Winnipeg, Canada. His expertise includes the subcellular and molecular basis of heart function in health and disease. He has been engaged in multidisciplinary research and education for promoting the scientific basis of cardiology, as well as training of professional manpower for combating heart disease for over 45 years.

Part I

Regulation of Proteases in Health and Disease

Physiological and Pathological Functions of Mitochondrial Proteases

1

Clea Bárcena, Pablo Mayoral, Pedro M. Quirós, and Carlos López-Otín

Abstract

Mitoproteases display an essential role in the preservation of mitochondrial homeostasis under regular and stress conditions. These enzymes perform tightly regulated proteolytic reactions by which they participate in mitochondrial protein trafficking, processing and activation of proteins, protein quality control, regulation of mitochondrial biogenesis, control of mitochondrial dynamics, mitophagy, and apoptosis. In this chapter, we have revised the physiological functions of the intrinsic mitochondrial proteases, analyzing their roles in the different compartments of this organelle and their connection to human pathology, primarily cancer, neurodegenerative disorders, and multisystemic diseases.

Keywords

Mitochondria • Mitoproteases • Mitochondrial dynamics • Cancer • Aging • Neurodegenerative disorders

1.1 Introduction

Due to their prokaryotic origin, mitochondria possess some structural characteristics that make them remarkably different from other organelles of eukaryotic cells. They have a double membrane with an intermembrane space, being the inner membrane expanded by the formation of numerous foldings named *cristae*. Enclosed among these *cristae*, we can find the mitochondrial matrix, where several copies of mitochondrial DNA genome are contained. Mitochondria play essential biological

C. Bárcena • P. Mayoral • P.M. Quirós • C. López-Otín (✉)
Departamento de Bioquímica y Biología Molecular, Facultad de Medicina, Instituto Universitario de Oncología (IUOPA), Universidad de Oviedo, 33006 Oviedo, Spain
e-mail: clo@uniovi.es

S. Chakraborti, N.S. Dhalla (eds.), *Proteases in Physiology and Pathology*,
DOI 10.1007/978-981-10-2513-6_1

functions, fundamental for the generation of most of the cell supply of adenosine triphosphate (ATP), the source of chemical energy within the cells.

The complex structure and the relevance of the mitochondrial function within the organisms justify the need to set an organization of the molecules that collaborate in the maintenance of the assembly and function of this energy machine, among which proteases are gaining an increasing attention. Proteases have been considered for many years just as performers of the catabolic reactions in the organisms; however, we are currently beholding a surprising and unexpected increment in the studies of these enzymes, positioning them as multifunctional molecules carrying essential functions in health and disease [1–3]. The study of proteases contained in the mitochondria, named *mitoproteases*, has also experienced an exciting expansion. Until very recently, they were seen only as members of the quality control system of mitochondria, in charge of the degradation of misfolded and damaged proteins or being responsible for the processing of proteins imported from cytosol into mitochondria [4]. Nevertheless, in recent years, we have witnessed an impressive progress in the knowledge of their functions, becoming increasingly evident that they are not merely actors of the catabolic functions of mitochondria. Instead, mitoproteases perform precise and tightly regulated proteolytic roles that determine time-specific functions of regulatory proteins.

Due to this increased knowledge and interest in the study of mitochondrial proteases, we have recently proposed the concept of *mitodegradome* to define the complete set of proteases and homologues that function in mitochondria from cells and tissues of an organism [5]. The mitodegradome would therefore modulate in an efficient and irreversible way the function of mitochondria in order to adapt these organelles to the diverse stress situations that they face, especially given the fact that there is an absence of de novo synthesis of mitochondria. This characteristic is of special relevance in post-mitotic tissues, endowing mitochondria and the mitodegradome an important role in health and aging. Consistent with this, malfunction or deregulation of mitochondrial proteases has been associated with numerous pathologies such as metabolic impairments, cancer, inflammatory diseases, and neurodegenerative disorders [6–8], as well as with the control of aging and longevity [9].

Mitoproteases have the common feature of their location in the mitochondria, which can be exclusive or shared with the cytosol. Beyond this characteristic, they form an assorted group attending to their proteolytic diversity. To set some clarity in this complex grouping, we have recently organized the mitoproteases according to their function, localization, and proteolytic nature in three groups: *intrinsic or resident mitoproteases*, which exert their function exclusively in this organelle regardless of the compartment in which they act; the *pseudo-mitoproteases*, which have a protease structure but are catalytically impaired; and *transient mitoproteases*, which are translocated into mitochondria only in some particular circumstances [5].

Among these three groups of defined mitoproteases, the *intrinsic* or *resident mitoproteases* are the most relevant, and it will be the main focus of this chapter. All the enzymes that belong to this group exert their function essentially in mitochondria, being considered as the bona fide mitochondrial proteases. In this group, we can find 20 enzymes, divided into 1 cysteine, 7 serine, and 12 metalloproteases.

The only member in the subgroup of cysteine proteases is the deubiquitinase USP30, which participates in the quality control system in the outer membrane [10]. By contrast, the metalloproteases are represented by members of seven different families, which exert a variety of functions in mitochondria. This subgroup includes the processing peptidases PMPCB, MIPEP, XPNPEP3, METAP1D, ATP23, and OMA1; the ATP-dependent proteases AFG3L2, SPG7, and YME1L1; the oligopeptidases NLN and PITRM1; and the relatively unknown enzyme OSGEPL1. Finally, the serine proteases are represented by beta-lactamase (LACTB); the processing peptidases IMMPL1, IMMPL2, and PARL; the ATP-dependent proteases LONP1 and CLPP; and the quality control protease HTRA2. Interestingly, many of these proteolytic enzymes are widely distributed from bacteria to mammals. In fact, there are bacterial orthologues of nearly all human mitochondrial proteases, including different oligopeptidases and aminopeptidases belonging to the M3 and M24 metalloprotease families; the highly conserved families of ATP-dependent proteases, such as Lon, ClpP, and FtsH proteases (orthologous of LONP1, CLPP, and the AAA proteases AFG3L2, SPG7, and YME1L1, respectively); the ATP-independent stress-response metalloprotease HtpX (orthologous of OMA1); and the serine proteases DegP and DegS (orthologous of HTRA2) (Table 1.1).

Functionally, the mitoproteases include the classical proteases involved in the import of proteins to mitochondria and in the protein quality control system. However, it is now recognized that mitoproteases exert a variety of functions within mitochondria, including mitochondrial protein trafficking, processing and activation of proteins, protein quality control, regulation of mitochondrial biogenesis, control of mitochondrial dynamics, mitophagy, and apoptosis [5].

The present chapter discusses the different proteolytic functions of mitoproteases in the mitochondrial compartments, focusing on recent advances of the study of this group of enzymes. Additionally, we present an overview of the role of these enzymes in human pathology.

1.2 Physiological Roles of Mitoproteases

1.2.1 Proteolysis in the Outer Membrane

The mitochondrial outer membrane plays essential roles in mitochondrial biogenesis, as well as in the control of mitochondrial dynamics and mitophagy. Since the amount of proteins in the outer membrane is lower compared with the inner membrane, the function of mitoproteases in the outer membrane is to regulate the fusion and fission machinery in order to maintain the integrity of the mitochondrial network. Apart from the intrinsic mitoproteases, transient proteases and the cytosolic ubiquitin proteasome system also collaborate in the regulation of these processes, maintaining the quality control in this membrane [11]. Actually, the only mitochondrial protease that exerts its biological function in the outer membrane is the deubiquitinating enzyme USP30. The role of this enzyme is to inhibit the function of mitofusin 1 and mitofusin 2 by specifically removing the non-degradative

Table 1.1 Intrinsic mitochondrial proteases

Symbol	Class	Localization	Mitochondrial function	Associated pathology in humans
OMA1	Metallo	Inner membrane	Mitochondrial dynamics, mitophagy, and apoptosis	Improves cisplatin chemosensitivity
PARL	Ser	Inner membrane	Mitophagy and apoptosis	Type 2 diabetes, Parkinson's-like disease, neuronal injury, and cerebral ischemia
AFG3L2	Metallo	Inner membrane	PQC and mitochondrial biogenesis	Dementia, ataxia spinocerebellar, spastic paraplegia
SPG7	Metallo	Inner membrane	PQC and mitochondrial biogenesis	Spastic paraplegia, ataxia, chronic ophthalmoplegia, type 2 diabetes, coronary artery disease
YME1L1	Metallo	Inner membrane	PQC and mitochondrial biogenesis	–
IMMP1L	Ser	Inner membrane	Protein import and activation	–
IMMP2L	Ser	Inner membrane	Protein import and activation	Thyroid cancer, Tourette syndrome
LACTB	Ser	Intermembrane space	Mitochondrial biogenesis	–
NLN	Metallo	Intermembrane space	PQC	–
ATP23	Metallo	Intermembrane space	PQC and processing peptidase	–
HTRA2	Ser	Intermembrane space	PQC, mitophagy, and apoptosis	Cancer, Parkinson's disease, essential tremor, neuronal injury, and cerebral ischemia
OSGEPL1	Metallo	Matrix	Mitochondrial biogenesis	–
CLPP	Ser	Matrix	PQC	Acute myeloid leukemias, Perrault syndrome
PITRM1	Metallo	Matrix	PQC	Alzheimer's disease
LONP	Ser	Matrix	PQC and mitochondrial biogenesis	Cancer, CODAS syndrome
METAP1D	Metallo	Matrix	Protein import and activation	Colon cancer
MIPEP	Metallo	Matrix	Protein import and activation	–

(continued)

Table 1.1 (continued)

Symbol	Class	Localization	Mitochondrial function	Associated pathology in humans
PMPCB	Metallo	Matrix	Protein import and activation	–
XPNPEP3	Metallo	Matrix	Protein import and activation	Nephropathy
USP30	Cys	Outer membrane	Mitochondrial dynamics and mitophagy	Parkinson's disease

PQC protein quality control

ubiquitylation marks [12]. Consequently, the inhibition or depletion of USP30 in human cells induces fusion, with a concomitant increment in interconnected mitochondria [10]. Recently, it has been observed that this effect is maintained in cells deficient in mitofusins [12], suggesting a regulation independent of these mitoproteins. Additionally, it has been described that USP30 also blocks parkin-mediated mitophagy through the removal of ubiquitin moieties from damaged mitochondria [13].

1.2.2 Proteolysis in the Intermembrane Space

Mitoproteases located at the intermembrane space mainly exert quality control functions. All the proteases associated with this region are ATP-independent enzymes, which is probably due to the compartmentalization of the intermembrane space, keeping it separated from the ATP production area. Among these proteases, we can find HTRA2 and ATP23, which participate in protein quality control; NLN, an oligopeptidase; and LACTB, whose function is still unknown.

HTRA2 is a trimeric serine protease with crucial roles in the degradation of oxidized proteins [14] and in the inhibition of mitophagy. This last function can be exerted under stress conditions through the degradation of parkin when liberated to the cytosol or by processing the intramitochondrial Mulan E3 ubiquitin ligase [15, 16]. Mouse cells deficient for this enzyme accumulate mutations in the mtDNA, which has led to suggest that this mitoprotease also collaborates in the maintenance of mtDNA integrity [17]. The decline in mitochondrial mass and membrane potential after loss of HTRA2, resulting in ATP depletion, is also remarkable [18]. HTRA2 also regulates cell death through different pathways. First, it seems to stimulate TNF-induced necroptosis as well as a secondary germ cell death pathway in *Drosophila melanogaster* [19, 20]. However, its most studied role in cell death is the promotion of apoptosis, through two different and somewhat opposing ways. Thus, some studies suggest that its activation by PARL avoids aggregation of pro-apoptotic proteins, as BAX, in the outer membrane [21]. Conversely, it has also been suggested that, during apoptosis, HTRA2 autoactivates itself before being released to the cytoplasm, where it cleaves and inactivates anti-apoptotic proteins, such as X-linked inhibitor of apoptosis protein (XIAP) and the tumor-suppressor WT1 [22, 23], which finally results in the promotion of apoptosis.

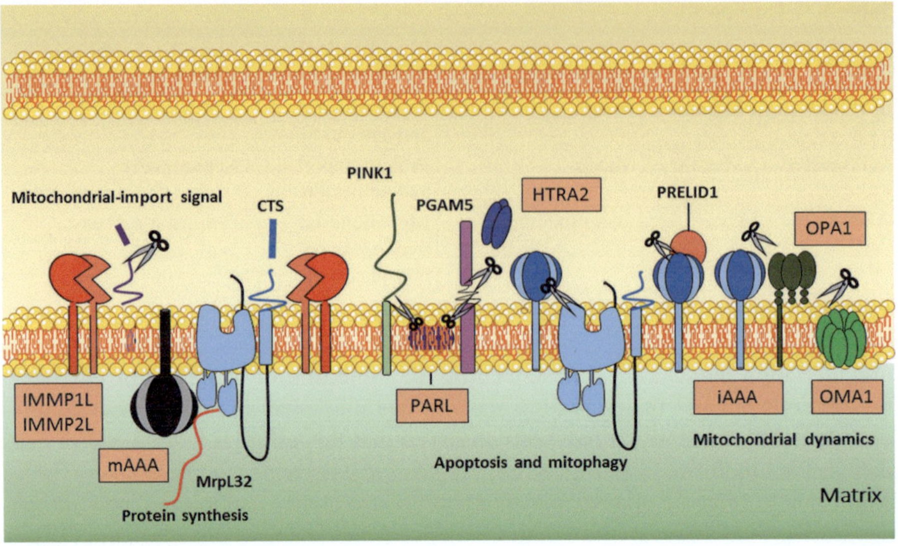

Fig. 1.1 Mitochondrial proteases in the inner membrane. Intrinsic mitochondrial proteases that exert their role in the inner membrane. The proteolytic activities of mitoproteases in this compartment are vastly diverse, participating in protein import, mitochondrial dynamics, and regulation of apoptosis and mitophagy

Among the quality control proteases, we can also find ATP23 (XRCC6BP1) [24, 25]. Although the function of this metalloprotease in mammals has not been confirmed yet, its yeast orthologue participates in the quality control of mitochondria through the degradation of lipid transfer proteins that are highly conserved in the course of evolution. The high functional conservation that is usually found among mitoproteases from yeast to mammals suggests that ATP23 might play a similar role in mammalian mitochondria [25].

NLN is supposed to be an oligopeptidase that functions in the intermembrane space. It is also known as *neurolysin* or *mitochondrial oligopeptidase M* and can be found both in the cytosol and inside of mitochondria [26].

Finally, among the mitoproteases from the intermembrane space, we can find LACTB, whose function has not been completely defined yet. However, as it is known to form internal filaments, it has been proposed to possess a structural function through the compartmentalization and structuration of the intramitochondrial membrane [27].

1.2.3 Proteolysis in the Inner Membrane

The mitoproteases located in the inner membrane are diverse and participate in several if not all mitochondrial processes that occur in this compartment (Fig. 1.1). Among them, we can find the ATP-dependent proteases mAAA (SPG7 y AFG3L2)

and iAAA (YME1L1), which participate in the quality control of the inner membrane, in the mitochondrial biogenesis, and in the regulation of mitochondrial dynamics. iAAA, and its only subunit YME1L1 protease, has its active site oriented to the intermembrane space, whereas mAAA, composed in humans of AFG3L2 and SPG7 (or paraplegin), has its active site oriented to the matrix [28, 29]. mAAA and iAAA proteases exert their quality control role through different mechanisms. First of all, these proteases degrade damaged or non-assembled subunits of the electron transport chain, so their absence provokes the malfunction of oxidative phosphorylation (OXPHOS) caused by the accumulation of defective complexes [29–31]. Besides, mAAA subunit AFG3L2 is critical to the maintenance of inner membrane integrity under aberrant protein accumulation caused by the loss of temporal and spatial coordination in the assembly of the oxidative phosphorylation complexes [32]. AAA proteases also have a role in mitochondrial biogenesis. Some years ago, it was observed that yeast cells without mAAA had a respiratory deficiency as a consequence of damaged mitochondrial translation as well as a defective processing of the ribosomal subunit MrpL32 [33]. It was later defined that this protease controls the assembly of mitochondrial ribosomes and the synthesis of mitochondrially encoded respiratory chain subunits [32]. mAAA protease, and in particular its subunit AFG3L2, also has a role in calcium homeostasis as it induces mitochondrial fragmentation through the processing of OPA1, thereby decreasing calcium uptake [34]. Also, in the case of iAAA (YME1L1), it degrades the translocase subunit TIM17A as a consequence of the stress response, decreasing protein import into mitochondria [35]. Additionally, YME1L1 protease also modulates cardiolipin levels and the resistance to apoptosis by degrading PRELID1 (known as Ups1 in yeast), a protein that prevents apoptosis by complexing to TRIAP1, a p53-regulated protein, and mediating this way the intramitochondrial transport of phosphatidic acid, necessary for cardiolipin synthesis and consequently apoptosis resistance [36].

Mitoproteases in the inner membrane also have a role in mitochondrial dynamics (Fig. 1.1). OMA1, an ATP-independent protease, is a stress-response protease that functions together with the AAA proteases. It is activated under different stress conditions, such as oxidative and heat stress, and membrane depolarization [37]. As a result, it carries the proteolytic cleavage of all long forms of OPA1, inhibiting this way the fusion process [38–40]. OMA1 negatively regulates itself by autoprocessing in order to control and limit the stress response [37]. The processing of the inner membrane fusion protein OPA1 by OMA1 and/or YME1L1, as well as the regulation of the abundance of mitofusins and DRP1 at the outer membrane, serves to regulate mitochondrial dynamics. In particular, the processing of OPA1 by YME1L1 seems to provide a connection between mitochondrial dynamics and OXPHOS function dependent on the metabolic state [41]. Actually, it has been recently reported that depolarization of the mitochondrial membrane leads to OMA1 activation, and depending on the energy status of the cell (ATP levels), YME1L1 will degrade OMA1 (high levels) or OMA1 will degrade YME1L1 (low levels), allowing cells to adapt mitochondrial dynamics to distinct cellular insults [42]. It has also been described that a third and unknown cysteine protease is able to cleave OPA1 in the C-terminal region, apparently an event that occurs in the liver after a meal.

The exact nature of this additional processing of OPA1 is still unknown; however, it seems to be dependent on mitofusin 2 and independent of OMA1 [43]. In consonance to this, cells from *Oma1*-deficient mice are unable to process and inactivate OPA1. For this reason, these cells are protected against mitochondrial fragmentation, and consequently they show an increase in highly connected mitochondria [39]. On the other hand, *Yme1l1*-deficient mice cells have constantly activated OMA1, so they show fragmented mitochondria [44]. As a consequence of this regulation of mitochondrial dynamics by OMA1 and AAA proteases, *Oma1*-null mice have an unbalanced OPA1 processing, being unable to adapt mitochondrial dynamics to stress conditions in the cells. As a result, these mutant mice are obese and have a defective thermogenesis, proving the connection between the control of metabolic homeostasis and the regulation of mitochondrial dynamics [39, 45]. Very recently, it has also been described that loss of Oma1 in yeast cells leads to elevated ROS levels and activation of stress survival responses in a TORC1-mediated way, linking mitochondrial quality control and TOR signaling in the response to stress stimuli [46].

Mitoproteases of the inner membrane also participate in the regulation of apoptosis by different mechanisms (Fig. 1.1). For example, mitochondrial fusion serves to maintain the integrity of cristae, avoiding this way the release of pro-apoptotic proteins and protecting against apoptosis. Probably for this reason, the absence of YME1L1 leads the processing of OPA1 by OMA1, causing a loss of the cristae integrity and mitochondrial fragmentation and eventually an increased susceptibility to apoptosis [44]. OMA1 also has a role in the regulation of mitophagy, as alteration of its activity stabilizes OPA1 and prevents mitochondrial fragmentation, a process that is required in order to elicit mitophagy [47]. On the contrary, it can be activated by the pro-apoptotic proteins BAX and BAK [48] and by other stress stimuli [37]. As a result of the degradation of OPA1, cytochrome c is released promoting apoptosis. Consequently, deficiency in this metalloprotease protects against apoptotic stimuli [39, 49].

PARL is a rhomboid intramembrane protease located in the inner membrane and with known roles in mitophagy and autophagy. In the first case, its function is related to the degradation of the mitochondrial kinase PINK1, which triggers the binding of parkin protein to depolarized mitochondria in order to induce mitophagy. Other mitoproteases participate in the processing of PINK1. Thus, under normal conditions, PINK1 enters the mitochondria, and once there, it is processed by MPP [50]. In addition, it is cleaved and consequently destabilized by PARL, leading to its release from mitochondria in order to be degraded by the ubiquitin-proteasome system in the cytosol [51]. Afterwards, PINK1 is recycled in a process still poorly understood involving other mitoproteases such as mAAA, CLPP, and LONP1 [50]. PARL also regulates mitophagy by participating in the degradation of PGAM5 [52], a phosphatase that activates mitophagy and necrosis under stress conditions. Its degradation by PARL would therefore prevent mitophagy [53, 54]. Finally, PARL also has a role in the regulation of apoptosis in a HTRA2-independent manner, through the remodeling of cristae and the control of cytochrome c release during apoptosis [55, 56]. Accordingly, mice deficient in PARL show an increased apoptosis that leads to a premature death [55].

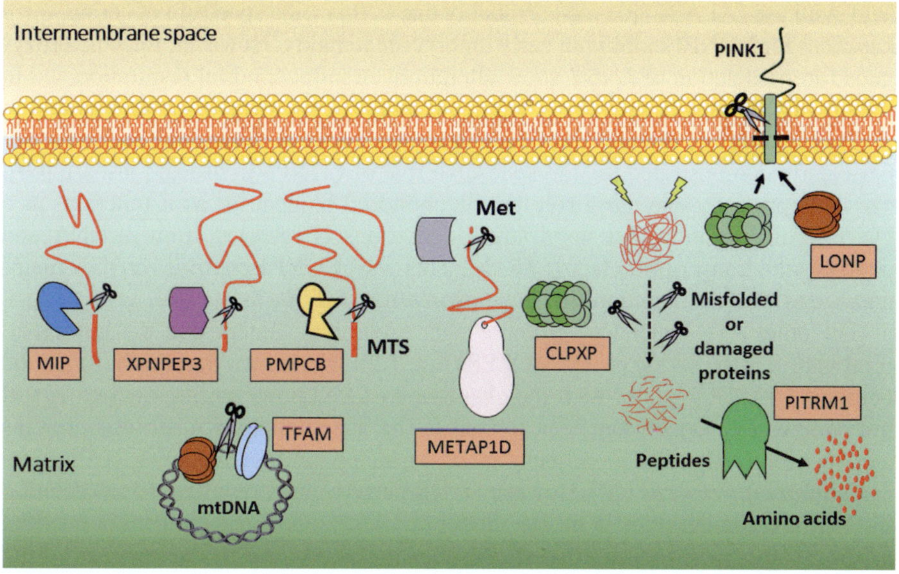

Fig. 1.2 Matrix mitochondrial proteases. Intrinsic mitochondrial proteases situated mainly in the matrix. Their roles in this compartment are predominantly the processing of imported peptides, maintenance of mtDNA, and degradation of misfolded or damaged proteins

The last mitoproteases from the inner membrane are IMMP1L and IMMP2L, processing peptidases that eliminate hydrophobic signals from proteins that have been imported into the intermembrane space after being processed by MPP [57]. These two mitoproteases also promote the assembly of yeast translocase inner membrane (TIM) complexes through the cleavage of specific carboxy-terminal subunits [58].

1.2.4 Proteolysis in the Mitochondrial Matrix

Among the ATP-dependent proteases that exert their function in the mitochondrial matrix, we can mention LONP1 and CLPP, which participate in quality control, mitochondrial biogenesis, and mitochondrial stress response (Fig. 1.2). LONP1 is a serine protease highly conserved through evolution. It participates in the quality control of the matrix by degrading oxidized, misfolded or mutated proteins, ensuring cell viability. Therefore, it can respond to different stress stimuli that can be potentially harmful [59, 60], as well as to normal conditions that require a reconditioning of the protein homeostasis in mitochondria. LONP1 has been shown to degrade several proteins in mammals, including succinate dehydrogenase subunit 5 (SDH5), aconitase, glutaminase C, cytochrome *c* oxidase isoform COX4–1, steroidogenic acute regulatory protein (StAR), mitochondrial transcription factor A (TFAM), cystathionine-β-synthase (CBS), heme oxygenase-1, and 5-aminolevulinic acid synthase (ALAS1) [61–68]. Probably due to this wide spectrum of action, mice *knockout* for LONP1 exhibit an early embryonic lethality. However, mice heterozygous for the deletion of *Lonp1* show alterations in mitochondrial respiration and in the OXPHOS system, probably causing an inability for metabolic reconversion in malignant cells and thereby showing a decreased tumoral susceptibility. This study has clearly demonstrated the indispensable role of LONP1 in life and disease [69]. This mitoprotease also has a role in mitochondrial biogenesis, as it functions as a DNA-binding protein that upon stress conditions is released from mtDNA and degrades the transcription factor TFAM. This way, LONP1 controls mtDNA maintenance as well as mitochondrial gene expression [68, 70, 71].

The other ATP-dependent protease with proteolytic function in the mitochondrial matrix is the serine protease CLPP (Fig. 1.2). It is known that CLPP forms the complex CLPXP together with the chaperone CLPX; however, its function is still not completely clear. It has been related to the degradation of misfolded proteins and, consequently, to the mitochondrial unfolded protein response (UPR^{mt}) in *Caenorhabditis elegans* [72]. However, it has been recently reported that its deletion in DARS2-deficient mice, a mouse model of UPR^{mt}, alleviates their mitochondrial cardiomyopathy, suggesting that CLPP does not participate in mammalian UPR^{mt} [73]. It has also been described that CLPP absence in mice provokes hearing loss, infertility, and growth retardation, probably through the accumulation of CLPX subunits and mtDNA [74].

Apart from these two ATP-dependent proteases, mitochondrial matrix also needs processing peptidases engaged in the import of proteins to mitochondria and oligopeptidases that deal with the peptides that result from the degradation of damaged proteins by LONP1, CLPP, and mAAA proteases (Fig. 1.2). Among them we find PITRM1, an oligopeptidase from the pitrilysin family. This metalloprotease also shows a presequence processing role that is critical for correct mitochondrial function [75–77]. PITRM1 degrades the mitochondrial amyloid β-protein in human cells [78] and, in yeast, it has also been observed that amyloid β-protein can in turn inhibit the yeast orthologue of PITRM1, impairing this way the processing activity of MPP. This would trigger the accumulation of preproteins in mitochondria, linking accumulation of amyloid β-protein to mitochondrial dysfunction [77].

As explained before, most of mitochondrial proteins are encoded in the nucleus DNA. In order to be translocated into the mitochondria after being synthesized in the cytosol, these proteins carry mitochondrial import signals that, once inside of the organelle, need to be removed. This action is performed by the so-called mitochondrial processing peptidases, a group of enzymes that includes PMPCB, MIPEP, METAP1D, and XPNPEP3 [79]. PMPCB (also known as β-MPP) is responsible for the processing of the majority of mitochondrial proteins, thus being the most important mitochondrial protein peptidase [80]. It forms a heterodimer with its non-protease homologue PMPCA (α-MPP), resulting in the MPP complex [81]. PMPCA, although without protease activity, facilitates the proteolytic processing of PMPCB by recognizing and binding to the mitochondrial targeting presequences from the imported proteins [82]. As already stated, PMPCB is the major mitochondrial processing peptidase. However, there are other processing peptidases with more defined targets and essential roles in the maturation and stabilization of the imported proteins. Among them, we can mention the mitochondrial intermediate peptidase or MIPEP (Oct1 in yeast) that cleaves octapeptides from cytosolic proteins imported into the mitochondria, whereas the aminopeptidase XPNPEP3 (also known as Icp55) cleaves one single amino acid from the amino terminus of MPP-processed substrates [83, 84]. Finally, the mitochondrial Met aminopeptidase METAP1D (also known as MAP1D) cleaves the Met residue from the amino terminus of some proteins encoded in the mitochondrial DNA, generating functionally active proteins [85].

Another proteolytic enzyme that functions in the mitochondrial matrix is OSGEPL1, whose role is still unknown although some studies suggest that it could participate in the mtDNA maintenance, as demonstrated for its orthologous proteins Kae1(OSGEP)/YgjD in multiple organisms [86].

1.3 Pathological Contributions of Mitoproteases

1.3.1 Mitochondrial Proteases in Cancer

The essential role of metabolism and bioenergetic remodeling in malignant processes [87, 88] has led to an increment in the studies that relate the function or malfunction of mitoproteases to different stages of cancer, from its first appearance to its progression and metastasis. Ten years ago, it was already observed that the expression of the aminopeptidase METAP1D was increased in colon cancer cells and that its downregulation was related to a reduction of the tumorigenic potential [89]. Nowadays, we know that several other mitoproteases are also implicated in the oncogenic process with antagonistic roles. For example, HTRA2 displays both pro- and anti-oncogenic roles as it degrades the tumor-suppressor protein WT1 [23], but also prevents cell invasion by modulating the cytoskeleton after being released to the cytosol following p53-dependent activation [90]. In this context, the mitoprotease OMA1 also responds to p53 stimulation, being upregulated and improving cisplatin chemosensitivity in gynecologic cancer cells by promoting mitochondrial fragmentation [91]. Also, a genetic study of risk of developing thyroid cancer has associated the presence of certain SNPs in the mitoprotease IMMP2L to this type of malignancy [92]. Furthermore, it was recently reported that CLPP is overexpressed in human acute myeloid leukemias and that its inactivation selectively kills the tumorigenic cells by inhibiting oxidative phosphorylation and mitochondrial metabolism [93]. Moreover, CLPP downregulation in human cancer cells seems to confer sensitivity to cisplatin therapy [94].

Besides this, LONP1 protease is the mitoprotease most studied for its relation to cancer. It has been consistently reported to promote tumorigenesis, as its upregulation protects malignant cells from different stresses, favoring their oncogenic potential [95]. Moreover, it has been observed that its upregulation in several tumors is related to a worsened prognostic. On the contrary, downregulation of LONP1 decreases the tumorigenic properties of cancer cells [69, 96–102]. Its implication in tumorigenesis might be explained by its role in the adaptation to the hypoxic

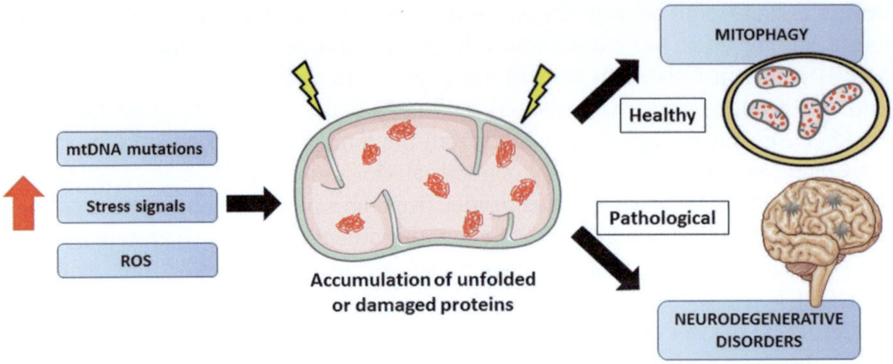

Fig. 1.3 **Mitochondrial proteases in neurodegenerative diseases**. Mitochondria need to adapt constantly to a changing environment in order to maintain their homeostasis. A deregulation of this status, such as an increment in reactive oxygen species (ROS), a disproportionate stress stimuli, or accumulation of mutations in mtDNA, might lead to malfunction of mitoproteases and the accumulation of an excess of misfolded or damaged proteins. In a healthy situation, this should be solved by activation of the mitophagy pathway. However, under pathological conditions or mutations in mitoproteases, mitophagy might be inhibited, causing different pathologies such as neurodegenerative disorders

conditions that are found in tumors in a HIF1α-mediated way [64]. More specifically, it has been observed that LONP1 induces the necessary metabolic switch in tumor cells and protects them from entering senescence [69]. Indeed, it was recently described that increase in LONP1 levels in tumor cells facilitates survival to apoptosis through the increase stability of the HSP60-mtHSP70 complex [103].

1.3.2 Mitochondrial Proteases in Neurodegenerative Diseases

Besides their implication in cancer, mitoproteases have been classically related to neurodegenerative diseases. The inability to solve the excessive accumulation of unfolded or oxidized proteins and deficiencies in the proteolytic function of these enzymes prompts a loss of proteostasis that can be added to defects in mitochondrial dynamics and mitophagy. These defects in mitochondrial function can be a consequence of certain mutations, a result of aging or a combination of both. Eventually, the inability to remove defective mitochondria or its malfunction causes neurodegenerative disorders such as Parkinson's or Alzheimer's disease (Fig. 1.3). Actually, mutations in HTRA2 have been reported to impair the clearance of α-synuclein deposits in mitochondria, participating in the pathogenesis of Parkinson's disease [104, 105]. Alterations in HTRA2 have also been proved to cause hereditary essential tremor in humans, which can be related to a later development of Parkinson's disease [106]. In fact, missense mutations in *Htra2* in mice cause a neuromuscular alteration, whereas its complete absence induces a neurodegenerative disorder that resembles Parkinson's disease [107, 108]. Mutations in PARL also cause a Parkinson's-like syndrome [109], probably due to its essential role in the processing of PINK1 and PGAM5. Finally, the deubiquitylase USP30 has also been implicated in Parkinson's disease by opposing to parkin-mediated autophagy, thereby promoting this neurodegenerative disorder [13].

Apart from their implication in Parkinson's disease, PARL and HTRA2 have also been linked to striatal neuronal injury cerebral ischemia [110]. Additionally, intragenic deletions in IMMP2L are known to cause Tourette syndrome, while SNPs and CNVs in this gene have been associated with several neurodevelopmental disorders as well as with autism [111–113]. Meanwhile, the already exposed role of PITRM1 in the degradation of amyloid β-protein accumulation involves an obvious implication in the pathology of Alzheimer's disease [78].

The ATP-dependent protease mAAA has also been implicated in neurological disorders by mutations in its two different subunits. Mutations in both AFG3L2 and SPG7 cause spastic paraplegia and ataxia disorders [114–116], whereas mutations in SPG7 have also been related to chronic ophthalmoplegia owing to the associated defects in mtDNA maintenance [117, 118]. Indeed, deletion of *Afg3l2* subunit in mice causes hyperphosphorylation of Tau protein and defective mitochondrial anterograde transport, leading to a neurological alteration that resembles axonal degeneration disorders observed in humans. Deletion in mice of the other mAAA subunit, *Spg7*, also causes axonopathy and abnormal mitochondria, developing a neurodegenerative disorder [119, 120]. More recent studies with mice deficient for the mAAA subunits have demonstrated their relevance for neurological function and their implication in cerebellar degeneration [121]. Apart from its relevance in correct axonal maintenance, AFG3L2 has also been implicated in dementia and neurodegeneration through its essential role in mitochondrial protein synthesis and Purkinje cells' survival [122–125]. Interestingly, antioxidants such as N-acetylcysteine or vitamin E have been observed to restore mitochondrial function in neurons with a depletion in AFG3L2, opening new horizons for therapeutic approaches in neurodegenerative disorders [122].

1.3.3 Mitochondrial Proteases in Multisystemic Diseases

Finally, the essential role of mitochondria in every tissue of the body implies that its incorrect function is related to several other metabolic and multisystemic diseases. For example, type 2 diabetes has been related to mutations or SNPs in SPG7 and PARL. Moreover, mutations in SPG7 have also been linked to coronary artery disease [126, 127]. Furthermore, it was recently reported that cardiac-specific ablation of *Yme1l1* in mice induced dilated cardiomyopathy and heart failure through an increase in mitochondrial fragmentation due to constant activation of OMA1 and consequent OPA1 proteolysis. Interestingly, deletion of *Oma1* rescued these alterations [128]. Additionally, the essential function of OMA1 in the adaptation of mitochondrial dynamics to different stress stimuli is consistent with the observation that mice deficient in this mitoprotease show an obesity phenotype when fed a high-fat diet. The observed obesity is also accompanied with an impaired thermogenesis, an increment in hepatic steatosis and a marked alteration of glucose metabolism [39, 45].

Conversely, absence of the oligopeptidase NLN in mice provokes an increased glucose tolerance, insulin sensitivity, and gluconeogenesis [129]. Similarly, transgenic mice for *Lactb* also show an obesity phenotype [130]. Moreover, mutations in XPNPEP3 in humans induce abnormal cilia in kidneys and a severe nephropathy as a consequence [131]. Furthermore, mutations in the ATP-dependent protease CLPP cause Perrault syndrome, a heterogeneous condition characterized by sensorineural hearing loss and ovarian failure [132]. Finally, it has been recently described that mutations in LONP1 are behind the genetic cause of CODAS syndrome, a multisystemic disease that is presented as a developmental disorder with anomalies at several levels, including cerebral, ocular, dental, auricular, and skeletal [133, 134].

1.4 Conclusions

The study of mitoproteases has experienced an enormous expansion in the last years owing to the discovery that their functions are much more complex and refined than the merely catabolic reactions involved in the quality control of mitochondria. In fact, mitoproteases exert essential roles in order to regulate the energetic responses to different internal and external stresses, as well as to control the trafficking between cytosol and mitochondria, the correct mitonuclear communication, and the tight coordination of apoptotic signals. The complexity that this area of knowledge was acquiring made it necessary to establish an organization of the proteases located in mitochondria. For this reason, we have recently classified the mitoproteases by their spatial action, distinguishing between *intrinsic* or *resident mitoproteases*, *pseudo-mitoproteases*, and *transient* or *roaming mitoproteases* [5]. In the first group, we have included those mitoproteases that exert their function mainly in mitochondria, even though some of them can be sporadically found in cytosol. The second group encloses molecules with a protease structure but with an impaired catalytic activity. Finally, the group of *transient* or *roaming mitoproteases* includes those mitoproteases that although their main function takes place in a different part of the cell, they translocate into mitochondria in some situations such as apoptosis or autophagy. In this chapter, we have revised the physiological function of intrinsic mitoproteases according to the compartment of mitochondria where they exert their main function (Table 1.1). This way, we have compiled the proteolytic activities that can be found in the outer membrane, in the intermembrane space, in the inner membrane, and in the matrix. In all compartments, we can detect mitoproteases with functions related to mitochondrial dynamics, trafficking and processing of other peptides, quality control, apoptosis or mitophagy.

Finally, we have conducted a revision of the implication of the different mitoproteases in pathology (Table 1.1). In this context, we have revised the implication of mitoproteases in cancer, where they can act as protumoral (METAP1D, CLPP, and LONP1) and as tumor suppressors (OMA1 and IMMP2L) or may exert a dual role (HTRA2). Also, we have studied the relation of mitoproteases in neurodegenerative disorders, including Parkinson's disease (HTRA2, PARL, and USP30), spastic paraplegia and ataxia (AFG3L2 and SPG7), dementia (AFG3L2), or Tourette syndrome (IMMP2L). At last, we have examined the literature connecting mitoproteases with a variety of multisystemic disorders, such as type 2 diabetes (SPG7 and PARL), cardiac disease (SPG7 and YME1L1), obesity (OMA1 and LACTB), abnormal kidney micromorphology (XPNPEP3), Perrault syndrome (CLPP), and CODAS syndrome (LONP1). This way, mitoproteases prove to be intimately related to the maintenance of the equilibrium between health and disease, not only in relation to energy homeostasis but also in the development of cancer, neurological syndromes, and multisystemic disorders.

Acknowledgments The work in our laboratory was supported by grants from Ministerio de Economía y Competitividad and Instituto de Salud Carlos III (RTICC). We also thank the generous support by J.I. Cabrera. The Instituto Universitario de Oncología is supported by Fundación Bancaria Caja de Ahorros de Asturias.

Databases Degradome: http://degradome.uniovi.es/diseases
Merops: http://merops.sanger.ac.uk/
Mitocarta: http://www.broadinstitute.org/pubs/MitoCarta/

References

1. Lopez-Otin C, Bond JS (2008) Proteases: multifunctional enzymes in life and disease. J Biol Chem 283:30433–30437
2. Lopez-Otin C, Hunter T (2010) The regulatory crosstalk between kinases and proteases in cancer. Nat Rev Cancer 10:278–292
3. Turk B, Turk D, Turk V (2012) Protease signalling: the cutting edge. EMBO J 31:1630–1643
4. Koppen M, Langer T (2007) Protein degradation within mitochondria: versatile activities of AAA proteases and other peptidases. Crit Rev Biochem Mol Biol 42:221–242
5. Quiros PM, Langer T, Lopez-Otin C (2015) New roles for mitochondrial proteases in health, ageing and disease. Nat Rev Mol Cell Biol 16:345–359
6. Bulteau AL, Bayot A (2011) Mitochondrial proteases and cancer. Biochim Biophys Acta 1807:595–601
7. Rugarli EI, Langer T (2012) Mitochondrial quality control: a matter of life and death for neurons. EMBO J 31:1336–1349
8. Goard CA, Schimmer AD (2014) Mitochondrial matrix proteases as novel therapeutic targets in malignancy. Oncogene 33:2690–2699
9. Lopez-Otin C, Blasco MA, Partridge L, Serrano M, Kroemer G (2013) The hallmarks of aging. Cell 153:1194–1217
10. Nakamura N, Hirose S (2008) Regulation of mitochondrial morphology by USP30, a deubiquitinating enzyme present in the mitochondrial outer membrane. Mol Biol Cell 19:1903–1911
11. Escobar-Henriques M, Langer T (2014) Dynamic survey of mitochondria by ubiquitin. EMBO Rep 15:231–243
12. Yue W, Chen Z, Liu H, Yan C, Chen M, Feng D, Yan C, Wu H, Du L, Wang Y, Liu J, Huang X, Xia L, Liu L, Wang X, Jin H, Wang J, Song Z, Hao X, Chen Q (2014) A small natural molecule promotes mitochondrial fusion through inhibition of the deubiquitinase USP30. Cell Res 24:482–496
13. Bingol B, Tea JS, Phu L, Reichelt M, Bakalarski CE, Song Q, Foreman O, Kirkpatrick DS, Sheng M (2014) The mitochondrial deubiquitinase USP30 opposes parkin-mediated mitophagy. Nature 510:370–375
14. Clausen T, Kaiser M, Huber R, Ehrmann M (2011) HTRA proteases: regulated proteolysis in protein quality control. Nat Rev Mol Cell Biol 12:152–162

15. Cilenti L, Ambivero CT, Ward N, Alnemri ES, Germain D, Zervos AS (2014) Inactivation of Omi/HtrA2 protease leads to the deregulation of mitochondrial Mulan E3 ubiquitin ligase and increased mitophagy. Biochim Biophys Acta 1843:1295–1307
16. Park HM, Kim GY, Nam MK, Seong GH, Han C, Chung KC, Kang S, Rhim H (2009) The serine protease HtrA2/Omi cleaves Parkin and irreversibly inactivates its E3 ubiquitin ligase activity. Biochem Biophys Res Commun 387:537–542
17. Goo HG, Jung MK, Han SS, Rhim H, Kang S (2013) HtrA2/Omi deficiency causes damage and mutation of mitochondrial DNA. Biochim Biophys Acta 1833:1866–1875
18. Plun-Favreau H, Burchell VS, Holmstrom KM, Yao Z, Deas E, Cain K, Fedele V, Moisoi N, Campanella M, Miguel Martins L, Wood NW, Gourine AV, Abramov AY (2012) HtrA2 deficiency causes mitochondrial uncoupling through the F(1)F(0)-ATP synthase and consequent ATP depletion. Cell Death Dis 3:e335
19. Sosna J, Voigt S, Mathieu S, Kabelitz D, Trad A, Janssen O, Meyer-Schwesinger C, Schutze S, Adam D (2013) The proteases HtrA2/Omi and UCH-L1 regulate TNF-induced necroptosis. Cell Commun Signal 11:76
20. Yacobi-Sharon K, Namdar Y, Arama E (2013) Alternative germ cell death pathway in Drosophila involves HtrA2/Omi, lysosomes, and a caspase-9 counterpart. Dev Cell 25:29–42
21. Chao JR, Parganas E, Boyd K, Hong CY, Opferman JT, Ihle JN (2008) Hax1-mediated processing of HtrA2 by Park allows survival of lymphocytes and neurons. Nature 452:98–102
22. Suzuki Y, Imai Y, Nakayama H, Takahashi K, Takio K, Takahashi R (2001) A serine protease, HtrA2, is released from the mitochondria and interacts with XIAP, inducing cell death. Mol Cell 8:613–621
23. Hartkamp J, Carpenter B, Roberts SG (2010) The Wilms' tumor suppressor protein WT1 is processed by the serine protease HtrA2/Omi. Mol Cell 37:159–171
24. Osman C, Wilmes C, Tatsuta T, Langer T (2007) Prohibitins interact genetically with Atp23, a novel processing peptidase and chaperone for the F1Fo-ATP synthase. Mol Biol Cell 18:627–635
25. Potting C, Wilmes C, Engmann T, Osman C, Langer T (2010) Regulation of mitochondrial phospholipids by Ups1/PRELI-like proteins depends on proteolysis and Mdm35. EMBO J 29:2888–2898
26. Serizawa A, Dando PM, Barrett AJ (1995) Characterization of a mitochondrial metallopeptidase reveals neurolysin as a homologue of thimet oligopeptidase. J Biol Chem 270:2092–2098
27. Polianskyte Z, Peitsaro N, Dapkunas A, Liobikas J, Soliymani R, Lalowski M, Speer O, Seitsonen J, Butcher S, Cereghetti GM, Linder MD, Merckel M, Thompson J, Eriksson O (2009) LACTB is a filament-forming protein localized in mitochondria. Proc Natl Acad Sci U S A 106:18960–18965
28. Koppen M, Metodiev MD, Casari G, Rugarli EI, Langer T (2007) Variable and tissue-specific subunit composition of mitochondrial m-AAA protease complexes linked to hereditary spastic paraplegia. Mol Cell Biol 27:758–767
29. Stiburek L, Cesnekova J, Kostkova O, Fornuskova D, Vinsova K, Wenchich L, Houstek J, Zeman J (2012) YME1L controls the accumulation of respiratory chain subunits and is required for apoptotic resistance, cristae morphogenesis, and cell proliferation. Mol Biol Cell 23:1010–1023
30. Hornig-Do HT, Tatsuta T, Buckermann A, Bust M, Kollberg G, Rotig A, Hellmich M, Nijtmans L, Wiesner RJ (2012) Nonsense mutations in the COX1 subunit impair the stability of respiratory chain complexes rather than their assembly. EMBO J 31:1293–1307
31. Zurita Rendon O, Shoubridge EA (2012) Early complex I assembly defects result in rapid turnover of the ND1 subunit. Hum Mol Genet 21:3815–3824
32. Richter U, Lahtinen T, Marttinen P, Suomi F, Battersby BJ (2015) Quality control of mitochondrial protein synthesis is required for membrane integrity and cell fitness. J Cell Biol 211:373–389
33. Nolden M, Ehses S, Koppen M, Bernacchia A, Rugarli EI, Langer T (2005) The m-AAA protease defective in hereditary spastic paraplegia controls ribosome assembly in mitochondria. Cell 123:277–289

34. Maltecca F, Baseggio E, Consolato F, Mazza D, Podini P, Young SM Jr, Drago I, Bahr BA, Puliti A, Codazzi F, Quattrini A, Casari G (2015) Purkinje neuron Ca2+ influx reduction rescues ataxia in SCA28 model. J Clin Invest 125:263–274
35. Rainbolt TK, Atanassova N, Genereux JC, Wiseman RL (2013) Stress-regulated translational attenuation adapts mitochondrial protein import through Tim17A degradation. Cell Metab 18:908–919
36. Potting C, Tatsuta T, Konig T, Haag M, Wai T, Aaltonen MJ, Langer T (2013) TRIAP1/PRELI complexes prevent apoptosis by mediating intramitochondrial transport of phosphatidic acid. Cell Metab 18:287–295
37. Baker MJ, Lampe PA, Stojanovski D, Korwitz A, Anand R, Tatsuta T, Langer T (2014) Stress-induced OMA1 activation and autocatalytic turnover regulate OPA1-dependent mitochondrial dynamics. EMBO J 33:578–593
38. Ehses S, Raschke I, Mancuso G, Bernacchia A, Geimer S, Tondera D, Martinou JC, Westermann B, Rugarli EI, Langer T (2009) Regulation of OPA1 processing and mitochondrial fusion by m-AAA protease isoenzymes and OMA1. J Cell Biol 187:1023–1036
39. Quiros PM, Ramsay AJ, Sala D, Fernandez-Vizarra E, Rodriguez F, Peinado JR, Fernandez-Garcia MS, Vega JA, Enriquez JA, Zorzano A, Lopez-Otin C (2012) Loss of mitochondrial protease OMA1 alters processing of the GTPase OPA1 and causes obesity and defective thermogenesis in mice. EMBO J 31:2117–2133
40. Head B, Griparic L, Amiri M, Gandre-Babbe S, van der Bliek AM (2009) Inducible proteolytic inactivation of OPA1 mediated by the OMA1 protease in mammalian cells. J Cell Biol 187:959–966
41. Mishra P, Carelli V, Manfredi G, Chan DC (2014) Proteolytic cleavage of Opa1 stimulates mitochondrial inner membrane fusion and couples fusion to oxidative phosphorylation. Cell Metab 19:630–641
42. Rainbolt TK, Lebeau J, Puchades C, Wiseman RL (2016) Reciprocal degradation of YME1L and OMA1 adapts mitochondrial proteolytic activity during stress. Cell Rep 14:2041–2049
43. Sood A, Jeyaraju DV, Prudent J, Caron A, Lemieux P, McBride HM, Laplante M, Toth K, Pellegrini L (2014) A Mitofusin-2-dependent inactivating cleavage of Opa1 links changes in mitochondria cristae and ER contacts in the postprandial liver. Proc Natl Acad Sci U S A 111:16017–16022
44. Anand R, Wai T, Baker MJ, Kladt N, Schauss AC, Rugarli E, Langer T (2014) The i-AAA protease YME1L and OMA1 cleave OPA1 to balance mitochondrial fusion and fission. J Cell Biol 204:919–929
45. Quiros PM, Ramsay AJ, Lopez-Otin C (2013) New roles for OMA1 metalloprotease: from mitochondrial proteostasis to metabolic homeostasis. Adipocytes 2:7–11
46. Bohovych I, Kastora S, Christianson S, Topil D, Kim H, Fangman T, Zhou YJ, Barrientos A, Lee J, Brown AJ, Khalimonchuk O (2016) Oma1 links mitochondrial protein quality control and TOR signaling to modulate physiological plasticity and cellular stress responses. Mol Cell Biol 36(17):2300–2312
47. MacVicar TD, Lane JD (2014) Impaired OMA1-dependent cleavage of OPA1 and reduced DRP1 fission activity combine to prevent mitophagy in cells that are dependent on oxidative phosphorylation. J Cell Sci 127:2313–2325
48. Jiang X, Jiang H, Shen Z, Wang X (2014) Activation of mitochondrial protease OMA1 by Bax and Bak promotes cytochrome c release during apoptosis. Proc Natl Acad Sci U S A 111:14782–14787
49. Xiao X, Hu Y, Quiros PM, Wei Q, Lopez-Otin C, Dong Z (2014) OMA1 mediates OPA1 proteolysis and mitochondrial fragmentation in experimental models of ischemic kidney injury. Am J Physiol Renal Physiol 306:F1318–F1326
50. Greene AW, Grenier K, Aguileta MA, Muise S, Farazifard R, Haque ME, McBride HM, Park DS, Fon EA (2012) Mitochondrial processing peptidase regulates PINK1 processing, import and Parkin recruitment. EMBO Rep 13:378–385

51. Thomas RE, Andrews LA, Burman JL, Lin WY, Pallanck LJ (2014) PINK1-Parkin pathway activity is regulated by degradation of PINK1 in the mitochondrial matrix. PLoS Genet 10:e1004279
52. Sekine S, Kanamaru Y, Koike M, Nishihara A, Okada M, Kinoshita H, Kamiyama M, Maruyama J, Uchiyama Y, Ishihara N, Takeda K, Ichijo H (2012) Rhomboid protease PARL mediates the mitochondrial membrane potential loss-induced cleavage of PGAM5. J Biol Chem 287:34635–34645
53. Chen G, Han Z, Feng D, Chen Y, Chen L, Wu H, Huang L, Zhou C, Cai X, Fu C, Duan L, Wang X, Liu L, Liu X, Shen Y, Zhu Y, Chen Q (2014) A regulatory signaling loop comprising the PGAM5 phosphatase and CK2 controls receptor-mediated mitophagy. Mol Cell 54:362–377
54. Wang Z, Jiang H, Chen S, Du F, Wang X (2012) The mitochondrial phosphatase PGAM5 functions at the convergence point of multiple necrotic death pathways. Cell 148:228–243
55. Cipolat S, Rudka T, Hartmann D, Costa V, Serneels L, Craessaerts K, Metzger K, Frezza C, Annaert W, D'Adamio L, Derks C, Dejaegere T, Pellegrini L, D'Hooge R, Scorrano L, De Strooper B (2006) Mitochondrial rhomboid PARL regulates cytochrome c release during apoptosis via OPA1-dependent cristae remodeling. Cell 126:163–175
56. Frezza C, Cipolat S, Martins de Brito O, Micaroni M, Beznoussenko GV, Rudka T, Bartoli D, Polishuck RS, Danial NN, De Strooper B, Scorrano L (2006) OPA1 controls apoptotic cristae remodeling independently from mitochondrial fusion. Cell 126:177–189
57. Nunnari J, Fox TD, Walter P (1993) A mitochondrial protease with two catalytic subunits of nonoverlapping specificities. Science 262:1997–2004
58. Ieva R, Heisswolf AK, Gebert M, Vogtle FN, Wollweber F, Mehnert CS, Oeljeklaus S, Warscheid B, Meisinger C, van der Laan M, Pfanner N (2013) Mitochondrial inner membrane protease promotes assembly of presequence translocase by removing a carboxy-terminal targeting sequence. Nat Commun 4:2853
59. Lu B, Liu T, Crosby JA, Thomas-Wohlever J, Lee I, Suzuki CK (2003) The ATP-dependent Lon protease of Mus musculus is a DNA-binding protein that is functionally conserved between yeast and mammals. Gene 306:45–55
60. Venkatesh S, Lee J, Singh K, Lee I, Suzuki CK (2012) Multitasking in the mitochondrion by the ATP-dependent Lon protease. Biochim Biophys Acta 1823:56–66
61. Teng H, Wu B, Zhao K, Yang G, Wu L, Wang R (2013) Oxygen-sensitive mitochondrial accumulation of cystathionine beta-synthase mediated by Lon protease. Proc Natl Acad Sci U S A 110:12679–12684
62. Bezawork-Geleta A, Saiyed T, Dougan DA, Truscott KN (2014) Mitochondrial matrix proteostasis is linked to hereditary paraganglioma: LON-mediated turnover of the human flavinylation factor SDH5 is regulated by its interaction with SDHA. FASEB J 28:1794–1804
63. Kita K, Suzuki T, Ochi T (2012) Diphenylarsinic acid promotes degradation of glutaminase C by mitochondrial Lon protease. J Biol Chem 287:18163–18172
64. Fukuda R, Zhang H, Kim JW, Shimoda L, Dang CV, Semenza GL (2007) HIF-1 regulates cytochrome oxidase subunits to optimize efficiency of respiration in hypoxic cells. Cell 129:111–122
65. Bota DA, Davies KJ (2002) Lon protease preferentially degrades oxidized mitochondrial aconitase by an ATP-stimulated mechanism. Nat Cell Biol 4:674–680
66. Granot Z, Kobiler O, Melamed-Book N, Eimerl S, Bahat A, Lu B, Braun S, Maurizi MR, Suzuki CK, Oppenheim AB, Orly J (2007) Turnover of mitochondrial steroidogenic acute regulatory (StAR) protein by Lon protease: the unexpected effect of proteasome inhibitors. Mol Endocrinol 21:2164–2177
67. Tian Q, Li T, Hou W, Zheng J, Schrum LW, Bonkovsky HL (2011) Lon peptidase 1 (LONP1)-dependent breakdown of mitochondrial 5-aminolevulinic acid synthase protein by heme in human liver cells. J Biol Chem 286:26424–26430
68. Matsushima Y, Goto Y, Kaguni LS (2010) Mitochondrial Lon protease regulates mitochondrial DNA copy number and transcription by selective degradation of mitochondrial transcription factor a (TFAM). Proc Natl Acad Sci U S A 107:18410–18415

69. Quiros PM, Espanol Y, Acin-Perez R, Rodriguez F, Barcena C, Watanabe K, Calvo E, Loureiro M, Fernandez-Garcia MS, Fueyo A, Vazquez J, Enriquez JA, Lopez-Otin C (2014) ATP-dependent Lon protease controls tumor bioenergetics by reprogramming mitochondrial activity. Cell Rep 8:542–556
70. Lu B, Lee J, Nie X, Li M, Morozov YI, Venkatesh S, Bogenhagen DF, Temiakov D, Suzuki CK (2013) Phosphorylation of human TFAM in mitochondria impairs DNA binding and promotes degradation by the AAA+ Lon protease. Mol Cell 49:121–132
71. Liu T, Lu B, Lee I, Ondrovicova G, Kutejova E, Suzuki CK (2004) DNA and RNA binding by the mitochondrial lon protease is regulated by nucleotide and protein substrate. J Biol Chem 279:13902–13910
72. Haynes CM, Petrova K, Benedetti C, Yang Y, Ron D (2007) ClpP mediates activation of a mitochondrial unfolded protein response in C. elegans. Dev Cell 13:467–480
73. Seiferling D, Szczepanowska K, Becker C, Senft K, Hermans S, Maiti P, Konig T, Kukat A, Trifunovic A (2016) Loss of CLPP alleviates mitochondrial cardiomyopathy without affecting the mammalian UPRmt. EMBO Rep 17(7):953–964
74. Gispert S, Parganlija D, Klinkenberg M, Drose S, Wittig I, Mittelbronn M, Grzmil P, Koob S, Hamann A, Walter M, Buchel F, Adler T, Hrabe de Angelis M, Busch DH, Zell A, Reichert AS, Brandt U, Osiewacz HD, Jendrach M, Auburger G (2013) Loss of mitochondrial peptidase Clpp leads to infertility, hearing loss plus growth retardation via accumulation of CLPX, mtDNA and inflammatory factors. Hum Mol Genet 22:4871–4887
75. Stahl A, Moberg P, Ytterberg J, Panfilov O, Brockenhuus Von Lowenhielm H, Nilsson F, Glaser E (2002) Isolation and identification of a novel mitochondrial metalloprotease (PreP) that degrades targeting presequences in plants. J Biol Chem 277:41931–41939
76. Kambacheld M, Augustin S, Tatsuta T, Muller S, Langer T (2005) Role of the novel metallopeptidase Mop112 and saccharolysin for the complete degradation of proteins residing in different subcompartments of mitochondria. J Biol Chem 280:20132–20139
77. Mossmann D, Vogtle FN, Taskin AA, Teixeira PF, Ring J, Burkhart JM, Burger N, Pinho CM, Tadic J, Loreth D, Graff C, Metzger F, Sickmann A, Kretz O, Wiedemann N, Zahedi RP, Madeo F, Glaser E, Meisinger C (2014) Amyloid-beta peptide induces mitochondrial dysfunction by inhibition of preprotein maturation. Cell Metab 20:662–669
78. Falkevall A, Alikhani N, Bhushan S, Pavlov PF, Busch K, Johnson KA, Eneqvist T, Tjernberg L, Ankarcrona M, Glaser E (2006) Degradation of the amyloid beta-protein by the novel mitochondrial peptidasome, PreP. J Biol Chem 281:29096–29104
79. Schmidt O, Pfanner N, Meisinger C (2010) Mitochondrial protein import: from proteomics to functional mechanisms. Nat Rev Mol Cell Biol 11:655–667
80. Gakh O, Cavadini P, Isaya G (2002) Mitochondrial processing peptidases. Biochim Biophys Acta 1592:63–77
81. Teixeira PF, Glaser E (2013) Processing peptidases in mitochondria and chloroplasts. Biochim Biophys Acta 1833:360–370
82. Dvorakova-Hola K, Matuskova A, Kubala M, Otyepka M, Kucera T, Vecer J, Herman P, Parkhomenko N, Kutejova E, Janata J (2010) Glycine-rich loop of mitochondrial processing peptidase alpha-subunit is responsible for substrate recognition by a mechanism analogous to mitochondrial receptor Tom20. J Mol Biol 396:1197–1210
83. Vogtle FN, Wortelkamp S, Zahedi RP, Becker D, Leidhold C, Gevaert K, Kellermann J, Voos W, Sickmann A, Pfanner N, Meisinger C (2009) Global analysis of the mitochondrial N-proteome identifies a processing peptidase critical for protein stability. Cell 139:428–439
84. Vogtle FN, Prinz C, Kellermann J, Lottspeich F, Pfanner N, Meisinger C (2011) Mitochondrial protein turnover: role of the precursor intermediate peptidase Oct1 in protein stabilization. Mol Biol Cell 22:2135–2143
85. Serero A, Giglione C, Sardini A, Martinez-Sanz J, Meinnel T (2003) An unusual peptide deformylase features in the human mitochondrial N-terminal methionine excision pathway. J Biol Chem 278:52953–52963

86. Oberto J, Breuil N, Hecker A, Farina F, Brochier-Armanet C, Culetto E, Forterre P (2009) Qri7/OSGEPL, the mitochondrial version of the universal Kae1/YgjD protein, is essential for mitochondrial genome maintenance. Nucleic Acids Res 37:5343–5352
87. Hanahan D, Weinberg RA (2011) Hallmarks of cancer: the next generation. Cell 144:646–674
88. Wallace DC (2012) Mitochondria and cancer. Nat Rev Cancer 12:685–698
89. Leszczyniecka M, Bhatia U, Cueto M, Nirmala NR, Towbin H, Vattay A, Wang B, Zabludoff S, Phillips PE (2006) MAP 1D, a novel methionine aminopeptidase family member is overexpressed in colon cancer. Oncogene 25:3471–3478
90. Yamauchi S, Hou YY, Guo AK, Hirata H, Nakajima W, Yip AK, Yu CH, Harada I, Chiam KH, Sawada Y, Tanaka N, Kawauchi K (2014) p53-mediated activation of the mitochondrial protease HtrA2/Omi prevents cell invasion. J Cell Biol 204:1191–1207
91. Kong B, Wang Q, Fung E, Xue K, Tsang BK (2014) p53 is required for cisplatin-induced processing of the mitochondrial fusion protein L-Opa1 that is mediated by the mitochondrial metallopeptidase Oma1 in gynecologic cancers. J Biol Chem 289:27134–27145
92. Kohler A, Chen B, Gemignani F, Elisei R, Romei C, Figlioli G, Cipollini M, Cristaudo A, Bambi F, Hoffmann P, Herms S, Kalemba M, Kula D, Harris S, Broderick P, Houlston R, Pastor S, Marcos R, Velazquez A, Jarzab B, Hemminki K, Landi S, Forsti A (2013) Genome-wide association study on differentiated thyroid cancer. J Clin Endocrinol Metab 98:E1674–E1681
93. Cole A, Wang Z, Coyaud E, Voisin V, Gronda M, Jitkova Y, Mattson R, Hurren R, Babovic S, Maclean N, Restall I, Wang X, Jeyaraju DV, Sukhai MA, Prabha S, Bashir S, Ramakrishnan A, Leung E, Qia YH, Zhang N, Combes KR, Ketela T, Lin F, Houry WA, Aman A, Al-Awar R, Zheng W, Wienholds E, Xu CJ, Dick J, Wang JC, Moffat J, Minden MD, Eaves CJ, Bader GD, Hao Z, Kornblau SM, Raught B, Schimmer AD (2015) Inhibition of the mitochondrial protease ClpP as a therapeutic strategy for human acute myeloid leukemia. Cancer Cell 27:864–876
94. Zhang Y, Maurizi MR (2016) Mitochondrial ClpP activity is required for cisplatin resistance in human cells. Biochim Biophys Acta 1862:252–264
95. Pinti M, Gibellini L, Liu Y, Xu S, Lu B, Cossarizza A (2015) Mitochondrial Lon protease at the crossroads of oxidative stress, ageing and cancer. Cell Mol Life Sci 72:4807–4824
96. Nie X, Li M, Lu B, Zhang Y, Lan L, Chen L, Lu J (2013) Down-regulating overexpressed human Lon in cervical cancer suppresses cell proliferation and bioenergetics. PLoS One 8:e81084
97. Cheng CW, Kuo CY, Fan CC, Fang WC, Jiang SS, Lo YK, Wang TY, Kao MC, Lee AY (2013) Overexpression of Lon contributes to survival and aggressive phenotype of cancer cells through mitochondrial complex I-mediated generation of reactive oxygen species. Cell Death Dis 4:e681
98. Gibellini L, Pinti M, Boraldi F, Giorgio V, Bernardi P, Bartolomeo R, Nasi M, De Biasi S, Missiroli S, Carnevale G, Losi L, Tesei A, Pinton P, Quaglino D, Cossarizza A (2014) Silencing of mitochondrial Lon protease deeply impairs mitochondrial proteome and function in colon cancer cells. FASEB J 28(12):5122–5135
99. Bayot A, Gareil M, Chavatte L, Hamon MP, L'Hermitte-Stead C, Beaumatin F, Priault M, Rustin P, Lombes A, Friguet B, Bulteau AL (2014) Effect of Lon protease knockdown on mitochondrial function in HeLa cells. Biochimie 100:38–47
100. Bernstein SH, Venkatesh S, Li M, Lee J, Lu B, Hilchey SP, Morse KM, Metcalfe HM, Skalska J, Andreeff M, Brookes PS, Suzuki CK (2012) The mitochondrial ATP-dependent Lon protease: a novel target in lymphoma death mediated by the synthetic triterpenoid CDDO and its derivatives. Blood 119:3321–3329
101. Quirós PM, Bárcena C, López-Otín C (2014) Lon protease: a key enzyme controlling mitochondrial bioenergetics in cancer. Mol Cell Oncol 1(4):e968505
102. Gibellini L, Pinti M, Bartolomeo R, De Biasi S, Cormio A, Musicco C, Carnevale G, Pecorini S, Nasi M, De Pol A, Cossarizza A (2015) Inhibition of Lon protease by triterpenoids alters mitochondria and is associated to cell death in human cancer cells. Oncotarget 6:25466–25483

103. Kao TY, Chiu YC, Fang WC, Cheng CW, Kuo CY, Juan HF, Wu SH, Lee AY (2015) Mitochondrial Lon regulates apoptosis through the association with Hsp60-mtHsp70 complex. Cell Death Dis 6:e1642
104. Bonifati V, Rizzu P, van Baren MJ, Schaap O, Breedveld GJ, Krieger E, Dekker MC, Squitieri F, Ibanez P, Joosse M, van Dongen JW, Vanacore N, van Swieten JC, Brice A, Meco G, van Duijn CM, Oostra BA, Heutink P (2003) Mutations in the DJ-1 gene associated with autosomal recessive early-onset parkinsonism. Science 299:256–259
105. Strauss KM, Martins LM, Plun-Favreau H, Marx FP, Kautzmann S, Berg D, Gasser T, Wszolek Z, Muller T, Bornemann A, Wolburg H, Downward J, Riess O, Schulz JB, Kruger R (2005) Loss of function mutations in the gene encoding Omi/HtrA2 in Parkinson's disease. Hum Mol Genet 14:2099–2111
106. Unal Gulsuner H, Gulsuner S, Mercan FN, Onat OE, Walsh T, Shahin H, Lee MK, Dogu O, Kansu T, Topaloglu H, Elibol B, Akbostanci C, King MC, Ozcelik T, Tekinay AB (2014) Mitochondrial serine protease HTRA2 p.G399S in a kindred with essential tremor and Parkinson disease. Proc Natl Acad Sci U S A 111:18285–18290
107. Jones JM, Datta P, Srinivasula SM, Ji W, Gupta S, Zhang Z, Davies E, Hajnoczky G, Saunders TL, Van Keuren ML, Fernandes-Alnemri T, Meisler MH, Alnemri ES (2003) Loss of Omi mitochondrial protease activity causes the neuromuscular disorder of mnd2 mutant mice. Nature 425:721–727
108. Martins LM, Morrison A, Klupsch K, Fedele V, Moisoi N, Teismann P, Abuin A, Grau E, Geppert M, Livi GP, Creasy CL, Martin A, Hargreaves I, Heales SJ, Okada H, Brandner S, Schulz JB, Mak T, Downward J (2004) Neuroprotective role of the reaper-related serine protease HtrA2/Omi revealed by targeted deletion in mice. Mol Cell Biol 24:9848–9862
109. Shi G, Lee JR, Grimes DA, Racacho L, Ye D, Yang H, Ross OA, Farrer M, McQuibban GA, Bulman DE (2011) Functional alteration of PARL contributes to mitochondrial dysregulation in Parkinson's disease. Hum Mol Genet 20:1966–1974
110. Yoshioka H, Katsu M, Sakata H, Okami N, Wakai T, Kinouchi H, Chan PH (2013) The role of PARL and HtrA2 in striatal neuronal injury after transient global cerebral ischemia. J Cereb Blood Flow Metab 33:1658–1665
111. Bertelsen B, Melchior L, Jensen LR, Groth C, Glenthoj B, Rizzo R, Debes NM, Skov L, Brondum-Nielsen K, Paschou P, Silahtaroglu A, Tumer Z (2014) Intragenic deletions affecting two alternative transcripts of the IMMP2L gene in patients with Tourette syndrome. Eur J Hum Genet 22:1283–1289
112. Casey JP, Magalhaes T, Conroy JM, Regan R, Shah N, Anney R, Shields DC, Abrahams BS, Almeida J, Bacchelli E, Bailey AJ, Baird G, Battaglia A, Berney T, Bolshakova N, Bolton PF, Bourgeron T, Brennan S, Cali P, Correia C, Corsello C, Coutanche M, Dawson G, de Jonge M, Delorme R, Duketis E, Duque F, Estes A, Farrar P, Fernandez BA, Folstein SE, Foley S, Fombonne E, Freitag CM, Gilbert J, Gillberg C, Glessner JT, Green J, Guter SJ, Hakonarson H, Holt R, Hughes G, Hus V, Igliozzi R, Kim C, Klauck SM, Kolevzon A, Lamb JA, Leboyer M, Le Couteur A, Leventhal BL, Lord C, Lund SC, Maestrini E, Mantoulan C, Marshall CR, McConachie H, McDougle CJ, McGrath J, McMahon WM, Merikangas A, Miller J, Minopoli F, Mirza GK, Munson J, Nelson SF, Nygren G, Oliveira G, Pagnamenta AT, Papanikolaou K, Parr JR, Parrini B, Pickles A, Pinto D, Piven J, Posey DJ, Poustka A, Poustka F, Ragoussis J, Roge B, Rutter ML, Sequeira AF, Soorya L, Sousa I, Sykes N, Stoppioni V, Tancredi R, Tauber M, Thompson AP, Thomson S, Tsiantis J, Van Engeland H, Vincent JB, Volkmar F, Vorstman JA, Wallace S, Wang K, Wassink TH, White K, Wing K, Wittemeyer K, Yaspan BL, Zwaigenbaum L, Betancur C, Buxbaum JD, Cantor RM, Cook EH, Coon H, Cuccaro ML, Geschwind DH, Haines JL, Hallmayer J, Monaco AP, Nurnberger JI Jr, Pericak-Vance MA, Schellenberg GD, Scherer SW, Sutcliffe JS, Szatmari P, Vieland VJ, Wijsman EM, Green A, Gill M, Gallagher L, Vicente A, Ennis S (2012) A novel approach of homozygous haplotype sharing identifies candidate genes in autism spectrum disorder. Hum Genet 131:565–579

113. Gimelli S, Capra V, Di Rocco M, Leoni M, Mirabelli-Badenier M, Schiaffino MC, Fiorio P, Cuoco C, Gimelli G, Tassano E (2014) Interstitial 7q31.1 copy number variations disrupting IMMP2L gene are associated with a wide spectrum of neurodevelopmental disorders. Mol Cytogenet 7:54
114. Di Bella D, Lazzaro F, Brusco A, Plumari M, Battaglia G, Pastore A, Finardi A, Cagnoli C, Tempia F, Frontali M, Veneziano L, Sacco T, Boda E, Brussino A, Bonn F, Castellotti B, Baratta S, Mariotti C, Gellera C, Fracasso V, Magri S, Langer T, Plevani P, Di Donato S, Muzi-Falconi M, Taroni F (2010) Mutations in the mitochondrial protease gene AFG3L2 cause dominant hereditary ataxia SCA28. Nat Genet 42:313–321
115. Pierson TM, Adams D, Bonn F, Martinelli P, Cherukuri PF, Teer JK, Hansen NF, Cruz P, Mullikin For The Nisc Comparative Sequencing Program JC, Blakesley RW, Golas G, Kwan J, Sandler A, Fuentes Fajardo K, Markello T, Tifft C, Blackstone C, Rugarli EI, Langer T, Gahl WA, Toro C (2011) Whole-exome sequencing identifies homozygous AFG3L2 mutations in a spastic ataxia-neuropathy syndrome linked to mitochondrial m-AAA proteases. PLoS Genet 7(10):e1002325
116. Casari G, De Fusco M, Ciarmatori S, Zeviani M, Mora M, Fernandez P, De Michele G, Filla A, Cocozza S, Marconi R, Durr A, Fontaine B, Ballabio A (1998) Spastic paraplegia and OXPHOS impairment caused by mutations in paraplegin, a nuclear-encoded mitochondrial metalloprotease. Cell 93:973–983
117. Pfeffer G, Gorman GS, Griffin H, Kurzawa-Akanbi M, Blakely EL, Wilson I, Sitarz K, Moore D, Murphy JL, Alston CL, Pyle A, Coxhead J, Payne B, Gorrie GH, Longman C, Hadjivassiliou M, McConville J, Dick D, Imam I, Hilton D, Norwood F, Baker MR, Jaiser SR, Yu-Wai-Man P, Farrell M, McCarthy A, Lynch T, McFarland R, Schaefer AM, Turnbull DM, Horvath R, Taylor RW, Chinnery PF (2014) Mutations in the SPG7 gene cause chronic progressive external ophthalmoplegia through disordered mitochondrial DNA maintenance. Brain 137:1323–1336
118. Wedding IM, Koht J, Tran GT, Misceo D, Selmer KK, Holmgren A, Frengen E, Bindoff L, Tallaksen CM, Tzoulis C (2014) Spastic paraplegia type 7 is associated with multiple mitochondrial DNA deletions. PLoS One 9:e86340
119. Ferreirinha F, Quattrini A, Pirozzi M, Valsecchi V, Dina G, Broccoli V, Auricchio A, Piemonte F, Tozzi G, Gaeta L, Casari G, Ballabio A, Rugarli EI (2004) Axonal degeneration in paraplegin-deficient mice is associated with abnormal mitochondria and impairment of axonal transport. J Clin Invest 113:231–242
120. Maltecca F, Aghaie A, Schroeder DG, Cassina L, Taylor BA, Phillips SJ, Malaguti M, Previtali S, Guenet JL, Quattrini A, Cox GA, Casari G (2008) The mitochondrial protease AFG3L2 is essential for axonal development. J Neurosci 28:2827–2836
121. Martinelli P, La Mattina V, Bernacchia A, Magnoni R, Cerri F, Cox G, Quattrini A, Casari G, Rugarli EI (2009) Genetic interaction between the m-AAA protease isoenzymes reveals novel roles in cerebellar degeneration. Hum Mol Genet 18:2001–2013
122. Kondadi AK, Wang S, Montagner S, Kladt N, Korwitz A, Martinelli P, Herholz D, Baker MJ, Schauss AC, Langer T, Rugarli EI (2014) Loss of the m-AAA protease subunit AFG(3)L(2) causes mitochondrial transport defects and tau hyperphosphorylation. EMBO J 33:1011–1026
123. Almajan ER, Richter R, Paeger L, Martinelli P, Barth E, Decker T, Larsson NG, Kloppenburg P, Langer T, Rugarli EI (2012) AFG3L2 supports mitochondrial protein synthesis and Purkinje cell survival. J Clin Invest 122:4048–4058
124. Maltecca F, Magnoni R, Cerri F, Cox GA, Quattrini A, Casari G (2009) Haploinsufficiency of AFG3L2, the gene responsible for spinocerebellar ataxia type 28, causes mitochondria-mediated Purkinje cell dark degeneration. J Neurosci 29:9244–9254
125. Maltecca F, Baseggio E, Consolato F, Mazza D, Podini P, Young SM Jr, Drago I, Bahr BA, Puliti A, Codazzi F, Quattrini A, Casari G (2014) Purkinje neuron Ca2+ influx reduction rescues ataxia in SCA28 model. J Clin Invest 125(1):263–274

126. Almontashiri NA, Chen HH, Mailloux RJ, Tatsuta T, Teng AC, Mahmoud AB, Ho T, Stewart NA, Rippstein P, Harper ME, Roberts R, Willenborg C, Erdmann J, Consortium CA, Pastore A, McBride HM, Langer T, Stewart AF (2014) SPG7 variant escapes phosphorylation-regulated processing by AFG3L2, elevates mitochondrial ROS, and is associated with multiple clinical phenotypes. Cell Rep 7:834–847
127. Civitarese AE, MacLean PS, Carling S, Kerr-Bayles L, McMillan RP, Pierce A, Becker TC, Moro C, Finlayson J, Lefort N, Newgard CB, Mandarino L, Cefalu W, Walder K, Collier GR, Hulver MW, Smith SR, Ravussin E (2010) Regulation of skeletal muscle oxidative capacity and insulin signaling by the mitochondrial rhomboid protease PARL. Cell Metab 11:412–426
128. Wai T, Garcia-Prieto J, Baker MJ, Merkwirth C, Benit P, Rustin P, Ruperez FJ, Barbas C, Ibanez B, Langer T (2015) Imbalanced OPA1 processing and mitochondrial fragmentation cause heart failure in mice. Science 350:aad0116
129. Cavalcanti DM, Castro LM, Rosa Neto JC, Seelaender M, Neves RX, Oliveira V, Forti FL, Iwai LK, Gozzo FC, Todiras M, Schadock I, Barros CC, Bader M, Ferro ES (2014) Neurolysin knockout mice generation and initial phenotype characterization. J Biol Chem 289:15426–15440
130. Chen Y, Zhu J, Lum PY, Yang X, Pinto S, MacNeil DJ, Zhang C, Lamb J, Edwards S, Sieberts SK, Leonardson A, Castellini LW, Wang S, Champy MF, Zhang B, Emilsson V, Doss S, Ghazalpour A, Horvath S, Drake TA, Lusis AJ, Schadt EE (2008) Variations in DNA elucidate molecular networks that cause disease. Nature 452:429–435
131. O'Toole JF, Liu Y, Davis EE, Westlake CJ, Attanasio M, Otto EA, Seelow D, Nurnberg G, Becker C, Nuutinen M, Karppa M, Ignatius J, Uusimaa J, Pakanen S, Jaakkola E, van den Heuvel LP, Fehrenbach H, Wiggins R, Goyal M, Zhou W, Wolf MT, Wise E, Helou J, Allen SJ, Murga-Zamalloa CA, Ashraf S, Chaki M, Heeringa S, Chernin G, Hoskins BE, Chaib H, Gleeson J, Kusakabe T, Suzuki T, Isaac RE, Quarmby LM, Tennant B, Fujioka H, Tuominen H, Hassinen I, Lohi H, van Houten JL, Rotig A, Sayer JA, Rolinski B, Freisinger P, Madhavan SM, Herzer M, Madignier F, Prokisch H, Nurnberg P, Jackson PK, Khanna H, Katsanis N, Hildebrandt F (2010) Individuals with mutations in XPNPEP3, which encodes a mitochondrial protein, develop a nephronophthisis-like nephropathy. J Clin Invest 120:791–802
132. Jenkinson EM, Rehman AU, Walsh T, Clayton-Smith J, Lee K, Morell RJ, Drummond MC, Khan SN, Naeem MA, Rauf B, Billington N, Schultz JM, Urquhart JE, Lee MK, Berry A, Hanley NA, Mehta S, Cilliers D, Clayton PE, Kingston H, Smith MJ, Warner TT, Black GC, Trump D, Davis JR, Ahmad W, Leal SM, Riazuddin S, King MC, Friedman TB, Newman WG (2013) Perrault syndrome is caused by recessive mutations in CLPP, encoding a mitochondrial ATP-dependent chambered protease. Am J Hum Genet 92:605–613
133. Strauss KA, Jinks RN, Puffenberger EG, Venkatesh S, Singh K, Cheng I, Mikita N, Thilagavathi J, Lee J, Sarafianos S, Benkert A, Koehler A, Zhu A, Trovillion V, McGlincy M, Morlet T, Deardorff M, Innes AM, Prasad C, Chudley AE, Lee IN, Suzuki CK (2015) CODAS syndrome is associated with mutations of LONP1, encoding mitochondrial AAA(+) Lon protease. Am J Hum Genet 96:121–135
134. Dikoglu E, Alfaiz A, Gorna M, Bertola D, Chae JH, Cho TJ, Derbent M, Alanay Y, Guran T, Kim OH, Llerenar JC Jr, Yamamoto G, Superti-Furga G, Reymond A, Xenarios I, Stevenson B, Campos-Xavier B, Bonafe L, Superti-Furga A, Unger S (2015) Mutations in LONP1, a mitochondrial matrix protease, cause CODAS syndrome. Am J Med Genet A 167:1501–1509

The Role of Matrix Metalloproteinase-2 and Metalloproteinase-9 in Embryonic Neural Crest Cells and Their Derivatives

2

Rotem Kalev-Altman, Efrat Monsonego-Ornan, and Dalit Sela-Donenfeld

Abstract

Neural crest cells (NCCs) are transient cell populations that are initially residing at the dorsal-most part of the neural tube of the developing vertebrate embryo. At well-defined time points, NCCs detach from the neural tube as they undergo epithelial-to-mesenchymal transition (EMT) and migrate in distinct pathways to their final destinations. There, this unique cell population differentiates into a great variety of cell types including bone and cartilage tissues of the head and face, connective tissue of the heart, skin melanocytes, adipocytes, enteric neurons, and most of the peripheral sensory neurons, glia, and Schwann cells. Matrix metalloproteinases (MMPs) are a large family of matrix-degrading enzymes, which are divided into several subfamilies according to their structure and substrate specificity. The gelatinases subfamily, which includes MMP-2 and MMP-9 solely, is the most investigated group. Both MMP-2 and MMP-9 were previously reported to be expressed in embryonic NCCs and to have a role in their EMT and

R. Kalev-Altman
Koret School of Veterinary Medicine, The Robert H. Smith Faculty of Agriculture, Food and Environment, The Hebrew University of Jerusalem, Rehovot 76100, Israel

The Institute of Biochemistry and Nutrition, The Robert H. Smith Faculty of Agriculture, Food and Environment, The Hebrew University of Jerusalem, P.O.B 12, Rehovot 76100, Israel

E. Monsonego-Ornan (✉)
The Institute of Biochemistry and Nutrition, The Robert H. Smith Faculty of Agriculture, Food and Environment, The Hebrew University of Jerusalem, P.O.B 12, Rehovot 76100, Israel
e-mail: efrat.mo@mail.huji.ac.il

D. Sela-Donenfeld (✉)
Koret School of Veterinary Medicine, The Robert H. Smith Faculty of Agriculture, Food and Environment, The Hebrew University of Jerusalem, Rehovot 76100, Israel
e-mail: dalit.seladon@mail.huji.ac.il

S. Chakraborti, N.S. Dhalla (eds.), *Proteases in Physiology and Pathology*, DOI 10.1007/978-981-10-2513-6_2

migration processes. In this review we present the most recent data regarding the role of MMP-2 and MMP-9 in embryonic NCCs and in their various derivatives in later embryonic stages and in adults.

Keywords
Neural crest cells • Matrix metalloproteinases • MMP-2 • MMP-9 • Gelatinases • Osteoblasts • Chondrocytes • Melanocytes • Adipocytes • Enteric nervous system • Glial cells • Schwann cells • Embryo • Epithelial-to-mesenchymal transition

2.1 Introduction

Neural crest cells (NCCs) are a transient cell population that resides at the dorsal-most part of the neural tube (NT) of the developing vertebrate embryo. The NCCs detach from the neuroepithelium in a gradual synchronized manner from head to tail and migrate in distinct well-characterized pathways throughout the embryonic axis to their final destinations [1–3]. There, this unique cell population differentiates into a great variety of cell types including most of the peripheral sensory neurons, glia and Schwann cells, smooth muscle cells, connective tissue of the heart, skin melanocytes, and the secreting cells of several endocrine tissues as the thymus, thyroid, and adrenal glands [2, 4, 5]. Moreover, cranial NCCs contribute also to the majority of bone and cartilage tissues of the head and face such as the frontal, nasal, premaxillary, maxillary, mandibular, and temporal bones; Meckel's cartilage; jaw, ear, and hyoid cartilages; and tooth dentin [4, 6–10].

Due to their fundamental contribution to this vast number of cell types and organs, their appropriate formation and migration are crucial for normal embryonic development. Defects in these processes will lead to various birth disorders and pathologies such as different heart defects, albinism, malformation of craniofacial structures including cleft palate/lip, mal-innervation of the digestive tract (which leads to Hirschsprung's disease), and several other congenital syndromes like DiGeorge, Treacher Collins, and Waardenburg [5, 9]. Furthermore, some of the most aggressive tumors derive from NCC origin (i.e., glioblastoma, melanoma, cranial chondrosarcoma); as such, the process of NCC migration together with their ability to differentiate into many derivatives is in common with metastatic cancer cells [11, 12].

In order to engage in migration, NCCs undergo epithelial-to-mesenchymal transition (EMT) after which they acquire motility. EMT is a major process involved in many physiological and pathological conditions in embryos and adults and requires fundamental changes in the cells and their surrounding environment, such as breakage of cell-cell interactions, rearrangement of cell cytoskeleton, and remodeling of the extracellular matrix (ECM) [13–15]. Therefore, EMT is mandatory for NCCs in order to transform from being immotile neuroectodermal cells with typical cell-cell and cell-matrix connections to mesenchymal-separated cells with the ability to migrate throughout various stereotypic pathways. The NCC EMT process involves degradation of the basal membrane around the dorsal NT, rearrangement of their cytoskeleton (i.e., via the modulation of RhoA/Rho-kinase, calponin, alpha-catenin), loss of adhesion molecules, and remodeling of the ECM [2, 16–22]. Major signaling factors such as BMP and Wnt, as well as several transcriptional regulators like Snail and Sox9, were also found to act as regulators of NCC EMT [23–30]. However, the putative executers of these processes to trigger the cell separation and migration are still much less investigated.

2.2 Matrix Metalloproteinases

Matrix metalloproteinases (MMPs) are a family of twenty eight currently known secreted or membrane-bound enzymes with an ability to degrade different components of the ECM and basal laminae of diverse cell types. These soluble proteases are mainly secreted in their latent form as proenzymes or inactive zymogens that require proteolytic activation by different proteases, including MMPs. They are active in neutral pH and require zinc and calcium ions for their proteolytic function. MMP's activity is also closely regulated by several endogenous inhibitors. In tissue fluids, the main MMP inhibitor is α2-macroglobulin, a large serum protein, which binds MMPs and creates an irreversibly complex. However, the most thoroughly studied MMP inhibitors are the tissue inhibitors of MMPs (TIMPs), small molecules of 21–28 kDa which bind MMPs and reversibly block their activity. TIMPs are anchored to the extracellular matrix or secreted extracellularly, and thus far, four human TIMPs have been characterized, which differentially bind and influence the various MMPs [31–37].

The MMP family can be divided into five subgroups according to their structure and main, but not exclusive, substrates: (1) matrilysins, which degrade non-collagen components in the ECM, such as fibronectin, laminin, and proteoglycans; (2) collagenases and stromelysins, which mainly degrade a number of collagen types; (3) gelatinases, comprise of MMP-2 and MMP-9, which degrade with high-affinity gelatin, the product of collagen degradation; (4) membrane-type MMPs (MT-MMPs), collagenases that differ from the other groups by being anchored to the cell membrane via a transmembrane domain; and (5) other MMPs, which do not fit to either group and are known to degrade several different substrates [31–34].

The structure of MMPs includes four domains: pre-domain, pro-domain, catalytic domain, and hemopexin domain. The pre-domain is located in the N-terminus of the MMP protein and responsible for targeting the protein to the endoplasmic reticulum in order to be secreted. The pro-domain has a conserved sequence which includes several cysteine residues that connect to the zinc ion which prevents catalytic activity; this domain is being cleaved after the secretion of MMPs for their activation. Activation of the zymogen is usually mediated by intracellular furin-like proteinases that target the furin recognition motif between the pro-domain and the catalytic domain. The catalytic domain binds three ions: a zinc and a calcium ion which stabilize its structure and a second zinc ion that is needed for the enzyme activity.

Specifically, the gelatinase catalytic site has a fibronectin domain which improves the degradation of denatured collagen (gelatin). Finally, the hemopexin domain, found in the c-terminal of MMPs, is attached through the hinge region to the catalytic domain and is folded to a structure of four-bladed β propeller which is involved in substrate recognition and TIMP binding. MT-MMPs (MMP-14, MMP-15, MMP-16, and MMP-24) have an additional domain that anchors them to the cell membrane as a transmembrane domain or glycosylphosphatidylinositol (GPI) [32, 33, 38].

Members of the MMP family are implicated in many physiological processes which involve proteolytic cleavage or remodeling of the ECM during embryogenesis, angiogenesis, and tissue repair, via their involvement in cell proliferation, differentiation, apoptosis, and migration [39, 40]. They also control key reproductive events such as ovulation, embryo implantation, breast involution, and more [32, 41]. On the other hand, imbalance between the expression/activity of MMPs and TIMPs is linked to different pathological conditions such as in tumor invasion and metastasis, arthritis, nephritis, fibrosis, endometriosis, and diseases such as multiple sclerosis and Alzheimer's [41–44].

2.3 MMP-2 and MMP-9 in the Onset of NCC Migration

The gelatinase subfamily of MMPs is composed of two members: MMP-2, a 72-kDa protein also known as gelatinase A, and MMP-9, a 92-kDa protein also known as gelatinase B. MMP-2 and MMP-9 are implicated in hydrolysis of gelatin and collagen type IV as well as in the cleavage of other ECM proteins such as elastin, laminin, fibronectin, and aggrecan [10, 45, 46]. Both gelatinases are secreted as zymogens and known to be processed extracellularly into their active form.

Studies from several labs, including ours, have revealed that MMP-2 and MMP-9 are required for the EMT process and migration onset of NCCs. A previous study from Duong and Erickson had shown that in the chick embryo, MMP-2 is expressed in the dorsal neuroepithelium at the time when trunk NCCs begin to undergo EMT, suggesting that MMP-2 is involved in their detachment [47]. To assess whether MMP-2 is necessary for this process, the researchers have used a general pharmacological inhibitor of many MMPs or a specific antisense morpholino oligomer designed to prevent the translation of MMP-2; both treatments prevented NCC dispersion from the trunk level of the NT, in *ex-vivo* explants and in the embryo *in-vivo*. However, none of these treatments affected the motility of the NCCs once they have undergone EMT, indicating that MMP-2 has a restricted role during the onset of NCC migration, while they detach from the neuroepithelium and undergo EMT [47]. Whether MMP-2 is also implicated in the migration of cranial NCCs was not addressed in this work, but our lab recently found that indeed cranial chick NCCs express MMP-2 during their detachment from the NT and early migration (Fig. 2.1A, A′).

Recent work from our lab has investigated the role of the other gelatinase, MMP-9, in promoting NCC delamination and migration, using the chick embryo as a model system [48]. In this study, MMP-9 was found to be expressed in NCCs prior to,

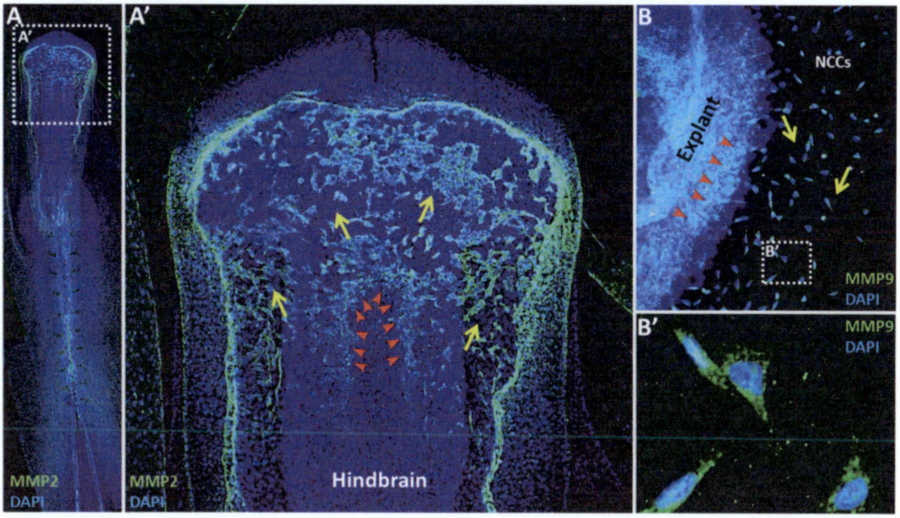

Fig. 2.1 The expression of MMP-2 and MMP-9 in migrating chick NCCs. (**A**, **A′**) Immunofluorescence labeling of MMP-2 protein in the cranial level of a 12-somite-stage whole-mount chick embryo, using MMP-2 antibody (*green*); MMP-2 is expressed in actively migrating NCCs as they detach from the NT (*arrowheads*) and during their migration (*arrows*). *Panel* **A′** is an enlargement of the boxed areas in *panel* **A**. (**B**) An *ex-vivo* explant of NCCs obtained from the hindbrain of 6–8-somite-stage embryo showing immunofluorescence labeling of MMP-9 protein, using MMP-9 antibody (*green*); MMP-9 is expressed in actively migrating NCCs as they detach from the explanted NT (*arrowheads*) and also during their migration as separated mesenchymal cells (*arrows*). *Panel* **B′** is an enlargement of the boxed areas in *panel* **B**. Blue staining (DAPI) represents cell nuclei

and during, their migration in the cranial and trunk axis levels (Fig. 2.1B, B′). The role of MMP-9 in executing NCC EMT and further migration was examined using loss and gain-of-function methods; addition of a specific MMP-9 pharmacological inhibitor or an antisense morpholino oligomer against MMP-9 *mRNA* inhibited the EMT and migration processes of NCCs, in explants and *in-vivo*. In contrast, addition of exogenous MMP-9 to NCC progenitors was sufficient to promote their migration ahead of time, as well as to rescue the effect of the MMP-9 inhibition on NCC migration. By analyzing possible substrates for MMP-9, our findings strongly suggested that MMP-9 executes NCC migration by degrading N-cadherin and laminin, two adhesion proteins, and that their degradation and/or downregulation in cell-cell and cell-basal lamina interactions, respectively, has been demonstrated to be necessary for NCC EMT [21, 48–50]. Together, these studies have shown that both gelatinases MMP-2 and MMP-9 play a role in executing NCC migration in the chick embryo.

Albeit the demonstrated role of MMP-2 and MMP-9 in early stages of NCC development in the chick embryo, their expression pattern or function in NCCs of other species, such as mouse, frog (*Xenopus laevis*), or teleost fish (*Zebra fish*), was not yet reported. Notably, in the zebra fish embryo, MMP-2 was found to be expressed from as early as one-cell stage through at least the 72 first hours of development, and its inhibition via injection of MMP-2 antisense morpholino oligomer greatly impaired zebra fish development [51]. In agreement with these data, Keow and colleagues have used a probe to tag the active form of MMP-2 and demonstrated the distribution of active MMP-2 in the developing zebra fish embryo [52]. The localization patterns of active MMP-2 that was presented in this study seemed reminiscent of the migration streams of NCCs. Therefore, similarly to the chick embryo, a role for MMP-2 in executing zebra fish NCC migration is possible although not directly tested. Furthermore, MMP-9 was also found to be expressed in the zebra fish embryo, throughout the first 5 days postfertilization [53]. In this study, the researchers showed that in 10-somite-stage old embryo, MMP-9 expression was localized to the anterior midline region of the embryonic mesoderm, whereas later on, MMP-9 was expressed around the head and eyes. These patterns of MMP-9 may implicate for its expression and role in the different stages of zebra fish NCC development.

2.4 MMP-2 and MMP-9 in NCC Derivatives in Embryos and Adults

Both MMP-2 and MMP-9 have been identified and characterized in different NCC derivatives including craniofacial cartilage and bone, cardiac cushion cells which contribute to the heart valves and septum, melanocytes, adipocytes, enteric neurons, and different neural supporting cells such as glia and Schwann cells. We present here the main findings according to gelatinases' tissue localization and function.

2.4.1 Cranial Cartilage and Bone

NCCs from the cranial axial level generate the majority of bone and cartilage craniofacial tissues including tooth dentin. The cranial NCC population migrates from different regions of the developing brain: the forebrain, midbrain, and hindbrain domains. NCCs that arise from the forebrain and rostral midbrain colonize the frontonasal and periocular regions, while caudal midbrain-derived NCCs populate the maxillary component of the first pharyngeal arch. Collectively, these NCCs give rise to the upper jaw, palatal mesenchyme, and extrinsic ocular muscles. The hindbrain is transiently partitioned into seven segments called rhombomeres. NCCs from these rhombomeres migrate in discrete segregated streams into the first through sixth pharyngeal arches, where they form the jaw, middle ear, hyoid, and thyroid cartilages [6, 8, 54, 55]. Different markers and signaling transducers were reported to be involved in the specification of cranial NCCs. This includes Sox10, Notch, Neuregulin1, BMP2, Neurogenin2, Hox genes, and more [56, 57]. For example, Hoxa2 was found to be essential for proper patterning of structures derived from the cranial NC since targeted inactivation of Hoxa2 resulted in lethality at birth and malformations of craniofacial elements derived from cranial NC [58–60].

Due to their important contribution to most of the craniofacial structures, most congenital craniofacial anomalies arise due to defects in NCC formation, proliferation, migration, and/or differentiation. For instance, inappropriate formation or migration of cranial NCCs has been described in detail as a major cause for Treacher Collins syndrome (TCS) [9, 61]. Cell-lineage tracing performed in mouse models of TCS revealed no migratory defects in cranial NCC migration [62]. Yet, fewer migrating NCCs were observed in TCS embryos compared with their wild-type littermates. This deficiency in NCC number resulted from extensive neuroepithelial cell apoptosis combined with reduced proliferation rate of the progenitor and migrating NCCs. Indeed, the protein treacle was found to play key role in ribosome maturation and therefore regulates NCC survival and proliferation [62, 63]. It was also demonstrated that the reduced proliferation rate results in nucleolar stress activation of p53 which in turn transcriptionally activates numerous pro-apoptotic effectors, which collectively are responsible for the high levels of NCC death observed in the pathogenesis of TCS [64].

In concomitant with the expression of MMP-2 and MMP-9 in early migrating NCCs, these gelatinases were also found in cranial NCC derivatives as craniofacial bones and cartilage. For example, MMP-2 and its activator MMP-14 (which will be discussed later), but not MMP-9, were found to be co-localized and to display gelatinolytic activity in the mesenchyme surrounding the palatal folds in E13.5–15.5 mouse embryos [65]. The results of this study suggested that MMP-2 has an important role not only in the process of palatal shelf fusion but also in their elevation, to create eventually normal palate. In addition, Chin and Werb have described previously that MMP-2 mRNA is strongly expressed in Meckel's cartilage and in mesenchymal areas of mandibles in E13–15 mouse embryos, whereas MMP-9 together with TIMP-1 and TIMP-2 was found mostly in the ossifying areas of the mandibles [66]. In line with these MMP-2 roles, defects in NCC development were shown to cause cleft lip/ plate [67–70]. Moreover, both MMP-2 and MMP-9 were recently found by our lab to be expressed in chondrocytes *in-vitro* and in rat chondrosarcoma cell line (Fig. 2.2A, B and C, D, respectively) [71, 72]. Yet, the direct link between NCCs and MMP-2 in this common congenital malformation awaits further research.

Furthermore, during mandibular arch development, MMP-9 was found in the cartilage tissues together with MMP-13 (a known activator of MMP-9), where a general inhibition of both resulted in dramatic defects in Meckel's cartilage [66, 73, 74]. Moreover, MMP-9 was reported to be expressed and to possibly have a role during rat odontogenesis, a process in which EMT, motility, and active remodeling of the dental primordia take place [75]. Notably, while cranial NCCs significantly contribute to tooth and mandible formation, these studies did not examine whether the described defects are linked to MMP's roles in NCCs.

Surprisingly, knockout (KO) mice for each gelatinase did not result in any marked cranial NCC-related defect [76, 77]. So far, most studies focused on the effect of the single KO of MMP-2 or MMP-9 on long-bone development in postnatal stages. For example, MMP-2-KO mice were reported to be significantly smaller at birth compared to control littermates and showed bone abnormalities such as loss

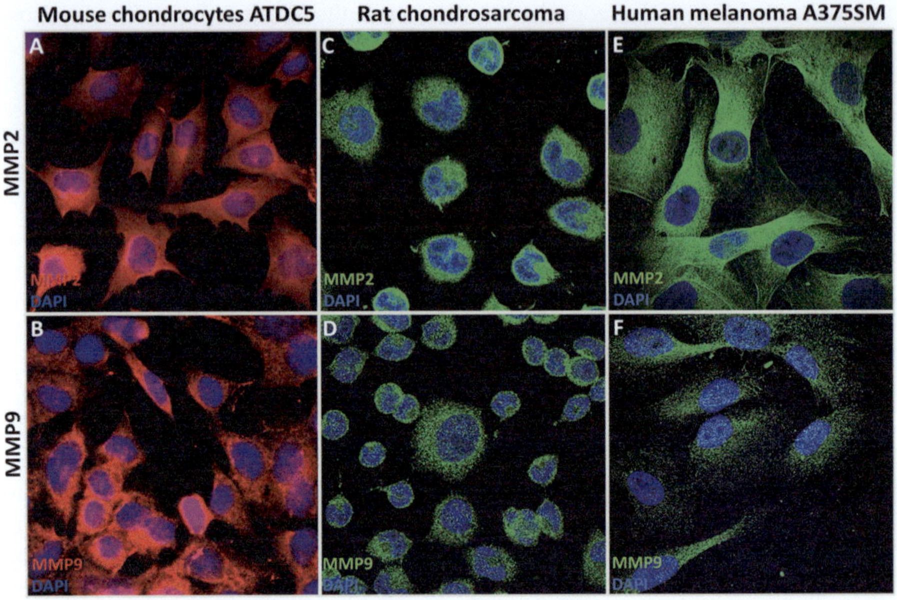

Fig. 2.2 The expression of MMP-2 and MMP-9 in mouse chondrocyte, rat chondrosarcoma, and human melanoma cell lines. Immunofluorescence labeling of MMP-2 and MMP-9 proteins using MMP-2 (**A**, **C**, **E**) or MMP-9 (**B**, **D**, **F**) antibodies, in mouse chondrocyte cell line ATDC5 (*red*) (**A**, **B**), rat chondrosarcoma cell line (*green*) (**C**, **D**), and human melanoma cell line A375SM (*green*) (**E**, **F**). Blue staining (DAPI) represents cell nuclei

of bone density. Yet, mild craniofacial defects have also been reported, such as shorter upper and lower jaws, nose, and overall skull length, which may imply on NCC developmental defects, although this has not been specifically examined [77, 78]. In support of this finding, a genetic missense mutation in the human MMP-2 gene has been described as causing inherited osteolytic/arthritic syndrome, which was suggested to be caused by the impairment of MMP-2 activity in bone and ECM remodeling [79]. Noticeably, patients with this syndrome suffer from several facial defects, which may also be due to the loss of MMP-2 in developing cranial NCCs. All of this data suggests a role of MMP-2 in the development of cranial NCCs. However, whether the MMP-2-KO has any effect on the early stages of NCC migration has not been examined yet.

In addition, Vu and coworkers found that MMP-9-KO mice presented shorter long bones, as the tibia and femurs, compared with those of wild-type littermates. These bones had a lengthened zone of hypertrophic cartilage in their growth plates due to inhibition of chondrocyte apoptosis which is normally regulated by MMP-9. Yet, similar to the MMP-2-KO mice, no clear cranial bone defect was reported [76]. One possible explanation for the normal NCC phenotypes in the mutants is the presence of the other gelatinase in NCC but not in the hypertrophic chondrocytes, which may compensate for the ablated gene and enables cell migration and ECM remodeling during NCC development. Therefore, future studies will be needed to investigate whether double KO for MMP-2 and MMP-9 genes affects cranial NCC migration and leads to major craniofacial defects in the mice. Surprisingly, such double KO line was not yet reported in the literature, which may indicate for its lethality.

The possible involvement of MMP-2 and MMP-9 in craniofacial development was also demonstrated in zebrafish embryos [80]. In this study, the researchers examined the mechanism by which antenatal usage of glucocorticoids (GC) induces teratogenic effects, as was previously described as causing fetal growth restriction and cleft palate [81]. Hillegass and colleagues have shown that GC treatment increases MMP-2, MMP-9, and MMP-13 mRNA levels and causes abnormal craniofacial development [80]. Although these studies have not directly examined the relation to NCC development, these data strongly suggest that GC-induced increase in MMPs affects cranial NCC ontogeny and therefore results in abnormal craniofacial development.

2.4.2 Heart

During cardiac development, NCCs that arise from the mid-otic up to the caudal end of the third somite level, termed "cardiac NCCs", migrate beneath the ectoderm into pharyngeal arches III, IV, and VI. Eventually, a subset of these cells invades the outflow tract of the developing heart and contributes to the formation of the heart valves and septum [5, 82–85]. The importance of NCCs to heart development was demonstrated in a series of studies in different embryonic models in which premigratory cardiac NCCs were ablated, resulting in a subsequent loss of NCC derivatives in the developing cardiovascular system leading to defects of the cardiac outflow tract, valve formation, and aortic arch arteries [86–90].

Yet, data on a possible link between cardiac NCC and MMP-2/MMP-9 activity is so far limited to avian embryos; MMP-2 mRNA was found to be expressed in cardiac NCCs of quail and chick embryos, and its activity levels were increased with the initiation of cushion NCC migration [91–93]. Moreover, Cai and Brauer have demonstrated how injection of a general MMP inhibitor (KB8301) to chick embryos at stages before cardiac NCCs begin to migrate caused major anomalies in cardiac morphogenesis, which was correlated with a decrease in MMP activity [94]. These data demonstrate that MMP-2 is expressed and acts during chick cardiac morphogenesis and suggests that perturbation of endogenous MMP activity may lead to NCC-related congenital defects. Further studies will be required in other embryonic models to analyze how conserved is the activity of MMP-2 in cardiac NCC, whether MMP-9 is also implicated in cardiac NCC development, and what is the mechanism of action of MMP-2 in heart morphogenesis.

2.4.3 Melanocytes

The NC is the major, if not the sole source of all pigment cells in the developing vertebrate, except those of the retina. The melanocyte progenitors reside in the ventral most part of the dorsal NT and are the last ones to emigrate in a dorsolateral migratory pathway [95, 96]. All pre-migratory NCCs in the dorsal NT express FoxD3, Sox10, and Pax3. The NCC-derived melanoblasts continue to express Sox10 and Pax3 during their delamination and migration, while they downregulate FoxD3 and upregulate Mitf when they exit the NT [2, 97, 98].

Many signaling cues and migration regulators have been reported to affect melanocyte specification and migration [95, 97, 99]. Yet, data regarding MMP-2/MMP-9 in embryonic melanocytes is limited. The expression of MMP-2 was so far reported in melanophores of frog embryos (*Xenopus laevis*), where it was also shown to promote melanophore migration [100]. This study used a chemical genomic screen and a functional approach to discover modulators of melanophore migration and identified one molecule, termed NSC 84093 that selectively inhibited melanophore migration via inhibition of MMP-2.

Much more knowledge exists regarding gelatinases in adult melanocytes. For instance, Simonetti et al. [101] analyzed the expression patterns of MMP-2 and MMP-9 in benign and invasive lesions of human melanomas. The results showed that invasive melanoma cells express MMP-2 and MMP-9 and that MMP-2 is present in both the melanoma cells and in the tumor-surrounding stromal or host cells [101]. In addition, MMP-9 was found to be variably expressed in the radial growth phase of primary melanoma cells, indicating that MMP-9 is correlated with early invasive stages of melanoma [102]. On the other hand, a different study argued that MMP-9 expression is absent during early stages of melanoma but present in advanced stages [103].

These data are in agreement with other studies which demonstrated MMP-2 and MMP-9 in melanoma cell lines or *in-vivo* [38, 104–106]. For example, both gelatinases were investigated for their involvement in canine cutaneous melanocytic tumors and were found to be expressed in both benign and malignant lesions. However, MMP-2 expression was significantly higher in benign melanocytic tumors than in malignant counterpart. In contrast, MMP-9 expression was elevated in malignant melanocytic tumors compared with benign tumors [38]. These results suggested that malignancy is associated with an increase of MMP-9 and a decrease of MMP-2 expression and that a switch in gelatinases' expression profile might occur during tumor progression, involving not only emergence of MMP expression but also its downregulation. Furthermore, our lab also found that both MMP-2 and MMP-9 are expressed in human melanoma cell line A375SM (Fig. 2.2E, F, respectively). The large amount of data regarding MMP-2 and MMP-9 in adult melanocytes as compared to embryos may result from a restricted activity of gelatinases only to pathological conditions. However, as melanoma invasion resembles in many instances to embryonic NCC migration [12], it is highly possible that the limited knowledge on expression/activity of gelatinases in embryonic melanocytes is due to insufficient research in this field.

2.4.4 Adipocytes

In contrast to other NCC derivatives, the adipocytes lineage of NCCs has not been carefully explored in the past. Billon et al. [107] have demonstrated that in the quail, both cephalic and trunk NCCs can differentiate into adipocytes *ex-vivo*, in an explant system [107]. These adipocytes had been shown to express the typical adipocyte differentiation markers such as CEBPα, PPARγ, and FABP4. Moreover, the researchers used Sox10-Cre transgenic mice to map NCC derivatives in order to genetically demonstrate that mice NCCs contribute to adipocytes also in normal development *in-vivo* [107].

Studies in adult tissues have shown that adipocytes express both gelatinases [108–111]. Human adipocytes and pre-adipocytes release MMP-2 and MMP-9, and this secretion is modulated during adipocyte differentiation. Moreover, inhibition of these MMP activities resulted in a blockage of adipocyte differentiation [108]. Secretion of both gelatinases was also increased during adipocyte differentiation of murine 3T3F442A pre-adipocyte cell line [108]. Differentiation of adipocytes leads to cell hyperplasia which is one of the changes in adipose tissue that leads to obesity, along with hypertrophy of the cells and increased angiogenesis [112, 113]. As different MMPs in general, and MMP-2 and MMP-9 in particular, have a central role in angiogenesis [76, 114–116], it was speculated that both gelatinases have a dual role in the development of adipose tissue and obesity by increasing the differentiation rate of pre-adipocytes to adipocytes on the one hand and by increasing angiogenesis on the other [108].

However, the role of MMP-2 in adipogenesis *in-vivo* remains elusive with controversial data. For instance, nude mice that were injected with MMP-2 knockdown pre-adipocytes showed small decrease in adipose tissue markers, but the tissue itself did not differ in weight or size from control [110]. In contrast, other studies showed that mice lacking MMP-2 or its activator MMP-14 presented impaired adipose tissue development [117–119].

Moreover, the expression of TIMP-1 was also found in pre-adipocytes and adipocytes and was strongly inhibited when pre-adipocytes entered differentiation [108, 120]. Furthermore, TIMP-1 expression in 3T3F442A pre-adipocyte cell line was also associated with enhanced gelatinase expression and activity, which may suggest its untraditional role as a gelatinases activator [120].

2.4.5 Enteric Neurons

The enteric nervous system which innervates the gastrointestinal tract consists of different neurons and glial cells that are distributed in two intramuscular plexuses along the gut and control the smooth muscle contractile activity [121, 122]. Back in 1954, Yntema and Hammond were the first to reveal that the neurons which innervate the gut are originated from vagal NCCs [123]. They demonstrated that upon ablation of the vagal region of the NC in avian embryos, enteric ganglia failed to form along the gut. Other studies also showed that vagal NCCs provide the majority of enteric precursors, whereas sacral NCCs arising caudal to the 28th somite contribute a smaller number of cells to the hindgut region only [124–126]. Vagal and sacral NCCs express different transcription regulators such as Sox10, endothelin receptor B (EndRb), Phox2b, receptor tyrosine kinas (RET), and low-affinity nerve growth receptor p75 [125, 127–129]. Different studies show that inappropriate NCC development and/or migration leads to mal-innervation of the gut and may also result in Hirschsprung disease, a congenital disease characterized by the absence of enteric ganglia (aganglionosis) along variable lengths of distal colon [2, 121, 122]. This congenital aganglionosis, which occurs in 1 in 5000 live births, is limited to the rectosigmoid colon in 80% of cases and most commonly presents with the failure of a newborn to pass meconium within 48 h of life [130, 131].

MMP-2/MMP-9 activity was found to be required for enteric NCC migration and network formation [125, 132]. At first, a broad-spectrum hydroxamate-based MMP inhibitor, named GM6001, was used in order to examine the possible role of MMPs in the migration of enteric mouse and rat NCCs and in the formation of the neural network within the developing gut [132]. Next, the researchers examined specifically the involvement of either MMP-3, MMP-8, or MMP-2/MMP-9 in these processes, using specific pharmacological inhibitors, and revealed that MMP-3 or MMP-8 inhibition had no effect on enteric NCC migration, whereas MMP-2/MMP-9 inhibition significantly decreased the distance that enteric NCCs migrated in the developing gut and the complexity of the neural network that the cells formed [125].Whether impaired activities of MMP-2/MMP-9 in enteric NCCs may be associated directly or indirectly with gut aganglionosis awaits further research.

2.4.6 Glial Cells

The NC is the source of all glial cells of the peripheral nervous system, including the peripheral nerves and the sensory, sympathetic, parasympathetic, and enteric ganglia [133]. NCCs that migrate from the NT in a ventral direction give rise to neurons in dorsal root sensory ganglia and to peripheral glial cells [134]. The peripheral glial cells include three types of cells, the satellite cells of the sensory and autonomic ganglia, the Schwann cells lining the peripheral nerves, and the enteric ganglia [135, 136].

MMPs in general, and MMP-2 and MMP-9 in particular, were reported previously in Schwann cells [137–139]. For example, Asundi et al. [137] have shown that syndecan-3 shedding in Schwann cells, a process that can have important consequences on cell adhesion, morphology, and migration, is being mediated by the proteases of the MMP family. The researchers also reported that analysis of MMP expression by gene microarray techniques has shown that Schwann cells express MMP-14 mRNA at very high levels. However, they mentioned that other mRNAs, including MMP-2 and MMP-9, were absent from Schwann cells [137]. Contradicting data was demonstrated in a different study, in which both MMP-2 and MMP-9 were found to increase the neurite-promoting potential of Schwann cell basal lamina and were also found to be upregulated in a degenerated nerve [139]. Furthermore,

Table 2.1 The expression and role of MMP-2 and MMP-9 in NCC derivatives. Summary of the data presented in this review regarding the expression and role of MMP-2 and MMP-9 in various NCC derivatives; cranial cartilage and bone, connective tissue of the heart, melanocytes, adipocytes, enteric neurons, and glial cells. Other NCC derivatives as the secreting cells of different endocrine glands as the thymus, thyroid, and adrenal are not discussed in this review

NCC derivative	Tissue/cell type	MMP-2	MMP-9	References
Cranial cartilage and bone	Palatal folds of E13.5–15.5 mouse embryo	+	–	[65]
	Meckel's cartilage and mandibles of E13–15 mouse embryos	+	+	[66, 73]
	Odontoblasts during odontogenesis in rats	–	+	[75]
	Zebra fish craniofacial structures	+	+	[80]
Heart	Quail and chick cushion cells	+	–	[91–94]
Melanocytes	Frog melanophores	+	–	[100]
	Benign and invasive human melanoma lesions	+	+	[101–106]
	Canine cutaneous melanocytic tumors	+	+	[38]
Adipocytes	Pre-adipocytes during their differentiation into adipocytes (human and murine)	+	+	[108, 118, 119]
Enteric neurons	Enteric nervous system formation in mouse embryos	+	+	[125, 132]
Glial cells	Schwann cells of rats, mice, and humans	+	+	[137–140]

MMP-2 immunoreactivity was also found in Schwann cells in human brain tissues and was suggested to play a role in antiproliferative activity which may balance between proteolytic and protease inhibitor activity; such a balance is crucial for both normal neuronal development and for neuronal response to injury [138].

Indeed, studies using damaged peripheral nerves were conducted to clarify the exact role of MMPs in Schwann cells. For example, Kobayashi et al. examined the role of MMPs in the development of mechanical allodynia through myelin-binding protein (MBP) degradation after rat L-5 spinal nerve crush (L-5 SNC) injury [140]. After conducting the L-5-SNC injury, the researchers found increase in MMP-9 but not in MMP-2 or MMP-7-mRNA levels, two other known regulators of MBP degradation. While they found no gelatinolytic activity of MMP-9 in uninjured nerves, the L-5-SNC injury caused major increased gelatinolytic activity of MMP-9. On the other hand, moderate gelatinolytic activity of MMP-2 was observed before the injury, and no change was noticed afterward [140].

Finally, MMP-3, which is a known activator of MMP-9 [141], was found to be expressed in microglial cells and ischemic neurons [142]. Another novel finding in this study was the presence of MMP-2 in normal astrocytes. These results suggested that MMP-2 is important in the repair process and could contribute to angiogenesis and glial scar formation (Table 2.1).

2.5 Activators and Repressors of MMP-2 and MMP-9 in NCCs

As was described before, MMP-2 and MMP-9 are secreted as latent zymogens and require an *in-situ* cleavage of the pro-domain for activation. This cleavage is often made by another already activated MMP or by several serine proteases. Two known activators of MMP-2 are MT1-MMP and MT3-MMP, which are also known as MMP-14 and MMP-16, respectively [143–146]. In a study conducted on mouse fibroblasts, the activation of MMP-2 by both MMP-14 and MMP-16 was found to form a ternary complex with the inhibitor TIMP-2, which is bound to the catalytic domain of the MT-MMP. Then, the pro-MMP-2 binds through its hemopexin domain to the C-terminus end of the same TIMP, and finally, a second-free MMP-14 molecule adjacent to the complex cleaves the pro-domain of the MMP-2 and activates it [146]. A different study demonstrated how MMP-2 expression is mediated by MMP-16. The researchers showed how MMP-16 suppression resulted in decrease in MMP-2-mRNA levels, whereas in parallel, TIMP-2 levels increased. In concomitant, an increase in MMP-2-mRNA levels and decrease in TIMP-2 levels were shown upon MMP-16-overexpression treatment, suggesting that MMP-16 positively regulates MMP-2 function [145, 147].

Despite the fact that the similarity between MMP-2 and MMP-9 regarding their activity and structure may suggest resemblance in their activation, no data has been demonstrated yet linking MMP-14 or MMP-16 directly to MMP-9 activation process. However, other mechanisms of MMP-9 activation are known. For example, MMP-9 is known to be activated indirectly via the MMP-14/MMP-2 axis; after MMP-14 activates pro-MMP-2 together with TIMP-2, the activated MMP-2 activates in turn the pro-MMP-9 [148, 149]. Another known mechanism in which pro-MMP-9 is being activated is via the MMP-14/MMP-13 cascade in which MMP-14 activates pro-MMP-13, and then activated MMP-13 cleaves directly pro-MMP-9 to its active form [150, 151]. Activation of pro-MMP-9 by other MMPs besides MMP-2 and MMP-13 has also been reported in several organs and cell types. This includes MMP-1, MMP-3, MMP-7, MMP-10, and MMP-26 [148–150, 152–154].

In consonant with the data gathered on MMP-2 and MMP-9 expression and roles in NCCs, both of their activators, MMP-14 and MMP-16, were also found to be expressed in migrating NCCs and in some of their derivatives. Data gathered from frog embryos show expression of MMP-14 at the timing of NCC migration as well as at later stages, and it was further demonstrated that MMP-2 expression is subsequent to the expression of MMP-14 in NCCs of these embryos [143, 155]. Moreover, MMP-14 expression was also found in migrating frog melanocytes [100]. Furthermore, MMP-16 was reported to be expressed in the frog embryo during development and to activate MMP-2 [145]. However, this expression has never been demonstrated to correlate with migrating NCCs or with NCC markers. In addition, very recently our lab found that MMP-16 is expressed and has a role in executing NCC migration in the avian embryo [156]. In this study, we demonstrated that MMP-16 is expressed in migrating cranial chick NCCs, and by using loss- and gain-of-function methods, we revealed that MMP-16 is required for normal NCC migration. Inhibition of MMP-16 attenuated NCC migration, and upon excess MMP-16 treatment, the NCC migration was enhanced and precocious. Furthermore, the effect of inhibiting MMP-16 was rescued by addition of MMP-9 which suggested that MMP-16 is upstream to MMP-9, which was previously shown to be required for NCC migration [48]. The mechanism by which MMP-16 promotes NCC migration was also described and, similarly to MMP-9, was found to involve N-cadherin and laminin, which leads to the detachment of the NCCs and to their migration [156].

Interestingly, studies in MMP-14-KO mice demonstrated several skeletal defects, as well as other malformations. These mice showed multiple craniofacial defects, which may imply on perturbed NCC development [157]. Surprisingly, MMP-16-KO mice had no clear phenotype besides small retardation of growth [158]; yet MMP-14 and MMP-16 double KO mice demonstrated an aggravation of the MMP-14-KO phenotype; some craniofacial bones did not ossify; all mice demonstrated severe cleft palate that led to perinatal lethality, as well as to several other major craniofacial defects, which may all be attributed to impaired NCC development. Also, these mice were about 30% smaller than wild types [158]. Notably, although MMP-14-KO mice exhibited some malformations, it is only the MMP-14/MMP-16 double KO that causes lethality. However, this study did not examine the KO effect on NCCs and whether the KO affects MMP-2/MMP-9 expression or activity. These findings, along with the similarities between both proteins, imply on resemblance of activity and on compensation that may occur in the single MT-MMP KO by the other MT-MMP [158]. Further studies will be needed to evaluate the effects of MMP-14/MMP-16 on different stages of NCC ontogeny.

2.6 Conclusion

The gelatinases, MMP-2 and MMP-9, are expressed and/or play important roles during NCC EMT and early migration in various embryonic species. Alternation in the expression and activity of these proteases in older embryos or adults in different NCC derivatives (i.e., cranial structures, melanocytes, heart septum and valves, adipocytes, enteric neurons, glial cells, etc.) leads to impaired development or malformations of various tissues. Yet, further investigations are needed in order to determine whether the role of these gelatinases in early stages of NCC development is linked to the later functioning of the different NCC derivatives or whether the early and late activities of MMP-2 and MMP-9 are two independent processes. Moreover, it is important to uncover the conservation in their roles in different species and the way they are regulated by their activators and inhibitors in the unique NCC population and descendants. Finally, other derivatives of the NCCs such as smooth muscle cells, the thymus and adrenal glands, and more were out of the scoop of this review, and the plausible expression and role of MMP-2 and MMP-9 in these cell types and tissues should be further examined.

References

1. Gammill LS, Bronner-Fraser M (2003) Neural crest specification: migrating into genomics. Nat Rev Neurosci 4:795–805
2. Le Douarin N, Kalcheim C (1999) The neural crest
3. Morales AV, Barbas JA, Nieto MA (2005) How to become neural crest: from segregation to delamination. Semin Cell Dev Biol 16:655–662
4. Le Lièvre CS, Le Douarin NM (1975) Mesenchymal derivatives of the neural crest: analysis of chimaeric quail and chick embryos. J Embryol Exp Morphol 34:125–154
5. Kirby ML, Waldo KL (1995) Neural crest and cardiovascular patterning. Circ Res 77:211–215
6. Gross JB, Hanken J (2008) Review of fate-mapping studies of osteogenic cranial neural crest in vertebrates. Dev Biol 317:389–400
7. McBratney-Owen B, Iseki S, Bamforth SD et al (2008) Development and tissue origins of the mammalian cranial base. Dev Biol 322:121–132
8. Noden DM (1983) The role of the neural crest in patterning of avian cranial skeletal, connective, and muscle tissues. Dev Biol 96:144–165
9. Trainor PA (2010) Craniofacial birth defects: the role of neural crest cells in the etiology and pathogenesis of Treacher Collins syndrome and the potential for prevention. Am J Med Genet Part A 152(A):2984–2994
10. Bar A, Shoval I, Monsonego-ornan E, Sela-donenfeld D (2014) Role of proteases in cellular dysfunction. 103–126
11. Ciccarone V, Spengler BA, Meyers MB et al (1989) Phenotypic diversification in human neuroblastoma cells: expression of distinct neural crest lineages. Cancer Res 49:219–225
12. Kulesa PM, Kasemeier-Kulesa JC, Teddy JM et al (2006) Reprogramming metastatic melanoma cells to assume a neural crest cell-like phenotype in an embryonic microenvironment. Proc Natl Acad Sci U S A 103:3752–3757
13. Thiery JP, Acloque H, Huang RYJ, Nieto MA (2009) Epithelial-mesenchymal transitions in development and disease. Cell 139:871–890
14. Kalluri R, Weinberg RA (2009) Review series the basics of epithelial-mesenchymal transition. J Clin Invest 119:1420–1428
15. Kerosuo L, Bronner-fraser M (2013) in Neural Crest. Development 23:320–332
16. Acloque H, Adams MS, Fishwick K et al (2009) Epithelial-mesenchymal transitions: the importance of changing cell state in development and disease. J Clin Invest 119:1438–1449
17. Barembaum M, Bronner-Fraser M (2005) Early steps in neural crest specification. Semin Cell Dev Biol 16:642–646
18. Kerosuo L, Bronner-Fraser M (2012) What is bad in cancer is good in the embryo: importance of EMT in neural crest development. Semin Cell Dev Biol 23:320–332
19. Trainor PA (2005) Specification of neural crest cell formation and migration in mouse embryos. Semin Cell Dev Biol 16:683–693
20. Ulmer B, Hagenlocher C, Schmalholz S et al (2013) Calponin 2 acts as an effector of noncanonical Wnt-mediated cell polarization during neural crest cell migration. Cell Rep 3:615–621
21. Shoval I, Kalcheim C (2012) Antagonistic activities of Rho and Rac GTPases underlie the transition from neural crest delamination to migration. Dev Dyn 241:1155–1168
22. Groysman M, Shoval I, Kalcheim C (2008) A negative modulatory role for Rho and Rho-associated kinase signaling in delamination of neural crest cells. Neural Dev 3:27
23. Sakai D, Wakamatsu Y (2005) Regulatory mechanisms for neural crest formation. Cells Tissues Organs 179:24–35
24. Burstyn-Cohen T, Stanleigh J, Sela-Donenfeld D, Kalcheim C (2004) Canonical Wnt activity regulates trunk neural crest delamination linking BMP/noggin signaling with G1/S transition. Development 131:5327–5339
25. García-Castro MI, Marcelle C, Bronner-Fraser M (2002) Ectodermal Wnt function as a neural crest inducer. Science 297:848–851

26. Sauka-Spengler T, Bronner-Fraser M (2008) A gene regulatory network orchestrates neural crest formation. Nat Rev Mol Cell Biol 9:557–568
27. Sela-Donenfeld D, Kalcheim C (1999) Regulation of the onset of neural crest migration by coordinated activity of BMP4 and Noggin in the dorsal neural tube. Development 126:4749–4762
28. Steventon B, Carmona-Fontaine C, Mayor R (2005) Genetic network during neural crest induction: from cell specification to cell survival. Semin Cell Dev Biol 16:647–654
29. Takashi S, Daisuke S, Noriko O et al (2006) Sox genes regulate type 2 collagen expression in avian neural crest cells. Develop Growth Differ 48:477–486
30. Sakai D, Suzuki T, Osumi N, Wakamatsu Y (2006) Cooperative action of Sox9, Snail2 and PKA signaling in early neural crest development. Development 133:1323–1333
31. Clark I, Swingler T, Sampieri C, Edwards D (2008) The regulation of matrix metalloproteinases and their inhibitors. Int J Biochem Cell Biol 40:1362–1378
32. Mannello F, Medda V (2012) Nuclear localization of matrix metalloproteinases. Prog Histochem Cytochem 47:27–58
33. Sternlicht M, Werb Z (2009) How matrix metalloproteinases regulate cell behavior. Annu Rev Cell Biol 17:463–516
34. Vu TH, Werb Z (2000) Matrix metalloproteinases: effectors of development and normal physiology. Genes Dev 14:2123–2133
35. Mannello F, Gazzanelli G (2001) Tissue inhibitors of metalloproteinases and programmed cell death: conundrums, controversies and potential implications. Apoptosis 6:479–482
36. Arbeláez LF, Bergmann U, Tuuttila A et al (1997) Interaction of matrix metalloproteinases-2 and -9 with pregnancy zone protein and alpha2-macroglobulin. Arch Biochem Biophys 347:62–68
37. Sottrup-Jensen L, Birkedal-Hansen H (1989) Human fibroblast collagenase-a-macroglobulin interactions. J Biol Chem 264:393–401
38. Pires I, Gomes J, Prada J et al (2013) MMP-2 and MMP-9 expression in canine cutaneous melanocytic tumours: evidence of a relationship with tumoural malignancy. Recent Adv Biol Ther Manag Melanoma:133–161
39. Zimowska M, Swierczynska M, Ciemerych M a. (2013) Nuclear MMP-9 role in the regulation of rat skeletal myoblasts proliferation. Biol Cell 105:334–344
40. Eguchi T, Kubota S, Kawata K et al (2008) Novel transcription-factor-like function of human matrix metalloproteinase 3 regulating the CTGF/CCN2 gene. Mol Cell Biol 28:2391–2413
41. Malemud CJ (2006) Matrix metalloproteinases (MMPs) in health and disease: an overview. Front Biosci 11:1696–1701
42. Szabo KA, Ablin RJ, Singh G (2004) Matrix metalloproteinases and the immune response. Clin Appl Immunol Rev 4:295–319
43. Biljana E, Boris V, Cena D, Veleska-stefkovska D (2011) Matrix metalloproteinases (with accent to collagenases). J Cell Anim Biol 5:113–120
44. Stamenkovic I (2003) Extracellular matrix remodelling: the role of matrix metalloproteinases. J Pathol 200:448–464
45. Quigley JP (1995) Matrix metalloproteinase-2 is an interstitial collagenase. J Biol Chem 270:5872–5876
46. Werb Z (1998) Chin J. Extracellular Matrix Remodeling during Morphogenesis 857:110–118
47. Duong TD, Erickson C a. (2004) MMP-2 plays an essential role in producing epithelial-mesenchymal transformations in the avian embryo. Dev Dyn 229:42–53
48. Monsonego-Ornan E, Kosonovsky J, Bar A et al (2012) Matrix metalloproteinase 9/gelatinase B is required for neural crest cell migration. Dev Biol 364:162–177
49. Taneyhill LA (2008) To adhere or not to adhere. Cell Adhes Migr 2:223–230
50. Shoval I, Ludwig A, Kalcheim C (2007) Antagonistic roles of full-length N-cadherin and its soluble BMP cleavage product in neural crest delamination. Development 134:491–501
51. Zhang J, Bai S, Zhang X et al (2003) The expression of gelatinase a (MMP-2) is required for normal development of zebrafish embryos. Dev Genes Evol 213:456–463

52. Keow JY, Pond ED, Cisar JS et al (2012) Activity-based labeling of matrix metalloproteinases in living vertebrate embryos. PLoS One 7:1–10
53. Yoong S, O'Connell B, Soanes A et al (2007) Characterization of the zebrafish matrix metalloproteinase 9 gene and its developmental expression pattern. Gene Expr Patterns 7:39–46
54. Minoux M, Rijli FM (2010) Molecular mechanisms of cranial neural crest cell migration and patterning in craniofacial development. Development 137:2605–2621
55. Trainor PA, Tam PP (1995) Cranial paraxial mesoderm and neural crest cells of the mouse embryo: co-distribution in the craniofacial mesenchyme but distinct segregation in branchial arches. Development 121:2569–2582
56. Morrison SJ, Perez SE, Qiao Z et al (2000) Transient notch activation initiates an irreversible switch from neurogenesis to gliogenesis by neural crest stem cells. Cell 101:499–510
57. Shah NM, Marchionni MA, Isaacs I et al (1994) Glial growth-factor restricts mammalian neural crest stem-cells to a glial fate. Cell 77:349–360
58. Rijli FM, Mark M, Lakkaraju S et al (1993) A homeotic transformation is generated in the rostral branchial region of the head by disruption of Hoxa-2, which acts as a selector gene. Cell 75:1333–1349
59. Trainor PA, Krumlauf R (2001) Hox genes, neural crest cells and branchial arch patterning. Curr Opin Cell Biol 13:698–705
60. Gendron-Maguire M, Mallo M, Zhang M, Gridley T (1993) Hoxa-2 mutant mice exhibit homeotic transformation of skeletal elements derived from cranial neural crest. Cell 75:1317–1331
61. Dixon J, Dixon J, Trainor P et al (2007) Treacher Collins syndrome. Orthod Craniofac Res 10:88–95
62. Dixon J, Jones NC, Sandell LL et al (2006) Tcof1/treacle is required for neural crest cell formation and proliferation deficiencies that cause craniofacial abnormalities. Proc Natl Acad Sci U S A 103:13403–13408
63. Valdez BC, Henning D, So RB et al (2004) The Treacher Collins syndrome (TCOF1) gene product is involved in ribosomal DNA gene transcription by interacting with upstream binding factor. Proc Natl Acad Sci U S A 101:10709–10714
64. Jones NC, Lynn ML, Gaudenz K et al (2008) Prevention of the neurocristopathy {Treacher} {Collins} syndrome through inhibition of p53 function. Nat Med 14:125–133
65. Gkantidis N, Blumer S, Katsaros C et al (2012) Site-specific expression of Gelatinolytic activity during morphogenesis of the secondary palate in the mouse embryo. PLoS One 7:1–14
66. Chin JR, Werb Z (1997) Matrix metalloproteinases regulate morphogenesis, migration and remodeling of epithelium, tongue skeletal muscle and cartilage in the mandibular arch. Development 124:1519–1530
67. Johnston MC, Bronsky PT (1991) Animal models for human craniofacial malformations. J Craniofac Genet Dev Biol 11:277–291
68. Mossey PA, Little J, Munger RG et al (2009) Cleft lip and palate. Lancet 374:1773–1785
69. Satokata I, Maas R (1994) Msx1 deficient mice exhibit cleft palate and abnormalities of craniofacial and tooth development. Nat Genet 6:348–356
70. Metzis V, Courtney AD, Kerr MC et al (2013) Patched1 is required in neural crest cells for the prevention of orofacial clefts. Hum Mol Genet 22:5026–5035
71. Simsa S, Hasdai A, Dan H, Ornan EM (2007) Differential regulation of MMPs and matrix assembly in chicken and turkey growth-plate chondrocytes. Am J Physiol Regul Integr Comp Physiol 292:R2216–R2224
72. Tong A, Reich A, Genin O, Pines M (2003) Expression of chicken 75-kDa gelatinase B-like enzyme in perivascular chondrocytes suggests its role in vascularization of the growth plate. 18:1443–1452
73. Miettinen PJ, Chin JR, Shum L et al (1999) Epidermal growth factor receptor function is necessary for normal craniofacial development and palate closure. Nat Genet 22:69–73
74. Shimo T, Kanyama M, Wu C et al (2004) Expression and roles of connective tissue growth factor in Meckel's cartilage development. Dev Dyn 231:136–147

75. Randall LE, Hall RC (2002) Temperospatial expression of matrix metalloproteinases 1, 2, 3, and 9 during early tooth development. Connect Tissue Res 43:205–211
76. Vu TH, Shipley JM, Bergers G et al (1998) MMP-9/gelatinase B is a key regulator of growth plate angiogenesis and apoptosis of hypertrophic chondrocytes. Cell 93:411–422
77. Itoh T, Ikeda T, Gomi H et al (1997) Unaltered secretion of β-amyloid precursor protein in gelatinase a (matrix metalloproteinase 2)-deficient mice. J Biol Chem 272:22389–22392
78. Mosig RA, Dowling O, DiFeo A et al (2007) Loss of MMP-2 disrupts skeletal and craniofacial development and results in decreased bone mineralization, joint erosion and defects in osteoblast and osteoclast growth. Hum Mol Genet 16:1113–1123
79. Martignetti JA, Aqeel AA, Sewairi WA et al (2001) Mutation of the matrix metalloproteinase 2 gene (MMP2) causes a multicentric osteolysis and arthritis syndrome. Nat Genet 28:261–265
80. Hillegass JM, Villano CM, Cooper KR, White LA (2008) Glucocorticoids alter craniofacial development and increase expression and activity of matrix metalloproteinases in developing Zebrafish (Danio rerio). Toxicol Sci 102:413–424
81. Abbott BD (1995) Review of the interaction between TCDD and glucocorticoids in embryonic palate. Toxicology 105:365–373
82. Waldo K, Zdanowicz M, Burch J et al (1999) A novel role for cardiac neural crest in heart development. J Clin Invest 103:1499–1507
83. Kirby ML, Waldo KL (1990) Role of neural crest in congenital heart disease. Circulation 82:332–340
84. Creazzo TL, Godt RE, Leatherbury L et al (1998) Role of cardiac neural crest cells in cardiovascular development. Annu Rev Physiol 60:267–286
85. Stoller JZ, Epstein JA (2005) Cardiac neural crest. Semin Cell Dev Biol 16:704–715
86. Jiang X, Rowitch DH, Soriano P et al (2000) Fate of the mammalian cardiac neural crest. Development 127:1607–1616
87. Kirby ML, Turnage KL, Hays BM (1985) Characterization of conotruncal malformations following ablation of "cardiac" neural crest. Anat Rec 213:87–93
88. Nakamura T, Colbert MC, Robbins J (2006) Neural crest cells retain multipotential characteristics in the developing valves and label the cardiac conduction system. Circ Res 98:1547–1554
89. Hutson MR, Kirby ML (2007) Model systems for the study of heart development and disease. Cardiac neural crest and conotruncal malformations. Semin Cell Dev Biol 18:101–110
90. Sato M, Yost HJ (2003) Cardiac neural crest contributes to cardiomyogenesis in zebrafish. Dev Biol 257:127–139
91. Alexander SM, Jackson KJ, Bushnell KM, McGuire PG (1997) Spatial and temporal expression of the 72-kDa type IV collagenase (MMP-8) correlates with development and differentiation of valves in the embryonic avian heart. Dev Dyn 209:261–268
92. Cai DH, Vollberg TM Sr, Hahn-Dantona E et al (2000) MMP-2 expression during early avian cardiac and neural crest morphogenesis. Anat Rec 259:168–179
93. Robbins JR, McGuire PG, Wehrle-Haller B, Rogers SL (1999) Diminished matrix metalloproteinase 2 (MMP-2) in Ectomesenchyme-derived tissues of the patch mutant mouse: regulation of MMP-2 by PDGF and effects on mesenchymal cell migration. Dev Biol 212:255–263
94. Cai DH, Brauer PR (2002) Synthetic matrix metalloproteinase inhibitor decreases early cardiac neural crest migration in chicken embryos. Dev Dyn 224:441–449
95. Nitzan E, Krispin S, Pfaltzgraff ER et al (2013) A dynamic code of dorsal neural tube genes regulates the segregation between neurogenic and melanogenic neural crest cells. Development 140:2269–2279
96. Erickson CA, Goins TL (1995) Avian neural crest cells can migrate in the dorsolateral path only if they are specified as melanocytes. Development 121:915–924
97. Thomas AJ, Erickson CA (2008) The making of a melanocyte: the specification of melanoblasts from the neural crest. Pigment Cell Melanoma Res 21:598–610

98. Krispin S, Nitzan E, Kassem Y, Kalcheim C (2010) Evidence for a dynamic spatiotemporal fate map and early fate restrictions of premigratory avian neural crest. Development 137:585–595
99. Simões-Costa M, Bronner ME (2015) Establishing neural crest identity: a gene regulatory recipe. Development 142:242–257
100. Tomlinson ML, Guan P, Morris RJ et al (2009) A chemical genomic approach identifies matrix metalloproteinases as playing an essential and specific role in Xenopus Melanophore migration. Chem Biol 16:93–104
101. Simonetti O, Lucarini G, Brancorsini D et al (2002) Immunohistochemical expression of vascular endothelial growth factor, matrix metalloproteinase 2, and matrix metalloproteinase 9 in cutaneous melanocytic lesions. Cancer 95:1963–1970
102. van den Oord JJ, Paemen L, Opdenakker G, de Wolf-Peeters C (1997) Expression of gelatinase B and the extracellular matrix metalloproteinase inducer EMMPRIN in benign and malignant pigment cell lesions of the skin. Am J Pathol 151:665–670
103. MacDougall JR, Bani MR, Lin Y, Muschel RJ, Kerbel RS (1999) "Proteolytic switching": opposite patterns of regulation of gelatinase B and its inhibitor TIMP-1 during human melanoma progression and consequences of gelatinase B overexpression. Br J Cancer 80:504–512
104. Redondo P, Lloret P, Idoate M, Inoges S (2005) Expression and serum levels of MMP-2 and MMP-9 during human melanoma progression. Clin Exp Dermatol 30:541–545
105. Hofmann UB, Westphal JR, Waas ET et al (1999) Matrix metalloproteinases in human melanoma cell lines and xenografts: increased expression of activated matrix metalloproteinase-2 (MMP-2) correlates with melanoma progression. Br J Cancer 81:774–782
106. Chen Y, Chen Y, Huang L, Yu J (2012) Evaluation of heparanase and matrix metalloproteinase-9 in patients with cutaneous malignant melanoma. J Dermatol 39:339–343
107. Billon N, Iannarelli P, Monteiro MC et al (2007) The generation of adipocytes by the neural crest. Development 134:2283–2292
108. Bouloumié A, Sengenès C, Portolan G et al (2001) Adipocyte produces matrix metalloproteinases 2 and 9 involvement in adipose differentiation. Diabetes 50:2080–2086
109. O'Hara A, Lim FL, Mazzatti DJ, Trayhurn P (2009) Microarray analysis identifies matrix metalloproteinases (MMPs) as key genes whose expression is up-regulated in human adipocytes by macrophage-conditioned medium. Pflugers Arch Eur J Physiol 458:1103–1114
110. Bauters D, Scroyen I, Van Hul M, Lijnen HR (2015) Gelatinase A (MMP-2) promotes murine adipogenesis. Biochim Biophys Acta - Gen Subj 1850:1449–1456
111. Christensen S, Purslow PP (2016) The role of matrix metalloproteinases in muscle and adipose tissue development and meat quality: a review. Meat Sci 119:138–146
112. Ailhaud G, Grimaldi P, Négrel R (1992) Cellular and molecular aspects of adipose tissue development. Annu Rev Nutr 12:207–233
113. Wang QA, Tao C, Gupta RK, Scherer PE (2013) Tracking adipogenesis during white adipose tissue development, expansion and regeneration. Nat Med 19:1338–1344
114. Sang QX (1998) Complex role of matrix metalloproteinases in angiogenesis. Cell Res 8:171–177
115. Nagase H, Woessner JF (1999) Matrix Metalloproteinases. Am Soc Biochem Mol Biol 274(3):21491–21494
116. Gonzalez-Villasana V, Fuentes-Mattei E, Ivan C et al (2015) Rac1/Pak1/p38/MMP-2 axis regulates angiogenesis in ovarian cancer. Clin Cancer Res 21:2127–2137
117. Van Hul M, Lijnen HR (2008) A functional role of gelatinase A in the development of nutritionally induced obesity in mice. J Thromb Haemost 6:1198–1206
118. Chun TH, Hotary KB, Sabeh F et al (2006) A pericellular collagenase directs the 3-dimensional development of white adipose tissue. Cell 125:577–591
119. Van Hul M, Lupu F, Dresselaers T et al (2012) Matrix metalloproteinase inhibition affects adipose tissue mass in obese mice. Clin Exp Pharmacol Physiol 39:544–550
120. Scroyen I, Cosemans L, Lijnen HR (2009) Effect of tissue inhibitor of matrix metalloproteinases-1 on in vitro and in vivo adipocyte differentiation. Thromb Res 124:578–583

121. Goldstein AM, Hofstra RMW, Burns AJ (2013) Building a brain in the gut: development of the enteric nervous system. Clin Genet 83:307–316
122. Sasselli V, Pachnis V, Burns AJ (2012) The enteric nervous system. Dev Biol 366:64–73
123. Yntema CL, Hammond WS (1954) The origin of intrinsic ganglia of trunk viscera from vagal neural crest in the chick embryo. J Comp Neurol 101:515–541
124. Le Douarin NM, Teillet MA (1973) The migration of neural crest cells to the wall of the digestive tract in avian embryo. J Embryol Exp Morphol 30:31–48
125. Anderson RB, Stewart AL, Young HM (2006) Phenotypes of neural-crest-derived cells in vagal and sacral pathways. Cell Tissue Res 323:11–25
126. Wang X, Chan AKK, Sham MH et al (2011) Analysis of the sacral neural crest cell contribution to the hindgut enteric nervous system in the mouse embryo. Gastroenterology 141:992–1002
127. Nataf V, Lecoin L, Eichmann A, Le Douarin NM (1996) Endothelin-B receptor is expressed by neural crest cells in the avian embryo. Proc Natl Acad Sci U S A 93:9645–9650
128. Pattyn A, Morin X, Cremer H et al (1999) The homeobox gene Phox2b is essential for the development of autonomic neural crest derivatives. Nature 399:366–370
129. Young HM, Hearn CJ, Ciampoli D et al (1998) A single rostrocaudal colonization of the rodent intestine by enteric neuron precursors is revealed by the expression of Phox2b, Ret, and p75 and by explants grown under the kidney capsule or in organ culture. Dev Biol 202:67–84
130. Kenny SE, Tam PKH, Garcia-Barcelo M (2010) Hirschsprung's disease. Semin Pediatr Surg 19:194–200
131. Amiel J, Sproat-Emison E, Garcia-Barcelo M et al (2008) Hirschsprung disease, associated syndromes and genetics: a review. J Med Genet 45:1–14
132. Anderson RB, Turner KN, Nikonenko AG et al (2006) The cell adhesion molecule L1 is required for chain migration of neural crest cells in the developing mouse gut. Gastroenterology 130:1221–1232
133. Ayer-Le Lievre CS, Le Douarin NM (1982) The early development of cranial sensory ganglia and the potentialities of their component cells studied in quail-chick chimeras. Dev Biol 94:291–310
134. Jessen KR, Mirsky R (2005) The origin and development of glial cells in peripheral nerves. Nat Rev Neurosci 6:671–682
135. Le Douarin N, Dulac C, Dupin E, Cameron-Curry P (1991) Glial cell lineages in the neural crest. Glia 4:175–184
136. Mirsky R, Jessen KR (1999) The neurobiology of Schwann cells. Brain Pathol 9:293–311
137. Asundi VK, Erdman R, Stahl RC, Carey DJ (2003) Matrix metalloproteinase-dependent shedding of syndecan-3, a transmembrane heparan sulfate proteoglycan, in Schwann cells. J Neurosci Res 73:593–602
138. Yamada T, Miyazaki K, Koshikawa N et al (1995) Selective localization of gelatinase A, an enzyme degrading β-amyloid protein, in white matter microglia and in Schwann cells. Acta Neuropathol 89:199–203
139. Ferguson TA, Muir D (2000) MMP-2 and MMP-9 increase the neurite-promoting potential of schwann cell basal laminae and are upregulated in degenerated nerve. Mol Cell Neurosci 16:157–167
140. Kobayashi H, Chattopadhyay S, Kato K et al (2008) MMPs initiate Schwann cell-mediated MBP degradation and mechanical nociception after nerve damage. Mol Cell Neurosci 39:619–627
141. Ramos-desimone N, Hahn-dantona E, Sipley J et al (1999) Activation of Matrix Metalloproteinase-9 (MMP-9) via a converging plasmin/Stromelysin-1 cascade enhances tumor cell invasion. Biochemistry 274:13066–13076
142. Rosenberg GA, Cunningham LA, Wallace J et al (2001) Immunohistochemistry of matrix metalloproteinases in reperfusion injury to rat brain: activation of MMP-9 linked to stromelysin-1 and microglia in cell cultures. Brain Res 893:104–112

143. Hasebe T, Hartman R, Fu L et al (2007) Evidence for a cooperative role of gelatinase A and membrane type-1 matrix metalloproteinase during Xenopus laevis development. Mech Dev 124:11–22
144. Itoh Y, Seiki M (2006) MT1-MMP: a potent modifier of pericellular microenvironment. J Cell Physiol 206:1–8
145. Walsh LA, Cooper CA, Damjanovski S (2007) Soluble membrane-type 3 matrix metalloproteinase causes changes in gene expression and increased gelatinase activity during Xenopus laevis development. Int J Dev Biol 51:389–395
146. Zhao H, Bernardo MM, Osenkowski P et al (2004) Differential inhibition of membrane type 3 (MT3)-matrix metalloproteinase (MMP) and MT1-MMP by tissue inhibitor of metalloproteinase (TIMP)-2 and TIMP-3 regulates pro-MMP-2 activation. J Biol Chem 279:8592–8601
147. Jalali S, Singh S, Agnihotri S et al (2015) A role for matrix remodelling proteins in invasive and malignant meningiomas. Neuropathol Appl Neurobiol 41:16–28
148. Toth M, Chvyrkova I, Bernardo MM et al (2003) Pro-MMP-9 activation by the MT1-MMP/MMP-2 axis and MMP-3: role of TIMP-2 and plasma membranes. Biochem Biophys Res Commun 308:386–395
149. Fridman R, Toth M, Pena D, Mobashery S (1995) Activation of progelatinase B (MMP-9) by gelatinase A (MMP-2). Cancer Res 55:2548–2555
150. Dreier R, Grässel S, Fuchs S et al (2004) Pro-MMP-9 is a specific macrophage product and is activated by osteoarthritic chondrocytes via MMP-3 or a MT1-MMP/MMP-13 cascade. Exp Cell Res 297:303–312
151. Knäuper V, Smith B, Lopez-Otin C, Murphy G (1997) Activation of progelatinase B (proMMP-9) by active collagenase-3 (MMP-13). Eur J Biochem 248:369–373
152. Sang Q, Birkedal-hansen H, Van Wart HE (1995) Proteolytic and non-proteolytic activation of human neutrophil progelatinase B. BBA 1251:99–108
153. Nakamura H, Fujii Y, Ohuchi E et al (1998) Activation of the precursor of human stromelysin 2 and its interactions with other matrix metalloproteinases. Eur J Biochem 253:67–75
154. Uria JA, Lopez-Otin C (2000) Matrilysin-2, a new matrix metalloproteinase expressed in human tumors and showing the minimal domain organization required for secretion, latency, and activity. Cancer Res 60:4745–4751
155. Harrison M, Abu-Elmagd M, Grocott T et al (2004) Matrix metalloproteinase genes in Xenopus development. Dev Dyn 231:214–220
156. Roth L, Kalev-Altman R, Monsonego-Ornan E, Sela-Donenfeld D (2017) A new role of the membrane-type matrix metalloproteinase 16 (MMP16/MT3-MMP) in neural crest cell migration. Int J Dev Biol 61(3–4–5):245–256
157. Holmbeck K, Bianco P, Caterina J et al (1999) MT1-MMP-deficient mice develop dwarfism, osteopenia, arthritis, and connective tissue disease due to inadequate collagen turnover. Cell 99(1):81–92
158. Shi J, Son MY, Yamada S et al (2008) Membrane-type MMPs enable extracellular matrix permissiveness and mesenchymal cell proliferation during embryogenesis. Dev Biol 313(1):196–209

3 The Matrix Metalloproteinase and Tissue Inhibitors of Metalloproteinase Balance in Physiological and Pathological Remodeling of Skeletal Muscles

Hala S. Alameddine

Abstract

Skeletal muscle is a highly plastic tissue that undergoes physiological or pathological remodeling in response to various stimuli such as exercise, immobilization, injury, disease, or aging. This remodeling process implies subtle or more profound changes to skeletal muscle structure and composition that involves extracellular matrix (ECM) degradation by matrix metalloproteinases. The balance between matrix metalloproteinases (MMPs) and their physiological inhibitors, the tissue inhibitors of metalloproteinases (TIMPs), regulates tissue homeostasis. Upregulation of MMPs and/or TIMPs correlates with vascular growth and enlargement in endurance-exercised individuals or with inflammation and regeneration of muscle fibers in injured or diseased muscles. They, further, contribute to the development of fibrosis by regulating cytokine/chemokine production and release/activation of growth factors. Those induce phenotypic transformation and favor the production of ECM components. It is, therefore, important to define the exact pattern of MMP/TIMP expression and regulation in normal and diseased muscles in order to identify potential targets for therapeutic approaches or biomarkers for specific disease entities and therapeutic follow-up.

Keywords

MMPs • TIMPs • Skeletal muscle remodeling • Muscle injury • Muscle diseases

H.S. Alameddine (✉)
Institut de Myologie, Bâtiment Joseph Babinski,
Groupe Hospitalier Pitié-Salpêtrière 47, Bd de l'Hôpital, 75651 Paris Cedex 13, France
e-mail: h.alameddine@institut-myologie.org

S. Chakraborti, N.S. Dhalla (eds.), *Proteases in Physiology and Pathology*,
DOI 10.1007/978-981-10-2513-6_3

3.1 Introduction

Skeletal muscle is the most abundant tissue in the human body and is prone to modifications in response to variation of functional demands, traumatic injury, aging, or disease. Development, hypertrophy, hyperplasia, atrophy, degeneration/regeneration, or fibrosis are all remodeling events that modify tissue composition, architecture, vascularization, or innervation. Variations of fiber size, gene expression, contractile parameters, or capillary density result from signaling events that influence cell-cell and/or cell-matrix interactions and modify the balance between MMPs and TIMPs.

Considered for a long time as a stable tissue, skeletal muscle has a remarkable capacity to adapt to different types of stimuli throughout life. Since their initial formation until the end of life in aging individuals, the development and adaptation of skeletal muscles are governed by an interactive cross talk between these muscles and their environment. The amplitude of cellular, molecular, or structural modifications induced by remodeling stimuli varies from subtle changes of gene expression and cell and tissue architecture to more profound modifications leading, for example, to total reconstitution of adult skeletal muscle tissue after severe trauma. The course of remodeling includes modifications to the ECM that surrounds muscle fibers, to the morphology of skeletal muscle fibers (formation, maturation, atrophy, or hypertrophy), or to muscle architecture (necrosis, regeneration, angiogenesis, fibrosis) and function (shift from slow to fast and vice versa, decrease or improvement of contractile parameters).

The ECM, formed by a complex and dynamic network of macromolecules, engulfs normal adult skeletal muscle fibers and serves as a support tissue. It is not a static structure. It contributes to structural integrity of muscle fibers and plays a role in signal transduction. It also has an essential role in key aspects of cell biology through the production, degradation, or remodeling of ECM components. Physical and biochemical properties of the ECM vary with the functional status of skeletal muscles. Modifications of ECM composition or physical properties such as rigidity, porosity, topography, and insolubility may have functional repercussions on anchorage-dependent biological functions such as cell division, cell migration, and tissue polarity. Defects in ECM components (laminin and type VI collagen), or other proteins participating to the molecular link between ECM and nucleus (laminin A-C or Emerin), via cytoskeletal elements (intermediate filaments, desmin), are responsible of several types of muscular diseases [1–7]. The MMP/TIMP balance regulates cellular microenvironment that affects myoblast proliferation, migration, and differentiation and potentially modifies cell fate. Cells exposed to cryptic fragments resulting from ECM hydrolysis, to growth factors liberated from the ECM or to cytokines or chemokines, respond to these stimuli by modifying their functional status and regulating MMP/TIMP expression [8, 9].

Enzymes of both serine and matrix metalloproteinase family degrade ECM components. Collectively, the different MMP family members degrade all ECM components despite a certain substrate specificity of each member [10]. Tissue homeostasis is characterized by the balance between basal levels of MMP expression/activity and their physiological inhibitors, the TIMPs. Transient increase of MMPs associates with various physiological or pathological remodeling situations such as ovulation, uterine resorption after pregnancy, bone development, angiogenesis, mammary involution, wound healing and exercise, immobilization, and injury or disease of skeletal muscles [11–24]. In normal situations, bursts of MMP expression underscore the tight regulation of MMP/TIMP expression and the essential role played by MMP/TIMP balance in tissue homeostasis. Strong evidence also points to the role played by MMPs in the pathogenic process in inflammatory diseases and other pathological conditions with focal or generalized tissue destruction. The degradation of certain ECM components (laminin, fibronectin) [25, 26], while others are preserved (laminin(s) that serves as a guidance cue for re-innervating axons) [27, 28] during the degeneration/regeneration process, indicates that selective regulation of MMP expression/activation probably occurs for the accomplishment of a specific task.

As far as we know from mice models of MMP deficiency, these enzymes are dispensable for embryogenesis probably because of functional redundancy, enzymatic compensation, or adaptive development. However, studies in animal models point to their essential role after birth as regulators of microenvironmental changes in remodeling tissues including development or disease [29]. MMPs are involved in cell migration by creating space for cells to migrate. They modulate cell functions by producing specific substrate-cleavage fragments with independent biological activity. They regulate tissue architecture through effects on the ECM and intercellular junctions and can activate, deactivate, or modify the activity of signaling molecules. MMP and related ADAM (a disintegrin and metalloproteases) and ADAM-TS (ADAMs with thrombospondin repeats) families are involved in shedding growth factors, which are synthesized as cell-membrane-bound precursor forms, or cell-surface-adhesion molecules such as syndecan-1. The wide range of MMP targets that range from peptide growth factors going through tyrosine kinase receptors, cell adhesion molecules, cytokines and chemokines, as well as other MMPs and unrelated proteases has considerably extended their role. MMPs have evolved from simple hydrolytic enzymes to regulators of signal transduction, of innate and adaptive immunity, and modifiers of cellular/molecular phenotype.

3.2 MMPs and TIMPs in Skeletal Muscles: Expression and Role

The involvement of a metallo-endopeptidase activity in myoblast fusion is reported for the first time by Couch and Strittmatter in the early 1980s [30, 31], but the identity of the enzyme(s) and its mode of action remained unsolved. A decade later, the role of Meltrin-alpha or ADAM-12, member of the metalloproteinase/disintegrin protein family, is identified in C2C12 cell fusion [32]. We know now that different members of the MMP family have, either directly or indirectly, a role in myogenic cell migration and fusion. Obviously, cell fusion occurs between neighboring cells, and MMPs favor myogenic cell migration by helping myogenic cells come into

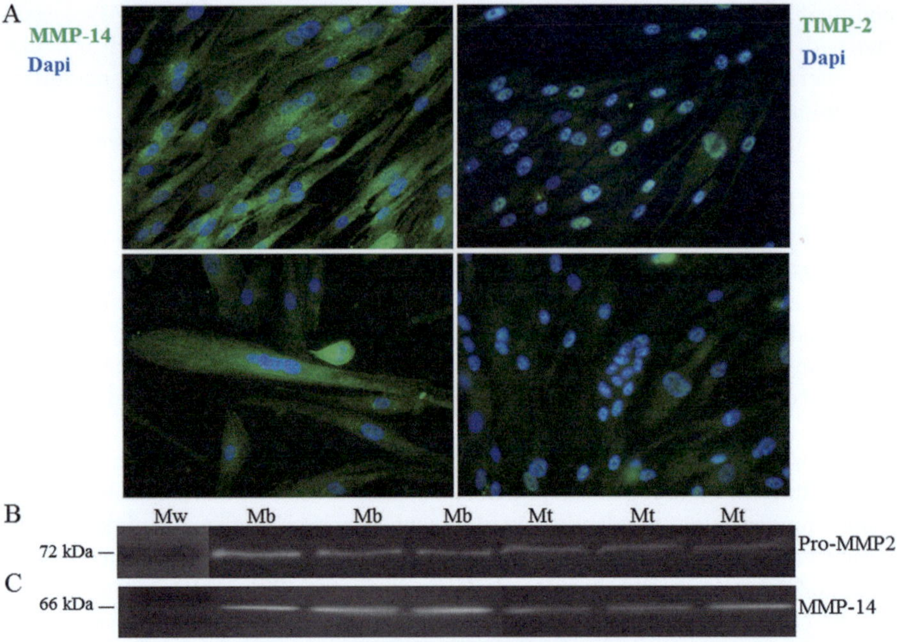

Fig. 3.1 Expression of MMP-2, MMP-14, and TIMP-2 in primary cultures of human myogenic cells, immunofluorescence (*panel* **A**), zymography (*Panel* **B**), and immunoblotting (*Panel* **C**) *Panel* **A**: immunolocalization of MMP-14 and TIMP-2 in human myogenic cells (myoblasts) and (myotubes). MMP-14 labeling is patchy and spreads along the surface of the cells, whereas TIMP-2 is detected in certain cell nuclei. *Panel* **B**: MMP-2 is secreted into conditioned medium of myoblasts and myotubes. *Panel* **C**: MMP-14 is detected in both myoblasts and myotubes. *MW* molecular weight, *Mb* myoblasts, *Mt* myotubes

close contact. These features, used in experimental settings of therapeutic trials, proved to have a relative success depending on the measured outcomes. On one hand, MMP overexpression is used to improve the efficiency of cell therapy for the treatment of animal models of Duchenne muscular dystrophy (DMD) [33, 34] and resolve fibrosis [35]. On the other hand, MMP-9 inhibition in young and old *mdx* mice models of DMD had opposite effect on the structure and function of muscles of dystrophic mice confirming the dual role played by MMP-9 in skeletal muscles [36–39].

3.2.1 MMPs /TIMPs in Skeletal Muscle Cells

Myogenic cells of various animal species ranging from mice to men express MMP-1, MMP-2, MMP-3, MMP-7, MMP-9, MMP-10, MMP-14, and MMP-16 [21, 40–47]. Human and murine myoblasts and myotubes constitutively express MMP-14, TIMP-2, and pro-MMP-2 (Fig. 3.1, unpublished results) but not MMP-9. However, myoblasts of different embryological origins may have different MMP expression patterns [34]. Myoblasts isolated from masseter have a constitutive MMP-9 expression, whereas myoblasts isolated from limb muscles do not [40, 44]. In the latter, MMP-9 is induced by TNF-α [48], phorbol ester [21, 43], and factors released from damaged muscles [21, 49]. Similarly, tumorigenic human rhabdomyosarcoma and C2C12 mouse myogenic cells express different MMP levels and variety that correlate with different invasive or fusion capacities [34, 50, 51]. Myoblasts also respond differently to cytokines and growth factors. Serum fibronectin, platelet-derived growth factor-BB (PDGF-BB), transforming growth factor-β (TGF-β), or insulin growth factor-1 (IGF-1) do not affect MMP-9 expression in adult mouse myoblasts, whereas tumor necrosis factor-α (TNF-α) and basic fibroblast growth factor (β-FGF) induce MMP-9 expression. However, none of these growth factors affects MMP-1 or MMP-2 expression [52].

In normal adult skeletal muscles, the exact panel of MMP/TIMP proteins/activity/mRNA expressed is still not clearly defined. However, one can assume that basal levels of MMP/TIMP protein/mRNA would be detected in the various cell types composing the muscle. Gelatinases A and B—both proteins and activity—are the most documented together with MMP-1 and MMP-14, but data about other MMPs remain elusive and often restricted to mRNA expression. In normal adult muscles, *in situ* zymography, which does not allow the distinction between MMP-2 and MMP-9, reveals a weak gelatinase activity in the sarcoplasm, endomysium, and perimysium [19, 33, 53]. Intracellularly, gelatinase activity localizes to mononucleated cells, presumably fibroblasts, endothelial or satellite cells, and patchy spots within the sarcoplasm [33]. It probably corresponds to MMP-2 which is the only gelatinase detected by gel zymography in normal muscle extracts [21]. Recent reports confirm intracellular localization of MMP-2 in type I and type II fibers where it is more prominent [54].

3.2.1.1 Role of MMPs in Myogenic Cells

Different and multiple MMPs and TIMPs play a role in myoblast migration and/or fusion and probably interfere in muscle fiber maturation, remodeling, and regeneration. Myogenesis *in vitro* involves myoblast cell cycle arrest, migration, and fusion to form multinucleated myotubes with centrally located nuclei that migrate to the periphery upon maturation of myofibers *in vivo*. Factors that affect either myogenic cell migration or fusion necessarily affect myogenesis/regeneration.

MMPs Contribute to Cell Migration

MMPs/TIMPs have been involved in myoblast migration in *ex vivo* or *in vivo* experiments. Upon culture of mandibular explants of young embryos in presence of protease inhibitors, morphogenesis is altered only when hydroxamic acid, a specific MMP inhibitor, is used [55]. Furthermore, MMP inhibition by large-spectrum inhibitors or siRNA decreases or inhibits cell migration *in vitro* [42, 46]. On the contrary, treatment of myoblasts by MMPs, by substances that induce MMP overexpression/activation, or, eventually, by gene transduction increases myogenic cell migration in all cell model systems and migration assays. MMP-1 [56], MMP-2, and MMP-7 [41, 42] as well as concanavalin that induces MT1-MMP overexpression and MMP-2 activation [57] or TNF-alpha that increases MMP-9 levels [48] all favor myogenic cell migration and dissemination far from the injection site after these cells are grafted at a single location. Data point to a specific role of MMP-9 in cell migration as neither MMP-2 overexpression nor higher overall gelatinolytic activity can compensate for MMP-9-induced cell migration [34]. Specific MMP-9 inhibition decreases the migration of human primary myoblasts [44], whereas the addition of exogenous MMP-9 disrupts the balance with TIMPs and improves migration [34].

MMPs Contribute to Cell Fusion

Besides ADAM-12, the enzyme required for myotube formation [32], other MMPs have been involved in myogenic cell fusion. Overexpression of various MMPs in myoblasts increases their propensity to form myotubes and generates more fibers than parental control, upon grafting into dystrophic muscles [41]. However, despite strong evidence for MMP/TIMP implication in the fusion process, it is difficult to ascertain whether one or several MMP/TIMP couple(s) operate in this process.

The first protein duo involved in cell fusion of masseter myoblasts is TIMP-1/MMP-9 because of inverse regulation during myogenesis *in vitro*. Upregulation of TIMP-1 concomitantly to MMP-9 downregulation suggested that mononucleated myogenic cells express MMP-9 to allow migration and then upregulate TIMP-1 to halt migration and authorize cell fusion [44]. However, this observation does not apply to all myogenic cell types. In C2C12 cell lines, frequently used as a model system for *in vitro* myogenesis, high TIMP-1 levels inhibited cell migration without affecting myoblast fusion [34].

The second set of MMP/TIMP proteins reported to function in cell fusion is MMP-2/MMP-14/TIMP-2 [45]. Pros and cons exist in favor and against each one of the partners. The conjunction of TIMP-2 upregulation during myogenic cell fusion [45] and decrease of muscle mass in TIMP-2 knockout mice led to suggest its involvement in cell fusion [58, 59]. Once more, TIMP-2 upregulation proved to be insufficient to induce myoblast fusion *in vitro* [34], and the atrophy of muscle fibers observed in TIMP-2-deficient mice results from abnormalities of muscle innervation [34]. The second partner, MMP-14, is presented as a multilateral regulator of muscle differentiation *in vitro* [47] and is thought to act at a specific stage of myoblast fusion. Inhibition of MT1-MMP by shMT1-MMP partially inhibits myoblast fusion, whereas treatment with BB94, a large-spectrum inhibitor, totally inhibits myotube formation [47]. Data indicate there is functional complementarity between MMP-2 and MMP-14. Stable transfection of these proteases in myoblasts favors cell fusion of the grafted cells with regenerating myofibers. It increases the number of myonuclei within regenerated myotubes without affecting the total number of formed myotubes. In contrast, MMP-14 deficiency does not prevent muscle fiber formation *in vivo*. MMP-14-deficient mice are born with severe generalized abnormalities of skeletal and extra-skeletal connective tissue, defective angiogenesis, and skeletal muscles with centrally located myonuclei indicative of defective maturation. These mice die prematurely by 3–12 weeks of age [47, 60]. Concerning the third member, MMP-2, the initial reports describe no major abnormalities of

MMP-2-deficient mice. They present as normal, fertile, but have slower growth rate [61]. Closer examination reveals marked age-related bone density loss, bone abnormalities, craniofacial defects, sclerotic cranial sutures, and articular cartilage destruction [62] resembling human Multicentric Osteolysis with Arthropathy (MOA) [63]. This explains slower growth rate that may have functional repercussion on skeletal muscle architecture. Indeed parallel alignment and skeletal muscle fiber organization depend on mechanical tension induced by bone growth [64]. MMP-2 and MMP-14 double mutant mice [65] die immediately after birth with respiratory failure, abnormal blood vessels, accumulation of ECM components, and small centrally nucleated muscle fibers that remind of myofiber immaturity in centronuclear myopathies. Central nucleation is a characteristic feature of incomplete maturation of myofibers during development. It can be due to gene defects or to the absence of appropriate vascularization. Besides the indirect effect these proteins can have on muscle fiber maturation through vascular defects, *in vitro* data indicate they have a complementary not an overlapping role on myotube formation [65].

The third set of proteins that influences myotube formation is TIMP-3/MMP-9. TIMP-3 regulates myogenesis via miR-206-TIMP3-TACE-TNF-α-p38 signaling pathway. It acts by regulating autocrine release of TNF-α, known to induce MMP-9 upregulation [48]. The perturbation of TIMP-3 downregulation, which occurs transiently during myoblast fusion or muscle regeneration, blocks TNF-α release, p38 MAPK activation, myogenic gene expression, and myotube formation. On the contrary, supplementation at physiological concentration of TNF-α rescues myoblast differentiation regulated by miR-206 that promotes myogenesis and mediates TIMP-3 downregulation [66]. The active role of these proteins in myogenic cell fusion is corroborated in regenerating soleus in which overexpression of TIMP-3 impairs the release of TNF-α, downregulates myogenic gene expression, and delays the formation of new fibers.

Finally, the addition of TIMP-1, TIMP-2, or TIMP-3 or overexpression of RECK, the membrane-localized MMP inhibitor, inhibits myotube formation [47]. Myogenic regulatory factors (MRFs), which determine myogenic cell fate, regulate RECK expression: MyoD suppresses promoter activity while MRF4 activates it [67].

3.3 MMPs in Remodeling Skeletal Muscles

Remodeling skeletal muscles undergo intra- and/or extracellular modifications that cover a wide range of adaptive behavior. Skeletal muscle fibers atrophy in response to denervation [20], immobilization [17], aging [68], and cancer [69] but improve their contractile performance in acute or long-term exercise that induce hypertrophy or hyperplasia [70]. Eccentric exercise, especially when the task is novel, produces muscle soreness and, possibly, strength loss as a secondary consequence of strain-induced muscle damage [71]. Chemicals, trauma, or disease cause also muscle damage and induce a regenerative response similar to wound healing [64]. These modifications imply the occurrence of interactive cross talk among several partners including MMPs and TIMPs. The cascade of regulatory mechanisms varies with the nature and time frame that govern MMP/TIMP induction/activation/regulation, in physiological and pathological remodeling situations [72].

3.3.1 Skeletal Muscle Plasticity

Physical and chemical signals allow the body to increase or decrease skeletal muscle size throughout adulthood. In adult mammals, an increase in muscle mass occurs primarily as an increase in muscle fiber size (hypertrophy) rather than muscle fiber number (hyperplasia) [70]. The most vigorous muscle hypertrophy is accompanied by ECM remodeling [73], in response to increase of anabolic hormonal signaling or strength training. Hemodynamic and mechanical stimuli activate satellite cells, increase angiogenesis, favor the proliferation of capillary-associated cells, and regulate MMP activity in the absence of any significant inflammation. Among the enzymes modified by physical activity are MMP-2 and MMP-14 [70, 74, 75]. MMP-2 is increased with high-intensity exercise and is particularly prominent in skeletal muscles with high percentage of fast muscle fibers [16]. MMP-9 is induced after a single bout of exercise [72, 76]. High-intensity exercise is required to promote the expression of MMP-2 in skeletal muscles, and the influence of exercise on MMP-2 expression is dominant in muscles containing a high percentage of fast fibers [16]. Visualization of gelatinase activity by high-resolution *in situ* zymography localized MMP-2 activity in myonuclei in endurance-trained rat soleus muscles, while activated satellite cells are identified as the source of MMP-9 upregulation that occurs within few hours of exercise [53]. Endurance training induces various biochemical modifications, hypertrophy, and transition of muscle fiber types accompanied by an increase of capillary density [77]. In this whole system setting, autocrine/paracrine networks may explain the induction of angio-myogenesis. In response to non-injurious exercise, moderately and highly trained athletes have systemic pro-inflammatory molecules produced in plasma [78], but there is no clear evidence indicating that exercise induces an inflammatory response in muscle tissue [79]. Induction of mild inflammatory response after a single bout of exercise is rapid and resolves within 24 h. With chronic loading, both muscle and tendon show adaptive changes with increased collagen production and MMP/TIMP changes indicative of matrix turnover [79]. Pro-inflammatory molecules and MMPs can mediate mobilization of stem cells [80] and angiogenesis [81, 82]. Peroxisome-proliferator-activated receptor (PPAR)-δ activation induces direct transcriptional activation of MMP-9 which, at turn, degrade insulin-like growth factor-binding protein (IGFBP)-3. This results in IGF-1 receptor (IGF-1R) activation in surrounding target cells, augmented angiogenesis, and improved regeneration [83]. Various other stimuli induce MMP-1, MMP-2, MMP-3, MMP-7, MMP-8, MMP-12, and MT1-MMP in vascular tissues [18].

In skeletal muscle, expression at basal levels of MMP-2, TIMP-1, IGF-1, and ciliary neurotrophic factor, conveys an advantage for aged persons to adapt to resistance training [84]. Specific MMP-1 and MMP-9 polymorphisms correlate with increased body mass and muscle strength in elderly exercised women. Increased cell body mass and maximal isometric strength correlate with G insertion in MMP-1 (G+/−1607), while homozygote for 21 or less CA repeats in MMP-9 (cytosine-adenine microsatellite) gains more isometric strength than carriers of longer microsatellites [85]. On the contrary, muscle fiber atrophy in response to unloading, immobilization, denervation, or aging involves MMP-2 activity [86, 87] although both MMP-2 and MMP-9 gelatinases are upregulated [88–92]. In such cases, MMP-2-deficient mice show preserved laminin and type IV collagen and less muscle atrophy than MMP-9-deficient mice [93]. Upregulation of MMP-2 activity is reported in denervated muscles after stretching associated or not to electrical stimulation [86], but this response seems to be restricted to atrophic muscles because single or repetitive stretching sessions fail to regulate MMP-2 levels in normal muscles [94].

3.3.2 Skeletal Muscle Degeneration and Regeneration

Muscle regeneration is the tissue-specific wound healing response to experimental, accidental, or disease-induced muscle injuries. It follows the typical stages of wound healing response during which degeneration/regeneration cycles recapitulate skeletal muscle fiber formation after an initial phase of necrosis and inflammation [64]. A successful regenerative process implies a perfect coordination of cellular and molecular interactions between the various tissue components and leads to tissue reconstitution after substantial remodeling of ECM components. Deregulation of this process by delayed vessel growth, perturbation of inflammatory reaction, and proliferation of fibrous tissue negatively affect the efficiency of muscle regeneration.

During muscle regeneration, the regulation of MMPs follows a time course that parallels biochemical, molecular, and structural modifications observed in various types and extent of muscle injuries [95]. Classically, transient upregulation of MMP-2 and MMP-9 occurs in experimental muscle injuries [12, 21, 96–98] and in patients with chronic critical limb ischemia (CLI) or with peripheral arterial occlusive disease (PAD) [99] with a time frame that depends on the amplitude of tissue destruction. Cardiotoxin injury of skeletal muscles induces massive tissue destruction and prolongation of MMP-2 and MMP-9 upregulation by comparison to models with restricted injuries. Within the first week, free gelatinase activity increases progressively and peaks at day 7 before returning to control values when skeletal muscle fibers have already regenerated. During this time interval, two consecutive regulations of gelatinase expression and activation take place: first, a rapid induction of MMP-9 in inflammatory cells and activated satellite cells [21] that upregulate adrenomedullin (ADM), insulin-like growth factor (IGF-I and IGF-II), MMP-9, and monocyte chemoattractant protein (MCP)-1 necessary for angiogenesis, tissue regeneration, and phagocyte recruitment after injury [49] and, second, a simultaneous decrease and activation of latent MMP-2 within the first 24–48 h followed by a progressive reconstitution of latent and active MMP-2 concomitantly to myofiber regeneration/maturation by the end of the first week [21, 97, 100]. Within this time frame, increased processing of MT1-MMP from latent MMP-14 (63 kDa) to the fully active soluble 50 kDa form, which retain its ability to process MMP-2, follows TIMP-2 reduction [12] suggesting the implication of MMP-2/MMP-14/TIMP-2 in either myoblast fusion or angiogenesis. Additionally, MMP-3 and TIMP-1 are upregulated 24 h after injury, and then TIMP-1 decreases 2 days later, whereas MMP-3 increase is followed by a decrease of active MMP-3 [101]. Other MMPs such as MMP-10 and MMP-13, the murine homologue of human MMP-1, are important for efficient muscle regeneration [102, 103]. MMP-13 expression is pronounced after myoblast fusion and myotube formation *in vitro* [103], and its levels correlate with the extent of tissue damage [104]. MMP-10 is also critical for skeletal muscle maintenance and regeneration [102]. MMP-10-deficient muscles display impaired recruitment of endothelial cells, reduced levels of ECM proteins, diminished collagen deposition, and decreased fiber size, which collectively contribute to delayed muscle regeneration. MMP-10 acts via a CXCR4/SDF1 signaling axis that proved essential for efficient skeletal muscle regeneration [105].

Damaging exercise induced by downhill running in human volunteers releases MMP-2, MMP-9, TIMP-1, and MMP-2/TIMP-2 complex into circulation [106]. The acute increase of TIMP-1, MMP-2, and TIMP-2 does not correlate with muscle damage, but the elevated levels of circulating MMP-9 correlate with the increase of the number of circulating leukocytes induced by exercise.

Finally, MMPs are largely involved in tendon degeneration, rupture, and healing [107, 108]. Diagnostic and therapeutic opportunities in orthopedics are provided by MMPs and TIMPs [109] with clear correlation between increase in MMPs and poor healing in rotator cuff muscles [110]. The increase of MMP-1 and MMP-13 in rotator cuff synovial fluid correlates with decreased healing in the tendon-bone repair site. On the contrary, blockade of MMPs improves tendon-to-bone healing by increasing fibrocartilage and collagen organization and improving the scar strength [111, 112].

3.4 Deregulation of MMPs in Skeletal Muscle Diseases

Deregulation of MMP/TIMP balance has been described in various neuromuscular diseases, but the exact role they play in these diseases is still largely unknown. Interestingly, their use as biomarkers for prognostic use or therapeutic follow-up has been reported in Duchenne muscular dystrophy (DMD) and amyotrophic lateral sclerosis (ALS).

3.4.1 Inflammatory Myopathies

Chronic muscle inflammation and muscle weakness characterize inflammatory myopathies [113]. They were the first neuromuscular diseases to be investigated for MMP involvement in the pathogenesis of the diseases particularly in light of accumulating evidence in favor of their role in modulating innate immunity and inflammation [24, 114]. This makes them good targets for the control of inflammatory conditions [115].

All inflammatory myopathies have in common the upregulation of MMP-9 in muscle tissue. High MMP-2 and MMP-9 levels are reported in muscle homogenates of polymyositis, dermatomyositis, and inclusion body myositis (IBM) patients [116–118], but distinct MMP expression patterns are revealed by immunolabeling. MMP-9 antibodies label atrophic myofibers [117, 119] although Choi et al. [118] report that labeling is restricted to $CD8^+$ cytotoxic T cells. MMP-2 labeling has similar distribution but weaker intensity, whereas MMP-7 antibodies strongly label myofibers invaded by inflammatory cells only in polymyositis [119]. Further, MMP-1 overexpression is localized in mesenchymal cells resembling fibroblasts and around the sarcolemma. MMP-1 and MMP-9 protein overexpression is confirmed by transcript upregulation in polymyositis, in dermatomyositis, and to a lesser extent in IBM [117]. Finally, muscle biopsies of sporadic inclusion body myositis (s-IBM) patients show a co-localization of MMP-2 and beta-APP labeling in rimmed vacuoles, indicating a potential involvement of MMP-2 in the formation of amyloid deposits [118]. TIMP levels remain unchanged and MMP levels in the serum are not modified [117].

3.4.2 Muscular Dystrophies

Muscular dystrophies (MD) are a group of myopathies that weaken the musculoskeletal system and hamper locomotion. Progressive skeletal muscle weakness, defects in muscle proteins, and loss of muscle cells and tissue usually characterize skeletal muscles of affected patients [120]. Muscular dystrophies do not have an available treatment or cure, but physiotherapy, aerobic exercise, and low-intensity catabolic steroids help to maintain muscle tone [121].

Duchenne muscular dystrophy (DMD) is the emblematic form of these diseases. It is the most frequent and most severe X-linked recessive muscle disorder as it affects 1 in 3500 newborn boys and leads to death in the second or third decade of life [122, 123]. DMD and its milder form Becker muscular dystrophies (BMD) are due to mutations in the dystrophin gene leading, respectively, to the absence or the expression of a partially functional protein. Dystrophin is a constitutive element of the "dystrophin-glycoprotein complex" that anchors the cytoskeleton to the ECM through the outer membrane of muscle cells. Thereby, modifications that affect these proteins result in defective scaffolding that leads to the disruption of the cell membrane during muscle contraction and eventual weakening and wasting of myofibers. Interestingly, a number of severe forms of hereditary neuromuscular diseases such as Duchenne or Fukuyama muscular dystrophy have secondary abnormalities of dystroglycan [124–127] identified as an MMP target. In patients with sarcoglycanopathy, MMPs process β-dystroglycan and liberate a 30 kDa proteolytic fragment [128]. The disruption of the link between extracellular matrix and cell membrane hence leads to deleterious evolution to which MMPs/TIMPs participate in DMD or other dystrophies raising important interrogations about their regulation

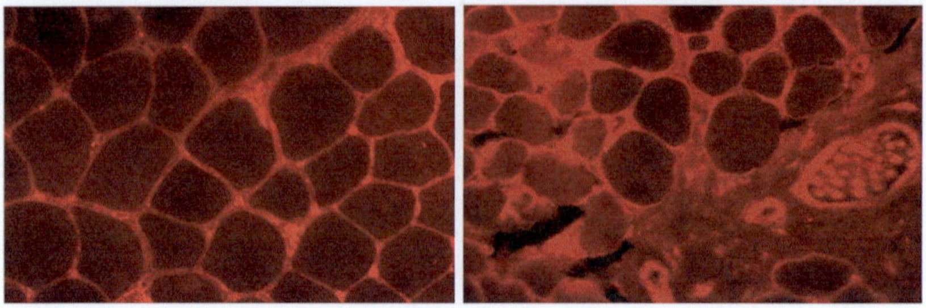

Fig. 3.2 TIMP-1 immunolocalization in normal (Control) and DMD muscles showing increased labeling of endomysial space with TIMP-1 antibody in DMD muscles by comparison to control muscles. Note the labeling of intramuscular vessels and nerves

and role in these diseases and their potential use as biomarkers or targets in these pathologies. Their role in pathological and physiological processes involving the dystrophin-glycoprotein complex has been reviewed recently [129].

In DMD patients, muscle fibers are lost and replaced by fibro-fatty tissue as a consequence of recurrent cycles of degeneration-regeneration cycles and exhaustion of regenerative capacity. Continuous necrosis-regeneration with its inflammatory component, cytokine production, generation of reactive oxygen species, release of growth factors, and production and activation of MMPs/TIMPs perpetuates constant remodeling of skeletal muscles. The restoration of dystrophin to the sarcolemma can be achieved by cell or gene therapy and result in the interruption of these vicious cycles [130–133].

Dystrophic muscles exhibit MMP-2 and MMP-9 in both x-linked muscular dystrophy (*mdx*) and canine x-linked muscular dystrophy (*CXMD*) [19, 21], whereas only MMP-2 is reported in normal muscles. MMP-9 is also upregulated in the serum of adult *mdx* mice [38] and is overexpressed throughout life span (Alameddine, unpublished results), and gelatinase activity is localized in necrotic fibers, inflammatory cells, and endomysium [33]. Differences in MMP expression patterns have been associated with different time course and amplitude of inflammation and regeneration in *mdx* muscles and experimental injuries [11]. Other MMP, ADAMTS, and TIMP genes are differently regulated in *mdx* muscles including MMP-3, MMP-8, MMP-9, MMP-10, MMP-12, MMP-14, MMP-15, Adamts2, and TIMP-1 mRNA or activity that is increased. On the contrary, MMP-11, Adamts1, Adamts5, Adamts8, TIMP-2, and TIMP-3 mRNA are downregulated [37]. *CXMD* muscles also evidenced upregulation of MMP regulatory molecules, MT1-MMP, TIMP-1, TIMP-2, and RECK [19].

In DMD muscles, TIMP-1, TIMP-2, and MMP-2 transcripts are upregulated, and MMP-2 activity is increased [134]. TIMP-1 levels are also elevated in the serum, plasma, and muscles of muscular dystrophy patients like in patients with fibrotic diseases (Fig.3.2 unpublished results). It correlates with TGF-β1 levels in DMD and congenital muscular dystrophy (CMD) but not with Becker muscular dystrophy patients [135]. TGF-β1 is released from decorin that is degraded by MMP-2.

TGF-β1 modulates the ability of cells to synthesize various ECM components and may modify the protein pattern produced by DMD fibroblasts upon their transformation to myofibroblasts. It increases MMP-7 that contributes to fibrosis [136–138].

3.4.3 Neurogenic Myopathies

Neurogenic myopathies are muscle diseases due to defects of peripheral nervous system with secondary muscle manifestations. Among them are Amyotrophic Lateral Sclerosis (ALS), Spinal Muscle Atrophy (SMA), Guillain-Barre Syndrome (GBS), and Chronic Axonal NeuroPathies (CANP) in which MMPs have been involved in the pathogenesis of the disease. Most of these pathologies have abnormalities of neuromuscular transmission, and despite evidence concerning the involvement of MMP-2 and MMP-9, we still ignore how other MMPs and principally MMP-3 are regulated. This enzyme has the particularity of degrading agrin that participates to the formation and stabilization of neuromuscular junctions (NMJs) [139–141], while agrin degradation leads to destabilization of NMJs and sarcopenia [142, 143]. Mice lacking MMP-3 exhibit higher levels of agrin and increased size and number of AchR in junctional folds [144–146]. Prolonged denervation preserves normal topography of NMJs and preserves agrin and musk at the denervated endplates [147] confirming the role of MMP-3 in the remodeling of NMJs.

As for the other MMPs, experimental evidence shows that initial MMP-9 increase in sciatic nerves and muscle tissue that occurs after axotomy is followed by its increase in the sera [148]. Transient or long-term denervation that results from nerve crush or axotomy is accompanied by remodeling of the nerve, muscle, and NMJs [149], increased turnover of ECM molecules, and proliferation of Schwann cells in the injured nerves. It does not modify MMP-2 and MMP-9 immunolocalization to NMJs, Schwann cells, and perineurium but differentially regulates expression levels of these enzymes. Denervated mouse muscles show persistent MMP-2 labeling at NMJs, but its activity and intensity are decreased in intramuscular nerves. On the contrary, MMP-9 labeling persists at NMJs but is enhanced in degenerated intramuscular nerves suggesting its involvement in axonal degeneration [20]. This is corroborated by high MMP-9 levels that correlate with selective vulnerability and denervation of fast muscles in ALS patients. In the presence of mutant SOD1, MMP-9 expressed by fast motor neurons themselves enhances the activation of endoplasmic reticulum (ER) stress and is sufficient to trigger axonal dieback [150]. On the contrary, reduction of MMP-9 function using gene ablation, viral gene therapy, or pharmacological inhibition significantly delayed muscle denervation.

In ALS patients, strong MMP-9 immunoreactivity reflects later or chronic stages of denervation and allows to differentiate between affected and non-affected patients [116]. Both pro and active forms of MMP-9 are elevated in the sera of ALS patients, whereas elevated TIMPs and MMP-9 activity are reported in postmortem ALS brain tissue, plasma, and CerebroSpinal Fluid (CSF) [151, 152]. Longitudinal monitoring of various MMP family members in the serum and CSF of ALS patients failed to establish a correlation between MMPs and TIMP-1 expression and the age of patients despite the elevation of MT1-MMP, MMP-2, MMP-9, and TIMP-1 in the serum of the majority of cases particularly in mild ALS patients. There is, however, a positive correlation between increased levels and disease duration [153]. The difference of MMP-2 and MMP-9 levels is particularly significant between mild and severe subgroups establishing MMP-9 as a candidate therapeutic target for ALS [150].

Monitoring MMP expression in transgenic mice models of ALS indicates they can be used as biomarkers of disease evolution and follow-up of treatments [154–157]. High gelatinase levels are the common factor linking pathology, response to oxidative stress and cytokine release in symptomatic mice [155]. MMP-9 levels are modulated throughout the course of disease progression in mice. MMP-9 elevation in the sera corresponds to an early stage of disease progression and declines at end stage [156], observation that reminds MMP-2 and MMP-9 increase in mild rather than in severe cases in human patients [153]. At last, in a mouse model of neuraminidase 1 deficiency which results in a neurodegenerative disease in human individuals, muscle biopsies show expansion of epimysial and perimysial spaces, extensive sarcolemmal invaginations, and infiltration of myofibers by fibroblast-like cells and ECM. MMP-2, MMP-9, and cathepsin increase in mice muscles associates with a "unique" pattern of muscle damage, progressive cytosolic fragmentation, and overt myofiber atrophy that may explain neuromuscular manifestations reported in patients with type II form of sialidosis [158].

3.5 MMPs/TIMPs Are Involved in Inflammation and Fibrosis of Diseased Skeletal Muscles

Normal wound healing involves transitory and highly organized response consisting of interrelated dynamic phases with overlapping time courses that lead to tissue replacement [159]. All successive stages, (1) hemostasis, (2) inflammation, (3) migration and proliferation, and (4) remodeling, necessitate high levels of extracellular proteolytic activity [160]. Interference with any of the inflammatory, proliferative, or regenerative phase jeopardizes the success of the repair process possibly leading to fibrosis. Building up of fibrosis results from deregulation between synthesis and hydrolysis of ECM components [161] normally accomplished by MMPs and enzymes of the plasminogen activation system [162–164]. Dysregulated MMP/TIMP balance [33, 162] and excessive or low-grade inflammation contribute to the development of fibrosis and exacerbation of pathological features or organ dysfunction [114, 165] in dystrophic or severely injured muscles.

3.5.1 MMP Involvement in Skeletal Muscle Inflammation

The initiation and progression of tissue remodeling involve a prior step of degradation and reorganization of the ECM scaffold to which inflammation is an indispensable contributor. However, if the inflammatory process is continuous, it may result in deleterious evolution of the disease with organ dysfunction and exacerbation of pathological features.

In a number of experimental injury models or different muscle pathologies, MMP overexpression correlates with the intensity of the inflammatory reaction or the persistence of low-grade inflammatory reaction [11, 21, 36, 38, 166]. Considering that white blood cells produce MMPs to facilitate their migration [167] and regulate their function [168, 169], one may predict that MMP levels increase concomitantly to inflammation. Indeed, MMP-9 is produced by inflammatory cells [170] and stored in granules [171–173] ready to be released during cell transmigration through vessel/capillary walls. Inflammatory cells also regulate MMP-9 production in a manner that resembles sequential variations of MMP-9 at the early stages of muscle regeneration [174]. Of the muscle diseases cited here-above, MMP elevation associates with inflammation in muscular dystrophies and inflammatory myopathies, but not in neurogenic myopathies in which it associates with tissue remodeling [116, 151–153].

During skeletal muscle degeneration/regeneration cycles in animal models or dystrophic patients, the presence of inflammatory cells [175–178] correlates with high MMP-9 in blood vessels, mononuclear cells, regenerating fibers [11, 19, 21, 38, 179], and the serum of dystrophic mice and DMD patients [166, 180]. MMP-1 signal and TIMP-1 and MMP-2 mRNAs localize in areas of degeneration/regeneration and high density of macrophages, whereas TIMP-2 transcripts distribute more homogeneously in mesenchymal fibroblasts [134, 179]. Similar to what is observed in dystrophic muscles, several MMPs including MMP-1, MMP-2, MMP-7, and MMP-9 are increased in correlation with the density of inflammatory cells in these myopathies [117, 119, 181].

Recent evidence indicates a potential benefit of controlling muscle inflammation or inhibiting MMP-9 activity by various natural or chemical compounds to improve skeletal muscle regeneration in dystrophic animal models [36–38, 182]. However, the potential use of MMP inhibition should be considered with great precaution as it may prove deleterious at the long term. The beneficial effect observed upon MMP-9 inhibition in young dystrophic mice [39] is completely reversed by its long-term inhibition which results in the development of fibroadipogenic tissue in old MMP-9-deficient *mdx* mice. Further, experimental data clearly indicate certain MMPs are necessary for the regenerative process [102, 105] and for improving myogenic cell engraftment in dystrophic skeletal muscles [34, 35].

3.5.2 MMP/TIMP Involvement in Skeletal Muscle Fibrosis

The robust regenerative ability of skeletal muscle can be impaired in case of severe injuries or in dystrophic muscles in which massive muscle loss yields a reparative, stabilizing fibrotic response that proceeds more rapidly than the growth of new muscle fibers. Recurrent wound healing response in dystrophic muscles generates, in cascade, molecular modifications that influence tissue reconstitution. In the first stage, a dysregulation between hydrolysis and synthesis of ECM components occurs in favor of hydrolysis and is followed by a resolution phase during which inhibitors are upregulated. Therefore, continuous cycles of necrosis-regeneration in dystrophic muscles may be associated to increase of both MMPs and TIMPs, as observed in DMD muscles [180].

Inflammatory cells contribute to the development of fibrosis by producing cytokines/chemokines [178, 183–185] that upregulate MMPs which, at turn, regulate the cytokines, their ligands, and receptors [186, 187]. MMPs also activate growth factors such as transforming growth factor beta (TGF-β) and connective tissue growth factor (CTGF) [188–190]. Each of these growth factors is reported to induce phenotypic transformation of myoblasts into myofibroblasts [191, 192], whereas CTGF also induces MMP upregulation [193]. Additionally, activated MMPs and a disintegrin and metalloproteinase (ADAM) signal through their receptors and downstream mitogen-activated kinases to activate the transcription of immediate-early genes that mediate fibrosis [194]. ADAM-17 (tumor necrosis factor-α converting enzyme, TACE, or MT4-MMP) and ADAM-12 are influential actors in the pathogenesis of inflammatory and fibrous connective tissue diseases. TACE overexpression and activation activate Epidermal Growth Factor Receptor (EGFR) and stimulate type I collagen expression [195]. It induces transcriptional regulation of MMP-2 and ADAM-12 that activates TGF-β signaling independently of its protease activity [188]. TNF-α release by TACE stimulates collagen synthesis in fibroblasts [196], whereas TNF-α inhibition significantly reduces necrosis [197, 198], the levels of TGF-β1, and type I collagen mRNA in *mdx* mice [199].

Serious indications, although not proven experimentally, indicate that TIMPs are involved in muscle fibrosis. Substantial proliferation of fibrotic tissue correlates with the elevation of TIMP-1 in the serum and endomysium of muscles of DMD patients and dystrophic mouse models [180] and (Fig. 3.2). In cells derived from heart muscles, all four TIMPs are reported to induce fibroblast cell proliferation, but only TIMP-2 upregulates collagen production [200]. This needs to be confirmed in skeletal muscle tissue.

3.6 Conclusion and Perspectives

A subtle balance between MMPs and TIMPs operates at the intracellular and extracellular levels to regulate homeostasis of skeletal muscle tissue. The role of certain MMPs has been clarified in skeletal muscle tissue remodeling. MMP-2 and its activator MMP-14 are linked to angiogenesis and vessel growth [201], and MMP-2 contributes to satellite cell activation by mediating HGF shedding from extracellular matrix in response to NO [202]. Furthermore, MMP-9 and MMP-10 are essential for muscle regeneration, and one may predict that other MMPs may prove essential for this process, particularly those involved in stem cell activation and recruitment or in efferocytosis and macrophage maturation (M1–M2 transition).

An important question that remains unsolved is to understand how two of the best-known and most explored MMPs, namely, MMP-2 and MMP-9, are involved in apparently two contradictory events. On one hand, MMP-2 is involved in muscle fiber atrophy [86, 87, 90, 93, 203] under denervation, immobilization, or unloading conditions. On the other hand, it is involved in muscle fiber hypertrophy and satellite cell activation upon functional overload [70]. A tentative explanation resides in the specific activation of intracellular or membrane-localized MMPs in response to a given stimulus. As for MMP-9, it is shown to be deleterious for skeletal muscle structure and function [36–38, 204, 205], whereas converging data argue in favor of its importance in skeletal muscle formation. Constitutive expression of active MMP-9 in skeletal muscle cells induces muscle fiber hypertrophy [206], and its inactivation decreases muscle and whole body mass and affects cross-sectional areas and fiber-type distribution [207]. Moreover, long-term inhibition leads to adverse effects and the development of fibroadipogenic tissue [39]. Such contradictory results are, essentially, based on serum or whole muscle extracts that neglect the complexity of skeletal muscles and the diversity of its cellular components. These cells may respond differently to the same stimulus, a possibility that underscores the need for in-depth investigation, by various means including immunolocalization and *in situ* zymography, of the nature of cells producing a given protease and how MMP production and activation are affected by the experimental or disease conditions. We also need to document MMP expression and regulation in the high number of diseases affecting skeletal muscles in order to identify potential targets or biomarkers of these pathologies. Circulating MMP-9 is already identified as a marker of DMD [180], and lately adamts5 has been identified as a treatment-responsive biomarker in dystrophic mice and validated in human patients [208].

Lastly, it is important to discuss the effects of MMP modulation in skeletal muscle remodeling. MMP overexpression or inhibition has been applied in animal models of muscle atrophy or muscular dystrophy to investigate their therapeutic benefit. Tetracycline therapy proved beneficial to limit muscle atrophy induced by immobilization [17], and chemical or genetic MMP-9 inhibition in *mdx* mice had beneficial effects reproduced by substances such as L-arginine, nitric oxide (NO) donors, doxycycline (Dox), and minocycline (Min). These substances decrease MMP-9 levels and orchestrate inflammation toward the "repair mode" favoring preservation of structural integrity and reducing fibrosis in different mouse models [205, 209–211]. However, long-term absence of MMP-9 in *mdx* mice impairs the development of the myogenic program, reduces muscle force, and leads to the development of fibroadipogenic tissue [39]. On the contrary, MMP-9 or MMP-1 overexpression by myogenic cells improves myoblast engraftment, favors gene complementation, and helps resolve fibrosis upon grafting into dystrophic muscles [34, 35, 212]. Such contradictory results raise legitimate interrogations about the opportunity to conduct clinical trials using MMP inhibition for the treatment of muscular dystrophies. This is particularly important to highlight that the inhibition of all the other MMPs investigated until now proved to have detrimental effect on skeletal muscle regeneration/maturation [102, 201].

Acknowledgments The author wishes to thank K.Mamchaoui and V. Mouly for generously providing primary human myogenic cells and the AFM-Telethon (Association Française contre les Myopathies) for financial support.

References

1. Allamand V, Briñas L, Richard P, Stojkovic T, Quijano-Roy S, Bonne G (2011) ColVI myopathies: where do we stand, where do we go? Skelet Muscle 1:1–14
2. Allamand V, Merlini L, Bushby K, Consortium for Collagen VI RM (2010) 166th ENMC international workshop on collagen type VI-related myopathies, 22-24 May 2009, Naarden, The Netherlands. Neuromuscul Disord 20:346–354
3. Bonne G, Mercuri E, Muchir A, Urtizberea A, Becane HM, Recan D et al (2000) Clinical and molecular genetic spectrum of autosomal dominant Emery-Dreifuss muscular dystrophy due to mutations of the lamin a/C gene. Ann Neurol 48:170–180
4. Bonne G, Di Barletta MR, Varnous S, Becane HM, Hammouda EH, Merlini L et al (1999) Mutations in the gene encoding lamin A/C cause autosomal dominant Emery-Dreifuss muscular dystrophy. Nat Genet 21:285–288
5. Demir E, Sabatelli P, Allamand V, Ferreiro A, Moghadaszadeh B, Makrelouf M et al (2002) Mutations in COL6A3 cause severe and mild phenotypes of Ullrich congenital muscular dystrophy. Am J Hum Genet 70:1446–1458
6. Helbling-Leclerc A, Zhang X, Topaloglu H, Cruaud C, Tesson F, Weissenbach J et al (1995) Mutations in the laminin alpha 2-chain gene (LAMA2) cause merosin-deficient congenital muscular dystrophy. Nat Genet 11:216–218
7. Muntoni F, Bonne G, Goldfarb LG, Mercuri E, Piercy RJ, Burke M et al (2006) Disease severity in dominant Emery Dreifuss is increased by mutations in both emerin and desmin proteins. Brain 129:1260–1268
8. Fortelny N, Cox JH, Kappelhoff R, Starr AE, Lange PF, Pavlidis P et al (2014) Network analyses reveal pervasive functional regulation between proteases in the human protease web. PLoS Biol 12:e1001869
9. Ricard-Blum S, Vallet SD (2016) Proteases decode the extracellular matrix cryptome. Biochimie 122:300–313
10. McCawley LJ, Matrisian LM (2001) Matrix metalloproteinases: they're not just for matrix anymore! Curr Opin Cell Biol 13:534–540
11. Bani C, Lagrota-Candido J, Pinheiro DF, Leite PE, Salimena MC, Henriques-Pons A et al (2008) Pattern of metalloprotease activity and myofiber regeneration in skeletal muscles of mdx mice. Muscle Nerve 37:583–592
12. Barnes BR, Szelenyi ER, Warren GL, Urso ML (2009) Alterations in mRNA and protein levels of metalloproteinases-2, -9, and -14 and tissue inhibitor of metalloproteinase-2 responses to traumatic skeletal muscle injury. Am J Physiol Cell Physiol 297:C1501–C1508
13. Bellayr I, Mu X, Li Y (2009) Biochemical insights into the role of matrix metalloproteinases in regeneration: challenges and recent developments. Future Med Chem 1:1095–1111
14. Brinckerhoff CE, Matrisian LM (2002) Matrix metalloproteinases: a tail of a frog that became a prince. Nat Rev Mol Cell Biol 3:207–214
15. Burrage PS, Mix KS, Brinckerhoff CE (2006) Matrix metalloproteinases: role in arthritis. Front Biosci 11:529–543

16. Carmeli E, Moas M, Lennon S, Powers SK (2005) High intensity exercise increases expression of matrix metalloproteinases in fast skeletal muscle fibres. Exp Physiol 90:613–619
17. Carmeli E, Kodesh E, Nemcovsky C (2009) Tetracycline therapy for muscle atrophy due to immobilization. J Musculoskelet Neuronal Interact 9:81–88
18. Chen Q, Jin M, Yang F, Zhu J, Xiao Q, Zhang L (2013) Matrix metalloproteinases: inflammatory regulators of cell behaviors in vascular formation and remodeling. Mediat Inflamm 2013:928315
19. Fukushima K, Nakamura A, Ueda H, Yuasa K, Yoshida K, Takeda S et al (2007) Activation and localization of matrix metalloproteinase-2 and -9 in the skeletal muscle of the muscular dystrophy dog (CXMDJ). BMC Musculoskelet Disord 8:54
20. Kherif S, Dehaupas M, Lafuma C, Fardeau M, Alameddine HS (1998) Matrix metalloproteinases MMP-2 and MMP-9 in denervated muscle and injured nerve. Neuropathol Appl Neurobiol 24:309–319
21. Kherif S, Lafuma C, Dehaupas M, Lachkar S, Fournier JG, Verdiere-Sahuque M et al (1999) Expression of matrix metalloproteinases 2 and 9 in regenerating skeletal muscle: a study in experimentally injured and mdx muscles. Dev Biol 205:158–170
22. Ohnishi J, Ohnishi E, Shibuya H, Takahashi T (2005) Functions for proteinases in the ovulatory process. Biochim Biophys Acta 1751:95–109
23. Paiva KBS, Granjeiro JM (2014) Bone tissue remodeling and development: focus on matrix metalloproteinase functions. Arch Biochem Biophys 561:74–87
24. Parks WC, Wilson CL, Lopez-Boado YS (2004) Matrix metalloproteinases as modulators of inflammation and innate immunity. Nat Rev Immunol 4:617–629
25. Gulati AK, Zalewski AA, Reddi AH (1983) An immunofluorescent study of the distribution of fibronectin and laminin during limb regeneration in the adult newt. Dev Biol 96:355–365
26. Gulati AK, Reddi AH, Zalewski AA (1983) Changes in the basement membrane zone components during skeletal muscle fiber degeneration and regeneration. J Cell Biol 97:957–962
27. Porter BE, Weis J, Sanes JR (1995) A motoneuron-selective stop signal in the synaptic protein S-laminin. Neuron 14:549–559
28. Sanes JR, Marshall LM, McMahan UJ (1978) Reinnervation of muscle fiber basal lamina after removal of myofibers. Differentiation of regenerating axons at original synaptic sites. J Cell Biol 78:176–198
29. Page-McCaw A, Ewald AJ, Werb Z (2007) Matrix metalloproteinases and the regulation of tissue remodelling. Nat Rev Mol Cell Biol 8:221–233
30. Couch CB, Strittmatter WJ (1983) Rat myoblast fusion requires metalloendoprotease activity. Cell 32:257–265
31. Couch CB, Strittmatter WJ (1984) Specific blockers of myoblast fusion inhibit a soluble and not the membrane-associated metalloendoprotease in myoblasts. J Biol Chem 259:5396–5399
32. Yagami-Hiromasa T, Sato T, Kurisaki T, Kamijo K, Nabeshima Y, Fujisawa-Sehara A (1995) A metalloprotease-disintegrin participating in myoblast fusion. Nature 377:652–656
33. Morgan J, Alameddine HS (2012). Stem cell based therapy for muscular dystrophies: cell types and environmental factors influencing their efficacy, muscular dystrophy, Dr. Madhuri Hegde (ed), InTech, doi:10.5772/30831. Available from: http://www.intechopen.com/books/muscular-dystrophy/stem-cell-based-therapy-for-muscular-dystrophies-cell-types-and-environmental-factors-influencing-th
34. Morgan J, Rouche A, Bausero P, Houssaïni A, Gross J, Fiszman MY et al (2010) MMP-9 overexpression improves myogenic cell migration and engraftment. Muscle Nerve 42:584–595
35. Pan H, Vojnits K, Liu TT, Meng F, Yang L, Wang Y et al (2015) MMP1 gene expression enhances myoblast migration and engraftment following implanting into mdx/SCID mice. Cell Adhes Migr 9:283–292
36. Hindi SM, Shin J, Ogura Y, Li H, Kumar A (2013) Matrix metalloproteinase-9 inhibition improves proliferation and engraftment of myogenic cells in dystrophic muscle of mdx mice. PLoS One 8:e72121

37. Kumar A, Bhatnagar S, Kumar A (2010) Matrix metalloproteinase inhibitor batimastat alleviates pathology and improves skeletal muscle function in dystrophin-deficient mdx mice. Am J Pathol 177:248–260
38. Li H, Mittal A, Makonchuk DY, Bhatnagar S, Kumar A (2009) Matrix metalloproteinase-9 inhibition ameliorates pathogenesis and improves skeletal muscle regeneration in muscular dystrophy. Hum Mol Genet 18:2584–2598
39. Shiba N, Miyazaki D, Yoshizawa T, Fukushima K, Shiba Y, Inaba Y et al (2015) Differential roles of MMP-9 in early and late stages of dystrophic muscles in a mouse model of Duchenne muscular dystrophy. Biochim Biophys Acta 1852:2170–2182
40. Balcerzak D, Querengesser L, Dixon WT, Baracos VE (2001) Coordinated expression of matrix-degrading proteinases and their activators and inhibitors in bovine skeletal muscle. J Anim Sci 79:94–107
41. Caron NJ, Asselin I, Morel G, Tremblay JP (1999) Increased myogenic potential and fusion of matrilysin-expressing myoblasts transplanted in mice. Cell Transplant 8:465–476
42. El Fahime E, Torrente Y, Caron NJ, Bresolin MD, Tremblay JP (2000) In vivo migration of transplanted myoblasts requires matrix metalloproteinase activity. Exp Cell Res 258:279–287
43. Guérin CW, Holland PC (1995) Synthesis and secretion of matrix-degrading metalloproteases by human skeletal muscle satellite cells. Dev Dyn 202:91–99
44. Lewis MP, Tippett HL, Sinanan AC, Morgan MJ, Hunt NP (2000) Gelatinase-B (matrix metalloproteinase-9; MMP-9) secretion is involved in the migratory phase of human and murine muscle cell cultures. J Muscle Res Cell Motil 21:223–233
45. Lluri G, Jaworski DM (2005) Regulation of TIMP-2, MT1-MMP, and MMP-2 expression during C2C12 differentiation. Muscle Nerve 32:492–499
46. Nishimura T, Nakamura K, Kishioka Y, Kato-Mori Y, Wakamatsu J, Hattori A (2008) Inhibition of matrix metalloproteinases suppresses the migration of skeletal muscle cells. J Muscle Res Cell Motil 29:37–44
47. Ohtake Y, Tojo H, Seiki M (2006) Multifunctional roles of MT1-MMP in myofiber formation and morphostatic maintenance of skeletal muscle. J Cell Sci 119:3822–3832
48. Torrente Y, El Fahime E, Caron NJ, Del Bo R, Belicchi M, Pisati F et al (2003) Tumor necrosis factor-alpha (TNF-alpha) stimulates chemotactic response in mouse myogenic cells. Cell Transplant 12:91–100
49. Dehne N, Kerkweg U, Flohe SB, Brune B, Fandrey J (2011) Activation of hypoxia-inducible factor 1 in skeletal muscle cells after exposure to damaged muscle cell debris. Shock 35:632–638
50. Diomedi-Camassei F, Boldrini R, Ravà L, Donfrancesco A, Boglino C, Messina E et al (2004) Different pattern of matrix metalloproteinases expression in alveolar versus embryonal rhabdomyosarcoma. J Ped Surg 39:1673–1679
51. Roomi MW, Kalinovsky T, Rath M, Niedzwiecki A (2014) In vitro modulation of MMP-2 and MMP-9 in pediatric human sarcoma cell lines by cytokines, inducers and inhibitors. Int J Oncol 44:27–34
52. Allen DL, Teitelbaum DH, Kurachi K (2003) Growth factor stimulation of matrix metalloproteinase expression and myoblast migration and invasion in vitro. Am J Physiol - Cell Physiol 284:C805–C815
53. Yeghiazaryan M, Zybura-Broda K, Cabaj A, Wlodarczyk J, Slawinska U, Rylski M et al (2012) Fine-structural distribution of MMP-2 and MMP-9 activities in the rat skeletal muscle upon training: a study by high-resolution in situ zymography. Histochem Cell Biol 138:75–87
54. Hadler-Olsen E, Solli AI, Hafstad A, Winberg JO, Uhlin-Hansen L (2015) Intracellular MMP-2 activity in skeletal muscle is associated with type II fibers. J Cell Physiol 230:160–169
55. Chin JR, Werb Z (1997) Matrix metalloproteinases regulate morphogenesis, migration and remodeling of epithelium, tongue skeletal muscle and cartilage in the mandibular arch. Development 124:1519–1530
56. Wang W, Pan H, Murray K, Jefferson BS, Li Y (2009) Matrix metalloproteinase-1 promotes muscle cell migration and differentiation. Am J Pathol 174:541–549

57. Ito H, Hallauer PL, Hastings KE, Tremblay JP (1998) Prior culture with concanavalin A increases intramuscular migration of transplanted myoblast. Muscle Nerve 21:291–297
58. Jaworski DM, Soloway P, Caterina J, Falls WA (2006) Tissue inhibitor of metalloproteinase-2(TIMP-2)-deficient mice display motor deficits. J Neurobiol 66:82–94
59. Lluri G, Langlois GD, Soloway PD, Jaworski DM (2008) Tissue inhibitor of metalloproteinase-2 (TIMP-2) regulates myogenesis and beta1 integrin expression in vitro. Exp Cell Res 314:11–24
60. Holmbeck K, Bianco P, Caterina J, Yamada S, Kromer M, Kuznetsov SA et al (1999) MT1-MMP-deficient mice develop dwarfism, osteopenia, arthritis, and connective tissue disease due to inadequate collagen turnover. Cell 99:81–92
61. Itoh T, Ikeda T, Gomi H, Nakao S, Suzuki T, Itohara S (1997) Unaltered secretion of beta-amyloid precursor protein in gelatinase A (matrix metalloproteinase 2)-deficient mice. J Biol Chem 272:22389–22392
62. Mosig RA, Dowling O, DiFeo A, Ramirez MC, Parker IC, Abe E et al (2007) Loss of MMP-2 disrupts skeletal and craniofacial development and results in decreased bone mineralization, joint erosion and defects in osteoblast and osteoclast growth. Hum Mol Genet 16:1113–1123
63. Martignetti JA, Aqeel AA, Sewairi WA, Boumah CE, Kambouris M, Mayouf SA et al (2001) Mutation of the matrix metalloproteinase 2 gene (MMP2) causes a multicentric osteolysis and arthritis syndrome. Nat Genet 28:261–265
64. Plaghki L (1985) Regeneration and myogenesis of striated muscle. J Physiol Paris 80:51–110
65. Oh J, Takahashi R, Adachi E, Kondo S, Kuratomi S, Noma A et al (2004) Mutations in two matrix metalloproteinase genes, MMP-2 and MT1-MMP, are synthetic lethal in mice. Oncogene 23:5041–5048
66. Liu H, Chen SE, Jin B, Carson JA, Niu A, Durham W et al (2010) TIMP3: a physiological regulator of adult myogenesis. J Cell Sci 123:2914–2921
67. Echizenya M, Kondo S, Takahashi R, Oh J, Kawashima S, Kitayama H et al (2005) The membrane-anchored MMP-regulator RECK is a target of myogenic regulatory factors. Oncogene 24:5850–5857
68. Fry CS, Lee JD, Mula J, Kirby TJ, Jackson JR, Liu F et al (2015) Inducible depletion of satellite cells in adult, sedentary mice impairs muscle regenerative capacity without affecting sarcopenia. Nat Med 21:76–80
69. Devine RD, Bicer S, Reiser PJ, Velten M, Wold LE (2015) Metalloproteinase expression is altered in cardiac and skeletal muscle in cancer cachexia. Am J Physiol Heart Circ Physiol 309:H685–H691
70. Zhang Q, Joshi SK, Lovett DH, Zhang B, Bodine S, Kim H et al (2014) Matrix metalloproteinase-2 plays a critical role in overload induced skeletal muscle hypertrophy. Muscles Ligaments Tendons J 4:362–370
71. Faulkner JA, Brooks SV, Opiteck JA (1993) Injury to skeletal muscle fibers during contractions: conditions of occurrence and prevention. Phys Ther 73:911–921
72. Rullman E, Norrbom J, Stromberg A, Wagsater D, Rundqvist H, Haas T et al (2009) Endurance exercise activates matrix metalloproteinases in human skeletal muscle. J Appl Physiol 106:804–812
73. Fry CS, Lee JD, Jackson JR, Kirby TJ, Stasko SA, Liu H et al (2014) Regulation of the muscle fiber microenvironment by activated satellite cells during hypertrophy. FASEB J 28:1654–1665
74. Haas TL, Milkiewicz M, Davis SJ, Zhou AL, Egginton S, Brown MD et al (2000) Matrix metalloproteinase activity is required for activity-induced angiogenesis in rat skeletal muscle. Am J Physiol Heart Circ Physiol 279:H1540–H1547
75. Brown MD, Hudlicka O (2003) Modulation of physiological angiogenesis in skeletal muscle by mechanical forces: involvement of VEGF and metalloproteinases. Angiogenesis 6:1–14
76. Rullman E, Rundqvist H, Wagsater D, Fischer H, Eriksson P, Sundberg CJ et al (2007) A single bout of exercise activates matrix metalloproteinase in human skeletal muscle. J Appl Physiol 102:2346–2351

77. Alameddine HS (2004) La plasticité du tissu musculaire. In: Didier JP (ed) La plasticité du tissu musculaire. Springer, Paris, France, pp 55–105
78. Reihmane D, Tretjakovs P, Kaupe J, Sars M, Valante R, Jurka A (2012) Systemic pro-inflammatory molecule response to acute submaximal exercise in moderately and highly trained athletes. Environmental and Experimental Biology 10:107–112
79. Rooney SI, Tobias JW, Bhatt PR, Kuntz AF, Soslowsky LJ (2015) Genetic response of rat supraspinatus tendon and muscle to exercise. PLoS One 10:e0139880
80. Klein G, Schmal O, Aicher WK (2015) Matrix metalloproteinases in stem cell mobilization. Matrix Biol 44–46:175–183
81. Raffeto JD, Khalil RA (2008) Matrix metalloproteinases and their inhibitors in vascular remodeling and vascular disease. Biochem Pharmacol 75:346–359
82. van Hinsbergh VWM, Koolwijk P (2008) Endothelial sprouting and angiogenesis: matrix metalloproteinases in the lead. Cardiovasc Res 78:203–212
83. Han JK, Kim HL, Jeon KH, Choi YE, Lee HS, Kwon YW et al (2013) Peroxisome proliferator-activated receptor-delta activates endothelial progenitor cells to induce angio-myogenesis through matrix metallo-proteinase-9-mediated insulin-like growth factor-1 paracrine networks. Eur Heart J 34:1755–1765
84. Dennis RA, Zhu H, Kortebein PM, Bush HM, Harvey JF, Sullivan DH et al (2009) Muscle expression of genes associated with inflammation, growth, and remodeling is strongly correlated in older adults with resistance training outcomes. Physiol Genomics 38:169–175
85. Fiotti N, Deiuri E, Altamura N, De Colle P, Moretti ME, Toigo G, et al. (2009) Body composition and muscular strength changes after moderate activity: association with matrix metalloproteinase polymorphisms. Arch Gerontol Geriatr 49(Suppl 1): 83–94
86. Peviani SM, Russo TL, Durigan JL, Vieira BS, Pinheiro CM, Galassi MS et al (2009) Stretching and electrical stimulation regulate the metalloproteinase-2 in rat denervated skeletal muscle. Neurol Res 32:891–896
87. Russo TL, Peviani SM, Durigan JL, Salvini TF (2008) Electrical stimulation increases matrix metalloproteinase-2 gene expression but does not change its activity in denervated rat muscle. Muscle Nerve 37:593–600
88. Reznick AZ, Menashe O, Bar-Shai M, Coleman R, Carmeli E (2003) Expression of matrix metalloproteinases, inhibitor, and acid phosphatase in muscles of immobilized hindlimbs of rats. Muscle Nerve 27:51–59
89. Berthon P, Duguez S, Favier FB, Amirouche A, Feasson L, Vico L et al (2007) Regulation of ubiquitin-proteasome system, caspase enzyme activities, and extracellular proteinases in rat soleus muscle in response to unloading. Pflugers Arch 454:625–633
90. Giannelli G, De Marzo A, Marinosci F, Antonaci S (2005) Matrix metalloproteinase imbalance in muscle disuse atrophy. Histol Histopathol 20:99–106
91. Wittwer M, Flück M, Hoppeler H, Müller S, Desplanches D, Billeter R (2002) Prolonged unloading of rat soleus muscle causes distinct adaptations of the gene profile. FASEB J 16:884–886
92. Stevenson EJ, Giresi PG, Koncarevic A, Kandarian SC (2003) Global analysis of gene expression patterns during disuse atrophy in rat skeletal muscle. J Physiol 551:33–48
93. Liu X, Lee DJ, Skittone LK, Natsuhara K, Kim HT (2010) Role of gelatinases in disuse-induced skeletal muscle atrophy. Muscle Nerve 41:174–178
94. Peviani SM, Gomes AR, Selistre de Araujo HS, Salvini TF (2009) MMP-2 is not altered by stretching in skeletal muscle. Int J Sports Med 30:550–554
95. Warren GL, Summan M, Gao X, Chapman R, Hulderman T, Simeonova PP (2007) Mechanisms of skeletal muscle injury and repair revealed by gene expression studies in mouse models. J Physiol 582:825–841
96. Ferre PJ, Liaubet L, Concordet D, SanCristobal M, Uro-Coste E, Tosser-Klopp G et al (2007) Longitudinal analysis of gene expression in porcine skeletal muscle after post-injection local injury. Pharmacol Res 24:1480–1489

97. Frisdal E, Teiger E, Lefaucheur JP, Adnot S, Planus E, Lafuma C et al (2000) Increased expression of gelatinases and alteration of basement membrane in rat soleus muscle following femoral artery ligation. Neuropathol Appl Neurobiol 26:11–21
98. Roach DM, Fitridge RA, Laws PE, Millard SH, Varelias A, Cowled PA (2002) Up-regulation of MMP-2 and MMP-9 leads to degradation of type IV collagen during skeletal muscle reperfusion injury; protection by the MMP inhibitor, Doxycycline. Eur J Vasc Endovasc Surg 23:260–269
99. Baum O, Ganster M, Baumgartner I, Nieselt K, Djonov V (2007) Basement membrane remodeling in skeletal muscles of patients with limb ischemia involves regulation of matrix metalloproteinases and tissue inhibitor of matrix metalloproteinases. J Vasc Res 44:202–213
100. Zimowska M, Brzoska E, Swierczynska M, Streminska W, Moraczewski J (2008) Distinct patterns of MMP-9 and MMP-2 activity in slow and fast twitch skeletal muscle regeneration in vivo. Int J Dev Biol 52:307–314
101. Urso ML, Szelenyi ER., Warren GL, Barnes BR (2010) Matrix metalloprotease-3 and tissue inhibitor of metalloprotease-1 mRNA and protein levels are altered in response to traumatic skeletal muscle injury. Eur J App Physiol 109:963–972
102. Bobadilla M, Sainz N, Rodriguez JA, Abizanda G, Orbe J, de Martino A et al (2014) MMP-10 is required for efficient muscle regeneration in mouse models of injury and muscular dystrophy. Stem Cells 32:447–461
103. Lei H, Leong D, Smith LR, Barton ER (2013) Matrix-metalloproteinase 13 (MMP-13) is a new contributor to skeletal muscle regeneration and critical for myoblast migration. Am J Physiol Cell Physiol 305:C529–C538
104. Wu N, Jansen ED, Davidson JM (2003) Comparison of mouse matrix metalloproteinase 13 expression in free-electron laser and scalpel incisions during wound healing. J Invest Dermatol 121:926–932
105. Bobadilla M, Sainz N, Abizanda G, Orbe J, Rodriguez JA, Paramo JA et al (2014) The CXCR4/SDF1 axis improves muscle regeneration through MMP-10 activity. Stem Cells Dev 23:1417–1427
106. Koskinen SO, Hoyhtya M, Turpeenniemi-Hujanen T, Martikkala V, Makinen TT, Oksa J et al (2001) Serum concentrations of collagen degrading enzymes and their inhibitors after downhill running. Scand J Med Sci Sports 11:9–15
107. Lo IK, Marchuk LL, Hollinshead R, Hart DA, Frank CB (2004) Matrix metalloproteinase and tissue inhibitor of matrix metalloproteinase mRNA levels are specifically altered in torn rotator cuff tendons. Am J Sports Med 32:1223–1229
108. Thornton GM, Shao X, Chung M, Sciore P, Boorman RS, Hart DA et al (2010) Changes in mechanical loading lead to tendonspecific alterations in MMP and TIMP expression: influence of stress deprivation and intermittent cyclic hydrostatic compression on rat supraspinatus and achilles tendons. Br J Sports Med 44:698–703
109. Pasternak B, Aspenberg P (2009) Metalloproteinases and their inhibitors-diagnostic and therapeutic opportunities in orthopedics. Acta Orthop 80:693–703
110. Reider B (2014) Big D. Am J Sports Med 42:25–26
111. Bedi A, Fox AJS, Kovacevic D, Deng X-h, Warren RF, Rodeo SA (2010) Doxycycline-mediated inhibition of matrix metalloproteinases improves healing after rotator cuff repair. Am J Sports Med 38:308–317
112. Bedi A, Kovacevic D, Hettrich C, Gulotta LV, Ehteshami JR, Warren RF et al (2010) The effect of matrix metalloproteinase inhibition on tendon-to-bone healing in a rotator cuff repair model. J Shoulder Elb Surg 19:384–391
113. Dalakas MC (2015) Inflammatory muscle diseases. N Engl J Med 372:1734–1747
114. Manicone AM, McGuire JK (2008) Matrix metalloproteinases as modulators of inflammation. Semin Cell Dev Biol 19:34–41
115. Clutterbuck AL, Asplin KE, Harris P, Allaway D, Mobasheri A (2009) Targeting matrix metalloproteinases in inflammatory conditions. Curr Drug Targets 10:1245–1254
116. Schoser BG, Blottner D (1999) Matrix metalloproteinases MMP-2, MMP-7 and MMP-9 in denervated human muscle. Neuroreport 10:2795–2797

117. Kieseier BC, Schneider C, Clements JM, Gearing AJ, Gold R, Toyka KV et al (2001) Expression of specific matrix metalloproteinases in inflammatory myopathies. Brain 124:341–351
118. Choi YC, Dalakas MC (2000) Expression of matrix metalloproteinases in the muscle of patients with inflammatory myopathies. Neurology 54:65–71
119. Schoser BGH, Blottner D, Stuerenburg HJ (2002) Matrix metalloproteinases in inflammatory myopathies: enhanced immunoreactivity near atrophic myofibers. Acta Neurol Scand 105:309–313
120. Emery AE (2002) The muscular dystrophies. Lancet 359:687–695
121. McAdam LC, Mayo AL, Alman BA, Biggar WD (2012) The Canadian experience with long-term deflazacort treatment in Duchenne muscular dystrophy. Acta Myol 31:16–20
122. Bushby K, Finkel R, Birnkrant DJ, Case LE, Clemens PR, Cripe L et al (2010) Diagnosis and management of Duchenne muscular dystrophy, part 1: diagnosis, and pharmacological and psychosocial management. Lancet Neurol 9:77–93
123. Bushby K, Finkel R, Birnkrant DJ, Case LE, Clemens PR, Cripe L et al (2010) Diagnosis and management of Duchenne muscular dystrophy, part 2: implementation of multidisciplinary care. Lancet Neurol 9:177–189
124. Hayashi YK, Ogawa M, Tagawa K, Noguchi S, Ishihara T, Nonaka I et al (2001) Selective deficiency of alpha-dystroglycan in Fukuyama-type congenital muscular dystrophy. Neurology 57:115–121
125. Jimenez-Mallebrera C, Torelli S, Brown SC, Feng L, Brockington M, Sewry CA et al (2003) Profound skeletal muscle depletion of alpha-dystroglycan in Walker-Warburg syndrome. Eur J Paediatr Neurol 7:129–137
126. Longman C, Brockington M, Torelli S, Jimenez-Mallebrera C, Kennedy C, Khalil N et al (2003) Mutations in the human LARGE gene cause MDC1D, a novel form of congenital muscular dystrophy with severe mental retardation and abnormal glycosylation of alpha-dystroglycan. Hum Mol Genet 12:2853–2861
127. Johnson EK, Li B, Yoon JH, Flanigan KM, Martin PT, Ervasti J et al (2013) Identification of new dystroglycan complexes in skeletal muscle. PLoS One 8:e73224
128. Yamada H, Saito F, Fukuta-Ohi H, Zhong D, Hase A, Arai K et al (2001) Processing of beta-dystroglycan by matrix metalloproteinase disrupts the link between the extracellular matrix and cell membrane via the dystroglycan complex. Hum Mol Genet 10:1563–1569
129. Bozzi M, Sciandra F, Brancaccio A (2015) Role of gelatinases in pathological and physiological processes involving the dystrophin–glycoprotein complex. Matrix Biol 44–46:130–137
130. Verhaart IE, Aartsma-Rus A (2012) Gene therapy for Duchenne muscular dystrophy. Curr Opin Neurol 25:588–596
131. Skuk D, Tremblay JP (2011) Intramuscular cell transplantation as a potential treatment of myopathies: clinical and preclinical relevant data. Expert Opin Biol Ther 11:359–374
132. Negroni E, Vallese D, Vilquin JT, Butler-Browne G, Mouly V, Trollet C (2011) Current advances in cell therapy strategies for muscular dystrophies. Expert Opin Biol Ther 11:157–176
133. Morgan JE, Partridge TA (1992) Cell transplantation and gene therapy in muscular dystrophy. BioEssays 14:641–645
134. Von Moers A, Zwirner A, Reinhold A, Bruckmann O, van Landeghem F, Stoltenburg-Didinger G et al (2005) Increased mRNA expression of tissue inhibitors of metalloproteinase-1 and -2 in Duchenne muscular dystrophy. Acta Neuropathol 109:285–293
135. Sun G, Haginoya K, Chiba Y, Uematsu M, Hino-Fukuyo N, Tanaka S et al (2010) Elevated plasma levels of tissue inhibitors of metalloproteinase-1 and their overexpression in muscle in human and mouse muscular dystrophy. J Neurol Sci 297:19–28
136. Fadic R, Mezzano V, Alvarez K, Cabrera D, Holmgren J, Brandan E (2006) Increase in decorin and biglycan in Duchenne muscular dystrophy: role of fibroblasts as cell source of these proteoglycans in the disease. J Cell Mol Med 10:758–769

137. Zanotti S, Gibertini S, Mora M (2010) Altered production of extra-cellular matrix components by muscle-derived Duchenne muscular dystrophy fibroblasts before and after TGF-β1 treatment. Cell Tissue Res 339:397–410
138. Zanotti S, Saredi S, Ruggieri A, Fabbri M, Blasevich F, Romaggi S et al (2007) Altered extracellular matrix transcript expression and protein modulation in primary Duchenne muscular dystrophy myotubes. Matrix Biol 26:615–624
139. Bolliger MF, Zurlinden A, Lüscher D, Bütikofer L, Shakhova O, Francolini M et al (2010) Specific proteolytic cleavage of agrin regulates maturation of the neuromuscular junction. J Cell Sci 123:3944–3955
140. Choi HY, Liu Y, Tennert C, Sugiura YI, A K, Kröger S, et al. (2013) APP interacts with LRP4 and agrin to coordinate the development of the neuromuscular junction in mice. elife 2013 2:e00220
141. Ruegg MA, Bixby JL (1998) Agrin orchestrates synaptic differentiation at the vertebrate neuromuscular junction. Trends Neurosci 21:22–27
142. Bütikofer L, Zurlinden A, Bolliger MF, Kunz B, Sonderegger P (2011) Destabilization of the neuromuscular junction by proteolytic cleavage of agrin results in precocious sarcopenia. FASEB J 25:4378–4393
143. Drey M, Sieber CC, Bauer JM, Uter W, Dahinden P, Fariello RG et al (2013) C-terminal Agrin fragment as a potential marker for sarcopenia caused by degeneration of the neuromuscular junction. Exp Gerontol 48:76–80
144. Van Saun M, Werle MJ (2000) Matrix metalloproteinase-3 removes agrin from synaptic basal lamina. J Neurobiol 43:140–149
145. VanSaun M, Herrera AA, Werle MJ (2003) Structural alterations at the neuromuscular junctions of matrix metalloproteinase 3 null mutant mice. J Neurocytol 32:1129–1142
146. Werle MJ, VanSaun M (2003) Activity dependent removal of agrin from synaptic basal lamina by matrix metalloproteinase 3. J Neurocytol 32:905–913
147. Chao T, Frump D, Lin M, Caiozzo VJ, Mozaffar T, Steward O et al (2013) Matrix metalloproteinase 3 deletion preserves denervated motor endplates after traumatic nerve injury. Ann Neurol 73:210–223
148. Demestre M, Parkin-Smith G, Petzold A, Pullen AH (2005) The pro and the active form of matrix metalloproteinase-9 is increased in serum of patients with amyotrophic lateral sclerosis. J Neuroimmunol 159:146–154
149. Ozawa J, Kurose T, Kawamata S, Kaneguchi A, Moriyama H, Kito N (2013) Regulation of connective tissue remodeling in the early phase of denervation in a rat skeletal muscle. Biomed Res 34:251–258
150. Kaplan A, Spiller Krista J, Towne C, Kanning Kevin C, Choe Ginn T, Geber A et al (2014) Neuronal matrix metalloproteinase-9 is a determinant of selective neurodegeneration. Neuron 81:333–348
151. Beuche W, Yushchenko M, Mader M, Maliszewska M, Felgenhauer K, Weber F (2000) Matrix metalloproteinase-9 is elevated in serum of patients with amyotrophic lateral sclerosis. Neuroreport 11:3419–3422
152. Lorenzl S, Albers DS, LeWitt PA, Chirichigno JW, Hilgenberg SL, Cudkowicz ME et al (2003) Tissue inhibitors of matrix metalloproteinases are elevated in cerebrospinal fluid of neurodegenerative diseases. J Neurol Sci 207:71–76
153. Niebroj-Dobosz I, Janik P, Sokołowska B, Kwiecinski H (2010) Matrix metalloproteinases and their tissue inhibitors in serum and cerebrospinal fluid of patients with amyotrophic lateral sclerosis. Eur J Neurol 17:226–231
154. Lorenzl S, Narr S, Angele B, Krell HW, Gregorio J, Kiaei M et al (2006) The matrix metalloproteinases inhibitor Ro 26-2853 extends survival in transgenic ALS mice. Exp Neurol 200:166–171
155. Fang L, Teuchert M, Huber-Abel F, Schattauer D, Hendrich C, Dorst J et al (2010) MMP-2 and MMP-9 are elevated in spinal cord and skin in a mouse model of ALS. J Neurol Sci 294:51–56

156. Soon CP, Crouch PJ, Turner BJ, McLean CA, Laughton KM, Atkin JD et al (2010) Serum matrix metalloproteinase-9 activity is dysregulated with disease progression in the mutant SOD1 transgenic mice. Neuromuscul Disord 20:260–266
157. Kriz J, Nguyen MD, Julien JP (2002) Minocycline slows disease progression in a mouse model of amyotrophic lateral sclerosis. Neurobiol Dis 10:268–278
158. Zanoteli E, van de Vlekkert D, Bonten EJ, Hu H, Mann L, Gomero EM et al (2010) Muscle degeneration in neuraminidase 1-deficient mice results from infiltration of the muscle fibers by expanded connective tissue. Biochim Biophys Acta 1802:659–672
159. Micallef L, Vedrenne N, Billet F, Coulomb B, Darby IA, Desmouliere A (2012) The myofibroblast, multiple origins for major roles in normal and pathological tissue repair. Fibrogenesis Tissue Repair 5(Suppl 1):S5
160. Mignatti P, Rifkin DB (1993) Biology and biochemistry of proteinases in tumor invasion. Physiol Rev 73:161–195
161. Serrano AL, Munoz-Canoves P (2010) Regulation and dysregulation of fibrosis in skeletal muscle. Exp Cell Res 316:3050–3058
162. Chen X, Li Y (2009) Role of matrix metalloproteinases in skeletal muscle: migration, differentiation, regeneration and fibrosis. Cell Adhes Migr 3:337–341
163. Lluis F, Roma J, Suelves M, Parra M, Aniorte G, Gallardo E et al (2001) Urokinase-dependent plasminogen activation is required for efficient skeletal muscle regeneration in vivo. Blood 97:1703–1711
164. Nagamine Y, Medcalf RL, Munoz-Canoves P (2005) Transcriptional and posttranscriptional regulation of the plasminogen activator system. Thromb Haemost 93:661–675
165. Sbardella D, Fasciglione GF, Gioia M, Ciaccio C, Tundo GR, Marini S et al (2012) Human matrix metalloproteinases: an ubiquitarian class of enzymes involved in several pathological processes. Mol Asp Med 33:119–208
166. Alameddine HS (2012) Matrix metalloproteinases in skeletal muscles: friends or foes? Neurobiol Dis 48:508–518
167. Goetzl EJ, Banda MJ, Leppert D (1996) Matrix metalloproteinases in immunity. J Immunol 156:1–4
168. Sopata I, Dancewicz AM (1974) Presence of a gelatin-specific proteinase and its latent form in human leucocytes. Biochim Biophys Acta 370:510–523
169. Mainardi CL, Hibbs MS, Hasty KA, Seyer JM (1984) Purification of a type V collagen degrading metalloproteinase from rabbit alveolar macrophages. Coll Relat Res 4:479–492
170. Opdenakker G, Van den Steen PE, Dubois B, Nelissen I, Van Coillie E, Masure S et al (2001) Gelatinase B functions as regulator and effector in leukocyte biology. J Leukoc Biol 69:851–859
171. Baram D, Vaday GG, Salamon P, Drucker I, Hershkoviz R, Mekori YA (2001) Human mast cells release metalloproteinase-9 on contact with activated T cells: juxtacrine regulation by TNF-alpha. J Immunol 167:4008–4016
172. Chakrabarti S, Zee JM, Patel KD (2006) Regulation of matrix metalloproteinase-9 (MMP-9) in TNF-stimulated neutrophils: novel pathways for tertiary granule release. J Leukoc Biol 79:214–222
173. Cowland JB, Borregaard N (1999) The individual regulation of granule protein mRNA levels during neutrophil maturation explains the heterogeneity of neutrophil granules. J Leukoc Biol 66:989–995
174. Kolaczkowska E, Arnold B, Opdenakker G (2008) Gelatinase B/MMP-9 as an inflammatory marker enzyme in mouse zymosan peritonitis: comparison of phase-specific and cell-specific production by mast cells, macrophages and neutrophils. Immunobiology 213:109–124
175. Gorospe JR, Tharp MD, Hinckley J, Kornegay JN, Hoffman EP (1994) A role for mast cells in the progression of Duchenne muscular dystrophy? Correlations in dystrophin-deficient humans, dogs, and mice. J Neurol Sci 122:44–56
176. Hodgetts S, Radley H, Davies M, Grounds MD (2006) Reduced necrosis of dystrophic muscle by depletion of host neutrophils, or blocking TNFalpha function with Etanercept in mdx mice. Neuromuscul Disord 16:591–602

177. Cai B, Spencer MJ, Nakamura G, Tseng-Ong L, Tidball JG (2000) Eosinophilia of dystrophin-deficient muscle is promoted by perforin-mediated cytotoxicity by T cell effectors. Am J Pathol 156:1789–1796
178. Spencer MJ, Montecino-Rodriguez E, Dorshkind K, Tidball JG (2001) Helper (CD4(+)) and cytotoxic (CD8(+)) T cells promote the pathology of dystrophin-deficient muscle. Clin Immunol 98:235–243
179. Zanotti S, Gibertini S, Di Blasi C, Cappelletti C, Bernasconi P, Mantegazza R et al (2011) Osteopontin is highly expressed in severely dystrophic muscle and seems to play a role in muscle regeneration and fibrosis. Histopathol 59:1215–1228
180. Nadarajah VD, van Putten M, Chaouch A, Garrood P, Straub V, Lochmuller H et al (2011) Serum matrix metalloproteinase-9 (MMP-9) as a biomarker for monitoring disease progression in Duchenne muscular dystrophy (DMD). Neuromuscul Disord 21:569–578
181. Rodolico C, Mazzeo A, Toscano A, Messina S, Aguennouz M, Gaeta M et al (2005) Specific matrix metalloproteinase expression in focal myositis: an immunopathological study. Acta Neurol Scand 112:173–177
182. Ogura Y, Tajrishi MM, Sato S, Hindi SM, Kumar A (2014) Therapeutic potential of matrix metalloproteinases in Duchenne muscular dystrophy. Front Cell Dev Biol 2:11
183. Kharraz Y, Guerra J, Mann CJ, Serrano AL, Munoz-Canoves P (2013) Macrophage plasticity and the role of inflammation in skeletal muscle repair. Mediat Inflamm 2013:491497
184. Mann CJ, Perdiguero E, Kharraz Y, Aguilar S, Pessina P, Serrano AL et al (2011) Aberrant repair and fibrosis development in skeletal muscle. Skelet Muscle 1:21
185. Morrison J, Lu QL, Pastoret C, Partridge T, Bou-Gharios G (2000) T-cell-dependent fibrosis in the mdx dystrophic mouse. Lab Investig 80:881–891
186. Ito A, Mukaiyama A, Itoh Y, Nagase H, Thogersen IB, Enghild JJ et al (1996) Degradation of interleukin 1beta by matrix metalloproteinases. J Biol Chem 271:14657–14660
187. Schonbeck U, Mach F, Libby P (1998) Generation of biologically active IL-1 beta by matrix metalloproteinases: a novel caspase-1-independent pathway of IL-1 beta processing. J Immunol 161:3340–3346
188. Atfi A, Dumont E, Colland F, Bonnier D, L'Helgoualc'h A, Prunier C et al (2007) The disintegrin and metalloproteinase ADAM12 contributes to TGF-beta signaling through interaction with the type II receptor. J Cell Biol 178:201–208
189. Hashimoto G, Inoki I, Fujii Y, Aoki T, Ikeda E, Okada Y (2002) Matrix metalloproteinases cleave connective tissue growth factor and reactivate angiogenic activity of vascular endothelial growth factor 165. J Biol Chem 277:36288–36295
190. Yu Q, Stamenkovic I (2000) Cell surface-localized matrix metalloproteinase-9 proteolytically activates TGF-beta and promotes tumor invasion and angiogenesis. Genes Dev 14:163–176
191. Li Y, Foster W, Deasy BM, Chan Y, Prisk V, Tang Y et al (2004) Transforming growth factor-beta1 induces the differentiation of myogenic cells into fibrotic cells in injured skeletal muscle: a key event in muscle fibrogenesis. Am J Pathol 164:1007–1019
192. Vial C, Zuniga LM, Cabello-Verrugio C, Canon P, Fadic R, Brandan E (2008) Skeletal muscle cells express the profibrotic cytokine connective tissue growth factor (CTGF/CCN2), which induces their dedifferentiation. J Cell Physiol 215:410–421
193. Yang M, Huang H, Li J, Huang W, Wang H (2007) Connective tissue growth factor increases matrix metalloproteinase-2 and suppresses tissue inhibitor of matrix metalloproteinase-2 production by cultured renal interstitial fibroblasts. Wound Repair Regen 15:817–824
194. Bhattacharyya S, Wu M, Fang F, Tourtellotte W, Feghali-Bostwick C, Varga J (2011) Early growth response transcription factors: key mediators of fibrosis and novel targets for anti-fibrotic therapy. Matrix Biol 30:235–242
195. Fukaya S, Matsui Y, Tomaru U, Kawakami A, Sogo S, Bohgaki T et al (2013) Overexpression of TNF-alpha-converting enzyme in fibroblasts augments dermal fibrosis after inflammation. Lab Investig 93:72–80
196. Lurton J, Soto H, Narayanan A, Raghu G (1999) Regulation of human lung fibroblast C1q-receptors by transforming growth factor-beta and tumor necrosis factor-alpha. Exp Lung Res 25:151–164

197. Piers AT, Lavin T, Radley-Crabb HG, Bakker AJ, Grounds MD, Pinniger GJ (2011) Blockade of TNF in vivo using cV1q antibody reduces contractile dysfunction of skeletal muscle in response to eccentric exercise in dystrophic mdx and normal mice. Neuromuscul Disord 21:132–141
198. Radley HG, Davies MJ, Grounds MD (2008) Reduced muscle necrosis and long-term benefits in dystrophic mdx mice after cV1q (blockade of TNF) treatment. Neuromuscul Disord 18:227–238
199. Gosselin LE, Martinez DA (2004) Impact of TNF-alpha blockade on TGF-beta1 and type I collagen mRNA expression in dystrophic muscle. Muscle Nerve 30:244–246
200. Lovelock JD, Baker AH, Gao F, Dong J-F, Bergeron AL, McPheat W et al (2005) Heterogeneous effects of tissue inhibitors of matrix metalloproteinases on cardiac fibroblasts. Am J Physiol Heart C 288:H461–H468
201. Miyazaki D, Nakamura A, Fukushima K, Yoshida K, Takeda S, Ikeda SI (2011) Matrix metalloproteinase-2 ablation in dystrophin-deficient mdx muscles reduces angiogenesis resulting in impaired growth of regenerated muscle fibers. Hum Mol Genet 9:1787–1799
202. Yamada M, Sankoda Y, Tatsumi R, Mizunoya W, Ikeuchi Y, Sunagawa K et al (2008) Matrix metalloproteinase-2 mediates stretch-induced activation of skeletal muscle satellite cells in a nitric oxide-dependent manner. Int J Biochem Cell B 40:2183–2191
203. Liu X (2011) Emerging ideas: matrix metalloproteinase-2 in muscle atrophy. Clin Orthop Relat R 469:1797–1799
204. Hnia K, Hugon G, Rivier F, Masmoudi A, Mercier J, Mornet D (2007) Modulation of p38 mitogen-activated protein kinase cascade and metalloproteinase activity in diaphragm muscle in response to free radical scavenger administration in dystrophin-deficient mdx mice. Am J Pathol 170:633–643
205. Hnia K, Gayraud J, Hugon G, Ramonatxo M, De La Porte S, Matecki S et al (2008) L-arginine decreases inflammation and modulates the nuclear factor-kappaB/matrix metalloproteinase cascade in mdx muscle fibers. Am J Pathol 172:1509–1519
206. Dahiya S, Bhatnagar S, Hindi SM, Jiang C, Paul PK, Kuang S et al (2011) Elevated levels of active matrix metalloproteinase-9 cause hypertrophy in skeletal muscle of normal and dystrophin-deficient mdx mice. Hum Mol Genet 20:4345–4359
207. Mehan R, Greybeck B, Emmons K, Byrnes W, Allen D (2011) Matrix metalloproteinase-9 deficiency results in decreased fiber cross-sectional area and alters fiber type distribution in mouse hindlimb skeletal muscle. Cells Tissues Organs 194:510–520
208. Coenen-Stass AM, McClorey G, Manzano R, Betts CA, Blain A, Saleh AF et al (2015) Identification of novel, therapy-responsive protein biomarkers in a mouse model of Duchenne muscular dystrophy by aptamer-based serum proteomics. Sci Rep 5:17014
209. Brunelli S, Sciorati C, D'Antona G, Innocenzi A, Covarello D, Galvez BG et al (2007) Nitric oxide release combined with nonsteroidal antiinflammatory activity prevents muscular dystrophy pathology and enhances stem cell therapy. Proc Natl Acad Sci U S A 104:264–269
210. Sciorati C, Buono R, Azzoni E, Casati S, Ciuffreda P, D'Angelo G et al (2010) Co-administration of ibuprofen and nitric oxide is an effective experimental therapy for muscular dystrophy, with immediate applicability to humans. Br J Pharmacol 160:1550–1560
211. Zordan P, Sciorati C, Campana L, Cottone L, Clementi E, Querini PR et al (2013) The nitric oxide-donor molsidomine modulates the innate inflammatory response in a mouse model of muscular dystrophy. Eur J Pharmacol 715:296–303
212. Bedair H, Liu TT, Kaar JL, Badlani S, Russell AJ, Li Y et al (2007) Matrix metalloproteinase-1 therapy improves muscle healing. J Appl Physiol 102:2338–2345

Role of BMP1/Tolloid like Proteases in Bone Morphogenesis and Tissue Remodeling

4

Sibani Chakraborty, Ankur Chaudhuri, and Asim K. Bera

Abstract

BMP-1/ tolloid like proteases collectively referred as BTPs are members of the astacin family of metalloproteases belonging to the metzincin clan. Four members of BTP are identified in humans; BMP-1, mTLD, mTLL1 and mTLL2. The BTPs are involved in several pathophysiological diseases including bone morphogenesis, fibrosis, tissue remodeling and tumor progression. BTPs are important regulators to activate several growth factors and helps to release anti-angiogenic fragments from parent proteins. Three dimensional structure of BTPs reveal the presence of a highly unusual disulphide bridge present within the cysteine-rich loop region in the active site. The activity of BTPs is controlled mainly by activators. The only endogenous inhibitor of mammalian BTPs is α2-macroglobulin. Several small molecular inhibitors of BTPs have been reported. Further studies will help to explore the full spectrum of activities of BTPs which will help in recognition of BTPs as new targets for future therapies.

Keywords

BMP1/Tolloid like Proteases (BTPs) • Morphogenesis • Tissue remodeling • Cancer activators • Inhibitors

S. Chakraborty (✉) • A. Chaudhuri • A.K. Bera
Department of Microbiology, West Bengal State University,
Barasat, Kolkata 700126, West Bengal, India
e-mail: sibani_chakraborty@yahoo.co.in; ankur.micro1986@gmail.com; akrbera@gmail.com

S. Chakraborti, N.S. Dhalla (eds.), *Proteases in Physiology and Pathology*,
DOI 10.1007/978-981-10-2513-6_4

4.1 Introduction

Bone morphogenic protein1/ tolloid like proteases commonly referred to as the BTPs [1] belong to astacin family of zinc metalloproteases and are involved in several pathophysiological diseases of bone morhogenesis, tissue repair and tumor progression. They are characterised having a unique 18 amino acid signature sequence (HExxHxxGxxHxxxRxDR) which is a part of an approximately 200 amino acid astacin-like protease/catalytic domain, followed by several CUB (for complement C1r/C1s, uEGF, BMP-1) and EGF domains (Fig. 4.1). The BTPs belongs to the metzincin clan [2] which includes astacins, the matrix metalloproteases (MMPs) and the ADAMs/ADAMTSs present in humans. In mammals there are four BMP-1-like proteases: bone morphogenetic protein-1 and mammalian tolloid are alternative spliced products of the same gene [3] and also mammalian tolloid-like1 and 2 [4]. The entire BMP-1/ tolloid like protease subfamily is referred to as the BTPs. Several disorders related to bone in vertebrates have been identified with this class of proteins.

BMP-1or Procollagen C-proteinase is a multi-domain, glycosylated, secretory, monomeric zinc endopeptidase. Structural studies indicate that the zinc atom is penta-coordinately bound to all the astacins except in bone morphogenetic protein-1 (BMP-1), where the tyrosine residue does not participate in binding to the active site zinc atom. Structurally catalytic domain of BTPs is closely related to the digestive enzyme astacin from the crayfish *Astacus astacus* with a sequence identity of above 35% (Table 4.1 and Fig. 4.2). Low resolution crystal structures available for the catalytic domain of BMP-1 indicates that the protease exist as a monomer in solution whereas both mTLD-1 and mTLL-1 exist as dimers [5]. Crystal structure of

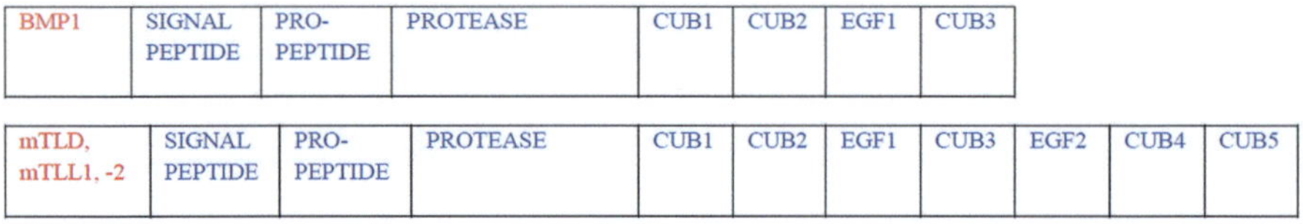

Fig. 4.1 Domain structure of BTPs. The BTPs consists of a signal peptide, pro-peptide, an protease/ catalytic domain followed by several CUB (for complement C1r/C1s, uEGF, BMP-1) and EGF (epidermal growth factor) domains

Table 4.1 BMP-1/ Tolloid-like Proteases in human

Protease	Species/tissue	UniProt ID	Total length	Length of protease domain	Identity (%)
BMP-1/ mTLD	Embryos, tissues	P13497	986	121–321 (201)	35.3
TLL1	Embryos, tissues	O43897	1031	148–348 (201)	37.1
TLL2	Embryos, tissues	Q9Y6L7	866	150–350 (201)	36.6

Note: Sequence identity of protease domain with astacin. Abbreviations: *BMP-1*-bone morphogenetic protein-1, *mTLD* mammalian tolloid, *TLL1* tolloid like 1, *TLL2* tolloid like 2

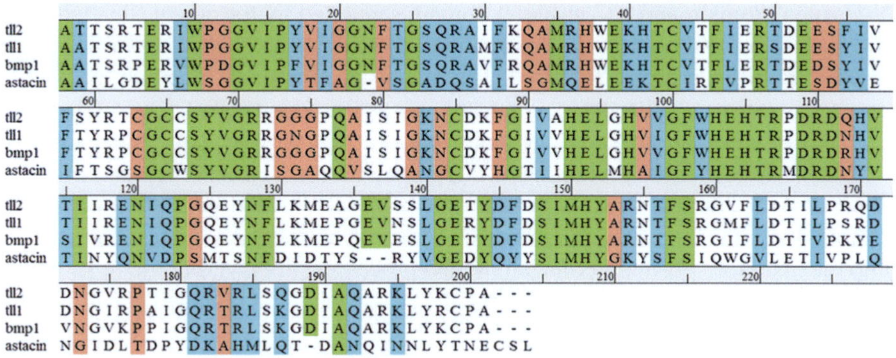

Fig. 4.2 Multiple Sequence alignment of protease domain of BTPs and astacin. Identical (conserved), strong and weak residues represented as green, cyan and orange colour respectively.The abbreviation for species are the same as used in Table 4.1

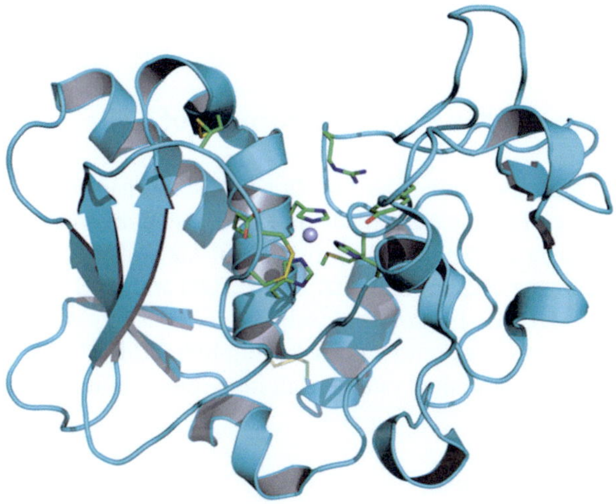

Fig. 4.3 X-ray structure of BMP-1 solved at 1.27 Å resolution. Disulphide bridges and important residues in substrate recognition involved at active site are represented as stick. Catalytic zinc is represented as CPK model

BMP-1 (Fig. 4.3) indicates it has a deep active site cleft within which is the three conserved His residues binding to the catalytic zinc. But unlike the prototypical protease astacin and other members of the family that a conserved tyrosine in BMP-1 is not involved in binding zinc at the active site [6]. BTPs differ from the other astacin like proteases in having a highly unusual additional disulphide bond in the active site region. This occurs in the cysteine-rich loop region and is believed to control the access of substrates through change in conformation [6]. Like other astacin like proteases, the catalytic domain of BMP-1 consist of a 100 residue

N-terminal domain and a C-terminal domain separated by the active site cleft. The N-terminal domain contains a five stranded β- sheet as also two long helices. The C-terminal domain contains few regular secondary structures. The two disulphide bonds present in the other astacin members is conserved in BMP-1 (Cys^{64}-Cys^{84} and Cys^{42}-Cys^{198}) with another additional disulphide bridge between Cys^{62} and Cys^{65} [7].

BMP-1/tolloid like proteinases (BTPs) has been detected in species from *Drosophila* to human. Procollagen C-proteinase (also called bone morphogenetic protein-1) of the astacin family cleaves procollagen and non-collagen substrates into the insoluble fibrillary collagen. It also processes prolysyl oxidase to its active form which is responsible for catalyzing the cross-linking of collagen fibrils thereby contributing to the structural stability of collagen [7]. BTPs are important regulators to activate several growth factors such as BMP-2/−4, TGF-β/1, GDF-8/−11 and IGFs and helps to release anti-angiogenic fragments from parent proteins [8]. This chapter summarises the recent advances on the structure, function and diverse role of BTPs of the astacin family and discusses their role as potential targets in therapeutic use. Table 4.1 describes the different members of the BTP's reported so far in human.

4.2 Structure of the Protease Domain of BTPs

The sequence similarity among the metalloprotease domain of several BTPs and astacin is indicative of the fact that the metalloprotease domain structure of BTPs is similar to that of astacin. Among the proteases of known three dimensional structures, digestive enzyme astacin (PDB ID: 1AST, 1QJJ and 1QJI share 35% sequence identity with TLL-1/BMP-1) is most closely related to BTPs. All BTPs are characterised by the presence of a highly unusual disulphide bridge between two adjacent cysteine residues present in the cysteine-rich loop region of the active site.

4.2.1 Primary Structure

The protease domain of BTPs containing about 200 amino acid residues consist of two sub-domains separated by a deep active site cleft with the catalytic zinc at its bottom.

4.2.2 X-ray Structure

Bone morphogenetic protein-1 (BMP-1) was initially found in bone extracts. It can induce bone formation at ectopic sites. BMP1 is a zinc-dependent metalloproteinase. The BMP-1 molecule contains an astacin-like catalytic domain, three CUB domains and an EGF (epidermal growth factor) domain which is located between the second and third CUB domains (Fig. 4.1). The CUB 1 domain is required for the secretion of the molecule. Glu483 on the β4-β5 loop of CUB2 is essential for the proteinase activity of BMP1. BMP-1, an astacin family member belongs to the metzincin clan which includes (in humans) the astacins, the meprins, the matrix metalloproteinases (MMPs), and ADAMs/ADAMTSs. BMP1 is the most active form. The isolated protease domain is relatively promiscuous, where specificity is conferred by the BMP-1 non-catalytic domains [9]. Procollagen C-proteinase (PCP) activity is one of the most important functional activities of BMP-1. Some residues are important for PCP activation such as Cys62, Cys65, Cys66, Lys86, Glu93 and Lys176 [10] and Cys84 is important for stability of BMP1 [11].

4.3 Substrate Specificity

Comparing the structure of catalytic domain of astacin from crayfish *Astacus astacus* it is found to be closely related to BMP-1/TLL-1 (Fig. 4.4).Three histidines, a tyrosine and a water molecule coordinated with Zn^{2+}ion at the active site is unique in astacins and serralysins enzyme family. Surprisingly these structural homologs have different substrate specificity. Astacins specificities are dominated by small aliphatic side chain in P1′ and a proline in P2′, whereas BMP-1/TLL-1 prefers strictly an Asp in P1′ and Asp/Glu in P2′ site [12]. Furthermore, BMP-1 crystal and molecular structure reveals a unique conformation of a small cysteine-rich loop within the active site cleft, which suggests that the flap movement is required to control substrate recognition. Therefore in comparison to astacin, BMP-1/TLL-1 protease's substrate specificity governed not only by S1' pocket but also registered by a unique cysteine-rich loop. Although they have conserved catalytic residues but the most structural differences between BMP-1 and astacin are around the substrate-binding pocket. A confirmed basic residue in S1′ fits well with the specificity of BMP-1 to counter opposite charge acidic amino acid, aspartate in the P1′ position.

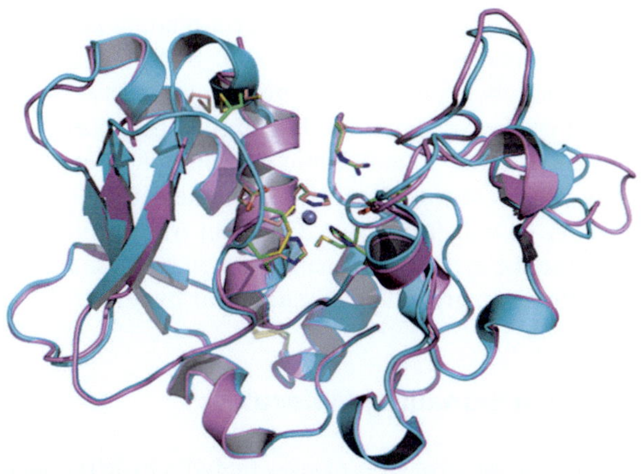

Fig. 4.4 Superimposition of TLL-1(pink) on BMP-1 (cyan). Important residues involved in substrate recognition are represented as stick. Catalytic zinc is represented as CPK model

BMP-1 S1' pocket is larger than astacin due to open conformational loop near Arg176.

In summary the cysteine-rich loop and the disulfide bonded network of BMP-1 clamps its substrate into a favourable position for catalysis [6].The unusual hydrophobic character and the flexibility of the cysteine-rich loop have not been observed for astacin or other related matrix metalloproteases. This loop controls substrate selectivity, therefore may guide in designing potent inhibitors of BMP-1/TLL-1 proteases against potential human metalloproteases anti-targets.

4.4 Biological Roles of BTPs

BTPs have involved as interesting therapeutic agents for treatment of healing bone fractures, periodontal bone defects, enhancing bone response around alloplastic materials implanted in bone and also in prevention of osteoporosis. In dentistry, bone morphogenetic proteins have been tested in periodontal, implant and restorative endodontic procedures [13].

4.4.1 Matrix Assembly

Properties of several connective tissues differ by tensile strength, elasticity and resistance to compressive loads. Composition and organization of matrix as also the cross linking which stabilizes the structural networks of extracellular matrix (ECM) proteins are responsible for properties of different connective tissues [14, 15]. The deamination of specific lysins by oxidation in the structure of collagen and elastin helps to trigger the formation of cross-links in the structure is carried out by the Lysyl oxidases. BTPs are the main enzymes for rate limiting step of fibril formation, in regulation of cell responses to tissue stiffness. In bone and dentin, mineralization causes a further degree of stiffness. Small integrin-binding ligand N-linked glycoproteins usually have increase glycosylation and phosphorylation sites and also bind calcium. The acidic domains of these proteins on release by proteolysis helps in the enhancement of their calcium binding ability. Substrates of BTPs such as dentin matrix protein-1 (DMP-1) and dentin sialophosphoprotein (DSPP) presents the above property which are well-described substrates of BTPs [16].

A direct effect on growth factors by BTPs can trigger simultaneously matrix assembly and increase the synthesis of matrix proteins. These functions are important related to development, growth and tissue repair by BTPs.

4.4.2 Activation of Signalling Molecules

The expression of a large number of proteases including BMP-1 and mTLD, protease inhibitors and growth factors can positively or negatively be modulated by TGF-β1 [17, 18]. The upregulation of some essential ECM matrix proteins like collagen and fibronectin can be induced by TGF-β1 [8]. The activation of TGF-β1 can take place both in a proteolytic or non-proteolytic manner. BMP1 plays an important role in the proteolytic pathway [19]. It is also known that BTPs contribute to the maintenance of high levels of TGF-β1 by stimulating the degradation of soluble betaglycan and CD109, two antagonists of TGF-β [16]. Several members of the TGF-β or IGF superfamilies can be triggered by BTPs that targets different pathophysiological processes such as morphogenesis, muscle growth, tumor progression and dorso-ventral patterning [20–22]. BTPs can also help to degrade several proteins that shows angiogenic properties. BTPs can convert prolactin and growth hormones into effective anti-angiogenic molecules [23] and remove the pro-metastatic potential of angiopoietin-like protein2 [24].

4.5 BTP's and Disease

The involvement of BTPs in a wide range of diseases in human have been reported. Since BMP-1 also known as Procollagen C-Proteinase have been shown to be responsible for a number of diseases in human, BMP-1 can be used as new target for drug therapy.

4.5.1 Fibrosis

Procollagen C-proteinases and procollagen C-proteinase enhancers (PCPEs) are important druggable targets for preventing excess deposition of collagen in fibrosis. In a rabbit skin model of hypertrophic scarring, the scar was reduced when a small molecule inhibitor of BTPs was injected into the rabbit [25]. This may be due to reduce collagen deposition followed by wounding. Transcription of PCPE-1 is up-regulated in hypertrophic and keloid scars [26]. BMP-1 and PCPE-1 was found to be expressed in a mouse model of corneal scarring [27]. Increasing the amount of circulating mTLD in a rat model resulted in renal fibrosis. PCPE-1 level was up-regulated in a rat model of liver fibrosis [28]. BTPs have been associated with cardiac fibrosis. Human samples with myocardial fibrosis showed an increased expression of PCPE-1 at protein levels linked to aortic valve stenosis [29]. The potentiality of PCPEs for anti-fibrotic therapies as new drug targets corroborate with the findings.

4.5.2 Bone Disorders

Bone disorders have been attributed to the role of BTPs. Missense mutations in the BTPs cleavage sites of procollagenα1 and α2 chains have been describes in osteogenesis imperfect (OI) [30]. This leads to delay in procollagen processing which results in increase in bone mineral density together with susceptibility of bone fracture. Recent studies have identified homozygous missense mutations in BMP-1/mTLD in recessive forms of OI [31–33]. This also results in defective processing of procollagen and perturbs fibril formation resulting in bone disorders.

4.5.3 Cancer

BMP-1 includes a group of signature extracellular matrix proteins linked with tumor metastasis has been confirmed by proteomic study [34]. BMP-1 and PCPE-1are desmoplastic markers associated with colorectal cancer [35]. microRNA and miR-194 downregulation of BMP-1 expression reduce metastasis of lung by reducing the level of TGF-β activation and MMP-2/−9 expression [36]. Angiogenesis is an important component of metastasis. PCPE-1 is one of the five proteins required for formation of endothelial cell lumen [37]. Results indicate that BMP-1 and PCPE-1 are noted components of tumor microenvironment. Future studies will unfold the inhibition of BMP-1 activity and its significant impact on metastasis and cancer.

4.6 Modulators of BTPs

4.6.1 Activators

Regulations of BTPs are controlled by different types of activators. Procollagen C proteinase enhancers (PCPE) are the first activators of BTPs. The two isomeric form of PCPEs are PCPE-1 and PCPE-2. Both isomers are found at high level in heart [38]. Periostin is substrate specific enhancer of BTPs, strongly expressed during development and matrix assembly [39]. Another two activators are fibronectin and chordin [40, 41].

4.6.2 Inhibitors

Inhibition of Procollagen C-Proteinases/BMP-1 may interfere in the progression of fibrosis and are thus important druggable targets as excessive fibrillogenesis of collagen can lead to diseases like keloids, formation of surgical adhesions and deep-seated fibrosis of organs. Some small molecules functioning as inhibitors of PCP have been reported. Small molecule inhibitors such as EDTA, 1,10-phenanthroline [42] N-ethylmaleimide (NAM) [43], UK3863 [44] and glutamic acid/diamino acid derived hydroxamates also inhibit BMP-1. Hydroxamate derivatives of diamino acids were reported to act as successful inhibitors of PCP [45]. Bone morphogenic protein 1 is inhibited by an exogenous inhibitor called sizzled protein which is a member of secreted frizzled-related protein (sFRP) family from ***Xenopus laevis*** shown to strongly inhibit other human tolloid isoforms mTLD and mTLL-1 [46]. Endogenous macromolecular inhibitor α-2 macroglobulin present in serum also inhibits BMP-1 [47].

Quantitative structure activity relationship (QSAR) describes a set of physiochemical properties or molecular descriptors to a response variable which could be a biological activity of the chemicals [48]. Some sulphonamide derivatives as PCP inhibitors were studied by QSAR which helps to design and synthesis of new PCP inhibitors that can be used for treatment of fibrotic condition [49].

4.7 Conclusion and Future Perspectives

The involvement of the BMP-1 /tolloid like proteases (BTPs) in several chronic fibrotic conditions like pulmonary, renal and liver fibrosis, scleroderma and muscle wasting indicates that these enzymes can be good therapeutic targets for the designing of potent, specific inhibitors at physiological condition. BTP activities are controlled by substrate specific activators rather than inhibitors. BTPs are therapeutic targets to prevent deposition of excess collagen in fibrosis. Cardiac fibrosis has been related to BTPs and associated proteins. BTPs are associated with bone genetic disorders. BMP-1 and PCPE-1 (Procollagen C-proteinase enhancers) are thus important components of the tumor microenvironment. Future studies will help to unveil if the inhibition of BMP-1and PCPE-1 can have an important role in tumor progression and metastasis.The designing of specific BTP inhibitors for the treatment of metalloprotease-associated diseases is an important goal for future disease management.

Acknowledgement The authors sincerely thank Dr. Indrani Sarkar of Narula Institute of Technology for her useful suggestions in preparation of the chapter.

References

1. Muir A and Greenspan DS (2011) Metalloproteinases in Drosophila to humans that are central players in developmental processes. J BiolChem286:41905–41911
2. Sterchi EE, Stocker Wand Bond JS (2008) Meprins, membrane-bound and secreted astacin metalloproteinases. Mol Asp Med 29:309–328
3. Takahara K, Lyons GE, Greenspan DS (1994) Bone morphogenetic protein-1 and a mammalian tolloid homologue (mTld) are encoded by alternatively spliced transcripts which are differentially expressed in some tissues. J BiolChem 269:32572–32578
4. Urist MR (1965) Bone: formation by autoinduction. Science 150:893–899
5. Constam DB, Robertson EJ (1999) Regulation of bone morphogenetic protein activity by prodomains and proprotein convertases. J Cell Biol 144:139–149
6. Mac Sweeney A, Gil-Parrado S, Vinzenz D, Bernardi A, Hein A, Bodendorf U, Erbel P, Logel C, Gerhartz B (2008) Structural basis for the substrate specificity of bone morphogenetic protein 1/tolloid-like metalloproteases. J MolBiol 384:228–239
7. Wozney JM, Rosen V, Celeste AJ et al (1988) Novel regulators of bone formation: molecular clones and activities. Science 242:1528–1534
8. Vadon-Le Goff S, Hulmes DJS, Moali C (2015) BMP-1/tolloid-like proteinases synchronize matrix assembly with growth factor activation to promote morphogenesis and tissue remodeling. Matrix Biol 44–46C:14–23

9. Wermter C, Howel M, Hintze V, Bombosch B, Aufenvenne K, Yiallouros I et al (2007) The protease domain of procollagen C proteinase (BMP1) lacks substrate selectivity, which isconferred by non-proteolytic domains. BiolChem 388:513–521
10. Garrigue-Antar L, Barker C, Kadler KE (2001) Identification of amino acid residues in bone morphogenetic protein-1 important for procollagen C-proteinase activity. J Biol Chem 276:26237–26242
11. Gaoxiang Ge and Neung-SeonSeo et al. (2004) Bone morphogenetic protein-1/Tolloid-related metalloproteinases process Osteoglycin and enhance its ability to regulate collagen Fibrillogenesis. J Biol Chem 279:41626–41633
12. Kessler E (2004) Procollagen C-endopeptidase. In: Barret AJ, Rawlings ND, Woessner JF (eds) Handbook of proteolytic enzymes. Elsevier Ltd, London, pp 609–617
13. King GN, King N, Hughes FJ (1998) Effect of two delivery systems for recombinant human bone morphogenetic protein-2 on periodontal regeneration in vivo. J Periodontal Res 33:226–236
14. Gopalakrishnan B, Wang WM, Greenspan DS (2004) Biosynthetic processing of the pro- alpha1(V)pro-alpha2(V)Proalpha3(V) procollagen heterotrimer. J Biol Chem 279:30904–30912
15. Imamura Y, Steiglitz BM, Greenspan DS (1998) Bone morphogenetic protein-1 processes the NH2-terminal propeptide, and a furin-like proprotein convertase processes the COO terminal propeptide of pro-alpha1(V) collagen. J Biol Chem 273:27511–27517
16. Delolme F, Anastasi C, Alcaraz LB, Mendoza V, Vadon-Le Goff S, Talantikite M et al (2015) Proteolytic control of TGF-beta coreceptor activity by BMP-1/tolloid-like proteases revealed by quantitative iTRAQ proteomics. Cell Mol Life Sci 72:1009–1027
17. Lee SB, Solow-Cordero DE, Kessler E, Takahara K, Greenspan DS (1997) Transforming growth factor-beta regulation of bone morphogenetic protein-1 procollagen C-proteinase and related proteins in fibrogenic cells and keratinocytes. J Biol Chem 272:19059–19066
18. Tovar-Vidales T, Fitzgerald AM, Clark AF, Wordinger RJ (2013) Transforming growth factor-beta2 induces expression of biologically active bone morphogenetic protein-1 in human trabecular meshwork cells. Invest Ophthalmol Vis Sci 54:4741–4748
19. Ge G, Greenspan DS (2006) BMP1 controls TGFbeta1 activation via cleavage of latent TGFbeta-binding protein. J Cell Biol 175:111–120
20. Marques G, Musacchio M, Shimell MJ, Wunnenberg-Stapleton K, KWY C, O'Connor MB (1997) Production of a DPP activity gradient in the early drosophila embryo through the opposing actions of the SOG and TLD proteins. Cell 91:417–426
21. Piccolo S, Agius E, Lu B, Goodman S, Dale L, De Robertis EM (1997) Cleavage of chordin by xolloid metalloproteinase suggests a role for proteolytic processing in the regulation of Spemann organizer activity. Cell 91:407–416
22. Kim B, Huang G, Ho WB, Greenspan DS (2011) Bone morpho- genetic protein-1 processes insulin-like growth factor- binding protein 3. J BiolChem 286:29014–29025
23. Ge G, Fernandez CA, Moses MA, Greenspan DS (2007) Bone morphogenetic protein 1 processes prolactin to a 17-kDa antiangiogenic factor. Proc Natl AcadSci U S A104:10010–10015
24. Odagiri H, Kadomatsu T, Endo M, Masuda T, Morioka MS, Fukuhara S et al (2014) The secreted protein ANGPTL2 promotes metastasis of osteosarcoma cells through integrin alpha5- beta1, p38 MAPK, and matrix metalloproteinases. Sci signal 7:ra7
25. Reid RR, Mogford JE, Butt R, deGiorgio-Miller A, Mustoe TA (2006) Inhibition of procollagen C-proteinase reduces scar hypertrophy in a rabbit model of cutaneous scarring. Wound Repair Regen 14:138–141
26. Wong VW, You F, Januszyk M, Gurtner GC, Kuang AA (2014) Transcriptional profiling of rapamycin-treated fibroblasts from hypertrophic and keloid scars. Ann PlastSurg 72:711–719
27. MalecazeF MD, Fournie P, Tricoire C, CassagneM MM et al (2014) Upregulation of bone morphogenetic protein-1/mammalian tolloid and procollagen C-proteinase enhancer-1 in corneal scarring. Invest Ophthalmol Vis Sci 55:6712–6721

28. Ogata I, Auster AS, Matsui A, Greenwel P, Geerts A, D'Amico T et al (1997) Up-regulation of type I procollagen Cproteinase enhancer protein messenger RNA in rats with CCl4-induced liver fibrosis. Hepatology 26:611–617
29. Beaumont J, Lopez B, Hermida N, Schroen B, San JG, Heymans S et al (2014) microRNA-122 down-regulation may play a role in severe myocardial fibrosis in human aortic stenosis through TGF-beta1 up-regulation. ClinSci (Lond) 126:497–506
30. Lindahl K, Barnes AM, Fratzl-Zelman N, Whyte MP, Hefferan TE, Makareeva E et al (2011) COL1 C-propeptide cleavage site mutations cause high bone mass osteogenesis imperfecta. Hum Mutat 32:598–609
31. Asharani PV, Keupp K, Semler O, Wang W, Li Y, Thiele H et al (2012) Attenuated BMP1 function compromises osteogenesis, leading to bone fragility in humans and zebrafish. Am J Hum Genet 90:661–674
32. Martinez-Glez V, Valencia M, Caparros-Martin JA, Aglan M, Temtamy S, Tenorio J et al (2012) Identification of a mutation causing deficient BMP1/mTLD proteolytic activity in autosomal recessive osteogenesis imperfecta. Hum Mutat 33:343–350
33. Valencia M, Caparros-Martin JA, Sirerol-Piquer MS, Garcia-Verdugo JM, Martinez-Glez V, Lapunzina P et al (2014) Report of a newly indentified patient with mutations in BMP1 and underlying pathogenetic aspects. Am J Med Genet A 164A:1143–1150
34. Naba A, Clauser KR, Lamar JM, Carr SA, Hynes RO (2014) Extracellularmatrix signatures of humanmammary carcinoma identify novel metastasis promoters. elife 3:e01308
35. Torres S, Bartolome RA, Mendes M, Barderas R, Fernandez-Acenero MJ, Pelaez-Garcia A et al (2013) Proteome profiling of cancer-associated fibroblasts identifies novel proinflammatory signatures and prognostic markers for colorectal cancer. Clin Cancer Res 19:6006–6019
36. Wu X, Liu T, Fang O, Leach LJ, Hu X, Luo Z (2014) miR-194 suppresses metastasis of non-small cell lung cancer through regulating expression of BMP1 and p27(kip1). Oncogene 33:1506–1514
37. Newman AC, Nakatsu MN, Chou W, Gershon PD, Hughes CC (2011) The requirement for fibroblasts in angiogenesis: fibroblast-derived matrix proteins are essential for endothelial cell lumen formation. MolBiol Cell 22:3791–3800
38. Steiglitz BM, Keene DR and Greenspan DS (2002) PCOLCE2 encodes a functional procollagen C-proteinase enhancer (PCPE2) that is a collagen-binding protein differing in distribution of expression and post-translational modifica- tion from the previously described PCPE1. J BiolChem277:49820–49830
39. Yamaguchi Y (2014) Periostin in skin tissue and skin-related diseases. AllergolInt 63:161–170
40. Huang G, Zhang Y, Kim B, Ge G, Annis DS, Mosher DF et al (2009) Fibronectin binds and enhances the activity of bone morphogenetic protein 1. J BiolChem 284:25879–25888
41. Inomata H, Haraguchi T, Sasai Y (2008) Robust stability of the embryonic axial pattern requires a secreted scaffold for chordin degradation. Cell 134:854–865
42. Leung MK, Fessler LI, Greenberg DB andFessler JH (1979) Separate amino and carboxyl procollagen peptidases in chick embryo tendon.J BiolChem254:224–232
43. Kessler E, Adar R, Goldberg B andNiece R (1986) Partial purification and characterization of a procollagen C-proteinase from the culture medium of mouse fibroblasts.CollRelat Res6:249–266
44. Allan GA, Gedge JI, Nedderman AN, Roffey SJ, Small HF, Webster R (2006) Pharmacokinetics and metabolism of UK-383,367 in rats and dogs: a rationale for long-lived plasma radioactivity. Xenobiotica 36:399–418
45. Robinson LA, Wilson DM, Delaet NG, Bradley EK, Dankwardt SM, Campbell JA, Martin RL, Van Wart HE, Walker KA, Sullivan RW (2003) Novel inhibitors of procollagen C-proteinase. Part 2: glutamic acid hydroxamates. Bioorg Med Chem Lett 13:2381–2384
46. Bijakowski C, Vadon-Le Goff S, Delolme F, Bourhis JM, Lécorché P, Ruggiero F, Becker-Pauly C, Yiallouros I, Stöcker W, Dive V, Hulmes DJS, Moali C (2012) Sizzled is unique among secreted frizzled-related proteins for its ability to specifically inhibit bone morphogenetic protein-1 (BMP-1)/tolloid-like proteinases. J BiolChem 287:33581–33593

47. Zhang Y, Ge G, Greenspan DS (2006) Inhibition of bone morphogenetic protein 1 by native and altered forms of alpha2-macroglobulin. J BiolChem 281:39096–39104
48. Hansch C, Kurup A, Garg R, Gao H (2001) Chem-bioinformatics and QSAR: a review of QSAR lacking positive hydrophobic terms. Chem Rev 101:619–672
49. Khazaei A, Sarmasti N, Seyf JY,Rostami Z, Zolfigol MA (2015) QSAR study of the non-peptidic inhibitors of procollagen C-proteinase based on multiple linear regression, principle component regression, and Partial Least Square Arabian J Chem

Role of Proteases in the Regulation of *N*-Myristoyltransferase

5

Sujeet Kumar, Umashankar Das, Jonathan R. Dimmock, and Rajendra K. Sharma

Abstract

N-Myristoyltransferase (NMT) catalyzes the amide-linked addition of the myristoyl (C14:0) moiety to the amino-terminal glycine of substrate molecules. The upregulation of NMT in cancerous states was originally reported by our group, and subsequently this protein has drawn significant interest as a therapeutic target. However, the regulation of this enzyme in various physiological states is still in a state of infancy. NMT possesses PEST sequences and is a target for *m*-calpain-mediated degradation. The two isoforms, NMT1 and NMT2, interact differentially with the proteases caspase-3 and *m*-calpain. Recently, we have shown that the interaction of these isoforms (i.e., NMT1 and NMT2) with caspases regulates co- and posttranslational protein myristoylation. In this chapter, we have discussed the findings of the regulation of NMT functions by various proteases and the effects of proteolytic processing of NMT on its biochemical behavior.

S. Kumar
Department of Pathology and Laboratory Medicine, College of Medicine, University of Saskatchewan, Saskatoon, SK S7N 5E5, Canada

Department of Chemistry and Biochemistry, University of the Sciences, Philadelphia, PA 19104, USA

U. Das • J.R. Dimmock
College of Pharmacy and Nutrition, University of Saskatchewan, Saskatoon, SK S7N 5C9, Canada

R.K. Sharma (✉)
Department of Pathology and Laboratory Medicine, College of Medicine, University of Saskatchewan, Saskatoon, SK S7N 5E5, Canada
e-mail: rajendra.sharma@usask.ca

S. Chakraborti, N.S. Dhalla (eds.), *Proteases in Physiology and Pathology*, DOI 10.1007/978-981-10-2513-6_5

Keywords
Caspase • Calpain • *N*-Myristoyltransferase • Lipidation • Protein myristoylation

5.1 Introduction

The process of protein lipidation acts as a critical regulator for proper targeting of numerous cellular proteins to their destined sub-cellular localizations i.e., to various membranes, membrane domains, and organelles [1, 2]. The lipid moiety attached in the prevalent forms of lipidic modification in eukaryotes includes (1) a 14-carbon myristoyl chain, (2) a 16-carbon palmitoyl chain, (3) isoprenoids (15-carbon farnesyl or 20-carbon geranylgeranyl), and (4) glycosylphosphatidylinositol (GPI) anchors [3]. The addition of the C14:0 (myristoyl) group to proteins, referred to as "myristoylation," is unique among these processes. A stable amide bond between the C14:0 group and the exposed amino-terminal glycine renders the half-life of the myristoyl moiety on the protein equivalent to the half-life of the polypeptide chain backbone [4, 5]. The catalytic subunit of cAMP-dependent protein kinase was the very first myristoylated protein discovered, blocked at the N-terminus by n-tetradecanoic acid, and a similar modification was soon reported in the β-subunit of calcineurin [6, 7]. Furthermore, isoform-specific protein myristoylation has been shown for many proteins to aid in their tissue-specific cellular roles [8, 9].

The process of protein myristoylation is catalyzed by the enzyme *N*-myristoyltransferase (NMT; EC 2.3.1.97) and involves the covalent transfer of a myristoyl group (C14:0) to the amino-terminal glycine moiety of substrate protein molecules [4, 5, 10–12]. The enzyme is found ubiquitously distributed in lower and higher eukaryotes [13]. The availability of an N-terminal exposed glycine moiety is an absolute requirement and the modification usually occurs on a general consensus motif of GXXXS/T (where X is any amino acid) [4, 14]. However, the availability of Ser/Thr at position 5, downstream of N-terminal acceptor glycine, is not a mandatory requirement. In such cases, Asp/Gln at the penultimate position to the acceptor glycine acts as a facilitator of myristoylation [9, 15, 16]. Protein myristoylation takes place in either of the following two ways: (i) in a co-translational manner, when the protein chain being synthesized on the ribosome (after the synthesis of up to 100 amino acids) is subjected to the removal of the initiator methionine, by the enzyme methionine aminopeptidase, thus exposing a penultimate N-terminal glycine, and (ii) in a post-translational fashion, when an internal glycine within a polypeptide is exposed following a proteolytic cleavage event, mostly in the apoptotic states [12, 17, 18].

The process of protein myristoylation follows the ordered reaction mechanism, the apoenzyme first binds to the ligand myristoyl-CoA (MYR) allowing to form a binary complex of NMT-MYR which subsequently binds to protein/peptide substrates. The catalytic transfer of myristoyl from the donor MYR to acceptor glycine occurs through direct nucleophilic addition-elimination reaction. After the

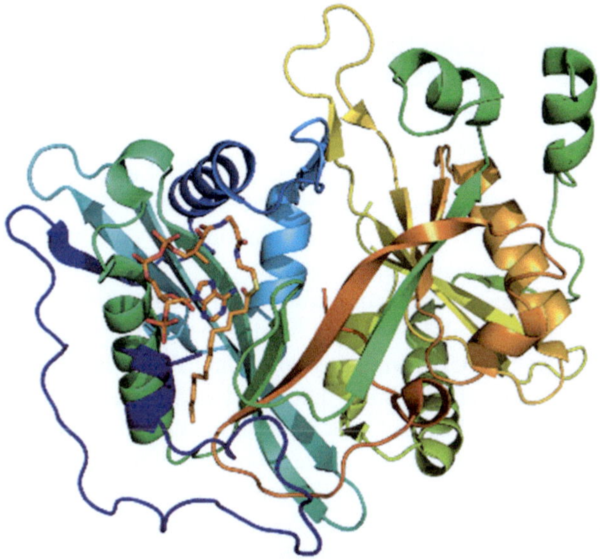

Fig. 5.1 N-Myristoyltransferase structure showing bound myristoyl-CoA in the N-terminal half of the enzyme

formation of an enzyme-product complex from the enzyme-substrate complex, the sequential release of CoA and myristoyl-peptide follows, thus allowing to complete the cycle [19, 20]. The enzyme is highly selective for MYR *in vitro* and *in vivo* [4, 14, 17]. NMTs have an architecture consisting of a saddle-shaped β-sheet flanked by α-helices and are members of the GNAT superfamily of enzymes [19, 20]. The regions corresponding to N- and C-terminal portions of the enzyme reflect a pseudo-twofold symmetry. The MYR binding site is located within the N-terminal portion, whereas the C-terminal segment provides for the occupancy site of the substrate-peptide/protein (Fig. 5.1). NMTs show very high common preferences for MYR but possess varying peptide substrate specificities [4, 21].

5.2 Differential Regulation of N-Myristoyltransferase Isoforms

NMT, in the mammalian species, exists in two major isoforms, NMT1 and NMT2 [12]. The two human isoforms (hNMT1 and hNMT2) reflect divergence only in the amino-terminal domains and share an overall ~77% identity [21]. These two isozymes, hNMT1 and hNMT2, are encoded by different genes, have been cloned and show differential specificities. The knockdown of hNMT1 is detrimental to tumor growth *in vivo* and is thus considered a potential novel cancer target [22–26]. The knockdown of hNMT1 also results in defective myelopoiesis in mice, thus indicating its essentiality in cellular development [27]. Among the two isoforms of human NMTs, hNMT1, but not hNMT2, exists in multiple isoforms. hNMT2 appears as a single 65 kDa protein, whereas four distinct isoforms (from 49 to 68 kDa) of hNMT1 are observed in cells [21]. The first report on the multiple isoforms of NMT came from studies on bovine brain NMT, and further, isozymes with varying size and tissue distribution have been reported from a variety of sources [21, 28–31]. Five isoforms of NMT exist in the murine leukemia cell line L1210, whereas two isoforms were reported in bovine brain cortex [29, 32]. Later it was reported that the bovine brain contains a heterogeneous mixture of NMT subunits [30]. To date it is not fully understood how the *in vivo* regulations are conferred on to the NMT isoforms.

N-Myristoylation lies upstream of multiple pro-proliferative and oncogenic pathways, and thus NMT is a validated target for therapeutic interventions in cancer [12, 23–26]. The selective knockdown using small interfering RNAs (siRNAs) against NMT isozymes in human cells shows that targeting NMT isozyme induces apoptosis, with depletion of hNMT2 having a 2.5-fold greater response than hNMT1, and that the loss of hNMT2 shifts the expression of the Bcl-family proteins toward apoptosis [24]. The silencing of hNMT1 inhibits cell replication associated with a loss of activation of c-Src and its target FAK. Also the signaling through the c-Raf/MEK/ERK/Elk pathway is reduced [24]. However, the two isoforms have been suggested only to have partially overlapping functions, and hNMT1 is more critical for tumor cell proliferation [24]. It is recently reported that in HeLa cells, upon NMT inhibition, the cell death occurs via apoptosis which is following, or concurrent with, accumulation in the G1 phase [33]. In response to NMT inhibition, the cell cycle regulation-associated proteins are downregulated, whereas the proteins involved in the endoplasmic reticulum stress and unfolded protein response are upregulated in the HeLa cells as well as breast (MCF-7, MDA-MB-231) and colon (HCT116) cancer cell lines [33].

This chapter discusses how the interaction of NMTs with various proteases regulates the biochemical behavior of *N*-myristoyltransferase and why cells need to allow for the proteolytic processing in demand to diversified physiological requirements.

5.3 Regulation of NMT Functions by Proteases

The sequence of both human NMT isoforms contains a higher percentage of Pro (**P**), Glu (**E**), Ser (**S**), and Thr (**T**) amino acid residues forming a **PEST** sequence which may act as an intramolecular signal for rapid proteolytic degradation [34]. Calpains, Ca^{2+}-dependent neutral proteases, recognize PEST signature and are responsible for the degradation of PEST-containing proteins [35]. An earlier study from our laboratory suggested that bovine cardiac NMT1 activity is completely abolished by *m*-calpain *in vitro* [36]. Degradation of NMT by *m*-calpain is shown to be inhibited by the calpain inhibitor, calpastatin [36]. Furthermore, we have reported earlier that in colorectal adenocarcinomas, the activity and expression of *m*-calpain is significantly higher [37]. The two isoforms, NMT1 and NMT2, in human colorectal adenocarcinoma tissues and human colon cancer cell lines (HCCLs) show

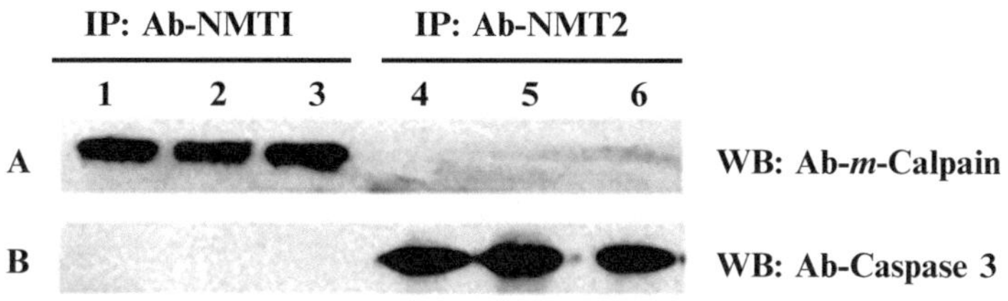

Fig. 5.2 Interaction between NMTs (NMT1 and NMT2) and proteases (*m*-calpain and caspase-3) by immunoprecipitation analysis in human colon cancer. *Lanes 1 and 4*, human normal colorectal sample; *lanes 2 and 5*, human adenocarcinoma samples; and *lanes 3 and 6*, HT29 colon cancer cell line (For details see Selvakumar et al. [38])

differential interactions with *m*-calpain and caspase-3 [38]. It was observed that NMT1 interacts with *m*-calpain in normal, adenocarcinoma, and HT29 colon cancer cell line (Fig. 5.2A, lanes 1–3), whereas NMT2 does not (Fig. 5.2A, lanes 4–6). These findings suggest that the two NMT isoforms may regulate cellular signaling differently and have been discussed elsewhere in details [39]. It is also observed that overexpression of calpastatin downregulates calpain activation but increases caspase-3-like activity and also accelerates the appearance of apoptotic nuclear morphology [40]. Besides being degraded by calpain, calpastatin may also be a target for degradation by caspases during apoptosis [41]. It is possible that calpain might be indirectly activated by caspase-3 via calpastatin degradation. Immunoprecipitation analysis shows that both in normal and cancerous samples, NMT2 interacts with caspase-3 (Fig. 5.2B, lanes 4–6), whereas NMT1 does not (Fig. 5.2B, lanes 1–3). These findings reveal that NMTs may be involved in the calpain/caspase-mediated pathway during the development of cancer [38, 39]. The differential interaction of the two isoforms of the NMT with *m*-calpain and caspase-3 is remarkable. *m*-Calpain showed to interact with NMT1 only, whereas NMT2 with caspase-3, indicating that a differential regulation may exist for NMT1 and NMT2 by *m*-calpain and caspase-3 [38] (Fig. 5.3).

A recent study has shown that both forms of human NMT (i.e., NMT1 and NMT2) are cleaved upon execution of apoptosis [42]. The timing of the cleavage of NMTs is in parallel with that of the apoptotic marker PARP-1, suggesting it is due to the action of caspases. The cleavages of NMT1 and NMT2 result in the decrease of molecular mass by ~11 and ~10 kDa, respectively. The general and caspase-8-specific inhibitors completely block the cleavage of NMT1, whereas caspase-3-specific inhibitor shows only partial protection against fragmentation. However, NMT2 cleavage is abrogated by all of these caspase inhibitors [42]. The findings suggest that NMT1 is a substrate for both caspase-3 and caspase-8 and NMT2 is likely a substrate of caspase-3 but not caspase-8. Identification of the amino-terminal residues of the cleaved fragments revealed that caspase-8-mediated NMT1 cleavage (~48 kDa) occurs mainly after Asp^{72} with minor cleavage at Asp^{38} [42]. However, it is suggested that the Asp^{38}-NMT1 fragment generated by purified caspase-8 might be nonphysiological. The caspase-3-mediated NMT1 cleavage results in a single ~48 kDa fragment corresponding to a cleavage at Asp^{72}. N-terminal

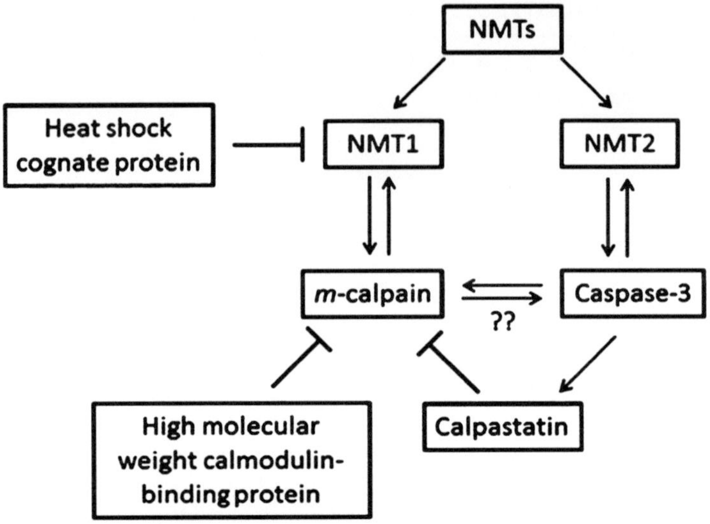

Fig. 5.3 Schematic illustration of interaction between NMTs (NMT1 and NMT2) with *m*-calpain and caspase-3. NMT exists in two major isoforms, NMT1 and NMT2, which interact differentially with calpain and caspase 3

sequencing of the caspase-3 cleavage products of NMT2 revealed that the processing of NMT2 was at both Asp^{25} and Asp^{67} [42]. The primary sequence information reveals that the caspase cleavage sites for both NMTs are located in the amino-terminus portion of the molecule. The caspase cleavage, either by caspase-3 or caspase-8, does not involve the carboxy-terminal catalytic domain of NMTs, and the caspase-cleaved enzymes retain their catalytic activity [42]. Of particular interest in the N-terminus of the enzymes, there is a presence of a poly-lysine block in the cleaved portion of the enzyme. This poly-lysine block is the part of the ribosomal targeting signal in consistence with a co-translational protein myristoylation model. The co-translational myristoylation requires the proximity of NMT within the protein synthetic complex (ribosomal complex) to timely capture its substrate when prompted by an appropriate *N*-myristoylation consensus signal available on a prospective nascent polypeptide [31]. The cleavage of the NMT1 leads to a cytosolic relocalization (>55%) from the predominantly membrane-bound form (64%), whereas NMT2 relocalizes to membranes (>80%) from the predominantly cytosolic (62%) when cleaved [31].

We have shown that the amino-terminal region of the catalytic module of NMT1 shows variability both in length and nature of the amino acids among the orthologous NMTs (Fig. 5.4) [43]. The minimal sequence length upstream of a $3_{10}\alpha A'$ conserved region is found in *Trypanosoma brucei* and *Trypanosoma cruzi*. The *T. brucei* and *T. cruzi* NMTs lack the sequence length parallel to the region corresponding to the first 28 amino acids of the catalytic domain of human NMT1 (Fig. 5.4). Using biochemical analyses on the purified recombinant proteins, we have shown that the deletion of the 28 residues at N-terminus enhances the

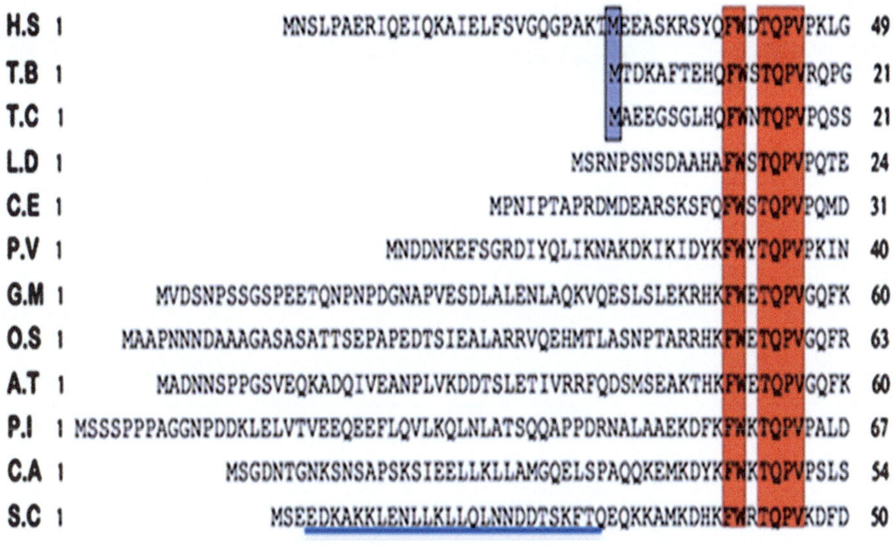

Fig. 5.4 Sequence comparisons of the N-terminal region orthologous NMTs from divergent species reflecting differences in the sequence lengths. The first conserved residues across all sequences are shown in *red* boxes (For details see Kumar and Sharma [43])

myristoyltransferase activity by ~2.3-fold without affecting the stability of the enzyme [43]. It has been previously observed that the differential processing in various apoptotic states results in diverse N-terminal regions which correspond to cleavage at positions 20 and 35 in the catalytic domain [44]. The observed physiological isoforms encountered in the apoptotic states closely serve the domain boundaries of the purified recombinant proteins employed in the biochemical analyses in our laboratory. The N-terminal truncation mutant in our study is closer in domain boundary to the physiological isoform generated by caspase-3-mediated cleavage at Asp^{72}. This deletion distinguishably enhances V_{max} by ~2-fold for MYR as compared to the catalytic module of NMT1; however, the affinity for MYR is reduced (Fig. 5.5a). On the other hand, the peptide binding affinity is enhanced with a concurrent increase in the V_{max} by ~1.7-fold (Fig. 5.5b). The truncation increases the enzymatic efficiency for peptide substrate without changing its efficiency toward MYR [43].

5.4 Summary

N-Myristoylation was primarily discovered as a co-translational modification that occurred on amino-terminal glycine residue of a translating nascent polypeptide chain bound to the ribosome. In consistence with this, the human NMTs reflected to possess a poly-lysine cluster facilitating it to tether to the ribosome. However, subsequently it has become widely documented that N-myristoylation occurs on many proteins in a post-translational fashion as well. The post-translational myristoylation

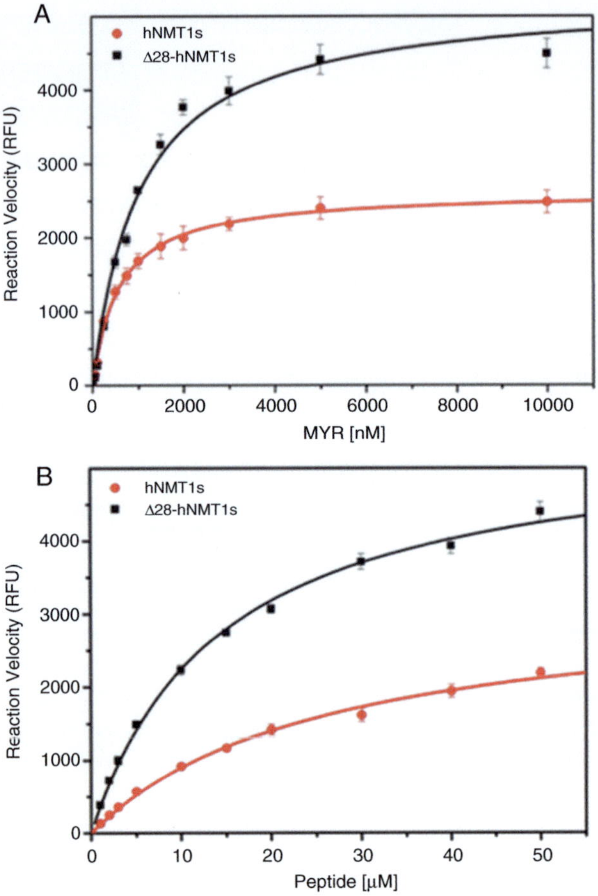

Fig. 5.5 Enzymatic activity of the catalytic domain (hNMT1s) and N-terminal truncation in catalytic domain (Δ28-hNMT1s) was measured under standard assay conditions as described (43). (**a**) Michaelis-Menten analysis of the myristoyl-CoA (MYR) and (**b**) Michaelis-Menten analysis of peptide substrate by the proteins used in study (For details see Kumar and Sharma [43])

usually occurs after a proteolytic cleavage event, which allows for the unmasking of a glycine residue at internal positions, mostly in the apoptotic states. The proteolytic cleavage of NMTs itself alters its sub-cellular localization either through the removal of ribosome-targeting domain containing the poly-lysine region in NMT1 or that of a negatively charged domain upstream of the poly-lysine domain in NMT2. The ribosome binding and/or membrane binding is mediated by poly-lysine domain and the caspase-mediated cleavage and of NMT1, and therefore the removal of the polybasic domain of hNMT1 translocates the caspase-cleaved NMT1 to cytosol from its primarily membrane-bound state. In cells undergoing apoptosis, a disproportionate number of substrates become quickly available following proteolytic cleavage and require being N-myristoylated to perform their physiological functions. This increases the demand for the myristoylation process, and to meet physiological requirements for myristoylation of the substrates, cells could undergo an increased synthesis of the NMT enzyme. However, the rate of polypeptide synthesis may not match the requirements within the rapid time scales of apoptosis. The removal of the ribosome-binding poly-lysine domain might represent the controlling features for the rapid relocalization of NMTs for the prompt post-translational myristoylation of substrate proteins available after the caspase cleavage events during apoptosis. Furthermore, the proteolytic processing as observed earlier may serve to meet the increased demands of myristoylation by removal of the N-terminal regulatory regions, and therefore enhancing the turnover rates of peptide substrates.

References

1. Resh MD (2016) Fatty acylation of proteins: the long and the short of it. Prog Lipid Res 63:120–131
2. Resh MD (2013) Covalent lipid modifications of proteins. Curr Biol 23:R431–R435
3. Hannoush RN, Sun J (2010) The chemical toolbox for monitoring protein fatty acylation and prenylation. Nat Chem Biol 6:498–506
4. Farazi TA, Waksman G, Gordon JI (2001) The biology and enzymology of protein N-myristoylation. J Biol Chem 276:39501–39504
5. Rajala RV, Datla RS, Moyana TN et al (2000) N-myristoyltransferase. Mol Cell Biochem 204:135–155
6. Aitken A, Cohen P, Santikarn S et al (1982) Identification of the NH2-terminal blocking group of calcineurin B as myristic acid. FEBS Lett 150:314–318
7. Carr SA, Biemann K, Shoji S et al (1982) N-Tetradecanoyl is the NH2-terminal blocking group of the catalytic subunit of cyclic AMP-dependent protein kinase from bovine cardiac muscle. Proc Natl Acad Sci U S A 79:6128–6131
8. Cebula M, Moolla N, Capovilla A et al (2013) The rare TXNRD1_v3 ("v3") splice variant of human thioredoxin reductase 1 protein is targeted to membrane rafts by N-acylation and induces filopodia independently of its redox active site integrity. J Biol Chem 288:10002–10011
9. Kumar S, Parameswaran S, Sharma RK (2015) Novel myristoylation of the sperm-specific hexokinase 1 isoform regulates its atypical localization. Biol Open 4:1679–1687
10. Kumar S, Sharma RK (2014) An improved method and cost effective strategy for soluble expression and purification of human N-myristoyltransferase 1 in E. coli. Mol Cell Biochem 392:175–186
11. Selvakumar P, Kumar S, Dimmock JR et al (2011) NMT1 (N-myristoyltransferase 1). Atlas Genet Cytogenet Oncol Haematol 15:570–575
12. Wright MH, Heal WP, Mann DJ et al (2010) Protein myristoylation in health and disease. J Chem Biol 3:19–35
13. Boutin JA (1997) Myristoylation. Cell Signal 9:15–35
14. Resh MD (1999) Fatty acylation of proteins: new insights into membrane targeting of myristoylated and palmitoylated proteins. Biochim Biophys Acta 1451:1–16
15. Maurer-Stroh S, Eisenhaber B, Eisenhaber F et al (2002) N-terminal N-myristoylation of proteins: refinement of the sequence motif and its taxon-specific differences. J Mol Biol 317:523–540
16. Utsumi T, Sato M, Nakano K et al (2001) Amino acid residue penultimate to the amino-terminal gly residue strongly affects two cotranslational protein modifications, N-myristoylation and N-acetylation. J Biol Chem 276:10505–10513
17. Wilcox C, Hu JS, Olson EN (1987) Acylation of proteins with myristic acid occurs cotranslationally. Science 238:1275–1278

18. Zha J, Weiler S, Oh KJ et al (2000) Posttranslational N-myristoylation of BID as a molecular switch for targeting mitochondria and apoptosis. Science 290:1761–1765
19. Bhatnagar RS, Futterer K, Farazi TA et al (1998) Structure of N-myristoyltransferase with bound myristoylCoA and peptide substrate analogs. Nat Struct Biol 5:1091–1097
20. Bhatnagar RS, Futterer K, Waksman G et al (1999) The structure of myristoyl-CoA:protein N-myristoyltransferase. Biochim Biophys Acta 1441:162–172
21. Giang DK, Cravatt BF (1998) A second mammalian N-myristoyltransferase. J Biol Chem 273:6595–6598
22. Bhandarkar SS, Bromberg J, Carrillo C et al (2008) Tris (dibenzylideneacetone) dipalladium, a N-myristoyltransferase-1 inhibitor, is effective against melanoma growth in vitro and in vivo. Clin Cancer Res 14:5743–5748
23. Das U, Kumar S, Dimmock JR et al (2012) Inhibition of protein N-myristoylation: a therapeutic protocol in developing anticancer agents. Curr Cancer Drug Targets 12:667–692
24. Ducker CE, Upson JJ, French KJ et al (2005) Two N-myristoyltransferase isozymes play unique roles in protein myristoylation, proliferation, and apoptosis. Mol Cancer Res 3:463–476
25. Kumar S, Dimmock JR, Sharma RK (2011) The potential use of N-myristoyltransferase as a biomarker in the early diagnosis of colon cancer. Cancers (Basel) 3:1372–1382
26. Selvakumar P, Lakshmikuttyamma A, Shrivastav A et al (2007) Potential role of N-myristoyltransferase in cancer. Prog Lipid Res 46:1–36
27. Yang SH, Shrivastav, A, Kosinski et al (2005) N-myristoyltransferase 1 is essential in early mouse development. J Biol Chem 280: 18990–18995
28. King MJ, Sharma RK (1992) Demonstration of multiple forms of bovine brain myristoyl CoA:protein N-myristoyl transferase. Mol Cell Biochem 113:77–81
29. Boutin JA, Ferry G, Ernould AP et al (1993) Myristoyl-CoA:protein N-myristoyltransferase activity in cancer cells. Purification and characterization of a cytosolic isoform from the murine leukemia cell line L1210. Eur J Biochem 214:853–867
30. Glover CJ, Felsted RL (1995) Identification and characterization of multiple forms of bovine brain N-myristoyltransferase. J Biol Chem 270:23226–23233
31. Glover CJ, Hartman KD, Felsted RL (1997) Human N-myristoyltransferase amino-terminal domain involved in targeting the enzyme to the ribosomal subcellular fraction. J Biol Chem 272:28680–28689
32. McIlhinney RA, McGlone K, Willis AC (1993) Purification and partial sequencing of myristoyl-CoA:protein N-myristoyltransferase from bovine brain. Biochem J 290:405–410
33. Thinon E, Morales-Sanfrutos J, Mann DJ et al (2016) N-Myristoyltransferase inhibition induces ER-stress, cell cycle arrest, and apoptosis in cancer cells. ACS Chem Biol http://www.ncbi.nlm.nih.gov/pubmed/?term=27267252
34. Rechsteiner M (1990) PEST sequences are signals for rapid intracellular proteolysis. Semin Cell Biol 1:433–440
35. Rechsteiner M, Rogers SW (1996) PEST sequences and regulation by proteolysis. Trends Biochem Sci 21:267–271
36. Raju RV, Kakkar R, Datla RS et al (1998) Myristoyl-coA:protein N-myristoyltransferase from bovine cardiac muscle: molecular cloning, kinetic analysis, and in vitro proteolytic cleavage by m-calpain. Exp Cell Res 241:23–35
37. Lakshmikuttyamma A, Selvakumar P, Kanthan R et al (2004) Overexpression of m-calpain in human colorectal adenocarcinomas. Cancer Epidemiol Biomark Prev 13:1604–1609
38. Selvakumar P, Smith-Windsor E, Bonham K et al (2006) N-myristoyltransferase 2 expression in human colon cancer: cross-talk between the calpain and caspase system. FEBS Lett 580:2021–2026
39. Sharma RK, Kumar S, Parameswaran S et al (2014) Regulation of N-myristoyltransferase by the calpain and caspase systems. Indian J Biochem Biophys 51:506–511
40. Neumar RW, Xu YA, Gada H et al (2003) Cross-talk between calpain and caspase proteolytic systems during neuronal apoptosis. J Biol Chem 278:14162–14167
41. Wang KK, Posmantur R, Nadimpalli R et al (1998) Caspase-mediated fragmentation of calpain inhibitor protein calpastatin during apoptosis. Arch Biochem Biophys 356:187–196

42. Perinpanayagam MA, Beauchamp E, Martin DD et al (2013) Regulation of co- and post-translational myristoylation of proteins during apoptosis: interplay of N-myristoyltransferases and caspases. FASEB J 27:811–821
43. Kumar S, Sharma RK (2015) N-terminal region of the catalytic domain of human N-myristoyltransferase 1 acts as an inhibitory module. PLoS One 10:e0127661
44. Mahrus S, Trinidad JC, Barkan DT et al (2008) Global sequencing of proteolytic cleavage sites in apoptosis by specific labeling of protein N termini. Cell 134:866–876

Role of Tissue Factor-FVIIa Blood Coagulation Initiation Complex in Cancer

6

Abhishek Roy, Ramesh Prasad, Anindita Bhattacharya, Kaushik Das, and Prosenjit Sen

Abstract

Since a century ago, an intricate relationship exists between cancer progression and thromboembolism. In various case studies, thromboembolic complications have been found to maintain an intricate relationship with the progression of various tumours like breast, lung, colon and glioblastoma. Moreover, coagulation factors have also been reported to be involved for metastatic augmentation complications in cancer patients with elevated levels of complication in cancer-associated thrombosis. Production and protease activity of various coagulation factors like thrombin and tissue factor (TF)-FVIIa complex affect tumour progression and propagation actively. TF exerts both coagulant as well as PAR2-dependent cancerous activity by eliciting various cell survival signalling pathways, like P42/44MAPK and PI3K/AKT. However, the molecular elucidation of the role of these coagulation factors in cancer-associated thrombosis and metastatic progression has not been understood till date.

Keywords

Blood coagulation • Cancer • Factor VIIa • Microvesicles • Macrophages • Signalling • Tissue factor

A. Roy • R. Prasad • A. Bhattacharya • K. Das • P. Sen (✉)
Department of Biological Chemistry, Indian Association for the Cultivation of Science, Jadavpur, Kolkata 700032, India
e-mail: bcps@iacs.res.in

S. Chakraborti, N.S. Dhalla (eds.), *Proteases in Physiology and Pathology*,
DOI 10.1007/978-981-10-2513-6_6

6.1 Introduction

The phenomenon of blood coagulation has been studied enormously in the last few decades. Although intrinsic factors involved in blood coagulation are not fully understood, its relation to various other pathophysiological and systemic disorders has been uncovered through several research studies. Blood coagulation is a complex series of enzymatic cascade reactions processed either by one or both the pathways: intrinsic and extrinsic. The concept of the coagulation cascade was first proposed in the year of 1964, which deals with the conversion of inactive protease precursors (zymogen) into their activated form. The extrinsic coagulation process is initiated by tissue factor (TF) on the cell surface [1]. In veins and arteries, coagulation factor VII (FVII) circulates as zymogen (inactive), and an endothelial layer forms a lining between the bloodstream and the extravascular layer where TF, the receptor for FVII, resides. Damage of the endothelial lining exposes TF to the blood, where it binds with its ligand FVII and form TF-FVIIa proteolytic active binary complex. This activated complex then activates FX to FXa, which binds with its cofactor FVa, and converts prothrombin into thrombin. Further, thrombin activates platelets and proteolytically activates soluble fibrinogen to insoluble fibrin, which eventually in conjugation with activated platelets forms mesh-like structure, known as blood clot [2]. In the intrinsic pathways, FVIIIa with its ligand FIXa forms tenase FVIIIa-FIXa complex in the presence of Ca^{2+} on membrane surface, which plays a crucial role to generate FXa and finally leads to activation of downstream cascade in a similar fashion that of extrinsic pathways. TF-FVIIa binary complex formation is the key step to initiate blood coagulation, where FVIIa acts as serine protease; on the other hand, TF acts as cofactor for FVIIa. Free FVIIa has very less proteolytic activity, but TF allosterically modifies FVIIa conformation in such a manner that FVIIa proteolytic activity gets enhanced several folds than that of free form. Structurally, FVII consists of four domains: GLA domain (contains gamma-carboxyglutamate-rich residues), two epidermal growth factor domains (EGF1 and EGF2) and serine protease (SP) domain as shown in Fig. 6.1. Although allosteric modifications imparted by TF on FVIIa are still under investigation, so far it is known that upon activation of FVIIa from FVII, a proteolytic cleavage occurs in the single peptide bond between Arg152 and Ile153$^{(16)}$ that results to the formation of two peptide chains (heavy chain and light chain) linked covalently by a single disulphide bond. The newly generated N-terminal Ile153$^{(16)}$ of FVIIa (heavy chain) makes a salt bridge with Asp343$^{(194)}$ by spontaneous insertion of its N-terminal tail into the cavity (the chymotrypsin numbering is denoted in superscript with parentheses). It has been reported that TF-binding promotes N-terminal insertion, which accounts for allosteric regulation of FVIIa [3]. Like FVII, human FX and FIX also belong to same chymotrypsin homology family, having the catalytic triad of residues, namely, His$^{(57)}$, Asp$^{(102)}$ and Ser$^{(195)}$, in the protease domain (heavy chain). FX circulates in blood as zymogen form; upon proteolytic cleavage between Arg194$^{(15)}$ and Ile195$^{(16)}$ by TF-FVIIa complex, it gets activated to form FXa, an active serine protease with 52-residue activation peptide release. FIX is also a single-chain protein, in which

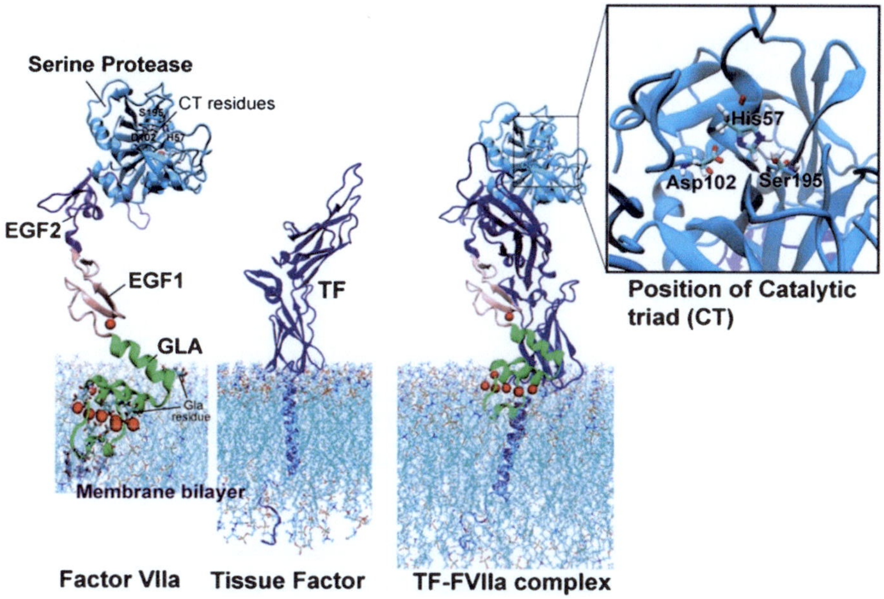

Fig. 6.1 Structure of TF-FVIIa with membrane lipid bilayer. In the left panel, cartoon diagram of FVIIa, showing four domains GLA (green), EGF1 (pink), EGF2 (blue) and serine protease (SP) (cyan) domain. Full length TF is shown in blue colour ribbon representation. The catalytic triad residues (His57, Ser195 and Asp102) are located in SP domain in TF-FVIIa complex, as shown in stick representation in the right panel. Bound calcium ions with Gla residues of GLA domain are shown in red colour ball representation

proteolytic cleavage occurs between two regions: Arg145-Ala146 and Arg180(15)-Val181(16) with a 35-residue activation peptide release upon activation of FIXa [4].

Recently, TF has converged the focus of attention due to its diverse role in different pathological and nonhemostatic conditions [5]. TF is a transmembrane glycoprotein (47-kDa molecular weight), having 263 amino acids in length, comprising of an extracellular (219 amino acids), single transmembrane-spanning domain (23 amino acids) and a cytosolic tail region (21 amino acids). TF contains post-translational modification in asparagine residues (11, 24, 137 and 261) by N-linked glycosylation [6]. At present the effect of N-linked glycosylation is not completely understood. In addition to post-translational modifications, extracellular part of TF contains two disulphide bonds, which is formed between Cys49-Cys57 and Cys186-Cys209. TF plays a crucial role in the initiation of blood coagulation [7]; it has also major contributions in maintaining haemostasis [8]. TF is constitutively present on cell surface of fibroblasts, pericytes, epithelial cells, etc., and in some abnormal conditions, TF can also get expressed in monocytes and endothelial cells [9–11]. TF is an essential factor for normal development of embryo, absence of which leads to a defective angiogenesis and causes embryonic lethality in mice [12, 13]. It is evident from various studies of TF in different cell types that TF exists into two populations on the cell surface: active TF (decryptic) and cryptic TF (inactive) [14, 15]. On cell surface, the majority of TF is non-functional and only small fraction of TF is functionally active [16]. Active TF binds to factor VIIa and cleaves FX, FIX and peptidyl substrate. However, cryptic TF can bind FVIIa but with reduced affinity and cleaves a peptidyl substrate but not FX. Recent studies suggest that TF also exists in soluble forms (sTF), which is generated either by alternative splicing (called alternatively spliced human tissue factor, asHTF) or proteolytic cleavage [17].

As mentioned earlier, apart from coagulation TF-FVIIa protease complex has diverse activities. This complex may affect the process of metastasis through various cellular signalling pathways, among which tumour angiogenesis at the site of secondary tumour growth will form. It also leads to the production of proteins that provide a favourable environment for metastasis and apoptosis inhibition, and majorly it acts as an adhesion molecule for the survival of cancer cells [18]. In this chapter, we are trying to highlight the role of TF-FVIIa proteolytic complex in the context of various cellular signalling pathways promoting the progression and propagation of cancer cell.

6.2 Tissue Factor-FVIIa Acts as a Signal Transducer

Till now the impact and contribution of coagulation proteases TF-FVIIa and its association with various cellular signalling pathways have been well accepted to get detail picture of this complex as a signal transducer [19–21]. During endothelial disruption, TF acts as a cofactor for FVIIa and forms TF-FVIIa protease-activated complex, which initiates various signalling pathways by cleaving and activating proteinase-activated receptor (PAR) family proteins [22, 23], which are seven transmembrane domain cellular receptors that get activated by self-ligation of the proteolytically cleaved end of extracellular amino terminus. TF-FVIIa activates directly by cleaving after arginine residue of sequence NH_2-SSKGRSLIGKV-COOH of PAR2 as shown in Fig. 6.2. PAR2 is generally present in human epithelial cells of neuronal, intestinal, airway and vascular tissues. Its expression level gets elevated in injured tissues or after inflammatory mediator treatment [24–26]. Here exposure of the tethered ligand that folds back to the second extracellular loop results to activation of PAR2. Unlike PAR2, PAR1 gets indirectly activated by TF-FVIIa proteolytic complex through thrombin and FXa, cleaving after arginine residue in the sequence NH_2-LDPRSFLLRN-COOH, as shown in Fig. 6.2. PAR1 is the prototypical receptor of thrombin, but is also activated by several other proteases, such as activated protein C, MMP-1 (matrix metalloproteinase) and plasmin. Activation of specific PAR subtypes is cancer dependent, for example, gene expression and their regulation evoked by TF-FVIIa through PAR2 in MDA-MB-231 cells occur in glioblastoma cell lines through thrombin-mediated activation of PAR1 [27]. TF-FVIIa, FXa, trypsin and tryptase activate PAR2, whereas thrombin and plasmin activate PAR4 [28]. In mouse, PAR3 has been found to serve as a cofactor for PAR4 [29], but in human it has been reported that PAR3 may also get directly activated by thrombin [30]. After activation, PARs interact to heterotrimeric G-proteins, which

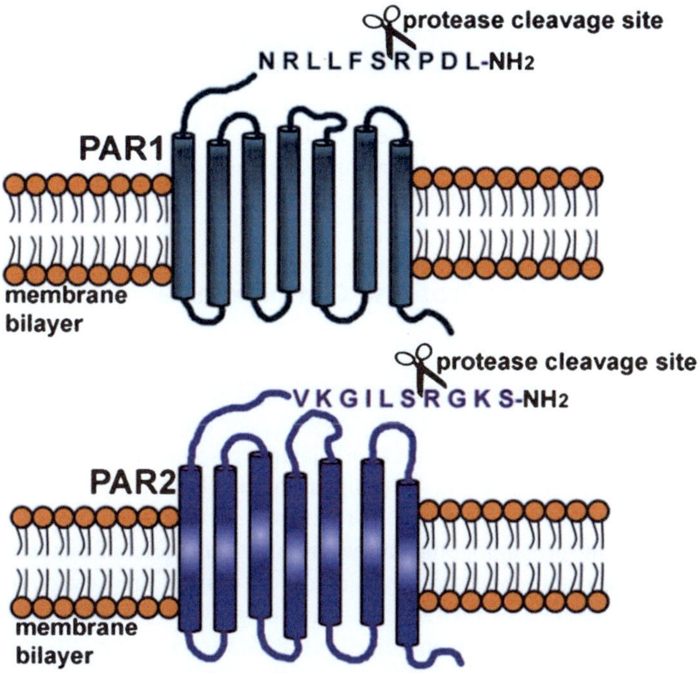

Fig. 6.2 Protein structures of proteinase-activated receptors (PAR1 and PAR2). Amino acid sequences and the protease cleavage site for receptor activation are shown in figure. Scissors symbol indicate the cleavage sites after arginine residue

leads to initiate further signalling events [28]. PARs expression level is usually found higher in various cancer types. Several evidences have revealed the intricate correlation between aggressive behaviour of cancer cells and PAR expression [31–34]. Ternary coagulation complex (TF-FVIIa-FXa) also cleaves and activates PAR2. TFPI (tissue factor pathway inhibitor) suppresses PAR2 activation by targeting ternary complex (involving FXa); however, the PAR2 signalling induced by TF-FVIIa binary complex is not affected by TFPI, because binary complex is independent of FXa.

Substantial evidences exist that TF plays pivotal role in the development of oncogenic processes and specifically tumour metastasis [35]. Most human epithelial cancers have high levels of TF, and the experimental studies clearly demonstrate that TF-driven tumour angiogenesis enhances tumour growth and promotes metastasis. It is important to note that TF could enhance cell migration through activation of PARs but may have the potential to initiate the migration in protease-independent manners like binding with integrins, decreasing the interactions with extracellular matrix by activating MMPs. Investigators reported that TF-dependent migratory ability of SW620 colorectal cancer cells [36], glioma cells [37] and MDA-MB231 breast cancer cells [35] appears to be dependent on both the proteolytic activity of FVIIa and the activation of PAR2. Experimentally it is evident that protein kinase C phosphorylates the serine residues of the cytoplasmic tail of TF [38]; however, the role of the cytoplasmic domain of TF in stimulating signalling and TF-FVIIa association in gene expression is not completely understood yet [39]. It is also reported that both intra- and extracellular domains of TF are essential for the prometastatic function of TF [40, 41]. More precisely both TF and FVIIa contribute to tumour progression through various signalling pathways and play a vital role in regulating the activity of various oncogenes.

6.3 Role of TF-FVIIa Protease Complex in Cancer Progression

Binding of FVIIa to TF results to a string of signalling events that regulates a broad range of cellular responses, such as cell survival, gene transcription and cytoskeletal changes, which are required for a cell to adequately respond to its local environment and malignant transformation [42–44]. Signalling of TF-FVIIa via PAR2 activation has been reported to be involved in the production of tumour-promoting molecules and tumour growth, proangiogenic to the invasive behaviour of cancer cells [45]. These findings are also supported by a study on murine breast cancer model in which genetic deficiency of PAR2 delays tumour growth and angiogenesis, and it was not observed in PAR1-deficient mice [46]. PAR2 activation elicits calcium transients and activation of the major members of the MAPK family, p42/44, P38 and JNK as shown in Fig. 6.3. In addition, Src-like kinase, PI3 kinase, the JAK/STAT pathway and Rho GTPases Rac1 and Cdc42 are activated, culminating in cell survival and cytoskeletal rearrangements [28]. Activation of PAR2 in tumour cells triggers the complex signalling mechanisms that affect both migratory and invasive properties which leads to the secretion of chemotactic and proangiogenic factors such as VEGF and IL-8 [47]. In colon cancer, PAR2 activation affects the decomposition of the extracellular matrix by enhancing the expression of MMP2 and MMP9, all of which promote metastasis and invasion [48]. In vitro and in vivo signalling also demonstrates that PAR1 activation by thrombin in tumour and endothelial cells also leads to proangiogenic behaviour that is involved with VEGF production and signalling as well as MMPs secretion [49–51].

As shown in Fig. 6.3. activation of both the MAPK and PI3 kinase pathways by PAR2 activation contributes to a promalignant transcriptional programme and stimulates the oncogenic protein synthesis. TF-FVIIa-mediated PAR2 activation also leads to the transcriptional activation and production of VEGF, CXCL1, Cyr61, CTGF, VEGF-C and IL-8 as well as of immunologic modulators, such as M-CSF and GM-CSF [27, 35, 52]. TF-dependent signalling through PAR2 enhances the inflammatory response, angiogenesis and cellular migration [53–56]. Being very similar to PAR2, various pro-tumoural responses may get evoked by PAR1 [57]. PAR1 expression in epithelial tumours is elevated by the transcription factor Egr-1, but inhibited by the tumour suppressor p53 protein [58]. Microarray analysis on cancer cell lines proclaims that activated PAR1 and PAR2 induce similar pro-tumoural responses [27]. Furthermore, PAR1 has been reported to induce oncogenic

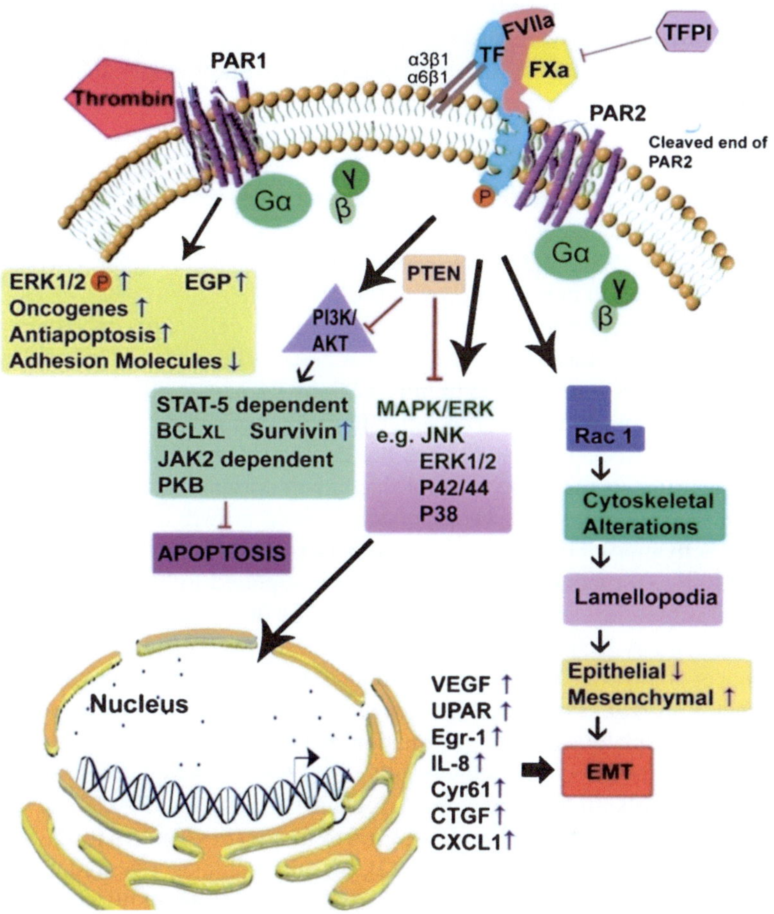

Fig. 6.3 Different protease-mediated factors (TF-FVIIa, FXa and thrombin) can activate various cellular signalling through the activation of PAR family. Some of the pathways involving PAR2 activation are PI3K/AKT, P42/44 and P38 MAPK which leads to the enhanced transcriptional activation of various genes (VEGF, UPAR, IL-8, etc.). In addition to this, cytoskeletal alteration through Rac-1 also contributes to cancer progression. Another receptor PAR1 (activated by thrombin) also contributes to tumour progression by increasing various antiapoptotic proteins

transformation in NIH3T3 cells [59]. Overexpression and activation of PAR1 in melanoma nonmetastatic cell lines stimulate the Akt/PKB signalling pathway, resulting to a reduction in both Bim and Bax expression. It also diminishes the levels of cleaved caspase-3 and caspase-9. In vivo experiment suggests that inhibition of PAR1 activity decreases tumour growth, confirming effects stimulated by this receptor [60].

Elevated PAR1 expression in MCF-7 cells promotes its in vivo tumour growth potential [61]. PAR1 also promotes angiogenesis through VEGF production in various melanoma models. In vivo study also demonstrated that blockade of both host and tumour PAR1 significantly decreases the tumour cell metastatic potential. Analogous observations were made in mouse models that support the PAR1 involvement in melanoma and breast cancer metastases [62]. Surprisingly, it has been mentioned that MMP-1 may act as a relevant PAR1 activator in the microenvironment of cancer cells [63]. Although PAR1 signalling induces a similar series of proteins in breast cancer cells, the activation of TF-FVIIa/PAR2 axis appears to elicit a more efficient production of these angiogenesis and immune regulators [27]. In mice harbouring a mammary tumour virus promoter-driven polyoma middle T antigen (PyMT) cassette, PAR2 deficiency resulted in delay of angiogenic switch and a concomitant reduction in tumour growth [46].

Lan et al. have reported that PAR2-induced cancer cell migration requires miR-125b involvement by targeting Gab2 and that NSun2- dependent RNA methylation contributes to the downregulation of miR-125b by PAR2 signalling. This study proposes the involvement of a novel epigenetic mechanism in which the altering expression of miRNA-125b modulates the cancer cell migratory ability [64]. All these studies suggest the importance of this binary complex as a key regulator during cancer progression. While TF-dependent tumour growth is critically dependent on PAR2 activation, TF-dependent metastasis is dependent on the formation of thrombin [65]. TF-mediated PAR2 signalling is reported to be dependent to some extent on β1-integrins, as PAR2 signalling could be diminished using a β1-integrin-inhibiting antibody [65].

Yokota et al. reported in a hyperthrombotic mouse model (thrombomodulin deficient) that a thrombin-mediated TF-dependent metastasis is associated with the hyperactivity of platelets and the formation of platelet leukocyte aggregates [66]. Activated PARs in blood cells, cancer and vessel wall cells by thrombin lead to the transcriptional activation of many proangiogenic genes such as VEGF and its receptor (VEGFR), MMP-2, basic fibroblast growth factor (bFGF), MAP, TF, angiopoietin-2 (Ang-2) and PI3 kinases [67–72]. In vitro studies reported that VEGF secretion from platelets and cancer cells occurs within few minutes of activation [69]. Furthermore, PAR activation by thrombin induces production of reactive oxygen species (ROS) via elevated expression of hypoxia-induced factor-1 (HIF-1) [73]. HIF-1 activates the transcription and expression of VEGF gene in response to fatty acid metabolism (arachidonic acid) [74]. These lines of experimental key findings altogether suggest that the presence of TF and blood coagulation enzymes in the microenvironment of tumour-associated cells plays a pivotal role in the neoplastic progression, mainly through the stimulation of PAR1 and PAR2 receptors.

6.4 Role of TF-FVIIa Protease Complex in Cancer Propagation

Apart from proliferation, the property of metastasis contributes immensely in the propagation and dissemination of cancer cells. The pivotal property of metastasis requires the cells to change their shape through alteration in cytoskeletal dynamics and the turnover of cell-cell as well as cell-matrix junctions to invade the

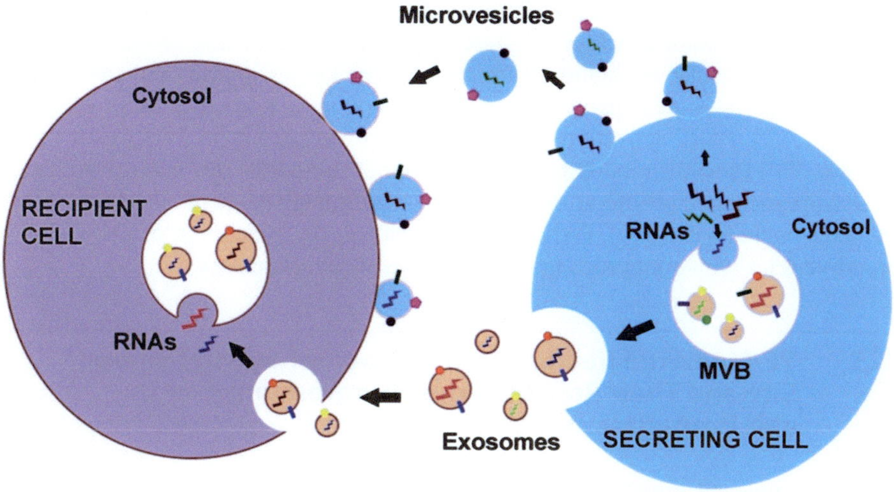

Fig. 6.4 Cancer cell-derived microvesicles generation. Cancer cells secreting eMVs containing proteins, RNAs and miRNAs fuse with the recipient cells to alter its phenotype

surrounding tissues [75]. This invasion is facilitated by the secretion of proteolytically active matrix metalloproteinases (MMPs) to degrade extracellular matrix (ECM) proteins and eventually promote tumour metastasis [76]. An important characteristic feature of tumour cells is to promote angiogenesis, the formation of new blood vessels from pre-existing vascular network. Although it is a normal physiological phenomenon, tumour cells show an elevated level of angiogenesis to get oxygen and nutrients and also to remove metabolic wastes [77]. Spreading not only involves the direct movement of cells but also happens through the secretion of pro-cancerous components from itself. It consists of cell-derived extracellular microvesicles (eMVs) and small signalling molecules. Extracellular microvesicles are cell-derived vesicles shed from almost all cell types and are primarily involved in transporting mRNA, miRNA and proteins between cells [78]. eMVs include (i) ectosomes (also called microvesicles, MVs, or microparticles, MPs) membranous structures ranging from 100 to 1000nm in diameter generated by simple outward budding of plasma membrane and can be pelleted down at 10,000 g and (ii) exosomes are endocytic origin and much smaller in size having diameter of 30–100 nm, collected by centrifugation at 1,00,000 g [79]. Microvesicles (MVs) are circulatory in nature, and growing evidence confirms their abundance in almost all body fluids as blood, urine, saliva, synovial fluid or even in interstitial spaces between cells and elsewhere [80]. TF-FVIIa-mediated microvesicle release has a profound contribution in the propagation of cancer cells and their metastasis. As shown in Fig. 6.4, a cancer cell (secreting cell) is shedding MVs to transform the neighbouring recipient cell to cancer cell. Previous studies have shown that various factors, like microRNA (miRNA), mRNA along with pro-cancerous protein content in the MVs and exosomes, contribute to the enhancement of cancer propagation [78, 81–86].

Over the past few years, emerging evidences indicate that these MVs act as an important contributor to various disease progression like atherosclerosis, liver diseases, cardiovascular anomalies, acute and chronic kidney injuries and cancer [87–90]. Microvesicles are capable of inducing antiapoptotic properties to the recipient cells [91, 92]. PAR2 activation in hypoxic cancer cells results in microvesicle generation [93]. Previous reports have already demonstrated the direct involvement of PAR2 in cancer cell proliferation and migration, although the detail mechanism still remains unelucidated [94]. PAR activation leads to the generation of extracellular microvesicles with both procoagulant and pro-cancerous activity.

6.5 TF-Mediated Recruitment of Macrophage to Tumour Site and Their Role in Cancer Progression and Propagation

Experimental and clinical experiments performed in the last 3–4 decades have established the vivid role of the blood clotting system in supporting tumour progression and metastasis. Gil et al. have shown that the macrophage recruitment in tumour microenvironment is mediated by TF-FVIIa-dependent clot formation in tumour cells [95]. They have further shown that these clot-mediated recruitments of tumour-associated macrophages (TAMs) are essential for the survival of tumour cells. These TAMs provide a suitable microenvironment for tumour growth, tumour survival, motility, tumour cell invasion, intravasation and angiogenesis [96]. Previous reports claim that ablation of TAMs from the tumour mass results in the inhibition of tumour progression and metastasis. Inflammation mediated by IL-6, TNF-α, IFN-γ, IL-4, TGF-β, arginase-1, etc. in the tumour mass induces the recruitment of macrophages [97, 98]. In turn these recruited macrophages secrete several growth factors and cytokines which positively regulate the tumour cell division and conversion of benign tumour to malignant one. Under the influence of CSF-1, TAMs promote angiogenesis via production of VEGF [99]. Macrophages are also found to promote neoangiogenesis in glioblastoma models [100]. Clot formations along with recruitment of functional macrophages are needed to establish the premetastatic niche in order to support tumour cell survival. TF expression is higher in cancer cells and the surrounding stromal macrophages which lead to improved tumour growth and metastasis. This is due to activation of signalling cascades via coagulation proteases (FVIIa, FXa and thrombin).

The microenvironment of a tumour mass is a complicated heterogenous system that supports the tumour cells and positively affects the process of cancer progression and propagation [101]. A number of leukocytes having role in both adaptive as well as innate immune responses get recruited to the tumour microenvironment in response to different stimuli. Among these recruited immune cell population, macrophages are the most prevalent and remain present in the tumour vicinity in all stages of tumour progression [102]. Monocytes are the universal precursors of macrophages. These precursor cells circulate in the blood for immune surveillance of the entire body. During any tissue damage, a number of various cytokines get

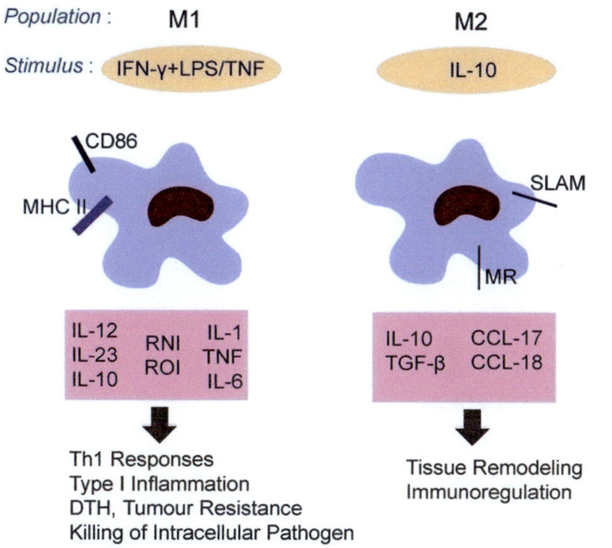

Fig. 6.5 Classification of macrophage (M1 and M2) depending on its inflammatory nature. M1 and M2 type of macrophage behave differently in the human body due to differential expression of different receptors and cytokines

released from the affected cells, and concomitantly a gradient of these chemoattractant gets established from the affected area to the blood. Sensing these chemicals, monocytes become activated and cross the endothelial barrier and finally reach to the affected area where they differentiate into macrophages and contribute to the wound healing procedure. The morphology of these differentiated macrophages is very much heterogenous, and moreover they perform diverse type of functions [103]. These macrophages can be subdivided into two categories: classically activated or M1 class and alternatively activated or M2 class [104], as shown in Fig. 6.5. The role of macrophages in the tumour microenvironment is ubiquitous. Being proinflammatory, M1 category of macrophage is a key player in immune system for the suppression of cancer. On contrary M2 macrophage is anti-inflammatory and proangiogenic in nature [105].

As shown in Fig. 6.6, due to TF-FVIIa proteolytic activity-mediated clot formation, TAMs get aligned along with the blood vessels to promote tumour cell intravasation into the circulation, which is a vital phenotype of malignancy [106]. TAM secretes epidermal growth factor (EGF) and other ligands of EGF family which form a paracrine loop by interacting with CSF-1, synthesized by the tumour cell to support directional tumour cell migration and invasion [107]. Abundant elevated coagulation factors expression and macrophage infiltration have been observed in various tumours. It was shown previously that FXIIa and TF-FVIIa treatment of monocytic cell line THP-1 transform their phenotypic expression to M2-like phenotype of interleukin (IL)-4^{high}, IL-10^{high}, tumour necrosis factor (TNF)-α^{high} and

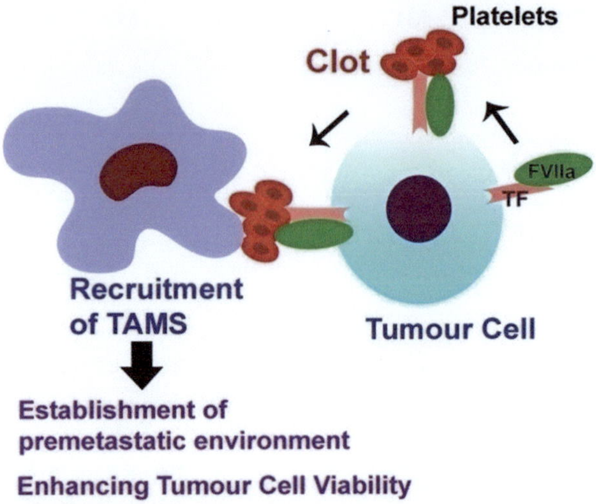

Fig. 6.6 TAMs-dependent tumour cell intravasation into the circulation. Interaction of tumour cells and its neighbouring TAMs through TF-FVIIa-mediated blood clot, leads to tumour progression and metastasis

transforming growth factor (TGF)-β^{high}. It has been elucidated recently that TAM activation by coagulation factors could induce VEGF/MMP-9 expression, which promotes the invasion of cancer cells. Studies assess that HUVEC cells cocultured with TAM (PMA-treated THP-1 macrophages cocultured with cancer cells) express higher levels of FXIIa [108]. In the current scenario, the understanding behind the metastatic progression of cancer cells by TAM has emerged a focus of attention in the field of cancer. Lastly, transformation of TAM-like cells by coagulation factors facilitates cancer cell migration and invasion.

6.6 Cancer-Associated Blood Coagulation Disorders

A proper equilibrium of coagulation inhibitors and coagulation factors must be maintained in the body to maintain the normal dynamics of blood coagulation. A reduced expression of natural coagulation prohibitors or their utilization may aggravate procoagulative activity in cancer patients. In healthy individuals coagulation inhibitors such as vitamin K-dependent coagulation inhibitors protein C and S and antithrombin III (AT-III) compensate the procoagulative activity. Whereas in cancer patients, lack of these coagulation inhibitors results in the enhanced clot formation [109]. AT-III, which is a natural thrombin inhibitor, acts irreversibly to inactivate thrombin, leading to the thrombin-antithrombin complex (TAT) formation. Higher levels of TAT are associated with acute lymphatic leukaemia (ALL) [110, 111]. Due to the lack of AT-III production, its concentration decreases in patients with disseminated malignancy. Protein C inhibits the initiation of coagulation on the surface of cancer cells and in small tumour blood vessels [112]. Activated form of protein

C (APC) also inhibits blood coagulation by neutralizing activated factor V and factor VIII [113, 114]. In spite of the presence of high amount of protein C, TF-mediated tumour-induced coagulation activation leads in hypercoagulability, mostly in dispersed malignancies.

Cancer patients often suffer from bleeding disorders in advanced malignant stage, and defect in coagulation is also associated with disseminated intravascular coagulation deep venous thrombosis (DVT) linked with malignancies like acute promyelocytic leukaemia and prostate cancer [115]. In thromboembolism, a large fraction of patients suffering from acute myelogenous leukaemia; renal, ovarian, pancreatic, gastric and lung cancer; and non-Hodgkin lymphoma also show enhanced level of thrombotic complications. As a consequence of abnormal blood coagulation, these patients have high risk of cerebrovascular disease, peripheral arterial/venous thrombosis and acute coronary syndrome-associated haemorrhagic complications. Another major consequence of abnormal coagulation disorder is disseminated intravascular coagulation (DIC), which is responsible for many malignant diseases in acute form. Cancer cells release high level of coagulation stimulatory factors like VEGF which activates the process of coagulation in damaged cancer blood vessels and surrounding tissues with concordant production of fibrinogen and fibrin degradation products. This local consumption of platelets and fibrinogen in fast-growing tumours lead to the systemic fibrinogen and platelet deficiency [116].

6.7 Conclusion

As a whole, proteolytic activity of coagulation factors especially TF-FVIIa complex, thrombin and FXa has significant contribution in cancer progression and propagation in various aspects. Several studies suggest that even picomolar concentration of FVIIa in the presence of TF can activate PAR2, which can regulate the switch point for initiating various signalling pathways [117]. We have also summarized that the signalling performed by TF/FVIIa/FXa ternary complex promotes metastasis which includes the enhanced development of blood vessels in the neighbourhood and suppression of apoptosis [118, 119]. The general mechanisms underlying cancer transformations in normal cells are directed by the signalling pathways, elicitated by coagulation factor activation and the recruitment of TAM substantiate TF-FVIIa complex proteolytic activity in tumour cells. In summary, these studies suggest that in therapeutic approach, the development of anticancer drugs with the potential to bind to TF-FVIIa complex or individual proteins may have beneficial effect in the treatment of such disease conditions. These inhibitors not only focus on tumour therapy but also contribute significantly in the treatment of coagulation disorders. Thus, specific inhibition of this binary complex functionally altered the TAM-dependent tumour viability and metastasis.

References

1. Versteeg HH, Peppelenbosch MP, Spek CA (2001) The pleiotropic effects of tissue factor: a possible role for factor VIIa-induced intracellular signalling? Thromb Haemost 86:1353–1359
2. Hoffman M (2003) A cell-based model of coagulation and the role of factor VIIa. Blood Rev 17:S1–S5
3. Higashi S, Nishimura H, Aita K, Iwanaga S (1994) Identification of regions of bovine factor VII essential for binding to tissue factor. J Biol Chem 269:18891–18898
4. Yoshitake S, Schach BG, Foster DC, Davie EW, Kurachi K (1985) Nucleotide sequence of the gene for human factor IX (antihemophilic factor B). Biochemistry 24:3736–3750
5. Mackman N (2009) The many faces of tissue factor. J Thromb Haemost 7(Suppl 1):136–139
6. Paborsky LR, Harris RJ (1990) Post-translational modifications of recombinant human tissue factor. Thromb Res 60:367–376
7. Butenas S, Orfeo T, Mann KG (2009) Tissue factor in coagulation: which? Where? When? Arterioscler Thromb Vasc Biol 29:1989–1996
8. Rao LVM, Pendurthi UR (2012) Regulation of tissue factor coagulant activity on cell surfaces. J Thromb Haemost 10:2242–2253
9. Pawlinski R, Mackman N (2010) Cellular sources of tissue factor in endotoxemia and sepsis. Thromb Res 125:S70–S73
10. Contrino J, Hair G, Kreutzer DL, Rickles FR (1996) In situ detection of tissue factor in vascular endothelial cells: correlation with the malignant phenotype of human breast disease. Nat Med 2:209–215
11. Drake TA, Morrissey JH, Edgington TS (1989) Selective cellular expression of tissue factor in human tissues. Implications for disorders of hemostasis and thrombosis. Am J Pathol 134:1087–1097
12. Bugge TH, Xiao Q, Kombrinck KW, Flick MJ, Holmbäck K, Danton MJ, Colbert MC, Witte DP, Fujikawa K, Davie EW, Degen JL (1996) Fatal embryonic bleeding events in mice lacking tissue factor, the cell-associated initiator of blood coagulation. Proc Natl Acad Sci U S A Nat Acad Sci 93:6258–6263
13. Carmeliet P, Mackman N, Moons L, Luther T, Gressens P, Van Vlaenderen I, Demunck H, Kasper M, Breier G, Evrard P, Müller M, Risau W, Edgington T, Collen D (1996) Role of tissue factor in embryonic blood vessel development. Nature 383:73–75
14. Chen VM, Hogg PJ (2013) Encryption and decryption of tissue factor. J Thromb Haemost 11(Suppl 1):277–284
15. Rao LVM, Kothari H, Pendurthi UR (2012) Tissue factor encryption and decryption: facts and controversies. Thromb Res 129(Suppl):S13–S17
16. Bach RR (2006) Tissue factor encryption. Arterioscler Thromb Vasc Biol 26:456–461
17. Bogdanov VY, Balasubramanian V, Hathcock J, Vele O, Lieb M, Nemerson Y (2003) Alternatively spliced human tissue factor: a circulating, soluble, thrombogenic protein. Nat Med 9:458–462
18. Versteeg HH, Spek CA, Peppelenbosch MP, Richel DJ (2004) Tissue factor and cancer metastasis: the role of intracellular and extracellular signaling pathways. Mol Med 10:6–11
19. Sallah S, Wan JY, Nguyen NP, Hanrahan LR, Sigounas G (2001) Disseminated intravascular coagulation in solid tumors: clinical and pathologic study. Thromb Haemost 86:828–833
20. Degen JL, Palumbo JS (2012) Hemostatic factors, innate immunity and malignancy. Thromb Res 129:S1–S5
21. Kocatürk B, Versteeg HH (2013) Tissue factor-integrin interactions in cancer and thrombosis: every Jack has his Jill. J Thromb Haemost 11(Suppl 1):285–293
22. Riewald M, Ruf W (2001) Mechanistic coupling of protease signaling and initiation of coagulation by tissue factor. Proc Natl Acad Sci U S A 98:7742–7747
23. Camerer E, Qazi AA, Duong DN, Cornelissen I, Advincula R, Coughlin SR (2004) Platelets, protease-activated receptors, and fibrinogen in hematogenous metastasis. Blood 104:397–401

24. Arora P, Ricks TK, Trejo J (2007) Protease-activated receptor signalling, endocytic sorting and dysregulation in cancer. J Cell Sci 120:921–928
25. Lin H, Liu AP, Smith TH, Trejo J (2013) Cofactoring and dimerization of proteinase-activated receptors. Pharmacol Rev 65:1198–1213
26. Ossovskaya VS, Bunnett NW (2004) Protease-activated receptors: contribution to physiology and disease. Physiol Rev 84:579–621
27. Albrektsen T, Sørensen BB, Hjortø GM, Fleckner J, Rao LVM, Petersen LC (2007) Transcriptional program induced by factor VIIa-tissue factor, PAR1 and PAR2 in MDA-MB-231 cells. J Thromb Haemost 5:1588–1597
28. Versteeg HH, Ruf W (2006) Emerging insights in tissue factor-dependent signaling events. Semin Thromb Hemost 32:24–32
29. Coughlin SR, Nakanishi-Matsui M, Zheng Y-W, Sulciner DJ, Weiss EJ, Ludeman MJ (2000) PAR3 is a cofactor for PAR4 activation by thrombin. Nature 404:609–613
30. Ostrowska E, Reiser G (2008) The protease-activated receptor-3 (PAR-3) can signal autonomously to induce interleukin-8 release. Cell Mol Life Sci 65:970–981
31. Veiga CDSB, Carneiro-Lobo TC, Coelho CJBP, Carvalho SMF, Maia RC, Vasconcelos FC, Abdelhay E, Mencalha AL, Ferreira AF, Castro FA, Monteiro RQ (2011) Increased expression of protease-activated receptor 1 (PAR-1) in human leukemias. Blood Cells Mol Dis 46:230–234
32. Ribeiro FS, Simão TA, Amoêdo ND, Andreollo NA, Lopes LR, Acatauassu R, Rumjanek FD, Albano RM, Pinto LFR, Monteiro RQ (2009) Evidence for increased expression of tissue factor and protease-activated receptor-1 in human esophageal cancer. Oncol Rep 21:1599–1604
33. Ikeda O, Egami H, Ishiko T, Ishikawa S, Kamohara H, Hidaka H, Mita S, Ogawa M (2003) Expression of proteinase-activated receptor-2 in human pancreatic cancer: a possible relation to cancer invasion and induction of fibrosis. Int J Oncol 22:295–300
34. Yin Y-J, Salah Z, Grisaru-Granovsky S, Cohen I, Even-Ram SC, Maoz M, Uziely B, Peretz T, Bar-Shavit R (2003) Human protease-activated receptor 1 expression in malignant epithelia: a role in invasiveness. Arterioscler Thromb Vasc Biol 23:940–944
35. Hjortoe GM, Petersen LC, Albrektsen T, Sorensen BB, Norby PL, Mandal SK, Pendurthi UR, Rao LVM (2004) Tissue factor-factor VIIa-specific up-regulation of IL-8 expression in MDA-MB-231 cells is mediated by PAR-2 and results in increased cell migration. Blood 103:3029–3037
36. Zhou H, Hu H, Shi W, Ling S, Wang T, Wang H (2008) The expression and the functional roles of tissue factor and protease-activated receptor-2 on SW620 cells. Oncol Rep 20:1069–1076
37. Gessler F, Voss V, Dützmann S, Seifert V, Gerlach R, Kögel D (2010) Inhibition of tissue factor/protease-activated receptor-2 signaling limits proliferation, migration and invasion of malignant glioma cells. Neuroscience 165:1312–1322
38. Zioncheck TF, Roy S, Vehar GA (1992) The cytoplasmic domain of tissue factor is phosphorylated by a protein kinase C-dependent mechanism. J Biol Chem 267:3561–3564
39. Camerer E, Gjernes E, Wiiger M, Pringle S, Prydz H (2000) Binding of factor VIIa to tissue factor on keratinocytes induces gene expression. J Biol Chem 275:6580–6585
40. Mueller BM, Ruf W (1998) Requirement for binding of catalytically active factor VIIa in tissue factor-dependent experimental metastasis. J Clin Invest 101:1372–1378
41. Bromberg ME, Konigsberg WH, Madison JF, Pawashe A, Garen A (1995) Tissue factor promotes melanoma metastasis by a pathway independent of blood coagulation. Proc Natl Acad Sci U S A 92:8205–8209
42. Antoniak S, Sparkenbaugh E, Pawlinski R (2014) Tissue factor, protease activated receptors and pathologic heart remodelling. Thromb Haemost 112:893–900
43. Åberg M, Eriksson O, Mokhtari D, Siegbahn A (2014) Tissue factor/factor VIIa induces cell survival and gene transcription by transactivation of the insulin-like growth factor 1 receptor. Thromb Haemost 111:748–760
44. Versteeg HH, Spek CA, Slofstra SH, Diks SH, Richel DJ, Peppelenbosch MP (2004) FVIIa:TF induces cell survival via G12/G13-dependent Jak/STAT activation and BclXL production. Circ Res 94:1032–1040

45. Ruf W, Disse J, Carneiro-Lobo TC, Yokota N, Schaffner F (2011) Tissue factor and cell signalling in cancer progression and thrombosis. J Thromb Haemost 9(Suppl 1):306–315
46. Versteeg HH, Schaffner F, Kerver M, Ellies LG, Andrade-Gordon P, Mueller BM, Ruf W (2008) Protease-activated receptor (PAR) 2, but not PAR1, signaling promotes the development of mammary adenocarcinoma in polyoma middle T mice. Cancer Res 68:7219–7227
47. Schaffner F, Ruf W (2008) Tissue factor and protease-activated receptor signaling in cancer. Semin Thromb Hemost 34:147–153
48. Tang J-Q, Fan Q, Wu W-H, Jia Z-C, Li H, Yang Y-M, Liu Y-C, Wan Y-L (2010) Extrahepatic synthesis of coagulation factor VII by colorectal cancer cells promotes tumor invasion and metastasis. Chin Med J 123:3559–3565
49. Maragoudakis ME, Tsopanoglou NE, Andriopoulou P, Maragoudakis M-EM (2000) Effects of thrombin/thrombosis in angiogenesis and tumour progression. Matrix Biol 19:345–351
50. Coughlin SR (2000) Thrombin signalling and protease-activated receptors. Nature 407:258–264
51. Nierodzik ML, Karpatkin S (2006) Thrombin induces tumor growth, metastasis, and angiogenesis: evidence for a thrombin-regulated dormant tumor phenotype. Cancer Cell 10:355–362
52. Liu Y, Mueller BM (2006) Protease-activated receptor-2 regulates vascular endothelial growth factor expression in MDA-MB-231 cells via MAPK pathways. Biochem Biophys Res Commun 344:1263–1270
53. Rao LVM, Pendurthi UR (2005) Tissue factor-factor VIIa signaling. Arterioscler Thromb Vasc Biol 25:47–56
54. Bluff JE, Brown NJ, Reed MWR, Staton CA (2008) Tissue factor, angiogenesis and tumour progression. Breast Cancer Res 10:204
55. Schaffner F, Ruf W (2009) Tissue factor and PAR2 signaling in the tumor microenvironment. Arterioscler Thromb Vasc Biol 29:1999–2004
56. Åberg M, Siegbahn A (2013) Tissue factor non-coagulant signaling - molecular mechanisms and biological consequences with a focus on cell migration and apoptosis. J Thromb Haemost 11:817–825
57. Villares GJ, Zigler M, Bar-Eli M (2011) The emerging role of the thrombin receptor (PAR-1) in melanoma metastasis--a possible therapeutic target. Oncotarget 2:8–17
58. Bar-Shavit R, Turm H, Salah Z, Maoz M, Cohen I, Weiss E, Uziely B, Grisaru-Granovsky S (2011) PAR1 plays a role in epithelial malignancies: transcriptional regulation and novel signaling pathway. IUBMB Life 63:397–402
59. Whitehead I, Kirk H, Kay R. (1995) Expression cloning of oncogenes by retroviral transfer of cDNA libraries. Mol Cell Biol 15:704–710
60. Zigler M, Kamiya T, Brantley EC, Villares GJ, Bar-Eli M (2011) PAR-1 and thrombin: the ties that bind the microenvironment to melanoma metastasis. Cancer Res 71:6561–6566
61. Boire A, Covic L, Agarwal A, Jacques S, Sherifi S, Kuliopulos A (2005) PAR1 is a matrix Metalloprotease-1 receptor that promotes invasion and tumorigenesis of breast cancer cells. Cell 120:303–313
62. Ünlü B, Versteeg HH (2014) Effects of tumor-expressed coagulation factors on cancer progression and venous thrombosis: is there a key factor? Thromb Res 133:S76–S84
63. Kim S-J, Shin J-Y, Lee K-D, Bae Y-K, Choi I-J, Park SH, Chun K-H (2011) Galectin-3 Facilitates Cell Motility in Gastric Cancer by Up-Regulating Protease-Activated Receptor-1(PAR-1) and Matrix Metalloproteinase-1(MMP-1). Vanacker J-M, editor. PLoS One 6:e25103
64. Yang L, Ma Y, Han W, Li W, Cui L, Zhao X, Tian Y, Zhou Z, Wang W, Wang H (2015) Proteinase-activated receptor 2 promotes cancer cell migration through RNA methylation-mediated repression of miR-125b *. Publ JBC Pap Press
65. Versteeg HH, Schaffner F, Kerver M, Petersen HH, Ahamed J, Felding-Habermann B, Takada Y, Mueller BM, Ruf W (2008) Inhibition of tissue factor signaling suppresses tumor growth. Blood 111:190–199

66. Yokota N, Zarpellon A, Chakrabarty S, Bogdanov VY, Gruber A, Castellino FJ, Mackman N, Ellies LG, Weiler H, Ruggeri ZM, Ruf W (2014) Contributions of thrombin targets to tissue factor-dependent metastasis in hyperthrombotic mice. J Thromb Haemost 12:71–81
67. Tsopanoglou NE, Maragoudakis ME (2004) Role of thrombin in angiogenesis and tumor progression. Semin Thromb Hemost 30:63–69.
68. Tang DG, Diglio CA, Honn KV (1993) 12(S)-HETE-induced microvascular endothelial cell retraction results from PKC-dependent rearrangement of cytoskeletal elements and alpha V beta 3 integrins. Prostaglandins 45:249–267
69. Tsopanoglou NE, Maragoudakis ME (1999) On the Mechanism of Thrombin-induced Angiogenesis: potentiation of vascular endothelial growth factor activity on endothelial cells by up-regulation of its receptors. J Biol Chem 274:23969–23976
70. Xie Q, Bao X, Chen ZH, Xu Y, Keep RF, Muraszko KM, Xi G, Hua Y (2016) Role of protease-activated receptor-1 in Glioma growth. Acta Neurochir Suppl 121:355–360
71. Yamahata H, Takeshima H, Kuratsu J-I, Sarker KP, Tanioka K, Wakimaru N, Nakata M, Kitajima I, Maruyama I (2002) The role of thrombin in the neo-vascularization of malignant gliomas: an intrinsic modulator for the up-regulation of vascular endothelial growth factor. Int J Oncol 20:921–928
72. Richard DE, Vouret-Craviari V, Pouysségur J (2001) Angiogenesis and G-protein-coupled receptors: signals that bridge the gap. Oncogene 20:1556–1562
73. Koizume S, Jin M-S, Miyagi E, Hirahara F, Nakamura Y, Piao J-H, Asai A, Yoshida A, Tsuchiya E, Ruf W, Miyagi Y (2006) Activation of cancer cell migration and invasion by ectopic synthesis of coagulation factor VII. Cancer Res 66:9453–9460
74. Krishnamoorthy S, Jin R, Cai Y, Maddipati KR, Nie D, Pagès G, Tucker SC, Honn KV (2010) 12-Lipoxygenase and the regulation of hypoxia-inducible factor in prostate cancer cells. Exp Cell Res 316:1706–1715
75. Friedl P, Alexander S (2011) Cancer invasion and the microenvironment: plasticity and reciprocity. Cell 147:992–1009
76. Stamenkovic I (2000) Matrix metalloproteinases in tumor invasion and metastasis. Semin Cancer Biol 10:415–433
77. Folkman J (2003) Fundamental concepts of the angiogenic process. Curr Mol Med 3:643–651
78. Balaj L, Lessard R, Dai L, Cho Y-J, Pomeroy SL, Breakefield XO, Skog J (2011) Tumour microvesicles contain retrotransposon elements and amplified oncogene sequences. Nat Commun 2:180
79. van Doormaal FF, Kleinjan A, Di Nisio M, Büller HR, Nieuwland R (2009) Cell-derived microvesicles and cancer. Neth J Med 67:266–273
80. Liu M-L, Williams KJ (2012) Microvesicles: potential markers and mediators of endothelial dysfunction. Curr Opin Endocrinol Diabetes Obes 19:121–127
81. Wang T, Gilkes DM, Takano N, Xiang L, Luo W, Bishop CJ, Chaturvedi P, Green JJ, Semenza GL (2014) Hypoxia-inducible factors and RAB22A mediate formation of microvesicles that stimulate breast cancer invasion and metastasis. Proc Natl Acad Sci 111:E3234–E3242
82. Janowska-Wieczorek A, Wysoczynski M, Kijowski J, Marquez-Curtis L, Machalinski B, Ratajczak J, Ratajczak MZ (2005) Microvesicles derived from activated platelets induce metastasis and angiogenesis in lung cancer. Int J Cancer 113:752–760
83. Wysoczynski M, Ratajczak MZ (2009) Lung cancer secreted microvesicles: underappreciated modulators of microenvironment in expanding tumors. Int J Cancer 125:1595–1603
84. Giusti I, D'Ascenzo S, Millimaggi D, Taraboletti G, Carta G, Franceschini N, Pavan A, Dolo V (2008) Cathepsin B mediates the pH-dependent proinvasive activity of tumor-shed microvesicles. Neoplasia 10:481–488
85. Li J, Zhang Y, Liu Y, Dai X, Li W, Cai X, Yin Y, Wang Q, Xue Y, Wang C, Li D, Hou D, Jiang X, Zhang J, Zen K, Chen X, Zhang C-Y (2013) Microvesicle-mediated Transfer of MicroRNA-150 from Monocytes to Endothelial Cells Promotes Angiogenesis. J Biol Chem 288:23586–23596

86. Hong BS, Cho J-H, Kim H, Choi E-J, Rho S, Kim J, Kim JH, Choi D-S, Kim Y-K, Hwang D, Gho YS (2009) Colorectal cancer cell-derived microvesicles are enriched in cell cycle-related mRNAs that promote proliferation of endothelial cells. BMC Genomics 10:556
87. Blum A (2009) The possible role of red blood cell microvesicles in atherosclerosis. Eur J Intern Med 20:101–105
88. Lemoinne S, Thabut D, Housset C, Moreau R, Valla D, Boulanger CM, Rautou P-E (2014) The emerging roles of microvesicles in liver diseases. Nat Rev Gastroenterol Hepatol 11:350–361
89. Gatti S, Bruno S, Deregibus MC, Sordi A, Cantaluppi V, Tetta C, Camussi G (2011) Microvesicles derived from human adult mesenchymal stem cells protect against ischaemia-reperfusion-induced acute and chronic kidney injury. Nephrol Dial Transplant 26:1474–1483
90. Muralidharan-Chari V, Clancy JW, Sedgwick A, D'Souza-Schorey C (2010) Microvesicles: mediators of extracellular communication during cancer progression. J Cell Sci 123:1603–1611
91. Bruno S, Grange C, Collino F, Deregibus MC, Cantaluppi V, Biancone L, Tetta C, Camussi G (2012) Microvesicles derived from mesenchymal stem cells enhance survival in a lethal model of acute kidney injury. PLoS One 7:e33115
92. Razmkhah F, Soleimani M, Mehrabani D, Karimi MH, Kafi-Abad SA (2015) Leukemia cell microvesicles promote survival in umbilical cord blood hematopoietic stem cells. EXCLI J 14:423–429
93. Svensson KJ, Kucharzewska P, Christianson HC, Sköld S, Löfstedt T, Johansson MC, Mörgelin M, Bengzon J, Ruf W, Belting M (2011) Hypoxia triggers a proangiogenic pathway involving cancer cell microvesicles and PAR-2-mediated heparin-binding EGF signaling in endothelial cells. Proc Natl Acad Sci U S A 108:13147–13152
94. Hu L, Xia L, Zhou H, Wu B, Mu Y, Wu Y, Yan J (2013) TF/FVIIa/PAR2 promotes cell proliferation and migration via PKCα and ERK-dependent c-Jun/AP-1 pathway in colon cancer cell line SW620. Tumour Biol 34:2573–2581
95. Gil-Bernabé AM, Ferjancic S, Tlalka M, Zhao L, Allen PD, Im JH, Watson K, Hill SA, Amirkhosravi A, Francis JL, Pollard JW, Ruf W, Muschel RJ (2012) Recruitment of monocytes/macrophages by tissue factor-mediated coagulation is essential for metastatic cell survival and premetastatic niche establishment in mice. Blood 119:3164–3175
96. Qian B-Z, Li J, Zhang H, Kitamura T, Zhang J, Campion LR, Kaiser EA, Snyder LA, Pollard JW (2011) CCL2 recruits inflammatory monocytes to facilitate breast-tumour metastasis. Nature 475:222–225
97. Grivennikov SI, Greten FR, Karin M (2010) Immunity, inflammation, and cancer. Cell 140:883–899
98. Brown D, Trowsdale J, Allen R (2004) The LILR family: modulators of innate and adaptive immune pathways in health and disease. Tissue Antigens 64:215–225
99. Yeo E-J, Cassetta L, Qian B-Z, Lewkowich I, Li J, Stefater JA, Smith AN, Wiechmann LS, Wang Y, Pollard JW, Lang RA (2014) Myeloid WNT7b mediates the angiogenic switch and metastasis in breast cancer. Cancer Res 74:2962–2973
100. Du R, Lu KV, Petritsch C, Liu P, Ganss R, Passegué E, Song H, Vandenberg S, Johnson RS, Werb Z, Bergers G (2008) HIF1alpha induces the recruitment of bone marrow-derived vascular modulatory cells to regulate tumor angiogenesis and invasion. Cancer Cell 13:206–220
101. Noy R, Pollard JW, Abraham D, Zins K, Sioud M, Lucas T, Schäfer R, Stanley ER, Aharinejad S, Adeegbe DO, Nishikawa H, Balkwill FR, Mantovani A, Balkwill F, Charles KA, Mantovani A, Belai EB, de Oliveira CE, Gasparoto TH, Ramos RN et al (2014) Tumor-associated macrophages: from mechanisms to therapy. Immunity 41:49–61
102. Chanmee T, Ontong P, Konno K, Itano N. (2014) Tumor-associated macrophages as major players in the tumor microenvironment. Cancers (Basel) 6:1670–1690
103. Gordon S, Taylor PR (2005) Monocyte and macrophage heterogeneity. Nat Rev Immunol 5:953–964
104. Martinez FO, Gordon S (2014) The M1 and M2 paradigm of macrophage activation: time for reassessment. F1000Prime Rep 6:13

105. Jetten N, Verbruggen S, Gijbels MJ, Post MJ, De Winther MPJ, Donners MMPC (2014) Anti-inflammatory M2, but not pro-inflammatory M1 macrophages promote angiogenesis in vivo. Angiogenesis 17:109–118
106. Wyckoff JB, Wang Y, Lin EY, Li J, Goswami S, Stanley ER, Segall JE, Pollard JW, Condeelis J (2007) Direct visualization of macrophage-assisted tumor cell intravasation in mammary tumors. Cancer Res 67:2649–2656
107. Condeelis J, Pollard JW (2006) Macrophages: obligate partners for tumor cell nigration, invasion, and metastasis. Cell 124:263–266
108. Ma Y-Y, He X-J, Wang H-J, Xia Y-J, Wang S-L, Ye Z-Y, Tao H-Q (2011) Interaction of coagulation factors and tumor-associated macrophages mediates migration and invasion of gastric cancer. Cancer Sci 102:336–342
109. Kvolik S, Jukic M, Matijevic M, Marjanovic K, Glavas-Obrovac L (2010) An overview of coagulation disorders in cancer patients. Surg Oncol 19:e33–e46
110. Chojnowski K, Wawrzyniak E, Treliński J, Niewiarowska J, Cierniewski C (1999) Assessment of coagulation disorders in patients with acute leukemia before and after cytostatic treatment. Leuk {&} lymphoma 36:77–84
111. Appel IM, Hop WCJ, van Kessel-Bakvis C, Stigter R, Pieters R (2008) L-Asparaginase and the effect of age on coagulation and fibrinolysis in childhood acute lymphoblastic leukemia. Thromb Haemost 100:330–337
112. Wojtukiewicz MZ, Sierko E, Zacharski LR, Zimnoch L, Kudryk B, Kisiel W (2003) Tissue factor-dependent coagulation activation and impaired fibrinolysis in situ in gastric cancer. Semin Thromb Hemost 29:291–300
113. Zangari M, Saghafifar F, Anaissie E, Badros A, Desikan R, Fassas A, Mehta P, Morris C, Toor A, Whitfield D, Siegel E, Barlogie B, Fink L, Tricot G (2002) Activated protein C resistance in the absence of factor V Leiden mutation is a common finding in multiple myeloma and is associated with an increased risk of thrombotic complications. Blood Coagul Fibrinolysis 13:187–192
114. Zanetta L, Marcus SG, Vasile J, Dobryansky M, Cohen H, Eng K, Shamamian P, Mignatti P (2000) Expression of Von Willebrand factor, an endothelial cell marker, is up-regulated by angiogenesis factors: a potential method for objective assessment of tumor angiogenesis. Int J Cancer 85:281–288
115. Munter G, Hershko C (2001) Increased warfarin sensitivity as an early manifestation of occult prostate cancer with chronic disseminated intravascular coagulation. Acta Haematol 105:97–99
116. Bernathova M, Jaschke W, Pechlahner C, Zelger B, Bodner G (2006) Primary angiosarcoma of the breast associated Kasabach-Merritt syndrome during pregnancy. Breast 15:255–258
117. Camerer E, Huang W, Coughlin SR (2000) Tissue factor- and factor X-dependent activation of protease-activated receptor 2 by factor VIIa. Proc Natl Acad Sci U S A 97:5255–5260
118. Jiang X, Guo YL, Bromberg ME (2006) Formation of tissue factor-factor VIIa-factor Xa complex prevents apoptosis in human breast cancer cells. Thromb Haemost 96:196–201
119. Goodsell DS (2006) The molecular perspective: tissue factor. Oncologist 11:849–850

Metalloproteases in Adaptative Cell Responses

7

Pavel Montes de Oca Balderas

Abstract

Throughout evolution, cells have acquired molecular mechanisms that enable them to interact with and adapt to the extracellular milieu. In the last decade, ectodomain shedding (ES) has emerged as a critical sensing mechanism of the environment that may remodel cell membrane molecular repertoire, eliciting dynamic intracellular responses. ES is the proteolytic release of the extracellular domain (ectodomain) from cell membrane molecules (CMM). This proteolysis is mediated mainly by matrix metalloproteases (MMPs) or disintegrin and metalloproteases (ADAMs). Virtually, all functional categories of CMM are subject of this proteolysis; therefore, ES is involved in most cellular processes including proliferation, apoptosis, migration, and differentiation or pathologies such as inflammation or cancer. ES releases membrane molecule's ectodomain to the extracellular space where it can play biological functions. ES of transmembrane molecules also generates membrane-attached terminal fragments comprising transmembrane and intracellular domains. These fragments may be further processed by intramembrane-cleaving proteases (i-CLiPs), a mechanism known as regulated intramembrane proteolysis (RIP), which releases molecule's intracellular domain (ICD). Contrary to the initial hypothesis, fragments that result from ES and/or RIP are not necessarily in the pathway of degradation. Instead, they may carry out specific functions that cannot be performed by full-length native molecules. Thus, ES has emerged as a switch that unmasks multifunctional activities of CMM. In this chapter, the general mechanism of ES is reviewed, and

P. Montes de Oca Balderas (✉)
Unit of Dynamic Neurobiology, Department of Neurochemistry, Instituto Nacional de Neurologia y Neurocirugia, Insurgentes Sur #3877; Col. La Fama, Ciudad de Mexico, CP 14269, Mexico
e-mail: pavel73@hotmail.com

S. Chakraborti, N.S. Dhalla (eds.), *Proteases in Physiology and Pathology*, DOI 10.1007/978-981-10-2513-6_7

some considerations are formulated in an effort to disentangle the complexity that this proteolytic mechanism and the processing of CMM clusters represent for the understanding of cell signaling.

Keywords
Metalloproteases • MMP • ADAM • "Ectodomain shedding" • Signaling • RIP • i-CLiPs • Transmembrane molecules

7.1 Introduction

The cell membrane has played a fundamental role in evolution because it is the structure that maintains vital differences between the extracellular space and the cytosol. However, this isolating role must also permit information exchange between outside and inside of the cell. The flow of information at membrane level is itself also critical for cell survival and its adaptation to the extracellular environment. This function is carried out mainly by cell membrane molecules (CMM) that sense extracellular space information and transduce it into intracellular signals that enable dynamic adaptative cell responses to its environment. The messengers, receptors, and intracellular transducers that mediate cell signaling have been the subject of study through decades in cell biology. The work on this subject has resulted in the identification and modus operandi of cell membrane receptors that belong to different families, including the G protein-coupled receptors, the tyrosine kinase family of receptors, or the tyrosine phosphatase family of receptors, to name a few. This knowledge has even been translated into specific therapies for certain diseases. However, and despite the advances in the field, the mechanisms that mediate cell responses to the environment are far from being fully understood. One of the reasons for this is the nonlinear, network-like complexity that cell signaling presents. In this regard, the mechanism of ectodomain shedding (ES) has reached the arena of cell signaling as an unexpected player that may help to understand the complexity of this cell function.

Since the 1970s, CMM shedding was considered a mechanism that mediates membrane molecule turnover, accounting it as an initial step in a degradative pathway that concomitantly enables molecular release from cell membrane [1]. By then, this mechanism was known to occur at higher rates in transformed, activated, or proliferating cells in comparison with resting cells [1]. Shedding was known to be an active metabolic process requiring respiration, protein synthesis, and energy, although the responsible molecular mechanism was unknown [1]. Proteases that mediate ES or sheddases started to be unmasked by the middle of the 1990s; since then its number has augmented considerably, but more importantly, the number of molecules that undergo ES in the cell has been growing incessantly [2–6]. Nowadays, taking into account that all sort of CMM are subject to ES, it is considered that virtually any CMM may undergo this proteolytic mechanism under certain circumstances. This assumption poses new challenges for the study of cell signaling and cell responses since sheddases are part of a web with high redundancy (one CMM may be cleaved by different sheddases) and promiscuity (one sheddase may cleave a large portfolio of CMM). Furthermore, since ES may yield functional molecular fragments capable of cell signaling, not only from receptors but from all sort of CMM, this mechanism has called for a reconsideration of the paradigms of cell signaling [7–9].

7.2 Ectodomain Shedding and Metalloproteases

Today, the process of ES is defined as the proteolytic release of CMM extracellular domain (ectodomain) from transmembrane molecules and those attached by different bindings such as glycophosphatidylinositol (GPI) anchoring [10] (Fig. 7.1). All functional groups of CMM include members that are subject to ES [2, 4, 6, 10, 11] (see below). For this reason, ES is involved in a diverse array of cell functions in which these membrane molecules participate including cell proliferation, apoptosis, migration, differentiation, and signal transmission among others [6, 8, 10–12]. ES is also involved in the regulation of relevant cell processes such as development and inflammation or pathologies such as cancer, Alzheimer disease (AD), prion disease, and degeneration [3, 6, 9, 11, 13–17].

ES is carried out mainly by two different families of metalloproteases, also called sheddases: the *m*atrix *m*etallo*p*roteases (MMPs) and the *A d*isintegrin *a*nd *m*etalloproteases (ADAMs), both from the superfamily of metzicins [6, 8, 10, 11]. However, some other proteases have been involved in ES such as plasmin, thrombin, or others [8, 18, 19]. MMPs were initially described by their ability to cleave extracellular matrix molecules; however, their catalog of substrates has diversified and now includes CMM or even intracellular molecules [2, 4, 16, 20–22]. ADAMs were first described as potential integrin-binding proteins involved in sperm-egg fusion [3]. MMPs and ADAMs are dependent upon Zn^{2+} and share similar catalytic domains and prodomains [8, 10, 23, 24], whereas MMPs also depend on Ca^{2+} [23]. Until today, close to 30 different MMPs have been described that may be soluble or attached to cell membrane by a transmembrane domain or GPI [11, 23]. On the other hand, more than 30 ADAMs have been described, all of them with transmembrane domains. From these, only 12 ADAMs have demonstrated or predicted active metalloprotease domain [6, 24].

MMP's and ADAM's activity are regulated by different mechanisms, the first being their gene expression that is under control of different kinds of stimuli [2, 6, 23, 24]. mRNA posttranscriptional regulation has also been described by its stability, translation efficiency, and microRNA activity [25–27]. A critical regulatory step of these proteases is their activation by proteolysis, since MMPs and ADAMs are synthesized as zymogens that require cleavage of their prodomain to become active [10, 23]. This activation step may be carried out by other already active metalloproteases or other proteases, a feature that has helped to forge the concept of the protease web [7, 28]. Their proteolytic activity may also be inhibited by their compartmentalization [23] or by the so-called tissue inhibitor of metalloproteases

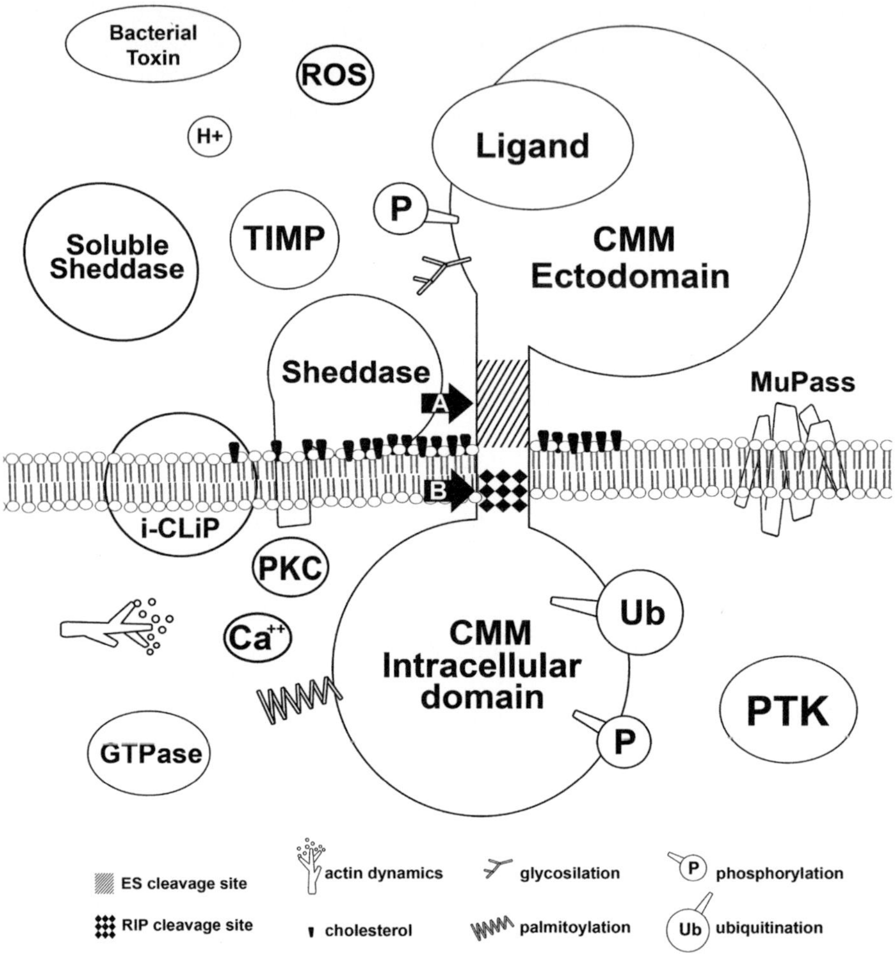

Fig. 7.1 General scheme of ectodomain shedding (ES) and regulated intramembrane proteolysis (RIP), players, and their regulators. Different classes of transmembrane molecules undergo ES releasing their ectodomain to the extracellular milieu. This proteolytic cleavage is mediated by sheddases that are soluble or attached to the cell membrane by a transmembrane domain or other anchoring. Sheddases' cleavage site (*arrow A*) is mainly located in the so-called stem region, <30 amino acids upstream the transmembrane domain. ES may be constitutive or induced by ligand binding, PKC activation, Ca^{++} dynamics, reactive oxygen species (*ROS*), pH, and bacterial toxins or inhibited by *TIMPs*. Target molecule posttranslational modifications such as ectodomain phosphorylation (*P*) or glycosylation, as well as its structural conformation, modulate its ES. Lipid rafts, actin dynamics, protein tyrosine kinases (*PTK*), and GTPase activity are also known to modulate ES. Target molecule ES generates NTF and CTF that remain associated to cell membrane that may further be subject to RIP mediated by an intramembrane-cleaving protease (*i-CLiP*). These proteases cleave within the transmembrane domain or few amino acids downstream it (*arrow B*). This second sequential proteolytic processing is regulated by intracellular posttranslational modifications such as ubiquitination (*Ub*), palmitoylation, or phosphorylation (*P*). RIP releases target molecule intracellular domain from cell membrane. Target molecule fragments generated by ES or RIP are known to carry out extracellular, intracellular, or intranuclear signaling functions. Some reports have demonstrated that multipass transmembrane molecules (*MuPass*) may also be sheddase or i-CLiP targets. See text for details and references (Modified from Ref. [4])

(TIMPs), four related proteins expressed by a wide array of cells [29]. In addition, it has been demonstrated that ADAM-17 transmembrane domain regulates its target specificity that can also be regulated by additional associated molecules or other proteases [30–32]. On the other hand, ADAM-10 intracellular domain (ICD) has been found to mediate its dimerization proposed to regulate its activity [33, 34] and control its constitutive activity [31]. Together, these regulatory mechanisms allow the spatiotemporal regulation of these enzymes and their activities. For more details on the expression, regulation, and activation of these proteases, please refer to the selected reviews and references therein [6, 11, 23, 24].

Although there is some consensus about the sequence specificity that some MMPs recognize, mainly a Pro residue at P3 position [35–37], there does not seem to be a sequence conservation in ADAM or MMP cleavage sites. In the same study by Turk et al. (2001), it was shown that MMPs have different amino acid preferences near P3 position. Nevertheless, in most cases of ES, cleavage occurs within transmembrane molecule stem region, that is, <30 amino acids upstream the transmembrane domain, a feature that has been considered more important than sequence itself [10, 38]. Sheddase loose sequence specificity provides their high promiscuity and redundancy and has some important implications for cell physiology (discussed below) [10]. It is credited that CMM are preferentially cleaved by a particular sheddase; however, it is well known that the same target molecule may be cleaved with different efficiencies by alternative sheddases [2, 3, 6]. This may become experimentally evident when the main associated sheddase is absent in a cell or may depend upon the cell model, tissue analyzed, or stimulation paradigm employed. The high promiscuity and diversity of sheddases, as well as their loose sequence specificity shared with the intramembrane-cleaving proteases (i-CLiPs) (see below), may be the evolutionary result of the fundamental role played by this proteolytic system in extracellular information sensing that conveys signals into the cell. The phylogenetic conservation from prokaryotic cells supports the notion of its fundamental role [39, 40].

ES may occur in a constitutive ligand-independent manner as has been described for γ-protocadherins, TrkA, interleukin-2 receptor beta (IL-2Rβ), or others [41–43] or in a ligand binding-dependent manner as described for Notch, ephrinB2, or others (Fig. 7.1) [44, 45]. As mentioned above, for some time it was believed that ES downregulated CMM function, since it was considered that it depended fully on the ectodomain [10]. Therefore, it was assumed that the remaining fragment attached to the cell membrane comprising the transmembrane and intracellular domains was nonfunctional; thus, ES was believed to be the initial step of a degradation pathway for molecular turnover [1]. However, studies on this proteolytic processing have demonstrated that this is not necessarily the case, since it has been found that ES enables extracellular (EC) and intracellular (IC) fragments to carry out additional functions beyond those of full-length native molecules, revealing their multifunctional nature [2, 4, 6]. Membrane-attached intracellular terminal fragments may be carboxy-terminal (CTF) or N-terminal (NTF) depending if the native molecule is a type I (with extracellular N-terminal) or type II (with intracellular N-terminal) transmembrane molecule. The activities of these fragments depend upon their IC nature. Interestingly, the work by Diaz-Rodriguez et al. [42] demonstrated that ES of receptor tyrosine kinase A (TrkA) is a ligand-independent cell mechanism regulated by the PKC pathway under control of different membrane receptors that generates IC fragments with tyrosine kinase activity. This clearly contradicts the notion that ES performs a receptor downregulatory function or the initial step in a degradation pathway of cell molecule turnover. Moreover, experiments with IL-2Rβ ES demonstrated that its CTF may potentiate the proliferative response to cytokine receptor activation, suggesting intracellular cross talk between full-length and shedded receptors, even if full-length receptor binds a different cytokine [43].

The full array or additional functions of CMM may depend or not upon the action of intracellular proteases that perform a second proteolytic process close to or within the transmembrane domain (Fig. 7.1) [4, 46–48]. This second proteolytic processing is known as regulated intramembrane proteolysis (RIP) and is performed by the intramembrane-cleaving proteases (i-CLiPs) [4, 46–50]. These include (1) presenilin (PS) from the γ-secretase (γ-sec) complex, (2) site 2 metalloproteases (S2P), (3) signal peptide peptidase (SPP) and aspartyl proteases, and (4) rhomboid serine proteases. The first three are believed to depend upon an initial ES to cleave target molecules, although it has been suggested that SPP may cleave without previous ES [51], whereas rhomboid proteases may cleave independently of it [46–50]. RIP enables the release into cytoplasm of intracellular domains (ICDs) from shedded molecules that may range from around a few amino acid residues (i.e., ≈40 from syndecan-3) up to some hundreds (i.e., ≈800 from neurogenic locus notch homolog protein [Notch]) [4, 46, 49]. As expected, length depends upon native molecule cytoplasmic domain size. Intracellular activities of some ICDs have already been studied and are known to regulate cell survival, gene transcription, or kinase activity among other functions [44, 52, 53]. However, for most CTFs, NTFs, and ICDs generated by ES and/or RIP, their intracellular functions have not been established or are suspected to be incomplete [53]. Notably, some ICDs have been observed to translocate into cell nucleus; therefore, it is inferred that they have a function within this cell compartment, as has been already demonstrated for Notch, ErbB1, and others that regulate gene transcription [4, 48, 51, 52, 54–56]. Interestingly, it has been reported that nuclear localization may also be achieved by membrane-tethered CTF, where it regulates gene transcription [57].

ES is known to take place in a constitutive or regulated manner by different stimuli that include ligand binding [44, 45], PKC activation [38, 52, 58–60], tyrosine kinase activity [38], Ca^{2+} dynamics [59, 61–65], oxidative stress [66–68], cell density [55, 69, 70], pH [71], or pathogen toxins [9]. Also, actin dynamics [72, 73], dynamin function [74], and GTPase activation [64, 75] have been shown to modulate ES. Posttranslational modifications of target molecule such as glycosylation [76, 77] and extracellular phosphorylation [78], as well as its structural conformation that precludes accessibility to sheddases [79], are also known to regulate ES (Fig. 7.1). It has been demonstrated that O-glycosylation not only downregulates ES, but in few cases it may upregulate it [77].

On the other hand, nowadays it is accepted that ES is a prerequisite and the main regulator for RIP, although some exceptions have been reported and rhomboid serine proteases are known to perform RIP without previous ES [46–48]. Similarly to ES, target molecule posttranslational modifications beyond proteolysis such as phosphorylation [44], palmitoylation [80], glycosylation [81], and ubiquitination [82] have been reported to regulate RIP (Fig. 7.1). Likewise, there is loose sequence specificity in cleavage site of those i-CLiPs analyzed, although it is commonly located within the transmembrane domain or few amino acids downstream where the only common motif are basic and hydrophobic amino acids [46, 47, 83]. Also, it has been found that helix destabilization at cleavage site facilitates i-CLiP's activity [46, 47, 83]. In the case of PS, their target molecules are supposed to require short (<30 amino acids) extracellular N-terminus tails generated after ES, apparently recognized by Nicastrin (also member of the γ-sec complex), although this is still disputed [46, 49, 50, 84, 85].

Cellular compartments where ES occurs have been investigated with different target molecules and include the pathway from the Golgi apparatus to the cell membrane and endosomes [2, 3, 6, 10, 86, 87]. In agreement with the accepted temporal sequence that exists between ES and RIP, the last has been reported to occur in the pathway from cell membrane to endosomes, and indeed the substrate traffic has been found to regulate its RIP [46, 47, 87, 88]. In addition, a role of lipid rafts as regulators or platforms where ES and RIP occur has been reported [46, 87, 89–91]. Also, ES has been found to occur in exosomes, where MPs have been found and are proposed to participate in the production of exosomes and in cell target activation through ES among other functions [92–94].

After examination of ES and RIP reports, a general scheme that includes recent findings of transmembrane molecule intracellular fate may be delineated (Fig. 7.2). In this scheme a considerable number of a given type of transmembrane molecule undergo ES in a constitutive or stimulated manner. From the total number of NTF or CTFs generated, some may carry out different intracellular functions including intracytoplasmic signaling [43] or intranuclear transcriptional regulation by a retrograde membrane trafficking pathway [57]. Concomitantly, or perhaps alternatively, NTFs or CTFs may follow the lysosomal pathway where they are degraded [43] or enter the exosome pathway through multivesicular bodies [92–94], whereas only a minor fraction undergoes RIP releasing their ICDs [55]. From endosomes, some transmembrane molecules may be recycled back to the cell membrane by recycling endosomes [95]. Seemingly, a large population of ICDs may follow the proteosomal pathway or other alternative undescribed degradative pathway [51, 96]. ICDs may directly regulate signal transduction pathways either by associating with kinases or regulating its RIP [44]. Ultimately, only a very small fraction of ICDs is translocated into the nucleus where it is involved in gene transcriptional regulation [48, 51]. Accordingly to the idea proposed by Anders et al., ICD's short life and low abundance reflect their important regulatory roles, as well as the low requirement of these fragments to fulfill their intracellular and intranuclear signaling functions [55].

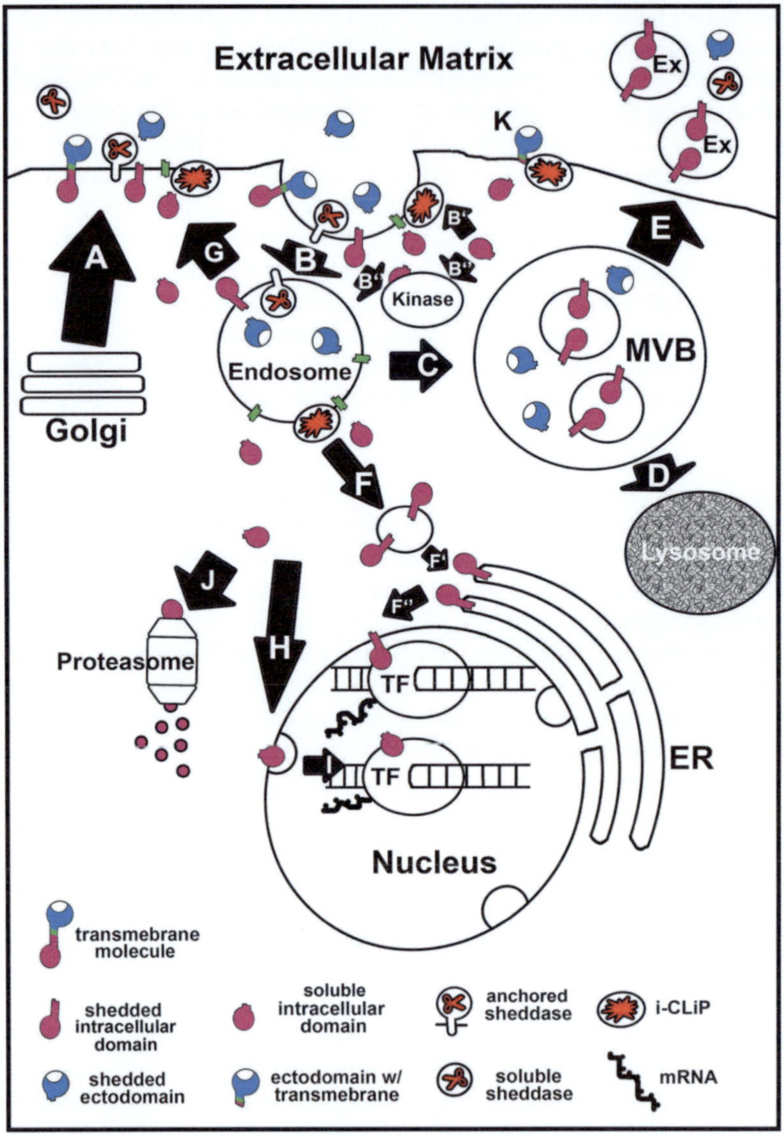

Fig. 7.2 Transmembrane molecule intracellular fate after ES and/or RIP. A large diversity of transmembrane molecules undergo ES and RIP in the cell membrane, although there are some reports that indicate sheddase activity in the pathway from Golgi to cell membrane (*arrow A*). In the cell membrane, these molecules are endocytosed and trafficked into the endosomal pathway that is known to regulate RIP (*arrow B*). NTF and CTF generated by ES, as well as ICD generated by RIP, may participate in cytoplasmic intracellular signaling by regulating i-CLiP (*arrow B'*) or kinase function (*arrow B''*), directly as kinases or phosphatases or regulating molecular localization. Concomitantly or perhaps alternatively, NTFs and CTFs may follow the multivesicular body pathway (*arrow C*) and then degraded in the lysosomal pathway (*arrow D*) or secreted (*arrow E*) in exosomes (*Ex*). From endosomes, some transmembrane molecules may be recycled back to the

7.3 Cell Membrane Molecules Subject to Ectodomain Shedding

As mentioned above, the spectra of CMM that undergo ES are wide and include all sort of functional molecules. Receptors, growth factors, cytokines, receptor ligands, cell adhesion molecules, and ion channels include members that have been reported to undergo ES. Either with transmembrane domains or other membrane anchoring, sheddases release extracellular soluble fragments from these molecules that in some cases have been reported to carry out functions associated with signal regulation in the EC milieu [4, 10]. Nevertheless, it must be considered that not all extracellular fragments are released to the EC milieu, despite they may lack transmembrane domain after processing. For instance, NMDAR subunit GluN1 extracellular fragment generated by MMP's activity was not found in the culture supernatant [97]. This observation suggests that after generation, it undergoes other fate, perhaps remaining attached to the extracellular matrix or cell membrane through some molecular interaction, or even internalized and degraded [98].

Despite the full number of molecules that undergo ER has not been evaluated, in a 2010 review [4] of the molecules expressed in the central nervous system (CNS) that are subject to ES and/or RIP (including only those with transmembrane domains), 110 molecules were found as target of ES. Of these, 9 were type II molecules, 93 were type I molecules, and only 2 were molecules with more than one transmembrane domain or multipass. These molecules belonged to different functional categories: 48 receptors, 21 receptor ligands, 25 involved in adhesion, 4 with adhesion-receptor function, 4 proteases, 2 involved in antigen presenting, 2 channel subunits, 1 with channel-adhesion function, 1 enzyme, and 2 without specific function described. In the receptor group, 11 belonged to the tyrosine protease kinase family, 3 to the protein tyrosine phosphatase family, 1 G protein-coupled receptor (GPCR), and 1 with guanylate cyclase activity. Despite these numbers are not updated, since new molecules have been discovered to undergo ER or other molecules have been found expressed in the CNS, the examples above are useful to substantiate the large diversity of molecules that are subject to ER, despite their function and biochemical or physicochemical properties.

Fig. 7.2 (continued) cell membrane by recycling endosomes (*arrow G*) or by a retrograde membrane trafficking pathway (*arrow F*), targeted to the endoplasmic reticulum (*ER, arrow F′*) from where they reach the nuclear envelope inner membrane (*arrow F″*) and participate in gene transcription regulation in association with transcription factors (*TF*). A minor fraction of RIP-generated ICDs is translocated into the nucleus through nuclear pores (*arrow H*) and carries out transcriptional regulatory functions (*arrow I*), whereas the major fraction follows the proteasomal pathway and is degraded (*arrow J*). Some work has suggested that transmembrane molecules may be cleaved without prior ES by presenilin or cleaved by ES-independent i-CLiPs (*K*) (See text for details and references. Modified from [4])

In a more recent review [16], the authors counted 182 molecules as MMP's targets. Nevertheless, this list included only 32 molecules reported as shedded membrane-bound substrates that included receptors, receptor ligands, adhesion molecules, and a plethora of molecules. Other categories of MMP's target molecules in this review included extracellular matrix molecules (38), soluble growth factors and cytokines (12), cryptic factors (29), chemokines and cytokines (18), immunity molecules (9), blood molecules (9), proteases and inhibitors (15), intracellular proteins (13), and multitask or multilocated molecules (7).

Notably, in both reviews just a few multipass transmembrane molecules were found to undergo ES, and none multipass transmembrane molecule has been reported to undergo both ES and RIP. Indeed, multipass transmembrane molecules ES and RIP are still under debate, despite some reports support this notion [6, 50]. These molecules are interesting in terms of ES and RIP because their processing would implicate additional complexity not observed with single-pass transmembrane domains, since these molecules could yield a larger diversity of fragments. This diversity and additional complexity would result from fragments with transmembrane domains with IC and EC loops, thus conferring the resulting fragments with biochemical and biophysical properties that could result in specific functional features different from those generated from single-pass transmembrane molecules. In this regard, ES has been demonstrated for GluN1 (NR1) subunit of NMDAR that is subject to activity-mediated ES in two different extracellular loops [97]. This subunit is also target of ES by an exogenous MMP that regulates its function [99]. Moreover, in our recent reports, we have observed the generation of GluN1 fragments that included different transmembrane domains and loops. Notably, in cultured astrocytes this correlates with a metabotropic-like flux-independent NMDAR, although causality was not established [100, 101]. In addition, GPCR PAR1 is a MMP-1 target that generates Ca^{2+}-dependent signals in breast cancer cells [102]. Also, GPCR brain angiogenesis inhibitor 1 (BAI1) has been reported to undergo ES generating the anti-angiogenic molecule vasculostatin [103]. More recently, GPCR's GPR37, GPR124, and GPR56 have also been reported as targets of ES [19, 104, 105]. In addition, in recent proteomic studies, other multipass membrane molecules have been identified as ES targets including GPCR, ion channel subunits, transporters, and others [106, 107]. On the other hand, GluR3 subunit from AMPAR is a multipass transmembrane molecule that apparently undergoes RIP independently of ES [108], whereas CXCR4 has also been reported to undergo γ-sec cleavage [50].

MPs have been found in the cytosol or intracellular organelles such as the nucleus, where their functions have been poorly studied and in some cases have been related to apoptosis [2, 16, 20–22]. Likewise, it has been reported that some nontraditional intracellular molecules are also targets of extracellular MPs, mainly metabolic enzymes, cytoskeletal proteins, or chaperones. Despite it has been argued that the extracellular presence of intracellular molecules is an artifact due to cell lysis, independent degradomic approaches have identified this kind of molecules as MMP's targets [2, 16, 20]. The review of these non-common MP's targets and locations is not the objective of the present chapter. For further information regarding these topics, please refer to the reviews above cited and references therein.

7.4 Considerations and Implications

ES influences most, if not all, CMM. Although initially considered a membrane molecule ectodomain delivery mechanism or an initial degradative step of membrane molecule turnover [1, 109], ES actually unveils the multifunctional nature of CMM enabling them to carry out their full array of intracellular and extracellular functions. These may further depend or not upon a second intracellular cleavage denominated RIP that releases cell membrane molecule ICD into cytosol and, in some cases, enables its translocation into the cell nucleus [4, 45, 48, 52, 57]. In these compartments, ICDs may act as enzymes and transcriptional or signal regulators, depending upon their intracellular domain nature [4, 42, 44, 45, 48, 52, 57]. Thus, intracellular signal transduction that was initially believed to be the intrinsic function only of full-length membrane receptors is indeed also carried out by a large diversity of CMM, including cell adhesion molecules, transmembrane receptor ligands, and proteases among others. This signal transduction mechanism elicited in receptor and membrane-bound receptor ligand-expressing cells has been termed reverse signaling or bidirectional signaling [41, 51, 110]. Considering the wide range of transmembrane molecules that are subject to ES, it seems plausible that this proteolytic system is indeed a general switch that regulates function of most transmembrane molecules, rather than an exceptional mechanism that some membrane molecules require for proper action in vivo [109], a cell surface molecule release mechanism [1], or sequential degradative steps of membrane molecule turnover that generate ICDs as by-products [1, 109].

Taking into account that ES and RIP of some transmembrane molecules may occur (1) constitutively or triggered by non-specific extracellular or intracellular stimuli such as oxidative stress, Ca^{2+} dynamics, or PKC activation and (2) within given intracellular compartments and/or membrane domains such as lipid rafts or endosomes and (3) that shedded receptors may cross talk with full-length receptors or perform intracellular kinase activities, then it is tempting to speculate that after a given stimulus and together with or independently from specific extracellular ligands, bundles or clusters of transmembrane molecules may concomitantly and stochastically be processed by ES and RIP (Fig. 7.3). This cluster processing would then yield bundles of NTFs and CTFs along with ICDs that coexist with full-length receptors and active MPs that collectively participate in intracellular signaling pathways (Fig. 7.3). These molecular ensembles or supramolecular entities could be transported to organelles and regulate their functions such as gene expression in the nucleus and ultimately determine adaptative cell responses. If true, this conjecture would confirm that cell signaling is not a linear phenomenon involving "one ligand = one receptor = signaling pathways." Instead, cell signaling would be an integrative response that results from the activation of canonical intracellular pathways accompanied by molecular bundles or ensembles of CMM that undergo ES and/or RIP and their resultant fragments. Together, these molecular ensembles would contain and carry the whole and specific information released from cell membrane or intracellular vesicles in the context of a particular cell type, tissue, and extracellular signal. In support of this idea, it is well known that CMM are not freely

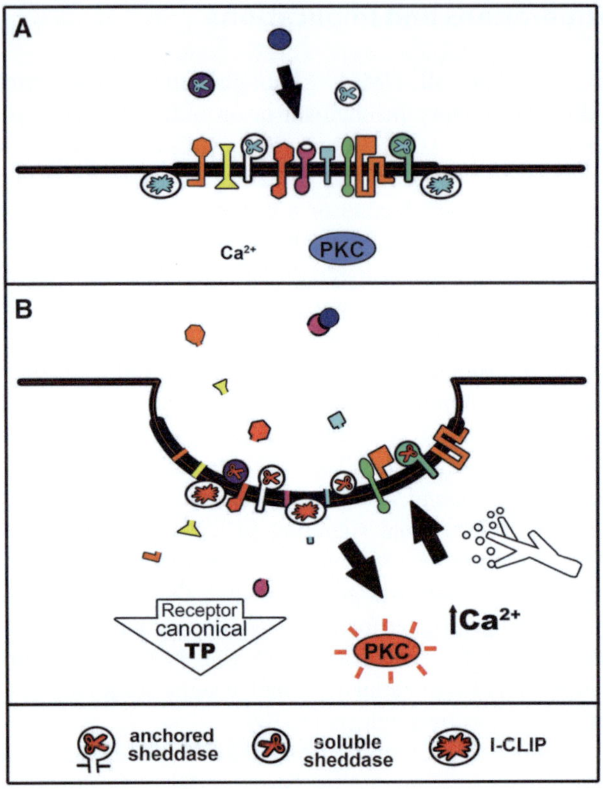

Fig. 7.3 The putative role of molecular bundles or ensembles processed by ES and/or RIP and their NTFs, CTFs, and ICDs in cell signaling. (**a**) Under basal conditions, bundles or ensembles of CMM in full-length form are organized in cell membrane regions such as lipid rafts (*thick black line* in the membrane). In these conditions, MPs and i-CLiPs are inactive, and ES/RIP does not occur or occurs in low levels. Under these conditions, cytoplasmic Ca^{2+} is also low and PKC is not activated. (**b**) After ligand binding to its receptor (*arrow panel A*) or some unspecific stimulus that rise cytoplasmic Ca^{2+} activates PKC and/or initiates actin dynamics (*black arrow* on the *left panel B*), MPs and i-CLiPs are activated (*black arrow* on the *right panel B*) and ES and RIP are initiated. Then bundles or ensembles of CMM are processed, generating CTFs, NTFs, and ICDs from them. The specific processing of these CMM depends upon their identity but also from the MPs and i-CLiPs expressed by the cell, helping to carve the specific intracellular response to a particular stimulus. Beyond the canonical IC transduction pathways (TP)(*open arrow panel B*) associated with a particular receptor (JAK/STAT, NFκB, etc.), the fragments generated by ES and/or RIP play a role in IC signaling, as explained in the text and Fig. 7.2. Together, IC TP and molecular fragments generated from CMM by ES and/or RIP and their IC activities induce the adaptative intracellular response to a given stimuli or condition that includes, for example, gene regulation within the nucleus (See text for details and references)

moving in the membrane bilayer. Instead, molecules are grouped together, and their movement is constrained by the interactions among them or by membrane biophysical properties that are given, for instance, by lipid rafts [111]. Moreover, this arrangement into supramolecular entities seems to be a common strategy in the cell that optimizes molecular interactions and therefore cellular responses [112–115]. Taking this conjecture further, the specific MPs and CMM expressed by each cell would then suppose a unique supramolecular signaling entity of proteases, receptors, their fragments, and activated signal transducers that would elicit specific intracellular responses (Fig. 7.3). This conjecture could help to understand the old fundamental question of cell biology regarding how different receptor-mediated intracellular cascades are able to generate specific cellular responses, despite the intracellular components are the same [114]. In addition, this may further account for some of the variability, diversity, and redundancy observed in cell biology models; however, experiments are needed to test this idea.

Interestingly, complexity of this model would be increased because cleavage of a given target molecule by more than one sheddase would open the possibility that different cleavage sites are recognized and consequently that resulting fragments generated have different N- or C- terminal sequences. These terminal sequence differences may yield important distinct results, as they occur with α- and β-secretase cleavage sites in amyloid precursor protein (APP) that differ by only two amino acids. This generates extracellular fragments with different biochemical characteristics that in one case result in pathological features. With this panorama, it is possible to conceive that this phenomenon may also occur with other shedded targets. This may also occur with intracellular fragments, as demonstrated by ADAM-10-CTFs generated by ADAM-9 or ADAM-15, which showed different intracellular traffics, thus suggesting different intracellular functions [56]. Similarly, complexity may still be increased since MP's biochemical properties modify their function. In this regard, it has been reported that the expression of membrane-tethered ADAM-9 or a soluble spliced variant has opposing functions in cancer cell migration [116]. At the same time, my unpublished observations suggest that intercellular contacts cast additional intricacy to the ES mechanism that could be related to cell density sensing, among other functions.

Thus, considering these observations, despite redundancy and promiscuity in MP's web, the set of sheddases expressed by a cell would be critical since shedded molecules and resulting fragments size would vary accordingly (Fig. 7.3). This would finally result into a specific response to a stimulus and/or localization and fate of the molecular cluster and fragments. Moreover, involvement of different sheddases in target molecule ES that depends upon cell type, tissue, or stimulus analyzed together with their specific glycosylation pattern indicates that despite their promiscuity and redundancy, sheddases display cell-type-specific specificity as hypothesized by Arribas and Borroto [10]. It is possible that this cell-type-specific specificity may provide the basis for the cell distinctive activation of intracellular pathways that results in cell specialization or differentiation. This idea is further supported by the spatial and temporal control in the expression of the sheddase, the substrate, and the regulatory factors [38]. Moreover, consistently with this idea, ES specificity for lipopolysaccharide (LPS) or tetradecanoyl phorbol acetate (TPA) activation has been recently demonstrated in immune-derived cells [117].

An additional important implication that rises from sheddases' high promiscuity and their self-proteolytic inter-regulatory interactions is the possibility that, when experimental lack of cleavage of a given target molecule is observed in the absence or inhibition of a particular sheddase *x*, this does not necessarily implies direct involvement of sheddase *x* in target processing. Instead, it may hint that sheddase *x* regulates indirectly through proteolytic processing of the sheddase involved directly in target molecule processing. Thus, appropriate experiments should be designed in order to conclude direct processing of a given target by a sheddase. Franzke et al. have previously discussed direct and indirect sheddase involvement as a putative reason that keeps sheddase identity of certain targets still under debate [61]. Also, since there are no sheddase exhaustive listings and some are better studied than others, this may lead to underestimate or overestimate their role. Finally, it is possible that overexpression of target molecules or sheddases in experimental cell models together with their high promiscuity may result in physiological nonrelevant cleavage by sheddases or i-CLiPs or other artifactual observations as previously discussed by Arribas and Borroto and Kirkin et al. [10, 51].

Other aspect to consider is that when CMM disappearance from cell membrane is studied, despite endocytosis may be the main mechanism involved, ES should also be considered, since intracellular mechanisms such as actin dynamics and GTPase activation among others are known to mediate both processes [64, 72, 75, 118, 119]. In this regard, to my knowledge, the proportion of CMM that undergoes ES or endocytosis in a cell has not been examined, although in some studies both have been reported to occur to the same molecule [120]. Also, despite it is known that some relationship exists between ES and endocytosis, this has been studied only partially [74, 82, 121, 122]. These aspects could be relevant for differential regulation of intracellular cell signaling pathways that are known to be closely associated with endocytosis [123–125] but also, as reviewed here, with ES. Moreover, it is possible that ES of certain molecules may regulate the mechanism of endocytosis itself. The rational for this is that the force exerted by the cytoskeleton and the endocytic machinery required to form an invagination and finally an endocytic vesicle depends upon the number of membrane molecular attachments with the extracellular matrix. Thus, when these molecular attachments are cleaved by ES, the force required is reduced and then endocytosis facilitated. However, to my knowledge, this has not been proved experimentally.

Intriguingly, it has been reported that ADAM-10-ICD translocates into cell nucleus only in 30% of transfected cells [56], in agreement with our unpublished observations made with an IL-2Rβ truncated construction mimicking a putative IL-2Rβ-ICD. This suggests that ICD's nuclear translocation occurs in a regulated manner, otherwise all ADAM-10-ICD or putative IL-2Rβ-ICD-transfected cells should show nuclear staining. This implication challenges the conception that any protein smaller than 40 kDa freely diffuses into the cell nucleus, in spite of lacking a nuclear localization signal (NLS) [126]. Nevertheless, considering these observations, it is possible to conceive that any protein smaller than 40 kDa without NLS enters the cell nucleus through nuclear pores under particular cell circumstances (present in 30% of cells in the experiments described above). Therefore, it seems that nuclear entrance of proteins below 40 kDa is mediated by an unspecific-regulated mechanism rather than a free diffusion mechanism. However, more experiments are required to explain these observations and test this possibility. Strikingly, Pcdh-γA3-ICD that is <200 aa (≈25 kDa) has a bipartite nuclear localization motif [96]. An ICD of such size would be expected to enter the nucleus freely and would not require a nuclear localization motif, unless it does so as part of a molecular complex that impedes such free entrance. Interestingly, when ICD size of the molecules that potentially undergo ES and RIP is analyzed, most of them are under the 40 kDa cutoff [4], opening the possibility that these fragments may freely enter the nucleus either by free diffusion or the putative unspecific-regulated mechanism.

7.5 Conclusions

The proteolytic system conformed by ES and RIP has emerged as an essential mechanism that mediates cell interaction with its environment [4, 7, 48]. This system acts as a regulatory switch of CMM that unveils their multifunctional identity and the full array of intracellular signals elicited by cell interactions with its extracellular milieu or neighboring cells that ultimately allow the adaptative cell responses.

The literature on ES indicates that all functional categories of CMM may be target of these proteolytic mechanisms, supporting the notion that under certain physiological circumstances, any CMM may be target of this proteolytic system. The stimuli that initiate CMM ES by sheddases are diverse and in some cases unspecific (Ca2+, PKC, ROS, etc.), opening the possibility that bundles or clusters of transmembrane CMM are processed yielding functional NTF, CTF, and ICD that participate in intracellular cell signaling (Fig. 7.3). This could also happen even with a specific messenger that initiates cell signaling, since intracellular Ca2+ rise or PKC activity induces ES of certain CMM. Thus, cell signaling would actually be the result of receptor activation by its ligand that results in its associated intracellular pathway activation, together with the actions of many CMM that undergo ES and yield functional NTF, CTF, and ICD that exert their function intracellularly (Fig. 7.3). Then, the receptors expressed and the pool of MPs activated by the stimulus together confer specificity to the response, as predicted earlier [10] and have been reported recently [117]. Therefore, cell signaling seems to be the integrative response of IC transduction pathways and CMM organized into supramolecular complexes or ensembles and their fragments that sense the EC milieu and elicit specific intracellular responses. This kind of signaling could help to understand the old fundamental question of cell biology regarding how different receptor-mediated intracellular cascades are able to generate specific cellular responses, despite the intracellular components are the same [114], and could account for the different responses observed among cell types to a given stimuli or even within the same population of cells [101].

Regarding the therapeutic use of metalloprotease inhibitors to treat different malignancies as has been proposed before [2, 6, 11, 14, 127], it is imperative that the sheddase-substrate web is well known but also that the putative supramolecular signaling entities and resulting fragments are considered. This would avoid undesired side effects, as suggested by other authors, further considering recent work reporting metalloprotease intracellular function [16, 21, 22, 128]. This could also be related with particular off-target or individual effects by certain signaling regulatory drugs. In this regard, several approaches have been employed trying to disentangle the degradome of MPs or the sheddome in certain cell types [20, 106, 107, 117, 129]. One limitation of some of these valuable efforts is that they have aimed to secreted proteins (the secretome), thus dismissing ES resultant fragments that are not released. An additional detail is that in most of these efforts, only one MP's inhibitor has been used to validate ES, and it is known that no inhibitor acts on all sheddases [127]. On the other hand, the same kind of analysis is necessary when PS or other i-CLiP's inhibitors are considered to treat different malignancies [46, 47, 130], due to the unexpected roles of these proteases and their products in intracellular signaling. Nevertheless, when the still not well-understood intracellular roles of CTFs, NTFs, and ICDs are appreciated, novel alternative therapeutic approaches will be envisaged for the future.

Acknowledgments Funding: National Council of Science and Technology (CONACyT) projects 138425 and 132706. I also thank Dr. Horacio Montes de Oca Balderas for his help with the review and edition of this manuscript.

References

1. Black PH (1980) Shedding from the cell surface of normal and cancer cells. Adv Cancer Res 32:75–199
2. Butler GS, Overall CM (2009) Updated biological roles for matrix metalloproteinases and new "intracellular" substrates revealed by degradomics. Biochemistry 48:10830–10845
3. Yang P, Baker KA, Hagg T (2006) The ADAMs family: coordinators of nervous system development, plasticity and repair. Prog Neurobiol 79:73–94
4. Montes de Oca-B P (2010) Ectodomain shedding and regulated intracellular proteolysis in the central nervous system. Cent Nerv Syst Agents Med Chem 10:337–359
5. Raucci A, Cugusi S, Antonelli A et al (2008) A soluble form of the receptor for advanced glycation endproducts (RAGE) is produced by proteolytic cleavage of the membrane-bound form by the sheddase a disintegrin and metalloprotease 10 (ADAM10). FASEB J 22:3716–3727
6. Reiss K, Saftig P (2009) The "A Disintegrin And Metalloprotease" (ADAM) family of sheddases: physiological and cellular functions. Semin Cell Dev Biol 20:126–137
7. Turk B, Turk D, Turk V (2012) Protease signalling: the cutting edge. EMBO J 31:1630–1643
8. Murphy G (2010) Fell-Muir lecture: Metalloproteinases: from demolition squad to master regulators. Int J Exp Pathol 91:303–313
9. Rodríguez D, Morrison CJ, Overall CM (2010) Matrix metalloproteinases: what do they not do? New substrates and biological roles identified by murine models and proteomics. Biochim Biophys Acta, Mol Cell Res 1803:39–54
10. Arribas J, Borroto A (2002) Protein ectodomain shedding. Chem Rev 102:4627–4638

11. Cauwe B, Van den Steen PE, Opdenakker G (2007) The biochemical, biological, and pathological kaleidoscope of cell surface substrates processed by matrix metalloproteinases. Crit Rev Biochem Mol Biol 42:113–185
12. Rivera S, Khrestchatisky M, Kaczmarek L et al (2010) Metzincin proteases and their inhibitors: foes or friends in nervous system physiology? J Neurosci 30:15337–15357
13. Agrawal SM, Lau L, Yong VW (2008) MMPs in the central nervous system: where the good guys go bad. Semin Cell Dev Biol 19:42–51
14. Rosenberg GA (2009) Matrix metalloproteinases and their multiple roles in neurodegenerative diseases. Lancet Neurol 8:205–216
15. Weber S, Saftig P (2012) Ectodomain shedding and ADAMs in development. Development 139:3693–3709
16. Butler GS (2000) Overall CM (2013) matrix metalloproteinase processing of signaling molecules to regulate inflammation. Periodontol 63:123–148
17. Gill SE, Parks WC (2008) Metalloproteinases and their inhibitors: regulators of wound healing. Int J Biochem Cell Biol 40:1334–1347
18. Bajou K, Peng H, Laug WE et al (2008) Plasminogen activator inhibitor-1 protects endothelial cells from FasL-mediated apoptosis. Cancer Cell 14:324–334
19. Vallon M, Aubele P, Janssen K-P, Essler M (2012) Thrombin-induced shedding of tumour endothelial marker 5 and exposure of its RGD motif are regulated by cell-surface protein disulfide-isomerase. Biochem J 441:937–944
20. Butler GS, Overall CM (2009) Proteomic identification of multitasking proteins in unexpected locations complicates drug targeting. Nat Rev Drug Discov 8:935–948
21. Mannello F, Medda V (2012) Nuclear localization of matrix metalloproteinases. Prog Histochem Cytochem 47:27–58
22. Strongin AY (2006) Mislocalization and unconventional functions of cellular MMPs in cancer. Cancer Metastasis Rev 25:87–98
23. Hadler-Olsen E, Fadnes B, Sylte I et al (2011) Regulation of matrix metalloproteinase activity in health and disease. FEBS J 278:28–45
24. Giebeler N, Zigrino P (2016) A Disintegrin and Metalloprotease (ADAM): historical overview of their functions. Toxins (Basel)
25. Jiang Y, Muschel RJ (2002) Regulation of matrix metalloproteinase-9 (MMP-9) by translational efficiency in murine prostate carcinoma cells. Cancer Res 62:1910–1914
26. Xia H, Qi Y, Ng SS et al (2009) microRNA-146b inhibits glioma cell migration and invasion by targeting MMPs. Brain Res 1269:158–165
27. Iyer V, Pumiglia K, DiPersio CM (2005) Alpha3beta1 integrin regulates MMP-9 mRNA stability in immortalized keratinocytes: a novel mechanism of integrin-mediated MMP gene expression. J cell Sci 118:1185–95
28. Krüger A (2009) Functional genetic mouse models: promising tools for investigation of the proteolytic internet. Biol Chem 390:91–97
29. Murphy G (2011) Tissue inhibitors of metalloproteinases. Genome Biol 12:233
30. Li X, Pérez L, Pan Z, Fan H (2007) The transmembrane domain of TACE regulates protein ectodomain shedding. Cell Res 17:985–998
31. Maretzky T, Evers A, Gall S Le, et al (2015) The cytoplasmic domain of a disintegrin and metalloproteinase 10 (ADAM10) regulates its constitutive activity but is dispensable for stimulated ADAM10-dependent shedding. J Biol Chem 290:7416–7425.
32. Maretzky T, McIlwain DR, Issuree PDA et al (2013) iRhom2 controls the substrate selectivity of stimulated ADAM17-dependent ectodomain shedding. Proc Natl Acad Sci U S A 110:11433–11438
33. Xu P, Liu J, Sakaki-Yumoto M, Derynck R (2012) TACE activation by MAPK-mediated regulation of cell surface dimerization and TIMP3 association. Sci signal 5:ra34.
34. Deng W, Cho S, Su P-C et al (2014) Membrane-enabled dimerization of the intrinsically disordered cytoplasmic domain of ADAM10. Proc Natl Acad Sci U S A 111:15987–15992
35. Turk BE, Huang LL, Piro ET, Cantley LC (2001) Determination of protease cleavage site motifs using mixture-based oriented peptide libraries. Nat Biotechnol 19:661–667

36. Nagase H, Fields GB (1996) Human matrix metalloproteinase specificity studies using collagen sequence-based synthetic peptides. Biopolymers 40:399–416
37. Nagase H (2001) Substrate specificity of MMPs. Matrix Met Inhib Cancer Ther:39–66
38. Hayashida K, Bartlett AH, Chen Y, Park PW (2010) Molecular and cellular mechanisms of ectodomain shedding. Anat Rec (Hoboken) 293:925–937
39. Brown MS, Ye J, Rawson RB, Goldstein JL (2000) Regulated intramembrane proteolysis: a control mechanism conserved from bacteria to humans. Cell 100:391–398
40. Li X, Wang B, Feng L et al (2009) Cleavage of RseA by RseP requires a carboxyl-terminal hydrophobic amino acid following DegS cleavage. Proc Natl Acad Sci 106:14837–14842
41. Reiss K, Maretzky T, Haas IG et al (2006) Regulated ADAM10-dependent ectodomain shedding of γ-protocadherin C3 modulates cell-cell adhesion. J Biol Chem 281:21735–21744
42. Díaz-Rodríguez E, Cabrera N, Esparís-Ogando A et al (1999) Cleavage of the TrkA neurotrophin receptor by multiple metalloproteases generates signalling-competent truncated forms. Eur J Neurosci 11:1421–1430
43. Montes de Oca P, Malarde V, Proust R et al (2010) Ectodomain shedding of interleukin-2 receptor beta and generation of an intracellular functional fragment. J Biol Chem 285:22050–22058
44. Georgakopoulos A, Litterst C, Ghersi E et al (2006) Metalloproteinase/Presenilin1 processing of ephrinB regulates EphB-induced Src phosphorylation and signaling. EMBO J 25:1242–1252
45. Mumm JS, Schroeter EH, Saxena MT et al (2000) A ligand-induced extracellular cleavage regulates gamma-secretase-like proteolytic activation of Notch1. Mol Cell 5:197–206
46. Beel AJ, Sanders CR (2008) Substrate specificity of gamma-secretase and other intramembrane proteases. Cell Mol Life Sci 65:1311–1334
47. McCarthy JV, Twomey C, Wujek P (2009) Presenilin-dependent regulated intramembrane proteolysis and γ-secretase activity. Cell Mol Life Sci 66:1534–1555
48. Lal M, Caplan M (2011) Regulated intramembrane proteolysis: signaling pathways and biological functions. Physiol 26:34–44
49. Wolfe MS (2009) Intramembrane-cleaving proteases. J Biol Chem 284:13969–13973
50. Boulton ME, Cai J, Grant MB (2008) γ-Secretase: a multifaceted regulator of angiogenesis: angiogenesis review series. J Cell Mol Med 12:781–795
51. Kirkin V, Cahuzac N, Guardiola-Serrano F et al (2007) The Fas ligand intracellular domain is released by ADAM10 and SPPL2a cleavage in T-cells. Cell Death Differ 14:1678–1687
52. Brou C, Logeat F, Gupta N et al (2000) A novel proteolytic cleavage involved in notch signaling: the role of the disintegrin-metalloprotease TACE. Mol Cell 5:207–216
53. Hambsch B, Grinevich V, Seeburg PH, Schwarz MK (2005) γ-Protocadherins, presenilin-mediated release of C-terminal fragment promotes locus expression. J Biol Chem 280:15888–15897
54. Lin SY, Makino K, Xia W et al (2001) Nuclear localization of EGF receptor and its potential new role as a transcription factor. Nat Cell Biol 3:802–808
55. Anders L, Mertins P, Lammich S et al (2006) Furin-, ADAM 10-, and gamma-secretase-mediated cleavage of a receptor tyrosine phosphatase and regulation of beta-catenin's transcriptional activity. Mol Cell Biol 26:3917–3934
56. Tousseyn T, Thathiah A, Jorissen E et al (2009) ADAM10, the rate-limiting protease of regulated intramembrane proteolysis of notch and other proteins, is processed by ADAMS-9, ADAMS-15, and the γ-secretase. J Biol Chem 284:11738–11747
57. Hieda M, Isokane M, Koizumi M et al (2008) Membrane-anchored growth factor, HB-EGF, on the cell surface targeted to the inner nuclear membrane. J Cell Biol 180:763–769
58. Aichem A, Masilamani M, Illges H (2006) Redox regulation of CD21 shedding involves signaling via PKC and indicates the formation of a juxtamembrane stalk. J Cell Sci 119:2892–2902
59. Zhang L, Bukulin M, Kojro E et al (2008) Receptor for advanced glycation end products is subjected to protein ectodomain shedding by metalloproteinases. J Biol Chem 283:35507–35516
60. Gooz M (2010) ADAM-17: the enzyme that does it all. Crit Rev Biochem Mol Biol 45:146–169

61. Franzke CW, Bruckner-Tuderman L, Blobel CP (2009) Shedding of collagen XVII/BP180 in skin depends on both ADAM10 and ADAM9. J Biol Chem 284:23386–23396
62. Litterst C, Georgakopoulos A, Shioi J et al (2007) Ligand binding and calcium influx induce distinct ectodomain/γ- secretase-processing pathways of EphB2 receptor. J Biol Chem 282:16155–16163
63. Higashiyama S, Iwabuki H, Morimoto C et al (2008) Membrane-anchored growth factors, the epidermal growth factor family: beyond receptor ligands. Cancer Sci 99:214–220
64. Nagano O, Murakami D, Hartmann D et al (2004) Cell-matrix interaction via CD44 is independently regulated by different metalloproteinases activated in response to extracellular Ca2+ influx and PKC activation. J Cell Biol 165:893–902
65. Boneberg E-M, Illges H, Legler DF, Fürstenberger G (2009) Soluble CD146 is generated by ectodomain shedding of membrane CD146 in a calcium-induced, matrix metalloprotease-dependent process. Microvasc Res 78:325–331
66. Kim J, Lin J, Adam RM et al (2005) An oxidative stress mechanism mediates chelerythrine-induced heparin-binding EGF-like growth factor ectodomain shedding. J Cell Biochem 94:39–49
67. Qin J, Goswami R, Dawson S, Dawson G (2008) Expression of the receptor for advanced glycation end products in oligodendrocytes in response to oxidative stress. J Neurosci Res 86:2414–2422
68. Hoefer MM, Illges H (2009) Ectodomain shedding and generation of two carboxy-terminal fragments of human complement receptor 2/CD21. Mol Immunol 46:2630–2639
69. Asher RA, Morgenstern DA, Properzi F et al (2005) Two separate metalloproteinase activities are responsible for the shedding and processing of the NG2 proteoglycan in vitro. Mol Cell Neurosci 29:82–96
70. Andersson CX, Fernandez-Rodriguez J, Laos S et al (2005) Shedding and gamma-secretase-mediated intramembrane proteolysis of the mucin-type molecule CD43. Biochem J 387:377–384
71. Choi SJ, Lee KH, Park HS et al (2005) Differential expression, shedding, cytokine regulation and function of TNFR1 and TNFR2 in human fetal astrocytes. Yonsei Med J 46:818–826
72. Furman MI, Krueger LA, Linden MD et al (2004) Release of soluble CD40L from platelets is regulated by glycoprotein IIb/IIIa and actin polymerization. J Am Coll Cardiol 43:2319–2325
73. Middelhoven PJ, van Buul JD, Kleijer M et al (1999) Actin polymerization induces shedding of FcgammaRIIIb (CD16) from human neutrophils. Biochem Biophys Res Commun 255:568–574
74. Carey RM, Balcz BA, Lopez-Coviella I, Slack BE (2005) Inhibition of dynamin-dependent endocytosis increases shedding of the amyloid precursor protein ectodomain and reduces generation of amyloid beta protein. BMC Cell Biol 6:30
75. Robert S, Maillet M, Morel E, et al (2005) Regulation of the amyloid precursor protein ectodomain shedding by the 5-HT4 receptor and Epac. FEBS Lett 579:1136–42
76. Eggert S, Paliga K, Soba P, Evin G, Masters CL, Weidemann A (2004) Beyreuther K. The proteolytic processing of the amyloid precursor protein gene family members APLP-1 and APLP-2 involves alpha-, beta-, gamma-, and epsilon-like cleavages: modulation of APLP-1 processing by n-glycosylation. J Biol Chem 279(18):18146–18156
77. Goth CK, Halim A, Khetarpal SA et al (2015) A systematic study of modulation of ADAM-mediated ectodomain shedding by site-specific O-glycosylation. Proc Natl Acad Sci U S A 112:14623–14628
78. Zimina EP, Fritsch A, Schermer B et al (2007) Extracellular phosphorylation of collagen XVII by ecto-casein kinase 2 inhibits ectodomain shedding. J Biol Chem 282:22737–22746
79. Asundi VK, Erdman R, Stahl RC, Carey DJ (2003) Matrix metalloproteinase-dependent shedding of syndecan-3, a transmembrane heparan sulfate proteoglycan, in Schwann cells. J Neurosci Res 73:593–602
80. Boutet P, Agu S, Atkinson S, et al (2008) Of the MHC class I-related chain B protein 1. J Immunol 8–12
81. May P, Bock HH, Nimpf J, Herz J (2003) Differential glycosylation regulates processing of lipoprotein receptors by γ-secretase. J Biol Chem 278:37386–37392

82. Gupta-Rossi N, Six E, LeBail O et al (2004) Monoubiquitination and endocytosis direct gamma-secretase cleavage of activated notch receptor. J Cell Biol 166:73–83
83. Sato T, Tang T-C, Reubins G et al (2009) A helix-to-coil transition at the epsilon-cut site in the transmembrane dimer of the amyloid precursor protein is required for proteolysis. Proc Natl Acad Sci U S A 106:1421–1426
84. Struhl G, Adachi A (2000) Requirements for presenilin-dependent cleavage of notch and other transmembrane proteins. Mol Cell 6:625–636
85. Xie T, Yan C, Zhou R et al (2014) Crystal structure of the γ-secretase component nicastrin. Proc Natl Acad Sci 111:13349–13354
86. Yokozeki T, Wakatsuki S, Hatsuzawa K et al (2007) Meltrin beta (ADAM19) mediates ectodomain shedding of Neuregulin beta1 in the Golgi apparatus: fluorescence correlation spectroscopic observation of the dynamics of ectodomain shedding in living cells. Genes Cells 12:329–343
87. Sannerud R, Annaert W (2009) Trafficking, a key player in regulated intramembrane proteolysis. Semin Cell Dev Biol 20:183–190
88. Urra S, Escudero CA, Ramos P et al (2007) TrkA receptor activation by nerve growth factor induces shedding of the p75 neurotrophin receptor followed by endosomal gamma-secretase-mediated release of the p75 intracellular domain. J Biol Chem 282:7606–7615
89. Gil C, Cubí R, Aguilera J (2007) Shedding of the p75NTR neurotrophin receptor is modulated by lipid rafts. FEBS Lett 581:1851–1858
90. Wakatsuki S, Kurisaki T, Sehara-Fujisawa A (2004) Lipid rafts identified as locations of ectodomain shedding mediated by Meltrin beta/ADAM19. J Neurochem 89:119–123
91. Toth M, Sohail A, Mobashery S, Fridman R (2006) MT1-MMP shedding involves an ADAM and is independent of its localization in lipid rafts. Biochem Biophys Res Commun 350:377–384
92. Sanderson MP, Keller S, Alonso A et al (2008) Generation of novel, secreted epidermal growth factor receptor (EGFR/ErbB1) isoforms via metalloprotease-dependent ectodomain shedding and exosome secretion. J Cell Biochem 103:1783–1797
93. Shimoda M, Khokha R (2013) Proteolytic factors in exosomes. Proteomics 13:1624–1636
94. Stoeck A, Keller S, Riedle S et al (2006) A role for exosomes in the constitutive and stimulus-induced ectodomain cleavage of L1 and CD44. Biochem J 393:609–618
95. Hémar A, Subtil A, Lieb M et al (1995) Endocytosis of interleukin 2 receptors in human T lymphocytes: distinct intracellular localization and fate of the receptor alpha, beta, and gamma chains. J Cell Biol 129:55–64
96. Haas IG, Frank M, Véron N, Kemler R (2005) Presenilin-dependent processing and nuclear function of γ-protocadherins. J Biol Chem 280:9313–9319
97. Pauly T, Ratliff M, Pietrowski E et al (2008) Activity-dependent shedding of the NMDA receptor glycine binding site by matrix metalloproteinase 3: a PUTATIVE mechanism of postsynaptic plasticity. PLoS One 3:e2681
98. Kato A, Rouach N, Nicoll RA, Bredt DS (2005) Activity-dependent NMDA receptor degradation mediated by retrotranslocation and ubiquitination. Proc Natl Acad Sci U S A 102:5600–5605
99. Szklarczyk A, Ewaleifoh O, Beique JC et al (2008) MMP-7 cleaves the NR1 NMDA receptor subunit and modifies NMDA receptor function. FASEB J 22:3757–3767
100. De Oca Balderas PM, Ospina GG, del Ángel AS (2013) Mitochondrial impairment induced by 3-nitropropionic acid is enhanced by endogenous metalloprotease activity inhibition in cultured rat striatal neurons. Neurosci Lett 546:16–20
101. Montes de Oca Balderas P, Aguilera P (2015) A metabotropic-like flux-independent NMDA receptor regulates Ca2+ exit from endoplasmic reticulum and mitochondrial membrane potential in cultured astrocytes. PLoS One 10:e0126314

102. Boire A, Covic L, Agarwal A et al (2005) PAR1 is a matrix metalloprotease-1 receptor that promotes invasion and tumorigenesis of breast cancer cells. Cell 120:303–313
103. Kaur B, Brat DJ, Devi NS, Van Meir EG (2005) Vasculostatin, a proteolytic fragment of brain angiogenesis inhibitor 1, is an antiangiogenic and antitumorigenic factor. Oncogene 24:3632–3642
104. Mattila SO, Tuusa JT, Peta ja-Repo UE (2016) The Parkinson's-disease-associated receptor GPR37 undergoes metalloproteinase-mediated N-terminal cleavage and ectodomain shedding. J Cell Sci 129:1366–1377
105. Paavola KJ, Stephenson JR, Ritter SL et al (2011) The N terminus of the adhesion G protein-coupled receptor GPR56 controls receptor signaling activity. J Biol Chem 286:28914–28921
106. Fong KP, Barry C, Tran AN et al (2011) Deciphering the human platelet sheddome. Blood 117:15–27
107. Sanz R, Ferraro GB, Fournier AE (2015) IgLON cell adhesion molecules are shed from the cell surface of cortical neurons to promote neuronal growth. J Biol Chem 290:4330–4342
108. Meyer EL, Strutz N, Gahring LC, Rogers SW (2003) Glutamate receptor subunit 3 is modified by site-specific limited proteolysis including cleavage by γ-secretase. J Biol Chem 278:23786–23796
109. Kopan R, Ilagan MXG (2004) Gamma-secretase: proteasome of the membrane? Nat Rev Mol Cell Biol 5:499–504
110. Voss M, Lettau M, Paulsen M, Janssen O (2008) Posttranslational regulation of Fas ligand function. Cell Commun Signal 6:11
111. Mayor S, Rao M (2004) Rafts: scale-dependent, active lipid organization at the cell surface. Traffic 5:231–240
112. Kagan JC, Magupalli VG, Wu H (2014) SMOCs: supramolecular organizing centres that control innate immunity. Nat Rev Immunol 14:821–826
113. Schroder K, Tschopp J (2010) The inflammasomes. Cell 140:821–832
114. Johenning FW, Ehrlich BE (2002) Signaling microdomains: InsP(3) receptor localization takes on new meaning. Neuron 34:173–175
115. Delmas P, Wanaverbecq N, Abogadie FC et al (2002) Signaling microdomains define the specificity of receptor-mediated InsP(3) pathways in neurons. Neuron 34:209–220
116. Fry JL, Toker A (2010) Secreted and membrane-bound isoforms of protease ADAM9 have opposing effects on breast cancer cell migration. Cancer Res 70:8187–8198
117. Shirakabe K, Shibagaki Y, Yoshimura A et al (2014) A proteomic approach for the elucidation of the specificity of ectodomain shedding. J Proteome 98:233–243
118. Middelhoven PJ, Van Buul JD, Hordijk PL, Roos D (2001) Different proteolytic mechanisms involved in fc gamma RIIIb shedding from human neutrophils. Clin Exp Immunol 125:169–175
119. Mayor S, Pagano RE (2007) Pathways of clathrin-independent endocytosis. Nat Rev Mol Cell Biol 8:603–612
120. Agüera-González S, Boutet P, Reyburn HT, Valés-Gómez M (2009) Brief residence at the plasma membrane of the MHC class I-related chain B is due to clathrin-mediated cholesterol-dependent endocytosis and shedding. J Immunol 182:4800–4808
121. Zhang RY, Li L, Wu J, Tang YQ (2012) Vascular endothelial-cadherin: a possible link between endocytosis and ectodomain shedding. J Allergy Clin Immunol 129:266
122. Fukuda S, Nishida-Fukuda H, Nakayama H et al (2012) Monoubiquitination of pro-amphiregulin regulates its endocytosis and ectodomain shedding. Biochem Biophys Res Commun 420:315–320
123. Cavalli V, Corti M, Gruenberg J (2001) Endocytosis and signaling cascades: a close encounter. FEBS Lett 498:190–196
124. Di Fiore PP, De Camilli P (2001) Endocytosis and signaling. An inseparable partnership. Cell 106:1–4

125. Ceresa BP, Schmid SL (2000) Regulation of signal transduction by endocytosis. Curr Opin Cell Biol 12:204–210
126. van der Aa MAEM, Mastrobattista E, Oosting RS et al (2006) The nuclear pore complex: the gateway to successful nonviral gene delivery. Pharm Res 23:447–459
127. Vandenbroucke RE, Libert C (2014) Is there new hope for therapeutic matrix metalloproteinase inhibition? Nat Rev Drug Discov 13:904–927
128. Eguchi T, Kubota S, Kawata K et al (2008) Novel transcription-factor-like function of human matrix metalloproteinase 3 regulating the CTGF/CCN2 gene. Mol Cell Biol 28:2391–2413
129. Schlage P, Egli FE, Nanni P et al (2014) Time-resolved analysis of the matrix metalloproteinase 10 substrate degradome. Mol Cell Proteomics 13:580–593
130. Wolfe MS (2009) γ-Secretase in biology and medicine. Semin Cell Dev Biol 20:219–224

Proteases from Protozoa and Their Role in Infection

8

Anupama Ghosh and Sanghamitra Raha

Abstract

One of the major classes of virulence factors acting in different host-pathogen interaction systems is comprised of proteases. Pathogen-secreted or membrane-associated proteases could be found to participate in different stages of establishment of infection. They are explored as candidate drug targets due to their key participation in the disease development process carried out by the pathogen. In this chapter we present an extensive review of the proteases of different protozoan parasites. Throughout the article we have made an effort to provide a comprehensive list of different proteases from various parasitic protozoa that have been demonstrated to execute major functions in the respective infection processes. Attempts have also been made to present their mode of action with respect to host invasion and disease development.

Keywords

Protozoa • Parasite • Protease

A. Ghosh (✉)
Division of Plant Biology, Bose Institute, Centenary Campus, Kolkata 700054, India
e-mail: ghosh.anupama@jcbose.ac.in; ghosh.anupama1982@gmail.com

S. Raha (✉)
Integrated Science Education and Research Center and Department of Biotechnology, Visva Bharati University, Shantiniketan 731235, India
e-mail: srr1987@gmail.com

S. Chakraborti, N.S. Dhalla (eds.), *Proteases in Physiology and Pathology*,
DOI 10.1007/978-981-10-2513-6_8

8.1 Introduction

Protozoan parasites constitute one of the most prevalent groups of disease-causing pathogens worldwide. Included among a myriad of human diseases that are caused by protozoan parasites are amoebiasis, leishmaniasis, malaria and trypanosomiasis. An attempt to understand the disease-causing mechanisms of these pathogens over the years has revealed a number of key components that are instrumental towards host invasion and in-host survival of the pathogen. These effectors of disease establishment by the pathogen are primarily proteins exhibiting varied enzymatic activities. Pathogen-derived proteases are among one of these effectors and possibly the most important ones. Protozoan proteases aid in their pathogenesis in many different ways. For instance, they play a role in degradation of extracellular matrix that enables the pathogen to invade host cells. Besides, they participate in direct cytolysis and phagocytosis of target cells. In some cases protozoan proteases help in immune evasion by the pathogen through modulation or degradation of host immune molecules (Fig. 8.1). This contribution of protozoan proteases towards disease development however is not limited to a particular class of enzyme; instead different classes of proteases have been documented to participate in the virulence of the pathogen. In this review we put forward an effort to present different aspects of protozoan proteases in relation to their roles in pathogenesis.

Proteases belonging to almost all the major classes have been found to participate in the virulence of the associated pathogen. Included among them are cysteine, serine and aspartyl proteases.

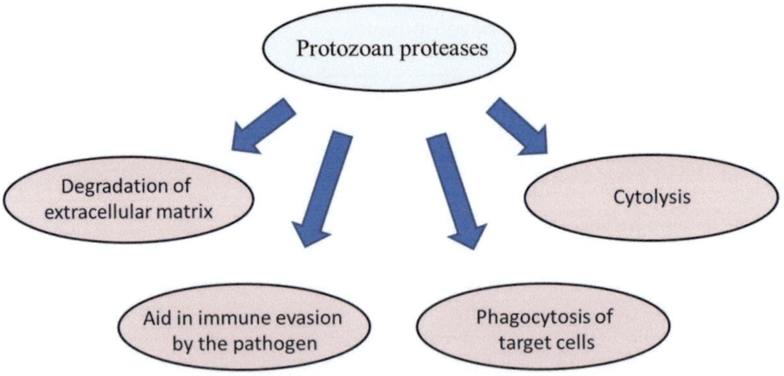

Fig. 8.1 Schematic representation of different roles played by protozoan proteases in parasite pathogenesis. Protozoan proteases play varied roles during host invasion by the parasite. Included among these is the degradation of the host extracellular matrix that aid in the passage of the parasite to the host cell surface. Besides, protozoan proteases also participate actively in the degradation of host immune molecules like immunoglobulins, thereby enabling the pathogen to evade host immune responses. Other activities of the protozoan protease involve cytolysis and phagocytosis of target cells

8.2 Protozoan Cysteine Proteases

Parasitic protozoan cysteine proteases have been found to be involved in all the major steps of disease establishment starting from invasion of host cells till immune evasion [1–3]. It is due to their key role in virulence of the associated pathogen that in some cases cysteine proteases have been investigated elaborately for their potential to become promising drug targets [4].

8.2.1 Malaria Parasite Cysteine Proteases

Among the best characterized cysteine proteases of the malaria parasite *Plasmodium* are the family C1 (papain family) of clan CA cysteine proteases. Genome sequence analysis of *Plasmodium* has revealed that the parasite harbours many members of this particular family of cysteine proteases that include falcipains, dipetidyl peptidases, proteins related to serine-rich antisera (SERA) and a calpain homologue [5].

There are four *falcipain* genes within the genome of the parasite, among which the proteins encoded by *falcipain*2 and *falcipain*3 have been found to catalyse hydrolysis of native haemoglobin and denatured globin [6]. Genetic studies involving RNAi-mediated downregulation of both *falcipain1* and *falcipain2* genes in *Plasmodium falciparum* showed inhibited development of erythrocytic parasites [7]. Haemoglobin hydrolysis is an important part of *Plasmodium* survival within the bloodstream. During rapid asexual multiplication of the parasite within the erythrocyte, it takes up the erythrocyte cytosol to an acidic food vacuole through a specialized organelle called cytostome. It is within this acidic vacuole that the haemoglobin is degraded [8]. Falcipains have a very intricate domain architecture that enables them to specifically target erythrocyte haemoglobins for hydrolysis within the acidic food vacuole. The primary structure of each of the four falcipains can be divided into two domains, the N-terminal prodomain and the C-terminal mature domain. While the N-terminal prodomain possesses a trafficking subdomain that guides the localization of the protease to the acidic food vacuole, the C-terminal mature domain comprises the haemoglobin-binding subdomain needed for substrate recognition [9]. Due to their participation in one of the vital functions of plasmodium blood stage cells that is absolutely essential for the survival of the pathogen within host erythrocyte, falcipains are treated as promising candidates for drug targets against malaria [10].

Among the other members of the clan CA cysteine proteases, besides falcipains, dipeptidyl aminopeptidases (DPAP) are the ones that also catalyse haemoglobin hydrolysis within the food vacuoles of host erythrocytes [11, 12]. *Plasmodium* genome codes for three *dpap* genes. DPAP1 has been shown to be absolutely essential for the survival of the pathogen since viable forms of *DPAP1* deletion strains could never be generated by applying different genetic approaches [12]. However *DPAP2* deletion studies involving the effect of DPAP1 and DPAP2 inhibitors ML4118S in both the human parasite *P. falciparum* and rodent parasite *P. berghei* indicated an additional role of *Plasmodium* dipeptidyl aminopeptidases in the transmission of the disease [13]. Besides, a forward chemical genetic study involving a library of 1200 covalent serine and cysteine protease inhibitors revealed a possible role of DPAPs in the erythrocyte rupture by the parasite in coordination with a subtilisin family serine protease PfSUB1 [14]. All these studies indicate a vital contribution of *Plasmodium* DPAPs in different stages of the pathogenic development of the malaria parasite. Suitable drugs that could target these aminopeptidases therefore pose great promise for controlling the spread of the disease. Another very important target for drug development against malaria could be the only calpain homologue of *Plasmodium* sp. Regulated knockdown studies of the respective gene in *Plasmodium falciparum* revealed its necessity in the cell cycle progression of the pathogen to maintain a proper disease cycle [15].

Serine-rich antigen (SERA) family of cysteine-like proteases is another group of *Plasmodium* proteases with possible roles in maintaining the disease cycle of the pathogen. *Plasmodium* genome codes for 9 *SERA* genes [16, 17], among which several codes for proteins that contain an atypical serine residue at the active site instead of a canonical cysteine residue. Studies have shown that SERA family members act as substrates of PfSUB1 [18] and possibly play a role in PfSUB1 induced release of *Plasmodium* cells from the erythrocytes (egress). Attempts to generate *SERA 1*, *SERA 4*, *SERA 5* and *SERA 9* deletion strains of *Plasmodium falciparum* revealed the lethal nature of Δ*SERA 5* strains indicating an important blood stage function of this particular SERA protein [19]. Moreover a twofold increase in expression of *SERA 5* in *SERA 4* null strains is in support of the common notion of functional redundancy among different members of SERA family of proteins.

8.2.2 *Entamoeba histolytica* Cysteine Proteases

Entamoeba histolytica is the causative agent of human amoebiasis. The parasite genome codes for 50 cysteine protease genes, most of which are expressed only under conditions of host invasion [20]. Studies have also shown that *E. histolytica*-derived secretory products contain a large proportion of cysteine proteases constituting important virulence factors [21]. During invasion of the host tissue, the parasite trophozoites first adhere themselves to the colonic epithelia. This initial interaction with the host tissue takes place through protein-protein interactions between parasite Gal/GalNAc-lectin and mucin glycoproteins of human colon [22, 23]. Subsequent stages involve rapid degradation of colonic mucus by the parasite secretory proteases. This leads to induction of cell death in colonic epithelial cells either through apoptosis or necrosis [24–26]. *Entamoeba* cysteine proteases constitute as a major contributor to all these molecular events. For instance, studies led by Kris Chadee and group demonstrated for the very first time the involvement of *E. histolytica*-secreted cysteine proteases in the disruption of MUC2 polymers of the intestinal mucous layer [27]. In a later study by the same group, the role of individual *E. histolytica*-secreted cysteine proteases in the degradation of the colonic mucus layer was identified. In this study, through antisense inhibition of *E. histolytica* cysteine protease 5 (EhCP5), it has been shown that cysteine protease-deficient trophozoites exhibited a reduced activity towards degradation of protective colonic mucus barrier. These cells however were not at all compromised in their ability to disrupt Chinese hamster ovary (CHO) monolayers devoid of a mucous layer when tested under ex vivo conditions [28]. Further research involving detailed study of molecular dynamics of *E. histolytica* cysteine proteases interacting with the extracellular matrix (ECM) of colonic epithelia demonstrated the perturbation of colonic ECM by *E. histolytica* cysteine protease A5 (CP-A5) [29]. These studies also demonstrated direct interaction of CP-A5 with the cell surface integrins of human colonic epithelia leading to secretion of proinflammatory cytokines [30]. Besides CP-A5 protease has also been found to bring its effect on remodelling of ECM through activation of host matrix metalloproteases (MMPs) [31]. Through these studies the role of cysteine proteases in disruption of cells beyond epithelia has also been evidenced. For instance, CP-A5 has been found to aid trophozoites to invade the layer of loose connective tissue, "lamina propria," beneath the colonic epithelium. Accordingly inhibition of the *CP-A5* expression is associated with a mark reduction in the invasive property of the amoebic trophozoites [32]. The importance of cysteine proteases as virulence factors of *E. histolytica* could further be accessed by their ability to restore pathogenicity in avirulent strains of the pathogen when expressed ectopically [33]. Besides EhCP5 and CP-A5, other cysteine proteases also play significant roles in defining the disease pathology associated with amoebiasis. At this point EhCP112 is definitely worth a mention. EhCP112 together with an adherence domain-containing protein, EhADh112, form a 112 KDa surface adhesin of *E. histolytica*. Named as EhCPADH, this surface adhesin plays a significant role in adhesion and subsequent phagocytosis of the host cells. Accordingly the protein complex was found to translocate from plasma membrane to phagocytic vacuoles during phagocytosis of the target cells [34]. Further studies to characterize the function of the said complex came up with the observation that the constituent cysteine protease (EhCP112) by itself is capable of disrupting cell monolayers through digestion of extracellular protein matrix [35, 36]. These studies involving recombinant EhCP112 also revealed the ability of the protein to execute peptidase function over a wide pH range. Further support in favour of EhCP112 being important for *E. histolytica* pathogenesis came from RNAi-mediated gene silencing studies within the pathogen. For instance, Mario A Rodriguez and his group demonstrated reduced virulence of *E. histolytica* trophozoites in response to RNAi-mediated silencing of *Ehcp112* gene [37].

8.2.3 *Leishmania* Cysteine Proteases

Genome data of *L. major* reveals a group of about 65 cysteine proteases encoded by the parasite that are grouped into 4 clans and 13 families [38]. Most of these cysteine proteases have the potential to serve as important virulence factors, thereby playing crucial roles in establishing host-pathogen interactions. Among these cysteine proteases, those belonging to clan CA, family C1, are the best studied in *Leishmania*, for example, cathepsin L-like CPA and CPB and cathepsin B-like CPC cysteine proteases. These cysteine proteases also show a stage-specific induced expression. CPA and CPB, for example, are found in higher levels in amastigotes than in stationary phase promastigotes [39]. Further evidence in support of the absolute necessity of cathepsin L-like cysteine proteases for the survival of the parasite within the macrophages is demonstrated by the pathogen retarded growth in the presence of cysteine protease inhibitors [40]. Later studies involving specific inhibitors of cathepsin B family proteases revealed a similar role of this class of proteases as well in the parasite survival within the host macrophages. A detailed investigation into the mechanism however showed an induced production of biologically active form of transforming growth factor β (TGF-β) within the infected host macrophages [41]. Although initially noted in *L. amazonensis* and *L. major*, later studies involving *L. chagasi* also demonstrated localized production of activated TGF-β1 at the site of maximum parasite invasion within liver tissue [42]. Studies led by Lashitew Gedamu could finally link the incidences of activated TGF-β induction, *Leishmania* infection and the parasite survival within host tissue. In this study Gedamu and his group showed that it is cathepsin B of *L. donovani* as well as *L. chagasi* that can cleave host TGF-β precursor into its active form [43]. Possibly this induced production of biologically active TGF-β is the basis of the parasite growth and survival within the host macrophages. It is because of this important role of *Leishmania* cysteine proteases in the virulence of the parasite that a lot of efforts have been made during the past years to develop cysteine protease inhibitor-based cure for the disease. As a part of this approach, several natural products including different flavones and quercetin have been screened for their ability to inhibit *L. Mexicana* cathepsin L-like cysteine protease CPB [44]. Besides, different peptidomimetic compounds have also been shown to have specific inhibitory activities towards cathepsin B-like cysteine protease such as CPC in *L. major* [45]. Cysteine proteases, specially the cathepsin L- and B-like proteases due to their tremendous contribution towards parasite growth and survival within host tissue, constitute some of the most popular targets to treat leishmaniasis. Nevertheless there always remains a question of specificity of the inhibitors towards parasite cysteine proteases in particular. This is because humans also contain cathepsin-like cysteine proteases, and thereby any cross-reactivity of the inhibitor with human proteases may lead to detrimental outcomes.

8.3 Serine Proteases from Parasitic Protozoa

Serine proteases are among the most abundant group of proteases in different organisms. This class of proteases has gained tremendous functional diversity during the course of evolution [46, 47]. Among the many biological functions exhibited by parasite serine proteases, involvement in defining the pathogenicity of concerned protozoan parasites is one. Serine proteases are found to play a significant role in the pathogenicity of several parasitic protozoans including the genera *Plasmodium* and *Entamoeba*.

8.3.1 Serine Proteases from *Plasmodium falciparum*

Plasmodium falciparum genome codes for proteases belonging to different clans of serine protease such as chymotrypsin-/trypsin-like, subtilisin-like and rhomboid proteases [48]. Among these, those belonging to subtilisin-like serine protease clans have been studied extensively over the years and have been found to play major roles in disease development by the pathogen. *P. falciparum* genome codes for three serine protease genes that code for proteases belonging to subtilisin-like serine protease family. These are named as PfSUB1, PfSUB2 and PfSUB3. These proteases specifically PfSUB1 and PfSUB2 have been found to play a role in both egress and invasion of host tissue during asexual blood stage lifecycle of the parasite [14, 18, 49]. In fact PfSUB1 was initially identified as an essential serine protease that is released from the parasite micronemes into the parasitophorous vacuolar space just before egress [18]. This release of PfSUB1 is found to be dependent on *Plasmodium falciparum* cGMP-dependent protein kinase G (PfPKG) activity and increased intracellular Ca^{2+} levels [50, 51]. Besides aiding in parasite egress, PfSUB1 also primes the outer surface of malaria merozoites for subsequent erythrocyte invasion. This priming step involves processing of three merozoite surface proteins MSP-1, MSP-6 and MSP-7 that is necessary for subsequent erythrocyte invasion. Accordingly inhibition of PfSUB1 activity has been found to be associated with increased accumulation of unprocessed MSPs on the merozoite surface [52]. Besides MSPs, PfSUB1 catalyses the proteolytic processing of a number of proteins from merozoite, parasitophorous vacuole or parasitophorous vacuolar membrane indicating multiple roles of the protein in the lifecycle of the parasite [53]. In addition to PfSUB1, PfSUB2 also participates in proteolytic maturation of merozoite MSPs and apical membrane antigen 1 (AMA1). For instance, MSP1 proteolytic processing needs both PfSUB1 and PfSUB2. While PfSUB1 catalyses primary processing of the protein just prior to egress, PfSUB2 catalyses secondary processing during erythrocyte invasion [54]. Expression of the gene coding for PfSUB2 is tightly regulated with the protein being available strictly during the merozoite differentiation stage. At this stage the protein is actually stored within the apical secretory organelle called microneme, and during merozoite release it is translocated to the posterior pole within the merozoite dense granules where it participates in the late stage of erythrocyte invasion by the parasite [49, 55]. This trafficking of PfSUB2 between merozoite micronemes and merozoite cell surfaces is regulated by autocatalytic protease activity of the protein. While the transmembrane domain of PfSUB2 is necessary for microneme targeting, the cytoplasmic domain is required for surface translocation of the protease following release from merozoite micronemes [56]. Unlike PfSUB1 and PfSUB2, very little is known about the molecular mechanism of action of the third subtilisin-like serine protease from *Plasmodium falciparum*, PfSUB3. Although not directly experimentally demonstrated, the observation that the expression of the protein is induced at the late asexual blood stage of the parasite indicates a possible involvement of the protease in the merozoite egress and invasion processes [57]. Besides, a recent study identified profilins as the substrate of PfSUB3 [58]. Since eukaryotic profilins are multifunctional proteins with a primary role in regulation of actin filament assembly, a probable function of the protease in *Plasmodium* motility, virulence and immune evasion has been envisaged. Another class of proteases that have been found to play a role in cleavage of microneme proteins thereby enabling host cell invasion by the pathogen are rhomboid proteases [59]. For example, a plasma membrane-localized *Plasmodium falciparum* rhomboid protease PfROM4 catalyses the cleavage of erythrocyte-binding antigen 175 (EBA175) expressed on the merozoite cell surface [60]. EBA175 aids in the initial binding of merozoites with the erythrocytes through their interactions with erythrocyte surface glycophorin A. Many of the *Plasmodium* surface adhesins other than EBA175 alone have been found to be processed by the parasite rhomboid proteases PfROM1 and PfROM4 during the disengagement of the merozoite from the erythrocyte binding for the subsequent host cell invasion process. Included among these surface adhesins are TRAP, CTRP, MTRAP, PFF0800c and others [61]. Besides a role in general shedding of *Plasmodium* cell surface adhesins, rhomboid proteases have also been found to have significant contribution towards proper modification of parasitophorous vacuole that aid in parasite development within the host [62]. Also in the case of *Plasmodium berghei* ROM3, PbROM3, a vital role in sporogony has been demonstrated [63].

8.3.2 Serine Proteases from Other Apicomplexan Parasites

Plasmodium spp. are not the only members of apicomplexan parasites that exhibit ample use of serine proteases in their virulence mechanisms. Similarly, significant contribution of serine proteases could be noticed in the molecular mechanisms of host cell invasion by *Toxoplasma gondii*. Being an obligate intracellular pathogen, *T. gondii* spends its entire lifecycle within a specialized parasitophorous vacuole formed in the cytoplasm of infected cells [64]. During its survival within the vacuole, the parasite produces many subtilisin-like serine proteases which play important roles in the establishment of infection by the pathogen like TgSUB1. TgSUB1 is processed within the secretory pathway of the parasite and finally secreted as smaller products by the microneme [65]. Secreted and processed TgSUB1 then participate in the processing of various micronemal proteins, thereby regulating adhesive properties of different cell surface macromolecular complexes that are involved in host cell invasion [66]. Besides TgSUB1, *T. gondii* also possess other subtilisin-like serine proteases like TgSUB2. A genetic approach to study the function of TgSUB2 revealed its indispensible nature with respect to pathogen survival. The protein undergoes autocatalytic processing during its passage through the pathogen secretory pathway and localizes to rhoptries where it associates with ROP1. TgSUB2 therefore functions as a rhoptry protein maturase [67]. Among serine proteases other than subtilisin-like family, rhomboid proteases of *T. gondii* also contribute significantly to the host cell invasion by the pathogen. Rhomboid proteases are intramembrane serine proteases. *T. gondii* genome codes for five nonmitochondrial rhomboid proteases. However the expression of each of these rhomboid proteases is dependent on the morphological form of the parasite. While TgROM1, TgROM4 and TgROM5 are expressed in the tachyzoite stage which is responsible for disease development by the parasite, TgROM2 and TgROM3 are expressed in the oocyst stage that is involved in transmission of the pathogen [68]. The key protease activity necessary for host cell invasion by *T. gondii* is provided by TgROM5 that catalyses the cleavage of MIC2 cell surface adhesin [68]. Although majority of the serine proteases from *T. gondii* are needed during host cell invasion which is the primary step of establishing infection, there are instances available in the literature that shows important roles of these proteases in the intracellular survival of the pathogen as well. For example, genetic studies revealed key functions of TgROM1 in the intracellular growth of *T. gondii* [69]. Besides, TgROM4 has been demonstrated to maintain the normal apical-posterior gradient of *T. gondii* cell surface adhesins which is a prerequisite for efficient cell motility and successful host cell invasion by the pathogen [70]. However despite their individual roles in adhesin cleavage, host cell invasion and regulation of intracellular growth of *T. gondii*, none of the rhomboid proteases are indispensible for the pathogen lifecycle [71]. This gives us a glimpse of the redundant pathways maintained by the pathogen comprising different classes of proteases to ensure host invasion by the pathogen and its subsequent survival within the host environment.

8.3.3 *Trypanosoma* Serine Proteases

Trypanosomes are unicellular flagellated protozoa, most of which are transmitted to vertebrate hosts by means of blood-feeding insects. These parasites cause various fatal diseases in human like Chagas disease caused by *Trypanosoma cruzi* and sleeping sickness caused by *Trypanosoma brucei*. One of the primary steps in the establishment of disease by these parasites involves colonization of suitable host cells. *T. cruzi*, for example, is capable of invading various types of mammalian cells upon release into the host bloodstream through the bite of insect vectors. Once in contact with a target cell, the parasite initiates invasion process that is found to be linked with recruitment and fusion of host cell lysosomes at the invasion site [72]. This lysosome-mediated host cell entry of trypanosomes is however dependent on increased intracellular calcium influxes within host cytoplasm [73]. One of the key parasite enzymes involved in regulation of these calcium-signalling events is a serine protease called oligopeptidase B. Accordingly *T. cruzi* mutants lacking *oligopeptidase B* gene were found to be defective in both host cell invasion and establishment of infection [74]. Homologues of this key serine protease have also been reported in *T. brucei*. But unlike *T. Cruzi* oligopeptidase B, *T. Brucei* oligopeptidase B exhibits trypsin-like enzyme specificity [75]. In addition to oligopeptidase B, another class of secreted serine oligopeptidase has been reported in *T. cruzi*. This oligopeptidase however is secreted into the extracellular milieu through flagellar pockets of the parasite. Moreover intracellular localization of the protein within the reservosomes which are the acidic organelles present in the posterior region of the parasite indicates a possible involvement of the protein in the general proteolysis activities of the organelle [76]. A signal peptide peptidase having serine protease activity further adds on to the list of serine proteases essential for the survival of the parasite. Genetic deletion strains of *T. brucei* for the abovementioned genes showed defective growth both in vivo and in vitro [77].

8.3.4 *Entamoeba* Serine Proteases

When it comes to the question of serine proteases involved in pathogenicity of *Entamoeba* sp., the predominant role played by the rhomboid proteases comes to the forefront. For instance, *E. histolytica* genome codes for four rhomboid proteases, among which only one possesses catalytic residues necessary for exhibiting proteolytic activity [78]. This ROM1 protein is usually localized within the parasite surface. Upon phagocytosis of erythrocytes, however, it relocalizes to internal vesicles. The localization is again altered during surface receptor capping when it could be found at the base of the cap. The protease has been found to catalyse the cleavage of the heavy subunit of Gal/GalNAclectin (Hgl) in vitro indicating a possibility of Hgl to be one of the physiological substrates of EhROM1 [78]. Accordingly *E. histolytica* strains that are silenced for *EhROM1* exhibited defects in both adhesion and phagocytosis. However no changes in either cap formation or complement resistance could be noticed in the said strain [79]. Although these observations were obtained primarily in the non-virulent strains of the parasite, genetic deletion studies of *ROM1* in virulent strains revealed even more novel functions. ROM1-deleted virulent strains of *E. histolytica* exhibited defective motility indicating a possible role of the protease in the amoebic motility [8].

8.4 Protozoan Aspartyl Proteases and Their Role in Pathogenesis

One of the best characterized and explored protozoan aspartyl proteases playing significant role in pathogen virulence includes plasmepsins from *Plasmodium* sp. Primarily, the protease aids haemoglobin degradation within the parasite food vacuole. The resulting degradation products that include mainly amino acids serve as both nutrient and energy sources for the survival of the pathogen within intra-erythrocytic environment [80]. Haemoglobin degradation function for plasmodial plasmepsins however has been successfully assigned for only four of the total ten members, namely, plasmepsins I, II, and IV and histoaspartic protease. Among them a detailed study on the mechanism of trafficking of cytosolic plasmepsin II to the acidic food vacuole of the parasite has been done. This particular study revealed an initial transport of the protein through the secretory system to the cytostomal vacuole. Within the cytostomal vacuole, the protease binds to its substrate haemoglobin from where it is carried to the food vacuole where the actual degradation of the haemoglobin takes place [81]. Moreover when it comes to the rest of the six plasmepsins, the functions are mostly uncharacterized. Nevertheless a very unique function could be demonstrated for one of the members, plasmepsin V by Goldberg

DE and colleagues [82]. According to their study, plasmepsin V plays a key role in the cleavage of the PEXEL motif within the *Plasmodium* exported proteins while they are still present within the parasite endoplasmic reticulum. PEXEL is an essential signature motif for plasmodial exported proteins [83]. These PEXEL proteins thus processed are then exported to the erythrocyte cytoplasm through an ATP-driven translocon channel. *Plasmodium* exportome thus translocated and deposited to erythrocyte cytosol plays a key role in suitable orchestration of the host environment for subsequent survival of the pathogen. Anything interfering in the normal protein export process of the parasite therefore might alter the virulence of the organism. Plasmepsin V hence can be considered to play a significant contribution towards the pathogenicity of *Plasmodium falciparum*.

8.5 Protozoan Proteases as Drug Targets

It is due to their very important roles in the disease development process of the parasitic protozoans that the proteases could be targeted for chemotherapy of the concerned disease. There are several instances available in the literature where one or more of the key virulence-related proteases of the parasite are targeted to control the respective disease. One such example of drug development against parasite proteases dates back to 1996. During this period Erickson JW and his group designed a series of low molecular weight compounds with potential to inhibit a key serine protease of *Plasmodium* plasmepsin I that is involved in haemoglobin degradation [84]. Further studies in this direction led to the discovery of an adaptive inhibitor for the entire plasmepsin group of serine proteases of *Plasmodium*. Adaptive inhibitors in general are targeted to and are specific to one of the members of a family of proteins but have the flexibility to inhibit other members as well although with less efficiency. This particular adaptive inhibitor against *Plasmodium* plasmepsin family is targeted to plasmepsin II but can also inhibit plasmepsins IV and I and HAP with almost equal efficiency [85]. Likewise aspartic acid proteases from other protozoan parasites have also been evaluated extensively for their potential to become competent drug targets [86]. Cysteine proteases and serine proteases also serve as attractive targets for drug discovery against protozoan infection. However the greatest hindrance in such drug development programmes involves specificity of the protease inhibitors used. In order to be accurate in imparting its inhibitory activities to parasite proteases and not to similar family of host proteases, a drug needs to be designed in accordance with the unique structure function relationship of the parasite protease [87].

8.6 Conclusion

Protozoan proteases play key roles in the process of disease development by pathogenic protozoa. These proteases participate in almost every essential step during the infection of the host. Their contribution in the disease process starting from the initial recognition of the host cell receptors to their final entry within the host cell is well substantiated in the literature. This class of protozoan proteases however is not limited to any specific family of protease, but almost all the major family of proteolytically active enzymes are involved. These proteases therefore serve as promising drug targets to control respective infection processes. However active research is ongoing to identify more and more unique protease targets in different protozoan parasite systems that can be used for development of suitable drugs against the associated diseases.

Acknowledgements A.G. acknowledges the financial support of the Dept. of Science and Technology, Govt. of India, for DST INSPIRE faculty fellowship.

References

1. Klemba M, Goldberg DE (2002) Biological roles of proteases in parasitic protozoa. Annu Rev Biochem 71:275–305
2. Mottram JC, Brooks DR, Coombs GH (1998) Roles of cysteine proteinases of trypanosomes and *Leishmania* in host-parasite interactions. Curr Opin Microbiol 1:455–460
3. Sajid M, McKerrow JH (2002) Cysteine proteases of parasitic organisms (2002). Mol Biochem Parasitol 121:159–159
4. McKerrow JH, Engel JC, Caffrey CR (1999) Cysteine protease inhibitors as chemotherapy for parasitic infections. Bioorg Med Chem 7:639–644
5. Wu YM, Wang XY, Liu X, Wang YF (2003) Data-mining approaches reveal hidden families of proteases in the genome of malaria parasite. Genome Res 13:601–616
6. Shenai BR, Rosenthal PJ (2002) Reducing requirements for hemoglobin hydrolysis by *Plasmodium falciparum* cysteine proteases. Mol Biochem Parasitol 122:99–104
7. Malhotra P, Dasaradhi PV, Kumar A, Mohmmed A, Agrawal N, Bhatnagar RK, Chauhan VS (2002) Double-stranded RNA-mediated gene silencing of cysteine proteases (falcipain-1 and -2) of *Plasmodium falciparum*. Mol Microbiol 45:1245–1254
8. Francis SE, Sullivan DJ, Goldberg DE (1997) Hemoglobin metabolism in the malaria parasite *Plasmodium falciparum*. Annu Rev Microbiol 51:97–123
9. Pandey KC, Dixit R (2012) Structure-function of falcipains: malarial cysteine proteases. Journal of Tropical Medicine 2012:345195
10. Rosenthal PJ, Sijwali PS, Singh A, Shenai BR (2002) Cysteine proteases of malaria parasites: targets for chemotherapy. Curr Pharm Des 8:1659–1672
11. Dalal S, Klemba M (2007) Roles for two aminopeptidases in vacuolar hemoglobin catabolism in *Plasmodium falciparum*. J Biol Chem 282:35978–35987
12. Klemba M, Gluzman I, Goldberg DE (2004) A *Plasmodium falciparum* dipeptidylaminopeptidase I participates in vacuolar hemoglobin degradation. J Biol Chem 279:43000–43007
13. Tanaka TQ, Deu E, Molina-Cruz A, Ashburne MJ, Ali O, Suri A, Kortagere S, Bogyo M, Williamsona KC (2013) *Plasmodium* Dipeptidyl Aminopeptidases as malaria transmission-blocking drug targets. Antimicrob Agents Chemother 57:4645–4652
14. Arastu-Kapur S, Ponder EL, Fonovic UP, Yeoh S, Yuan F, Fonovic M, Grainger M, Phillips CI, Powers JC, Bogyo M (2008) Identification of proteases that regulate erythrocyte rupture by the malaria parasite *Plasmodium falciparum*. Nat Chem Biol 4:203–213
15. Russo I, Oksman A, Vaupel B, Goldberg DE (2009) A calpain unique to alveolates is essential in *Plasmodium falciparum* and its knockdown reveals an involvement in pre-S-phase development. Proc Natl Acad Sci U S A 106:1554–1559

16. Le Roch KG, Zhou Y, Blair PL, Grainger M, Moch JK, Haynes JD, De La Vega P, Holder AA, Batalov S, Carucci DJ, Winzeler EA (2003) Discovery of gene function by expression profiling of the malaria parasite life cycle. Science 301:1503–1508
17. Bozdech Z, Llinas M, Pulliam BL, Wong ED, Zhu J, DeRisi JL (2003) The transcriptome of the intraerythrocytic developmental cycle of *Plasmodium falciparum*. PLoS Biol 1:E5
18. Yeoh S, O'Donnell RA, Koussis K, Dluzewski AR, Ansell KH, Osborne SA, Hackett F, Withers-Martinez C, Mitchell GH, Bannister LH, Bryans JS, Kettleborough CA, Blackman MJ (2007) Subcellular discharge of a serine protease mediates release of invasive malaria parasites from host erythrocytes. Cell 131:1072–1083
19. McCoubrie JE, Miller SK, Sargeant T, Good RT, Hodder AN, Speed TP, de Koning-Ward TF, Crabb BS (2007) Evidence for a common role for the serine-type *Plasmodium falciparum* serine repeat antigen proteases: implications for vaccine and drug design. Infect Immun 75:5565–5574
20. Tillack M, Biller L, Irmer H, Freitas M, Gomes MA, Tannich E, Bruchhaus I (2007) The *Entamoeba histolytica* genome: primary structure and expression of proteolytic enzymes. BMC Genomics 8:170
21. Gadasi H, Kobiler D (1983) *Entamoeba histolytica*: correlation between virulence and content of proteolytic enzymes. Exp Parasitol 55:105–110
22. Chadee K, Ndarathi C, Keller K (1990) Binding of proteolytically-degraded human colonic mucin glycoproteins to the gal/GalNAc adherence lectin of *Entamoeba histolytica*. Gut 31:890–895
23. McCoy JJ, Mann BJ, Petri WA Jr (1994) Adherence and cytotoxicity of *Entamoeba histolytica* or how lectins let parasites stick around. Infect Immun 62:3045–3050
24. Ragland BD, Ashley LS, Vaux DL, Petri WA Jr (1994) *Entamoeba histolytica*: target cells killed by trophozoites undergo DNA fragmentation which is not blocked by Bcl-2. Exp Parasitol 79:460–467
25. Huston CD, Boettner DR, Miller-Sims V, Petri WA Jr (2003) Apoptotic killing and phagocytosis of host cells by the parasite *Entamoeba histolytica*. Infect Immun 71:964–972
26. Berninghausen O, Leippe M (1997) Necrosis versus apoptosis as the mechanism of target cell death induced by *Entamoeba histolytica*. Infect Immun 65:3615–3621
27. Moncada D, Keller K, Chadee K (2003) *Entamoeba histolytica* Cysteine proteinases disrupt the polymeric structure of colonic mucin and alter its protective function. Infect Immun 71:838–844
28. Moncada D, Keller K, Ankri S, Mirelman D, Chadee K (2006) Antisense inhibition of *Entamoeba histolytica* cysteine proteases inhibits colonic mucus degradation. Gastroenterology 130:721–730
29. Thibeaux R, Dufour A, Roux P, Bernier M, Baglin AC, Frileux P, Olivo-Marin JC, Guillen N, Labruyere E (2012) Newly visualized fibrillar collagen scaffolds dictate *Entamoeba histolytica* invasion route in the human colon. Cell Microbiol 14:609–621
30. Hou Y, Mortimer L, Chadee K (2010) *Entamoeba histolytica* Cysteine proteinase 5 binds integrin on colonic cells and stimulates NFkappaB-mediated pro-inflammatory responses. J Biol Chem 285:35497–35504
31. Thibeaux R, Ave P, Bernier M, Morcelet M, Frileux P, Guillen N, Labruyere E (2014) The parasite *Entamoeba histolytica* exploits the activities of human matrix metalloproteinases to invade colonic tissue. Nat Commun 5:5142
32. Bansal D, Ave P, Kerneis S, Frileux P, Boche O, Baglin AC, Dubost G, Leguern AS, Prevost MC, Bracha R, Mirelman D, Guillen N, Labruyere E (2009) An ex-vivo human intestinal model to study *Entamoeba histolytica* pathogenesis. PLoS Negl Trop Dis 3:e551
33. Matthiesen J, Bar AK, Bartels AK, Marien D, Ofori S, Biller L, Tannich E, Lotter H, Bruchhaus I (2013) Overexpression of specific cysteine peptidases confers pathogenicity to a non-pathogenic *Entamoeba histolytica* clone. MBio 4:e00072–e00013
34. Garcia-Rivera G, Rodriguez MA, Ocadiz R, Martinez-Lopez MC, Arroyo R, Gonzalez-Robles A, Orozco E (1999) *Entamoeba histolytica*: a novel cysteine protease and an adhesin form the 112kDa surface protein. Mol Microbiol 33:556–568

35. Ocadiz R, Orozco E, Carrillo E, Quintas LI, Ortega-Lopez J, Garcia-Perez RM, Sanchez T, Castillo-Juarez BA, Garcia-Rivera G, Rodriguez MA (2005) EhCP112 is an *Entamoeba histolytica* secreted cysteine protease that may be involved in the parasite-virulence. Cell Microbiol 7:221–232
36. Quintas-Granados LI, Orozco E, Brieba LG, Arroyo R, Ortega-Lopez J (2009) Purification, refolding and autoactivation of the recombinant cysteine proteinase EhCP112 from *Entamoeba histolytica*. Protein Expr Purif 63:26–32
37. Ocadiz-Ruiz R, Fonseca W, Martinez MB, Ocadiz-Quintanar R, Orozco E, Rodriguez MA (2013) Effect of the silencing of the Ehcp112 gene on the in vitro virulence of *Entamoeba histolytica*. Parasit Vectors 6:248–256
38. Ivens AC, Peacock CS, Worthey EA, Murphy L, Aggarwal G, Berriman M, Sisk E, Rajandream MA, Adlem E, Aert R, Anupama A, Apostolou Z, Attipoe P, Bason N, Bauser C, Beck A, Beverley SM, Bianchettin G, Borzym K, Bothe G, Bruschi CV, Collins M, Cadag E, Ciarloni L, Clayton C, Coulson RMR, Cronin A, Cruz AK, Davies RM, De Gaudenzi J, Dobson DE, Duesterhoeft A, Fazelina G, Fosker N, Frasch AC, Fraser A, Fuchs M, Gabel C, Goble A, Goffeau A, Harris D, Hertz-Fowler C, Hilbert H, Horn D, Huang YT, Klages S, Knights A, Kube M, Larke N, Litvin L, Lord A, Louie T, Marra M, Masuy D, Matthews K, Michaeli S, Mottram JC, Muller-Auer S, Munden H, Nelson S, Norbertczak H, Oliver K, O'Neil S, Pentony M, Pohl TM, Price C, Purnelle B, Quail MA, Rabbinowitsch E, Reinhardt R, Rieger M, Rinta J, Robben J, Robertson L, Ruiz JC, Rutter S, Saunders D, Schafer M, Schein J, Schwartz DC, Seeger K, Seyler A, Sharp S, Shin H, Sivam D, Squares R, Squares S, Tosato V, Vogt C, Volckaert G, Wambutt R, Warren T, Wedler H, Woodward J, Zhou SG, Zimmermann W, Smith DF, Blackwell JM, Stuart KD, Barrell B, Myler PJ (2005) The genome of the kinetoplastid parasite, *Leishmania major*. Science 309:436–442
39. Rafati S, Salmanian AH, Hashemi K, Schaff C, Belli S, Fasel N (2001) Identification of *Leishmania major* cysteine proteinases as targets of the immune response in humans. Mol Biochem Parasitol 113:35–43
40. Mottram JC, Souza AE, Hutchison JE, Carter R, Frame MJ, Coombs GH (1996) Evidence from disruption of the lmcpb gene array of *Leishmania mexicana* that cysteine proteinases are virulence factors. Proc Natl Acad Sci U S A 93:6008–6013
41. Barralnetto M, Barral A, Brownell CE, Skeiky YAW, Ellingsworth LR, Twardzik DR, Reed SG (1992) Transforming growth-factor-Beta in Leishmanial infection - a parasite escape mechanism. Science 257:545–548
42. Wilson ME, Young BM, Davidson BL, Mente KA, McGowan SE (1998) The importance of TGF-beta in murine visceral leishmaniasis. J Immunol 161:6148–6155
43. Somanna A, Mundodi V, Gedamu L (2002) Functional analysis of cathepsin B-like cysteine proteases from *Leishmania donovani* complex - evidence for the activation of latent transforming growth factor beta. J Biol Chem 277:25305–25312
44. de Sousa LRF, Wu HM, Nebo L, Fernandes JB, da Silva MFDF, Kiefer W, Schirmeister T, Vieira PC (2015) Natural products as inhibitors of recombinant cathepsin L of *Leishmania mexicana*. Exp Parasitol 156:42–48
45. Schad C, Baum U, Frank B, Dietzel U, Mattern F, Gomes C, Ponte-Sucre A, Moll H, Schurigt U, Schirmeister T (2016) Development of a new Antileishmanial Aziridine-2,3-Dicarboxylate-based inhibitor with high selectivity for parasite cysteine proteases. Antimicrob Agents Chemother 60:797–805
46. Page MJ, Di Cera E (2008) Serine peptidases: classification, structure and function. Cell Mol Life Sci 65:1220–1236
47. Page MJ, Di Cera E (2008) Evolution of peptidase diversity. J Biol Chem 283:30010–30014
48. Alam A (2014) Serine proteases of malaria parasite *Plasmodium falciparum*: potential as antimalarial drug targets. Interdisciplinary Perspectives on Infectious Diseases 2014:453186
49. Harris PK, Yeoh S, Dluzewski AR, O'Donnell RA, Withers-Martinez C, Hackett F, Bannister LH, Mitchell GH, Blackman MJ (2005) Molecular identification of a malaria merozoite surface sheddase. PLoS Pathog 1:241–251

50. Collins CR, Hackett F, Strath M, Penzo M, Withers-Martinez C, Baker DA, Blackman MJ (2013) Malaria parasite cGMP-dependent protein kinase regulates blood stage merozoite secretory organelle discharge and egress. PLoS Pathog 9:e1003344
51. Agarwal S, Singh MK, Garg S, Chitnis CE, Singh S (2013) Ca(2+) -mediated exocytosis of subtilisin-like protease 1: a key step in egress of *Plasmodium falciparum* merozoites. Cell Microbiol 15:910–921
52. Koussis K, Withers-Martinez C, Yeoh S, Child M, Hackett F, Knuepfer E, Juliano L, Woehlbier U, Bujard H, Blackman MJ (2009) A multifunctional serine protease primes the malaria parasite for red blood cell invasion. EMBO J 28:725–735
53. Silmon de Monerri NC, Flynn HR, Campos MG, Hackett F, Koussis K, Withers-Martinez C, Skehel JM, Blackman MJ (2011) Global identification of multiple substrates for *Plasmodium falciparum* SUB1, an essential malarial processing protease. Infect Immun 79:1086–1097
54. Child MA, Epp C, Bujard H, Blackman MJ (2010) Regulated maturation of malaria merozoite surface protein-1 is essential for parasite growth. Mol Microbiol 78:187–202
55. Barale JC, Blisnick T, Fujioka H, Alzari PM, Aikawa M, Braun-Breton C, Langsley G (1999) *Plasmodium falciparum* subtilisin-like protease 2, a merozoite candidate for the merozoite surface protein 1-42 maturase. Proc Natl Acad Sci U S A 96:6445–6450
56. Child MA, Harris PK, Collins CR, Withers-Martinez C, Yeoh S, Blackman MJ (2013) Molecular determinants for subcellular trafficking of the malarial sheddase PfSUB2. Traffic 14:1053–1064
57. Alam A, Bhatnagar RK, Chauhan VS (2012) Expression and characterization of catalytic domain of *Plasmodium falciparum* subtilisin-like protease 3. Mol Biochem Parasitol 183:84–89
58. Alam A, Bhatnagar RK, Relan U, Mukherjee P, Chauhan VS (2013) Proteolytic activity of *Plasmodium falciparum* subtilisin-like protease 3 on parasite profilin, a multifunctional protein. Mol Biochem Parasitol 191:58–62
59. Dowse TJ, Pascall JC, Brown KD, Soldati D (2005) Apicomplexan rhomboids have a potential role in microneme protein cleavage during host cell invasion. Int J Parasitol 35:747–756
60. O'Donnell RA, Hackett F, Howell SA, Treeck M, Struck N, Krnajski Z, Withers-Martinez C, Gilberger TW, Blackman MJ (2006) Intramembrane proteolysis mediates shedding of a key adhesin during erythrocyte invasion by the malaria parasite. J Cell Biol 174:1023–1033
61. Baker RP, Wijetilaka R, Urban S (2006) Two plasmodium rhomboid proteases preferentially cleave different adhesins implicated in all invasive stages of malaria. PLoS Pathog 2:922–932
62. Vera IM, Beatty WL, Sinnis P, Kim K (2011) Plasmodium protease ROM1 is important for proper formation of the Parasitophorous vacuole. PLoS Pathog 7:e1002197
63. Lin JW, Meireles P, Prudencio M, Engelmann S, Annoura T, Sajid M, Chevalley-Maurel S, Ramesar J, Nahar C, Avramut CMC, Koster AJ, Matuschewski K, Waters AP, Janse CJ, Mair GR, Khan SM (2013) Loss-of-function analyses defines vital and redundant functions of the plasmodium rhomboid protease family. Mol Microbiol 88:318–338
64. Kim K (2004) Role of proteases in host cell invasion by *Toxoplasma gondii* and other Apicomplexa. Acta Trop 91:69–81
65. Miller SA, Binder EM, Blackman MJ, Carruthers VB, Kim K (2001) A conserved subtilisin-like protein TgSUB1 in microneme organelles of *Toxoplasma gondii.* J Biol Chem 276:45341–45348
66. Lagal V, Binder EM, Huynh MH, Kafsack BFC, Harris PK, Diez R, Chen D, Cole RN, Carruthers VB, Kim K (2010) *Toxoplasma gondii* protease TgSUB1 is required for cell surface processing of micronemal adhesive complexes and efficient adhesion of tachyzoites. Cell Microbiol 12:1792–1808
67. Miller SA, Thathy V, Ajioka JW, Blackman MJ, Kim K (2003) TgSUB2 is a *Toxoplasma gondii* rhoptry organelle processing proteinase. Mol Microbiol 49:883–894
68. Brossier F, Jewett TJ, Sibley LD, Urban S (2005) A spatially localized rhomboid protease cleaves cell surface adhesins essential for invasion by *Toxoplasma.* Proc Natl Acad Sci U S A 102:4146–4151

69. Brossier F, Starnes GL, Beatty WL, Sibley LD (2008) Microneme rhomboid protease TgROM1 is required for efficient intracellular growth of *Toxoplasma gondii*. Eukaryot Cell 7:664–674
70. Buguliskis JS, Brossier F, Shuman J, Sibley LD (2010) Rhomboid 4 (ROM4) affects the processing of surface Adhesins and facilitates host cell invasion by *Toxoplasma gondii*. PLoS Pathog 6:e1000858
71. Shen B, Buguliskis JS, Lee TD, Sibley LD (2014) Functional analysis of rhomboid proteases during *Toxoplasma* invasion. MBio 5:e01795–e01714
72. Rodriguez A, Samoff E, Rioult MG, Chung A, Andrews NW (1996) Host cell invasion by trypanosomes requires lysosomes and microtubule/kinesin-mediated transport. J Cell Biol 134:349–362
73. Rodriguez A, Rioult MG, Ora A, Andrews NW (1995) A trypanosome-soluble factor induces Ip3 formation, intracellular Ca2+ mobilization and microfilament rearrangement in host-cells. J Cell Biol 129:1263–1273
74. Caler EV, de Avalos SV, Haynes PA, Andrews NW, Burleigh BA (1998) Oligopeptidase B-dependent signaling mediates host cell invasion by *Trypanosoma cruzi*. EMBO J 17:4975–4986
75. Morty RE, Lonsdale-Eccles JD, Morehead J, Caler EV, Mentele R, Auerswald EA, Coetzer THT, Andrews NW, Burleigh BA (1999) Oligopeptidase B from *Trypanosoma brucei*, a new member of an emerging subgroup of serine oligopeptidases. J Biol Chem 274:26149–26156
76. da Silva-Lopeza RE, Morgado-Diaz JA, dos Santos PT, Giovanni-De-Simone S (2008) Purification and subcellular localization of a secreted 75 kDa *Trypanosoma cruzi* serine oligopeptidase. Acta Trop 107:159–167
77. Moss CX, Brown E, Hamilton A, Van der Veken P, Augustyns K, Mottram JC (2015) An essential signal peptide peptidase identified in an RNAi screen of serine peptidases of *Trypanosoma brucei*. PLoS One 10:e0123241
78. Baxt LA, Baker RP, Singh U, Urban S (2008) An *Entamoeba histolytica* rhomboid protease with atypical specificity cleaves a surface lectin involved in phagocytosis and immune evasion. Genes Dev 22:1636–1646
79. Baxt LA, Rastew E, Bracha R, Mirelman D, Singh U (2010) Downregulation of an *Entamoeba histolytica* rhomboid protease reveals roles in regulating parasite adhesion and phagocytosis. Eukaryot Cell 9:1283–1293
80. Rastew E, Morf L, Singh U (2015) *Entamoeba histolytica* Rhomboid protease 1 has a role in migration and motility as validated by two independent genetic approaches. Exp Parasitol 154:33–42
81. Klemba M, Beatty W, Gluzman I, Goldberg DE (2004) Trafficking of plasmepsin II to the food vacuole of the malaria parasite *Plasmodium falciparum*. J Cell Biol 164:47–56
82. Russo I, Babbitt S, Muralidharan V, Butler T, Oksman A, Goldberg DE (2010) Plasmepsin V licenses *Plasmodium* proteins for export into the host erythrocyte. Nature 463:632–636
83. Horrocks P, Muhia D (2005) Pexel/VTS: a protein-export motif in erythrocytes infected with malaria parasites. Trends Parasitol 21:396–399
84. Silva AM, Lee AY, Gulnik SV, Majer P, Collins J, Bhat TN, Collins PJ, Cachau RE, Luker KE, Gluzman IY, Francis SE, Oksman A, Goldberg DE, Erickson JW (1996) Structure and inhibition of plasmepsin II, a hemoglobin-degrading enzyme from *Plasmodium falciparum*. Proc Natl Acad Sci U S A 93:10034–10039
85. Nezami A, Kimura T, Hidaka K, Kiso A, Liu J, Kiso Y, Goldberg DE, Freire E (2003) High-affinity inhibition of a family of *Plasmodium falciparum* proteases by a designed adaptive inhibitor. Biochemistry 42:8459–8464
86. Coombs GH, Goldberg DE, Klemba M, Berry C, Kay J, Mottram JC (2001) Aspartic proteases of *Plasmodium falciparum* and other parasitic protozoa as drug targets. Trends Parasitol 17:532–537
87. McKerrow JH, Rosenthal PJ, Swenerton R, Doyle P (2008) Development of protease inhibitors for protozoan infections. Curr Opin Infect Dis 21:668–672

Regulation of Extracellular Matrix Remodeling and Epithelial-Mesenchymal Transition by Matrix Metalloproteinases: Decisive Candidates in Tumor Progression

9

Y. Rajesh and Mahitosh Mandal

Abstract

Tumor biology is intricate and multifaceted. The genetic and epigenetic alterations accelerate normal cells to transform into aggressive malignant phenotype. Molecular principles of invasion and metastasis are indeed indispensable for profound understanding of tumorigenesis. The seeding pioneer cells from growing tumor eventually discharge from the original clump of mutant cells, invading adjacent tissues and mobilizing to distant sites. This attribute of cancer cells reduces patient's survival rate and prognosis. Inquisition of mechanistic approach for metastasis is bestowed by two processes—extracellular matrix (ECM) remodeling and epithelial-mesenchymal transition (EMT). Proteases pave the way for invaders by breaking down the ECM and releasing pro-invasive factors from cell surface and ECM. Indeed, highly conserved EMT program leads to dissemination of single tumor cells from primary tumors. The zinc-dependent matrix MMPs are the most important effectors in these processes and frequently overexpressed in most of the tumors. Besides proteolysis, by activating or deactivating several growth factors, MMPs affect tumor neoangiogenesis and proliferation. The tissue inhibitors of metalloproteinases (TIMPs) play a central role in complex regulation of MMPs. An apt equilibrium between TIMPs and MMPs is significant in cell invasion and metastasis. These concepts are encouraged for pursuing MMPs as a signature for predicting metastasis and also as therapeutic target. A comprehensive understanding regarding enzyme-substrate interactions and regulation and specific MMPs' functionality in cancer addresses that MMP inhibitors (MMPIs) should be specific in terms of MMP or degrading definite

Y. Rajesh • M. Mandal (✉)
School of Medical Science and Technology, Indian Institute of Technology Kharagpur, Kharagpur 721302, India
e-mail: mahitosh@smst.iitkgp.ernet.in

S. Chakraborti, N.S. Dhalla (eds.), *Proteases in Physiology and Pathology*,
DOI 10.1007/978-981-10-2513-6_9

substrates. The scientific and clinical drive for second-generation MMPIs through the development of pharmaceutical reagents and clinical trials determining the therapeutic benefit to cancer patients should be geared up.

Keywords
Invasion • Metastasis • Extracellular matrix (ECM) • Epithelial-mesenchymal transition (EMT) • Matrix metalloproteinases (MMPs)

9.1 Introduction

The malfunctioning in cellular activities that are pivotal for growth, differentiation, and tissue integrity leads to cancer. Failure in growth control results in amassing cells and producing tumor. The key risk and underlying reason behind tumor-related deaths are not primary tumors, but the secondary tumors, i.e., metastasis. The phenotypic and biochemical alterations in relation to growth factors, cell-cell adherence, and genetic expression occur during metamorphosis of a normal cell into invasive cell [1].The attributes acquired by a healthy cell in the course of transformation to malignant one are cell division in the absence of external growth stimulatory signals, growth in spite of exogenous growth inhibitory signals, evading apoptosis, neoangiogenesis, potent immortalization, and invasion and metastasis [2]. The molecular principles of invasion and metastasis are indispensable for profound understanding of tumorigenesis. Furthermore, the issue is immensely significant since 90% of all cancer deaths are attributed to metastasis. The conventional therapeutic approaches target rapidly proliferating cells. New insights toward molecular progression of invasion and metastasis might pave the way for new, highly specific and potential tumor management strategies. The prerequisite for this resolution is further research in the field, for the better comprehension of these processes.

9.2 Invasion and Metastasis: The Critical Players of Cancer Progression

9.2.1 Invasion

Cellular invasion is a cohesive procedure involving pathological processes (developing embryo, repairing tissue, healing wound, and immunity check) that are well choreographed in the body. Basically, the chemo signals (hormones, growth factors/metabolites), physical signals (tissue stiffness, cell density/cellular pattern and organization), and physicochemical proceedings (diffusion, cell activation and deactivation) influence the migrating cells in a tissue. Mutational alterations in cellular invasion signaling lead to arthritis, atherosclerosis, aneurism, multiple sclerosis, chronic obstructive pulmonary disease (COPD), and cancer. Ninety percent of cancer-related deaths are attributed to this only. Cell invasion occurs as single cells or as collection of cells in sheets or clusters on the basis of cell type and host tissue matrix. In cancer, invasion occurs with less homogeneity. However, in leukemia, lymphoma, sarcomas, and glioma, cells invade heterogeneously in pattern of single cells. But in tumors from epithelial origin, collective cell configurations infiltrate poorly into structured clusters or sheets. They expand, dedifferentiate (epithelial-mesenchymal transformation (EMT)), and disseminate as single cells, ensuing metastasis and poor prognosis [3]. We will be focusing on single cell invasion (key mode of invasion in cancer). Cellular invasion is associated with immunity, angiogenesis, and metastasis.

9.2.1.1 Immune Response

Immune cell invasion is a chief component, essential for infiltrative potential. Normally, in immune response against infection, cells infiltrate the disrupted sites with the help of various growth factors and cytokines' released form blood clot comprising cross-linked fibrin and extracellular matrix (ECM) proteins. The ECM deposition is carried out by neutrophils, monocytes, and lymphocytes. Subsequently, fibroblasts' invasion offers contractile force for wound closure. Implication of invasion signaling in immune cell migration during tissue repair is also associated with disease progression such as in cancer [4]. The type of immune cells prevailing in tumor microenvironment even aids as prognostic factor. It has been also proposed that macrophages and mast cells maintain tumor inflammation, tumor growth promotion, and tumor growth management by lymphocytes [5]. Hence, invading the potential of macrophages and mast cell inhibition might attribute significantly toward anticancer drive.

9.2.1.2 Angiogenesis

New vasculature sprouts by penetrating the tissue matrix and provides nutrients to tissue in morphogenesis and regeneration. Angiogenesis in cancer occurs due to deficit of nutrient diffusion and oxygen exchange [6], and this erratic signaling forms new blood vessels with altered structure. The tumorigenic abnormalities leading to enhanced permeability and retention (EPR) effect are poorly aligned, and irregular-shaped endothelial cells result in large fenestrations, leaky vasculature, and deficient lymphatic drainage. This EPR effect could be exploited for delivering macromolecular drugs [7]. Many anti-angiogenic strategies targeting endothelial cell invasion are under clinical evaluation. The combinational approaches inhibiting both endothelial and tumor cells' invasion are on the horizon [8].

9.2.1.3 Lymphangiogenesis

Lymphangiogenesis refers to sprouting lymph vasculature for draining waste, during morphogenesis and regeneration. It has a huge role in tumor progression and metastasis and in tumors lacking sufficient lymphatic vessels causing EPR effect. However, in lymph node metastasis (breast, colon, and prostate), the primary route is lymphatic vasculature [9], and in some tumors, pro-lymphangiogenesis factors promote lymph node metastasis [10]. Hence, lymphangiogenesis inhibitors might effectively target tumor metastasis.

9.2.1.4 Cancer Metastasis

The features that are involved in cancer metastasis are invading tumor cells to blood/lymph vessel, intravasation, extravasation, and forming a secondary tumor by invading into the tissue. In brain tumors, the cancer cells infiltrate in the organ of origin via cellular invasion signaling rather than metastasizing to other organs. This results in poor survival outcome. Therefore, inhibiting cancer cells invading potential would aid in improvising the therapeutic outcome. Likewise, poor survival outcome is found in breast cancer metastasis. In breast cancer, cells metastasize to the lungs and bone marrow [11].

9.2.2 Metastasis

In cancer, metastasis is the foremost cause of mortality in patient. The scientific and clinical drive has to be geared up to unravel the poorly understood mechanisms of metastasis. The enhanced knowledge in genetic/cellular behavior and biological proceedings in cancer progression has added some new prospects in the diagnosis, prognosis, and treatment of metastasizing diseases. Fine understanding regarding the barrier's role and paracellular permeability allowed formulating a new path toward regulation of trespassing cancer cells and invading cells. The EMT biomarker offers new opportunity in the field of both prognostic methods and therapeutic target for the metastatic prospective of a primary tumor. Angiogenesis has been already established as a significant area in cancer therapy. A method directed toward the detection of organ-specific spreading of solid tumors may grant a new approach for targeting metastatic tumor cells. The genetic and epigenetic basis of metastasis and the acquisition of ability to complete a series of steps involved in metastasis during emergence of secondary tumors has to be revealed. Enormous challenges have to be sorted out to anticipate these lines of research into clinical practice [11].

9.2.2.1 Mechanism Underlying Metastasis

Cell migration bypasses the physical resistance of 3D tissue networks involving different strategies depending upon the tumor type and surrounding tissue. The different patterns of invasion are:

- In squamous cell esophageal cancers, invasion occurs through cone-like arrangements.
- In breast cancers (lobular), cells migrate through ECM as single-file patterns.
- In thyroid cancers (anaplastic), invasion is carried out by single and sparse cells.

Single-cell migration involves isolated and dispersed tumor cells in an adjacent tissue, whereas in collective cell invasion, healthy adjacent cells are being displaced by moving cancerous tissue. Actually, cancer cells migrate and invade through the ECM as single cell in fibroblast-/leukocyte-like fashion.

For establishing secondary sites of tumor growth, cancer cells leave the primary tumor by losing adhering potential and gaining migratory and invasive capability, to disseminate to distant organs. This cascade is convoyed by variations in gene expression and functions (loosing epithelial markers and gaining mesenchymal markers), permeation of basement membrane, invasion of surrounding tissues, and accession to blood and lymph vasculature. After intravasation, survival, and dissemination, target organs are recognized for further extravasation to develop as secondary tumors [12].

Invasion and Cell Migration

Cells with migratory potential invade tissues and vessels through extending cell membrane protrusions under the influence of cyclic actin polymerization and depolymerization. Initially, cells adhere to ECM via integrin- and FAK-containing complexes. Then cells contract through actin-myosin 2 followed by the disruption of cellular adhesion at the trailing edge. It has been reported that ECM remodeling and degradation facilitate invasion through proteases/integrins and other adhesion receptors on the cell surface. Cadherins and other cell-cell adhesion molecules in migrating cell sheets or clusters assist intercellular adhesion. In the absence of EMT, collective tumor cell migrates through Podoplanin (small transmembrane glycoprotein) via actin reorganization involving RhoA/ROCK and ezrin pathway. Single-cell migration occurs in slow, "mesenchymal" or in fast, "amoeboid" (requires no proteolytic ECM remodeling) fashion. The adhesion and signaling molecules involved in migration and invasion are integrins, CD44, and IgCAMs [13]. The Friedl group highlighted that upon blocking protease function cells switch from mesenchymal to amoeboid fashion of migration [14].

Tumor Cell Dissemination and Epithelial-Mesenchymal Transition

During invasive progression, epithelial tumor cells breach the basement membrane and form rigid sheet organized by adjacent belts of cell-cell adhesion molecules. This underlying process involves EMT [15, 16]. Reduced cell polarity and epithelial protein's expression (E-cadherin, occludin, claudins, cytokeratins, or catenin proteins), spindle-shaped morphology, increased migratory potential and mesenchymal proteins (N-cadherin, vimentin, tenascin C, laminin-β1 or collagen type VI α) are the characteristic features of EMT [17]. The pathways involved here are RTKs, TGFβ, WNT, NOTCH, hedgehog [18, 19], and NF-κB [20]. The transcription factors regulating EMT transcriptome program are the Snail family (SNAI1/Snail, SNAI2/slug), ZEB family (ZEB1, ZEB2), and TWIST1, TWIST2, and E12/E47 [19].

EMT promotes metastasis by [21]:

- Loss of cell-cell adhesion in invading tumor cells, as shown in E-cadherin knockout mouse models).
- Protein-degrading enzymes (matrix metalloproteinases (MMPs) that aided tissue/vessel invasion are overexpressed in the tumor stroma).
- Exposed cryptic sites by cleaving ECM components (laminin 5/collagen IV) stimulate migration/angiogenesis.

- Activity of small GTPase is affected by the released δ-catenin from dissolved E-cadherin complex.
- E-cadherin activity modulates RTK signaling by stimulating/repressing the EGFR activity.
- E-cadherin modulates β-catenin signaling prominently in colon cancer.
- The Snail and Twist family inhibits apoptosis. Twist interferes with the cellular differentiation and oncogene-induced senescence.

9.2.2.2 Organ-Specific Metastasis

Cancerous cells from a primary tumor eventually develop into a secondary tumor through a series of interrelated and sequential steps. During metastasis maximum circulating tumor cells fail to grow at distant sites. Latent period ranging from couple of years (as in breast cancer) to few months (as in lung cancer) exists between infiltrating cancer cells at distant site, colonization, and progression to a secondary tumor. The organs generally assailed by metastases are the bone, lung, liver, brain, and adrenal medulla. Lungs are the common site of metastasis for many primary tumors as they only first filter the tumor cells spreading through blood circulation, whose venous output directly flows into the lungs. The incidence of pulmonary metastasis is highest in testis, melanoma, osteosarcoma, and head and neck tumors. During metastasis to solid organs, the liver is one of the most common sites. In case of lung, breast, melanoma, renal, and colorectal tumor patients, cells frequently metastasize to the brain. Bone metastasis is mostly seen in prostate, breast, and lung cancer [22–24].

9.2.3 Regulatory Mechanisms in Invasion and Metastasis

The protein families that play a role in invasion and metastasis are enlisted in Table 9.1, where some proteins are products of promoter/suppressor genes or targets of proteins encoded by these genes. For instance, DNA-binding HIF fails to degrade upon tumor suppressor von Hippel-Lindau (VHL) protein's mutation; subsequently, CXCR4 gene (encoding motility factor receptor attracting metastatic cells) gets constitutively activated. The key players in distinct activities of invasive cells are the members of these protein families. The proteins crucial in homotypic and heterotypic cell-cell adhesion that counteract primary invasion and stimulate metastasis are cadherins and IgCAMs [26]. Apart from mechanistic role, they extensively participate in signal transduction through their association with cytoplasmic components and undergo ectodomain shedding for regulating invasion by their soluble fragments. Integrin receptors and their ECM protein ligands mechanistically regulate cell-matrix adhesion and de-adhesion from cell to matrix and vice versa. This interaction arrests cells in the matrix, assists migration and motility factors + receptors, and stimulates locomotory machinery of cancer cells through their invasion pathways [27]. Proteases pave the way for invaders by breaking down the ECM and releasing pro-invasive factors from cell surface and ECM. Proteases with inactive precursors get activated by other proteases, whereas active forms are neutralized by

Table 9.1 Protein families influencing invasion and metastasis [25]

Gene type	Gene involved	Role of protein	Type of progression
Promoter genes	ERBB2 (HER2, Neu)	Receptor tyrosine kinase	Metastasis
	FGF3	Heparin-binding growth factor	Invasion and metastasis
	$KRAS_2$, Hras	Small GTPases	Invasion
	$MADH_2$	Transcription factor	Metastasis
	MYC	HLH transcription factor	Invasion
	PI3K-γ	Protein and lipid tyrosine kinase	Invasion
	S100A4	Calcium binding	Metastasis
	SNAI1	Zinc finger transcription factor	Invasion
	SRC	Non-receptor tyrosine kinase	Invasion
Suppressor genes	CDH_1	Calcium-dependent cadherins	Invasion and metastasis
	KAI_1	TCR/CD3 tetraspanin coreceptor	Metastasis
	$MADH_4$	Transcription factor	Invasion
	MAP_2K_4	Serine/threonine kinase	Metastasis
	PTEN	Protein and lipid phosphatase	Invasion
	$TIMP_2$	Protease inhibitor	Invasion
	TP53	Regulator of transcription; growth arrest and apoptosis	Invasion
	TXNIP	Thioredoxin-binding protein	Metastasis
	VHL	Proteasomal degradation; Transcription regulation	Invasion

specific inhibitors. Normally, cells are anchorage dependent, and when they are released from their substratum, they undergo apoptosis. But, invasive cancer cells evade apoptosis via activation of growth and survival pathways and inactivation of death pathways. The abovementioned proteins participate in integrated invasion programs by forming multi-protein complexes, such as β-catenin assist as invasion suppressor in E-cadherin/catenin complex and as invasion promoter in APC/GSK-3β complex. These types of networks mediate positive and negative invasion signaling pathways [28] in terms of invasion factors binding to specific receptors, implicated in invasion and metastasis.

9.3 Processes Involved in Inquisition of Mechanistic Approach of Tumor Progression

The cancer biology is intricate and multifaceted. During genetic and epigenetic changes, normal cells transform into aggressive malignant phenotype. The manipulation of migrating behavior of tumor generally affects proliferation/apoptosis or both. Rather than regulating the migrating behavior of diffuse tumor, envision of interventions that specifically target the invasive phenotype should be addressed. The distant settlements of cancer cells are a bad news, with significantly reduced survival rate and poor prognosis. However, cancer cells depending on feedback and paracrine signaling from other tumor cells and stromal cells have a profound influence on carcinogenesis and metastasis. It is well established that hallmarks of cancer include cell growth and metastasis facilitated by MMPs and TGF-β, which remodel the ECM. Indeed, highly conserved EMT program gives rise to dissemination of single tumor cells from primary tumors. Thus, it would be valid saying that inquisition of the mechanistic approach of distant settlements/metastasis is being bestowed by two processes—ECM and EMT.

9.3.1 Extracellular Matrix (ECM) Remodeling in Tumor Progression

In cancer development, local microenvironment/niche of a cell has an important role to play. The major components of niche are composite grid of ECM macromolecules with distinct physical, biochemical, and biomechanical properties. It has a significant and regulatory role during embryonic development, tissue development, and organ homeostasis, but its deregulation results in cancer progression. The components of ECM are proteins and polysaccharide macromolecules (collagen, elastin, fibronectin, and laminin) [29]. The macromolecules constituting ECM glycosaminoglycans (hyaluronan, chondroitin and dermatan sulfate, heparan sulfate, keratan sulfate), proteoglycans, and fibrous proteins are produced by the adjacent cells attached to ECM.

- Hyaluronan resists compressive forces, creates cell-free pockets, and acts as lubricating agent [30].
- Chondroitin level increases during brain injury, contributes in the regeneration of damaged neurons, and limits the production of new neurites [31].
- In coagulation cascade, dermatan sulfate binds to thrombin and increases the activity of active protein C [32]. These are majorly regulated by TGF-β [33].
- During skeletal muscle regeneration, the concentration of heparin sulfate glycosaminoglycans increases. This influences the hedgehog, wingless, and other developmental pathways [34].
- The glycosaminoglycans perform as cofactors, coreceptors, stabilizers (for growth factors/cytokines/chemokines), enzyme activity regulators, signaling molecules (in wound healing/tumorigenesis/infections), and targets (for pathogen binding/invading cells) [32].

ECM serves as scaffold (tissue organization), biochemical and biomechanical cues (cell growth/survival/migration/differentiation), and modulator (vascular development/immune functioning). It aids in stem cell regulation and prevention of tumor cells' invasion by conserving cellular polarity and architecture [35]. In cancer progression, abnormal ECM directly promotes cellular transformation and metastasis. ECM deregulation varies the stromal cells' activity and activation of tumor-associated angiogenesis/inflammation, resulting in a tumorigenic microenvironment.

9.3.1.1 Deregulated ECM Dynamics: A Player in Cancer Initiation and Progression

The salient feature in tissue fibrosis is excess ECM production or reduced ECM turnover. The prognostic indicators of breast cancer treatment are breast density, reflecting elevated collagen and proteoglycan levels. Enhanced deposition, reduced remodeling, or amplified posttranslational modifications (crosslinking of certain ECM components) reflect increased breast density and high collagen content. The remarkable changes noted in the architecture of tumor-associated ECM are highly linearized collagen I in breast tumors and either adjacently oriented to epithelium or projected perpendicularly into the tissue. Consistently, many ECM components are frequently overexpressed in cancer, and deregulated biomechanical properties can be oncogenic. Upregulated integrin signaling promotes cell survival and proliferation, upon increased collagen deposition or ECM stiffness owing to LOX overproduction. Even deregulation of ECM remodeling evades apoptosis. Tissue invasion is promoted by MMPs, which removes the physical barrier posed by basement membrane [35, 36]. The role of ECM remodeling in cancer initiation and progression has been depicted in Fig. 9.1. However, ECM of tumor basement membrane is porous and leaky and even promotes tumor cell metastasis during cancer progression.

9.3.1.2 Disruption of the Basement Membrane and ECM

Basement membrane (BM) is an integral contributor to epithelial structure, providing physical boundary as well as a signaling substrate orienting cells via integrin-based adhesions. In developing state, the BM of epithelial tumors acts as a barrier to the invading transformed cells. Tumor cells that proteolytically disrupt the BM attain the potential for metastasizing and malignant progression. Usually, the activity of ECM proteases is under tight control governed by specific localization, autoinhibition, and secreted tissue inhibitors. But in cancerous condition, this tight regulation is disrupted by diverse mechanisms and proteolytic activities on basement membrane and interstitial extracellular matrices. Additionally, a diverse array of bioactive cleaved peptides generated by extracellular proteases facilitate tumor invasion and modulate migration, cancer cell proliferation and survival, and tumor angiogenesis [37]. During different stages of cancer progression, deciphering activities of pro- and antimetastatic components of extracellular proteases separately will aid in designing clinically effective generation of protease inhibitors.

Malignant transformation is facilitated by actin-rich protrusions, termed invadopodia, under the influence of integrin-mediated adhesion and focal adhesion formation, and MMP-mediated matrix degradation aided tumor cell invasion, followed by tumor cell migration (via elevated Rho and Rac GTPase activity), actin assemblage, and actomyosin-dependent cell tension. Thereafter, migratory phenotype is dictated by dominant Rho family GTPase activity; mesenchymal migratory phenotype is dictated by Rac GTPase activity, while amoeboid migration is favored by high RhoA GTPase activity. Regularly, Ras oncogene stimulates Rho activity (for promoting amoeboid migratory phenotype), whereas p53 reduces RhoA activity (for inhibiting tumor cell migration). During metastatic cascade, EMT favors metastasis of transformed cells. This is nurtured by TGF-β secreted by infiltrating immune cells or

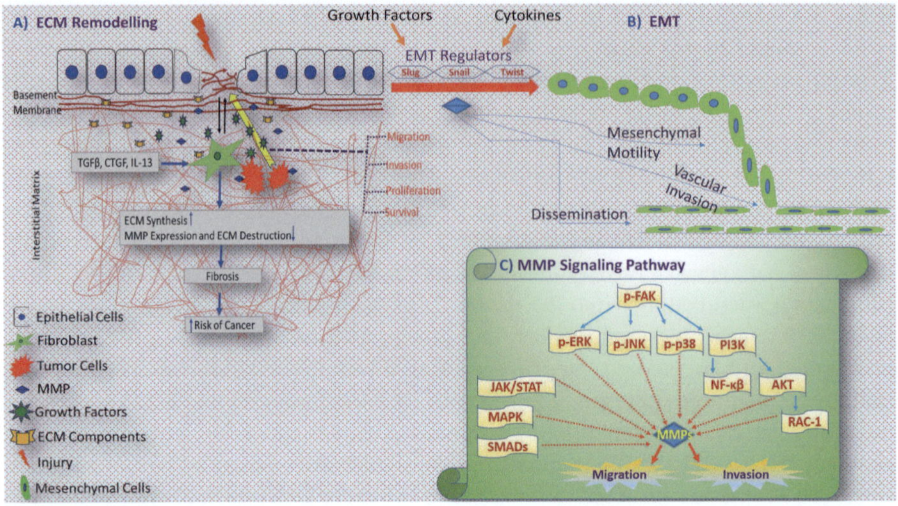

Fig. 9.1 (**a**) ECM Remodeling – The chronic inflammation/ tissue injury, TGFβ, connective tissue growth factor (CTGF), IL13 and other factors stimulate the chief ECM producer (fibroblasts and myofibroblasts) for more ECM production. This ECM contributes towards positive feedback loop by further stimulating fibroblasts for continuous production of ECM. Resultant fibrosis is a major risk for developing cancer. Tumor cells induce fibroblast activation, endothelial cell proliferation and leucocyte recruitment; in contact with stromal cells and effect of growth factors. Activated fibroblasts and endothelial cells express MMPs and secrete growth factors/ ECM components. The leucocytesdisplays activated phenotype under influence of MMPs, cytokines and chemokines. The MMP derived from both the tumor and stromal cells hastens ECM degradation and growth factor release, increasing stromal cell activation and remodeling process. The tumor cells promotecell migration, invasion, survival, and proliferation. (**b**) EMT – Under the influence of EMT regulators (Slug/Snail/Twist) stimulated growth factors and cytokines, the epithelial cells undergo mesenchymal transition. This enhances the mesenchymal motility of the cells. They even attain the ability of invading the vascular regions. The expression levels of mesenchymal markers (N Cadherin/ Vimentin/ Fibronectin) also get upregulated. (**c**) MMP Signaling Pathway – Various signals integrate towards MMP's promoter activation. This ultimately results in elevation in MMP's expression level. Thus, influencing the cell's migratory and invasive potential

through ECM degradation. Cancer metastasis is stimulated by ECM stiffness promoting TGF-β-induced EMT and a basal phenotype. Conversely, metastases could be prevented by inhibiting collagen crosslinking and reducing matrix stiffening [38].

9.3.2 Epithelial-Mesenchymal Transition (EMT) in Tumor Progression

EMT plays a critical role in the embryonic development, formation of body plan, differentiation, and tissue repair. It is a highly conserved cellular program allowing polarized, immotile epithelial cells to convert into motile mesenchymal cells. Adversely, it promotes invasion and metastasis, induces stemness, averts apoptosis and senescence, and contributes to immunosuppression.

Basically, EMT is a biologic process allowing the interaction of a polarized epithelial cell with BM. It endures multiple biochemical changes, assuming mesenchymal phenotype. The molecular processes involved in EMT are activation of transcription factors, expression and reorganization of specific cell-surface proteins and cytoskeletal proteins, production of ECM-degrading enzymes, and varied specific microRNA expression. Consequently, the involved factors are being exploited as EMT biomarkers [39].

9.3.2.1 Correlation of EMT with Cancer Progression and Metastasis

The hallmarks of primary epithelial cancers are elevated epithelial cell proliferation and angiogenesis [40]. The EMT activation in acquisition of invasive potential and eventual metastatic dissemination is the critical mechanism involved in malignant transformation [16]. In vitro and in vivo studies demonstrate that carcinoma cells acquire mesenchymal phenotype and express mesenchymal markers (SMA, FSP1, vimentin, and desmin) [41]. These typical cells are seen at the invasive front of primary tumors eventually entering into the invasion-metastasis cascade [16]. The correlation of EMT with cancer progression has been illustrated in Fig. 9.1. The EMT-inducing signals stemming from the tumor-associated stroma include HGF, EGF, PDGF, and TGF-β, which subsequently induce EMT transcription factors (Snail, Slug, zinc finger E-box binding homeobox 1 (ZEB1), Twist, Goosecoid, and FOXC2) [16, 42]. These transcription factors, once expressed and activated, pleiotropically choreograph the EMT drive under the influence of intracellular signaling like ERK, MAPK, PI3K, Akt, Smads, RhoB, β-catenin, lymphoid enhancer-binding factor (LEF), Ras, c-Fos, and integrins. The EMT program is mediated by the disruption of cell-cell adherence junctions and integrin-driven cell-ECM adhesions. Intense research also intensifies that TGF-β regulates tumor progression and metastasis [39]. The signaling pathways involved in TGF-β-induced EMT are:

- Smad proteins mediated TGF-β-induced EMTs via ALK-5 receptor facilitating motility [43]. Signaling pathways mediating β-catenin and LEF cooperation with Smads induce EMT [44, 45]. Studies also demonstrate that TGF-β/Smad/LEF/PDGF axis induces EMT phenotype.
- p38 MAPK and RhoA-mediated TGF-β induce EMT under integrin influence. Fibulin-5 augments TGF-β-induced EMT in a MAPK-dependent mechanism. TGF-β induces EMT in Ras-transformed cells (via MAPK) [46, 47].
- Raf mediates TGF-β-induced EMT promoting invasiveness [48].
- COX-2 inactivates Smad signaling and enhances TGF-β-induced EMT through PGE2-dependent mechanism [49].

The link between loss of E-cadherin expression and EMT has been well established [50]. Furthermore, in EMT, upon ectopic expression of E-cadherin containing β-catenin binding site, epithelial cell adhesion complexes reorganize suppressing cell proliferation. Such cells lose mesenchymal phenotype [45]. The mutations in E-cadherin gene make EMT cells more susceptible to EMT and metastasis [51]. The actions of EMT-inducing transcription factors facilitating acquisition of mesenchymal phenotype illustrate the central role played by E-cadherin loss [52]. A correlation between loss of E-cadherin and Wnt signaling or high expression of Snail in nucleus has been reported. The expression of *Snail* and E-cadherin is inversely correlated with the prognosis of breast cancer and oral cancer patients [53, 54]. Some labs also report that MMP-3 facilitates genomic instability via Rac1b- and ROS-induced EMT [55] and noncoding microRNA regulation of EMT program (miR200 and miR205 inhibiting the E-cadherin expression repressors, ZEB1 and ZEB2 maintaining epithelial cell phenotype) [56, 57]. A loss of miR200 in breast cancer is correlated with increased expression of vimentin and decreased levels of E-cadherin [58], but miR21 upregulation facilitates TGF-β-induced EMT [59].

Here, it is also important to record the association between proteases and ECM network. Invasion is favored by Snail and Zeb that induce the metalloprotease expression for basement membrane's degradation. By increasing the ROS cellular levels, MMP-3 triggers EMT by inducing Snail1 expression. FGF1-induced MMP-13 and Eplysin via TGF-β also trigger EMT. In colon cancer, metastasis is being promoted by overexpression of transmembrane serine protease TMPRSS4-induced EMT through Zeb transcription and E-cadherin downregulation. EMT, invasion, and metastasis are also being promoted by an ECM protein "Periostin" secreted by osteoblasts via PI3K/Akt interaction with integrins [60].

9.4 Aspect of Extracellular Proteases in Tumor Progression

The tumor cells invade and metastasize by breaching ECM and several tissue layers. For this, activation of proteolytic enzymes is a must, and apart from ECM proteins, other components such as glycosaminoglycans are also degraded. Perhaps zinc-dependent MMPs are vital in this process and are frequently overexpressed in most of the tumors. The endopeptidases are secreted by macrophages, mast cells, and fibroblasts. The cellular substrates involved in MMP's degradation are fibronectin, collagen, laminin, and proteoglycans. Actually in ECM, soluble MMPs are secreted as inactive proenzymes which are activated by other enzymes [61]. By activating or deactivating several growth factors, MMPs also affect tumor neoangiogenesis and proliferation. The tissue inhibitors of metalloproteinases (TIMPs) play a central role in complex regulation of MMPs. An apt equilibrium between TIMPs and MMPs is of essential relevance in cell invasion and metastasis.

9.4.1 Role of Matrix Metalloproteinases in Cancer

Advanced cancer research dealing with intended role of proteolysis in tumor invasion and metastasis analyzed the role of MMPs in tumor progression through identified members of MMP family (as secreted enzymes + ECM components as substrate) and elevated expression of MMPs. This motivated many laboratories for designing proof-of-principle experiments. Initial experiments justified that the endogenous MMP inhibitors modulated the MMP activity by manipulating the levels of TIMPs.

These advances encouraged for pursuing MMPs as therapeutic targets, but unlikely the number of MMPs expanded to more than 20. Their expression pattern determines cell- and tissue-specificity. The expansion of MMPs' role affecting angiogenesis and growth of both benign and malignant tumors, substantially the range of potential relevant substrates, got broadened. The development of pharmaceutical reagents and initiation of clinical trials targeting MMPs enhance the therapeutic benefit to cancer patients [1]. Further understanding of MMP biology, their regulation, expression pattern in different types of tumor, role in invasion, metastasis, ECM remodeling, EMT and tumor growth, and biomarkers will finely tune our knowledge in developing anticancer therapeutics by potentially inhibiting MMPs.

9.4.2 MMP Biology

MMPs are zinc-dependent endopeptidases of metzincin family of enzymes coding highly conserved zinc-binding motif. They are also known as matrixins degrading all kinds of ECM proteins and acting upon various bioactive molecules. MMPs cleave cell-surface receptors, release apoptotic ligands (like FAS ligand), and mediate chemokine and cytokine activation or inactivation [62]. MMPs regularly affect cellular proliferation, differentiation, migration, apoptosis, angiogenesis, and host defense. The MMPs' physiological and cellular function includes cell migration through ECM degradation, changes in cellular behavior, and modification in the activity of biologically active molecules by direct cleavage/their inhibitors/their release. They are released as inactive proenzymes, activated by factors and TIMP. Brew et al. reported that imbalance in the MMP and TIMP levels might lead to pathological conditions [63]. Various reports validate correlation between overexpression of MMPs and inflammatory, malignant, and degenerative diseases [64, 65].

The structural sketch of MMPs defines three domains, namely:

- *Propeptide domain*—keeps the enzyme inactive. The conserved cysteine residue interacts with zinc in the active site to inhibit binding and cleavage of the substrate. The enzyme gets activated upon proteolytic cleavage intracellularly (by furin) or extracellularly (by other MMPs/serine proteinases such as plasmin) of this domain [66].
- *Catalytic domain*—zinc-binding motif is its structural signature. The active site comprises of Zn^{2+} ion bound by three histidine residues forming a shallow groove that binds to the substrate.
- *Hinge region*—the 75 amino acids' long linker region that connects the catalytic domain to the C-terminal domain is essential for enzyme's stability.
- *Hemopexin-like C-terminal domain*—its polypeptide chain organizes into four β-sheets that symmetrically arranges around a central channel, resulting into four-bladed β-propeller structure. The structural flat surface is involved in interactions between proteins determining substrate specificity (e.g., TIMP).

An interface between cysteine-sulfhydryl group (propeptide domain) and zinc ion (bound to catalytic domain) keeps the enzyme in inactive form. Actually, MMPs synthesized as inactive zymogens require proteolytic removal of propeptide domain for activation. Major MMPs are activated extracellularly under the influence of serine proteinases or other activated MMPs. Some MMPs (MMP-11, MMP-28, and MT-MMPs) are intracellularly activated by furin-like serine proteinases [66]. MMP-2 gets activated at the cell surface involving MMP-14 (MT1- MMP) and TIMP-2 via unique multistep pathway (TIMP-2 binds to MMP-14 at amino terminal and pro-MMP-2 at carboxy terminal, and then cleavage of non-inhibited MMP-14 occurs from bound pro-MMP-2, and for MMP-2 activation, the removal of residual portion of MMP-2 propeptide is necessary) [67]. Thrombospondin-1 shows inhibitory action on activated MMP-2 and MMP-9 by binding to pro-MMP-2 and pro-MMP-9 [68]. The well-established endogenous MMP inhibitors are TIMP-1, TIMP-2, TIMP-3, and TIMP-4 reversibly inhibiting MMPs in 1:1 stoichiometric fashion [69].

Proteolytic degradation of ECM's structural components by MMPs enhances the cell migratory potential (such as cleavage of laminin 5 and collagen type IV results in the "cryptic sites" that enable migration [70, 71] and cleavage of IGF-BP and perlecan releases IGFs and FGFs [72, 73]. However, MMP-2 and MMP-9 influence the release of TGF-β from an inactive extracellular complex [74]. MMPs even target cell adhesion molecules such as E-cadherin and CD44 resulting in the release of extracellular domain fragments. This mechanism increases the invasive potential of the cells [75, 76]. It has been also observed that MMP-14 cleaves α-v integrin subunit precursor. This enhances the migratory potential of cancer cells. Additionally, MMPs cleave proteinase inhibitors (serpins) and even other MMPs [66].

9.4.3 MMP Expression in Tumor and Underlying Mechanism

Since long it is understood that proteinase activity is essential for tumor cells to invade and metastasize to distant sites, where the potentially invasive cells first attach to BM via cell-surface receptors (integrin) and administer extracellular proteolytic action, and cellular locomotive action is initiated, depending on chemotactic factors. Since proteolysis of BM and ECM components is considered to be an essential step in cancer invasion and metastasis, tumor proteases are considered to be accessible targets for therapeutic intervention. It is evident that MMP activity contributes to early-stage tumorigenesis, angiogenesis, and later events of invasion and metastasis. Let us now overview the role played by MMPs in tumor progression.

The MMPs are abundantly expressed in malignant disease. A survey on tumor-associated MMPs is shown in Table 9.2. Literature findings reveal that a positive correlation exists between level of MMPs expression and tumor grade. Alongside, research on levels of endogenously produced inhibitors reveals that elevated MMP levels and reduced TIMP levels define the aggressiveness of a tumor [81–84]. These findings led to a concept stating that aggressive, invasive, and metastatic potential of

Table 9.2 Survey on tumor-associated MMPs [77–80]

MMP	Common name	Structural class	Role	Type of cancer	MMP inhibition: effect on tumors
MMP-1	Collagenase-1	Simple hemopexin domain	(1) Degrade major component of bone ECM	Breast	–
			(2) It is expressed by fibroblasts, keratinocytes, endothelial cells, monocytes, and macrophages	Colon	
			(3) MMP-1 was significantly downregulated, while TIMP-1 levels were increased, in a time- and pressure-dependent manner in a smooth muscle cell	Gastrointestinal	
				Head and neck	
				Prostate	
				Esophagus	
MMP-2	Gelatinase A, 72-kDa type IV collagenase	Gelatin binding	(1) It is widely expressed in embryonic CNS	Glioma	Reduction in tumor-induced angiogenesis and in experimental metastasis
			(2) Expression of MMP-2 and β-catenin loss has a role in the pathogenesis and progression of ESC	Bladder	
			(3) Decreased E-cadherin has an important role in the development of both ESC and EEC	Breast	
				Cervical	
				Colon	
				Lung	
				Melanoma	
				Pancreas	
				Prostate	
				Skin	
				Stomach	

(continued)

Table 9.2 (continued)

MMP	Common name	Structural class	Role	Type of cancer	MMP inhibition: effect on tumors
MMP-3	Stromelysin-1, proteoglycanase	Simple hemopexin domain	It limits plaque growth and enhances plaque stability	Glioma	Enhanced tumor growth following carcinogen treatment
				Breast	
				Colon	
				Lung	
				Pancreas	
				Prostate	
MMP-7	Matrilysin	Minimal domain	Major role in invasion and metastasis of cancer	Glioma	Reduction in tumor formation in min mice
				Breast	
				Lung	
				Prostate	
				Stomach	
MMP-9	Gelatinase B, 92-kDa type IV collagenase	Gelatin binding	Favors differentiation and reconstitution of the stem/progenitor cell pool	Glioma	Aberrant angiogenesis and apoptosis in the developing bone
				Bone	
				Breast	
				Lung	
				Lymphoma	
				Myeloma	
				Ovary	
				Pancreas	
				Prostate	
				Skin	
MMP-10	Stromelysin-2, Transin-2 MMP-11 Furin activated	Simple hemopexin domain	Is associated with aggressiveness of cancer cell	Head and neck	–

MMP-13	Collagenase-3	Simple hemopexin domain	(1) Critical role in cartilage destruction	Breast	–
			(2) Expressed by chondrocytes (degrade their matrix, when stimulated by retinoic acid)	Head and neck	
				Skin	
MMP-14	MT1-MMP, MT-MMP1	Transmembrane	(1) Major activator of pro-MMP-2 and is essential for skeletal development	Breast	–
			(2)Generated in vitro by cleavage of membrane-bound native MT1-MMP with several recombinant MMPs (both active MT1-MMP and MMP-2)	Cervical	
				Colon	
				Head and neck	
				Liver	
				Lung	
				Ovary	
				Pancreas	
				Stomach	

a tumor is influenced by proteolytic degradation, owing to the imbalance between proteases and their inhibitors. The protein level data reveals that MMPs are mostly produced by stroma surrounding the tumors in response to signals from tumor cells through soluble factors or by cellular contact. The MT1-MMP acts as receptor and activator of gelatinase A (produced by surrounding stromal cells). It has been also observed that advanced tumor cells undergoing EMT express a range of elevated MMPs. The MMP levels are usually assayed by zymography, immunoblotting, ELISA, and IHC.

In tumorigenesis, MMPs potentially affect both cellular proliferation and apoptosis by altering BM and ECM substrates [85, 86]. Apart from BM and ECM components being the potential substrates for MMPs, growth and apoptotic factors also provide the basis for MMPs' effect on cellular processes, ultimately leading to the establishment and growth of tumors. Such potentiating role of MMPs in processing or releasing of such factors has to be certainly assessed. The various signaling pathways integrate for promoting MMP activity and subsequent cell migration and invasion. This network has been illustrated in Fig. 9.1. The MMPI inhibition targeting growth factors or their receptors has been reported [87]. The gelatinase A activity can accomplish the cleavage of FGF type 1 receptor [88]. Such an event maintains its ligand-binding ability, which may further modulate growth and angiogenesis-allied activities of FGF. Other reported MMP substrates are EGFR ligands such as amphiregulin, TGF-α, and HB-EGF [87, 89]. TNF-α is also processed to its soluble form by several MMPs in vitro [90]. The growth factor's activity can also be synchronized by sequestering or by binding proteins (BPs), for instance, IGF-BPs regulate IGF's bioavailability by checking its interaction with receptors. Accumulating evidences demonstrate potential cleavage of IGF-BPs by various MMPs [91, 92]; hence, free IGF increases tumor cell's proliferation rate. Some ECM proteins even sequester growth factors [93]. For instance, an ECM protein "decorin" binds to TGF-β, being substrate of matrilysin, stromelysin-l, and gelatinase A upon enzymatic cleavage release of TGF-β [94]. The action of MMPs on ECM proteins results in death signals transmitted to cells through integrin, a cell-matrix adhesion molecule [95]. E-cadherin controls tumor growth, and loss of this aids in overcoming the normal contact inhibition of cellular proliferation [96].

9.4.4 MMPs in Tumor Invasion and Metastasis

The invasiveness of tumor cells in correlation with metastasis can be evaluated quantitatively in vitro using embryonic chick heart invasion assay [97], amnion invasion assay [98], and matrigel invasion assay [99, 100]. Such assays showcase a correlation between invasive ability and MMP's expression. The examination of increasing metastatic variant of murine melanoma cells expressing type IV collagenase [101] involved a collagen degradation assay. The study implicated that collagenase activity increases with the metastatic potential of cells. In case of natural MMP inhibitors—TIMPs [98]—protease inhibitors blocked tumor cell proliferation, attachment to amnion, and migration through noncoated filters. Schultz and colleagues using recombinant TIMP-l demonstrated that invasive murine melanoma cells depend on metalloproteinase activity for invasion [102]. Some studies also demonstrate that sufficient level of MMP's expression increases invasion. The implantation of DU-145 cells in nude mice generated an invasive phenotype through the expression of matrilysin [103]. Moreover, gelatinase A activator MT1-MMP increased the invasive potential in three different tumor cell lines [104].

The introduction of recombinant or transfected TIMPs modified MMP activity, demonstrating the role of MMPs in in vivo metastatic models [105]. The administration of recombinant TIMP (rTIMP) to mice bearing B16-FlO melanoma cells significantly reduced the number but not the size of metastasis. This study indicates that inhibition of proteinase activity does not affect the growth rate of tumor cells [102]. Thc TIMP-2 distinctly reduced the tumor growth rate and moderately suppressed hematogenous metastasis in mice bearing TIMP-2-transfected cell line. The tumor growth rate was regulated by inhibiting tumor mass expansion and subsequent suppression of local invasion. Of late TIMP and TIMP-4 have been reported as a metastasis inhibitor in a breast cancer mouse model [105]. The gelatinase B-positive cells generated metastatic lesions in nude mice. The MMPI, batimastat [BB-94], has significantly reduced the number of lung metastasis [106]. Radiolabeling of tumor cells before injecting serves the purpose of monitoring the distribution of cells to organs and site of arrest. BB-94 was not able to arrest in the lungs but prevented retention by blocking extravasation. In an ovarian cancer model, it reduced the tumor size and increased the survival time [107]. CT1746, orally active, and gelatinase A/B, stromelysin-l specific MMPI, prolonged survival time (51 to 78 days), reduced primary tumor growth (by 32%), and significantly reduced total spread and tumor metastasis [108]. Multiple studies highlight that tumor invasion and resulting metastasis mediated through MMPs could be controlled by both natural and synthetic metalloproteinase inhibitors.

9.4.5 MMPs: Regulator Protein Family of ECM Remodeling and EMT

As we all know, MMPs belong to metalloendopeptidase family which cleaves the ECM protein components and thereby plays a fundamental role in tissue remodeling. Since long MMPs were thought to function principally as ECM composition's regulators facilitating cell migration by removing barriers like collagen. However, the role of MMPs in the regulation of growth factors and their receptors, cytokines, and chemokines, adhesion receptors and cell-surface proteoglycans, and a variety of enzymes is well implicated. Therefore, MMPs play a significant role in controlling cellular interactions in response to the environment. The proteolytic activity of MMPs leads to important insight. On the basis of specificity for ECM proteins, MMPs are broadly classified into collagenases, gelatinases, stromelysins, and matrilysins [109].

Table 9.3 EMT traits and corresponding MMP expression in different cancer

Cancer model	Specific EMT traits	MMPs expressed	References
Bladder cancer	E-cadherin reorganization	MMP-2	[111]
Breast cancer	Lack of E-cadherin, vimentin and invasive potential	Activation of MT1-MMP, MMP-2	[112]
Bronchial cancer	Altered vimentin expression and migratory potential	MMP-3, MMP-9, and MMP-11	[113, 114]
Prostate cancer	Loss of E-cadherin, vimentin, and invasive potential	MT-MMP	[115]
Cervical cancer	Loss of E-cadherin, vimentin, and invasive potential	Activation of MT1-MMP, MMP-2	[116]
Squamous cell carcinoma	Loss of E-cadherin, vimentin	MMP-2	[117]

EMT is a fundamental biological process in tumor metastasis. During which transcription factors from zinc finger family (like Snail, Slug, Twist, and MMPs) are upregulated. A highly invasive A431-III tumor subline displayed the correlation between MMP levels and EMT promotion. Treatment with a broad-spectrum MMP inhibitor (GM6001) reduced vimentin and fibronectin in A431-P and A431-III cells. This indicates that the MMP-9 induced EMT in association to elevated invasion and metastasis. Reports highlight that the prevalence of MMP-2, MMP-3, and MMP-9 leads to the disruption of cell adhesion by processing the cell-cell and cell-ECM contact components and by interfering E-cadherin's function. MMP's procession of E-cadherin initiates EMT and detaches tumor cells and transfers into the stroma; stationary epithelial cells attain motility [110]. The EMT traits and corresponding MMP expression in different cancer have been listed in Table 9.3. Hence, EMT-associated MMPs are also a promising therapeutic target. The correlation among MMPs and ECM remodeling/EMT in cancer progression has been also finely depicted in Fig. 9.1.

9.4.6 MMPs as Biomarkers in Cancer

Clinically, several biomolecules assisting diagnostic decision-making for cancer patients have been developed. A biomolecule has to be sensitive and specific enough to employ as a diagnostic marker. They can be detected in the blood, serum, saliva, or urine. In cancer patients, body fluids with elevated levels of numerous MMPs have been reported. This aroused the question whether MMP profiling of body fluids can be employed as a cancer diagnostic marker. It is very appealing since it unlocks the way for quick, noninvasive test convenient for screening large populations with increased risk of cancer. A study involving 300 colon cancer cases demonstrated high serum level of MMP-9 in malignant and premalignant lesions in comparison with benign lesions. The selected threshold value showed sensitivity up to 99% and specificity up to 63% [118]. A quick and noninvasive analysis predicting the suspicion of cancer would be effective in terms of both time and resources. The serum MMP-9 levels are currently being employed as an accurate test for colon cancer patients [119]. There are also evidences displaying the presence of MMP-2 and MMP-9 as a marker of bladder and prostate cancer [120, 121]. Additionally, studies involving tissue fluids of different cancers assessed the diagnostic value of MMP-2, MMP-7, and MMP-9 and/or TIMP-1 and TIMP-2 [77]. However, these tests fail to demark the patients with malignant tumors and benign tumors or an inflammatory disease. Similar genre of MMPs are upregulated in different types of cancer as well as in inflammatory diseases, hence IHC is often employed in clinical pathology as a differentiating tool for benign and malignant tumors and also among different type of cancers. Recently, research has established MMP-11 to be more effective in this context [122, 123].

The standard treatment procedure involves surgical abscission of tumor followed by adjuvant therapy like radio/chemo/hormonal therapy or angiogenic/kinase inhibitors. The treatment procedure implemented might pose adverse effects, and inability to identify patients with low risk of tumor recurrence might be overtreated. Recent development of biomarkers predicting the chances of relapse has contributed toward treatment stratification. There are accumulated evidences addressing the potential of MMPs and TIMPs as prognostic markers in different cancer. However, owing to conflicting results, it is not that easy to draw general conclusions concerning prognostic value of MMPs/TIMPs in cancer. The possible reasons behind this are [77]:

- Since MMPs are multifunctional in nature and their specific role depends on the acting substrate in a biological situation such as variation between patients, organs, phases of tumorigenesis, and progression.
- TIMPs are multifaceted proteins and their MMP-independent roles are in starting phase.
- Variations in the parameters considered for studying such as:
  - MMP/TIMP level in blood/urine or tissue samples/extracts
  - Enzymatic activity or total expression at transcription/translation level
  - Separating active enzymes and proenzymes or MMP/TIMP expressing cell types

Therefore, a confirmatory approach for validating the findings to establish MMPs/TIMPs as a prognostic marker in the field of cancer has to be generated.

9.5 MMPs: Therapeutic Intervention

Since long, MMPs were considered to be matrix-degrading enzymes and MMPIs to downregulate the invasive and metastatic potential. Clinically, the drugs were not able to offer survival benefit to patients and sometimes reduced the survival rate, and severe adverse effects were also reported. The reasons behind the therapeutic failure which helped in modifying the therapeutic strategy involving MMPs are:

- The patients enrolled in clinical trials were in advanced stage, and MMPs are involved in early stages of tumorigenesis, and the drugs targeting MMPs might be effective if applied in nonmetastatic patients.
- Preclinical mouse experiments involving MMPIs were generally successful, as they were administered before developing metastasis.
- The first-generation MMPIs come under broad-spectrum inhibitors inhibiting both tumor-promoting and tumor-repressing MMP activities.
- Level of MMP expression in patients enrolled for clinical trials should be checked and administered as adjuvant therapy with conventional cytostatic drugs or radiation [124].

These findings indicate that new MMPIs developed should be specific in terms of MMP or degradation of certain substrates. This is challenging because the active sites of MMPs are very similar among MMP family members. But different MMPs have different subsites/pockets in active site clefts. The substrate specificity is defined by the ability of the cleaving substrate to fit into these pockets [125]. Instead of targeting the active site Zn ion, such subsites should be focused for designing more specific MMPIs [126]. Additionally, targeting exosites or noncatalytic sites in MMPs may inhibit detrimental effects of MMP, as it is known that large protein substrates require cross talk between active site and noncatalytic domains for efficient cleavage [127]. An example displaying exosite's requirement is collagenases (MMP-1) which aided cleavage of triple-helical collagen. The large triple-helical cord is processed (unwinding hemopexin domain of collagenases acts as unhelicase) to fit into the cleft of active site and hydrolyzed [128, 129]. Targeting exosites of defined substrates might aid in inhibiting specific MMP functions only, not their entire activity. However, comprehensive understanding regarding enzyme-substrate interactions and regulation and role of specific MMPs in different cancers will certainly aid in designing such drugs. Another possibility is targeting MMPs at expression level [130, 131] or generating specific cytostatic drugs as prodrugs by exploiting the advantage of elevated MMP expression in tumors. Prodrug approach would release higher concentrations of active drug in the tumor environment after being processed by cancer-associated MMP and successively reduce the adverse effects in other tissues. The MMP synthesis could be usually inhibited by agents which avert them from associating with the molecules mediating their activities to the cell surface or impeding their enzymatic activity. The different approaches inhibiting MMP gene transcription, targeting extracellular factor signal transduction pathways or nuclear factors that stimulate genetic expression and downregulate MMP production, have been illustrated in Table 9.4.

Table 9.4 Different approaches inhibiting MMP gene transcription, targeting extracellular factors, signal transduction pathways, or nuclear factors that stimulate genetic expression and downregulate MMP's production

MMP inhibition strategy	Approach involved	Molecules	Experimental model	Observations	References
Inhibiting MMP synthesis	Antisense mRNA/oligonucleotide-transfected cells	–	Mouse models	Reduced tumor burden or metastasis by downregulating MMP-7 or MMP-9	[132–134]
	Ribozyme-targeted mRNA directly inhibits MMP synthesis				
Inhibiting signal transduction pathways inducing MMP transcription	Inhibit tyrosine kinase receptor signaling	Halofuginone (Coccidiostat)	Chickens	Regulating MMP gene expression and experimental cancer cell metastasis	[135]
Inhibiting MMP and other protein's interaction	Inhibiting binding of MMP-2 to α-v-$\beta3$ integrin	–	Animal models	Specifically targets cancer-promoting function	[136]
MMP activity's manipulation	Fuses to MMP cleavage site, upon activation by MMP cleavage at cell surface internalized to the cell	Recombinant proteins containing anthrax toxin	–	Cell death	[137]
				Tumor treatment	

(continued)

Table 9.4 (continued)

MMP inhibition strategy	Approach involved	Molecules	Experimental model	Observations	References
MMP blockage	Collagen peptidomimetics (mimic the cleavage sites of MMP substrates)	Batimastat and Marimastat	Phase III trials	–	[138–142]
	Collagen non-peptidomimetics (based on confirmation of MMP active site)	BAY 12-9566, Prinomastat/AG3340, BMS-275291, and CGS 27023A/MMI270	Phase II/III trails	–	
	Tetracycline derivatives	Col-3 (Metastat)	Phase II trials	Inhibits both the activity and synthesis of MMPs	
	Small peptides	–	Animal models	Inhibits MMP-2 and MMP-9 enzymatic activity	
	Bisphosphonates	–	–	Inhibits MMP enzymatic activity	
	Unconventional inhibition	AE-941(Neovastat)	Phase III trials	Inhibits MMPs	
	Green tea component	–	Phase III trials	Inhibits MMP-2 and MMP-9	
	Acetylsalicylic acid	–	–	Reduces risk of colon cancer by inhibiting MMP-2 activity	
Targeting extracellular factors		IFN-γ, IFN-β, and IFN-α	Diverse human cancer cells	Inhibit transcription of several MMPs via transcription factor STAT1	[143–148]
	Blocking IL-1 or EGF receptor			Abolishing MMP's production	
	Blockade of TGF-β	Soluble TGF-β receptor antagonist	Breast cancer mouse model	Inhibit tumor metastasis and production of active MMP-2 and MMP-9	

Targeting signal transduction pathways	Interfere with the TGF-β signaling pathway	Halofuginone	Bladder cancer metastasis	Blocking MMP2 expression	[149]
	Selective inhibition of p38 MAPK activity	SB203580	Transformed keratinocytes and squamous cell carcinoma cells	Abolish MMP-1, MMP-9, and MMP-13	[150, 151]
		Malolactomycin D	NIH3T3 cells	Transcription inhibition under control of RAS-responsive element suppressing several MMP expression	[152]
	RAS farnesyltransferase inhibitor	Manumycin A	Lung cancer cells	Block hyaluronan-mediated MMP-2 secretion	[153]
Targeting nuclear factors	Interact with the AP-1 binding site	Glucocorticoids	–	Prevent the upregulation of MMPs	[154–158]
	Inhibit AP-1 binding activity	Nobiletin	Fibrosarcoma cells	Suppress the production of MMP-1 and MMP-9 and invasive potential	
	Interfere with AP-1-induced transcription	Curcuminoids		Inhibit MMP-9 expression	
	Block the degradation of inhibitor of κB (IκB)	PS-341	Multiple myeloma	Maintain NF-κB in an inactive status	
	Adenoviral delivery of wild-type p53	–	Squamous cell carcinoma cells	Carry mutant forms of p53 and inhibit MMP expression and invasive properties (independent of pro-apoptotic effect of p53)	
Blocking pro-MMP activation	–	Green tea catechins	–	Blocks MT1-MMP-dependent activation of pro-MMPs	[159]
Inhibiting proprotein convertase	Reduces processing of pro-MMP-2 and prevents MT1-MMP activation	α1-PDX	–	Prevents tumor growth and invasive potential	[160, 161]

(continued)

Table 9.4 (continued)

MMP inhibition strategy	Approach involved	Molecules	Experimental model	Observations	References
Natural MMP inhibitors	–	Neovastat	–	Display anti-angiogenic and antimetastatic activity of these effects depending on the inhibition of MMP enzymatic activity and VEGF inhibition	[162]
		Genistein		Confer tumor inhibition growth and invasion by interfering with the expression ratio and activity of several MMPs and TIMPs	
Marine source MMP inhibitors	Marine saccharoid MMPIs			Inhibit MMP by direct downregulation of MMP-9 transcription or via inhibition of activator protein-1 (AP-1) pathway or nuclear factor-κB (NF-κB) pathway	[163]
	Marine flavonoid and polyphenol MMPIs	Flavonoid glycosides, isorhamnetin-3-O-b-D-glucosides, and quercetin-3-O-b-Dglucoside	Human fibrosarcoma cells	Inhibited MMP-2 and MMP-9	
	Marine fatty acid MMPIs	Oleic acid and elaidic acid	–	Inhibit MMP-2 and MMP-9 by binding to neutrophil elastase	[164]
		Eicosapentaenoic (EPA) and docosahexaenoic acid (DHA)	Colon cancer model	Inhibited lung metastasis with reduced MMP-9 activity	
	Anticoagulant and antiproliferative agent	Sulfated polysaccharide from Ecklonia cava (brown	HL-60 and U-937 cells	Antiproliferative effect	
		algae)			

		Fucoidan extract from cladosiphon (seaweed)	HT1080 cells	Reduced cell invasion by suppressing MMP-2 and MMP-9 activity	[165]
				Suppress the expression and secretion of VEGF, thereby inhibiting invasion and angiogenesis	
	Inhibiting both NF-κB and AP-1 reporter	*Eisenia bicyclis*, *Ecklonia cava*, and *Ecklonia stolonifera* extracts	–	Reduced MMP-1 expression	
	LPS-induced production of nitric oxide, prostaglandin E2, inducible nitric oxide synthase, and COX-2 suppression	Dieckol	Murine BV2 microglia	–	
	Interference with the transcription factor AP-1	Flavonoid glycoside from *E. cava*	HT1080 cells	Inhibited the expression of MMP-2 and MMP-9 and elevated TIMP-1 expression	
	–	Ageladine A	–	Inhibit MMP-1, MMP-2, MMP-8, MMP-9, MMP-12, and MMP-13	[166]
Miscellaneous	A succinyl hydroxamic acid bearing close structural similarity (substrate-based design)	Actinonin	–	–	[167]

9.6 Conclusion

It is immensely significant to observe that 90% of cancer deaths are attributed to metastasis. The underlying molecular principles of invasion and metastasis would aid in profound understanding of tumorigenesis. New insights toward molecular progression of invasion and metastasis might pave the way for new, highly specific, and potential tumor management strategies. The prerequisite for this resolution is further research. Cancer cells leave the primary tumor to disseminate to distant organs. This cascade is convoyed by variations in gene expression and functions, such as loss of epithelial markers and gain of mesenchymal markers. The genes determining these activities are defined as metastasis initiation genes, promoting cell motility, EMT, ECM degradation, angiogenesis, or evasion of immune system. ECM deregulation varies the stromal cell behavior, switch on tumor-associated angiogenesis, and inflammation. EMT is implicated in promoting carcinoma invasion and metastasis. Apart from endowing cells with migratory and invasive properties, EMT induces stemness, averts apoptosis and senescence, and contributes to immunosuppression. The EMT program is mediated by the disruption of cell-cell adherence junctions and integrin-driven cell-ECM adhesions.

Proteases pave the way for invaders by breaking down the ECM and releasing pro-invasive factors from cell surface and ECM. A diverse array of bioactive cleaved peptides generated by extracellular proteases facilitate tumor invasion and modulate migration, cancer cell proliferation/survival, and tumor angiogenesis. During different stages of cancer progression, deciphering activities of pro- and antimetastatic components of extracellular proteases will aid in designing clinically effective generation of protease inhibitors. It is also important to record the association between proteases and ECM network. Tumor cells breach through ECM via activation of proteolytic enzymes. The MMP family member's functionality (as secreted enzymes + ECM components as substrate) and elevated expression of MMP motivate several laboratories for designing principle experiments and scrutinizing the MMPs' role in tumor progression. These advances are encouraged for pursuing MMPs as therapeutic targets. Further understanding of MMP biology, their regulation, expression pattern in different types of tumor, role in invasion, metastasis, ECM remodeling, EMT and tumor growth, and biomarkers will finely tune our knowledge in developing anticancer therapeutics (by potentially inhibiting MMPs). The second-generation MMPIs developed should be specific in terms of MMP or degradation of certain substrates. MMPs should be targeted at expression, transcription, extracellular factors, signal transduction pathways, or nuclear factor level. The development of pharmaceutical reagents and initiation of clinical trials via strategic inhibition of MMPs would aid in paradigm shift in cancer therapy.

Acknowledgments We gratefully acknowledge the Department of Science and Technology (DST)—INSPIRE—India.

References

1. Leber MF, Efferth T (2009) Molecular principles of cancer invasion and metastasis. Int J Oncol 34:881–895
2. Hahn WC, Weinberg RA (2002) Rules for making human tumor cell. N Engl J Med 347:1593–1603
3. Veiseh O, Kievit FM, Ellenbogen RG, Zhang M (2011) Cancer cell invasion: treatment and monitoring opportunities in nanomedicine. Adv Drug Deliv Rev 63(8):582–596
4. Coussens LM, Werb Z (2002) Inflammation and cancer. Nature 420:860–867
5. Pages F, Galon J, Dieu-Nosjean MC, Tartour E, Sautes-Fridman C, Fridman WH (2010) Immune infiltration in human tumors: a prognostic factor that should not be ignored. Oncogene 29:1093–1102
6. Friedl P, Gilmour D (2009) Collective cell migration in morphogenesis, regeneration and cancer. Nat Rev Mol Cell Biol 10:445–457
7. Maeda H, Sawa T, Konno T (2001) Mechanism of tumor-targeted delivery of macromolecular drugs, including the EPR effect in solid tumor and clinical overview of the prototype polymeric drug SMANCS. J Control Release 74:47–61
8. Cook KM, Figg WD (2010) Angiogenesis inhibitors: current strategies and future prospects. CA Cancer J Clin 60:222–243
9. Tammela T, Alitalo K (2010) Lymphangiogenesis: molecular mechanisms and future promise, cell 140 (2010) 460–476. Cell 140:460–476
10. Alitalo K, Tammela T, TVP P (2005) Lymphangiogenesis in development and human disease. Nature 438:946–953
11. Martin TA, Ye L, Sanders AJ, Lane J, Jiang WG (2000) Cancer invasion and Metastasis: molecular and cellular perspective. Madame Curie Biosci Database:1–34
12. Yilmaz M, Christofori G, Lehembre F (2007) Distinct mechanisms of tumor invasion and metastasis. Trends Mol Med 13(12):535–541
13. Van Zijl F, Krupitza G, Mikulits W (2011) Initial steps of metastasis: cell invasion and endothelial transmigration. Mutat Res Rev Mutat Res 728(1–2):23–34
14. Sabeh F, Shimizu-Hirota R, Weiss SJ (2009) Protease-dependent versus -independent cancer cell invasion programs: three-dimensional amoeboid movement revisited. J Cell Biol 185:11–19
15. Christofori G (2006) New signals from the invasive front. Nature 441:444–450
16. Thiery JP (2002) Epithelial–mesenchymal transitions in tumour progression. Nat Rev Cancer 2:442–454
17. Jechlinger M, Grunert S, Tamir IH, Janda E, Ludemann S, Waerner T, Seither P, Weith A, Beug H, Kraut N (2003) Expression profiling of epithelial plasticity in tumor progression. Oncogene 22:7155–7169
18. Massague J (2008) TGFbeta in cancer. Cell 134:215–230
19. Huber MA, Kraut N, Beug H (2005) Molecular requirements for epithelial–mesenchymal transition during tumor progression. Curr Opin Cell Biol 17:548–558
20. Huber MA, Azoitei N, Baumann B, Grunert S, Sommer A, Pehamberger H, Kraut N, Beug H, Wirth T (2004) NF-kappaB is essential for epithelial–mesenchymal transition and metastasis in a model of breast cancer progression. J Clin Invest 114:569–581
21. Geiger TR, Peeper DS (2009) Metastasis mechanisms. Biochim Biophys Acta Rev Cancer 1796(2):293–308
22. Debois JM (2002) TXNXM1: the anatomy and clinics of metastatic cancer. Kluwer Academic Publisher, Dordrecht
23. Schouten LJ, Rutten J, Huveneers HA TA (2002) Incidence of brain metastases in a cohort of patients with carcinoma of the breast, colon, kidney, and lung and melanoma. Cancer 95:2698–2705
24. GR M (2002) Metastasis to bone: causes, consequences and therapeutic opportunities. Nat Rev Cancer 2:584–593

25. Martin TA, Ye L, Sanders A et al (2000) Cancer invasion and metastasis: molecular and cellular perspective. 1–7
26. Birchmeier W BJ (1994) Cadherin expression in carcinomas: role in the formation of cell junctions and the prevention of invasiveness. Biochim Biophys Acta 1198:11–26
27. Trusolino L, Bertotti ACP (2001) A signalling adapter function for alpha6β4 integrin in the control of HGF-dependent invasive growth. Cell 107:643–654
28. Mareel M LA (2003) Clinical, cellular, and molecular aspects of cancer invasion. Physiol Rev 83:337–376
29. Arakaki PA, Marques MR, MCS (2009) MMP-1 polymorphism and its relationship to pathological processes. J Biosci 34(2):313–320
30. Alberts B (2008) Molecular biology of the cell, 5th edn. Garland Science, New York
31. Friedlander DR et al (1994) The neuronal chondroitin sulfate proteoglycan neurocan binds to the neural cell adhesion molecules Ng-CAM/L1/NILE and N-CAM, and inhibits neuronal adhesion and neurite outgrowth. J Cell Biol 125(3):669–680
32. Trowbridge JM, RLG (2002) Dermatan sulfate: new functions from an old glycosaminoglycan. Glycobiology 12(9):117R–125R
33. Bassols A, JM (1988) Transforming growth factor beta regulates the expression and structure of extracellular matrix chondroitin/dermatan sulfate proteoglycans. J Biol Chem 263(6):3039–3045
34. Hacker U, Nybakken K, NP (2005) Heparan sulphate proteoglycans: the sweet side of development. Nat Rev Mol Cell Biol 6(7):530–541
35. Lu P, Weaver VM, Werb Z (2012) The extracellular matrix: a dynamic niche in cancer progression. J Cell Biol 196(4):395–406
36. Lu P, Takai K, Weaver VM, Werb Z (2011) Extracellular matrix degradation and remodeling in development and disease. Cold Spring Harb Perspect Biol 3(12):1–24
37. Pickup MW, Mouw JK, Weaver VM (2014) The extracellular matrix modulates the hallmarks of cancer. EMBO Rep 15(12):1243–1253
38. Gupta GP, Massagu J (2006) Cancer metastasis: building a framework. Cell 127(4):679–695
39. Kalluri R, Weinberg RA (2009) Review series the basics of epithelial-mesenchymal transition. J Clin Invest 119(6):1420–1428
40. Hanahan D, Weinberg RA (2000) The hallmarks of cancer. Cell 100:57–70
41. Yang J, Weinberg R (2008) Epithelial mesenchymal transition: at the crossroads of development and tumor metastasis. Dev Cell 14:818–829
42. Jechlinger M, Grunert S, Beug H (2002) Mechanisms in epithelial plasticity and metastasis: insights from 3D cultures and expression profiling. J Mammary Gland Biol Neoplasia 7:415–432
43. Zeisberg M et al (2003) BMP-7 counteracts TGFbeta1- induced epithelial-to-mesenchymal transition and reverses chronic renal injury. Nat Med 9:964–968
44. Kim K, Lu Z, Hay ED (2002) Direct evidence for a role of beta-catenin/LEF-1 signaling pathway in induction of EMT. Cell Biol Int 26:463–476
45. Eger A, Stockinger A, Schaffhauser B, Beug HA, Foisner R (2000) Epithelial mesenchymal transition by c-Fos estrogen receptor activation involves nuclear translocation of beta-catenin and upregulation of beta-catenin/lymphoid enhancer binding factor-1 transcriptional activity. J Cell Biol 148:173–188
46. Bhowmick NA et al (2001) Transforming growth factor-beta1 mediates epithelial to mesenchymal transdifferentiation through a RhoA-dependent mechanism. Mol Biol Cell 12(12):27–36
47. Bhowmick NA, Zent R, Ghiassi MM, M, Moses HL (2001) Integrin beta 1 signaling is necessary for transforming growth factor-beta activation of p38MAPK and epithelial plasticity. J Biol Chem 276:46707–46713
48. Watanabe T et al (2001) Molecular predictors of survival after adjuvant chemotherapy for colon cancer. N Engl J Med 344:1196–1206

49. Neil JR, Johnson KM, Nemenoff RA, Schiemann WP (2008) Cox-2 inactivates Smad signaling and enhances EMT stimulated by TGF-beta through a PGE2-dependent mechanisms. Carcinogenesis 29:2227–2235
50. Tepass U, Truong K, Godt D, Ikura MA, Peifer M (2000) Cadherins in embryonic and neural morphogenesis. Nat Rev Mol Cell Biol 1:91–100
51. Muta H et al (1996) E-cadherin gene mutations in signet ring cell carcinoma of the stomach. Jpn J Cancer Res 87:843–848
52. Medici D, Hay ED, Olsen BR (2008) Snail and slug promote epithelial-mesenchymal transition through beta-catenin-T-cell factor-4-dependent expression of transforming growth factorbeta3. Mol Biol Cell 19:4875–4887
53. Blanco MJ et al (2002) Correlation of snail expression with histological grade and lymph node status in breast carcinomas. Oncogene 21:3241–3246
54. Yokoyama K et al (2001) Reverse correlation of Ecadherin and snail expression in oral squamous cell carcinoma cells in vitro. Oral Oncol 37:65–71
55. Radisky DC et al (2005) Rac1b and reactive oxygen species mediate MMP-3-induced EMT and genomic instability. Nature 436:123–127
56. Korpal M, Lee ES, Hu G, Kang Y (2008) The miR-200 family inhibits epithelial-mesenchymal transition and cancer cell migration by direct targeting of E-cadherin transcriptional repressors ZEB1 and ZEB2. J Biol Chem 283:14910–14914
57. Gregory PA et al (2008) The miR-200 family and miR-205 regulate epithelial to mesenchymal transition by targeting ZEB1 and SIP1. Nat Cell Biol 10:593–601
58. Park SM, Gaur AB, Lengyel E, Peter ME (2008) The miR-200 family determines the epithelial phenotype of cancer cells by targeting the E-cadherin repressors ZEB1 and ZEB2. Genes Dev 22:894–907
59. Zavadil J, Narasimhan M, Blumenberg MA, Schneider RJ (2007) Transforming growth factorbeta and microRNA: mRNA regulatory networks in epithelial plasticity. Cells Tissues Organs 185:157–161
60. Thiery JP, Acloque H, Huang RYJ, Nieto MA (2009) Epithelial-mesenchymal transitions in development and disease. Cell 139(5):871–890
61. Weinberg RA (2006) The biology of cancer. Garland Science, New York
62. Van Lint PLC (2007) Chemokine and cytokine processing by matrix metalloproteinases and its effect on leukocyte migration and inflammation. J Leukoc Biol 82:1375–1381
63. Brew KNH (2010) The tissue inhibitors of metalloproteinases (TIMPs): an ancient family with structural and functional diversity. Biochim Biophys Acta 1803:55–71
64. Johnson LL, Dyer RHD (1998) Matrix metalloproteinases. Curr Opin Chem Biol 2:466–471
65. Massova I, Kotra LP, Fridman RMS (1998) Matrix metalloproteinases: structures, evolution, and diversification. FASEB J 12:1075–1095
66. Sternlicht MDWZ (2001) How matrix metalloproteinases regulate cell behavior. Annu Rev Cell Dev Biol 17:463–516
67. Deryugina EI, Ratnikov B, Monosov E, Postnova TI, DiScipio R et al (2001) MT1-MMP initiates activation of pro-MMP-2 and integrin alphavbeta3 promotes maturation of MMP-2 in breast carcinoma cells. Exp Cell Res 263:209–223
68. Rodriguez-Manzaneque JC, Lane TF, Ortega MA, Hynes RO, Lawler J et al (2001) Thrombospondin-1 suppresses spontaneous tumor growth and inhibits activation of matrix metalloproteinase-9 and mobilization of vascular endothelial growth factor. Proc Natl Acad Sci U S A 98:12485–12490
69. DR E (2001) Matrix metalloproteinase inhibitors in cancer therapy. In: Clende. Humana Press, Totowa
70. Giannelli G, Falk-Marzillier J, Schiraldi O, Stetler-Stevenson WG Q V (1997) Induction of cell migration by matrix metalloprotease-2 cleavage of laminin-5. Science (80–)277:225–228
71. Xu J, Rodriguez D, Petitclerc E, Kim JJ, Hangai M et al (2001) Proteolytic exposure of a cryptic site within collagen type IV is required for angiogenesis and tumor growth in vivo. J Cell Biol 154:1069–1079

72. Mañes S, Mira E, Barbacid MM, Ciprés A, Fernández-Resa P et al (1997) Identification of insulin-like growth factor-binding protein-1 as a potential physiological substrate for human stromelysin-3. J Biol Chem 272:25706–25712
73. Whitelock JM, Murdoch AD, Iozzo RVUP (1996) The degradation of human endothelial cell-derived perlecan and release of bound basic fibroblast growth factor by stromelysin, collagenase, plasmin, and heparanases. J Biol Chem 271:10079–10086
74. Yu QSI (2000) Cell surface-localized matrix metalloproteinase-9 proteolytically activates TGF-beta and promotes tumor invasion and angiogenesis. Genes Dev 14:163–176
75. Noë V, Fingleton B, Jacobs K, Crawford HC, Vermeulen S et al (2001) Release of an invasion promoter E-cadherin fragment by matrilysin and stromelysin-1. J Cell Sci 114:111–118
76. Kajita M, Itoh Y, Chiba T, Mori HOA (2001) Membrane-type 1 matrix metalloproteinase cleaves CD44 and promotes cell migration. J Cell Biol 153:893–904
77. Hadler-Olsen E, Winberg JO, Uhlin-Hansen L (2013) Matrix metalloproteinases in cancer: their value as diagnostic and prognostic markers and therapeutic targets. Tumor Biol 34(4):2041–2051
78. Coussens LM, Fingleton B, Matrisian LM (2002) Matrix metalloproteinase inhibitors and cancer–trials and tribulations. Science 295(80):2387–2392
79. Anfinsen CB Christian B (1995) Adv Prot Chem 47. (Academic Press)
80. Nagase H, Visse R, Murphy G (2006) Structure and function of matrix metalloproteinases and TIMPs. Cardiovasc Res 69(3):562–573
81. Davies B, Mi1es DW, Happerfield LC, Naylor MS, Bobrow LG, Rubens RD, Balkwill FR (1993) Activity of type IV collagenases in benign and malignant breast disease. Brit J Cancer 67:1126–1131
82. Davies B, Waxman J, Wasan H et al (1993) Levels of matrix metalloproteases in bladder cancer correlate with tumor grade and invasion. Cancer Res 53:5365–5369
83. Jung K, Nowak L, Lein M, Priem F, Schnorr D, Loening SA (1997) Matrix metalloproteinases I and 3, tissue inhibitor of metalloproteinase-l and the comp1ex of metalloproteinase- I tissue inhibitor in plasma of patients with prostate cancer. Int J Cancer 74:220–223
84. Halaka A, Bunning R, Bird C, Gibson M, Reynolds J (1983) Production of collagenase and inhibitor (TIMP) by intracranial tumors and dura in vitro. J Neurosurg 59:444–461
85. Boudreau N, Sympson C, Werb Z, Bissell M (1995) Suppression ofiCE and apoptosis in mammary epithelial cells by extracellular matrix. Seience 267:891–893
86. Alexander CM, Howard E w, BisselI MJ, Werb Z (1996) Rescue of mammary epithelial cell apoptosis and entactin degradation by a tissue inhibitor of metalloproteinase- l transgene. J Cell Biol 135:1667–1677
87. Arribas J, Coodly L, Vollmer P, Kishimoto TK, Rosejohn S, Massagué J (1996) Diverse cell surface protein ectodomains are shed by a system sensitive to metalloprotease inhibitors. J Biol Chem 271:11376–11382
88. Levi E, Fridman R, Miao H, Ma Y, Yayon A, Vlodavsky I et al (1996) Matrix metalloproteinase 2 releases active soluble ectodomain of fibroblast growth factor receptor. Proc Natl Acad Sci U S A 93:7069–7074
89. Suzuki M, Raab G, Moses M, Femandez C, Klagsbrun M (1997) Matrix metalloproteinase- 3 releases active heparin-binding EGF-like growth factor by cleavage at a specific juxtamembrane site. J Bio Chem 272:31730–31737
90. Gearing AJ, Beckett P, Christodoulou M, Churchill M, Clements JDAH et al (1994) Processing of tumour necrosis factor-a precursor by metalloproteinases. Nature 370:555–557
91. Thrailkill KM, QuarIes LD, Nagase H, Suzuki K, Serra DM, Fowlkes JL (1995) Characterization of insulin-like growth factor-binding protein 5-degrading proteases produced throughout murine osteoblast differentiation. Endocrinology 136:3527–3533
92. Rajah R, Nunn SE, Herrick DJ, Grunstein MM, Cohen P (1996) Leukotriene d-4 induces MMP-l, which functions as an IGFBP protease in human airway smooth muscle cells. Am J Phys 15:1014–L1022

93. Vlodavsky I, Korner G, Ishai-Michaeli R, Bashkin P, Bar-Shavit R, Fuks Z (1990) Extracellular matrix-resident growth factors and enzymes: possible involvement in tumor metastasis and angiogenesis. Cancer Metast Rev 9:203–226
94. Imai K, Hiramatsu A, Fukushima D, Pierschbacher MD, Okada Y (1997) Degradation of decorin by matrix metalloproteinases: identification of the cleavage sites, kinetic analyses and transforming growth factor-betal release. Biochem J 322:809–814
95. Frisch SM, Francis H (1994) Disruption of epithelial cell-matrix interactions induces apoptosis. J Cell Biol 124:619–626
96. StCroix B, Sheehan C, Rak J, Florenes V, Slingerland J, Kerbel R (1998) E-cadherin-dependent growth suppression is mediated by the cyclin-dependent kinase inhibitor p27Kip1. J Cell Biol 142:557–571
97. Mareei M, Kint J, Meyvisch C (1979) Methods of study of the invasion of malignant C3H mouse fibroblasts into embryonic chick heart in vitro. Virchows Arch B 30:95–111
98. Thorgeirsson U, Liotta L, Kalebic T, Margulies I, Thomas K, Rios-Candelore ME et al (1982) Effect of natural protease inhibitors and a chemoattractant on tumor cell invasion in vitro. J Natl Cancer Inst 69:1049–1054
99. Repesh L (1989) A new in vitro assay for quantitating tumor ceH invasion. Invas Metast 9:192–208
100. Hendrix M, Seftor E, Seftor R, Fidler I (1987) A simple quantitative assay for studying the invasive potential of high and low metastatic variants. Cancer Lett 38:137–147
101. Liotta LA, Tryggvason K, Garbisa S, Hart I, Foltz CM, Shafie S (1980) Metastatic potential correlates with enzymatic degradation of basement membrane collagen. Nature 284:67–68
102. Schultz, R., Silberman, S., Persky, B., Bajkowski, A., and Carmichael D (1988) Inhibition by human recombinant tissue inhibitor of metalloproteinases of human amnion invasion and lung colonization by murine BI6-FlO melanoma cells. Cancer Res 48:5539–5545
103. Powell W c, Knox JD, Navre M, Grogan TM, Kittelson J, Nagle RB et al (1993) Expression of the metalloproteinase matrilysin in DU-145 cells increases their invasive potential in severe combined immunodeficient mice. Cancer Res 53:417–422
104. Deryugina E, Luo G, Reisfeld R, Bourdon M, Strongin A (1997) Tumor cell invasion through matrigel is regulated by activated matrix metalloproteinase-2. Anticancer Res 17:3201–3210
105. Gomez D, Alonso D, Yoshiji H, Thorgeirsson U (1997) Tissue inhibitors of metalloproteinases: structure, regulation and biological function. Eur J Cell Biol 74:111–122
106. Chirivi RGS, Garofalo A, Crimmin MJ, Bawden LJ, Stoppacciaro AB, PD et al (1994) Inhibition of the metastatic spread and growth of BI6-BL6 murine melanoma by a synthetic matrix metalloproteinase inhibitor. Int J Cancer 58:460–464
107. Davies B, Brown PD, East N, Crimmin MJ, Balkwill FR (1993) A synthetic matrix metalloproteinase inhibitor decreases tumor burden and prolongs survival of mice bearing human ovarian carcinoma xenografts. Cancer Res 53:2087–2091
108. An ZL, Wang XE, Willmott N, Chander SK, Tickle S, Docherty AJP et al (1997) Conversion of highly malignant colon cancer from an aggressive to a controlled disease by oral administration of a metalloproteinase inhibitor. Clin Exp Metastas 15:184–195
109. Stamenkovic I (2003) Extracellular matrix remodelling: the role of matrix metalloproteinases. J Pathol 200:448–464
110. Lin CY et al (2011) Matrix metalloproteinase-9 cooperates with transcription factor snail to induce epithelial-mesenchymal transition. Cancer Sci 102(4):815–827
111. Gavrilovic J, Moens GTJ et al (1990) Expression of transfected transforming growth factor alpha induces a motile fibroblast-like phenotype with extracellular matrix-degrading potential in a rat bladder carcinoma cell line. Cell Regul 1(13):1003–1014
112. Gilles C, Polette MCC et al (2001) Contribution of MT1-MMP and of human laminin-5 gamma2 chain degradation to mammary epithelial cell migration. J Cell Sci 114(16):2967–2976
113. Buisson AC, Zahm JMPM et al (1996) Gelatinase B is involved in the in vitro wound repair of human respiratory epithelium. J Cell Physiol 166(2):413–426

114. Legrand C, Gilles CZJ et al (1999) Airway epithelial cell migration dynamics. MMP-9 role in cell-extracellular matrix remodeling. J Cell Biol 146(2):517–529
115. Daja MM, Niu XZZ et al (2003) Characterization of expression of matrix metalloproteinases and tissue inhibitors of metalloproteinases in prostate cancer cell lines. Prostate Cancer Prostatic Dis 6(1):15–26
116. Gilles C, Polette MPJ et al (1994) Epithelial-to-mesenchymal transition in HPV-33-transfected cervical keratinocytes is associated with increased invasiveness and expression of gelatinase A. Int J Cancer 59(5):661–666
117. Yokoyama K, Kamata NFR et al (2003) Increased invasion and matrix metalloproteinase-2 expression by Snailinduced mesenchymal transition in squamous cell carcinomas. Int J Oncol 22(4):891–898
118. Hurst NG, Stocken DD, Wilson S, Keh C, Wakelam MJIT (2007) Elevated serum matrix metalloproteinase 9 (MMP-9) concentration predicts the presence of colorectal neoplasia in symptomatic patients. Br J Cancer 97(7):971–977
119. Wilson S, Wakelam MJ, Hobbs RF, Ryan AV, Dunn JA R, VD et al. (2006) Evaluation of the accuracy of serum MMP-9 as a test for colorectal cancer in a primary care population. BMC Cancer 6(258)
120. Eissa S, Ali-Labib R, Swellam M, Bassiony M, Tash FE-Z, TM. (2007) Noninvasive diagnosis of bladder cancer by detection of matrix metalloproteinases (MMP-2 and MMP-9) and their inhibitor (TIMP-2) in urine. Eur Urol 52(5):1388–1396
121. Roy R, Louis G, Loughlin KR, Wiederschain D, Kilroy SM L, CC et al (2008) Tumor-specific urinary matrix metalloproteinase fingerprinting: identification of high molecular weight urinary matrix metalloproteinase species. Clin Cancer Res 14(20):6610–6617
122. Cribier B, Noacco G, Peltre BGE (2002) Stromelysin 3 expression: a useful marker for the differential diagnosis dermatofibroma versus dermatofibrosarcoma protuberans. J Am Acad Dermatol 46(3):408–413
123. Kim HJ, Lee JY, Kim SH, Seo YJ, Lee JH, Park JK et al (2007) Stromelysin-3 expression in the differential diagnosis of dermatofibroma and dermatofibrosarcoma protuberans: comparison with factor XIIIa and CD34. Br J Dermatol 157(2):319–324
124. Morrison CJ, Butler GS, Rodriguez DOC (2009) Matrix metalloproteinase proteomics: substrates, targets, and therapy. Curr Opin Cell Biol 21(5):645–653
125. Terp GE, Cruciani G, Christensen ITJF (2002) Structural differences of matrix metalloproteinases with potential implications for inhibitor selectivity examined by the GRID/CPCA approach. J Med Chem 45(13):2675–2684
126. Cuniasse P, Devel L, Makaritis A, Beau F, Georgiadis D M, M et al. (2005) Future challenges facing the development of specific active-site-directed synthetic inhibitors of MMPs. Biochimie 87(3–4):393–402.
127. Chung L, Shimokawa K, Dinakarpandian D, Grams F FG, H N (2000) Identification of the (183)RWTNNFREY(191) region as a critical segment of matrix metalloproteinase 1 for the expression of collagenolytic activity. J Biol Chem 275(38):29610–29617
128. Chung L, Dinakarpandian D, Yoshida N, Lauer-Fields JL F, GB Visse R et al (2004) Collagenase unwinds triple-helical collagen prior to peptide bond hydrolysis. EMBO J 23(15):3020–3030
129. Manka SW, Carafoli F, Visse R, Bihan D, Raynal N F, RW et al (2012) Structural insights into triple-helical collagen cleavage by matrix metalloproteinase 1. Proc Natl Acad Sci U S A 109(31):12461–12466
130. Kousidou OC, Mitropoulou TN, Roussidis AEKD, Theocharis ADKN (2005) Genistein suppresses the invasive potential of human breast cancer cells through transcriptional regulation of metalloproteinases and their tissue inhibitors. Int J Oncol 26(4):1101–1109
131. Mitropoulou TN, Tzanakakis GN, Kletsas D KH, NK K (2003) Letrozole as a potent inhibitor of cell proliferation and expression of metalloproteinases (MMP-2 and MMP-9) by human epithelial breast cancer cells. Int J Cancer J Int du Cancer 104(2):155–160
132. Hua J MR (1996) Inhibition of matrix metalloproteinase 9 expression by a ribozyme blocks metastasis in a rat sarcoma model system. Cancer Res 56:5279–5284 56:5279–5284

133. Yonemura Y, Endo Y, Fujita H, Kimura K, Sugiyama K et al (2001) Inhibition of peritoneal dissemination in human gastric cancer by MMP-7-specific antisense oligonucleotide. J Exp Clin Cancer Res 20:205–212
134. Kondraganti S, Mohanam S, Chintala SK, Kin Y, Jasti SL et al. (2000) Selective suppression of matrix metalloproteinase-9 in human glioblastoma cells by antisense gene transfer impairs glioblastoma cell invasion. Cancer Res 60:6851–6855, 60:6851–6868
135. Elkin M, Reich R, Nagler A, Aingorn E, Pines M et al (1999) Inhibition of matrix metalloproteinase-2 expression and bladder carcinoma metastasis by halofuginone. Clin Cancer Res 5:1982–1988, 5:1982–1988
136. Silletti S, Kessler T, Goldberg J, Boger DL C DA (2001) Disruption of matrix metalloproteinase 2 binding to integrin alpha vbeta 3 by an organic molecule inhibits angiogenesis and tumor growth in vivo. Proc Natl Acad Sci U S A 98:119–124
137. Liu S, Netzel-Arnett S, Birkedal-Hansen HLS (2000) Tumor cell-selective cytotoxicity of matrix metalloproteinase-activated anthrax toxin. Cancer Res 60:6061–6067
138. Hidalgo MES (2001) Development of matrix metalloproteinase inhibitors in cancer therapy. J Natl Cancer Inst 93:178–193
139. Koivunen E, Arap W, Valtanen H, Rainisalo A, Medina OP et al (1999) Tumor targeting with a selective gelatinase inhibitor. Nat Biotechnol 17:768–774
140. Falardeau P, Champagne P, Poyet P, Hariton C DE (2001) Neovastat, a naturally occurring multifunctional antiangiogenic drug, in phase III clinical trials. Semin Oncol 28:620–625
141. Garbisa S, Biggin S, Cavallarin N, Sartor L, Benelli R et al (1999) Tumor invasion: molecular shears blunted by green tea. Nat Med 5:1216
142. Jiang MC, Liao CFLP (2001) Aspirin inhibits matrix metalloproteinase-2 activity, increases E-cadherin production, and inhibits in vitro invasion of tumor cells. Biochem Biophys Res Commun 282:671–677
143. Ala-aho R, Johansson N, Grénman R, Fusenig NE, López-Otín C et al (2000) Inhibition of collagenase-3 (MMP-13) expression in transformed human keratinocytes by interferon-gamma is associated with activation of extracellular signal-regulated kinase-1,2 and STAT1. Oncogene 19:248–257
144. Ma Z, Qin HBE (2001) Transcriptional suppression of matrix metalloproteinase-9 gene expression by IFN-gamma and IFN-beta: critical role of STAT-1alpha. J Immunol 167:5150–5159
145. Slaton JW, Karashima T, Perrotte P, Inoue K, Kim SJ et al (2001) Treatment with low-dose interferon-alpha restores the balance between matrix metalloproteinase-9 and E-cadherin expression in human transitional cell carcinoma of the bladder. Clin Cancer Res 7:2840–2853
146. Mengshol JA, Mix KSBC (2002) Matrix metalloproteinases as therapeutic targets in arthritic diseases: bull's-eye or missing the mark? Arthritis Rheum 46:13–20
147. Lal A, Glazer CA, Martinson HM, Friedman HS, Archer GE et al (2002) Mutant epidermal growth factor receptor up-regulates molecular effectors of tumor invasion. Cancer Res 62:3335–3339
148. Muraoka RS, Dumont N, Ritter CA, Dugger TC, Brantley DM et al (2002) Blockade of TGF-beta inhibits mammary tumor cell viability, migration, and metastases. J Clin Invest 109:1551–1559
149. McGaha TL, Phelps RG, Spiera HBC (2002) Halofuginone, an inhibitor of type-I collagen synthesis and skin sclerosis, blocks transforming growth- factor-beta-mediated Smad3 activation in fibroblasts. J Invest Dermatol 118:461–470
150. Simon C, Goepfert HBD (1998) Inhibition of the p38 mitogen-activated protein kinase by SB 203580 blocks PMA-induced Mr 92,000 type IV collagenase secretion and in vitro invasion. Cancer Res 58:1135–1139
151. Johansson N, Ala-aho R, Uitto V, Grénman R, Fusenig NE et al (2000) Expression of collagenase-3 (MMP-13) and collagenase-1 (MMP-1) by transformed keratinocytes is dependent on the activity of p38 mitogen-activated protein kinase. J Cell Sci 113:227–235

152. Futamura M, Kamiya S, Tsukamoto M, Hirano A, Monden Y et al (2001) Malolactomycin D, a potent inhibitor of transcription controlled by the Ras responsive element, inhibits Ras-mediated transformation activity with suppression of MMP-1 and MMP-9 in NIH3T3 cells. Oncogene 20:6724–6730
153. Zhang Y, Thant AA, Machida K, Ichigotani Y, Naito Y et al (2002) Hyaluronan-CD44s signaling regulates matrix metalloproteinase-2 secretion in a human lung carcinoma cell line QG90. Cancer Res 62:3962–3965
154. Karin MCL (2001) AP-1--glucocorticoid receptor crosstalk taken to a higher level. J Endocrinol 169:447–451
155. Al. ST et al (2002) Inhibition of activator protein-1 binding activity and phosphatidylinositol 3-kinase pathway by nobiletin, a polymethoxy flavonoid, results in augmentation of tissue inhibitor of metalloproteinases-1 production and suppression of production of matrix meta. Cancer Res 62:1025–1029
156. Mohan R, Sivak J, Ashton P, Russo LA, Pham BQ et al (2000) Curcuminoids inhibit the angiogenic response stimulated by fibroblast growth factor-2, including expression of matrix metalloproteinase gelatinase B. J Biol Chem 275:10405–10412
157. Adams J, Palombella VJ, Sausville EA, Johnson J, Destree A et al (1999) Proteasome inhibitors: a novel class of potent and effective antitumor agents. Cancer Res 59(59):2615–2622
158. Ala-aho R, Grénman R, Seth PKV (2002) Adenoviral delivery of p53 gene suppresses expression of collagenase-3 (MMP-13) in squamous carcinoma cells. Oncogene 21:1187–1195
159. Annabi B, Lachambre MP, Bousquet-Gagnon N, Page M, Gingras D et al (2002) Green tea polyphenol (−)-epigallocatechin 3-gallate inhibits MMP-2 secretion and MT1-MMP-driven migration in glioblastoma cells. Biochim Biophys Acta 1542:209–220
160. Bassi DE, Lopez De Cicco R, Mahloogi H, Zucker S, Thomas G et al (2001) Furin inhibition results in absent or decreased invasiveness and tumorigenicity of human cancer cells. Proc Natl Acad Sci U S A 98:10326–10331
161. Khatib AM, Siegfried G, Chrétien M, Metrakos PSN (2002) Proprotein convertases in tumor progression and malignancy: novel targets in cancer therapy. Am J Pathol 160:1921–1935
162. Huang X, Chen S, Xu L, Liu Y, Deb DK et al (2005) Genistein inhibits p38 map kinase activation, matrix metalloproteinase type 2, and cell invasion in human prostate epithelial cells. Cancer Res 65:3470–3478
163. Wang S, Cheng Y, Wang F, Sun L, Liu C et al (2008) Inhibition activity of sulfated polysaccharide of Sepiella maindroni ink on matrix metalloproteinase (MMP)-2. Biomed Pharmacother 62:297–302
164. Berton A, Rigot V, Huet E, Decarme M, Eeckhout Y et al (2001) Involvement of fibronectin type II repeats in the efficient inhibition of gelatinases a and B by long-chain unsaturated fatty acids. J Biol Chem 276:20458–20465
165. Huxley-Jones J, Clarke TK, Beck C, Toubaris G, Robertson DL et al (2007) The evolution of the vertebrate metzincins; insights from Ciona intestinalis and Danio rerio. BMC Evol Biol 7:63
166. Fujita M, Nakao Y, Matsunaga S, Seiki M, Itoh Y et al (2003) Ageladine a: an antiangiogenic matrix metalloproteinase inhibitor from the marine sponge Agelas Nakamurai. J Am Chem Soc 125:15700–15701
167. Lelièvre Y, Bouboutou R, Boiziau JCT (1989) Inhibition of synovial collagenase by actinonin. Study of structure/activity relationship. Pathol Biol 37:43–46

Proteases and Protease Inhibitors in Male Reproduction

10

V.S. Gurupriya and Sudhir C. Roy

Abstract

From the development of the spermatozoa within the seminiferous tubule of testis to the fertilization events that occur in female reproductive tract, all the reproductive processes in mammals are regulated by a highly orchestrated and integrative mechanism. Several proteases and their protease inhibitors form an important part of this mechanism. So far, vast arrays of proteases have been identified in the reproductive system of mammals playing critical role in the major events associated with the several male reproductive processes. Several endogenous inhibitors of these proteases are also produced in the male reproductive tissues/fluids that cater to the role of regulating the protease production/degradation, activation/inactivation, etc. Thus, there exist a fine balance between the production of these proteases and their regulators for maintaining the blood-testes barrier and the gamete development. A disturbance in this equilibrium leads to progression of reproductive failures including azoospermia, impaired sperm functions, low fertilizing efficiency, etc. and culminates in infertility cases. This chapter focuses on an account of such proteases and the protease inhibitors with their role in mammalian male reproduction.

Keywords

Proteases • Protease inhibitors • Male • Reproduction • Fertility • Infertility

V.S. Gurupriya • S.C. Roy (✉)
Molecular Biology Laboratory, ICAR-National Institute of Animal Nutrition & Physiology, Hosur Road, Adugodi P.O, Banglaore 560030, India
e-mail: scroy67@gmail.com

S. Chakraborti, N.S. Dhalla (eds.), *Proteases in Physiology and Pathology*,
DOI 10.1007/978-981-10-2513-6_10

10.1 Introduction

The productive function of mammalian male reproductive system is dependent on the outcome of an integrated network of reproductive hormones, nervous system, cytokines, growth factors, and several other known and unknown biomolecules. The system acts in a coordinated way to produce, store, and transport spermatozoa and seminal plasma and finally to release the semen into the female reproductive tract where it can fertilize the ova to produce an offspring. Before fertilization, mammalian spermatozoa must undergo a defined physiological process known as "capacitation" in the female reproductive tract, which is associated with hyperactivated sperm motility and protein tyrosine phosphorylation. During acrosome reaction, the release of acrosomal enzymes including proteases at proper time and place aids in sperm penetration to the zona pellucida and fusion with the oocyte plasma membrane. Proteases play important roles during the initial stages of testes development, spermatogenesis, and epididymal maturation until the fertilization events that occur in female reproductive tract. An array of proteases including serine proteases, adamalysin-related proteinases (ADAMs), matrix metalloproteinases (MMPs), and protease inhibitors such as serine protease inhibitors (SERPINs), serine protease inhibitors Kazal type (SPINKs), Kazal inhibitors, and tissue inhibitor of metalloproteinases (TIMPs) are found to play important roles in the most of these male reproductive events. However, in an organism, a control of proteolysis by the proteases is accomplished through a balance of production, degradation, and inactivation of proteases, via interaction with endogenous protease inhibitors with the intention that protease action should be precise in terms of time and place. In addition to the protease inhibitors, the expressions of proteases are also regulated by nervous system, sex hormones, and growth factors accountable for maintaining the male reproductive system. When one of these mechanisms of regulation fails, it can result in the onset or progression of reproductive failures that may end up with infertility. Hence, a proper understanding of the protease/protease inhibitor system in mammalian male reproduction may pave the way to counterbalance the disturbance in equilibrium of the proteases and their regulators that predispose to male reproductive disorders.

10.2 Proteases and Protease Inhibitors in Mammals

Proteases are proteolytic enzymes catalyzing cleavage or hydrolysis of peptide bonds in proteins through a mechanism called proteolysis. Proteolysis is the key mechanism for controlling the activity of many proteins including shedding of cell surface protein domains; activation or inactivation of cytokines, hormones, and growth factors; exposure of hidden protein domains for exhibiting functional roles entirely different from the parent molecule; and degradation of multiple extracellular matrix (ECM) components of basement membrane facilitating cell migration, invasion, etc. [1]. As illustrated in Fig. 10.1, proteases are categorized based on their hydrolysis mechanism as exopeptidases and endopeptidases. The exopeptidases

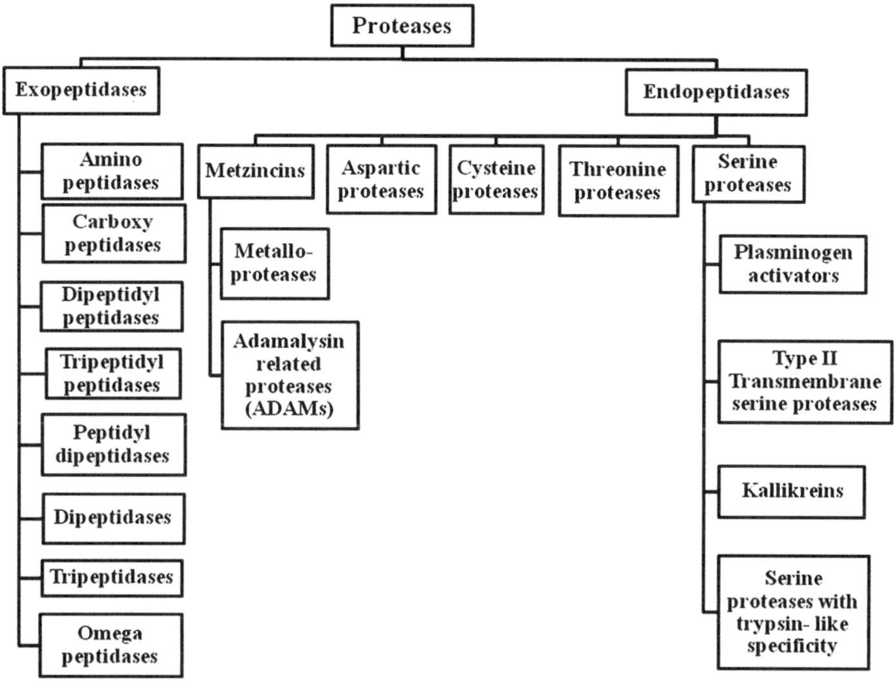

Fig. 10.1 Classification of proteases: The proteases are classified based on the hydrolysis of peptide bonds in proteins as exopeptidases and endopeptidases. The exopeptidases are further classified based on the identity of the liberated fragment as aminopeptidases, carboxypeptidases, dipeptidyl peptidases, tripeptidyl peptidases, peptidyl dipeptidases, dipeptidases, tripeptidases, omega peptidases, etc. and the endopeptidases based on the catalytic site present as serine proteases, metzincins, threonine, cysteine, and aspartic proteases

attack only the peptide bonds confined at/or close to the amino- or carboxy-terminal portion of peptide chains. The endopeptidases attack the internal peptide bonds in the polypeptides. The exopeptidases can be classified further based on the size or identity of the liberated fragment as aminopeptidases, carboxypeptidases, dipeptidyl peptidases, tripeptidyl peptidases, peptidyl dipeptidases, dipeptidases, tripeptidases, omega peptidases, etc. Dipeptidyl-peptidase IV purified from porcine seminal plasma and carboxypeptidase C identified from human seminal plasma are examples for exopeptidases.

The endoproteases are divided into five subclasses depending on their catalytic sites as metzincins, aspartic, cysteine, threonine, and serine proteases (Fig. 10.1). In metzincins, zinc is present at their catalytic sites. They also exploit an activated water molecule to attack the peptide bond in the substrate. Metzincins are subdivided into four distinct families: matrixins, adamalysins, astacins, and serralysins. Matrixins and adamalysin-related proteases play fundamental role in many physiological processes. Matrix metalloproteinases (MMPs) or matrixins are extracellular matrix (ECM)-digesting enzymes and are Ca^{2+}- and Zn^{2+}-dependent endopeptidases active at neutral pH. They use a metal ion to polarize a water molecule to hydrolyze the peptide bond in the substrate [2]. MMPs are secreted as latent forms which can be activated by chaotropic agents or by cleavage of the inhibitory propeptide by MMP family of proteases or the plasminogen activator of the urokinase type. The active MMPs have a relative molecular mass of about 10 kDa less than the latent/pro-forms. MMPs are also involved in the release and activation of growth factors and cytokines [3]. MMPs can be broadly classified into four groups: (i) collagenases that are active against native collagen, (ii) gelatinases that have high activity against gelatin and denatured and type IV collagens, (iii) stromelysins that degrade noncollagen ECM components, and (iv) membrane-type MMPs (MT-MMPs) that are transmembrane molecules mainly cleaving ECM components at the same time activating other MMPs also [4]. Adamalysin-related proteinases also known as ADAMs (a disintegrin and metalloprotinease) contain a disintegrin domain having specific role in cell adhesion and proteolytic processing. In mice and human, respectively, at least 34 and 26 ADAM genes have been identified so far [5]. Unlike mammals, avian genome lacks ADAM1–7 and ADAM30 genes [6]. ADAMTS (a disintegrin and metalloproteinase with thrombospondin type 1 motifs) are the soluble counterparts of the ADAMs. They contain thrombospondin type 1 motifs that help in ECM association, inflammation, angiogenesis, etc.

In aspartic proteases, an aspartate is present at the catalytic site to attack the peptide bond linkage in the substrate. The digestive proteases like pepsin and cathepsin H of the spermatozoa are some of the examples of aspartic proteases. Cysteine proteases have at their catalytic sites Cys molecules that act as nucleophiles [7]. Cysteine proteases are common in plants and animals, act as lysosomal enzymes, and exhibit tissue-specific expression for bone growth and lung function [8, 9]. Many of the cathepsins, calpains I and II in seminal fluid belong to the class of cysteine proteases [10, 11]. Threonine proteases are classes of proteases having the catalytic site with Cys, Ser, or Thr to act as a nucleophile [7].

Serine proteases are the most common protease in the both insects and mammals [12]. Serine proteases have a conserved catalytic triad of a His, Ser, and Asp to coordinate a water molecule. Serine protease family can be subdivided into 16 subfamilies including plasminogen activators (PAs), type II transmembrane serine proteases (TTSPs), kallikreins, and serine protease with trypsin-like specificity [7]. Plasminogen activators (PA) are trypsin-like proteases that help in the conversion of plasminogen to plasmin and digestion of fibronectin, laminin, vitronectin, etc. [13]. They are physiological activators of pro-/latent metalloproteases (MMPs) for collagen degradation [14]. Two classes of PAs are known in mammals: the tissue-type plasminogen activator (t-PA) and the urokinase-type plasminogen activator (u-PA). Both catalyze the activation of plasminogens. The u-PA is associated with the physiological and pathological tissue remodeling, whereas t-PA is mainly involved in thrombolysis and neurobiology. Type II transmembrane serine proteases (TTSPs) interact with the cell surface and soluble or secreted proteins, cell matrix components, and proteins on surrounding cells. TTSPs are synthesized as zymogens and are activated by cleavage of arginine or lysine present at the highly conserved activation motif and remain as membrane bound after activation. Kallikrein family of

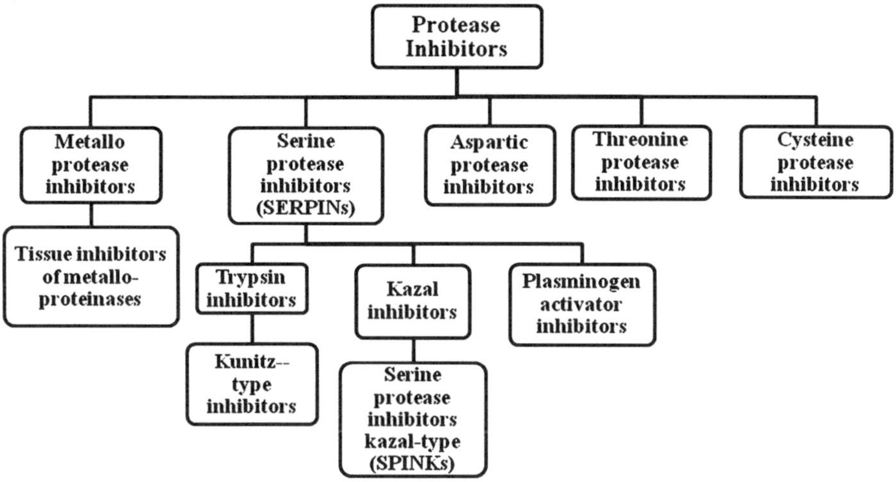

Fig. 10.2 Classification of major protease inhibitors: Protease inhibitors are classified based on the type of protease they inhibit as metalloproteinase inhibitors, serine protease inhibitors, aspartic protease inhibitors, threonine protease inhibitors, and cysteine protease inhibitors. The serine protease inhibitors include major proteases such as trypsin inhibitors, Kazal inhibitors, plasminogen activator inhibitors, etc.

proteases is present in many animal species including human, rat, mouse, etc. and found to express in tissues including the prostate, breast, ovary, and testis. Plasma and tissue kallikreins are the two categories coming under kallikrein family. The kallikreins have a main role in regulation of blood pressure and semen liquefaction [15]. Serine protease with trypsin-like specificity includes acrosin- and testes-specific serine proteases [16]. Acrosin has an enzymatic activity for the limited proteolysis of the zona pellucida and a lectin-like carbohydrate-binding activity for binding of acrosome-reacted sperm to the zona [17, 18]. Testes-specific serine proteases play different roles in spermatogenesis and are required for germ cell survival during meiosis. They also help in sperm-oocyte interaction and penetration of the ZP.

An overactivity of proteases may cause premature activation of pathways in tissues leading to tissue damage. Their action needs to be controlled by protease inhibitors by keeping the proteases in its pro-form or zymogen form so as to maintain the tissue integrity, cell migration, cell signaling, cell surface and tissue remodeling, tissue support, repair, growth and development, etc. [19]. More than 2% of human genes belong to the class of proteases or protease inhibitors [7]. It shows the importance of balanced mechanism of proteolysis in mammalian system. The protease inhibitors are broadly classified into five groups based on the type of protease they inhibit: (i) metalloproteinase inhibitors, (ii) serine protease inhibitors, (iii) aspartic protease inhibitors, (iv) threonine protease inhibitors, and (v) cysteine protease inhibitors (Fig. 10.2). Among these, the metalloproteinase inhibitors and the serine protease inhibitors are most commonly associated with the reproductive functions of mammals. Tissue inhibitors of metalloproteinases (TIMPs) are natural inhibitors of matrix metalloproteinases (MMPs). TIMPs are 21–34 kDa proteins possessing N-terminal MMP inhibitory domain. The C-terminal domain is involved in formation of complexes with the proenzymes, thus regulating the MMP activation process. They are mostly resistant to heat denaturation and proteolytic degradation. They inhibit the proteolytic activity of MMPs up to the gene expression levels by cleavage of the latent forms or by inhibition of active MMPs [4, 20]. The TIMP-1, TIMP-2, and TIMP-4 are secreted, whereas TIMP-3 is ECM associated. TIMPs vary in their solubility characters, interaction with proMMPs, and expression pattern. The serine protease inhibitors, most commonly known as SERPINs, inhibit the activity of many classes of serine proteases including plasminogen activators. SERPINs include the trypsin inhibitors such as Kunitz-type inhibitors, Kazal inhibitors, plasminogen activator inhibitors, etc. The soya bean trypsin inhibitor (STI) of the pancreas and ovomucoid is an example of trypsin inhibitors. Serine protease inhibitors Kazal type are popularly known as SPINKs, and several members of this group, e.g., SPINK1–2, SPINK4–9, SPINK13–14, have been identified from different mammalian species [21–23]. The primary function of SPINK family of protease inhibitors is regulation of serine protease activities to prevent uncontrolled proteolysis [24]. Imbalance between SPINKs and serine proteases causes diseases, such as pancreatitis, skin barrier defects, and cancer [25]. PA inhibitors are generally known as PAIs which includes the clades-A, clades-B, and clades-E.

10.3 Role of Proteases and Protease Inhibitors in Male Reproduction

Proteases play an important role in several processes of male reproduction including initial testes development, spermatogenesis, epididymal sperm maturation, and spermatozoa-associated fertilization events that occur in female reproductive tract. Before fertilization, a mammalian spermatozoon must spend some finite time in the female reproductive tract to undergo a defined physiological and biochemical process known as "capacitation" [26–29]. The capacitation process is characterized by two landmark events such as hyperactivated motility of sperm and protein tyrosine phosphorylation, and these lead to acrosome reaction of spermatozoa before fertilization [30–32]. Acrosome reaction is the fusion of the sperm plasma membrane with the outer acrosomal membrane, uncovering the inner acrosomal membrane resulting in vesiculation and the discharge of acrosomal enzymes including proteases at the proper time and place that can aid in sperm penetration to the zona pellucida and oocyte plasma membrane. Protease activity must be strictly controlled by protease inhibitors by keeping the sperm proteases in its zymogen form to maintain the tissue integrity and reproductive roles of spermatozoa [33]. Equilibrium between proteases and their inhibitors should be maintained for conserving blood-testis barrier integrity and gamete development [34]. A disturbance in this equilibrium leads to progression of reproductive failures including azoospermia, impaired sperm functions, low fertilizing efficiency, etc. that may culminate in infertility cases. The schematic representation of the role of proteases and their inhibitors, the importance

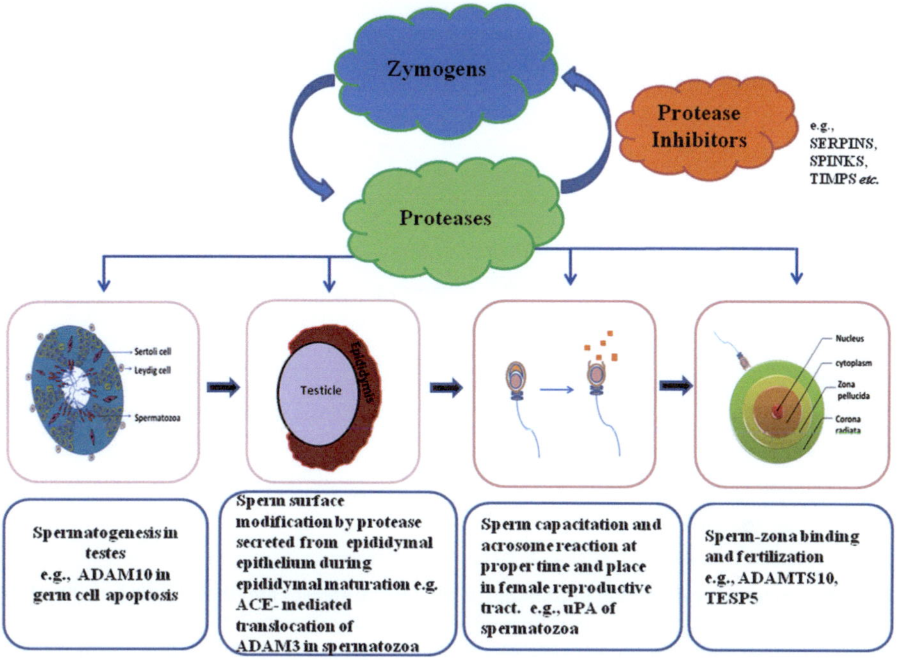

Fig. 10.3 Schematic depiction of the role of proteases and protease inhibitors: Different classes of proteases are involved in several male reproductive processes like spermatogenesis, sperm maturation, capacitation, acrosome reaction, and fertilization. Their action must be strictly controlled by protease inhibitors by keeping the sperm proteases in its pro-form or zymogen form to maintain the tissue integrity and reproductive roles of spermatozoa. Some of the examples of proteases/protease inhibitors involved in various male reproductive processes have been indicated in boxes

of maintaining the equilibrium between proteases, and their regulators are depicted in Fig. 10.3 and Fig. 10.4.

In mammals, MMPs play vital roles in several physiological processes that are associated with male reproduction. MMP-2 and MMP-9 have been demonstrated in mouse fetal testes [35]. In human and mouse testes, the expressions of MMP-18, MMP-23, MMP-26, and MMP-28 were detected [36–39]. In dogs, the expression of metalloproteinase and semen parameters are positively correlated [20]. Latent forms of the matrix metalloproteinases in semen are inversely correlated with sperm quality trait and ejaculate volume in dogs and humans [20, 40]. Both latent and active forms of MMP-9 and MMP-2 are present in canine seminal fluid with predominant expression of latent forms [41]. Expressions of MMP-2, MMP-9, TIMP-1, and TIMP-2 were detected in the testes, epididymis, and ejaculated semen of dogs, and MMP-2 was present in spermatozoa from all the developmental stages as well as in ejaculated semen, and thus MMP-2 could be considered as a marker for predicting the quality of semen [42]. MMP-2 has been detected in the inner acrosomal membrane of bull, human, and mouse spermatozoa [43]. In mouse spermatozoa, MMP-2

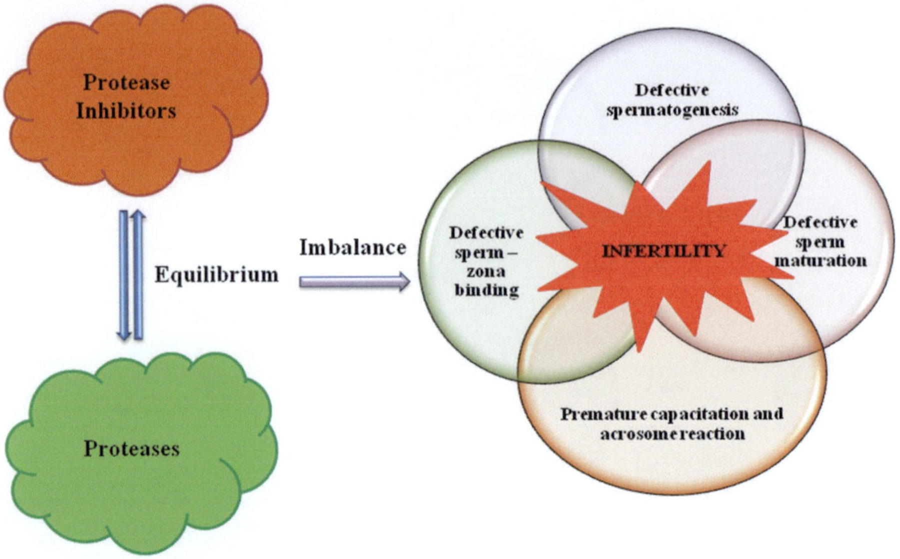

Fig. 10.4 Schematic depiction of imbalance between proteases and their inhibitors: The loss of equilibrium between proteases and their inhibitors may cause premature activation of pathways or tissue damage affecting blood-testes barrier, gamete development, epididymal sperm maturation, sperm-egg binding during fertilization, etc. leading to reproductive disorders that culminate in infertility cases

inhibition demonstrated decreased fertilization in vitro [43]. Using zymography, in ram epididymal fluids, at least five, and in boar and stallion at least seven, gelatinolytic bands were detected [44]. Of these, two (54–66 kDa) gelatinolytic bands were found to be metalloproteinases. In turkey seminal plasma, two metalloproteinases of 58 and 66 kDa were detected [45]. In buffaloes, metalloproteinases of 42, 95, 52, and 33 kDa were detected in spermatozoa, and those of 78, 68, 62, and 98 kDa were detected in seminal plasma [46]. Similarly, in cattle, metalloproteinases of 62, 42, 85, and 78 kDa were detected in spermatozoa, and those of 68, 78, and 75 kDa were detected in seminal plasma [46]. MMP-12 is solely produced by macrophages, and its presence in leucocytospermic human seminal plasma revealed its role in predicting the inflammatory conditions in male genital tract [47].

ADAMTS (a disintegrin and metalloproteinase with thrombospondin type 1 motifs) are zinc-dependent metalloproteinases and lack a transmembrane domain but instead contain a thrombospondin motif that helps in extracellular membrane association, inflammation, angiogenesis, etc. Of 19 ADAMTSs, only 2 ADAMTSs, viz., ADAMTS2 and ADAMTS20, display testis-specific expressions [6]. In human, the levels of ADAMTS1 and ADAMTS5 were negatively correlated with sperm count and motility [48]. In mouse, ADAMTS2 null mutation demonstrated increased spermatogonia and decreased mature spermatozoa in testes [49]. ADAM2 is absent from human spermatozoa but present in testes. It is present in both spermatozoa and testes of mouse and monkey [6]. In mouse, the expression of ADAM31 gene was demonstrated in Leydig cells [50]. ADAMTSs are largely involved in sperm-egg recognition [51]. In mouse, ADAMTS10 is secreted from the testis during the later stages of spermatogenesis and incorporated into the acrosomal domain of developing spermatids [52]. Testicular tissue extracts demonstrated ADAMTS10 of 65 kDa, whereas spermatozoa from different segments of epididymis displayed ADAMTS10 of 50 kDa. During epididymal maturation, this particular protease is processed before being expressed on the surface of the peri-acrosomal region of the spermatids and mature spermatozoa. Like ADAM2, the ADAMTS10 also plays an important role in sperm adhesion to the zona pellucida in mouse and lost following acrosomal exocytosis. The use of galardin, a broad-spectrum inhibitor of metalloproteinases or anti-ADAMTS10 antibody, demonstrated reduced sperm-zona interaction. Expression of ADAMTS12 has been demonstrated in developing testicular chords at day 7 in chicken embryo [53]. In rat, expression of ADAMTS16 has been demonstrated in embryonic and adult testes. ADAMTS16 null rat models showed loss of sperm progenitor cells, azoospermia, cryptorchidism, and infertility [54]. A 70 kDa ADAMTS20 has been characterized from human and mouse testis during prenatal developmental stages [55]. Predominant expression of bovine ADAMTSL3 has been demonstrated in testis compared to other tissues [56].

Of the proteases, metzincins, serine- and cysteine proteases are the most abundant proteolytic enzymes in rat, mouse, and human genome [6]. Plasminogen activators are present in the sperm head and seminal plasma of various mammalian species [57, 58]. They are vital for the drive of spermatocytes across the blood-testes barrier, binding, and penetration of the spermatozoon through the layers surrounding the egg [59]. The prostate glands synthesize the t-PA in human [60]. The Sertoli cells control spermatogenesis by producing PA [61]. The u-PA receptor (u-PAR) synthesized by germ cells is present on the surface of spermatids and mature spermatozoa whereas plasminogen and u-PA in the seminiferous tubules. The PAs from seminal plasma bind to the u-PAR present in the cell membrane of the spermatozoon and enhance local generation of plasmin. This locally generated plasmin-induced proteolysis may be an essential step for stimulating sperm motility toward the lumen of the seminiferous tubule, initiation of sperm capacitation, hyperactivation, and fertilizing capacity [62, 63]. The pachytene and diakinetic spermatocytes express t-PA [64]. The t-PA and u-PA, located on the outer acrosomal membrane and plasma membrane of human spermatozoa, may participate in sperm maturation, capacitation, acrosome reaction, and binding of spermatozoa to the zona pellucida [65]. The spermatozoa stored in the vas deferens retain almost no u-PA activity, whereas the ejaculated spermatozoa have precise u-PA activity on their cell surface, mostly around the head region. Hence, during ejaculation the u-PA secretion from the accessory glands may be activated and may bind to the surface of the spermatozoa via its specific receptors [66]. Moreover, the u-PA is one of the enzymes involved in sperm surface modification during sperm maturation [66, 67]. The u-PA-mediated proteolysis enables spermatozoa to move toward the ampulla of the fallopian tubes and prevents adhesion of spermatozoa to fibrin deposits on the tubal mucosa and promotes capacitation and fertilization at the ampulla of the fallopian tubes. Sertoli cells are mainly considered the main source of t-PA in rodent testis [68]. Urokinase PA is detected in the stereo-ciliated epithelial cells and in the lumen of the vas deferens of mice [66]. There are evidences for urokinase PA expression in the seminal vesicles and epididymis of mice and monkeys and t-PA mRNA in the epididymis of monkeys [59]. Rat and monkey epididymal epithelial cells also secrete both t-PA and u-PA, and the PA activities (PAA) seemed to be diminished toward the cauda epididymis [59]. This confirms the role of the PA in sperm maturation process.

The kallikreins have important role in motility, penetration, and migration of sperm through the cervix and uterus [69]. Leydig cells are the unique site of expression of kallikrein-21, kallikrein-24, and kallikrein-27 in the testis [7, 70]. Kallikrein-like proteases (KLKs) such as KLK-2, KLK-3, KLK-4, KLK-5, KLK-8, KLK-11, KLK-12, KLK-14, and KLK-15 are also secreted by the prostate and help in semen liquefaction [71]. KLK-14 has been recognized as a potential activator of KLK-3 (also known as prostate-specific antigen, PSA) from prostatic fluid. Further, its ability to degrade seminogelin I and II from seminal vesicle causes release of trapped spermatozoa from the seminal clot/semen coagulum thereby allow sperm motility and capacitation [72]. The semen liquefaction is activated within 5–20 min post-ejaculation and regulated by the action of eppin (an epididymal protease inhibitor), KLK-3, and other KLKs (kallikrein-like proteases) in the prostatic fluid. Eppin (or SPINLW1) is a member of Kunitz-type and the whey acidic protein (WAP)-type protease inhibitor gene family [73]. Human-ejaculated spermatozoa are coated with eppin over the head and tail regions [73–75]. The C-terminal region of eppin (aa 75–133) binds a fragment of semenogelin (Sg, aa 164–283) secreted from seminal vesicles and prevents the action of KLK-3 on Seminogelin [76]. Human epididymis protein-4, commonly known as HE-4, is an epididymal protease inhibitor. HE-4 has been detected in human seminal fluid [77]. The inhibitory nature of HE-4 on different serine, aspartyl, and cysteine proteases has demonstrated it as a cross class protease inhibitor [77].

Acrosin is a serine protease present in acrosome of spermatozoa and is involved in the recognition, binding, and penetration of the spermatozoa of/into the zona pellucida of the oocyte. It is the most widely studied and well-characterized acrosomal enzyme [78]. In mouse, rat, human, and swine, acrosin gene is localized on chromosomes 15, 7, 22, and 5, respectively [79–81]. Acrosin is present as an inactive precursor, namely, proacrosin in acrosome of freshly ejaculated sperm [82]. Upon acrosome reaction, proacrosin is autoproteolyzed into most stable active forms by an intrazymogen mechanism, and later it is delivered to the extracellular milieu [83, 84]. Acrosin is conserved among birds and mammals [85]. The presence of proacrosin/acrosin system was demonstrated in turkey spermatozoa [86]. Apart from acrosin, several other enzymes with properties similar to acrosin have been identified in mammalian spermatozoa [87]. Acrosin II of molecular mass 30.9 kDa is an acrosin-like protease similar to other avian acrosins [88]. The presence of trypsin-like activity in turkey and chicken spermatozoa extract has been demonstrated [89]. Kotlowska et al. [90] detected the sperm amidase activity and acrosomal serine proteases activity in turkey spermatozoa. Serine proteases of molecular weight 29 and 88 kDa have been detected in turkey seminal plasma [45]. In fresh and frozen/thawed dog spermatozoa, proacrosin, alpha-acrosin, and beta-acrosin with 40, 32, and 27 kDa bands, respectively, are identified by in vitro capacitation studies [91, 92]. The proacrosin to acrosin conversion during sperm-zona interaction has been demonstrated in boar spermatozoa [93, 94]. The three molecular forms of 64, 38, and 25 kDa human acrosin have been identified [95]. A direct correlation between the levels of proacrosin/acrosin activity with the fertilizing potential was observed in human spermatozoa [96]. The amino acid sequence of mouse proacrosin has a high degree of homology with that of porcine, human, and rat [97–100]. Serine protease of mouse sperm is different from other rodents and mammals. The amount of acrosin and gelatinase activity has been shown to be less for mouse sperm when compared with rat and hamster sperm [101]. The mouse proacrosin has an extra Cys residue at positions 143 and 144 and may cause incorrect formation of a disulfide bridge and prevent the acrosin activity. The crystal structures of pig and ram acrosins have been identified [16]. The presence of acrosin in ram epididymal fluids suggested that it may be released from dead spermatozoa [44]. Both proacrosin and acrosin have been purified from porcine, guinea pig, and human cauda epididymal and ejaculated spermatozoa [102–104]. In porcine spermatozoa, proacrosin with molecular weight 55 and 53 kDa and active forms with molecular sizes of 49, 43, and 35 kDa have been identified [93, 98, 105]. Analysis N- and C-terminal sequence of the five forms of boar proacrosin demonstrated that conversion of 55 kDa proacrosin to the 35 kDa mature acrosin occurs by removal of three C-terminal segments and cleavage of a peptide bond near the N-terminus [98, 105]. This peptide bond cleavage is responsible for the initiation of the protease activity of acrosin [98, 105]. Genomic and cDNA sequences of acrosins were characterized from human, bovine, boar, and mouse [99, 106, 107]. Two forms of proacrosin with different molecular masses (55 and 53 kDa) and three different forms of acrosin, named α (49 kDa), β (35 kDa), and γ (25 kDa), have been identified in ejaculated boar spermatozoa [108]. It has been suggested that the acrosin activity can be used as an index for assessing the fertility check in males [109]. Serine proteases other than acrosins have also been detected in mouse sperm acrosome [110].

The murine testis-specific serine proteases (TESP) are located as a gene cluster on chromosome 9F2-F3. Prss42/Tessp-2 is expressed in sperm membrane and cytoplasm, Prss43/Tessp-3 in the sperm membranes, and Prss44/Tessp-4 in the cytoplasm of the secondary spermatocytes and spermatids. Thus, these serine proteases play different roles in spermatogenesis and are required for germ cell survival during meiosis [111]. In mouse spermatozoa, the testes-specific serine proteases, namely, TESP1, TESP2, and TESP4, were identified in the acrosome, TESP3 from the spermatogenic cells of the testis, and TESP5 on the cauda epididymal sperm membranes. The 42 and 41 kDa isoforms of TESP5 were identified in mouse sperm extract [112]. Since these isoforms are glycosylphosphatidylinositol (GPI) anchored and are located in the lipid rafts of mouse sperm membranes, TESP5 is suggested to play important role in signal transduction at the sperm surface that help in sperm-zona binding [113, 114].

Two serine proteases, namely, BSp66 and its dimeric form BSp120, have been identified in bovine and hamster spermatozoa [87, 115]. In hamster spermatozoa, BSp66 is involved in sperm-oocyte interaction [87]. Proteases such as aspartate aminotransferase, hyaluronidase aminotransferase, and lactic dehydrogenase have been identified in buffalo and cattle spermatozoa [116]. Collagenase-like peptidase has been detected in human semen, rat, and bull epididymal spermatozoa and testes. A 110 kDa collagenase has been purified from rat testes [117]. Cathepsin-D proteases were identified from mouse testes [118]. Dipeptidyl protease-II was detected in guinea pig sperm acrosome [119]. Calpain II of 80 kDa has been detected in porcine sperm acrosome [120]. Aryl amidases were detected in bull spermatozoa [121].

Two proteases, namely, seminin and seminal pepsin, were identified in human seminal plasma with optimum pH 7.5 and 2.5–3.5, respectively [122]. Seminin-like protease was also identified in dog, rabbit, and bull seminal plasma [123]. Basic arginine esterase activity has been detected in human seminal plasma [124]. The molecular forms of 29–88 kDa, 37 kDa, 38–44 kDa, and 28–32 kDa turkey seminal plasma enzymes (TSPE) were identified in turkey seminal plasma [125]. Amidases were detected in vas deferens and testicular/epididymal fluids of turkey, guinea fowl, and chicken [125]. More protease activity was observed in vas deferens of turkey compared to testicular and epididymal fluids of guinea fowl and chicken. A 32–34 kDa prostate-specific antigen was identified in human seminal fluid [126]. In human seminal plasma, prostate-specific antigen is secreted from the prostate gland [126]. TMPRSS2, a type II transmembrane serine protease (TTSP), was found to express in epithelium of prostate gland and ejaculated semen in human [127].

TIMP-1 and TIMP-2 have been detected in human and bovine seminal plasma [128–130]. In bovine, TIMP-2 is suggested to play important role in fertility [130]. Expression of TIMP-1–3 has been demonstrated in mouse and human fetal testis [35, 131]. TIMP-2 was detected in Leydig cells during mouse gonad developmental stages [132]. TIMP-2 has been detected in caput fluid of ram and boar epididymis and in caput and caudal fluids of stallion epididymis [44].

The Kazal inhibitors are present in seminal plasma and spermatozoa of different mammalian species. The Kazal inhibitors include acrosin inhibitors, acrosin-trypsin inhibitors, ovoinhibitors, etc. Three isoforms of acrosin inhibitors were detected in boar-ejaculated spermatozoa. In boar, these acrosin inhibitors were secreted from seminal vesicle to the ejaculated semen [133]. Human acrosin-trypsin inhibitor (HUSI-II) was identified in human seminal plasma [134]. Acrosin inhibitor of 6.1 kDa has been identified in chicken seminal plasma [135]. The acrosin inhibitors protect the reproductive tissues, seminal plasma proteins, and viable spermatozoa from acrosin released from dead or damaged spermatozoa [136]. An ovoinhibitor has been identified and characterized from turkey seminal plasma, epithelium of ductus deferens, and epididymis [137]. The ovoinhibitor is demonstrated to have inhibitory action on subtilisin, trypsin, and elastase and antibacterial activity against *Bacillus subtilis* and *Staphylococcus aureus* in vitro [137]. The ovoinhibitor is suggested to have important role in maintaining the microenvironment for spermatozoa in the epididymis and ductus deferens in turkey [137].

Among the SPINKs that are identified from mammals, some are expressed in the epithelial cells of the male reproductive and some are involved in the process of fertilization [23]. The expression of SPINKL (SPINK-like) has been demonstrated in mouse seminal vesicle [138]. SPINKL was demonstrated to inhibit sperm capacitation and sperm motility in vitro [138]. SPINKL-bound spermatozoa were detected from the uterus but not from the oviduct. In mouse, the SPINKL was demonstrated to inhibit the capacitation-related signals such as cholesterol efflux, calcium influx, and cAMPi in uterine spermatozoa [139]. In mouse, SPINK2 expression has been demonstrated in spermatogenic cells and mature sperms in testes [140]. Mutation in SPINK2 locus demonstrated impaired spermatogenesis, loss of testicular integrity, and reduced sperm number and serine protease-mediated germ cell apoptosis. SPINK3 expression has been demonstrated in seminal vesicle and plasma membrane of apical hook of spermatozoa [141]. It is also known as Caltrin or calcium transport inhibitor. In mouse, the SPINK3- spermatozoa binding reduced $[Ca^{2+}]_i$ in the head and suppressed the acrosome reaction in spermatozoa before encountering egg. Release of sperm-bound SPINK3 by SPINK3-inhibiting trypsin-like activity (SITA) in uterine luminal fluid demonstrated restoration of sperm fertilization ability [141]. Expressions of SPINK8, SPINK10, SPINK11, and SPINK12 have been demonstrated in mouse epididymis [142]. In rat, specific expression of an androgen-responsive SPINK13 has been demonstrated in epididymal tissues and spermatozoa [23]. RNAi knockdown of SPINK13 gene in rat has demonstrated enhanced acrosomal exocytosis and fertility defects both in vivo and in vitro. SPINK13 is involved in proteolytic processing of epididymal proteins required for sperm maturation.

SERPIN-E1 is commonly known as plasminogen activators inhibitors (PAIs). PAIs play crucial role in spermatogenesis, spermatozoa capacitation, and fertilization [62]. PAI-1 has been identified from rat, mouse, and monkey testes [68]. In rat, the highest expression of PAI-1 mRNA has been detected in the germinal cells of testes [143]. Sertoli cells have also demonstrated to secrete PAI-1 to the adluminal compartment of seminiferous tubules [144]. PAI-1 is suggested to protect the Sertoli cell barrier during passage of pre-leptotene spermatocytes across the blood-testes barrier and prevents the release of developing spermatid in spermiation process. In monkey, the increased expression of PAI-1 has been demonstrated in caput epithelial cells compared to that of initial and caudal regions of the epididymis [59]. PAI-1 is suggested to play major role in regulation of epididymal sperm maturation. PAIs were also detected in human-ejaculated spermatozoa and seminal plasma [145, 146]. PAI-1 was detected in the head, middle piece, and tail regions of human and rhesus monkey spermatozoa [65].

SERPIN-E2 is generally known as protease nexin-1. SERPIN-E2 was identified from mouse seminal vesicle. The mouse deficient in SERPIN-E2 genes demonstrated fertility defects [147]. In mouse, SERPIN-E2 was detected in seminal vesicular secretions, epithelium of the seminal vesicle, epididymis, and vas deferens [148]. SERPIN-E2 was also detected on the acrosomal cap of testicular and epididymal spermatozoa. In ejaculated spermatozoa, SERPIN-E2 was detected mostly from the uncapacitated spermatozoa and was found to inhibit the sperm capacitation signals and sperm-zona binding in vitro. The presence of SERPIN-E2 in uncapacitated mouse spermatozoa suggested that it may act as a decapacitation factor in fertilization [148].

SERPIN-A5 is also known as protein C inhibitor (PCI) or PAI-3. Presences of SERPIN-b6b and SERPIN-b6c have been identified from mouse germ cells. In mouse, SERPIN-b6 is suggested to play role in testes development, spermatogenesis, and fertilization process [149]. Three SERPINs were electrophoretically detected in turkey seminal plasma [45]. Of these, two were from testes and epididymis, and the third was from ductus deferens and seminal plasma.

Some partially characterized protease inhibitors, namely, inhibitor-I and inhibitor-II of molecular weight 8.7 kDa and 6.8 kDa, were detected in seminal plasma and ejaculated spermatozoa of bulls [150]. The inhibitor-I was found to inhibit bull sperm acrosin, bovine trypsin, chymotrypsin, and porcine plasmin. The inhibitor-II has been shown to inhibit bovine trypsin, chymotrypsin, porcine plasmin, and pancreatic and urinary kallikreins. Kunitz-type proteinase inhibitor of 9.5–12 kDa has been detected from porcine seminal vesicular fluid [151]. Protease inhibitors of 86 and 26 kD have been demonstrated in both buffalo and cattle seminal plasma [46]. However, between the two species, the cattle seminal plasma demonstrated comparatively higher protease inhibitor activities than buffaloes.

10.4 Conclusions

A vast array of proteases and protease inhibitors have been identified in the male reproductive organs of mammals playing critical role in the major events associated with the several male reproductive processes including spermatogenesis, epididymal sperm maturation, capacitation, and fertilization. Protease inhibitors have important role for regulating the protease production/degradation, activation/inactivation, etc. There exists a fine balance between the protease and their regulators for maintaining the above male reproductive processes including tissue integrity, blood-testes barrier, and the gamete development. A disturbance in this equilibrium leads to progression of reproductive failures such as azoospermia, impaired sperm functions, low fertilizing efficiency, etc. that culminate in infertility cases. Although there have been considerable knowledge in many aspects of mammalian male reproductive health, a better understanding of the protease and protease inhibitors may in turn help in developing the accurate methodologies or specific approaches for improving the sperm function and male fertility.

Acknowledgments The authors thank the Director, National Institute of Animal Nutrition and Physiology (NIANP), Bangalore, India for providing funds and facilities for carrying out the present study. The first author was supported by Junior Research Fellowship from Indian Council of Medical Research, New Delhi. The authors also thank the Director, Indian Veterinary Research Institute (IVRI), Bareilly, India for providing an opportunity to the first author for conducting the present study at NIANP, Bangalore.

References

1. Hulboy DL, Rudolph LA, Matrisian LM (1997) Matrix metallo-proteinases as mediators of reproductive function. Mol Hum Reprod 3:27–45
2. Polgar L (1989) Mechanisms of protease action. CRC Press, Boca Raton, FL
3. Fowlkes JL, Winkler MK (2002) Exploring the interface between metallo-proteinase activity and growth factor and cytokine bioavailability. Cytokine Growth Factor Rev 13:277–287
4. Nagase H, Woessne JF Jr (1999) Matrix metalloproteinases. J Biol Chem 274:21491–21499
5. Choi H, Jin S, Kwon JT, Kim J, Jeong J, Kim J et al (2016) Characterization of mammalian ADAM2 and its absence from human sperm. PLoS One 11:e0158321. doi:10.1371/journal.pone.0158321
6. Le Magueresse-Battistoni B (2000) Proteases and their cognate inhibitors of the serine and metalloprotease subclasses in testicular physiology. In: Madame curie bioscience database [internet] Austin (TX) Landes Bioscience; 2000–2013. Source: http://wwwncbinlmnihgov/books/NBK6476
7. Puente XS, López-Otín C (2004) A genomic analysis of rat proteases and protease inhibitors. Genome Res 14:609–622
8. Domsalla A, Melzig MF (2008) Occurrence and properties of proteases in plant lattices. Planta Med 74:699–711
9. Chapman HA, Riese RJ, Shi GP (1997) Emerging roles for cysteine proteases in human biology. Annu Rev Physiol 59:63–88
10. Rojas FJ (1999) Calpain-calpastatin: a novel, complete calcium-dependent protease system in human spermatozoa. Mol Hum Reprod 5:520–526
11. Turk V, Stoka V, Vasiljeva O, Renko M, Sun T, Turk B, Turk D (2012) Cysteine cathepsins: from structure, function and regulation to new frontiers. Biochim Biophys Acta 1824:68–88
12. Heutinck KM, ten Berge IJM, Hack CE, Hamann J, Rowshani AT (2010) Serine proteases of the human immune system in health and disease. Mol Immunol 47:1943–1955
13. Dano K, Andreasen PA, GrondahlHansen J (1985) Plasminogen activators, tissue degradation, and cancer. Adv Cancer Res 44:139–266
14. Mazzieri R, Masiero L, Zanetta L (1997) Control of type IV collagenase activity by components of the urokinaseplasmin system: a regulatory mechanism with cell-bound reactants. EMBO J 16:2319–2332
15. Karakosta TD, Soosaipillai A, Diamandis EP, Batruch I, Drabovich AP (2016) Quantification of human kallikrein-related peptidases in biological fluids by multi-platform targeted mass spectrometry assays. Mol Cell Proteomics. doi:10.1074/mcp.M115.057695
16. Tranter R, Read JA, Jones R, Brady RL (2000) Effector sites in the three-dimensional structure of mammalian sperm β-acrosin. Structure 8:1179–1188
17. Bedford JM (1998) Mammalian fertilization misread? Sperm penetration of the eutherian zona pellucida is unlikely to be a lytic event. Biol Reprod 59:1275–1287. doi:10.1095/biolreprod59.6.1275
18. De los Reyes M, Barros C (2000) Immunolocalization of proacrosin/ acrosin in bovine sperm and sperm penetration through the zona pellucida. Anim Reprod Sci 58:215-228
19. Woessner JF (1991) Matrix metalloproteinases and their inhibitors in connective tissue remodeling. FASEB J 5:2145–2154
20. Tentes I, Asimakopoulos B, Mourvati E, Diedrich K, Al-Hasani S, Nikolettos N (2007) Matrix metalloproteinase (MMP)-2 and MMP-9 in seminal plasma. J Assist Reprod Gen 24:278–281
21. Kazal LA, Spicer DS, Brahinsky RA (1948) Isolation of a crystalline trypsin inhibitor-anticoagulant protein from pancreas. J Am Chem Soc 70:3034–3040
22. Van Hoef V, Breugelmans B, Spit J, Simonet G, Zels S, Vanden Broeck J (2013) Phylogenetic distribution of protease inhibitors of the Kazal-family within the arthropoda. Peptides 41:59–65

23. Ma L, Yu H, Ni Z, Hu S, Ma W, Chu C, Zhang Y (2013) SPINK13, an epididymis-specific gene of the Kazal-type serine protease inhibitor (SPINK) family, is essential for the acrosomal integrity and male fertility. J Biol Chem 288:10154–10165
24. Lu SM, Lu W, Qasim MA, Anderson S et al (2001) Predicting the reactivity of proteins from their sequence alone Kazal family of protein inhibitors of serine proteinases. Proc Natl Acad Sci 98:1410–1415
25. Stenman UH (2011) Role of the tumor-associated trypsin inhibitor SPINK1 in cancer development. Asian J Androl 13:628–629
26. Austin CR (1951) Observations on the penetration of sperm into the mammalian egg. Aust J Sci Res 4:581–596
27. Austin CR (1952) The "capacitation" of the mammalian sperm. Nature 170:326
28. Chang MC (1951) Fertilizing capacity of spermatozoa deposited into the fallopian tubes. Nature 168:697–698
29. Chang MC (1955) Development of fertilizing capacity of rabbit spermatozoa in the uterus. Nature 175:1036–1037
30. Yanagimachi R (1969) In vitro acrosome reaction and capacitation of golden hamster spermatozoa, bovine follicular fluid and its fractions. J Exp Zool 179:269–280
31. Yanagimachi R (1994) Mammalian fertilization. In: Knobil E, Neill JD (eds) The physiology of reproduction. Raven Press, New York, pp 189–317
32. Visconti PE, Bailey JL, Moore GD, Pan D, Olds-Clake P, Kopf GS (1995) Capacitation of mouse spermatozoa. I. Correlation between the capacitation state and protein tyrosine phosphorylation. Development 121:1129–1137
33. Zheng X, Geiger M, Ecke S, Bielek E, Donner P, Eberspacher U, Schleuning WD, Binder BR (1994) Inhibition of acrosin by protein C inhibitor and localization of protein C inhibitor to spermatozoa. Am J Phys 267:C466–C472
34. Rudolph-Owen LA, Cannon P, Matrisian LM (1998) Overexpression of the matrix metalloproteinase matrilysin results in premature mammary gland differentiation and male infertility. Mol Biol Cell 9:421–435
35. Guyot R, Magre S, Leduque P (2003) Differential expression of tissue inhibitor of metalloproteinases type 1 (TIMP-1) during mouse gonad development. Dev Dyn 227:357–366
36. Cossins J, Dudgeon TJ, Catlin G (1996) Identification of MMP18, a putative novel human matrix metalloproteinase. Biochem Biophys Res Commun 228:494–498
37. Velasco G, Pendas AM, Fueyo A (1999) Cloning and characterization of human MMP-23, a new matrix metalloproteinase predominantly expressed in reproductive tissues and lacking conserved domains in other family members. J Biol Chem 274:4570–4576
38. Lohi J, Wilson CL, Roby JD (2001) Epilysin, a novel human matrix metalloproteinase (MMP-28) expressed in testis and keratinocytes and in response to injury. J Biol Chem 276:10134–10144
39. Nuttall RK, Sampieri CL, Pennington CJ (2004) Expression analysis of the entire MMP and IMP gene families during mouse tissue development. FEBS Lett 563:129–134
40. Shimokawa K, Katayama M, Matsuda Y, Takahashi H, Hara I, Sato H (2002) Matrix metalloproteinase (MMP)-2 and MMP-9 activities in human seminal plasma. Mol Hum Reprod 8:32–36
41. Saengsoi W, Shia WY, Shyu CL, Wu JT, Warinrak C, Lee WM, Cheng FP (2011) Detection of matrix metalloproteinase (MMP)- 2 and MMP-9 in canine seminal plasma. Anim Reprod Sci 127:114–119
42. Warinrak C, Wu J, Hsu W, Liao J, Chang S, Cheng F (2014) Expression of matrix metalloproteinases (MMP-2, MMP-9) and their inhibitors (TIMP-1, TIMP-2) in canine testis, epididymis and semen. Reprod Domest Anim 50:48–57
43. Ferrer M, Rodriguez H, Zara L, Yu Y, Xu W, Oko R (2012) MMP2 and acrosin are major proteinases associated with the inner acrosomal membrane and may cooperate in sperm penetration of the zona pellucida during fertilization. Cell Tissue Res 349:881–895. doi:10.1007/s00441-012-1429-1

44. Metayer S, Dacheux F, Dacheux JL, Gatti JL (2002) Comparison, characterization and identification of protease inhibitors in epididymal fluids of domestic mammals. Matrix metalloproteinases are major fluid gelatinases. Biol Reprod 66:1219–1229
45. Kotlowska M, Kowalski R, Glogowski J, Jankowski J, Ciereszko A (2005) Gelatinases and serine proteinase inhibitors of seminal plasma and the reproductive tract of turkey (Meleagris gallopavo). Theriogenology 63:1667–1681
46. Gurupriya VS, Divyashree BC, Roy SC (2014) Cryogenic changes in proteases and antiprotease activities of buffalo (Bubalus bubalis) and cattle (Bos taurus) semen. Theriogenology 81:396–402
47. Urbschat A, Paulus P, Wiegratz I, Beschmann H, Hadji P, Hofmann R, Ochsendorf F (2014) Macrophage metalloelastase-12 is detectable in human seminal plasma and represents a predictor for inflammatory processes in the male genital tract. Andrologia 47:153–159
48. Aydos SE, Yukselten Y, Sunguroglu A, Demircan K, Aydos K (2016) Role of ADAMTS1 and ADAMTS5 in male infertility. Andrologia. doi:10.1111/and.12547
49. Li SW, Arita M, Fertala A, Bao Y, Kopen GC, Långsjö TK et al (2001) Transgenic mice with inactive alleles for procollagen Nproteinase (ADAMTS-2) develop fragile skin and male sterility. Biochem J 355:271–278
50. Liu L, Smith JW (2000) Identification of ADAM 31: a protein expressed in leydig cells and specialized epithelia. Endocrinology 141:2033–2042
51. Porter S, Clark IM, Kevorkian L (2005) The ADAMTS metalloproteinases. Biochem J 386:15–27
52. Dun MD, Anderson AL, Bromfield EG, Asquith KL, Emmett B, McLaughlin EA, Nixon B (2012) Investigation of the expression and functional significance of the novel mouse sperm protein, a disintegrin and metalloprotease with thrombospondin type 1 motifs number 10 (ADAMTS10). Int J Androl 35:572–589
53. Carre GA, Couty I, Hennequet-Antier C, Govoroun MS (2011) Gene expression profiling reveals new potential players of gonad differentiation in the chicken embryo. PLoS One 6:e23959
54. Abdul-Majeed S, Mell B, Nauli SM, Joe B (2014) Cryptorchidism and infertility in rats with targeted disruption of the Adamts16 locus. PLoS One 9:e100967
55. Llamazares M, Cal S, Quesada V, Lopez-Otin C (2003) Identification and characterization of ADAMTS-20 defines a novel subfamily of metalloproteinases-disintegrins with multiple thrombospondin-1 repeats and a unique GON domain. J Biol Chem 278:13382–13389
56. Liu Y, Zan L, Zhao S, Xin Y, Jiao Y, Li K (2012) Molecular characterization, expression pattern, polymorphism and association analysis of bovine ADAMTSL3 gene. Mol Biol Rep 39:1551–1560
57. Zaneveld LJD, De Jonge CJ (1991) Mammalian sperm acrosomal enzymes and the acrosome reaction. In: Dunbar BS, O'Rand MG (eds) A comparative overview of mammalian fertilization. Plenum Press, New York, pp 63–79
58. Zervos IA, Lavrentiadou SN, Tsantarliotou MP, Georgiadis MP, Kokolis NA, Taitzoglou IA (2010) Seasonal variation of plasminogen activator activity in spermatozoa and seminal plasma of boar, buck, bull and stallion. Reprod Dom Anim 45:e440–e446. doi:10.1111/j.1439-0531.2010.01597.x
59. Zhang T, Guo CX, Hu ZY, Liu YX (1997a) Localization of plasminogen activator and inhibitor, LH and androgen receptors and inhibin subunits in monkey epididymis. Mol Hum Reprod 3:945–952
60. Reese JH, McNeal JE, Redwine EA, Stamey TA, Freiha FS (1988) Tissue type plasminogen activator as a marker for functional zones, within the human prostate gland. Prostate 12:47–53
61. Zhang T, Zhou HM, Liu YX (1997b) Expression of plasminogen activator and inhibitor, urokinase receptor and inhibin subunits in rhesus monkey testes. Mol Hum Reprod 3:223–231
62. Liu YX (2007) Involvement of plasminogen activator and plasminogen activator inhibitor type 1 in spermatogenesis, sperm capacitation, and fertilization. Semin Thromb Hemost 33:29–40

63. Ebisch I, Steegers-Theunissen R, Sweep F, Zielhuis G, Geurts- Moespot A, Thomas C (2007) Possible role of the plasminogen activation system in human subfertility. Fertil Steril 87: 619–626
64. Le Magueresse-Battistoni B (2007) Serine proteases and serine protease inhibitors in testicular physiology: the plasminogen activation system. Reproduction 134:721–729
65. Liu K, Liu YX, Qun D, Zhou HM, Lin X, Hu ZY (1996) Preliminary studies on the role of plasminogen activator in seminal plasma of human and rhesus monkey. Mol Hum Reprod 2:99–104
66. Huarte J, Belin D, Bosco D, Sappino AP, Vassalli JD (1987) Plasminogen activator and mouse spermatozoa: urokinase synthesis in the male genital tract and binding of the enzyme to the sperm cell surface. J Cell Biol 104:1281–1289
67. Tulsiani DR, NagDas SK, Skudlarek MD, Orgebin-Crist MC (1995) Rat sperm plasma membrane mannosidase: localization and evidence for proteolytic processing during epididymal maturation. Dev Biol 167:584–597
68. Lacroix M, Smith FE, Fritz IB (1977) Secretion of plasminogen activator by sertoli cell enriched cultures. Mol Cell Endocrinol 9:227–236
69. Schill WB (1975) Significance of proteolytic sperm enzymes for the fertility. Hautarzt Oct 26:514–523
70. Matsui H, Takahashi T (2001) Mouse testicular leydig cells express Klk21, a tissue kallikrein that cleaves fibronectin and IGF-binding protein-3. Endocrinology 142:4918–4929
71. Yousef GM, Diamandis EP (2001) The new human tissue kallikrein gene family: structure, function, and association to disease. Endocr Rev 22:184–204
72. Emami N, Diamandis EP (2008) Human kallikrein-related peptidase 14 (KLK14) is a new activator component of the KLK proteolytic cascade. Possible function in seminal plasma and skin J Biol Chem 283:3031–3041
73. Richardson RT, Sivashanmugam P, Hall SH, Hamil KG, Moore PA (2001) Cloning and sequencing of human Eppin: a novel family of protease inhibitors expressed in the epididymis and testis. Gene 270:93–102
74. Wang Z, Widgren EE, Sivashanmugam P, O'Rand MG, Richardson RT (2005) Association of eppin with semenogelin on human spermatozoa. Biol Reprod 72:1064–1070
75. Sivashanmugam P, Hall SH, Hamil KG, French FS, O'Rand MG, Richardson RT (2003) Characterization of mouse Eppin and a gene cluster of similar protease inhibitors on mouse chromosome 2. Gene 312:125–134
76. Wang ZJ, Zhang W, Feng NH, Song NH, Wu HF (2008) Molecular mechanism of epididymal protease inhibitor modulating the liquefaction of human semen. Asian J Androl 10:770–775
77. Chhikara N, Saraswat M, Tomar AK, Dey S, Singh S, Yadav S (2012) Human epididymis protein4 (HE4): a novel cross class protease inhibitor. PLoS ONE 7:E47672E47672
78. Klemm U, MuÈller-Esterl W, Engel W (1991b) Acrosin, the peculiar sperm-specific serine protease. Hum Genet 87:635–641
79. Kremling H, Flake A, Adham IM, Radtke J, Engel W (1991a) Exon intron structure and nucleotide sequence of the rat proacrosin gene. DNA Seq 2:57–60
80. Adham IM, Szpirer C, Kremling H, Keime S, Szpirer J, Levan G, Engel W (1991) Chromosomal assignment of four rat genes coding for the spermatid-specific proteins proacrosin (ACR), transition proteins 1 (TNP1) and 2 (TNP2), and protamine 1 (PRM1). Cytogenet Cell Genet 57:47–50
81. Yasue H, Hisamatsu N, Awata T, Wada Y, Kusamoto H (1999) Clarification of the order of acrosin and aconitase 2 genes on the physical and linkage maps of porcine chromosome 5. Anim Genet 30:161–162
82. Gaboriau D, Howes EA, Clark J, Jones J (2007) Binding of sperm proacrosin/β-acrosin to zona pellucida glycoproteins is sulfate and stereo dependent synthesis of a novel fertilization inhibitor. Dev Biol 306:646–657
83. Urch UA (1991) Biochemistry and function of acrosin. In: Wassarman PM (ed) The biology and chemistry of mammalian fertilization. CRC Press, Chicago, pp 233–248

84. Nuzzo NA, Anderson RA, Zaneveld LJ (1990) Proacrosin activation and acrosin release during the guinea pig acrosome reaction. Mol Reprod Dev 25:52–60
85. Berlin S, Qu L, Ellegren HJ (2008) Adaptive evolution of gamete recognition proteins in birds. J Mol Evol 67:488–496
86. Slowinska M, Olczak M, Liszewska E, Wątorek W, Ciereszko (2010) Isolation, characterization and cDNA sequencing of acrosin from turkey spermatozoa. Comp Biochem Physiol 157:127–136
87. Cesari A, Katunar MR, Monclus MA, Vincenti A, De Rosas JC, Fornes MW (2005) Serine protease activity, bovine sperm protease, 66 kDa (BSp66), is present in hamster sperm and is involved in sperm–zona interaction. Reproduction 129:291–298
88. Slowinska A, Ciereszko A (2012) Identification of the second form of acrosin in turkey spermatozoa. Reprod Dom Anim 47:849–855
89. Froman DP (1990) Chicken acrosin: extraction and purification. Poult Sci 69:812–817
90. Kotlowska M, Dietrich G, Wojtczak M, Karol H, Ciereszko A (2007) Effects of liquid storage on amidase activity, DNA fragmentation and motility of turkey spermatozoa. Theriogenology 67:276–286
91. De los Reyes M, Medina G, Palomino J (2009) Western blot analysis of proacrosin/acrosin in frozen dog sperm during in vitro capacitation. Reprod Domestic Anim 44:350–353
92. De los Reyes M, Palomino J, Martínez V, Aretio C, Gutiérrez M (2011) Acrosin release and acrosin activity during incubation in capacitating media using fresh and frozen-thawed dog sperm. Biol Res 44:139–144
93. Baba T, Kashiwabara S, Watanabe K, Itoh H, Michikawa Y, Kimura K, Takada M, Fukamizu A, Arai Y (1989a) Activation and maturation mechanisms of boar acrosin zymogen based on the deduced primary structure. J Biol Chem 264:11920–11927
94. Topfer-Petersen E, Cechova D (1990) Zona pellucida induces conversion of proacrosin to acrosin. Int J Androl 13:190–196
95. Schleuning WD, Hell R, Fritz H (1976) Multiple forms of human acrosin: isolation and properties. Hoppe-Seyler's Zeitschrift für physiologische. Chemie 357:855–866
96. Shimizu Y, Kodama H, Fukuda J, Tanaka T (1997) Evidence of proacrosin molecule abnormality as a possible cause of low acrosin activity and unexplained failure of fertilization in vitro. J Androl 18:281–288
97. Kashiwabara S, Baba T, Takada M, Watanabe K, Yano Y, Arai Y (1990) Primary structure of mouse proacrosin deduced from the cDNA sequence and its gene expression during spermatogenesis. J Biochem 108:785–791
98. Baba T, Michikawa Y, Kawakura K, Arai Y (1989c) Activation of boar proacrosin is effected by processing at both N- and C-terminal portions of the zymogen molecule. FEBS Lett 244:132–136
99. Baba T, Watanabe K, Kashiwabara S, Arai Y (1989d) Primary structure of human proacrosin deduced from its cDNA sequence. FEBS Lett 244:296–300
100. Klemm U, Flake A, Engel W (1991a) Rat sperm acrosin: cDNA sequence, derived primary structure and phylogenetic origin. Biochim Biophys Acta 1090:270–272
101. Yamagata K, Honda A, Kashiwabara S, Baba T (1999) Difference of acrosomal serine protease system between mouse and other rodent sperm. Dev Genet 25:115–122
102. MuÈller-Esterl W, Fritz H (1981) Sperm acrosin. Methods Enzymol 80:621–632
103. Siegel MS, Bechtold DS, Kopta CI, Polakoski KL (1986) The rapid purification and partial characterization of human sperm proacrosin using an automated fast protein liquid chromatography (FPLC) system. Biochem Biophys Acta 883:567–573
104. Hardy DM, Oda MN, Friend DS, Huang TIT (1991) A mechanism for differential release of acrosomal enzymes during the acrosome reaction. Biochem J 275:759–766
105. Baba T, Michikawa Y, Kashiwabara S, Arai Y (1989b) Proacrosin activation in the presence of a 32-kDa protein from boar spermatozoa. Biochem Biophys Res Commun 160:1026–1032
106. Adham IM, Kremling H, Nieter S, Zimmermann S, Hummel M, Schroeter U, Engel W (1996) The structures of the bovine and porcine proacrosin genes and their conservation among mammals. Biol Chem Hoppe Seyler 377:261–265

107. Kremling H, Keime S, Wilhelm K, Adham IM, Hameister H, Engel W (1991b) Mouse proacrosin gene: nucleotide sequence, diploid expression, and chromosomal localization. Genomics 11:828–834
108. Polakoski KL, Parrish RF (1977) Boar proacrosin purification and preliminary activation studies of proacrosin isolated from ejaculated boar sperm. J Biol Chem 252:1888–1894
109. Cui YH, Zhao RL, Wang Q, Zhang ZY (2000) Determination of sperm acrosin activity for evaluation of male fertility. Asian J Androl 2:229–232
110. Honda A, Siruntawineti J, Baba T (2002a) Role of acrosomal matrix proteases in sperm-zona pellucida interactions. Hum Reprod Update 8:405–412
111. Yoneda R, Takahashi T, Matsui H, Takano N, Hasebe Y, Ogiwara K, Kimura AP (2013) Three testis-specific paralogous serine proteases play different roles in murine spermatogenesis and are involved in germ cell survival during meiosis. Biol Reprod 88:118–118
112. Yamagata K, Murayama K, Kohno N, Kashiwabara S, Baba T (1998) P-Aminobenzamidine-sensitive acrosomal protease(s) other than acrosin serves the sperm penetration of the egg zona pellucida in mouse. Zygote 6:311–319
113. Honda A, Yamagata K, Sugiura S (2002b) A mouse serine protease TESP5 is selectively included into lipid rafts of sperm membrane presumably as a glycosylphosphatidylinositol-anchored protein. J Biol Chem 277:16976–16984
114. KraÈmer EM, Klein C, Koch T, Boytinck M, Trotter J (1999) Compartmentation of Fyn kinase with glycosylphosphatidylinositol-anchored molecules in oligodendrocytes facilitates kinase activation during myelination. J Biol Chem 274:29042–29049
115. Cesari A, Cacciato CS, De Castro RE, Sanchez JJ (2003) Low temperature-induced dimerization of the bovine sperm serine protease BSp66. J Cell Biochem 88:1057–1065
116. Dhami AJ, Sahni KL (1994) Comparative appraisal of physicomorphological and enzymatic attributes of semen and their interrelationships in ox and buffalo bulls. J Appl Anim Res 5:13–20
117. Koren E, Milkovic S (1973) Collaginase like peptide in human rat and bull spermatozoa. J Rep Fertil 32:349–356
118. Erickson RP, Martin SR (1974) The relationship of mouse spermatozoa to mouse testicular cathepsins. Arch Biochem Biophys 165:114–120
119. Talbot P, Dicarlantonio G (1985) Cytochemical localization of dipeptidyl peptidaseII (DPPII) in mature guinea pig sperm. J Histochem Cytochem 33:1169–1172
120. Schollmeyer JE (1986) Identification of calpain II in porcine sperm. Biol Reprod 34:721–731
121. Meizel S, Cotham J (1972) Partial characterization of sperm bull arylamidases. J Rep Fertil 28:303–307
122. Ruenwongsa P (1974) Chulavatnatol M. A new acidic protease in human seminal plasma Biochem Biophys Res Commun 59:44–50
123. Morton DB (1977) The occurrence and function of proteolytic enzymes in the reproductive tract of mammals. In: Barret AJ (ed) proteinases in mammalian cells and tissues. Elsevier North Holland Biomedical Press, Amsterdam, pp 1977:445–500
124. Kobayashi T, Park JY, Matsuda Y, Hara I, Kaneki S, Oshio S, Akihama S, Fujimoto Y (1991) Basic arginine esterase from human seminal plasma: purification and some and some properties. Arch Androl 27:197–206
125. Thurston R, Korn N, Froman DP, Bodine AB (1993) Proteolytic enzymes in seminal plasma of domestic turkey (Meleagris gallopavo). Biol Reprod 48:393–402
126. Waheed A, Hassan MI, Van Etten RL, Ahmad F (2008) Human seminal protease and prostate-specific antigens are the same protein. J Biosci 133:195–207
127. Chen Y, Lee M, Lucht A, Chou F, Huang W, Havighurst T et al (2010) TMPRSS2, a serine protease expressed in the prostate on the apical surface of luminal epithelial cells and released into semen in prostasomes, is misregulated in prostate cancer cells. Am J Pathol 176:2986–2996. doi:10.2353/ajpath.2010.090665
128. Baumgart E, Lenk SV, Loening SA, Jung K (2002) Quantitative differences in matrix metalloproteinase (MMP)-2, but not in MMP-9, tissue inhibitor of metalloproteinase (TIMP)-1 or TIMP-2, in seminal plasma of normozoospermic and azoospermic patients. Hum Reprod 17:2919–2923

129. Shimokawa K, Katayama M, Matsuda Y, Takahashi H, Hara I, Sato H (2003) Complexes of gelatinases and tissue inhibitor of metalloproteinases in human seminal plasma. J Androl 24:73–77
130. McCauley TC, Zhang HM, Bellin ME (2001) Identification of a heparin-binding protein in bovine seminal fluid as tissue inhibitor of metalloproteinases-2. Mol Reprod Dev 58:336–341
131. Robinson LL, Sznajder NA, Riley SC, Anderson RA (2001) Matrix metalloproteinases and tissue inhibitors of metalloproteinases in human fetal testis and ovary. Mol Hum Reprod 7:641–648
132. Blavier L, DeClerck YA (1997) Tissue inhibitor of metalloproteinase-2 is expressed in the interstitial matrix in adult mouse organs and during embryonic development. Mol Biol Cell 8:1513–1527
133. Jonakova V, Calvete JJ, Mann K, Schafer W, Schmid ER, Topfer-Petersen E (1992) The complete primary structure of three isoforms of a boar sperm-associated acrosin inhibitor. FEBS Lett 297:147–150
134. Fink E, Hehlein-Fink C, Eulitz M (1990) Amino acid sequence elucidation of human acrosin-trypsin inhibitor (HUSI-II) reveals that Kazal- type proteinase inhibitors are structurally related to -subunits of glycoprotein hormones. FEBS Lett 270:222–224
135. Lessley BA, Brown KI (1978) Purification and properties of a proteinase inhibitor from chicken seminal plasma. Biol Reprod 19:223–234
136. Laskowski M Jr, Kato I (1980) Protein inhibitors of proteinases. Annu Rev Biochem 49:593–626
137. Slowinska M, Liszewska E, Nynca J, Bukowska J, Hejmej A, Biliska B et al (2014) Isolation and characterization of an ovoinhibitor, a multidomain Kazal-like inhibitor from turkey (Meleagris gallopavo) seminal plasma. Biol Reprod 91:108–109
138. Lin M, Lee R, Hwu Y, Lu C, Chu S, Chen Y, Chang W, Li S (2008) SPINKL, a Kazal-type serine protease inhibitor-like protein purified from mouse seminal vesicle fluid, is able to inhibit sperm capacitation. Reproduction 136:559–571
139. Tseng HC, Lee RK, Hwu YM, Lu CH, Lin MH, Li SH (2013) Mechanisms underlying the inhibition of murine sperm capacitation by the seminal protein, SPINKL. J Cell Biochem 114:888–898. doi:10.1002/jcb.24428
140. Lee B, Park I, Jin S, Choi H, Kwon JT, Kim J, Jeong J, Cho B-N, Eddy EM, Cho C (2011) Impaired spermatogenesis and fertility in mice carrying a mutation in the SPINK2 gene expressed predominantly in testes. J Biol Chem 286:29108–29117
141. Ou CM, Tang JB, Huang MS, Sudhakar Gandhi PS, Geetha S, Li SH, Chen YH (2012) The mode of reproductive-derived SPINK (serine protease inhibitor Kazal-type) action in the modulation of mammalian sperm activity. Int J Androl 35:52–62
142. Jalkanen J, Kotimäki M, Huhtaniemi I, Poutanen M (2006) Novel epididymal protease inhibitors with Kazal or WAP family domain. Biochem Biophys Res Commun 349:245–254
143. Zhou HM, Liu YX (1996) Localization of tissue-type plasminogen activator (tPA) and plasminogen activator inhibitor type-1 (PAI-1) messenger RNA (mRNA) in rat testis. Chin Sci Bull 41:455–458
144. Le Magueresse-Battistoni B (1998) Plasminogen activator inhibitor-1 is expressed in cultured rat sertoli cells. Biol Reprod 59:591–598
145. Smokovitis A, Kokolis N, Alexopoulos C, Alexaki E, Eleftheriou E (1987) Plasminogen activator activity, plasminogen activator inhibition and plasmin inhibition in spermatozoa and seminal plasma of man and various animal species-effect of plasmin on sperm motility. Fibrinolysis 1:253–257
146. Smokovitis A, Kokolis N, Taitzoglou I, Rekkas C (1992) Plasminogen activator: the identification of an additional proteinase at the outer acrosomal membrane of human and boar spermatozoa. Int J Fertil 37:308–314
147. Murer V, Spetz JF, Hengst U, Altrogge LM, de Agostini A, Monard D (2001) Male fertility defects in mice lacking the serine protease inhibitor protease nexin-1. Proc Natl Acad Sci 98:3029–3033

148. Lu CH, Lee RKK, Hwu YM, Chu SL, Chen YJ, Chang WC, Lin SP, Li SH (2011) SERPINE2, a serine protease inhibitor extensively expressed in adult male mouse reproductive tissues, may serve as a murine sperm decapacitation factor. Biol Reprod 84:514–525
149. Charron Y, Madani R, Nef S (2006) Expression of Serpinb6 serpins in germ and somatic cells of mouse gonads. Mol Reprod Dev 73:9–19
150. Cechova D, Fritz H (1976) Characterization of the proteinase inhibitors from bull seminal plasma and spermatozoa. Hoppe Seylers Z Physiol Chem 357:401–408
151. Veselsky L, Jonakova VC, Echova D (1985) A Kunitz type of proteinase inhibitor isolated from boar seminal vesicle fluid. Andrologia 17:352–358

Physiological and Pathological Functions of Cysteine Cathepsins

11

Mansi Manchanda, Nishat Fatima, and Shyam Singh Chauhan

Abstract

Cysteine cathepsins are lysosomal hydrolases that belong to the papain family of cysteine proteases. This group comprises of 11 members and a majority of them are endo-proteases. They are initially synthesized as inactive zymogens, which are then processed into their active forms in the acidic and reducing environment of the lysosomes. The most striking element of cysteine cathepsins is their active site that contains a catalytic triad of a cysteine, histidine, and an asparagine residue. Originally, turnover and degradation of intracellular proteins was considered the only function of cysteine cathepsins. However, substantial evidences accumulated over the years have established their role in several physiological and pathological processes. Tissue-specific distribution and gene knockout analysis of these housekeeping proteases established their several physiological functions including antigen presentation, bone and tissue remodeling, keratinocyte differentiation, extracellular matrix degradation, cell cycle regulation, and death. Expression and activity of these proteases are tightly regulated, and their deregulation has been reported in a variety of pathological conditions such as cancer, lung diseases, metabolic disorders, atherosclerosis, cardiomyopathy, rheumatoid arthritis, osteoporosis, etc. These proteases have been proposed to be potential drug targets, and some of their inhibitors are under phase I clinical trial. This chapter presents an overview of the structure, synthesis, mode of action, regulation of expression and activity, and physiological as well as pathological role of lysosomal cysteine cathepsins.

M. Manchanda • N. Fatima • S.S. Chauhan (✉)
Department of Biochemistry, All India Institute of Medical Sciences,
Ansari Nagar, New Delhi, India
e-mail: s_s_chauhan@hotmail.com

S. Chakraborti, N.S. Dhalla (eds.), *Proteases in Physiology and Pathology*,
DOI 10.1007/978-981-10-2513-6_11

Keywords
Cathepsins • Antigen presentation • Cell death • Angiogenesis • Extracellular matrix degradation • Cancer • Lung disorders • Cardiovascular diseases • Metabolic syndrome

Abbreviations

&: and
APCs: Antigen-presenting cells
BALF: Bronchoalveolar lavage fluid
bFGF: Basic fibroblast growth factor
CLIP: Class II-associated invariant chain
COPD: Chronic obstructive pulmonary disease
cTEC: Cortical thymus epithelial cells
DPPI: Dipeptidyl aminopeptidase I
ECM: Extracellular matrix
EMT: Epithelial-mesenchymal transition
ER: Endoplasmic reticulum
Ets: Erythroblast transformation-specific transcription factor
GAGs: Glycosaminoglycans
hCATL: human cathepsin L
HMWK: High-molecular-weight kininogens
IFN: Interferon
IL: Interleukin
IR: Insulin receptor
IRES: Internal ribosomal entry site
L: Left-hand domain
LMP: Lysosomal membrane permeabilization
LMWK: Low-molecular-weight kininogens
M6PR: Mannose-6-phosphate receptor
MMPs: Matrix metalloproteinases
MOMP: Mitochondrial outer membrane permeabilization
PAI: Plasminogen activator inhibitor
R: Right-hand domain
ROS: Reactive oxygen species
SMA: Smooth muscle actin
TGF: Transforming growth factor
TGN: Trans-Golgi network
TIMP: Tissue inhibitors of metalloproteinases
TNF: Tumor necrosis factor
UTR: Untranslated regions
VEGF: Vascular endothelial growth factor
XIAP: X-chromosome-linked inhibitor of apoptosis

11.1 Introduction

Proteolysis (protein degradation) is essential for several physiological functions such as angiogenesis, tissue remodeling, wound healing, senescence, immune response, protein turnover as well as trafficking, cell survival, proliferation, migration, signaling, and death [1–3]. The most abundant and ubiquitously present class of enzymes called proteases catalyzes this process. These enzymes irreversibly hydrolyze proteins and polypeptides into their constituent peptides and eventually into amino acids. They cleave peptide bonds either from the terminal ends (exopeptidases) or within the peptide chain (endopeptidases). Their actions determine the fate of target proteins, which result in rapid and efficient amplification of an organism's response to a physiological signal [4].

The complete repertoire of human proteolytic enzymes termed as degradome comprises of approximately 600 proteases/peptidases [5]. Based on their catalytic mechanism(s), mammalian proteases have been classified into five different subgroups such as matrix metalloproteinases (MMPs) and serine, cysteine, threonine, and aspartic proteases as shown in Fig. 11.1 [6]. While these proteases play vital role(s) in multiple cellular processes, their uncontrolled activity can be detrimental. Therefore, specific endogenous inhibitors for each class of proteases are present in the cell to safeguard against the potential damage caused by the undesired proteolytic activity [7]. MMPs and cysteine and serine proteases are inhibited by tissue inhibitors of metalloproteinases (TIMPs), cystatins, and serpins, respectively [8]. The biological activity of proteases is also influenced by both their spatial and temporal expression along with the cellular microenvironment (e.g., pH conditions, calcium ion concentration, and redox potential).

Proteases, their endogenous substrates, products, activators, and inhibitors form the proteolytic networks [9]. Majority of the proteases are expressed as inactive zymogens and require processing by other proteases amplifying the number of targets at each successive step [10]. Decades ago, proteases came into the limelight for their pathological role in cancer invasion and metastasis [11]. Aberrant proteolysis is implicated in many pathological conditions such as inflammation, cancer, vascular diseases, rheumatoid arthritis, neurodegenerative processes, and liver diseases like fibrosis [10].

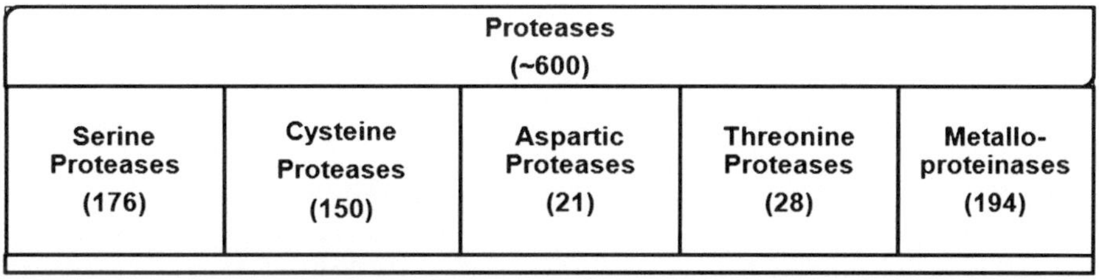

Fig. 11.1 Classification of human proteases. Human degradome includes five distinct classes of proteases. The number of members in each class is indicated in *parenthesis*

This chapter will focus on properties, localization, and regulation of cysteine cathepsins. In addition, an attempt has been made to delineate their role in various physiological and pathological processes such as antigen presentation, neovascularization, protein turnover and degradation, cell death, tumor invasion, and metastasis as well in cardiovascular disorders.

11.2 Cysteine Cathepsins

Cathepsins are a large and diverse group of enzymes ubiquitously present in all organisms ranging from prokaryotes to mammals. These peptidases were first discovered in the acidic gastric juice by Willstätter and Bamann in the year 1929 and were named as "cathepsin" which means *to digest*. Later in the year 1955, Christian de Duve identified the membranous sacs (lysosomes) enriched in acid hydrolases [12]. Lysosomal enzymes were characterized in the due course of time, and it was discovered that cysteine cathepsins play a major role in the degradation of proteins internalized by endocytosis/phagocytosis and in autophagy. However, under certain physiological and pathological conditions, cathepsins can also be secreted into the cytosolic and extracellular compartments and retain their proteolytic activity [13]. Cellular function and general properties of cysteine cathepsins have been summarized in Table 11.1.

Table 11.1 General properties and functions of cathepsins as revealed by knockout analysis

Cathepsin	Catalytic activity	Location	Phenotype of null mice	References
Cathepsin B	Endopeptidase, carboxydipeptidase	Ubiquitous	Impaired prohormone thyroglobulin processing and thyroid liberation, trypsinogen activation in pancreatitis	[224, 225]
Cathepsin C (dipeptidyl peptidase I)	Aminodipeptidase	Ubiquitous	Defects in activation of granzymes and serine proteases	[226]
Cathepsin F	Endopeptidase	Ubiquitous	Accumulation of lipofuscin in neurons and neurodegeneration	[227]
Cathepsin H	Endopeptidase, aminopeptidase	Ubiquitous	Impaired pulmonary surfactant protein B	[228]
Cathepsin K (cathepsin O2)	Endopeptidase	Mainly in osteoclasts, ovary	Osteopetrosis, deposition of bone matrix, exhibits clinical features of pycnodysostosis, reduced thyroglobulin processing	[91, 224]

(continued)

Table 11.1 (continued)

Cathepsin	Catalytic activity	Location	Phenotype of null mice	References
Cathepsin L	Endopeptidase	Ubiquitous	Reduced CD4+ T cell, epidermal hyperplasia, periodic hair loss, acanthosis, hyperkeratosis, dilated cardiomyopathy, impaired enkephalin processing, reduced thyroglobulin processing, abnormal bone development	[13, 224, 229, 230]
Cathepsin O	Endopeptidase	Ubiquitous	Normal	[131]
Cathepsin S	Endopeptidase	Mainly in APC	Impaired MHC class I and II and CD 1 antigen presentation	[131]
Cathepsin V (cathepsin L2)	Endopeptidase	Thymus and testis, cornea, keratinocytes	Normal	[13]
Cathepsin X (cathepsin Z)	Endopeptidase, carboxymonopeptidase	Ubiquitous	Not reported	[131]
Cathepsin W (lymphomain)	Endopeptidase	Cytotoxic lymphocytes	Immune defects	[231]

11.3 Structure and Mode of Action

The primary structure of all cysteine cathepsins consists of a signal peptide, propeptide, and catalytically active mature functional enzyme [3] (Fig. 11.2). Cathepsins are synthesized on rough endoplasmic reticulum (ER) as pre-proenzymes and contain a 10–20 amino acid-long signal peptide required for their entry into the lumen of ER. The signal sequence is then proteolytically cleaved by the signal peptidases in ER followed by the glycosylation of proenzymes in ER trans-Golgi network (TGN) [3]. Proteolytically inactive procathepsins possess a variable N-terminal pro-region which is 36-amino acid long in cathepsin X to 251 amino acids for cathepsin F. Propeptide acts as a potent reversible auto-inhibitor of the mature enzyme and prevents inappropriate activation of the catalytically functional cathepsins [14]. Maturation of zymogens can be a pH-dependent autocatalytic event or requires processing by other proteases, namely, pepsin, legumain, and cathepsin D occurring in acidic endo-/lysosomal compartments [15]. These released propeptides retain their inhibitory function and are degraded by surrounding proteases after performing their task [16]. Moreover, the propeptide acquires its structural confirmation prior to the remaining part of the enzyme and thus acts as a chaperone for the proper folding of active cathepsin. In addition, the prodomain also facilitates the endosomal/

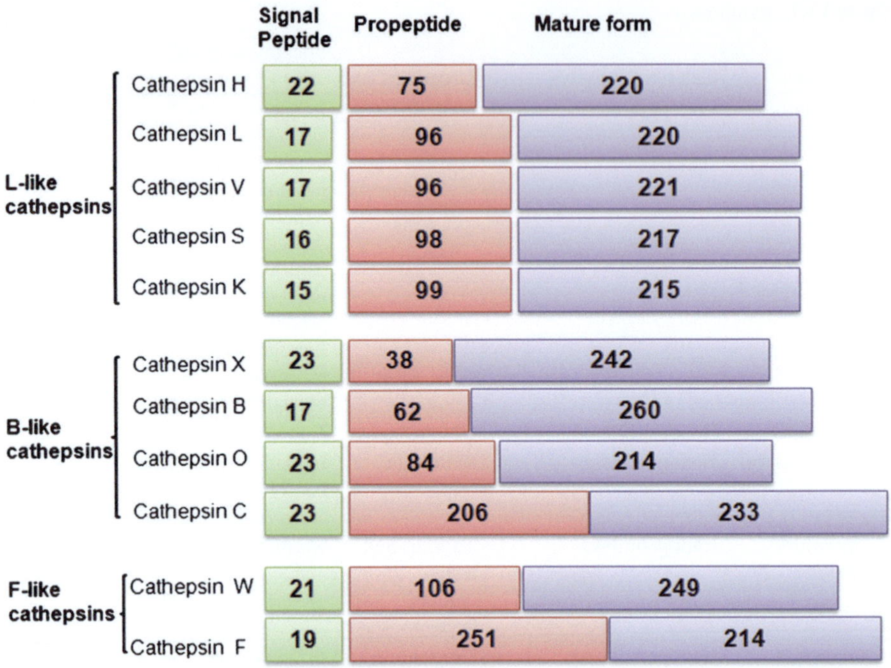

Fig. 11.2 **Primary structure of cysteine cathepsins.** The primary structure of all cysteine cathepsins consists of a signal peptide, propeptide, and catalytically active mature functional enzyme. Cysteine cathepsins have been classified in three categories ("L like," "B like," and "F like"). The members in each category have been arranged in the increasing order of the length of their proregion

lysosomal targeting of the proenzyme through the mannose-6-phosphate receptor (M6PR) pathway. The mature form of these enzymes contains disulfide-linked heavy and light chain subunits with molecular weights ranging from 20 to 35 kDa except for homotetrameric cathepsin C with a molecular weight of 200 kDa [17].

The crystal structure of cathepsins B, H, L, K, S, X, and C has been determined. All cysteine cathepsin has a characteristic papain-like fold, which is constituted by two different sub-domains: the helical core of the left-hand (L) and ß-barrel of the right-hand (R) domains [18]. These secondary structural elements along with active site are highly conserved in all the members. N-terminal L-domain contains three α-helices with the longest being the vertical or central helix. R-domain β-sheets form a coiled structure which is enclosed by α-helix at the bottom and is situated at the C-terminal [16]. Highly conserved V-shaped active-site cleft is situated at the center of both domains. The active site of cysteine proteases is occupied by the open structural confirmations of the substrate which makes an alternate contact with the backbone and side-chain atoms of both L- and R-domain amino acids [19]. Key catalytic residues of the active site include cysteine (Cys25) residue and histidine (His159) residue each located in different domain. The N-terminus of the central helix of L-domain contains the reactive-site cysteine residue, whereas histidine is

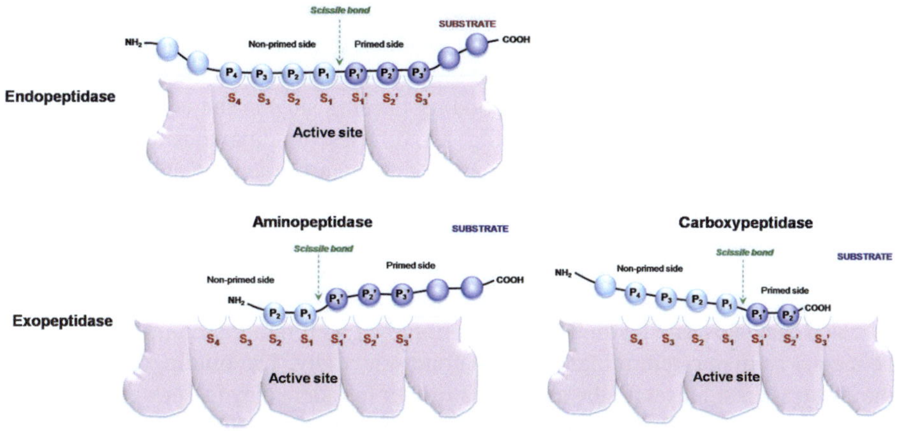

Fig. 11.3 Schematic representation of substrate-binding subsites in the active-site cleft and catalytic activities of different cysteine cathepsins. Structural features determine the enzyme-substrate interaction and hence the catalytic activity of cysteine cathepsins. Substrate-binding sites of endopeptidase cathepsins F, O, S, K, V, L, and W (*top*) and exopeptidase cathepsins B, C, H, and X (*left*, a diaminopeptidase and, *right*, a dicarboxypeptidase) have been diagrammatically represented. Peptide-binding subsites within the active site of cathepsins have been denoted as S (S4 to S3′). Each of these subsites interacts with seven different residues of the peptide/substrate, designated as P (P4 to P3′). Substrate-binding subsites N- terminus to the scissile peptide bond have been labeled as S1–S4 (non-primed sites) and those located on the C-terminal side as S1′–S3′ (primed sites). Amino acid binding to these sites has been designated as P1–P4 and P1′–P3′, respectively. In endopeptidases, all the subsites of active site accommodate the substrate amino acid residues and are cleaved at the scissile bond marked by an *arrow*. Additional structural features in exopeptidases (occluding loop of cathepsin B, mini-loop in cathepsin X, mini-chain in cathepsin H, and exclusion domain of cathepsin C) restrict the access of the substrate to C- (carboxypeptidases) and N- terminal (aminopeptidases) subsites, respectively

positioned within the R-domain [16]. Catalytic triad consists of negatively charged thiolate ion (Cys25), positively charged imidazolium ion (His159), and an asparagine or aspartic acid residue (Asn175) required for the activation of imidazolium ring [19]. All cysteine proteases perform proton-transfer catalysis using Cys residue as a nucleophile and His residue as a proton donor [16]. Nucleophilic cysteine attacks the carbonyl carbon atom of the peptide bond that generates the tetrahedral thioester transition intermediate which bears the negative charge on the carbonyl oxygen (oxyanion) [18]. Negatively charged ion pair is also stabilized by helix microdipole formed by catalytic Cys25 located at the N-terminus of α-helix and preceding Glu19. These two residues also stabilize the transient tetrahedral intermediate by H-bond interactions and form an oxyanion hole. After the collapse of tetrahedral intermediate, a proton is transferred from the positively charged His159 to the amino group of the cleaved peptide bond which results in the release of amine components [19].

Catalytic activity of cysteine cathepsins varies depending upon the enzyme-substrate interaction. As shown in Fig. 11.3, the substrate-binding pocket of these enzymes contains seven subsites, four on one side of the Cys25 and the remaining three on its other side. Amino acid residue of the substrate at the N-terminal of the scissile bond and the subsites of enzyme to which they bind are referred to as non-primed, while the amino acid residue of the substrate located on the C-terminal side of the scissile bond and their corresponding binding sites in the active cleft of the enzyme are termed as primed sites (Fig. 11.3). These binding subsites are denoted as S (S4 to S3′) extending over a 25°A long domain of the protease that interacts with seven residues of the peptide/substrate, designated as P (P4 to P3') [20]. Structural analysis revealed that S2, S1, and S1′ are the well-characterized binding sites in cysteine cathepsin where S2 site situated in the groove is occupied by hydrophobic and aromatic amino acids. Insights on the enzyme-inhibitor complexes revealed that substrate residues P2, P1, and P1' fit well into these binding sites and hence act as major determinants of substrate selectivity. The binding area between the substrate residues with the enzyme is further broadened by the presence of additional subsites S4, S3, S2′, and S3′ [21]. The substrate-binding site of endopeptidases along with two loops of L-domain binds specifically with the amino acid residues at P3, P1, and P2 position of the peptide targeted for degradation. Also residues at P2 and P1' positions of the substrate interact with amino acid residues of R-domain loops of the enzyme [22]. The amino acid residue at the P2 position of the substrate binds with both L- and R-domains and S2 pocket of the enzyme through hydrogen bonding and therefore determines its ability to bind the active site [16].

Cysteine cathepsins have a highly structured extended propeptide which is held in an opposite orientation over the substrate-binding cleft and thereby, physically blocks enzyme-substrate interaction [19]. This propeptide interacts non-covalently with the active site of all cysteine cathepsins except cathepsin X, where a disulfide linkage between the cysteine residue in the active site and the proregion holds propeptide over the substrate-binding pocket of the proenzyme [13]. Analysis of crystal structures of endopeptidase cathepsins K, L, S, F, V, O, and W revealed that active-site cleft is extended along the interface of L- and R- domains [21]. Substrate access to the active-site cleft of exopeptidases is restricted due to the additional structural elements (Fig. 11.3) such as a mini-loop of cathepsin X [23] and occluding loop of cathepsin B [24]. These loops block the binding of substrate at the C-terminus (primed side) of the active-site cleft limiting the access of substrate residues [25]. Dipeptidyl cathepsin B and monopeptidyl cathepsin X cleave the amino acid residues from the C-terminal of the substrate by using side-chain histidine residue(s) to position the negatively charged carboxylic group of the peptidyl substrate at the cleavage site. Cathepsin B contains an insertion of an 18-residue-long occluding loop (Pro107, Asp124) and utilizes the adjacent imidazolium rings of His110 and His111 to bind the carboxyl group of the C-terminal residue of the substrate [24]. However, the occluding loop of cathepsin B is very flexible as it gets displaced from the substrate-binding site under the acidic pH, and therefore, it also exhibits endopeptidase activity [22]. The proregion of monocarboxypeptidase cathepsin X contains three-residue-long "mini-loops" (between His23 and Tyr27) that extend toward the active-site cleft and influence the access of substrate to the S' binding region of the protease. Binding of the carboxylate group of P1' residue is favored by H-bond interactions with His23, Tyr27, and Trp202 in the S1′ region of active-site cleft [23].

The active site of aminopeptidase cathepsins H and C is blocked by the parts of their propeptide on the non-primed sides so that it can only be occupied by one or two substrate residues [26]. Propeptide of cathepsin H contains an octapeptide mini-chain, covalently linked by the disulfide bond to the enzyme that binds in the substrate orientation to the active-site cleft [27]. Active site (S2 site) of cathepsin H is occupied by the negatively charged carboxylic group of Thr83P at the C-terminus of the mini-chain and anchors the positively charged N-terminus of a substrate to the cleavage site. Attachment of mini-chain to the enzyme surface is enhanced by the insertion of four residues (Lys155A, Asp155D) and glycosylation of Asn111 [28]. Aminodipeptidase cathepsin C (dipeptidyl peptidase I, DPPI) exists as a tetramer and contains four identical active sites on the external face. The structure of each cathepsin C functional monomer is constituted by three domains, two domains of the papain-like structure and an "exclusion domain." Exclusion domain folds into an eight-stranded β-barrel and bears no homology with any member of papain family of proteases. Exclusion domains aid in tetrahedral structural arrangement and act as an extension for the active-site cleft, hence assisting DPPI activity. In case of cathepsin C, protruding β-hairpin (Lys82, Tyr93) and residues at N-terminus of the massive exclusion domain obstruct the entry of substrate into the active site. Hindering N-terminal of this exclusion domain of one dimer blocks the active-site cleft of the next, while the C-terminus of R-domain binds to the β-barrel of the adjoining exclusion domain of its neighboring partner [28]. The carboxylate group of Asp1 side chain oriented toward the substrate-binding site controls access of N-terminal amino group of the substrate into the S2 binding pocket. In addition, glycosylation of Asn5 residue in the N-terminal region of exclusion domain improves the jamming of active-site cleft [26, 28]. Interestingly, attachment of carbohydrate rings to key residues in these blocking elements plays a pivotal role in stabilizing their structure and simultaneously controls the access to active-site cleft [16]. Therefore, cysteine cathepsins have no definite substrate selectivity as a result of multiple interactions adding to the challenges faced in designing specific inhibitors targeting these proteases.

11.4 Classification

Cathepsins have been classified on the basis of their structure, active site, and substrate specificity into three different categories: (1) aspartic proteases (cathepsins D and E), (2) serine proteases (cathepsins A and G), and (3) cysteine proteases consisting of 11 lysosomal cathepsins (B, C, F, H, L, K, O, S, V, X, and W) [29]. Papain-like proteases are the most abundant among the cysteine proteases belonging to the C1A (clan CA) family [14]. Members of cysteine cathepsins exhibit diverse catalytic activity, and a majority of them are endopeptidases (F, O, S, K, V, L, and W) with wide variations in their substrate specificity. Additionally, cathepsins H and B possess both endo- and exopeptidase activities, whereas cathepsins C and X are exclusively amino- and carboxypeptidase, respectively [30]. Based on the length and amino acid sequence of their proregion cysteine, cathepsins have been further divided into "cathepsin L-like" (cathepsins L, K, S, H, and V) and "cathepsin B-like" (cathepsins B, C, O, and X) groups. With the exception of cathepsin H, cathepsin L subfamily proregion is approximately 100 (96–99) amino acid residues and contains consensus sequences ERF(X)NIN and GxNxFxD, whereas proregion of "B-like" cathepsins only contains GxNxFxD motif and exhibits a wide variation (38–206 amino acids) in its length [31]. Subsequent studies identified two more cysteine cathepsins (F and W) and based on phylogenetic analysis placed them in yet another group "F-like cathepsins." The propeptide region of this group of cathepsins contains a unique ERFNAQ motif [32].

11.5 Biosynthesis, Trafficking, and Localization

Cysteine cathepsins are synthesized as pre-proenzymes having N-terminal signal peptide that directs the enzyme into the lumen of endoplasmic reticulum (ER) (Fig. 11.4). The signal peptide is cleaved, and the inactive precursor undergoes N-linked glycosylation subsequent to its entry into the ER [16]. High mannose oligosaccharide on the protein is then phosphorylated in the trans-Golgi network (TGN) to mannose-6-phosphate for its further targeting to the lysosomes. Cathepsins are transported to the endosomal/lysosomal compartment by mannose-6-phosphate receptor (M6PR) pathway [3]. Activation of cathepsins occurs either by the low pH of the endosomes or through proteolytic processing by other protease such as pepsin, legumin, or cathepsin D [33]. Dissociation of the proregion is an autocatalytic event for endopeptidases such as the cathepsins B, H, L, S, and K, whereas the exopeptidases including cathepsins C and X are processed by other endopeptidases, such as the cathepsins L and S [34]. In the acidic environment of lysosomes, the mature enzyme exists as single- or double-chain forms linked by disulfide bonds [16]. Cysteine cathepsins being intracellular proteases are usually but not exclusively localized in the acidic and reducing endo-lysosomal vesicles. Under certain circumstances, they are targeted to the plasma membrane or even secreted [13]. Cell-matrix interactions have been suggested to influence their localization and activity [3]. Various negatively charged molecules such as glycosaminoglycans (GAGs) play a major role in autocatalytic activation of cysteine cathepsins [30, 35]. GAG binding induces a conformational change in the zymogen structure thus, weakening the interaction between the propeptide and active domain of the enzyme, thereby exposing the active site for processing [36]. Negatively charged GAG and polysaccharide (dextran sulfate) binding to the cathepsins facilitate the autocatalytic activation and confer protection against pH-induced changes under various pathological conditions [35]. Mature cathepsins may also be secreted into the extracellular space from late endosomes or lysosomes by Ca^{2+}-mediated fusion of lysosomes with the cell membrane [37]. Relatively small but a significant portion (approximately 5%) of cathepsins is secreted as proenzyme from the TGN into the extracellular milieu instead of them being targeted to the lysosomes. Particularly in keratinocytes and thyrocytes, mature cysteine cathepsin from late endosomes/lysosomes enters the retrograde transport vesicles for their extracellular secretion [38].

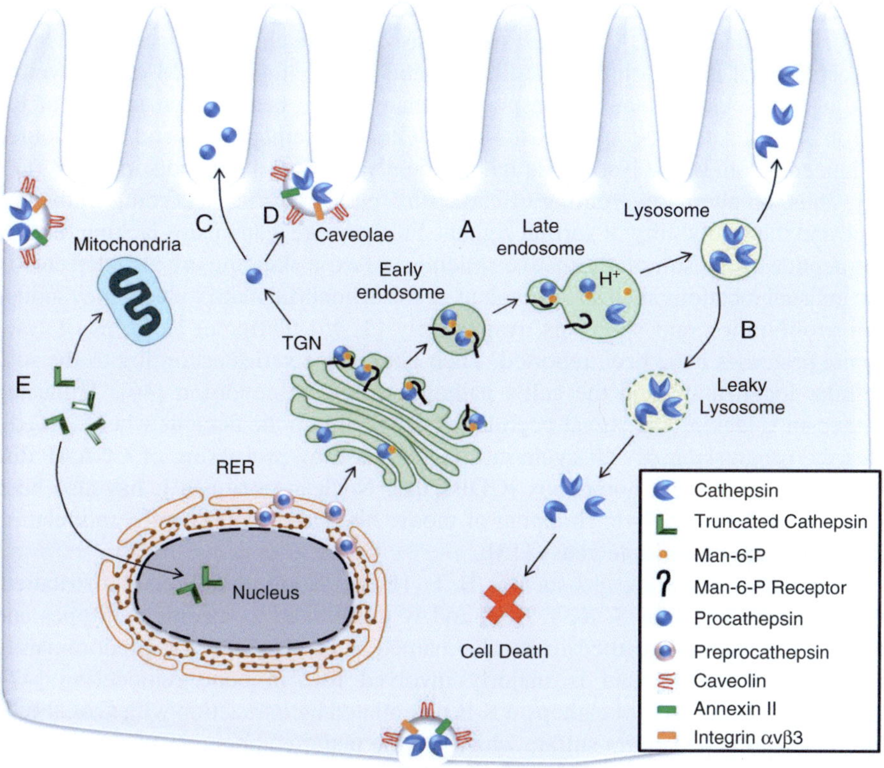

Fig. 11.4 Synthesis, sorting, and subcellular localization of cathepsins. Cysteine cathepsins are synthesized as pre-proenzymes having N-terminal signal peptide that guides entry of the polypeptide chain into the ER. Signal peptide is removed and three-dimensional structural confirmation is attained with the help of the proregion in the ER. Procathepsin undergoes disulfide bond formation and glycosylation with high mannose glycans that are later targeted to the TGN network. Procathepsins are then tagged with M6P, which is used to target the protein to the endosomal/lysosomal compartment through M6PR. Activation of cathepsins occurs either by the low pH of the endosomes or through proteolytic processing by other protease and is transformed into disulfide-linked heavy and light chains in the lysosomes (**A**). Leakage of lysosomal proteases can trigger the cell death pathways (**B**). A portion of the cathepsins that escape the phosphorylation of mannose residues pass through the exocytosis into extracellular compartments. Cysteine cathepsins being intracellular proteases are usually but not exclusively localized in the acidic and reducing endo-lysosomal vesicles. Relatively small but a significant portion (approximately 5%) of cathepsins is secreted as proenzyme from the TGN into the extracellular milieu (**C**). Cathepsins are either expressed by the transformed cells on their cell surface (caveolae) or secreted into the extracellular locations for tumor invasion and metastasis (**D**). Truncated cathepsins lacking the signal peptide as a result of alternative splicing and exon skipping are also detected in the unusual locations such as the nucleus or mitochondrial matrix (**E**)

Secretion of active cathepsin from epithelial cells is required for proenzyme processing and ECM remodeling in the pericellular environment [39, 40]. Cathepsins are either expressed by the transformed cells on their cell surface or secreted into the extracellular locations for tumor invasion and metastasis. Localization of cathepsin

B has been extensively demonstrated in the microdomains of the plasma membrane such as caveolae [41–43]. This cell surface-associated cathepsin B is involved in the intraepidermal migration of keratinocytes and remodeling of ECM during wound healing [44]. An alternative transport mechanism has been shown in case of the fibroblasts, hepatocytes, and breast tumors where procathepsins B and H are routed to late endosomes and lysosomes independently of M6P-mediated sorting at TGN [3]. Thus, an alternative routing of cathepsins perhaps varies according to the cell type and the availability of sorting signals. Furthermore, cathepsins lacking the signal peptide as a result of alternative splicing and exon skipping are also detected in the unusual locations such as the nucleus or mitochondrial matrix where they induce cell proliferation and apoptosis, respectively [3, 45]. Different isoforms of lysosomal proteases have been reported. Their abundance varies according to the subcellular localization and the cell's pathophysiological condition [46]. Truncated cathepsin L lacking the signal peptide is trafficked into the nucleus where it mediates the progression of cell cycle into the S phase by processing of CCAAT displacement protein/cut homeobox (CDP/Cux). Nuclear cathepsin L has also been reported in the proteolytic chopping of mouse histone H3 tail, thereby modulating the expression of multiple genes [13].

Some cysteine cathepsins such as B, L, H, and X are ubiquitously distributed, whereas others such as S, K, V, F, C, and W are limited to specific cell types and tissues. For example, cathepsin K (also named as cathepsin O2) is predominantly found in osteoclasts and is majorly involved role in bone remodeling [47]. Collagenolytic activity of cathepsin K is potentiated by interaction with GAGs such as chondroitin and keratin sulfate which are the major components of cartilage tissues. Similarly, cathepsin W is expressed in CD8-positive cells [48]; cathepsin S by various antigen-presenting cells such as macrophages, dendritic cells, and lymphocytes [49]; and cathepsin V (also termed as L2) which shares similarities to cathepsin L is majorly expressed in the testis and thymus [50]. Cathepsin O, on the contrary, is highly abundant in colon cancer cell [51]. Increased secretion of cathepsins from tumor cells and activated immune cells, either as inactive form (procathepsins B, L, and X) or as active enzymes (cathepsins B, H, K, L, S, and X), suggests that their functions are controlled by their cellular localization [52].

11.6 Regulation of Expression and Activity of Cysteine Cathepsins

Cathepsins are intracellular proteases which function optimally at acidic pH and reducing environment of the lysosomes. These peptidases have also been described in the cell nucleus, plasma membrane, and cytoplasm, and many of them are secreted extracellularly in various physiological and pathological conditions. Proteolytic activity of these enzymes can be regulated in several ways including pH and redox status of the pericellular microenvironment, procathepsin activation, and inhibition by endogenous inhibitors [21] . Furthermore, expression of cathepsins like any other protein is also regulated at the level of transcription, mRNA splicing, translation, posttranslational modifications, and protein trafficking [52]. Due to the importance of these proteases in various pathologies and change in expression profile at various stages of disease development, their regulation has gained significant attention in the recent past.

11.6.1 Transcriptional and Translational Regulation

The rate of transcription which plays a key role in the regulation of gene expression is determined by the strength of the promoter. Promoters of both cathepsins L and B have high G + C content, lack canonical TATA box, and contain binding sites for several transcription factors including multiple Sp1 binding sites which resemble the features of housekeeping genes [52–55]. Both these cathepsins are overexpressed in mouse and human tumors as a result of binding of Sp1 and Sp3 to the GC-rich regions of their promoter [53, 55–57]. Furthermore, the binding sites for NF-Y, Sp1, and Sp3 are essential for the transcription of the gene encoding human cathepsin L in melanoma cells [58]. The expression of cathepsin L is upregulated by several growth factors, proinflammatory cytokines, oncogenes, and tumor promoters [59–61]. Angiogenesis-promoting vascular endothelial growth factor (VEGF) also upregulates cathepsin L expression in glioblastoma cells. A 47 base pair region containing Sp1 and AP4 motifs plays a critical role in conferring VEGF responsiveness to cathepsin L promoter [60]. Interestingly, treatment of peritoneal macrophages with interferon (IFN)-α decreases the expression of cathepsin L without affecting the levels of cathepsin S [62]. The regulatory region of human cathepsin L contains multiple CpG islands which are methylated in lymphoma cells, thereby silencing its expression in this malignancy [58]. Consistent with this observation, promoter methylation downregulates the expression of cathepsin L in chronic myelogenous leukemia patients during accelerated phase/blast crisis [63]. Erythroblast transformation-specific (Ets) family transcription factors regulate the transcription of cathepsins K and C during bone and breast cancer progression [52].

Cathepsins are encoded by multiple mRNAs that contain variable length of 5′ and 3′ untranslated regions (UTRs) as a result of alternate splicing [64]. For instance, human cathepsin L (hCATL) is encoded by at least five mRNA species, namely, hCATL A, AI, AII, and AIII and B [65]. Among them, the first four are generated by the alternate splicing of the same primary transcript. The full length of exon I (280 nucleotides) is retained in the transcript variant hCATL A, whereas 27, 90, and 145 nucleotides are spliced out from the 3' end of this exon to generate AI, AII, and AIII variants, respectively. HCATL AIII, the shortest variant, is most efficiently translated, whereas the longer mRNA species exhibit lower translational efficiencies [45, 65]. The fifth mRNA species hCATL B is transcribed from an alternate promoter located in the first intron of human cathepsin L gene. The alternate promoter like its proximal counterpart also lacks a canonical TATA box. However, it contains putative binding sites for several transcription factors like AP1, AP4, GATA-1, Lmo2, NF-kappa B, etc. [66]. The most efficiently translated cathepsin L transcript AIII is abundantly expressed in cancers which may explain the elevated levels of the protease in malignancy [65]. In line with this, splice variant of cathepsin B lacking Alu sequence containing exon 2 is translated more efficiently compared to other variant(s) which may be responsible for overexpression of cathepsin B in malignancies of the breast, colon, prostate, and brain [52, 67, 68]. Thus, alternative splicing of cathepsin L and B pre-mRNA results in differential expression of these proteases in different cell types. Hence, understanding of these molecular mechanisms is essential to delineate the role of different mRNA variants in tumors.

Interestingly, the longest human cathepsin L splice variant A contains a functional internal ribosomal entry site (IRES) which is involved in overexpression of cathepsin L by hypoxia [69–71]. This IRES besides providing a translational control may also account for discrete mitochondrial and nuclear localization of the cathepsin L which lacks the signal peptide at the N-terminus [13].

11.6.2 Zymogen Activation

As outlined above, N-terminal proregion of cathepsin precursor is essential for the proper folding and targeting of precursor to the endocytic compartments. Autoinhibitory proregion blocks the contact of substrate with active-site cleft by binding the enzyme in opposite orientation and extended confirmation [13]. In majority of cathepsins, this proregion forms the loose contact with intact zymogen structure by non-covalent interactions such as salt bridges and hydrogen bonding [62]. Cathepsin X exhibits an exceptional structure in which proregion is covalently bonded by disulfide linkage with the active-site cysteine residue, thus avoiding the autocatalytic activation [72]. Endopeptidases such as cathepsins B, H, L and K are initially autocatalytically activated at mildly acidic pH followed by proteolytic processing of the remaining procathepsin molecules [36].

11.6.3 pH

Lysosomal cathepsins function optimally at low pH (pH < 4.5) and have narrow functional pH, ranging from 4 to 6 [62]. Acidic pH of lysosomal compartments (pH < 5) destabilizes the interaction of the inhibitory propeptide and the active site and thus facilitates the movement of propeptide from the normal position [14]. Furthermore, proenzyme after conformational change in propeptide structure exhibits very less catalytic activity that is just enough for the activation of other procathepsins and hence initiates the bimolecular chain reaction [73]. Cathepsins not only exhibit enzymatic activity at acidic pH but several of them retain significant activity over a wide range of pH [18]. For example, cathepsin S is stable at pH ranging from 4.5 to 8.0 with pH optima of 6.0. This perhaps facilitates its role in antigen processing and presentation within less acidic compartments [74]. Ubiquitously expressed cathepsin B also exhibits stability at neutral pH 7 with half-life of 15 min and optima in the range of 4–6 presumably accounting for its presence in distinct locations [75], while other cathepsins such as L, H, K, V, and F retain only partial activity at neutral pH and, therefore, are less active outside lysosomes [76].

11.6.4 Inhibitors of Cysteine Proteases

The most critical and preeminent control of cathepsins that has been spilled into the cytosol is accomplished by their endogenous protein inhibitors including cystatins, thyropins, and serpins [16]. These inhibitors based on their binding potencies and physiological role have been divided into emergency and regulatory inhibitors [77]. Emergency inhibitors exhibit competitive and reversible binding and rapidly form a complex with enzyme that remains associated for longer duration. This class of inhibitors is more abundant than the enzymes and is localized in a distinct compartment from that of the target enzyme [78]. Certain delayed inhibitors such as blood plasma serpins are converted to emergency inhibitor by heparin. In contrast, regulatory type of inhibitors not only blocks but also modulates the proteolytic activity [79]. Cystatins were first found in chicken egg white and exhibited their inhibitory effects against only papain-related proteases [78]. These proteins have been well characterized and belong to the MEROPS family I25. In humans, there are 12 members in cystatin superfamily that have been classified into three families including stefins (type I), cystatins (type II), and kininogen (type III) [15]. Cystatins are low-molecular-weight (10–13 kDa) competitive inhibitors that interact reversibly with the lysosomal cysteine proteases and prevent the inadvertent tissue damage by regulating the misplaced cathepsins. Cystatins are not very selective, and the picomolar amount is enough to inhibit the endopeptidases [80]. An inhibitory domain of cystatin is composed of five antiparallel β-sheets enfolded around a central α-helix. The binding groove of papain is blocked by the wedge shape formed by flexible N-terminal and two β-hairpin loops of cystatin. The two hairpin loops of cystatins dock with the “prime” subsites of the substrate-binding sites, whereas the elongated N-terminus interacts with the “unprimed” subsites [78].

11.6.4.1 Stefins

Stefins are present in the cytosolic compartments along with the low physiological concentration in serum and are potent intracellular inhibitors due to their stability in a wide pH range [16]. They exist as non-glycosylated single chain of approximately 100 amino acid residues [77]. In humans, this class comprises of two members, cystatins A and B. Human cystatin A has been localized in the skin epithelium and blood cells. On the contrary, human cystatin B is widely distributed mainly in the cell cytoplasm. Interestingly, this inhibitor has also been detected in the nucleus where it regulates the cathepsin L-mediated processing of histone proteins [15].

11.6.4.2 Cystatins

Type II cystatins are the members of MEROPS subfamily I25B consisting of cystatins C, D, E/M, F, G, S, SN, and SA [15]. Unlike the stefins, these inhibitors are synthesized as pro-inhibitors with 20–26 amino acid residue signal sequence required for their secretion into extracellular milieu [79]. Similar to type I cystatins, they are usually non-glycosylated (except cystatins F and E/M), single-chain polypeptides but larger in size having 120 amino acid residues. The most salient feature is the presence of Pro-Trp and two conserved disulfide bonds at the

C-terminal segment where the respective cysteine residues are held 10–20 amino acids apart. The cystatins C, D, S, SA, and SN share >50% sequence homology, while cystatins E/M, F, and G are only <35% identical in sequence [81]. They are present in most biological fluids [77]. Cystatins S and SA are found in the saliva and seminal fluids; cystatin SN is present in the saliva and tears [81]. Human cystatin E is present in the amniotic fluid [15], and cystatin F also called as "leukocystatin" is primarily expressed by the immune cells and the spleen [77]. Cystatin F-deficient mice exhibit altered regulation of cysteine proteases due to impaired granule biogenesis in eosinophils resulting in defective immune response to combat the pathogens [82]. Cystatin G is mainly expressed in epididymal and spermatogenic cells [15]. Cystatin C (also known as post-γ-globulin) is one of the most thoroughly studied human cystatins. It is secreted into all body fluids, and its concentration is particularly high in seminal plasma and cerebrospinal fluid [81]. Cystatin C displays broad-spectrum selectivity against all the papain-like proteases and cannot distinguish endo- or exopeptidases. Human cystatin C is a potent "emergency inhibitor" which rapidly neutralizes the activity of lysosomal cathepsins that escaped in the extracellular milieu [15].

11.6.4.3 Kininogens

Type III cystatins belong to I25C subfamily and comprise of three members including high (HMWK, approximately 120 kDa) and low (LMWK, approximately 60 kDa) molecular weight kininogens in humans [83]. However, kininogen T has only been described in rats [81]. These glycosylated inhibitors are synthesized as preproteins. Both HMWK and LMWK are synthesized in the liver and are encoded by splice variants generated as a result of alternate splicing of the same primary transcript. Human kininogens contain N-terminal heavy and C-terminal light chain connected by a disulfide bond. Three cystatin-like domains (D1–D3) are present in their heavy chain. However, only pentapeptide (QVVAG)-containing domains D2 and D3 are able to inhibit cysteine proteases [15]. Cathepsins L, S, and H are strongly inhibited by type III human cystatins, whereas cathepsin B is only weakly inhibited. Higher concentration of kininogen is found in hepatocytes, spleen, and dermatophytes [81]. Both human kininogens are present in equal amount in the blood and thus act as a potent inhibitor of cysteine proteases in circulation [77].

11.7 Physiological Functions of Cysteine Cathepsins

Assigning the specific roles to the cathepsin has been a challenging task owing to the redundancy in their functions. However, lysosomal cysteine cathepsins perform various important physiological functions as their deficiency in mice results in various hereditary disorders. Traditionally, these peptidases were only thought to participate in nonspecific proteolysis within the lysosome. However, this view is rapidly changing, and these proteases are now being implicated in specific biological roles.

11.7.1 Extracellular Matrix (ECM) Degradation

ECM occupies the void space between the cells and provides the meshwork for holding the cells embedded within the tissues. This noncellular component is also an important mediator for cross talk between the cells, angiogenesis, wound healing, bone remodeling, and many other physiological processes. ECM is mainly composed of fibrous proteins (elastin, collagen, laminins, and fibronectins), proteoglycans, water, and minerals. Deregulated synthesis and remodeling of ECM have been attributed to fibro-proliferative disorders and cancers affecting almost every organ of the body. In healthy tissue, ECM homeostasis is mainly monitored by matrix-degrading proteases like MMPs, serine proteases, and cysteine cathepsins [84]. ECM degradation is not necessarily restricted to the extracellular milieu, but a number of ECM proteins are also acquired intracellularly by endocytosis [14]. ECM proteins such as fibronectin; laminin; elastin collagen types I, IV, and XVII; and tenascin C are well-characterized substrates for cysteine cathepsins [85]. Earlier, the responsibility of bulk degradation of matrix proteins was solely assigned to MMPs. However, this concept was challenged as treatment with MMP inhibitor does not confer protection against ECM-related pathologies such as cancer, atherosclerosis, and many other fibrotic conditions. Interestingly, pan-inhibitor E64D was able to overcome the bone resorption by inhibiting cartilage degradation in osteoporosis [14] and thus highlighted the specialized role of cathepsins in the maintenance of tissue architecture. Collagen is the most abundant and viscoelastic structural component of ECM that exists in right-handed triple superhelical confirmation. Highly abundant forms of collagen, i.e., types I and II, are hydrolyzed only by specific proteases such as MMP-1, MMP-8, MMP-13 and the cysteine protease, cathepsin K [21]. The presence of additional C-terminal hemopexin domain in MMPs equips them with the ability to unwind the collagen helical structure and hydrolyzes the peptide bond in collagen helix into three-fourths and one-fourth telopeptides which are in turn cleaved by other tissue proteinases [14]. In contrast, cathepsins lacking such special structural features can only cleave the non-helical telopeptide regions of collagens except for cathepsin K [86]. Cathepsin K contains additional GAG binding site located opposite to the substrate-binding site which forms an oligomeric complex with ECM. Cathepsin K interaction with chondroitin sulfate resembles "beads on a string-like arrangement" [87]. It cleaves within the collagen helix at multiple sites as opposed to MMPs that specifically attack the peptide bond between 775 and 776 amino acid residues [14]. Cathepsin V, on the other hand, has the ability to form the complex with GAGs but lacks the collagenolytic activity. Hence, the collagenolytic activity cannot alone be assigned to the formation of cathepsin-GAG complexes but to its unique ability to accommodate proline residues in the P1 and P2 positions [88]. However, cathepsin K-like variant of cathepsin L generated by mutating S2 subsite possessed the similar affinity for type I collagen but lacked the collagenolytic activity supporting the critical role of both active-site and GAG complex formation [89]. Cathepsin K along with cathepsins B, L, and S degrades the telopeptide region of the collagen into monomers [21]. Cathepsin K is overexpressed in osteoarthritis and rheumatoid arthritis patients, and its specific inhibitors display therapeutic effect [21]. As a result, it has been implicated as a major player in the pathogenesis of osteoarthritis [14]. Consistent with this view, genetic loss of cathepsin K in humans leads to pycnodysostosis characterized by osteopetrosis, acroosteolysis, spondylolysis, and bone fragility [90]. Cathepsin K knockout mice also develop osteopetrosis due to reduced bone resorption and increased bone formation [91]. In addition, cathepsin K degrades other bone matrix-associated proteins such as osteocalcin and osteonectin, playing a major role in the proteolytic degradation of the cartilage [14]. Other cathepsins such as B, L, and S also increase in the synovial fluids of rheumatoid arthritis which explains their function in collagen degradation and bone remodeling [92]. Genetic deficiency of cathepsin L in the mouse model of arthritis abrogated the bone inflammation and cartilage destruction confirming its role in joint destruction [93]. Elastin, another element of the ECM, is required for the elasticity and flexibility of the tissues. Human cysteine cathepsins V, S, and K exhibit significant elastin-degrading activities, whereas cathepsin L displays minimal elastolytic ability. Cathepsin K possesses elastin-degrading potential which exceeds that of all other mammalian elastases; as a result, inflammatory cells may use this protease to degrade elastic lamina [94]. In the absence of cathepsin K, elastin degradation is compensated by cathepsins L and S [21].

Proteoglycans are the complex macromolecules found in interstitial matrix and basement membrane of the tissues. These molecules consisting of core proteins and covalently attached GAGs have also been listed among the various substrates degraded by cathepsins [21]. GAGs form the complex with the cathepsins and thereby regulate the matrix degradation and turnover. For example, collagenase activity of cathepsin K is negatively regulated by dermatan, heparan sulfate, and heparin [95]. Binding of chondroitin sulfate inhibits the cathepsin S-mediated collagen degradation [96] and the elastolytic activity of cathepsins V and K [97]. Additionally, protein core of proteoglycans is also degraded by cathepsins. Aggrecan, the major proteoglycan present in the cartilage tissue, is cleaved by cathepsins such as B, K, L, and S [98]. Apoptotic endothelial cells secrete cathepsin L which proteolytically processes basement membrane proteoglycan perlecan releasing antiapoptotic endorepellin (LG3) from its C-terminal end [99]. Interestingly, the generation of neuroprotective LG3 peptide was attributed to the cathepsin B activity rather than that of cathepsin L during cerebral ischemia [100]. Moreover, the degradation of basement membrane constituents such as nidogen 1 and nidogen 2 is mediated by cathepsin S expressed by keratinocytes [101]. Cathepsins L and B are also capable of degrading laminin present in the basal membrane [21]. Cathepsin K was also suggested to play a crucial role in basement membrane remodeling as mice lacking this protease showed the elevated levels of collagen IV, laminin, E-cadherin, and occludin in the colon [102]. Cell adhesion protein such as E-cadherin and ß-2 integrins is also processed by the cathepsins B, L, and S secreted by tumor cells [103]. Henceforth, regulation of this network of proteases is vital for the ECM homeostasis.

11.7.2 Mediators of Cell Death

Cell death is a biological process by which unwanted cells are eliminated through apoptosis, necrosis, necroptosis, and autophagy [104]. All these pathways sometimes work simultaneously or in tandem to decide the cell fate [105]. Lysosomal damage and subsequent drainage of its components in the cytoplasm can trigger the cell death pathway [106, 107]. Lysosomal membrane permeabilization (LMP) can progress to apoptosis, necrosis, autophagy, or necroptosis depending upon the type of cellular injury, leakage of the cathepsins, as well as the expression of their inhibitors [80, 108]. LMP is triggered by various agents such as reactive oxygen species (ROS), lipid metabolites, lysosomotropic compounds, as well as by proapoptotic factors such as Bax [109]. ROS such as hydrogen peroxide enter the lysosomes and get transformed into highly reactive hydrogen oxide free radical in a step that is facilitated by lysosomal iron that disrupts the lysosomal cell membrane by lipid peroxidation [108].

Apoptosis (derived from the Greek word meaning fall off) is a programmed cell death characterized by ATP-dependent biochemical pathways, defined morphological changes, and activation of executioner caspases in the dying cell. The apoptotic response is induced by either an intrinsic or extrinsic pathway based upon the source of cellular stress. Microinjecting cathepsins into the cytoplasm can experimentally induce apoptosis [110]. LMP can either be an initiating event leading to the caspase signaling cascade or it can just be a supportive event amplifying the death signals independent of caspase [111]. Release of cathepsins as a result of LMP triggers the further downstream events of an intrinsic apoptotic pathway. Cathepsins B, H, L, S, and K have been shown to activate the Bcl-2 family member proapoptotic Bid to truncated (t-Bid) form which then facilitates the oligomerization of Bax and Bak proteins [80]. This proapoptotic complex then translocates to mitochondria and induces mitochondrial outer membrane permeabilization (MOMP) by forming the pores [108, 112]. Cysteine cathepsins further facilitate apoptosis by degrading anti-apoptotic proteins such as Bcl-2 and Bcl-xL. They can also directly activate caspase-3, caspase-7, and caspase-9 as well as degrade X-chromosome-linked inhibitor of apoptosis (XIAP) and thereby expand the downstream apoptotic cascade [113]. The proapoptotic role of cathepsin B is further supported by the fact that genetic or pharmacological inhibition of this protease reduced apoptosis in various experimental models of liver injury including tumor necrosis factor (TNF)-α-mediated hepatocyte apoptosis [114–117], obstructive cholestasis [118], hepatic ischemia-reperfusion injury [119], and lipotoxicity [120]. Treatment with cholestasis causing toxic bile salt, glycochenodeoxycholate, induces LMP, cathepsin B translocation, caspase activation, and cell death in rat hepatocytes. Caspase inhibitors and overexpression of the cathepsin inhibitor cystatin A reduced this toxicity [121]. Cathepsin B translocation from lysosomes to the cytosol causes mitochondrial dysfunction and cytochrome c release which in turn induce apoptosis [120]. Redistribution of cathepsins to cytosol may activate classical mitochondrial pathway of apoptosis. This redistribution may also induce caspase-dependent or caspase-independent apoptosis [80]. LMP occurs as an early event followed by the release of cathepsin B into cytosol which induces caspase-independent apoptosis in response to microtubule-destabilizing drugs in non-small cell lung cancer cells [122]. However, the proapoptotic effect of cathepsin B is exerted through trypsinogen activation specifically in the case of the pancreas [123]. Cathepsin L contributes in palmitic acid-induced lipotoxicity in neuronal cells (PC12) as its pharmacological inhibition attenuates LMP, MMPs, and apoptosis [124]. In contrast, cathepsin L also exhibits antiapoptotic role in cancer cells because its downregulation sensitizes these cells to apoptosis in response to chemotherapeutic agents [125, 126].

Mitochondrial, lysosomal, and plasma membrane permeabilization are critical features of necrosis. LMP is a delayed event in H_2O_2-mediated necrosis, while it is an early step in TNF-α-induced necroptosis [108]. Cathepsin L cleaves DNA topoisomerase I during necrosis in diffuse systemic sclerosis. This is further corroborated by the fact that L929 cells undergoing necrosis exhibit higher levels of cathepsin L diffused into the cytoplasmic and nuclear compartments [127]. Similarly, ionophore toxin nigericin induces cathepsin B redistribution causing caspase-1 activation and interleukin (IL)-18 generation followed by necrosis in monocytic THP-1 cells [128]. Different cathepsins are involved in different types of adjuvant-mediated cell necrosis. For example, cathepsin C is involved in leucyl-leucine methyl ester-triggered necrosis, whereas cathepsins B and S mediate alum-associated cytotoxic effects [129]. Intracellular proteins are targeted to lysosomes by autophagy for cathepsin-mediated degradation [111].

11.7.3 Antigen Presentation

Various cysteine cathepsins play important roles in MHC class II antigen presentation by degrading the antigenic peptides and processing of invariant chain (Ii) [130]. Specific cathepsins comprising L, S, F, and V are expressed by different types of antigen-presenting cells (APCs) for the maturation of their antigenic complexes [16]. Cathepsin S is expressed by most of the antigen-presenting cells including dendritic cells, macrophages, and B cells. Spleen, lymph nodes, and vascular smooth muscle cells also express high levels of cathepsin S [74]. It is the most potent protease involved in immune response due to its activity over the broad pH range. Interestingly, cathepsin S-null mice exhibit impairment in the invariant chain (Ii) processing leading to deposition of MHC II complexes in the endosomes [2]. High levels of cathepsin S are secreted by the macrophages during various pathological conditions like rheumatoid arthritis, atherosclerosis, and bronchial asthma [62]. Due to the specific role of this protease in modulating the immune response, it is considered a major therapeutic target, and its commercial inhibitor "celera" has reached phase I clinical trial for the treatment of psoriasis [131]. Loss of cathepsin L in mice results in impaired Ii processing to the class II-associated invariant chain (CLIP) in the cortical thymic epithelial cells (cTEC), thereby confirming the role of this protease in antigen presentation [2]. This incomplete processing of Ii fragment in cathepsin L-deficient mice results in accumulation of I-Ab-bound p-12 and p18–22 Ii fragments leading to the defect in thymic selection of CD4+ T cells, whereas the cathepsin L-deficient cTECs and splenic APCs do not show any such accumulation of MHC II-bound Ii fragments [132]. However, cathepsin V that is exclusively present in the thymus and testis performs the same function in humans [16]. The p41 isoform of Ii specifically inhibits cathepsin L in APC where cathepsin S is the major protease involved in antigen presentation [131]. Cathepsins X, B, and H may also participate in antigen presentation, but they are not essential [133].

11.8 Cysteine Cathepsins and Human Pathologies

Given the role of cysteine cathepsins in maintenance of cellular homeostasis, alterations in their expression, localization, and activity have been associated with the development and progression of disorders like cancer [134], arthritis [92], neurodegenerative diseases [135], cardiomyopathies [136], obesity [137], liver fibrosis [138, 139], lung and autoimmune disorders [15, 140], as well as in viral and parasitic pathogenesis [141, 142]. However, in this chapter, the role of cysteine cathepsins only in the pathogenesis of cancer, cardiomyopathy, and lung and metabolic disorders has been discussed.

11.8.1 Cancer

Cancer is a multistage disease characterized by in situ development and proliferation of cancerous cells followed by dissemination of these cells to regional and distant organs by "metastasis." These metastatic cells leave the parent tumor and colonize the other tissues to form a secondary tumor. Central to this process of invasion and metastasis is the proteolysis of tissue scaffold holding the cells. As described earlier, cysteine cathepsins can degrade the constituents of epithelial basement membrane, cell-cell junctions, and ECM, which further facilitate intravasation and extravasation of the cancer cells. They also have additional specialized roles in various pro-tumorigenic processes like uncontrolled cell proliferation, signaling, angiogenesis, loss of cell contacts, migration that subsequently influences the tumor aggressiveness, and therapeutic resistance.

Elevated levels of cysteine cathepsins have been reported in cancers of the colon, brain, bladder, prostrate, breast, lung, ovary, head and neck, pancreas, skin, and bone [134, 143].Tumor-specific overexpression of various cysteine cathepsins is summarized in Table 11.2. High levels of both cathepsins L and B serve as prognostic markers in breast cancer and exhibit inverse correlation with disease-free and overall survival of these patients [144]. Prognostic significance of cathepsin B in patients with lymph node-negative disease is also documented [145]. Breast cancer cells that metastasize to different organs exhibit different expression pattern of these cathepsins. For example, the breast cancer cells that metastasize to brain express high levels of cathepsins B, C, S, and L, while those metastasizing to the lungs express high levels of cathepsins C, B, and L [146]. Interestingly, cathepsin L has been used to predict the outcome of systemic adjuvant hormone therapy in patients

Table 11.2 Tissue-specific overexpression of cathepsins in cancer

Disease	Cathepsin levels							References
	B	C	H	K	L	S	Z	
Colorectal cancer	✓				✓	✓		[232–235]
Breast cancer	✓	✓			✓	✓		[144]; [148]
Lung cancer	✓		✓		✓	✓		[156]; [143]; [157]
Pancreatic cancer	✓				✓			[134]
Islet cell tumor	✓				✓		✓	[103, 236]
Bone cancer				✓				[237]
Ovarian cancer	✓				✓			[150, 238]
Liver cancer	✓				✓			[139]
Skin cancer				✓				[158]
Pediatric acute myeloid leukemia	✓				✓			[153, 154]
Gallbladder cancer	✓				✓			[134]

with hormone receptor-positive breast cancer, and elevated levels of this protease suggest the poor outcome of the disease [147]. Levels of cathepsin H are also high in the serum and tissues of breast cancer patients [148]. Overexpression of cathepsins B and L has been reported in the serum and tissues of patients suffering from ovarian cancer [149, 150]. Similarly, cathepsins B and L are also increased in atypical invasive and aggressive meningiomas [151], gliomas [152], and hepatocellular carcinoma [139] compared to their benign counterparts and may serve as potential diagnostic markers. Expression of cathepsins L and B increases in parallel with histological grade of pancreatic adenocarcinoma and correlates with short overall survival after the surgical resection [134]. Increased levels of cathepsins L and B in pediatric acute myeloid leukemia patients are strong markers for poor prognosis of the disease [153, 154]. However, cathepsins B and H and plasminogen activator inhibitor (PAI)-1 are considered more sensitive biomarkers and have a major prognostic value in colorectal cancer [155]. As in other cancers, cathepsins B, L, H, and S are also upregulated in lung cancer patients [143]. Enhanced levels of cathepsin B correlate with hematogenous and intrapulmonary metastases of lung cancer cells [156]. Moreover, tumors and tumor cell-infiltrated lymph nodes have high cathepsin B activity which may be used as the predictor of poor prognosis in lung carcinoma [157]. Similarly, stromal fibroblasts in squamous cell carcinoma of the skin overexpress cathepsin K, which in turn promotes tumor invasion and metastasis by ECM degradation and vascularization [158]. Overexpression of cathepsins B and L is also reported in gall bladder cancer patients [159].

Formation of new blood vessels enhances the tumor vascularization and helps tumor cells to reach the bloodstream and metastasize to the secondary sites. During angiogenesis, endothelial cells proliferate, migrate, and invade the surrounding perivascular stroma, forming tube-like structures that give rise to neocapillary network. Various cysteine cathepsins such as B, L, S, and X have been implicated in angiogenesis [19, 134, 160]. Several pro-angiogenic factors and inhibitors are also the substrates for these cathepsins [14]. The use of the broad-spectrum inhibitor of cysteine cathepsins in the Rip1-Tag2 mouse model established the role of these proteases in angiogenesis, tumor growth, and invasiveness within pancreatic islet tumors. In these mice, genetic inhibition of cystatin C increases the formation of vascular networks, while cathepsin S deficiency leads to impaired tumor angiogenesis and invasion [103]. Cathepsin S-deficient mice display defective microvessel development despite high levels of VEGF and basic fibroblast growth factor (bFGF) and promote angiogenesis by the degradation of anti-angiogenic peptides canstatin and arresten [14]. Si-RNA-mediated knockdown of cathepsin S expression reduces cell proliferation, invasion, and angiogenesis in human hepatocellular carcinoma [161]. Similar to the findings in Rip1-Tag2 pancreatic cancer mouse model, cathepsin S overexpressed and secreted by both tumor and tumor-associated cells mediates tumor growth and vascularization in the syngeneic model of colorectal cancer. Loss of cathepsin S in these mice abrogates the formation of new blood vessels, cell growth, and viability and thus further establishes the pro-tumorigenic role of this protease [162]. Compared to benign tumors, increased cathepsin B levels are diffusely distributed in microvessel of neoplastic prostate cancer. Cell surface-associated overexpression of this protease is seen in highly metastatic prostate cancer [163]. Similarly, downregulation of cathepsin B reduces the aggressiveness and angiogenesis in gliomas [164, 165]. Cathepsin H is required for the development of blood vessels, tumor growth, and invasion in the mouse model of pancreatic islet cancer [166]. Although cathepsin L expressed by endothelial progenitor cell is required for ischemia-mediated neovascularization, its role in tumor angiogenesis is still not clear as deletion of this protease had no significant impact on the angiogenesis in Rip-Tag2 mice [52]. However, cathepsins L and S favor the generation of laminin-derived pro-angiogenic factor-gamma 2 and induce the neoplastic progression [156, 167]. Anti-angiogenic effects of cathepsins L and S are mediated by cleaving C-terminal of collagen XVIII that leads to endostatin formation [168]. Cathepsin B also contributes to angiogenesis by degrading the TIMPs which leads to increase in the angiogenesis-promoting MMP activity [169].

Cathepsin activity and expression are mainly localized at the invasive edges of the tumors which have been attributed to the ectodomain shedding of E-cadherin, transmembrane proteins, and other cell surface-associated molecules enhancing the invasion and migration of cancer cells [170]. Invasiveness of highly metastatic melanoma cells is assigned to cathepsins B and L, and their downregulation and pharmacological inhibition impair the invasive potential of human melanoma cells in matrigel invasion assays [143]. Furthermore, overexpression of cathepsin L confers highly invasive phenotype to nonmetastatic melanoma cells. Similarly, reduction in cathepsins B, L, H, and S expression/activity lowers the invasive ability of glioblastoma cells [143]. Likewise, cathepsin B facilitates the invasion of esophageal cancer fibroblasts, and cathepsin H performs the same role in prostate cancer [160]. It has been proposed that cathepsin H mediates the processing of talin (actin- and β-integrin tail-binding protein) which promotes activation of integrins and consequently migration of prostate cancer cells [171]. Cathepsin X removes C-terminal Tyr139 of profilin1, which abolishes its tumor-suppressor function. This cleavage also abrogates its ability to bind clathrin and enhance prostate cancer cell migration and invasion [172]. Coronin-3, a protein involved in the regulation of cytoskeletal dynamics, facilitates gastric cancer cell migration and invasion by increasing the expression of MMP-9 and cathepsin K [173]. Cancer cells that metastasize to the bone secrete large amount of cathepsin K [174]. The invasive ability of breast cancer cells is associated with the proteolytic activity of cathepsin B at the tumor cell surface [175]. Also, cell surface-associated clusters of proteases with cathepsin B detected in caveolae of human colorectal carcinoma cells degrade type IV collagen, thereby augmenting their invasive and migratory potential [176]. High cathepsin B activity is also implicated in the pathogenesis of invasive and metastatic thyroid carcinomas [160].

Cysteine cathepsins such as cathepsins L and Z are also involved in epithelial-mesenchymal transition (EMT), another vital feature of tumorigenesis. Cathepsin L plays an important role in transforming growth factor (TGF)-β1-mediated EMT, and its downregulation reduces the migration and invasion of epithelial cancer cells [177]. Similarly, its upregulation induces EMT in gastric cancer [178]. However, cathepsin Z stimulates EMT in metastatic hepatocellular carcinoma by upregulating the mesenchymal markers (fibronectin and vimentin) and downregulating the epithelial markers (E-cadherin and α-catenin). This protease also exerts its metastatic effect by influencing the ECM degradation through activation of other proteases such as MMP-2, MMP-3, and MMP-9 [179].

Tumor metastasis relies on its surrounding microenvironment, which is a rich source of proteases aiding the dissemination of the cancerous cells. Some proteases directly impact the tumor growth and invasion, while others indirectly regulate the expression of tumor- promoting molecules and signaling. Each cancer depending upon its origin expresses a distinct set of proteases which promotes tumor progression. Therefore, in-depth understanding of these proteases and their behavior in the tumor niche may prove useful in designing strategies for cancer therapeutics and management.

11.8.2 Lung Diseases

Despite the crucial role of cysteine cathepsins in the maintenance of lung microenvironment, their deregulated expression and activity have been implicated in several lung pathologies such as fibrosis, asthma, bronchopulmonary dysplasia, chronic obstructive pulmonary disease (COPD), and silicosis [15]. Cysteine cathepsins display distinct immunostaining patterns in normal human lung tissue. Bronchial and alveolar epithelial cells display intense staining for cathepsin K [180]. Cathepsin S is located on the surface of ciliated cells and may favor the motility of cilia by preventing unspecific binding with circulating plasma-derived proteins [181]. Cathepsins B and L are predominantly expressed in bronchial epithelial cells and protect against airborne foreign particles and microbes. However, significant amounts of cathepsin H are detected in macrophages, bronchial epithelial cells, and type II pneumocytes [180]. Type II alveolar epithelial cells also express cathepsin C, while X and S are mainly present in macrophages [182].

In lung fibrosis, the balance between repair and inflammatory pathway is regulated by multifaceted cross talk between the cells and surrounding ECM [15]. Inflammatory cells (neutrophils and macrophages) in the airway get activated in response to the lung damage and secrete proteases along with their inhibitors which then disturb ECM homeostasis and alter the lung architecture [15]. Fibroblasts from patients diagnosed with pulmonary fibrosis display higher-level expression and activity of cathepsin K than the normal lung specimens. In line with this observation, cathepsin K expression is temporally upregulated in bleomycin-induced lung fibrosis mouse model. Cathepsin K-null mice after bleomycin treatment display aggravated fibrosis due to increase in the ECM deposition, upregulation of α-SMA and vimentin, and decrease in collagen degradation by fibroblasts from these mice [182]. Similarly, exposure of experimental animals to crystalline silica particles induces pulmonary expression of cathepsin K [15]. The role of cathepsin K in preventing silicosis is further strengthened by the observation that in response to silica particles, silica-sensitive mouse strain (C57BL/6) expresses low levels of cathepsin K mRNA in comparison with the resistant strain (BALB/c) [15]. The anti-fibrotic role of cathepsin K is further confirmed by its ability to degrade fibrogenic cytokine TGF-β1 and thus diminish the ECM accumulation [183]. Bronchoalveolar lavage fluid (BALF) from silicosis patients has high levels of active cysteine cathepsins such as B, H, K, L, and S [184]. Interestingly, loss of cathepsins B and L lowers the expression of α-smooth muscle actin (SMA) in the fibroblasts from idiopathic pulmonary fibrosis patients. In addition, TGF-β1-mediated transdifferentiation of fibroblast is prolonged after treatment with cathepsin B inhibitor. This finding is further confirmed by the observed increase in levels of cystatin C as opposed to cathepsin B during TGF-β1-dependent differentiation of fibroblasts. Elevated cystatin C levels inhibit cathepsin B and facilitate TGF-β1-mediated pulmonary fibrosis [185]. Cathepsin B is also upregulated in models and patients with interstitial lung disease. Inflammation and progression of experimental pulmonary fibrosis can be attenuated by the use of CA-074 Me, a specific inhibitor of cathepsin B [186]. On the contrary, reduction in the extent of bleomycin-induced lung fibrosis by curcumin is associated with the induction of cathepsins K and L expression [185].

COPD includes emphysema and chronic bronchitis mainly caused by cigarette smoking and inhalation of particulate pollutants. Levels of cathepsins B, L, and S as well as their endogenous inhibitor cystatin C are increased in smokers afflicted with COPD [187]. The lungs of emphysema patients express high levels of cathepsin K [15]. Similarly, cathepsins B, L, H, K, and S are upregulated in IL-13 and IFN-γ transgenic mouse models of emphysema. Treatment of these transgenic mice with a common inhibitor of cysteine cathepsins attenuates lung inflammation and emphysema and hence establishes the pathogenic role of these proteases in COPD [180]. Ozone-induced hyperresponsiveness and inflammation in BALB/c mice are associated with increased BAL levels of cathepsin S. Treatment of these mice with cathepsin S inhibitor decreased the levels of proinflammatory cytokines IL-6 and TNF-α. These results further confirmed the role of cathepsin S in the oxidative stress-induced airway hyperresponsiveness and suggest its utility as a potential therapeutic target [188]. It is possible that cathepsin S mediates proteolysis of pulmonary ECM, basement membrane, and secretory leukocyte peptidase inhibitor (SLPI). Degradation of SLPI, the endogenous inhibitor of human neutrophil elastase, tilts the balance toward the breakdown of elastin causing loss of lung elasticity and hence emphysema [189].

Involvement of cysteine cathepsins in the pathogenesis of cystic fibrosis is extensively documented. High levels of active cathepsins B, L, and S have been detected in BALF of cystic fibrosis patients compared to healthy subjects [15]. Cathepsins B and S have been proposed as sputum markers of inflammation due to their remarkable correlation with IL-8 and neutrophil elastase [190]. Cathepsins worsen cystic fibrosis by degrading the antimicrobial molecules such as surfactant protein A, lactoferrin, SLPI, and human β-defensin-2 and β-defensin-3, thus increasing the vulnerability of the lungs to infection [184]. Furthermore, uncontrolled proteolysis is enhanced as a result of hydrolysis of anti-proteinase such as human α_1-proteinase inhibitor by cathepsin L and kininogens by cathepsin B [15, 184]. Cathepsin S has also been proposed as a biomarker for asthma pathogenesis [191]. It is upregulated in experimental model of ovalbumin-induced allergic inflammation where the use of selective inhibitor against this protease abrogated the inflammatory response [192]. In addition, cathepsin F has also been linked to the heightened immunoreactivity during asthma [193].

11.8.3 Cardiovascular Disorder

Cardiovascular dysfunctions such as hypertension, hypertrophic cardiomyopathy, dilated cardiomyopathy, diabetic cardiomyopathy, myocardial infarction, atherosclerosis, aortic aneurysm, neointima formation, and neovascularization are characterized by extensive ECM degradation and remodeling, a process in which the involvement of various cysteine cathepsins has been discussed earlier [194]. The major cysteine proteases implicated in cardiac dysfunctions are cathepsins B, L, S, and K [195]. The role of these proteases in the cardiovascular disorders has been established by studies in mouse model, patients, and cultured cardiac cells [196]. Angiotensin II, superoxide radicals, and cytokines stimulate the levels of these cathepsins, and their overexpression in the heart correlates with disorders like hypertension, coronary artery disease, and atherosclerosis [197].

Cathepsin L knockout mice exhibit abnormal heart rhythms and develop features resembling human dilated cardiomyopathy characterized by interstitial myocardial fibrosis and the appearance of pleomorphic nuclei in cardiomyocytes at 1 year of age [198]. Cardiac fibrosis in cathepsin L-deficient mice results from decreased ECM degradation due to lack of cardiac fibroblast-derived cathepsin L [198]. Pressure overload in cathepsin L knockout mice leads to activation of cellular stress pathways that aggravate cardiac hypertrophy and dysfunction [199]. In contrast to cathepsin L−/− mice, knockout of cathepsin K prevents contractile dysfunction and cardiac hypertrophy [196]. Cardiac hypertrophy is caused due to inflammation, fibrosis, and apoptosis, which engage Akt/GSK-3β pathways that are inactivated by cathepsin L [199]. The balance between cysteine cathepsins and their endogenous inhibitor cystatin C plays a vital role in the normal myocardial ECM remodeling. Consistent with this view, hypoxia-induced cardiac failure is associated with elevation of cystatin C levels, inhibition of cathepsin B activity, and myocardial deposition of collagen and fibronectin [200]. However, cathepsin B overexpressed in myocardial necrotic zone has been implicated in cardiac cell death [201]. In contrast to results in experimental models, levels of cathepsins B, L, S, and K are elevated in human dilated and hypertrophic cardiomyopathies [13]. Elevated expression of cathepsin L is seen in human abdominal aortic aneurysm and atheromata. High serum levels of this cathepsin show strong positive correlation with arterial stenosis suggesting its involvement in vascular diseases [202].

Atherosclerotic lesions involving arterial wall remodeling are also associated with high activities of elastase, collagenase, and gelatinase [203]. Cysteine cathepsins due to their potent elastinolytic and collagenolytic activities lead to the genesis and rupture of atherosclerotic plaques [136]. Apo E-deficient murine atherosclerotic lesions exhibit high levels of cathepsins B, L, and S [194]. Cathepsins K, S, and L are overexpressed in human atherosclerotic plaques. These proteases are mainly localized in the macrophages, smooth muscle cells, and fibrotic and lipid-rich areas of the plaque [137]. Expression of cathepsins K and S in the vascular wall facilitates the elastin proteolysis which aids in smooth muscle cell migration and collagen degradation of the fibrous cap [136]. Cathepsin K and S expression in the endothelium of atherosclerotic coronary arteries positively correlates with the breaks in elastic lamina, thus substantiating their role in neovascularization of atherosclerotic lesions [204, 205]. Pro-inflammatory cytokines such as TGF-β, IFN-γ, and IL-1β induce cathepsin L and S expression in human aortic smooth muscle cells and macrophages [137, 202, 205]. The role of cysteine proteases in the development of atherosclerosis has been studied in various knockout mouse models. Phenotypes associated with the knockout of these proteases in the proatherogenic genetic background have been summarized in Table 11.3. These findings highlight the pathogenic role of cysteine cathepsins in atherosclerosis.

Table 11.3 Role of cathepsins in atherosclerosis elucidated using experimental mouse model

Genotype of mouse model		Vascular phenotype					
Cathepsin deletion	Genetic defect leading to atherosclerosis	Plaque size	Plaque progression	Collagen content	Elastin breaks	Macrophages	Reference
K	Apo E	↓	↓	↑	↓	↑	[239–241]
K	LDLR	=		↓	↓	↑	[242]
L	LDLR	↓		↓	↓	↓	[243]
S	LDLR	↓	↓	↓	↓	↓	[241]
S	Apo E	↓	↓		↓		[244]

11.8.4 Metabolic Disorders

Diabetes and obesity are the most common human metabolic disorders. Apart from other factors, cysteine proteases such as cathepsins L, S, and K have been implicated in their pathogenesis [206–208]. Diabetes-prone C57BL/6J mice demonstrated differential expression of cathepsin L in their muscle tissues compared to the resistant strain of mice. Cathepsin L gene expression in muscle tissues inversely correlates with plasma glucose in these mice implying its association with glucose intolerance [207]. Interestingly, basal cathepsin L mRNA levels were found to be comparable in the muscle biopsies from monozygotic twin pairs discordant for type 2 diabetes and control subjects. However, cathepsin L mRNA levels were reduced by post-insulin clamp in diabetic twins that correlated with insulin-mediated glucose transport [207]. Cathepsin L deficiency confers protection against the insulitis and autoimmune type 1 diabetes (T1D) in diabetes-susceptible NOD mice [209]. Protective effects of cathepsin L deficiency seen in NOD mice have been attributed to increase in the ratio of T regulatory (Treg) cells that attenuate autoimmune response [208]. Similarly, deletion of cathepsins S and B conferred only partial protection against T1D [209]. In addition to protection by Treg cells, cathepsin L inhibition curtails the CD8+ T cells cytotoxicity by inhibiting granzyme B and hinders the development of T1D [210]. Cathepsin L degrades fibronectin, insulin receptor (IR), and insulin-like growth factor-1 receptor and plays an important role in fat deposition and glucose tolerance [211]. Cathepsin L knockout NOD mice have leaner phenotype with lower levels of serum glucose and insulin but accumulate IR-β subunits, glucose transporter, and fibronectin in their muscles accounting for enhanced insulin sensitivity and glucose utilization [211]. High cathepsin L levels in obese and diabetic patients further supported its involvement in the metabolic disorders [211]. Inhibition of the cysteine cathepsins B and L results in impaired autophagy, and accumulation of cathepsins B, D, and L pro-forms and triggers the caspase-dependent β-cell apoptosis cultured in hyperglycemic conditions [212].

Cathepsin S and H levels strongly correlate with pro-inflammatory cytokines in the tear glands of NOD mouse model of Sjögren's syndrome [213]. Compared to healthy subjects, patients with type 2 diabetes have increased serum cathepsin S levels [214]. This finding has been further corroborated by Jobs and coworkers, who reported that elevation in serum cathepsin S levels leads to decreased insulin sensitivity and higher susceptibility to develop type 2 diabetes [215]. Cathepsin S is overexpressed in adipose tissue of obese patients and therefore may serve as a biomarker for adiposity [216]. Its role as a biomarker is further corroborated by the observation that weight loss and amelioration of glycemic status following bariatric surgery lead to reduced serum cathepsin S levels [217]. Diet-induced obesity in cathepsin S −/− mice is associated with low blood and hepatic glucose and better glucose tolerance compared to wild-type littermates [218]. Therefore, cathepsin S inhibition may be of relevance in delaying the onset of diabetic phenotype.

Cathepsin K also is implicated in adipocyte differentiation and glucose metabolism [219]. White adipose tissue of obese db/db mice expresses higher cathepsin K levels as compared to the wild-type littermates. Like cathepsin S, levels of cathepsin K in white adipose tissue also display a decrease in obese mice undergoing weight loss, and hence it has also been proposed to be a marker and therapeutic target for obesity [220]. Subsequent studies suggest the involvement of cathepsin K in adipocyte differentiation and pathogenesis of obesity. Blocking of cathepsin K using specific inhibitors can impair conversion of pre-adipocyte into adipocytes during the early phase of cell differentiation [221]. Genetic ablation or pharmacological inhibition of cathepsin K leads to significant reduction in high-fat diet-induced weight gain and serum insulin and glucose levels with increase in fat and muscle fibronectin [222]. Upregulation of type I collagen as a result of cathepsin K inhibition abrogates pre-adipocyte differentiation [223]. Cathepsin K deficiency in mice also attenuates the high-fat diet-induced cardiac hypertrophy and pumping defects and alters subcellular distribution of intracellular Ca2+ in cardiac muscles [196]. Similar to reports in mice, cathepsin K levels are significantly higher in obese individuals and correlate with body mass index [221]. Thus, these consistent findings in mouse models and human subjects confirmed the role of cathepsin K in obesity, and its inhibition along with cathepsins S and L may potentially be used for treating this morbid disease.

11.9 Future Perspectives

Cathepsins were long thought to remain functional only within the confines of acidic lysosomal compartments, which implied that their role was limited to the intracellular protein degradation and turnover. On the contrary, a number of studies emphasized the role of cathepsin-mediated proteolysis in extracellular matrix degradation. Rampant expression, activity, or cytosolic escape of the lysosomal enzymes results in number of human pathologies. Cathepsins are usually overexpressed in cancer, and a majority of them are either secreted into the extracellular tumor microenvironment or targeted to the surface of malignant cells. Cathepsins B, L, and S have been used for diagnosis and predicting the outcome of chemotherapy in various malignancies. Several clinical conditions such as disorders of the lungs, heart, and kidney are caused due the overexpression and dysregulated ECM degradation. Specific inhibitors of these proteases may potentially be used to protect against the detrimental effects of their nonspecific proteolysis. However, targeting them to the specific site remains a challenge.

Disclosure of Potential Conflicts of Interest The authors declared no potential conflicts of interest.

References

1. Turk V, Turk B, Turk D (2001) Lysosomal cysteine proteases: facts and opportunities. EMBO J 20:4629–4633
2. Vasiljeva O, Reinheckel T, Peters C et al (2007) Emerging roles of cysteine cathepsins in disease and their potential as drug targets. Curr Pharm Des 13:387–403

3. Brix K, Dunkhorst A, Mayer K et al (2008) Cysteine cathepsins: cellular roadmap to different functions. Biochimie 90:194–207
4. Lopez-Otin C, Matrisian LM (2007) Emerging roles of proteases in tumour suppression. Nat Rev Cancer 7:800–808
5. Quesada V, Ordonez GR, Sanchez LM et al (2009) The Degradome database: mammalian proteases and diseases of proteolysis. Nucleic Acids Res 37:D239–D243
6. Lopez-Otin C, Bond JS (2008) Proteases: multifunctional enzymes in life and disease. J Biol Chem 283:30433–30437
7. Puente XS, Sanchez LM, Overall CM et al (2003) Human and mouse proteases: a comparative genomic approach. Nat Rev Genet 4:544–558
8. Rawlings ND, Waller M, Barrett AJ et al (2014) MEROPS: the database of proteolytic enzymes, their substrates and inhibitors. Nucleic Acids Res 42:D503–D509
9. Klingler D, Hardt M (2012) Profiling protease activities by dynamic proteomics workflows. Proteomics 12:587–596
10. Sevenich L, Joyce JA (2014) Pericellular proteolysis in cancer. Genes Dev 28:2331–2347
11. Fischer A (1946) Mechanism of the proteolytic activity of malignant tissue cells. Nature 157:442
12. de Duve C, Pressman BC, Gianetto R et al (1955) Tissue fractionation studies. 6. Intracellular distribution patterns of enzymes in rat-liver tissue. Biochem J 60:604–617
13. Reiser J, Adair B, Reinheckel T (2010) Specialized roles for cysteine cathepsins in health and disease. J Clin Invest 120:3421–3431
14. Brömme D, Wilson S (2011) Role of cysteine cathepsins in extracellular proteolysis. In: Extracellular matrix degradation. Springer, Berlin, pp 23–51
15. Lalmanach G, Saidi A, Marchand-Adam S et al (2015) Cysteine cathepsins and cystatins: from ancillary tasks to prominent status in lung diseases. Biol Chem 396:111–130
16. Turk V, Stoka V, Vasiljeva O et al (2012) Cysteine cathepsins: from structure, function and regulation to new frontiers. Biochim Biophys Acta 1824:68–88
17. Lecaille F, Kaleta J, Bromme D (2002) Human and parasitic papain-like cysteine proteases: their role in physiology and pathology and recent developments in inhibitor design. Chem Rev 102:4459–4488
18. Verma S, Dixit R, Pandey KC (2016) Cysteine proteases: modes of activation and future prospects as pharmacological targets. Front Pharmacol 7:107
19. Löser R, Pietzsch J (2015) Cysteine cathepsins: their role in tumor progression and recent trends in the development of imaging probes. Front Chem 3:37
20. Schechter I, Berger A (1967) On the size of the active site in proteases. I. Papain. Biochem Biophys Res Commun 27:157–162
21. Fonovic M, Turk B (2014) Cysteine cathepsins and extracellular matrix degradation. Biochim Biophys Acta 1840:2560–2570
22. Stoka V, Turk B, Turk V (2005) Lysosomal cysteine proteases: structural features and their role in apoptosis. IUBMB Life 57:347–353
23. Sivaraman J, Nagler DK, Zhang R et al (2000) Crystal structure of human procathepsin X: a cysteine protease with the proregion covalently linked to the active site cysteine. J Mol Biol 295:939–951
24. Musil D, Zucic D, Turk D et al (1991) The refined 2.15 a X-ray crystal structure of human liver cathepsin B: the structural basis for its specificity. EMBO J 10:2321–2330
25. Renko M, Požgan U, Majera D et al (2010) Stefin a displaces the occluding loop of cathepsin B only by as much as required to bind to the active site cleft. FEBS J 277:4338–4345
26. Turk D, Janjic V, Stern I et al (2001) Structure of human dipeptidyl peptidase I (cathepsin C): exclusion domain added to an endopeptidase framework creates the machine for activation of granular serine proteases. EMBO J 20:6570–6582
27. Guncar G, Podobnik M, Pungercar J et al (1998) Crystal structure of porcine cathepsin H determined at 2.1 a resolution: location of the mini-chain C-terminal carboxyl group defines cathepsin H aminopeptidase function. Structure 6:51–61

28. Turk D, Turk B, Turk V (2003) Papain-like lysosomal cysteine proteases and their inhibitors: drug discovery targets? In: Biochemical Society symposia. Portland Press Limited, London, pp 15–30
29. Rawlings ND, Barrett AJ, Bateman A (2012) MEROPS: the database of proteolytic enzymes, their substrates and inhibitors. Nucleic Acids Res 40:343–350
30. Fonović, M.Turk, B. (2014) Cysteine cathepsins and extracellular matrix degradation. Biochimica et Biophysica Acta (BBA)-general subjects 1840:2560-2570.
31. Karrer KM, Peiffer SL, DiTomas ME (1993) Two distinct gene subfamilies within the family of cysteine protease genes. Proc Natl Acad Sci 90:3063–3067
32. Wex T, Wex H, Brömme D (1999) The human Cathepsin F Gene a fusion product between an ancestral Cathepsin and Cystatin Gene. Biol Chem 380:1439–1442
33. Ishidoh K, Kominami E (2002) Processing and activation of lysosomal proteinases. Biol Chem 383:1827–1831
34. Dahl SW, Halkier T, Lauritzen C et al (2001) Human recombinant pro-dipeptidyl peptidase I (cathepsin C) can be activated by cathepsins L and S but not by autocatalytic processing. Biochemistry 40:1671–1678
35. Vasiljeva O, Dolinar M, Pungerčar JR et al (2005) Recombinant human procathepsin S is capable of autocatalytic processing at neutral pH in the presence of glycosaminoglycans. FEBS Lett 579:1285–1290
36. Caglic D, Pungercar JR, Pejler G et al (2007) Glycosaminoglycans facilitate procathepsin B activation through disruption of propeptide-mature enzyme interactions. J Biol Chem 282:33076–33085
37. Reddy A, Caler EV, Andrews NW (2001) Plasma membrane repair is mediated by ca(2+)-regulated exocytosis of lysosomes. Cell 106:157–169
38. Linke M, Herzog V, Brix K (2002) Trafficking of lysosomal cathepsin B-green fluorescent protein to the surface of thyroid epithelial cells involves the endosomal/lysosomal compartment. J Cell Sci 115:4877–4889
39. Guinec N, Dalet-Fumeron V, Pagano M (1993) "In vitro" study of basement membrane degradation by the cysteine proteinases, cathepsins B, B-like and L. digestion of collagen IV, laminin, fibronectin, and release of gelatinase activities from basement membrane fibronectin. Biol Chem Hoppe Seyler 374:1135–1146
40. Brix K, Herzog V (1994) Extrathyroidal release of thyroid hormones from thyroglobulin by J774 mouse macrophages. J Clin Invest 93:1388–1396
41. Arkona C, Wiederanders B (1996) Expression, subcellular distribution and plasma membrane binding of cathepsin B and gelatinases in bone metastatic tissue. Biol Chem 377:695–702
42. Mai J, Finley RL Jr, Waisman DM et al (2000) Human procathepsin B interacts with the annexin II tetramer on the surface of tumor cells. J Biol Chem 275:12806–12812
43. Cavallo-Medved D, Sloane BF (2003) Cell-surface cathepsin B: understanding its functional significance. Curr Top Dev Biol 54:313–341
44. Buth H, Wolters B, Hartwig B et al (2004) HaCaT keratinocytes secrete lysosomal cysteine proteinases during migration. Eur J Cell Biol 83:781–795
45. Abudula A, Rommerskirch W, Weber E et al (2001) Splice variants of human cathepsin L mRNA show different expression rates. Biol Chem 382:1583–1591
46. Mehtani S, Gong Q, Panella J et al (1998) In vivo expression of an alternatively spliced human tumor message that encodes a truncated form of cathepsin B. Subcellular distribution of the truncated enzyme in COS cells J Biol Chem 273:13236–13244
47. Drake FH, Dodds RA, James IE et al (1996) Cathepsin K, but not cathepsins B, L, or S, is abundantly expressed in human osteoclasts. J Biol Chem 271:12511–12516
48. Brown J, Matutes E, Singleton A et al (1998) Lymphopain, a cytotoxic T and natural killer cell-associated cysteine proteinase. Leukemia 12:1771–1781
49. Shi GP, Webb AC, Foster KE et al (1994) Human cathepsin S: chromosomal localization, gene structure, and tissue distribution. J Biol Chem 269:11530–11536

50. Bromme D, Li Z, Barnes M et al (1999) Human cathepsin V functional expression, tissue distribution, electrostatic surface potential, enzymatic characterization, and chromosomal localization. Biochemistry 38:2377–2385
51. Velasco G, Ferrando AA, Puente XS et al (1994) Human cathepsin O. Molecular cloning from a breast carcinoma, production of the active enzyme in Escherichia coli, and expression analysis in human tissues. J Biol Chem 269:27136–27142
52. Mohamed MM, Sloane BF (2006) Cysteine cathepsins: multifunctional enzymes in cancer. Nat Rev Cancer 6:764–775
53. Bakhshi R, Goel A, Seth P et al (2001) Cloning and characterization of human cathepsin L promoter. Gene 275:93–101
54. Qian F, Frankfater A, Chan SJ et al (1991) The structure of the mouse cathepsin B gene and its putative promoter. DNA Cell Biol 10:159–168
55. Jean D, Guillaume N, Frade R (2002) Characterization of human cathepsin L promoter and identification of binding sites for NF-Y, Sp1 and Sp3 that are essential for its activity. Biochem J 361:173–184
56. Sitabkhan Y, Frankfater A (2007) Differences in the expression of cathepsin B in B16 melanoma metastatic variants depend on transcription factor Sp1. DNA Cell Biol 26:673–682
57. Berquin IM, Sloane BF (1996) Cathepsin B expression in human tumors. In: Suzuki K, Bond JS (eds) Intracellular protein catabolis. Springer US, Berlin, pp 281–294
58. Jean D, Rousselet N, Frade R (2005) Expression of cathepsin L in human tumor cells is under the control of distinct regulatory mechanisms. Oncogene 25:1474–1484
59. Asanuma K, Shirato I, Ishidoh K et al (2002) Selective modulation of the secretion of proteinases and their inhibitors by growth factors in cultured differentiated podocytes. Kidney Int 62:822–831
60. Keerthivasan S, Keerthivasan G, Mittal S et al (2007) Transcriptional upregulation of human cathepsin L by VEGF in glioblastoma cells. Gene 399:129–136
61. Gerber A, Welte T, Ansorge S et al (2002) Expression of cathepsins B and L in human lung epithelial cells is regulated by cytokines. In: Cellular peptidases in immune functions and diseases, vol 2. Springer, New York, pp 287–292
62. Guha S, Padh H (2008) Cathepsins: fundamental effectors of endolysosomal proteolysis. Indian J Biochem Biophys 45:75–90
63. Samaiya M, Bakhshi S, Shukla AA et al (2011) Epigenetic regulation of cathepsin L expression in chronic myeloid leukaemia. J Cell Mol Med 15:2189–2199
64. Hook G, Jacobsen JS, Grabstein K et al (2015) Cathepsin B is a new drug target for traumatic brain injury therapeutics: evidence for E64d as a promising lead drug candidate. Front Neurol 6:178
65. Arora S, Chauhan SS (2002) Identification and characterization of a novel human cathepsin L splice variant. Gene 293:123–131
66. Seth P, Mahajan VS, Chauhan SS (2003) Transcription of human cathepsin L mRNA species hCATL B from a novel alternative promoter in the first intron of its gene. Gene 321:83–91
67. Gong Q, Chan SJ, BAJKOWSKI AS et al (1993) Characterization of the cathepsin B gene and multiple mRNAs in human tissues: evidence for alternative splicing of cathepsin B pre-mRNA. DNA Cell Biol 12:299–309
68. CANDANHIZEL MF, CURE H, PEZET D et al (1998) Evaluation of the 51 spliced form of human cathepsin B mRNA in colorectal mucosa and tumors. Oncol Rep 5:31–34
69. Jean D, Rousselet N, Frade R (2008) Cathepsin L expression is up-regulated by hypoxia in human melanoma cells: role of its 5′-untranslated region. Biochem J 413:125–134
70. Mittal S, Mir RA, Chauhan SS (2011) Post-transcriptional regulation of human cathepsin L expression. Biol Chem 392:405–413
71. Tholen M, Wolanski J, Stolze B et al (2015) Stress-resistant translation of cathepsin L mRNA in breast cancer progression. J Biol Chem 290:15758–15769
72. Nägler DK, Ménard R (1998) Human cathepsin X: a novel cysteine protease of the papain family with a very short proregion and unique insertions. FEBS Lett 434:135–139

73. Pungercar JR, Caglic D, Sajid M et al (2009) Autocatalytic processing of procathepsin B is triggered by proenzyme activity. FEBS J 276:660–668
74. Hsing LC, Rudensky AY (2005) The lysosomal cysteine proteases in MHC class II antigen presentation. Immunol Rev 207:229–241
75. Jordans S, Jenko-Kokalj S, Kühl NM et al (2009) Monitoring compartment-specific substrate cleavage by cathepsins B, K, L, and S at physiological pH and redox conditions. BMC Biochem 10:1
76. Turk B, Turk D, Turk V (2000) Lysosomal cysteine proteases: more than scavengers. Biochim Biophys Acta 1477:98–111
77. Turk V, Stoka V, Turk D (2008) Cystatins: biochemical and structural properties, and medical relevance. Front Biosci 13:5406–5420
78. Rzychon M, Chmiel D, Stec-Niemczyk J (2004) Modes of inhibition of cysteine proteases. Acta Biochim Pol 51:861–873
79. Turk B, Turk D, Salvesen GS (2002) Regulating cysteine protease activity: essential role of protease inhibitors as guardians and regulators. Curr Pharm Des 8:1623–1637
80. Boya P, Kroemer G (2008) Lysosomal membrane permeabilization in cell death. Oncogene 27:6434–6451
81. Wallin H, Bjarnadottir M, Vogel LK et al (2010) Cystatins–extra-and intracellular cysteine protease inhibitors: high-level secretion and uptake of cystatin C in human neuroblastoma cells. Biochimie 92:1625–1634
82. Matthews SP, McMillan SJ, Colbert JD et al (2016) Cystatin F ensures eosinophil survival by regulating granule biogenesis. Immunity 44:795–806
83. Grzonka Z, Jankowska E, Kasprzykowski F et al (2001) Structural studies of cysteine proteases and their inhibitors. Acta Biochim Pol 48:1–20
84. Lu P, Takai K, Weaver VM et al (2011) Extracellular matrix degradation and remodeling in development and disease. Cold Spring Harb Perspect Biol 3:a005058
85. Obermajer N, Jevnikar Z, Doljak B et al (2008) Role of cysteine cathepsins in matrix degradation and cell signalling. Connect Tissue Res 49:193–196
86. Chung L, Dinakarpandian D, Yoshida N et al (2004) Collagenase unwinds triple-helical collagen prior to peptide bond hydrolysis. EMBO J 23:3020–3030
87. Li Z, Kienetz M, Cherney MM et al (2008) The crystal and molecular structures of a cathepsin K: chondroitin sulfate complex. J Mol Biol 383:78–91
88. Choe Y, Leonetti F, Greenbaum DC et al (2006) Substrate profiling of cysteine proteases using a combinatorial peptide library identifies functionally unique specificities. J Biol Chem 281:12824–12832
89. Lecaille F, Chowdhury S, Purisima E et al (2007) The S2 subsites of cathepsins K and L and their contribution to collagen degradation. Protein Sci 16:662–670
90. Lotinun S, Kiviranta R, Matsubara T et al (2013) Osteoclast-specific cathepsin K deletion stimulates S1P-dependent bone formation. J Clin Invest 123:666–681
91. Gowen M, Lazner F, Dodds R et al (1999) Cathepsin K knockout mice develop osteopetrosis due to a deficit in matrix degradation but not demineralization. J Bone Miner Res 14:1654–1663
92. Huber L, Distler O, Tarner I et al (2006) Synovial fibroblasts: key players in rheumatoid arthritis. Rheumatology 45:669–675
93. Schurigt U (2013) Role of cysteine cathepsins in joint inflammation and destruction in human rheumatoid arthritis and associated animal models. INTECH Open Access Publisher, Rijeka
94. Novinec M, Grass RN, Stark WJ et al (2007) Interaction between human Cathepsins K, L, and S and Elastins MECHANISM OF ELASTINOLYSIS AND INHIBITION BY MACROMOLECULAR INHIBITORS. J Biol Chem 282:7893–7902
95. Li Z, Yasuda Y, Li W et al (2004) Regulation of collagenase activities of human cathepsins by glycosaminoglycans. J Biol Chem 279:5470–5479
96. Sage J, Mallevre F, Barbarin-Costes F et al (2013) Binding of chondroitin 4-sulfate to cathepsin S regulates its enzymatic activity. Biochemistry 52:6487–6498

97. Yasuda Y, Li Z, Greenbaum D et al (2004) Cathepsin V, a novel and potent elastolytic activity expressed in activated macrophages. J Biol Chem 279:36761–36770
98. Bastow ER, Last K, Golub S et al (2012) Evidence for lysosomal exocytosis and release of aggrecan-degrading hydrolases from hypertrophic chondrocytes, in vitro and in vivo. Biol Open 1:318–328
99. Cailhier J-F, Sirois I, Laplante P et al (2008) Caspase-3 activation triggers extracellular Cathepsin L release and Endorepellin proteolysis. J Biol Chem 283:27220–27229
100. Saini MG, Bix GJ (2012) Oxygen-glucose deprivation (OGD) and interleukin-1 (IL-1) differentially modulate Cathepsin B/L mediated generation of Neuroprotective Perlecan LG3 by neurons. Brain Res 1438:65–74
101. Sage J, Leblanc-Noblesse E, Nizard C et al (2012) Cleavage of Nidogen-1 by Cathepsin S impairs its binding to basement membrane partners. PLoS One 7:e43494
102. Arampatzidou M, Schütte A, Hansson GC et al (2012) Effects of cathepsin K deficiency on intercellular junction proteins, luminal mucus layers, and extracellular matrix constituents in the mouse colon. Biol Chem 393:1391–1403
103. Gocheva V, Zeng W, Ke D et al (2006) Distinct roles for cysteine cathepsin genes in multistage tumorigenesis. Genes Dev 20:543–556
104. Galluzzi L, Vitale I, Abrams JM et al (2012) Molecular definitions of cell death subroutines: recommendations of the nomenclature committee on cell death 2012. Cell Death Differ 19:107–120
105. Saeed WK, Jun DW (2014) Necroptosis: an emerging type of cell death in liver diseases. World J Gastroenterol 20:12526–12532
106. Kroemer G, Jäättelä M (2005) Lysosomes and autophagy in cell death control. Nat Rev Cancer 5:886–897
107. Kirkegaard T, Jäättelä M (2009) Lysosomal involvement in cell death and cancer. Biochimica et Biophysica Acta (BBA)-molecular. Cell Res 1793:746–754
108. Repnik U, Hafner Cesen M, Turk B (2014) Lysosomal membrane permeabilization in cell death: concepts and challenges. Mitochondrion 19(Pt A):49–57
109. Rodriguez-Muela N, Hernandez-Pinto AM, Serrano-Puebla A et al (2015) Lysosomal membrane permeabilization and autophagy blockade contribute to photoreceptor cell death in a mouse model of retinitis pigmentosa. Cell Death Differ 22:476–487
110. Bivik CA, Larsson PK, Kågedal KM et al (2006) UVA/B-induced apoptosis in human melanocytes involves translocation of cathepsins and Bcl-2 family members. J Investig Dermatol 126:1119–1127
111. Appelqvist H, Waster P, Kagedal K et al (2013) The lysosome: from waste bag to potential therapeutic target. J Mol Cell Biol 5:214–226
112. Chwieralski C, Welte T, Bühling F (2006) Cathepsin-regulated apoptosis. Apoptosis 11:143–149
113. Repnik U, Cesen MH, Turk B (2013) The endolysosomal system in cell death and survival. Cold Spring Harb Perspect Biol 5:a008755
114. Guicciardi ME, Miyoshi H, Bronk SF et al (2001) Cathepsin B knockout mice are resistant to tumor necrosis factor-alpha-mediated hepatocyte apoptosis and liver injury: implications for therapeutic applications. Am J Pathol 159:2045–2054
115. Werneburg NW, Bronk SF, Guicciardi ME et al (2012) Tumor necrosis factor-related apoptosis-inducing ligand (TRAIL) protein-induced lysosomal translocation of proapoptotic effectors is mediated by phosphofurin acidic cluster sorting protein-2 (PACS-2). J Biol Chem 287:24427–24437
116. Werneburg NW, Guicciardi ME, Bronk SF et al (2002) Tumor necrosis factor-alpha-associated lysosomal permeabilization is cathepsin B dependent. Am J Physiol Gastrointest Liver Physiol 283:G947–G956
117. Werneburg NW, Guicciardi ME, Bronk SF et al (2007) Tumor necrosis factor-related apoptosis-inducing ligand activates a lysosomal pathway of apoptosis that is regulated by Bcl-2 proteins. J Biol Chem 282:28960–28970

118. Canbay A, Guicciardi ME, Higuchi H et al (2003) Cathepsin B inactivation attenuates hepatic injury and fibrosis during cholestasis. J Clin Invest 112:152–159
119. Baskin-Bey ES, Canbay A, Bronk SF et al (2005) Cathepsin B inactivation attenuates hepatocyte apoptosis and liver damage in steatotic livers after cold ischemia-warm reperfusion injury. Am J Physiol Gastrointest Liver Physiol 288:G396–G402
120. Feldstein AE, Werneburg NW, Canbay A et al (2004) Free fatty acids promote hepatic lipotoxicity by stimulating TNF-alpha expression via a lysosomal pathway. Hepatology 40:185–194
121. Jones B, Roberts PJ, Faubion WA et al (1998) Cystatin a expression reduces bile salt-induced apoptosis in a rat hepatoma cell line. Am J Phys 275:G723–G730
122. Kirkegaard T, Jaattela M (2009) Lysosomal involvement in cell death and cancer. Biochim Biophys Acta 1793:746–754
123. Sendler M, Maertin S, John D et al (2016) Cathepsin-B activity initiates apoptosis via digestive protease activation in pancreatic acinar cells and experimental pancreatitis. J Biol Chem 291(28):14717–14731
124. Almaguel FG, Liu J-W, Pacheco FJ et al (2010) Lipotoxicity mediated cell dysfunction and death involves Lysosomal membrane Permeabilization and Cathepsin L activity. Brain Res 1318C:133–143
125. Zhang H, Zhang L, Wei L et al (2016) Knockdown of cathepsin L sensitizes ovarian cancer cells to chemotherapy. Oncol Lett 11:4235–4239
126. Cui F, Wang W, Wu D et al (2016) Overexpression of Cathepsin L is associated with gefitinib resistance in non-small cell lung cancer. Clin Transl Oncol 18:722–727
127. Pacheco FJ, Servin J, Dang D et al (2005) Involvement of lysosomal cathepsins in the cleavage of DNA topoisomerase I during necrotic cell death. Arthritis Rheum 52:2133–2145
128. Hentze H, Lin XY, Choi MS et al (2003) Critical role for cathepsin B in mediating caspase-1-dependent interleukin-18 maturation and caspase-1-independent necrosis triggered by the microbial toxin nigericin. Cell Death Differ 10:956–968
129. Jacobson LS, Lima H, Goldberg MF et al (2013) Cathepsin-mediated necrosis controls the adaptive immune response by Th2 (T helper type 2)-associated adjuvants. J Biol Chem 288:7481–7491
130. Honey K, Rudensky AY (2003) Lysosomal cysteine proteases regulate antigen presentation. Nat Rev Immunol 3:472–482
131. Zavasnik-Bergant T, Turk B (2006) Cysteine cathepsins in the immune response. Tissue Antigens 67:349–355
132. Honey K, Nakagawa T, Peters C et al (2002) Cathepsin L regulates CD4+ T cell selection independently of its effect on invariant chain: a role in the generation of positively selecting peptide ligands. J Exp Med 195:1349–1358
133. Riese RJ, Chapman HA (2000) Cathepsins and compartmentalization in antigen presentation. Curr Opin Immunol 12:107–113
134. Olson OC, Joyce JA (2015) Cysteine cathepsin proteases: regulators of cancer progression and therapeutic response. Nat Rev Cancer 15:712–729
135. Artal-Sanz M, Tavernarakis N (2005) Proteolytic mechanisms in necrotic cell death and neurodegeneration. FEBS Lett 579:3287–3296
136. Cheng XW, Shi GP, Kuzuya M et al (2012) Role for cysteine protease cathepsins in heart disease: focus on biology and mechanisms with clinical implication. Circulation 125:1551–1562
137. Lafarge JC, Naour N, Clement K et al (2010) Cathepsins and cystatin C in atherosclerosis and obesity. Biochimie 92:1580–1586
138. Manchanda M, Roeb E, Roderfield M, et al (2015) P0109: elevation of cathepsin L and B expression in liver fibrosis: a study in mice models and patients. J Hepatol 62:S341–S342
139. Leto G, Tumminello FM, Pizzolanti G et al (1997) Lysosomal cathepsins B and L and Stefin a blood levels in patients with hepatocellular carcinoma and/or liver cirrhosis: potential clinical implications. Oncology 54:79–83
140. Conus S, Simon H-U (2010) Cathepsins and their involvement in immune responses. Swiss Med Wkly 140:w13042

141. McKerrow JH, Caffrey C, Kelly B et al (2006) Proteases in parasitic diseases. Annu Rev Pathol Mech Dis 1:497–536
142. Simmons G, Gosalia DN, Rennekamp AJ et al (2005) Inhibitors of cathepsin L prevent severe acute respiratory syndrome coronavirus entry. Proc Natl Acad Sci U S A 102:11876–11881
143. Jedeszko C, Sloane BF (2004) Cysteine cathepsins in human cancer. Biol Chem 385:1017–1027
144. Thomssen C, Schmitt M, Goretzki L et al (1995) Prognostic value of the cysteine proteases cathepsins B and cathepsin L in human breast cancer. American Association for Cancer Research 1:741–746
145. Paik S, Shak S, Tang G et al (2004) A multigene assay to predict recurrence of tamoxifen-treated, node-negative breast cancer. N Engl J Med 351:2817–2826
146. Lah TT, Čerček M, Blejec A et al (2000) Cathepsin B, a prognostic indicator in lymph node-negative breast carcinoma patients: comparison with cathepsin D, cathepsin L, and other clinical indicators. Clin Cancer Res 6:578–584
147. Paik S, Tang G, Shak S et al (2006) Gene expression and benefit of chemotherapy in women with node-negative, estrogen receptor–positive breast cancer. J Clin Oncol 24:3726–3734
148. Berdowska I (2004) Cysteine proteases as disease markers. Clin Chim Acta 342:41–69
149. Zhang W, Wang S, Wang Q et al (2014) Overexpression of cysteine cathepsin L is a marker of invasion and metastasis in ovarian cancer. Oncol Rep 31:1334–1342
150. Scorilas A, Fotiou S, Tsiambas E et al (2002) Determination of cathepsin B expression may offer additional prognostic information for ovarian cancer patients. Biol Chem 383:1297–1303
151. Strojnik T, Kos J, Lah TT (2001) Cathepsins B and L are markers for clinically invasive types of meningiomas. Neurosurgery 48:598–605
152. Nakada M, Nakada S, Demuth T et al (2007) Molecular targets of glioma invasion. Cell Mol Life Sci 64:458–478
153. Jain M, Bakhshi S, Shukla AA et al (2010) Cathepsins B and L in peripheral blood mononuclear cells of pediatric acute myeloid leukemia: potential poor prognostic markers. Ann Hematol 89:1223–1232
154. Pandey G, Bakhshi S, Singh R et al (2014) Clinical significance of cathepsin L and B expression in pediatric acute myeloid leukemia. Cancer Res 74:1868–1868
155. Herszenyi L, Farinati F, Cardin R et al (2008) Tumor marker utility and prognostic relevance of cathepsin B, cathepsin L, urokinase-type plasminogen activator, plasminogen activator inhibitor type-1, CEA and CA 19-9 in colorectal cancer. BMC Cancer 8:194
156. Fujise N, Nanashim A, Taniguchi Y et al (2000) Prognostic impact of cathepsin B and matrix metalloproteinase-9 in pulmonary adenocarcinomas by immunohistochemical study. Lung Cancer 27:19–26
157. Werle B, Kraft C, Lah TT et al (2000) Cathepsin B in infiltrated lymph nodes is of prognostic significance for patients with nonsmall cell lung carcinoma. Cancer 89:2282–2291
158. Yan X, Takahara M, Xie L et al (2011) Stromal expression of cathepsin K in squamous cell carcinoma. J Eur Acad Dermatol Venereol 25:362–365
159. MEHRA S, KUMAR M, PANWAR R et al (2016) Abstract 3986: diagnostic significance of cathepsin L and cathepsin B expression in human gallbladder cancer - a pilot study. Cancer Res 76:3986–3986
160. Tan GJ, Peng ZK, Lu JP et al (2013) Cathepsins mediate tumor metastasis. World J Biol Chem 4:91–101
161. Fan Q, Wang X, Zhang H et al (2012) Silencing cathepsin S gene expression inhibits growth, invasion and angiogenesis of human hepatocellular carcinoma in vitro. Biochem Biophys Res Commun 425:703–710
162. Small DM, Burden RE, Jaworski J et al (2013) Cathepsin S from both tumor and tumor-associated cells promote cancer growth and neovascularization. Int J Cancer 133:2102–2112
163. Sinha AA, Gleason DF, Staley NA et al (1995) Cathepsin B in angiogenesis of human prostate: an immunohistochemical and immunoelectron microscopic analysis. Anat Rec 241:353–362

164. Mohanam S, Jasti SL, Kondraganti SR et al (2001) Down-regulation of cathepsin B expression impairs the invasive and tumorigenic potential of human glioblastoma cells. Oncogene 20:3665–3673
165. Yanamandra N, Gumidyala KV, Waldron KG et al (2004) Blockade of cathepsin B expression in human glioblastoma cells is associated with suppression of angiogenesis. Oncogene 23:2224–2230
166. Gocheva V, Chen X, Peters C et al (2010) Deletion of cathepsin H perturbs angiogenic switching, vascularization and growth of tumors in a mouse model of pancreatic islet cell cancer. Biol Chem 391:937–945
167. Wang B, Sun J, Kitamoto S et al (2006) Cathepsin S controls angiogenesis and tumor growth via matrix-derived angiogenic factors. J Biol Chem 281:6020–6029
168. Veillard F, Saidi A, Burden RE et al (2011) Cysteine cathepsins S and L modulate anti-angiogenic activities of human endostatin. J Biol Chem 286:37158–37167
169. Kostoulas G, Lang A, Nagase H et al (1999) Stimulation of angiogenesis through cathepsin B inactivation of the tissue inhibitors of matrix metalloproteinases. FEBS Lett 455:286–290
170. Sobotic B, Vizovisek M, Vidmar R et al (2015) Proteomic identification of cysteine Cathepsin substrates shed from the surface of cancer cells. Mol Cell Proteomics 14:2213–2228
171. Jevnikar Z, Rojnik M, Jamnik P et al (2013) Cathepsin H mediates the processing of Talin and regulates migration of prostate cancer cells. J Biol Chem 288:2201–2209
172. Pecar F, Kos U, J. (2015) Cathepsin X cleaves profilin 1 C-terminal Tyr139 and influences Clathrin-mediated endocytosis. PLoS One 10:e0137217
173. Ren G, Tian Q, An Y et al (2012) Coronin 3 promotes gastric cancer metastasis via the up-regulation of MMP-9 and cathepsin K. Mol Cancer 11:67–67
174. Le Gall C, Bonnelye E, Clezardin P (2008) Cathepsin K inhibitors as treatment of bone metastasis. Curr Opin Support Palliat Care 2:218–222
175. Victor BC, Anbalagan A, Mohamed MM et al (2011) Inhibition of cathepsin B activity attenuates extracellular matrix degradation and inflammatory breast cancer invasion. Breast Cancer Res 13:R115
176. Cavallo-Medved D, Mai J, Dosescu J et al (2005) Caveolin-1 mediates the expression and localization of cathepsin B, pro-urokinase plasminogen activator and their cell-surface receptors in human colorectal carcinoma cells. J Cell Sci 118:1493–1503
177. Zhang Q, Han M, Wang W et al (2015) Downregulation of cathepsin L suppresses cancer invasion and migration by inhibiting transforming growth factor beta mediated epithelial-mesenchymal transition. Oncol Rep 33:1851–1859
178. Yu S, Yu Y, Zhang W, et al (2016) FOXO3a promotes gastric cancer cell migration and invasion through the induction of cathepsin L. Oncotarget 7(23):34773–34784
179. Wang J, Chen L, Li Y et al (2011) Overexpression of Cathepsin Z contributes to tumor metastasis by inducing epithelial-mesenchymal transition in hepatocellular carcinoma. PLoS One 6:e24967
180. Lecaille F, Bromme D, Lalmanach G (2008) Biochemical properties and regulation of cathepsin K activity. Biochimie 90:208–226
181. Chapman HA, Riese RJ, Shi GP (1997) Emerging roles for cysteine proteases in human biology. Annu Rev Physiol 59:63–88
182. Buhling F, Waldburg N, Reisenauer A et al (2004) Lysosomal cysteine proteases in the lung: role in protein processing and immunoregulation. Eur Respir J 23:620–628
183. Zhang D, Leung N, Weber E et al (2011) The effect of cathepsin K deficiency on airway development and TGF-beta1 degradation. Respir Res 12:72
184. Lecaille F, Lalmanach G, Andrault PM (2016) Antimicrobial proteins and peptides in human lung diseases: a friend and foe partnership with host proteases. Biochimie 122:151–168
185. Kasabova M, Joulin-Giet A, Lecaille F et al (2014) Regulation of TGF-β1-driven differentiation of human lung fibroblasts: EMERGING ROLES OF CATHEPSIN B AND CYSTATIN C. J Biol Chem 289:16239–16251

186. Zhang L, Fu XH, Yu Y et al (2015) Treatment with CA-074Me, a Cathepsin B inhibitor, reduces lung interstitial inflammation and fibrosis in a rat model of polymyositis. Lab Investig 95:65–77
187. Takeyabu K, Betsuyaku T, Nishimura M et al (1998) Cysteine proteinases and cystatin C in bronchoalveolar lavage fluid from subjects with subclinical emphysema. Eur Respir J 12:1033–1039
188. Williams AS, Eynott PR, Leung SY et al (2009) Role of cathepsin S in ozone-induced airway hyperresponsiveness and inflammation. Pulm Pharmacol Ther 22:27–32
189. Geraghty P, Rogan MP, Greene CM et al (2008) Alpha-1-antitrypsin aerosolised augmentation abrogates neutrophil elastase-induced expression of cathepsin B and matrix metalloprotease 2 in vivo and in vitro. Thorax 63:621–626
190. Martin SL, Moffitt KL, McDowell A et al (2010) Association of airway cathepsin B and S with inflammation in cystic fibrosis. Pediatr Pulmonol 45:860–868
191. Cimerman N, Brguljan PM, Krasovec M et al (2001) Circadian and concentration profile of cathepsin S in sera from healthy subjects and asthmatic patients. Pflugers Arch 442:R204–R206
192. Fajardo I, Svensson L, Bucht A et al (2004) Increased levels of hypoxia-sensitive proteins in allergic airway inflammation. Am J Respir Crit Care Med 170:477–484
193. Somoza JR, Palmer JT, Ho JD (2002) The crystal structure of human cathepsin F and its implications for the development of novel immunomodulators. J Mol Biol 322:559–568
194. Lutgens SP, Cleutjens KB, Daemen MJ et al (2007) Cathepsin cysteine proteases in cardiovascular disease. FASEB J 21:3029–3041
195. Blondelle J, Lange S, Greenberg BH et al (2015) Cathepsins in heart disease–chewing on the heartache? Am J Physiol Heart Circ Physiol 308:H974–H976
196. Hua Y, Zhang Y, Dolence J et al (2013) Cathepsin K knockout mitigates high-fat diet-induced cardiac hypertrophy and contractile dysfunction. Diabetes 62:498–509
197. Cheng XW, Murohara T, Kuzuya M et al (2008) Superoxide-dependent cathepsin activation is associated with hypertensive myocardial remodeling and represents a target for angiotensin II type 1 receptor blocker treatment. Am J Pathol 173:358–369
198. Stypmann J, Glaser K, Roth W et al (2002) Dilated cardiomyopathy in mice deficient for the lysosomal cysteine peptidase cathepsin L. Proc Natl Acad Sci U S A 99:6234–6239
199. Sun M, Ouzounian M, de Couto G et al (2013) Cathepsin-L ameliorates cardiac hypertrophy through activation of the autophagy–Lysosomal dependent protein processing pathways. Journal of the American Heart Association: Cardiovascular and Cerebrovascular Disease 2:e000191
200. Xie L, Terrand J, Xu B et al (2010) Cystatin C increases in cardiac injury: a role in extracellular matrix protein modulation. Cardiovasc Res 87:628–635
201. Sehl PD, Tai JT, Hillan KJ et al (2000) Application of cDNA microarrays in determining molecular phenotype in cardiac growth, development, and response to injury. Circulation 101:1990–1999
202. Liu J, Sukhova GK, Yang JT et al (2006) Cathepsin L expression and regulation in human abdominal aortic aneurysm, atherosclerosis, and vascular cells. Atherosclerosis 184:302–311
203. Liu J, Sukhova GK, Sun JS et al (2004) Lysosomal cysteine proteases in atherosclerosis. Arterioscler Thromb Vasc Biol 24:1359–1366
204. Platt MO, Ankeny RF, Shi G-P et al (2007) Expression of cathepsin K is regulated by shear stress in cultured endothelial cells and is increased in endothelium in human atherosclerosis. Am J Physiol Heart Circ Physiol 292:H1479–H1486
205. Sukhova GK, Shi GP, Simon DI et al (1998) Expression of the elastolytic cathepsins S and K in human atheroma and regulation of their production in smooth muscle cells. J Clin Investig 102:576–583
206. Wilczynski C, Samarasinghe S, Emanuele M et al (2015) Cathepsins K and S: role in bone, adipocytes, and glucose regulation. Clinical Reviews in Bone and Mineral Metabolism 13:2–10

207. Huang X, Vaag A, Carlsson E et al (2003) Impaired Cathepsin L Gene expression in skeletal muscle is associated with Type 2 diabetes. Diabetes 52:2411–2418
208. Maehr R, Mintern JD, Herman AE et al (2005) Cathepsin L is essential for onset of autoimmune diabetes in NOD mice. J Clin Invest 115:2934–2943
209. Hsing LC, Kirk EA, McMillen TS et al (2010) Roles for cathepsins S, L, and B in insulitis and diabetes in the NOD mouse. J Autoimmun 34:96–104
210. Yamada A, Ishimaru N, Arakaki R et al (2010) Cathepsin L inhibition prevents murine autoimmune diabetes via suppression of CD8(+) T cell activity. PLoS One 5:e12894
211. Yang M, Zhang Y, Pan J et al (2007) Cathepsin L activity controls adipogenesis and glucose tolerance. Nat Cell Biol 9:970–977
212. Jung M, Lee J, Seo HY et al (2015) Cathepsin inhibition-induced lysosomal dysfunction enhances pancreatic beta-cell apoptosis in high glucose. PLoS One 10:e0116972
213. Li X, Wu K, Edman M et al (2010) Increased expression of cathepsins and obesity-induced proinflammatory cytokines in lacrimal glands of male NOD mouse. Invest Ophthalmol Vis Sci 51:5019–5029
214. Chen RP, Ren A, Ye SD (2013) Correlation between serum cathepsin S and insulin resistance in type 2 diabetes. Exp Ther Med 6:1237–1242
215. Jobs E, Risérus U, Ingelsson E et al (2013) Serum Cathepsin S is associated with decreased insulin sensitivity and the development of type 2 diabetes in a community-based cohort of elderly men. Diabetes Care 36:163–165
216. Taleb S, Lacasa D, Bastard JP et al (2005) Cathepsin S, a novel biomarker of adiposity: relevance to atherogenesis. FASEB J 19:1540–1542
217. Naour N, Rouault C, Fellahi S et al (2010) Cathepsins in human obesity: changes in energy balance predominantly affect cathepsin s in adipose tissue and in circulation. J Clin Endocrinol Metab 95:1861–1868
218. Lafarge JC, Pini M, Pelloux V et al (2014) Cathepsin S inhibition lowers blood glucose levels in mice. Diabetologia 57:1674–1683
219. Hua Y, Nair S (2015) Proteases in cardiometabolic diseases: pathophysiology, molecular mechanisms and clinical applications. Biochim Biophys Acta 1852:195–208
220. Chiellini C, Costa M, Novelli SE et al (2003) Identification of cathepsin K as a novel marker of adiposity in white adipose tissue. J Cell Physiol 195:309–321
221. Xiao Y, Junfeng H, Tianhong L et al (2006) Cathepsin K in adipocyte differentiation and its potential role in the pathogenesis of obesity. J Clin Endocrinol Metab 91:4520–4527
222. Yang M, Sun J, Zhang T et al (2008) Deficiency and inhibition of Cathepsin K reduce body weight gain and increase glucose metabolism in mice. Arterioscler Thromb Vasc Biol 28:2202–2208
223. Han J, Luo T, Gu Y et al (2009) Cathepsin K regulates adipocyte differentiation: possible involvement of type I collagen degradation. Endocr J 56:55–63
224. Friedrichs B, Tepel C, Reinheckel T et al (2003) Thyroid functions of mouse cathepsins B, K, and L. J Clin Investig 111:1733–1745
225. Halangk W, Lerch MM, Brandt-Nedelev B et al (2000) Role of cathepsin B in intracellular trypsinogen activation and the onset of acute pancreatitis. J Clin Invest 106:773–781
226. Pham CTN, Ley TJ (1999) Dipeptidyl peptidase I is required for the processing and activation of granzymes a and B in vivo. Proc Natl Acad Sci U S A 96:8627–8632
227. Tang C-H, Lee J-W, Galvez MG et al (2006) Murine Cathepsin F deficiency causes neuronal Lipofuscinosis and late-onset neurological disease. Mol Cell Biol 26:2309–2316
228. Bühling F, Kouadio M, Chwieralski CE et al (2011) Gene targeting of the cysteine peptidase Cathepsin H impairs lung surfactant in mice. PLoS One 6:e26247
229. ROTH W, DEUSSING J, BOTCHKAREV VA et al (2000) Cathepsin L deficiency as molecular defect of furless: hyperproliferation of keratinocytes and perturbation of hair follicle cycling. FASEB J 14:2075–2086
230. Potts W, Bowyer J, Jones H et al (2004) Cathepsin L-deficient mice exhibit abnormal skin and bone development and show increased resistance to osteoporosis following ovariectomy. Int J Exp Pathol 85:85–96

231. Ondr JK, Pham CTN (2004) Characterization of murine Cathepsin W and its role in cell-mediated cytotoxicity. J Biol Chem 279:27525–27533
232. Adenis A, Huet G, Zerimech F et al (1995) Cathepsin-B, Cathepsin-L, and Cathepsin-D activities in colorectal carcinomas - relationship with Clinicopathological parameters. Cancer Lett 96:267–275
233. Gormley JA, Hegarty SM, O'Grady A et al (2011) The role of Cathepsin S as a marker of prognosis and predictor of chemotherapy benefit in adjuvant CRC: a pilot study. Br J Cancer 105:1487–1494
234. Herszenyi L, Plebani M, Carraro P et al (1999) The role of cysteine and serine proteases in colorectal carcinoma. Cancer 86:1135–1142
235. Troy AM, Sheahan K, Mulcahy HE et al (2004) Expression of Cathepsin B and L antigen and activity is associated with early colorectal cancer progression. Eur J Cancer 40:1610–1616
236. Akkari L, Gocheva V, Kester JC et al (2014) Distinct functions of macrophage-derived and cancer cell-derived cathepsin Z combine to promote tumor malignancy via interactions with the extracellular matrix. Genes Dev 28:2134–2150
237. Husmann K, Muff R, Bolander ME et al (2008) Cathepsins and osteosarcoma: expression analysis identifies cathepsin K as an indicator of metastasis. Mol Carcinog 47:66–73
238. Liu JP, Guo Q, Chen BX et al (2006) Cathepsin B and its interacting proteins, bikunin and TSRC1, correlate with TNF-induced apoptosis of ovarian cancer cells OV-90. FEBS Lett 580:245–250
239. Lutgens E, Lutgens SP, Faber BC et al (2006) Disruption of the cathepsin K gene reduces atherosclerosis progression and induces plaque fibrosis but accelerates macrophage foam cell formation. Circulation 113:98–107
240. Lutgens SP, Kisters N, Lutgens E et al (2006) Gene profiling of cathepsin K deficiency in atherogenesis: profibrotic but lipogenic. J Pathol 210:334–343
241. Samokhin AO, Wong A, Saftig P et al (2008) Role of cathepsin K in structural changes in brachiocephalic artery during progression of atherosclerosis in apoE-deficient mice. Atherosclerosis 200:58–68
242. Guo J, Bot I, de Nooijer R et al (2009) Leucocyte cathepsin K affects atherosclerotic lesion composition and bone mineral density in low-density lipoprotein receptor deficient mice. Cardiovasc Res 81:278–285
243. Kitamoto S, Sukhova GK, Sun J et al (2007) Cathepsin L deficiency reduces diet-induced atherosclerosis in low-density lipoprotein receptor-knockout mice. Circulation 115:2065–2075
244. Rodgers KJ, Watkins DJ, Miller AL et al (2006) Destabilizing role of cathepsin S in murine atherosclerotic plaques. Arterioscler Thromb Vasc Biol 26:851–856

Role of Serine Proteases and Inhibitors in Cancer

12

Nitesh Kumar Poddar, Sanjeev Kumar Maurya, and Vanshika Saxena

Abstract

Serine proteases, the largest human protease family, are found in many key developmental and physiological processes in the biological system. Protease signalling pathways are stringently controlled, and deregulation of proteolytic activity results in the degradation of extracellular matrix which plays a major role in cancer progression. The Type II transmembrane serine protease, hepsin, matriptase-2 and TMPRSS4, and secreted serine protease, urokinase plasminogen activator (uPA), kallikreins and HtrA, are closely related to cancer-associated proteases and also involved in perturbation of uPA plasminogen system, matrix metalloproteases (MMPs), upregulation of adhesion molecules like integrin family, activation of intracellular signalling cascade, inhibition of apoptosis pathway in various types of cancers which causes cell proliferation, invasion and metastasis. Serpin, an endogenous serine protease inhibitor, regulates the homeostasis by maintaining a delicate balance with the serine protease and prevents the process of invasion and metastasis of cancer cells thus inhibiting tumour growth. This chapter focuses on the role of serine proteases and their inhibitors in different types of tumours associated with cancer prognostication and therapy.

Keywords

Type II transmembrane serine protease • Extracellular matrix • Matrix metalloproteases (MMPs) • Secreted serine proteases • Therapy • Tumour • Serpin • Urokinase plasminogen activator (uPA) • Plasminogen system

N.K. Poddar (✉) • S.K. Maurya • V. Saxena
Department of Biotechnology, Invertis University, Bareilly -243123, Uttar Pradesh, India
e-mail: nitesh.p@invertis.org; niteshpoddar@gmail.com

S. Chakraborti, N.S. Dhalla (eds.), *Proteases in Physiology and Pathology*,
DOI 10.1007/978-981-10-2513-6_12

12.1 Introduction

Proteases occupy a pivotal position among biological molecules required for the physiological roles in living systems and commercial biotechnology markets and medical fields. They are called proteolytic enzymes or systemic enzymes, and their catalytic function is to hydrolyse the peptide bond that links amino acids together in a polypeptide chain. These are also called peptidase or proteinase (Fig. 12.1).

A large variety of proteases are found in intracellular or extracellular space in all eukaryotic and prokaryotic cells. They are mainly located in different organelles of eukaryotic cells such as the cytosol, mitochondria, vacuoles, lysosomes and endoplasmic reticulum. These intracellular proteases are involved in many important functions such as regulating synthesis, activation and proteolysis of proteins. The extracellular proteases are mostly secreted in the gastrointestinal tract of animals or involve in the blood coagulation and complement cascade events. Consequently, different organisms or different tissues have different sets of proteases.

12.1.1 Cellular and Physiological Functions of Protease and Their Industrial Applications

Proteases exhibit many important cascades such as homeostasis and inflammation which control the dynamics of protein turnover in various hierarchical levels of

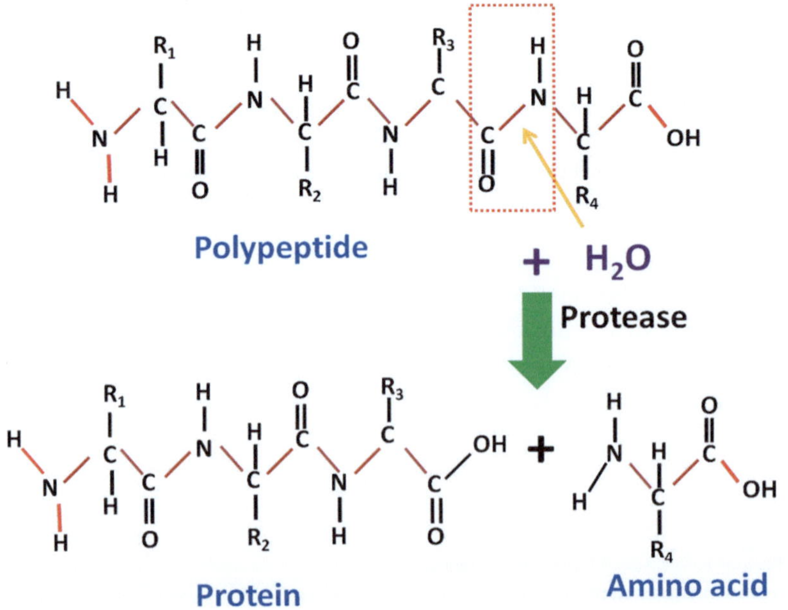

Fig. 12.1 Hydrolysis of peptide bond of protein by protease in the presence of water

biological organisation. In thermodynamics, the hydrolysis of peptide bond is energetically favourable, for example, the equilibrium constant, $K_{eq} = 10^5$, which indicates that proteolysis is irreversible and biological switches must be strictly controlled.

Proteases involve in different biological roles such as signal transduction through proteolysis of IkB-α: it is an inhibitory protein to release nuclear factor (NF-kB, a transcription factor) that enters from cytoplasm to nucleus [1], has defensive role in blood coagulation [2], displays the hydrolysed foreign proteins through major histocompatibility complex (MHC class I) in immune system [3], acts as a development process such as fertilisation [4] and, last but not the least, is useful for the proliferation programme in cell system with the help of cyclin degradation and programmed cell death and controls the homeostasis of biological system [5, 6].

Proteases have also been utilised in the field of food processing such as manufacturing of sauces, aroma formation for the milk products, tenderisation of meat and cold stabilisation of beer. These proteases are commonly used as a hypoallergenic food for digesting milk proteins into small peptides to protect the babies from developing milk allergies.

12.1.2 Classification of Protease

A well-known database, MEROPS (http://merops.sanger.ac.uk), was first developed by Barret et al. for the classification of proteases, their substrates and inhibitors on the basis of their homologies of their significant sequences and structures [7]. This database has hierarchical classification in which proteases are grouped into families and clans.

Furthermore, proteases can also be broadly categorised into two major types, exopeptidase and endopeptidases, characterised by their site of action of the peptide bond. Usually, exopeptidases break the peptide bond nearer to the amino- or carboxyl-termini of the substrate, while endopeptidases break peptide bonds distant from to the amino- or carboxyl-termini of the substrate (within a protein molecule).

On the basis of functional group/conserved amino acids found in the active site, proteases are also categorised into four major groups as shown in Table 12.1.

12.2 Serine Proteases

According to the MEROPS database, about 33% are serine proteases which are categorised into 40 families and 13 clans in both eukaryotes and prokaryotes [7, 8].

Usually, the family name is derived from the nucleophilic Ser present in the active site of the enzyme. The Ser amino acid of the active site cleaves the carbonyl terminus of the peptide bond to form an acyl-enzyme intermediate [9].

Thus, serine proteases (or serine endopeptidases) prefers serine at the active site for the hydrolysis of the peptide bond in proteins.

Table 12.1. Classification of proteases

Class	Mechanism	Location	Examples
Serine/threonine	Endopeptidases have active centres of serine/threonine in the catalytic core	Soluble	Trypsin, chymotrypsin, subtilisin, elastase, coagulation factor X, proteasome, g-glutamyl transpeptidase, proteasome
		Membrane	Rhomboid family
Aspartic type	The active site of an enzyme contains two highly conserved aspartate residues bonded with activated water in network fashion	Soluble	Pepsin, cathepsin, renin, HIV protease
		Membrane	Presenilins, signal peptide peptidase
Cysteine type	Carboxypeptidase use a cysteine in the catalytic core	Soluble	Bromelain, papain, cathepsins, caspases, calpain
		Membrane	–
Metallo type	Carboxypeptidase use a metal ion in the catalytic core	Soluble	Thermolysin, angiotensin-converting enzyme
		Membrane	Matrix metalloproteases

Usually, they are found in the form of zymogens (digestive enzymes are released in inactive forms) to regulate the enzyme activities by controlling the specific activation of proteolysis.

The main division of serine proteases is based on the site of cleavage of specific amino acids of the peptide bonds:

1. Trypsin such as serine peptidases prefers to cleave the peptide bonds which have lysine and arginine at the cleavage sites.
2. Chymotrypsin such as serine peptidases prefers aromatic amino acids (phenylalanine, tyrosine or tryptophan) at the cleavage site for the digestion of the peptide bond.
3. Elastase such as serine peptidases prefers to cleave amino acids with short side chain groups such as alanine in their cleavage site.

12.2.1 The Catalytic Mechanism

The prime contributors of amino acids for the catalytic mechanism of serine protease classes such as chymotrypsin (in eukaryote) and subtilisin (in prokaryote) enzymes are their catalytic triad (Fig. 12.2). This triad is found in the active site of enzyme and conserved in all serine proteases. The triad comprises of three amino acids, namely, His57, Ser195 and Asp102, bonded in a network fashion (Fig. 12.2). The position of each amino acid of the triad is far from one another in the primary structure of the protein, but once folded, they will be in close proximity to the

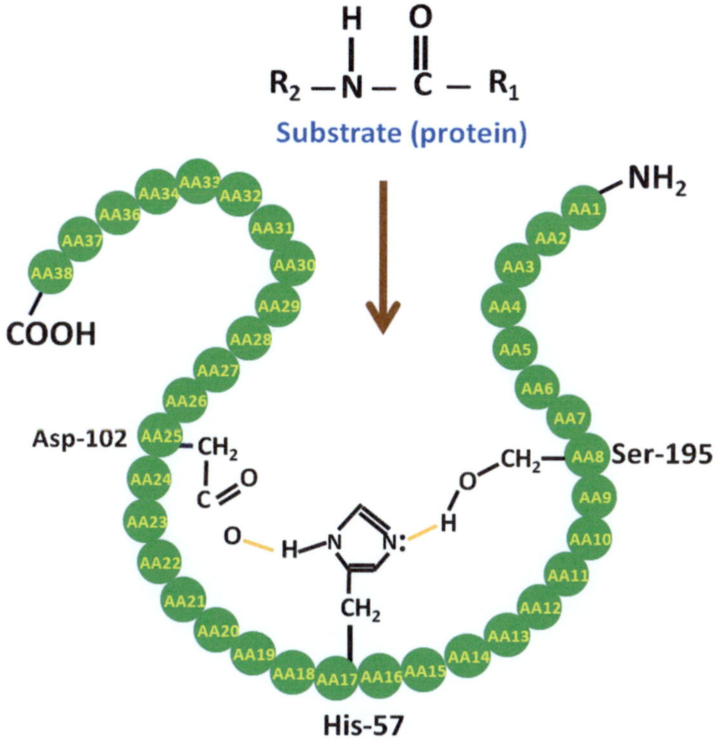

Fig. 12.2 Catalytic triad of serine protease consists of aspartic acid (Asp-102), histidine (His-57) and serine (Ser-195)

enzyme. This explicit tertiary structure of the triad members is vital for the specific catalysis of the enzyme.

Serine proteases follow ping-pong catalysis mechanism, and this involves formation of unstable enzyme-peptide intermediate by covalent catalysis mechanism, and finally the intermediate is stabilised, and consequently the peptide fragment is released [10].

The serine protease mechanism can be summarised in the following two steps: acylation followed by de-acylation process in which a nucleophilic attack takes place on the intermediate by water, which leads to the hydrolysis of the protein (Fig. 12.3). The overall process of the reaction mechanism utilises the catalytic triad (Asp-102-His-57-Ser-195) of serine protease. In this process, the serine-OH acts as a nucleophile, while histidine-NH acts as a base catalyst to activate the serine but later on it acts as an acid catalyst, whereas aspartate plays a supportive role by stabilising the histidine in the whole reactions.

The detailed process is given in the following steps:

1. The peptide binds to the active site of the enzyme, in such a way that the sessile bond of the protein (indicated by –N - C-) is placed into the active site (catalytic

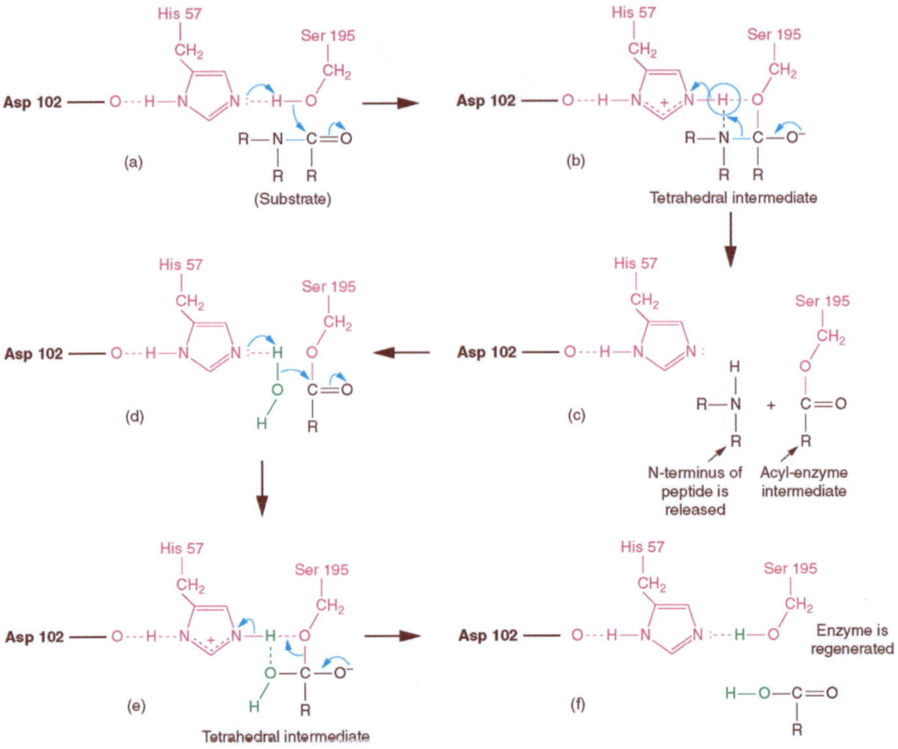

Fig. 12.3 Mechanism of a serine protease. In the acylation step, (**a**) substrate binds to active site of enzyme. (**b**) Tetrahedral intermediate is formed due to nucleophilic attack of serine on carbonyl part of peptide. (**c**) Acyl-enzyme intermediate is formed by breakage of peptide bond of the substrate. In the de-acylation step, (**d–e**) water acting as a nucleophile stabilises the cleavage peptide of carbonyl carbon and gives rise to a new tetrahedral intermediate with the nitrogen of the histidine. (**f**) Regeneration of the active site by releasing the product (Redrawn based on figure by Pratt CW, Cornely K (2012) Essential Biochemistry, 3rd edn. Wiley, New York, p 170)

triad) of the enzyme and the carbonyl-C part of peptide is present close to the nucleophilic serine [Fig. 12.3a].

2. First, the electron-rich –N atom of the histidine activates the serine residue of the catalytic triad by extracting the –H atom from serine-OH and make it more nucleophilic, and thus the serine-OH is more likely to attack the electron sink of the carbonyl-C of the peptide. Consequently, a tetrahedral intermediate is generated in which a newly covalent bond is formed between the carbonyl-C of the peptide and –O atom of serine, whereas the hydrogen part of serine is covalently bond to –N atom of histidine as well as a pair of electrons from the double bond of the –C = O moves to the –O atom of carbonyl part of peptide, creating a negative charge on the –O atom, and this causes an unstable carbonyl anion of the peptide [Fig. 12.3b]. Moreover, the histidine residue carries a positive charge due to the newly covalent bond with the serine-OH which is stabilised by the hydrogen bond of aspartic acid of the catalytic triad.

3. Because of the positive charge in the histidine, the histidine donate the –H atom to the –N atom of scissile bond of the peptide which results in the breaking of the sessile bond in which the sessile bond is now covalently bond with the –H atom of histidine. The negative charge on the oxygen atom that formed previously on the –C = O moves back to recreate a double bond. Thus, the peptide bond break results in the release of N-terminus part of the peptide, and C-terminus part of the peptide is covalently attached to serine residue, generating an acyl-enzyme intermediate. Now, the histidine residue of catalytic triad is back into the original form as an acid catalyst [Fig. 12.3c].
4. After that, water comes to play an active role in this catalysis reaction. The electron-rich nitrogen atom of histidine residue acts as a nucleophile and extracts the proton of water, and this allows the –OH part of water to act as a nucleophile, and because of this, it attacks the electron sink of –C = O part of the peptide. This is exactly the same step as in 1. Now, the new -N-H bond is formed, and histidine again carries a positive charge which is stabilised by the hydrogen bond of aspartic acid of the catalytic triad. Once again, the electron pairs from the –C = O of the substrate move back to the oxygen making it negative charge, as the bond between the –OH of water and the carbonyl-C of the substrate is formed. Overall, this generates other tetrahedral intermediate results in an unstable carbonyl anion of peptide [Fig. 12.3d, e].
5. Finally, in order to neutralise the positive charge of histidine, the covalent bond of -N-H of histidine residue is now breaking, and new covalent bond is formed between -H atom of -N-H bond of histidine and –O atom of serine residue by breaking the bond between the carbonyl-C of the peptide and oxygen atom of serine. Now, the electron-deficient carbonyl carbon of the peptide regains the previous double bond with the oxygen. Consequently the C-terminus of the peptide is now released along the formation of new –OH group of water [Fig. 12.3f]. In a nutshell, the peptide is hydrolysed with the help of protease by adding –H atom to N-terminus, and –OH atom is attached to C-terminus of the peptide bond of the substrate.

In mammals, serine proteases take part in multiple functions of living organisms such as protein digestion, blood coagulation, complement system, differentiation and development [11, 12]. Serine protease can be broadly classified into the following two broad classes based on their localisation within the extracellular space:

1. Secreted type
2. Membrane-anchored type

The secreted serine proteases are the well-characterised members of S1 family of serine proteases which are produced from secretory vesicles into the extracellular environment. Chymotrypsin, trypsin and thrombin are the prototype members of the S1 family (Fig. 12.4). Other examples of secreted serine proteases such as uPA and

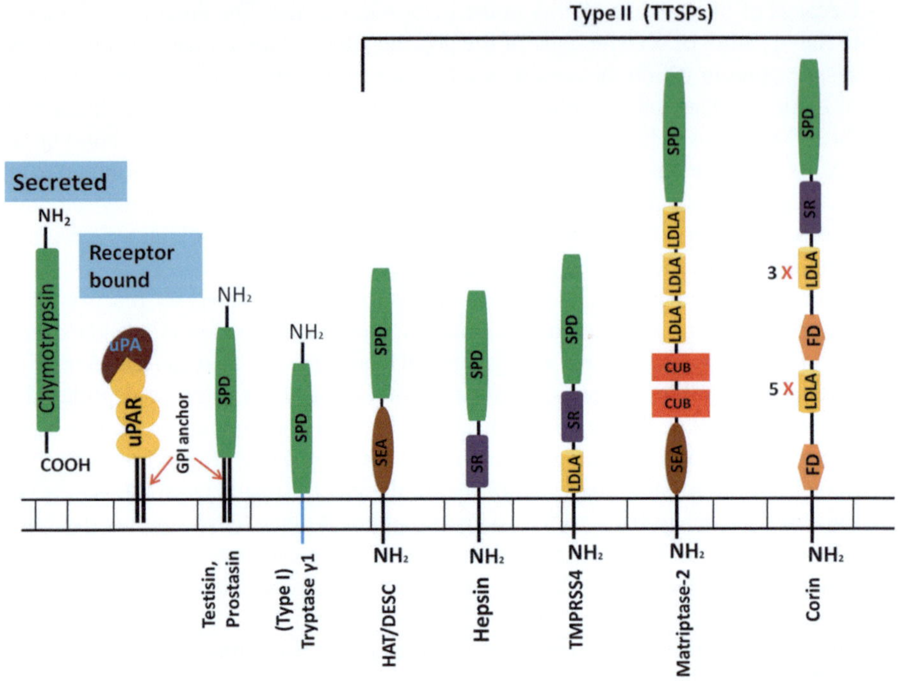

Fig. 12.4 Classification of serine proteases: (1) secreted serine protease type ((i) chymotrypsin type, (ii) role of uPA enzyme in pericellular proteolysis by binding to specific cell-surface receptors uPAR (GPI-anchored type)) and (2) membrane-anchored type: (i) The human GPI-anchored serine proteases, prostasin and testisin, (ii) Type 1 transmembrane serine protease, Tryptase γ1, (iii) The Type II transmembrane serine proteases (TTSPs): (a) the human airway trypsin-like protease expressed in squamous cell carcinoma (HAT/DESC) subfamily; (b) hepsin/TMPRSS subfamily which consists of SR, SEA and LDLA domains; (c) Matriptase subfamily, particularly, Matriptase-2, which consists of SEA, two CUB and three LDLA domains; and (d) the corin subfamily which consists of two FD domains, eight LDRA domains and one SR domain (abbreviations: *SR* scavenger receptor domain (group A), *SPD* serine protease domain consist of three catalytic residues histidine, aspartate and serine, *FD* Frizzled domain, *LDLA* LDL receptor class A)

kallikrein involve in pericellular proteolysis by either activating zymogen forms of other substrates or binding with the co-receptors (Fig. 12.4). These secreted serine proteases exhibit various biological events such as tissue repair, immunity and nutrient uptake [11].

In a recent decade, a structurally and functionally unique subgroup of S1 serine proteases, termed broadly as the membrane-anchored serine proteases, has been reported which are found to be directly anchored to the plasma membrane through its amino- or carboxy-terminal domains [13] (Fig. 12.4). In compare with secreted serine proteases, the membrane-anchored serine proteases are involved in a diverse array of physiological functions such as epithelial barrier, fertilisation, cell signalling, embryo development and tissue morphogenesis [11].

On the basis of their structural features, they can be divided into subgroups and are anchored to the membrane by three ways: (1) a carboxyl-terminal transmembrane domain through a GPI (glycosyl phosphatidylinositol) linkage which is added post-translationally, (2) a carboxy-terminal transmembrane domain (Type I) and (3) an amino-terminal transmembrane domain with a cytoplasmic extension (Type II transmembrane serine proteases – TTSPs) [13, 14]. Type I serine proteases, Tryptase γ1 and the GPI-anchored serine proteases, prostasin and testisin contain a carboxy-terminal hydrophobic extension that serves as a transmembrane domain (ranging from 310 to 370 amino acids). GPI anchors have been known to modify C-terminal domain of prostatin and testisin post-transcriptionally [15–17] (Fig. 12.4).

TTSPs are the group of membrane-anchored serine proteases with 17 members and 19 members of humans and mice, respectively. TTSPs are trypsin-like (family S1) proteases and have the potential to be linked to cellular membranes via a hydrophobic stretch at their amino terminus. These proteases have two parts, one with a cytoplasmic amino-terminal signal anchor of variable length (20 to 160 amino acids) and the other with a catalytic serine protease domain at the carboxyl-terminus. All the membrane-anchored serine proteases are structurally conserved catalytic domains and belong to S1 peptidase family. These serine proteases usually exist as zymogens (inactive form); and their autoactivation cleavage occurs after a basic amino acid residues present in a highly conserved activation motif producing a two-chain form with their chains bonded by a disulphide bridge, eventually, separating the pro- and catalytic domains with the catalytic domain remaining membrane bound. Some examples of TTSPs are TMPRSS2, matriptase, hepsin and TMPRSS4 [18]. Therefore, they represent enzymes whose peptide bond cleaving activities are specifically targeted to cellular membranes. They have been phylogenetically categorised into four subfamilies on the basis of C-terminal transmembrane domain: (1) the human airway trypsin-like (HAT)/differentially expressed in squamous cell carcinoma gene (DESC) subfamily, (2) the hepsin/transmembrane protease serine (TMPRSS) subfamily, (3) the matriptase subfamily, and (4) the corin subfamily (Fig. 12.4). In compare with the GPI-anchored and Type I serine proteases, which consist of SPD (serine protease domain) and membrane anchor, the TTSPs possess a stem region which is C-terminal to SPD having a variety of modular structural accessory domains (SEA, CUB, FD, SR) that are involved in protease activation, localisation and substrate recognition to maintain the homeostasis of pericellular microenvironment (Fig. 12.4).

Dysregulation of pericellular and extracellular proteolysis that involve the membrane-anchored and secreted serine proteases, respectively, are the hallmarks in various clinical disorders. Reports have shown that proteolytic breakdown of the extracellular matrix (ECM) is the key step in spreading tumour cell [19, 20]. The series of activities of proteases involved in tumour progression is collectively called as the cancer 'degredom'. A positive cooperativity between the aggressiveness of tumour and the overexpression of many proteases has been detected [20]. In the series of events in cancer progression, serine proteases may be involved in any of the fundamental processes of tumorigenesis with unique specifications [13]. In normal physiological conditions, an endogenous anti-serine protease system known as serpins regulate the serine protease activity and maintain the balance between proteases and their inhibitors in the organism. An imbalance between the proteolytic and antiproteolytic may be of major significance in the cancer development. For example, hepsin, a cell surface serine protease, and maspin, a serine protease inhibitor, are both showing highly upregulated and downregulated, respectively, in prostate cancer, and this causes an imbalance in cellular homeostasis which are believed to promote tumour growth, invasion and metastasis. This shows that the improper function of serine protease leads to cancer which will suggest the need of therapeutic agents against the serine protease to prevent tumour progression and metastasis. As a tumour biomarker, serine proteases are important in detecting certain cancers at an earlier stage. For example, determination of coagulation factor levels and serum prostate-specific antigen can be used for detecting thrombotic and prostate cancer patients. Furthermore, targeting and modulation of overexpressed proteases are the efficient selective approaches for the development of antitumour therapies [21]. Due to the ever-increasing, newly found roles of serine proteases in cancer, there has been increasing attention in the specific roles of members of serine proteases and their inhibitors in array of diverse cancer progression. In this chapter, a review of the role of these members of serine proteases (secreted type and membrane anchored of TTSPs family) and their inhibitors in tumour progression has been discussed in order to understand their therapeutic applications.

12.2.2 Secreted Serine Proteases and Its Role in Cancer

There are about 175 predicted serine proteases in humans. Most of them are found to be secretory in nature that has major roles in a multiple metabolic functions in maintaining the tissues homeostasis. For example, the uPA (urokinase plasminogen activator) and kallikrein system participate in a range of physiological function from cell growth, cell signalling to tissue remodelling process. However, dysregulation expression of serine proteases leads to tumour invasion and cancer. In this section, the main focus is on particular secreted serine proteases, which are reported to be the vital causes of cancer progression and metastasis.

12.2.2.1 Urokinase Plasminogen Activator

At physiological condition, the uPA system is linked with various tissues remodelling processes, immune system and inflammation, fibrinolysis, embryogenesis, angiogenesis, cell migration and activation and differentiation of white blood cells. The active form of uPA is mostly synthesised by cells, in tissues and extracellular fluids with mild intrinsic activity [22].

The uPA system belongs to a serine protease family, playing an important function in tumour invasion and metastasis in cancer. The plasminogen activator (PA) system comprises the two serine proteases, uPA and tissue plasminogen activator (tPA), the two serpin inhibitors, plasminogen activator inhibitor-1 (PAI-1) and plasminogen activator inhibitor-2 (PAI-2) and the glycolipid-anchored uPA receptor (uPAR). Both uPA and tPA catalyse the formation of active protease plasmin from the inactive zymogen, plasminogen, which can break down most extracellular proteins. However, tPA mainly acts as a fibrin-dependent pathway for blood clot dissolution process [23]. While uPA performs through fibrin-independent pathway and largely acts on the cell surface receptor-bound plasminogen activator like uPAR which controls the pericellular proteolysis of the system, this involves in the degradation of ECM and causes invasion and cancer metastasis [24] (Fig. 12.4). uPA and uPAR are observed to be highly expressed in various human cancers in contrast to the corresponding normal tissue. In this regard, uPAR is a highly glycosylated cell surface protein which do not contain transmembrane and intracellular domains but are attached to the cell membrane by a GPI anchor (Fig. 12.4).

uPA is a small trypsin-like protease having molecular weight of 53 kDa. It performs the catalysis of zymogen, plasminogen, into its active form plasmin that facilitates the degradation of various ECM proteins such as fibronectin (FN), vitronectin (VN) and fibrin which results in the loss of interactions between cells, leading to the invasion of cancer cells [25]. In addition, it is also able to activate the inactivated forms of various metalloproteases (MMPs) [26]. In this regard, uPA in combination with uPAR plays a pivotal role in inducing the proteolytic cascade reactions that promote tumour growth through the process of metastasis of cancer cells [27]. The effect of uPAR on cancer cell migration is characterised on the basis of whether it is protease dependent or not. The protease-dependent function is catalysed by uPAR-bound uPA. Since uPAR has no transmembrane structure, its non-protease function depends on the interaction with VN, integrin family, G-protein-coupled receptors and growth factor receptors to relay its downstream signals [28]. Signalling through uPAR activates Tyr kinases, Src, the serine kinase Raf, FAK and extracellular signal-regulated kinase (ERK)/mitogen-activated protein kinase (MAPK) pathway which results in the broad modulation of cell proliferation, metastasis and cell-cell interactions [29]. The universal functions of uPAR are proteolytic extracellular matrix degradation for the progression of cancer, angiogenesis, modulation of cAMP levels for downstream signalling and cell interaction with integrins, tyrosine kinases and serine/threonine kinases [30].

uPA, uPAR and also PAI-1 are constitutively expressed in human breast cancer. In most of the cancers especially breast cancer, high levels of uPA and uPAR are overexpressed, proposing the enhanced role in ECM degradation, migration and adhesion and cancer invasion [31]. PAI-1 and uPA are the first novel tumour biological predictive factors found, in evidence with their clinical utility for breast cancer [32]. uPA has also been observed to be a predictive marker in many types of organ cancers such as cancers of the lung [33], bladder [34], stomach [35], etc.

Several studies suggest that uPA binds with uPAR that facilitate the cell migration process through diverse cell signalling pathways. In this regard, integrins are crucial uPAR signalling co-receptors, and activation of integrin stimulates the focal adhesion kinase (FAK) and, thereby also, activates Src/MEK/ERK-dependent signalling pathways, resulting in transcriptional activation of the uPA promoter, which promotes tumour cell proliferation and tumour invasion (Fig. 12.5). Similarly, the p38 MAPK and myosin light-chain kinase (MLCK) pathways are involved in

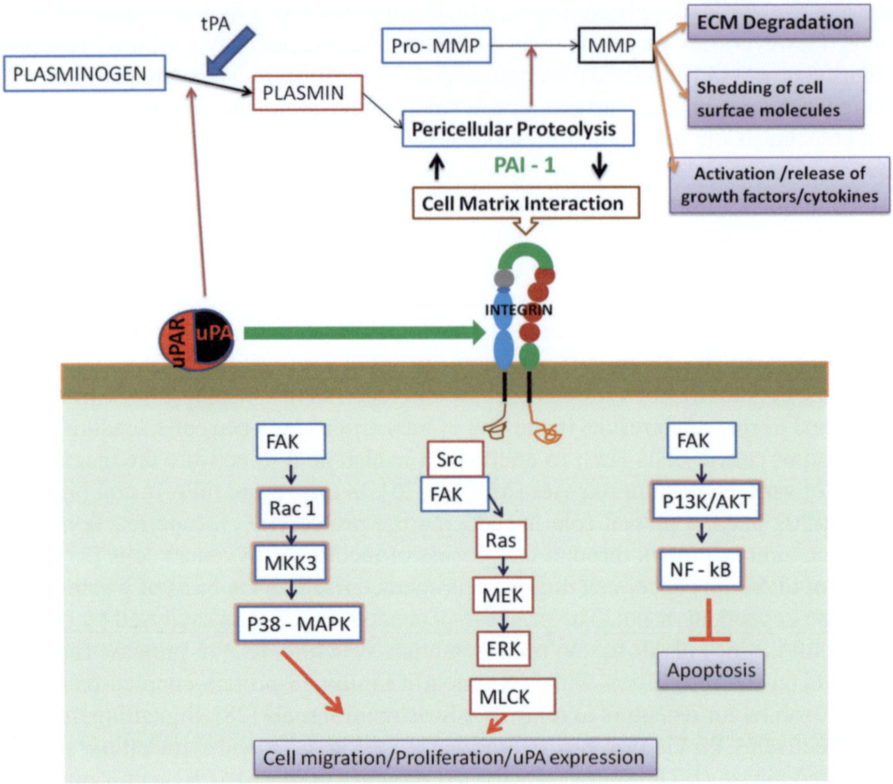

Fig. 12.5 uPA system in cancer malignancy. In the plasma membrane, uPA binds to uPAR which promotes uPA activation resulting in the catalysis of plasminogen into active form, plasmin. Plasmin can subsequently activate MMPs in the pericellular environment that results in invasion and metastasis via ECM degradation. Intracellular activation of uPA/uPAR along with integrins activates FAK, PI3/AKT and p38-MAPK signalling pathways, which finally leads to pathophysiological events, such as metastasis and inhibition of apoptosis pathway

uPA-promoted cell migration through MEK/ERK, PI3K/AKT and Ras/ERK signalling pathways, respectively [36] (Fig. 12.5).

12.2.2.2 Kallikreins

In 1930, Kraut and colleagues coined the term kallikrein (*kallikreas* is the Greek word for pancreas) from an identified substance (human kallikrein 1) that was present at significant concentration in the pancreas. Human tissue kallikreins (hKs) are secreted serine proteases that convert high molecular weight proteins into biologically active peptides known as kinins. There are two families of kallikreins, the tissue and plasma kallikreins. Human plasma kallikrein cleaves high molecular weight kininogen into a bradykinin which is a potent vasodilator nonapeptide (Fig. 12.6). The only enzyme which has been found with appreciable kallikrein activity is kallikrein I (hK1, pancreatic-renal kallikrein) in human tissue kallikreins. There are a

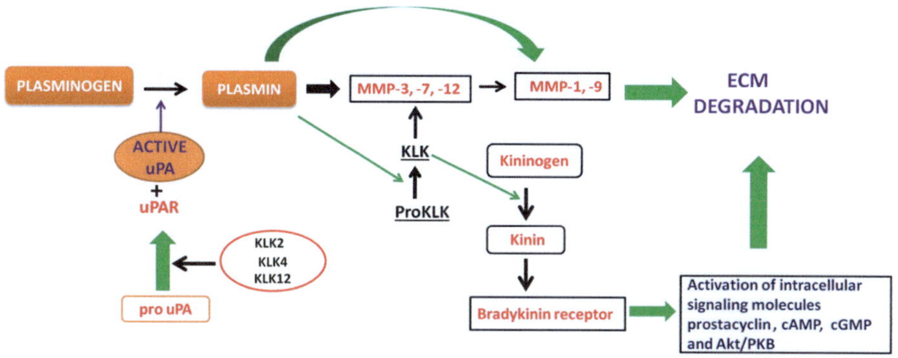

Fig. 12.6 Kallikrein's signalling pathway system. Kallikreins activate the uPA system resulting in the catalysis of plasminogen into active plasmin which in turn activates the pro-KLK proteins leading to the breakdown of various downstream targets, such as latent MMPs. Kallikreins take part in ECM remodelling directly and/or indirectly via activation of pro-MMPs. Active kallikrein is involved in the conversion of kininogen into kinin fragment which induces angiogenesis and other pathological processes via activation of the cAMP, Akt/PKB and VEGF pathway

total of 15 tissue kallikreins genes named as KLK1 to KLK15 encoding hK1 to hK15 which are mostly regulated by steroid hormones [37]. Most of the kallikreins are expressed in endocrine-related organs, such as the breast, ovary, prostrate and testis. These are secreted by extracellular matrix. It has been known for decades that the ECM degradation performs a crucial function in tumour metastasis through extracellular proteolytic activity. The ECM maintains its own structural integrity which involves various growth factors and signalling molecules. Therefore, imbalances created by the activity of extracellular proteases modify the microenvironment of the ECM which either directly or indirectly poses an impact on the number of cell activity processes such as apoptosis, angiogenesis and metastasis via the breakdown of ECM and non-ECM components [38]. This breakdown of ECM results in the alteration of cell-cell and cell-ECM interactions which in turn perturb the activity of growth factors and growth factor receptors and finally leading to either tumour promoting or tumour-suppressive effects. This shows that it is a very complex process and contains many factors. For example, the proteolytic activity of kallikrein is found to be deregulated in tumours such as adenocarcinomas, and it is also used in patient prognosis [38]. Similarly, several studies reported the overexpression of 12 *KLK* genes in ovarian carcinoma associated with steroid-hormone-regulated cancer [39]. Interestingly, kallikreins are normally found to be downregulated in breast, prostate and testicular tumours. Apart from steroid-hormone-regulated cancers, kallikreins are deregulated in various tumour types such as lung adenocarcinomas, pancreatic cancer and acute lymphoblastic leukaemia [38].

The current perception is that pericellular cascade is not only regulated by the serine protease system of uPA, uPAR and plasminogen, but it has adverse impacts on activation of MMPs, which is associated with the extracellular proteolysis in tumorigenesis. Despite it, the activation of the uPA-uPAR-MMP proteolytic cascade by various hK-family members further widens their various routes in cancer progression. Thus, many members of proteolytic network especially kallikrein family is involved in cascading reactions of tumour progression [38].

Angiogenesis is a process of differentiation of new capillary blood vessels from pre-existing vessels. Angiogenesis is mainly controlled by the ratio of pro- and anti-angiogenic growth factors exist in the blood. The increased ratio of pro-angiogenic stimuli and inhibitory regulators activates or switches on the angiogenic process.

Human tissue kallikrein possesses potent angiogenic effects by processing many elements of the extracellular matrix. It has been classified as a pleiotropic angiogenic agent which catalyses the inactive form of kininogen into active form of kinin peptides which in turn activate cAMP, Akt/PKB and VEGF (vascular endothelial growth factor) pathways, and this promotes the process of angiogenesis [40] (Fig. 12.6). KLKs may also participate in remodelling of ECM indirectly through the MMPs, uPA and kinin signalling pathways [41–43]. The kallikrein family such as KLK2, KLK4 and KLK12 activates the uPA system, resulting in plasmin formation, and this activated plasmin causes the breakdown of a number of ECM proteins, for example, fibronectin, proteoglycans and fibrin [28, 44–46] (Fig. 12.6). Similarly, KLK1 and KLK12 catalyse the conversion of kininogen to active kinin peptides and bradykinin, and this promotes angiogenesis and metastasis through the activation of downstream signalling pathways, for example, basic fibroblast growth factor (bFGF) cAMP, Akt/PKB and VEGF pathways [45, 47, 48] (Fig. 12.6).

12.2.2.3 PSA/hK3 (Prostate-Specific Antigen)

PSA consists of a 240-amino acids long glycoprotein, and it comes under the category of human glandular kallikrein family (hK3, a 33 kDa serine protease) [49]. PSA is mainly synthesised in prostrate ductal and acinal epithelium and is secreted into seminal plasma. PSA plays an important role in semen liquefaction by hydrolysing semenogelin I and II in the seminal coagulum [50]. It has chymotrypsin-like activity, does not hydrolyse synthetic substrates for plasmin and displays a weak interaction with aprotinin, a plasmin inhibitor [51]. This suggests that PSA primarily acts independently as a protease in protein degradation, and not via plasmin, like uPA. PSA is organ-specific and is characteristically expressed in prostatic epithelial cells, and its expression is regulated by androgens [52]. It has been observed that proteolytic cascade pathways may also exist in the absence of MMPs but through plasmin-dependent pathway which involves the degradation of type IV collagen, an essential part of a basement membrane [53]. In this regard, urokinase performs a pivotal role in the proteolytic cascade pathway in prostate cancer invasion. Furthermore, kallikreins such as PSA can activate pro-urokinase to its active form, and subsequently, uPA activates plasmin and, which in turn, can recruit collagenases from pro-collagenases which can cause massive degradation of ECM [54]. It has been shown that the dissolution of ECM involve direct degradation of fibronectin by uPA and degradation of fibronectin and laminin by plasmin before the degradation of collagen matrix [54].

12.2.2.4 HtrA1 (Prss11 or IGFBP-5)

HtrA1 (also known as Prss11 or IGFBP-5 or DegP) belongs to a family of high-temperature requirement factor A (HtrA) of oxidative stress-response proteases. It is a heat shock-induced envelope-associated serine protease and performs as a chaperone which is crucial for the survival of bacteria at elevated temperature [55, 56]. They are widely distributed from prokaryotes to eukaryotes. Evolutionarily, these serine proteases have independent ATP conserved sequences and are believed to act as a defence mechanism against cellular stresses including the proteolysis of the misfolded proteins to maintain the homeostasis of the cell [57]. There are four human HtrAs: HtrA1 [58], HtrA2 [59], HtrA3 (pregnancy-related serine protease, PRSP) [60] and HtrA4 [61]. They carry out a number of biological functions such as mitochondrial homeostasis, apoptosis and cell signalling, and their improper functions lead to various clinical disorders [62, 63].

HtrA1 is the first reported member of the human HtrA protein family isolated from a normal fibroblast cell [64]. HtrA1 is downregulated in a variety of cancers such as melanoma [65], glioma [66], ovarian tumours [67, 68], endometrial cancer [68, 69], lung cancers [70], etc. Interestingly, studies have reported that overexpression of HtrA1 functions as a tumour suppressor either by inhibiting the cancer cells or through the apoptosis of cancer cells [65]. Despite it, the mechanism of HtrA1 involved in cancer is still unexplored [71]. In cancer development, it was also proven that HtrA1 and HtrA3 are the inhibitors of growth factor systems such as transforming growth factor β (Tgfβ) family members which are the key regulators for cell growth and differentiation in different tissues [72].

One of the most promising approaches in the discovery of drug cancer is to rationally identify such type of therapeutic agent which regulates the apoptotic process [73]. The biochemical events in apoptosis is regulated by pro-apoptotic and anti-apoptotic proteins which come under the category of Bcl-2 family anti-apoptotic survival proteins such as inhibitors of apoptosis protein (IAP) family and caspases. It was found that HtrA2/Omi functions as a promoter in apoptotic cell death [74]. The mature HtrA2/Omi is capable of inducing apoptosis in human cells and functions as a caspase-independent system through its proteolytic activity and in a caspase-dependent manner through the degradation of IAPs [75]. The function of HtrA2 in tumorigenesis is not yet fully understood; however, its increased levels in the cell upon apoptotic stimuli might prevent the cells from apoptosis which act as a defence mechanism to malignancy. This suggested that HtrA proteins can be used as a novel approach in cancer therapy.

12.2.3 Membrane-Anchored Serine Proteases

The membrane-anchored Type II serine proteases are identified as essential part of the human degradome, and they function in the conversion of precursor molecules into active molecules in the pericellular microenvironment, playing absolute functions in tissues homeostasis and cancer [76]. The TTSP family is the recently known protease family, and much is still to be explored. TTSPs have diverse roles in mammalian system, and their structural homology does not linked to a common biochemical function. They are mostly participating in either hormone or growth factor activation or in the initiation of proteolytic cascades. This suggests that they are maintaining basic homeostasis by activating or deactivating the signalling molecules involve in the biochemical reactions.

Recently, great attention has been paid to the members of TTSP family such as hepsin, matriptase-2 and TMPRSS4 for their vital physiological roles, role in tumourigenic activity and distinctive regulatory mechanisms. These TTSPs are being increasingly documented for their important roles in regulating the pericellular microenvironment and thus providing new insights of their mechanisms of mammalian health and diseases.

12.2.3.1 Membrane-Anchored Type II Serine Proteases (TTSP) and Its Role in Cancer

Hepsin

In the United States, about 2 lakhs of new cases of prostate cancer of adult men and 40,000 deaths were observed in 1995. The occurrence of prostate cancer is more prevalent in the later age of 60 years and above, and about 80% of prostate cancers are diagnosed of this age group [77].

Hepsin (TMPRSS1) is one of the members of Type II transmembrane serine proteases and is expressed in prostate cancer [78]. Hepsin is mostly present in the liver but is also found at trace amount in tissues of the stomach, kidney, prostate, thyroid and inner ear. This subfamily is composed of seven members in human and mice. It possesses only one additional domain in its stem region, a group A scavenger receptor domain (SR) in addition to serine protease domain (SPD) (Fig. 12.4).

It can cleave and activate pro-uPA, pro-HGF, Laminin332 and pro-MSP [78]. It activates pro-hepatocyte growth factor and is inhibited by hepatocyte growth factor activator inhibitor-1B (HAI-1B) and hepatocyte growth factor activator inhibitor-2 (HAI-2) [79]. It is involved in the activation of various proteolytic cascades especially the activation of non-active proteases which leads to breakdown of the extracellular matrix proteins. Moreover, this membrane-associated serine protease helps in the blood coagulation pathway by converting factor VII to VIIa resulting in the formation of thrombin, deposition of pericellular fibrin as well as the activation of PAR-1 (protease activated receptor) [80]. Hepsin was also to be highly expressed in ovarian cancers [81–83].

The TMPRSS2 and TMPRSS4 are the Type II serine proteases that are highly expressed in prostate and pancreatic cancers, respectively, and activate proteolytic cascades which lead to metastasis events [84].

TMPRSS4

TMPRSS4 is a member of Type II transmembrane serine protease and is found to be overexpressed mostly in the pancreas, the thyroid and cancer tissues. As compared to hepsin, it possesses an additional low-density lipoprotein receptor class A (LDLA) domain which is N-terminal to the SR and SPD domains (Fig. 12.4). The

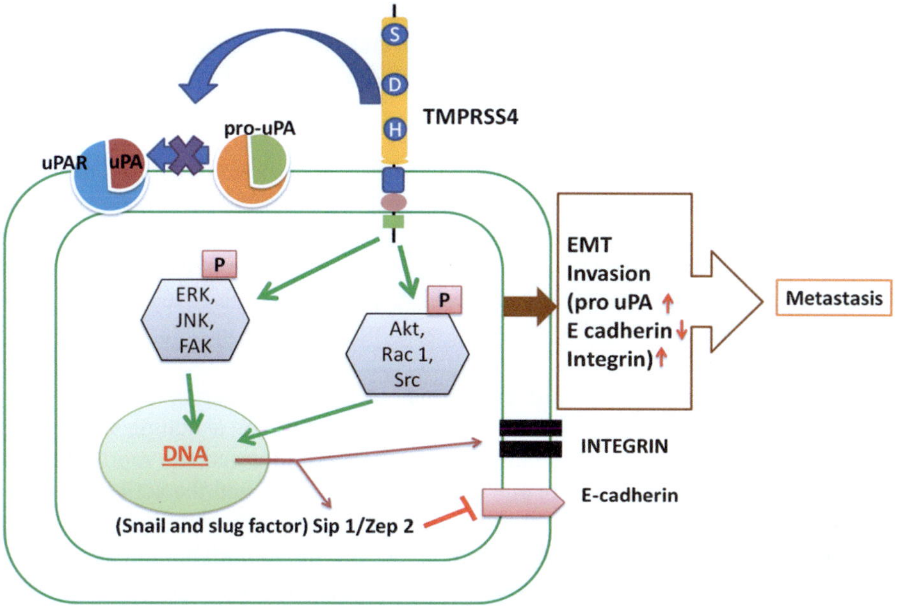

Fig. 12.7 Effect of overexpression of TMPRSS4 in cancer malignancy. It helps in the activation of the intracellular pathways through phosphorylation of ERK, JNK, Akt, Src, FAK and Rac1 which in turn upregulates integrin (ITG-α5) and transcription factors such as Sip1/Zeb2 (a repressor of E-cadherin) resulting in invasiveness and EMT. In the cell membrane, TMPRSS4 converts the precursor of uPA (pro-uPA) to its active form which accelerate the invasiveness

molecular mechanism of TMPRSS4 for metastasis of cancer cells is still unclear. However, it promotes the cancer progression by activating the loss of E-cadherin-mediated cell-cell adhesion and facilitating the epithelial-mesenchymal transition (EMT). The EMT is a process of conversion of epithelial cells into motile mesenchymal cells characterised by the change in the polarity of epithelial cells, cell-cell adhesion, enhanced proteolytic activity, migratory capacity and invasiveness resulting in increased production of ECM components [85].

One of the factors that contribute to metastasis is the downregulation of E-cadherin through E-cadherin transcriptional repressors/EMT-inducing transcription factors, including the snail superfamily consisting snail and slug factors, and this leads to EMT events in human epithelial cancer cells. In colon cancer, TMPRSS4 significantly promoted FAK signalling pathway activation that includes FAK, ERK1/2, Akt, Src and Rac1 activation which in turn stimulate the transcription factors SIP1/ZEB2, resulting in E-cadherin loss, a major event found in EMT (Fig. 12.7) [86–88]. Furthermore, interestingly, TMPRSS4 downregulates the expression of RECK, an inhibitor of tumour angiogenesis, via the activation of ERK1/2 pathway [89–91]. The overexpression of TMPRSS4 is the major event in hepatocellular carcinoma (HCC) progression and can be used as a good predictive biomarker for HCC.

TMPRSS4 activates uPA by two ways, one through increased gene expression (JNK and transcription factors Sp1 and Sp3 and AP-1 pathway) and another by activating pro-uPA, and this leads to enhanced invasion [92] (Fig. 12.7).

Matriptase-2

Matriptase-2 or TMPRSS6 (80–90-kDa cell surface glycoprotein) belongs to a Type II transmembrane serine protease family [93, 94]. Matriptase-2 comprises of a short N-terminal cytoplasmic tail, a transmembrane domain, an extracellular stem region containing a SEA domain (a single sea urchin sperm protein), two CUB domains (urchin embryonic growth factor), three LDLA repeats and a C-terminal trypsin-like SPD domain [93, 94] (Fig. 12.4). It was mostly found in breast and prostate cancers [95, 96].

Matriptase-2 is a hepatic membrane serine protease and is expressed as zymogen on the cell surface, and this inactive proenzyme undergoes shedding to a single chain form followed by autoactivation by cleavage at conserved site represented as RIVGG between the pro-domain and the catalytic domain, and the activated protease domain fragment remains on the membrane via a single disulphide bond linking the pro- and catalytic domains [97, 98]. Matriptase-2 shows high homology in terms of structure as well as its function with matriptase-1 [99], which is found to be overexpressed in epithelial cells, and in various cancers [100]. Matriptase-2 is primarily found to be expressed in human liver that shows connection with the dissolving of extracellular matrix proteins including laminin and fibronectin [94]. It was established that the degradome components such as hepsin, MTSP1, MMP26, plasminogen activator inhibitor-1 (PAI-1), uPAR, MMP15, TIMP3, TIMP4, maspin and RECK are associated with cancer progression in human prostatic tissues [101].

The protease activity can be controlled by its pericellular environment in various ways. In our living system, it was found that several proteases can be activated in an acidic environment such as cathepsins in lysosomes and pepsinogen in the stomach [102–104]. It was also observed that the activity of matriptase is firmly controlled by the chemical environment of the cell [105]. Like other secreted or lysosomal proteases which are activated by an acidic pH, matriptase is also activated in the same way but is unique in the sense that it is attached onto the surface of cells [106]. Matriptase is released as a zymogen and its autoactivation activity depends on intrinsic activity of matriptase zymogen, non-catalytic domains of the enzyme and post-translational modifications [107, 108].

This protease is mainly co-expressed with hepatocyte growth factor activator inhibitor-1 (HAI-1) in the normal epithelial components of tissues, suggesting that the protease activity of matriptase is tightly regulated [100, 109, 110].

Reports suggested that an imbalance of matriptase and HAI-1 ratio is the key factor for the indication of a cancer-related proteolytic events. It is being shown that the ratio of matriptase and HAI-1 has been increased in many cancers such as in breast and prostate [111, 112]. Although the proper mechanism of the dysregulation of matriptase activity is still not known, it may directly affect the cellular microenvironment via the activation or inactivation of downstream signalling molecules leading to the breakdown of ECM components and cell-cell adhesion [113].

12.3 Serpins for Diagnosis and Therapy in Cancer

The significance of regulated membrane-anchored and secreted serine proteases to maintain homeostasis and its relation with these enzymes and cancer reflects that these enzymes must be strictly controlled in normal physiological conditions. Therefore, enzymatic breakdown of serine proteases is considered to be one of the important regulators for maintaining cellular homeostasis. However, excessive enzymatic activity is often an adverse effect on the cellular processes, and this can also be associated with cancer. In conjunction with evolutionary development of proteases, regulators for proteases have also been developed. These anti-regulators of cellular serine protease are known as serpin. Selective serpins which are thought to be correlated with progression or remission of selected cancers have been selected for the critical reviews so that they can be used for diagnosis and therapy in cancer.

12.3.1 Serpin

SERPIN (an acronym of SERine Protease INhibitors) is a protein superfamily representing a core structure of 370–390 conserved amino acids residues with three β sheets (A, B, C) and seven to nine α-helixes. In humans, plasma serpins comprise 2%–10% of all proteins in the blood circulation and perform a crucial role in regulation of a various types of biological functions.

In the serpin, a reactive centre loop (RCL) is found to be involved in the inhibition of proteases target. This RCL is about 20–24 residues long and is present in the extended conformation above the body of the serpin scaffold. Serpins use S (stressed) which are in the native to R (relaxed) transition forms for inhibition of serine proteases. During this transition, the long, flexible RCL of serpin interacts with target protease by inserting itself into the centre of β-sheet A to form an extra strand that locks it into a canonical (key-like) conformation via a non-covalent, reversible mechanism [114]. This results into the distortion of the active site of protease which causes an irreversible covalent serpin-enzyme complex formation. This mechanism is also known as suicide substrate mechanism.

The serpin suicide inhibitors such as α-antitrypsin, α-antichymotrypsin, antithrombin and PAI-1 regulate coagulation pathway, neurotrophic factors, hormone transport, inflammation, angiogenesis, hormone transport, blood pressure and various biological processes. Surprisingly, not all the serpins are acting as protease inhibitors but few of them are found to inhibit other types of proteases whereas others are found to be non-inhibitors. For example, antigen-1 (SCCA-1) inhibits cysteinyl proteases of the papain family. Non-inhibitory serpins exhibit various important functions, including roles as chaperones, for example, the 47-kD heat shock protein (HSP47) and hormone transportation like cortisol-binding globulin [115]. Serpins such as PAI-1, maspin, neuroserpin, PEDF and SPINK1 have been selected to understand further of their antitumour mechanisms in various type of cancers.

12.3.1.1 Plasminogen Activator Inhibitor-1

PAI-1 consists of 400 amino acid residues long glycoprotein, with molecular weight varying from 38 to 70 kDa, on the basis of their degree of glycosylation and functions in a wide variety of clinical and non-clinical conditions [116].

PAI-1 has a dual role in biological system. It inhibits uPA and tPA to prevent plasminogen cleavage into active plasmin, and this results in the inhibition of the process of carcinogenesis [117]. PAI-1 binding to the uPA/uPAR complex triggers the internalisation of uPA/uPAR through low-density receptor-related protein-1 (LRP-1) via endocytosis, and this results in de-adhesion of plasma membrane matrix which facilitate tumour growth and dissemination [118]. All forms of PAI (activated, latent and cleaved) interact directly with LRP1 and enhance cell motility via activation of the JAK/Stat 1 pathway. Studies have shown that in many cancer patients, there were contradictory reports of having positive association between high levels of PAI-1 in tumours and blood with poor clinical outcome. This contradictory effect of PAI-1 has been elucidated by its pro-angiogenic activity (angiogenic activity at low concentration and anti-angiogenic activity at high concentration) and its anti-apoptotic of cells. The pro-angiogenesis activity of PAI-1 is postulated to be associated with PAI-1 inhibition of plasmin-mediated cleavage of FAS-ligand preventing the apoptosis of the endothelial cells [119]. Similarly, reports have shown that a PAI-1 deficiency in mice and cancer cells has the ability to promote the apoptosis process and also inhibit angiogenesis [120]. Nishioka et al. reported that the deletion of PAI-1 in gastric cancer cells decreased down the tumourigenicity [121]. These results revealed that PAI-1 can be used as a good therapeutic agent for cancer.

12.3.1.2 Maspin (SERPINB5)

Maspin, a 42 kDa mammary serpin, was first reported as a class II tumour suppressor in human breast cancer. It comes under the category of non-inhibitory serpin that promotes the tumour cell towards apoptosis and inhibits invasion and metastasis, and thus, maspin plays a vital role against tumour growth [122]. It is located in the cytoplasm but is also secreted to the cell surface, where it has been postulated to prevent angiogenesis and reduce the migration of many cell types in different experimental models [123, 124]. Maspin in contrast to PAI-1 consists of a relatively short, non-conserved, hydrophobic RCL, and therefore it is incapable of conversion of stressed to relaxed transition form for inhibition. Furthermore, it is incapable to inhibit either tPA or uPA as their postulated targets [125]. Because of these properties, maspin is considered as non-inhibitory category of serpin superfamily. However, recently, it was shown that maspin has inhibitory effect against plasminogen activators uPA and tPA, but they work only when these proteases are bound to macromolecular cofactors, that is, tPA bound to fibrin and uPA on the cell surface [126–128]. The expression of maspin gene is controlled at the transcription level and is found downregulated with the degree of malignancy. For example, the concentration levels of maspin are relatively very low in breast and prostate cancer cells as compared to normal cells [129]. Many cancer studies have shown that the involvement of cytosine methylation and chromatin condensation are associated with the deregulation of maspin expression during cancer progression [130]. This suggested that an epigenetic mechanism which is involved in cytosine methylation, histone, deacetylation and chromatin condensation inhibits and thus regulating the expression of maspin. Since, maspin is an inhibitor of angiogenesis, it regulates adhesion-mediated cell signalling pathway through extracellular and cell-cell contact adhesion molecules. For example, Maspin enhances the endothelial cell adhesion to FN, laminin, collagen and vitronectin, leading to the activation of integrin family and FAK signal transduction pathway. These cause the modulation of focal adhesion and cytoskeleton reorganisation which finally prevent the degradation of EC components and migration of tumorigenic cells [131, 132].

Numerous studies have reported that maspin suppresses tumour cells through induction of apoptosis pathway. For example, mammary carcinoma cells transfected with maspin gene provide the evidence of inhibition of invasion and metastasis in nude mice [133]. Maspin was hypothesised to induce tumour cell apoptosis by modulating mitochondrial permeability transition and initiating apoptotic death [134]. Thus, such discoveries of molecular mechanisms regarding maspin-mediated apoptosis paved a new pathway for the treatment of cancer.

Reports have shown that maspin expression and ubiquitin-proteasome pathway are inversely correlated with each other, where expression of maspin reduces with the increase in chymotrypsin-like activity of the proteasome [135]. As the ubiquitin-proteasome pathway modulates several biochemical events through protein regulation, it is postulated that deregulation of proteasome function is an important factor responsible for the malignancy of tumours [136]. Thus, the establishment of a new distinct relationship between maspin and the ubiquitin-proteasome pathway also provides an important clue for the suppression of a multitude of processes of tumour and metastasis.

Recently, the use of maspin alone or in association with mammaglobin B (a secretoglobin) is exploited as two biomarkers at different stages (cell proliferation and pathological stage) of the detection of the breast cancer [137]. In context with the epigenetic regulation of maspin, it was observed that in a pregnant woman, the *maspin* gene promoter was unmethylated in foetus with respect to maternal blood cell, and this opened a new avenue for developing further biomarkers for prenatal diagnosis [138]. The established anti-tumorigenic/anti-metastatic characteristic of maspin in cancer provides useful information regarding the development of therapeutic agents [139, 140]. In a nutshell, maspin can be exploited as an antitumour agent in different cellular events such as actin cytoskeleton, apoptosis, proteasome function, oxidative stress for the inhibition of cell invasion and angiogenesis.

12.3.1.3 Neuroserpin (SERPINI1)

Neuroserpin (NSP), a protease inhibitor of 46–55 kDa glycoprotein, was first recognised as a secreted protein from cultured chicken neuronal axons and is predominantly present in both central and peripheral nervous systems [141, 142]. Neuroserpin is a trypsin-type protease, preferentially inhibits tPA and to a minor extent uPA plasmin, but shows no inhibition towards thrombin [143].

Ischemic stroke is a single largest cause of stroke and accounts to be the second largest contributor to mortality in the world [144]. It is due to the obstruction of a certain cerebral artery resulting in an absence of blood flow to artery and brain tissue, and this could induce an energy metabolism disorder which in turn perturbs the ion gradients and an excessive release of excitotoxic neurotransmitters such as dopamine and glutamate which ultimately leads to neuronal death [78, 145]. The effective treatment for acute ischemic stroke is the administration of tPA within 3 hours on the onset of stroke. Meanwhile, tPA is capable of activating matrix MMPs and converting plasminogen to plasmin which results in the blood brain degradation [145, 146]. This extra administration of extravascular tPA beyond its therapeutic window (hours) causes a more deleterious effect on the brain. However, the adjuvant treatment with neuroserpin along with tPA is found to increase the therapeutic window, and this could have better treatment for cerebral ischemia. Thus, the neuroprotective effect of neuroserpin for the treatment of cerebral ischemia is dependent on the balanced expression of tPA which in turn regulates the recanalisation of the occluded vessel [147]. It was observed that *neuroserpin* gene is a cancer-associated gene and acts as a tissue-specific tumour suppressor gene in the brain [148]. Like other members of serpin family such as maspin and pancpin, this tissue-specific tumour-suppressive gene is found to be downregulated in brain tumour and even absent in brain cancer cells. As mentioned before, tPA converts inactive plasminogen into active plasmin which results in the breakdown of the extracellular matrix that facilitates the invasion of cancer cells and enables tumour migration. It was observed that neuroserpin functions in the inhibitory process of tPA, and its absence in the CNS causes brain tumorigenesis. Based on this finding, it may be suggested that neuroserpin may represent a new approach for cancer therapy. Additionally, neuroserpin is a tPA-independent mediator of neurite such as outgrowth, cell-cell adhesion and N-cadherin and NFκB expression. The tPA-independent regulatory effect of neuroserpin participates in tumorigenesis as well as emotional and cognitive processes. These new finding of the multifaceted roles of neuroserpin and its polymers will be helpful in designing better methods for treating cancer-like diseases.

12.3.1.4 PEDF (SERPINF1)

Pigment epithelium-derived factor (PEDF) is a 50 kDa secreted glycoprotein consist of 418 amino acids, and this was first described and purified from cultured human foetal retinal pigment epithelium cells. It comes under the category of a non-inhibitory member of the superfamily of serpin [149]. PEDF is a multifunctional member and is widely present in foetal and adult tissues. It is a well-known protein and plays important functions in many physiological and pathological processes [150, 151]. PEDF exhibits a protective mechanism against tumour and represents as a biomarker for prostate cancer patients. For example, it was recently found that the PEDF level in the venous blood patients was significantly high. PEDF can decrease tumour growth either indirectly through the inhibition of angiogenesis or directly through the activation of cell apoptosis and/or differentiation process which inhibits the process of invasion and metastasis. PEDF also works as a protective factor for neuronal components of the eye as well as an important inhibitor of the growth of ocular blood vessels. It is a very selective inhibitor of angiogenesis, and it works only on new blood-forming vessels, while the old cells have no such inhibitory effect. It is also found to be a reversible process [152]. Dawson et al. were the first who proposed that PEDF can regulate the blood vessel growth for angiogenesis at hypoxia condition as it was found in tumours [153]. PEDF mediate the anti-angiogenesis activity by effectively blocking the VEGF-driven vascular permeability through internalisation and degradation of VEGF receptors, VEGFR-1 or VEGFR-2, in VEGF-stimulated endothelial cells [154]. This shows that PEDF prevents the angiogenesis process by nullifying the VEGF activity and inhibits the tumour growth from becoming more malignant.

The PEDF acts as a regulation of cell proliferation, and invasion is gradually lost with the aggressiveness of metastatic melanoma. It has been found that the expression of PEDF is reduced in many tumour grade of various forms of cancers such as prostate adenocarcinoma, pancreatic adenocarcinoma and hepatocellular carcinoma [155]. The molecular mechanism of PEDF to regulate the metastasis of cancer is yet not fully understood, but few studies suggest that the VEGF/PEDF ratio and regulation of MMPs function by PEDF are the key events that prevent cell invasion and tumour dissemination.

Recent studies have shown that PEDF can also manifest its anti-angiogenesis activity by selectively inducing the apoptosis of endothelial cell. In this regard, Guan et al. observed that apoptotic cells are much higher in overexpressed PEDF in prostate cancer cells as referenced to control [156]. PEDF induces apoptosis of endothelial and tumour cells mainly by extrinsic and intrinsic pathways. The extrinsic pathway which is a cell surface death receptor-mediated pathway depends on the activation of Fas/FasL death pathway, whereas in the intrinsic pathway, also called as mitochondrial pathway, the apoptosis of cells is governed by mitochondrial permeability, manifested by Bcl-2 family proteins and caspases.

Another facet of PEDF is that it may exhibit antitumour activity by its ability to promote tumour cell differentiation. Crawford et al. showed that the intratumoural injection of rPEDF in primitive neuroblastomas which were grown in athymic mice results in tumour cell differentiation, evidenced by less malignant appearing cells histologically and immunohistochemical staining for neurofilament [157]. Filleur et al. suggested that PEDF functions in prostate neuroendocrine differentiation through feedforward mechanism [158]. Although very few studies have been done on this aspect, this added ability to prevent tumour cell growth by differentiation of malignant cells into normal phenotype is indeed promising and warrants further investigation.

Nowadays, PEDF has gained wide attention for the approach of making a potential endogenous agent for treating cancers. Therefore, drug delivery systems system such as gene therapy route in the form of a viral vector, systemic administration of naked PEDF (free, unmodified) or using nanoparticles for controlled release of drugs are needed for the smart delivery of PEDF, to neoplastic sites which not only results in tumour regression but also protects from any side effect [159]. For example, a new approach was used in enhancing PEDF expression through specific platinum-based chemotherapeutic phosphaplatin drugs [160].

12.3.1.5 SPINK1 (Kazal Type 1)

The serine protease inhibitor Kazal type 1 (SPINK1) was first purified from bovine pancreas as a pancreatic secretory trypsin inhibitor (PSTI) in 1948 by Kazal and colleagues [161]. Similarly, Stenman et al. (1982) also isolated SPINK1 from the urine of patients which were diagnosed with ovarian cancer, but they described it as a trypsin inhibitor (TATI) which is known to be associated with tumour [162]. This inhibitor consists of 56 amino acid residues containing three disulphide bonds and a trypsin-specific binding site formed by Lys-Ile. The primary function of SPINK1 is to inhibit pancreatic and small intestinal serine proteases. SPINK1 is produced in the acinal cells of pancreas where it prevents from autophagy of pancreas cells by inhibiting the trypsin activity in acinar cells. This is because trypsinogen, a precursor of trypsin, is also produced in the same cells and packed together with SPINK1 in zymogen granule [163]. Under normal physiological condition, the conversion of trypsinogen to trypsin is under strict control, and a balanced level of trypsinogen and trypsin is needed in acinar cells. So, it serves as a significant role in preventing the onset of pancreatitis [164].

SPINK1 is overexpressed in various human organ cancers found in the colon, breast, liver and urinary bladder [165, 166]. It has been reported that SPINK1 is found to be overexpressed in the prostrate, and its expression is directly proportional to the tumour grade [162]. The mechanism of SPINK1 involved in prostate cancer and tumour progression is not yet clear. It was observed that tumours producing SPINK1 also produce trypsin, and this trypsin can activate MMPs of the matrix leading metastasis of cancer. This shows that the imbalance secretion of SPINK1 and trypsin is linked with the adverse prognosis in cancer [167]. Thus, SPINK1 can be used as a good target for prostate cancer treatment.

It was proposed that SPINK1 can act as a growth factor because there are a lot of structural similarities, and 50% amino acid homology was found in between SPINK1 and epidermal growth factor (EGF). Thus, SPINK1 is also thought to be involved as a growth factor for tissue repair in inflammatory sites, and if it was prolonged, then it acts as a booster for cancerous cell [168]. Recently, it was reported that SPINK1 induces EMT through activating epidermal growth factor receptor (EGFR), causing proliferation of pancreatic and breast cancer cells [169].

For the specific therapy of cancer patients, it is necessary that specific or activated pathway, for that specific tumour should be targeted. Accordingly, small-molecule inhibitors should be discovered which interfere with specific signalling networks inside the cells. In this way, SPINK1 can be an excellent 'druggable' target [167, 170].

References

1. Lin YC et al (1995) Activation of NF-kappa B requires proteolysis of the inhibitor I kappa B-alpha: signal-induced phosphorylation of I kappa B-alpha alone does not release active NF-kappa B. Proc Natl Acad Sci U S A 92:552–556
2. Walsh PN, Ahmad SS (2002) Proteases in blood clotting. Essays Biochem 38:95–111

3. Borissenko L, Groll M (2007) Diversity of proteasomal missions: fine tuning of the immune response. Biol Chem 388:947–955
4. Roth S (2003) The origin of dorsoventral polarity in drosophila. Philos Trans R Soc Lond Ser B Biol Sci 358:1317–1329. discussion 1329
5. Bastians H et al (1999) Cell cycle-regulated proteolysis of mitotic target proteins. Mol Biol Cell 10:3927–3941
6. Turk B, Stoka V (2007) Protease signalling in cell death: caspases versus cysteine cathepsins. FEBS Lett 581:2761–2767
7. Rawlings ND et al (2008) MEROPS: the peptidase database. Nucleic Acids Res 36:D320–D325
8. Di Cera E (2009) Serine proteases. IUBMB Life 61:510–515
9. Hedstrom L (2002) Serine protease mechanism and specificity. Chem Rev 102:4501–4524
10. Fastrez J, Fersht AR (1973) Demonstration of the acyl-enzyme mechanism for the hydrolysis of peptides and anilides by chymotrypsin. Biochemistry 12:2025–2034
11. Puente XS et al (2005) A genomic view of the complexity of mammalian proteolytic systems. Biochem Soc Trans 33:331–334
12. Stroud RM (1974) A family of protein-cutting proteins. Sci Am 231:74–88
13. Hooper JD et al (2001) Type II transmembrane serine proteases. Insights into an emerging class of cell surface proteolytic enzymes J Biol Chem 276:857–860
14. Netzel-Arnett S et al (2003) Membrane anchored serine proteases: a rapidly expanding group of cell surface proteolytic enzymes with potential roles in cancer. Cancer Metastasis Rev 22:237–258
15. Chen LM et al (2001) Prostasin is a glycosylphosphatidylinositol-anchored active serine protease. J Biol Chem 276:21434–21442
16. Verghese GM et al (2006) Prostasin regulates epithelial monolayer function: cell-specific Gpld1-mediated secretion and functional role for GPI anchor. Am J Physiol Cell Physiol 291:C1258–C1270
17. Hooper JD et al (1999) Testisin, a new human serine proteinase expressed by premeiotic testicular germ cells and lost in testicular germ cell tumors. Cancer Res 59:3199–3205
18. Szabo R, Bugge TH (2008) Type II transmembrane serine proteases in development and disease. Int J Biochem Cell Biol 40:1297–1316
19. Lu P et al. (2011) Extracellular matrix degradation and remodeling in development and disease. Cold Spring Harb Perspect Biol 3: pii:a005058
20. Lopez-Otin C, Matrisian LM (2007) Emerging roles of proteases in tumour suppression. Nat Rev Cancer 7:800–808
21. Choi KY et al (2012) Protease-activated drug development. Theranostics 2:156–178
22. Hildenbrand R et al (2008) The urokinase-system--role of cell proliferation and apoptosis. Histol Histopathol 23:227–236
23. Collen D, Lijnen HR (1991) Basic and clinical aspects of fibrinolysis and thrombolysis. Blood 78:3114–3124
24. Blasi F (1997) uPA, uPAR, PAI-1: key intersection of proteolytic, adhesive and chemotactic highways? Immunol Today 18:415–417
25. Gondi CS et al (2007) Down-regulation of uPAR and uPA activates caspase-mediated apoptosis and inhibits the PI3K/AKT pathway. Int J Oncol 31:19–27
26. Andreasen PA et al (1997) The urokinase-type plasminogen activator system in cancer metastasis: a review. Int J Cancer 72:1–22
27. Fisher JL et al (2001) The expression of the urokinase plasminogen activator system in metastatic murine osteosarcoma: an in vivo mouse model. Clin Cancer Res 7:1654–1660
28. Sidenius N, Blasi F (2003) The urokinase plasminogen activator system in cancer: recent advances and implication for prognosis and therapy. Cancer Metastasis Rev 22:205–222
29. Aguirre Ghiso JA et al (1999) Tumor dormancy induced by downregulation of urokinase receptor in human carcinoma involves integrin and MAPK signaling. J Cell Biol 147:89–104
30. Preissner KT et al (2000) Urokinase receptor: a molecular organizer in cellular communication. Curr Opin Cell Biol 12:621–628

31. Stillfried GE et al (2007) Plasminogen binding and activation at the breast cancer cell surface: the integral role of urokinase activity. Breast Cancer Res 9:R14
32. Harbeck N et al (2002) Clinical utility of urokinase-type plasminogen activator and plasminogen activator inhibitor-1 determination in primary breast cancer tissue for individualized therapy concepts. Clin Breast Cancer 3:196–200
33. Oka T et al (1991) Immunohistochemical evidence of urokinase-type plasminogen activator in primary and metastatic tumors of pulmonary adenocarcinoma. Cancer Res 51:3522–3525
34. Hasui Y et al (1992) The content of urokinase-type plasminogen activator antigen as a prognostic factor in urinary bladder cancer. Int J Cancer 50:871–873
35. Nekarda H et al (1994) Tumour-associated proteolytic factors uPA and PAI-1 and survival in totally resected gastric cancer. Lancet 343:117
36. Han Q et al (2002) Rac1-MKK3-p38-MAPKAPK2 pathway promotes urokinase plasminogen activator mRNA stability in invasive breast cancer cells. J Biol Chem 277:48379–48385
37. Borgono CA et al (2004) Human tissue kallikreins: physiologic roles and applications in cancer. Mol Cancer Res 2:257–280
38. Borgono CA, Diamandis EP (2004) The emerging roles of human tissue kallikreins in cancer. Nat Rev Cancer 4:876–890
39. Schmitt M et al (2013) Emerging clinical importance of the cancer biomarkers kallikrein-related peptidases (KLK) in female and male reproductive organ malignancies. Radiol Oncol 47:319–329
40. Milkiewicz M et al (2006) Regulators of angiogenesis and strategies for their therapeutic manipulation. Int J Biochem Cell Biol 38:333–357
41. Desrivieres S et al (1993) Activation of the 92 kDa type IV collagenase by tissue kallikrein. J Cell Physiol 157:587–593
42. Menashi S et al (1994) Regulation of 92-kDa gelatinase B activity in the extracellular matrix by tissue kallikrein. Ann N Y Acad Sci 732:466–468
43. Saunders WB et al (2005) MMP-1 activation by serine proteases and MMP-10 induces human capillary tubular network collapse and regression in 3D collagen matrices. J Cell Sci 118:2325–2340
44. Takayama TK et al (2001) Characterization of hK4 (prostase), a prostate-specific serine protease: activation of the precursor of prostate specific antigen (pro-PSA) and single-chain urokinase-type plasminogen activator and degradation of prostatic acid phosphatase. Biochemistry 40:15341–15348
45. Giusti B et al (2005) The antiangiogenic tissue kallikrein pattern of endothelial cells in systemic sclerosis. Arthritis Rheum 52:3618–3628
46. Frenette G et al (1997) Prostatic kallikrein hK2, but not prostate-specific antigen (hK3), activates single-chain urokinase-type plasminogen activator. Int J Cancer 71:897–899
47. Colman RW (2006) Regulation of angiogenesis by the kallikrein-kinin system. Curr Pharm Des 12:2599–2607
48. Emanueli C, Madeddu P (2001) Targeting kinin receptors for the treatment of tissue ischaemia. Trends Pharmacol Sci 22:478–484
49. Watt KW et al (1986) Human prostate-specific antigen: structural and functional similarity with serine proteases. Proc Natl Acad Sci U S A 83:3166–3170
50. Peter A et al (1998) Semenogelin I and semenogelin II, the major gel-forming proteins in human semen, are substrates for transglutaminase. Eur J Biochem 252:216–221
51. Christensson A et al (1990) Enzymatic activity of prostate-specific antigen and its reactions with extracellular serine proteinase inhibitors. Eur J Biochem 194:755–763
52. Christensson A, Lilja H (1994) Complex formation between protein C inhibitor and prostate-specific antigen in vitro and in human semen. Eur J Biochem 220:45–53
53. Mackay AR et al (1990) Basement membrane type IV collagen degradation: evidence for the involvement of a proteolytic cascade independent of metalloproteinases. Cancer Res 50:5997–6001
54. Webber MM, Waghray A (1995) Urokinase-mediated extracellular matrix degradation by human prostatic carcinoma cells and its inhibition by retinoic acid. Clin Cancer Res 1:755–761

55. Lipinska B et al (1990) The HtrA (DegP) protein, essential for Escherichia coli survival at high temperatures, is an endopeptidase. J Bacteriol 172:1791–1797
56. Spiess C et al (1999) A temperature-dependent switch from chaperone to protease in a widely conserved heat shock protein. Cell 97:339–347
57. Clausen T et al (2002) The HtrA family of proteases: implications for protein composition and cell fate. Mol Cell 10:443–455
58. Hu SI et al (1998) Human HtrA, an evolutionarily conserved serine protease identified as a differentially expressed gene product in osteoarthritic cartilage. J Biol Chem 273:34406–34412
59. Gray CW et al (2000) Characterization of human HtrA2, a novel serine protease involved in the mammalian cellular stress response. Eur J Biochem 267:5699–5710
60. Nie G et al (2006) Serine peptidase HTRA3 is closely associated with human placental development and is elevated in pregnancy serum. Biol Reprod 74:366–374
61. Inagaki A et al (2012) Upregulation of HtrA4 in the placentas of patients with severe pre-eclampsia. Placenta 33:919–926
62. Zurawa-Janicka D et al (2013) Temperature-induced changes of HtrA2(Omi) protease activity and structure. Cell Stress Chaperones 18:35–51
63. Canfield AE et al (2007) HtrA1: a novel regulator of physiological and pathological matrix mineralization? Biochem Soc Trans 35:669–671
64. Zumbrunn J, Trueb B (1996) Primary structure of a putative serine protease specific for IGF-binding proteins. FEBS Lett 398:187–192
65. Baldi A et al (2002) The HtrA1 serine protease is down-regulated during human melanoma progression and represses growth of metastatic melanoma cells. Oncogene 21:6684–6688
66. Kotliarov Y et al (2006) High-resolution global genomic survey of 178 gliomas reveals novel regions of copy number alteration and allelic imbalances. Cancer Res 66:9428–9436
67. Chien J et al (2004) A candidate tumor suppressor HtrA1 is downregulated in ovarian cancer. Oncogene 23:1636–1644
68. Narkiewicz J et al (2009) Expression of human HtrA1, HtrA2, HtrA3 and TGF-beta1 genes in primary endometrial cancer. Oncol Rep 21:1529–1537
69. Bowden MA et al (2006) Serine proteases HTRA1 and HTRA3 are down-regulated with increasing grades of human endometrial cancer. Gynecol Oncol 103:253–260
70. Esposito V et al (2006) Analysis of HtrA1 serine protease expression in human lung cancer. Anticancer Res 26:3455–3459
71. Chien J et al (2006) Serine protease HtrA1 modulates chemotherapy-induced cytotoxicity. J Clin Invest 116:1994–2004
72. Oka C et al (2004) HtrA1 serine protease inhibits signaling mediated by Tgfbeta family proteins. Development 131:1041–1053
73. Fesik SW (2005) Promoting apoptosis as a strategy for cancer drug discovery. Nat Rev Cancer 5:876–885
74. Verhagen AM et al (2002) HtrA2 promotes cell death through its serine protease activity and its ability to antagonize inhibitor of apoptosis proteins. J Biol Chem 277:445–454
75. Suzuki Y et al (2001) A serine protease, HtrA2, is released from the mitochondria and interacts with XIAP, inducing cell death. Mol Cell 8:613–621
76. Antalis TM et al (2010) The cutting edge: membrane-anchored serine protease activities in the pericellular microenvironment. Biochem J 428:325–346
77. Carter BS et al (1990) Epidemiologic evidence regarding predisposing factors to prostate cancer. Prostate 16:187–197
78. Moran P et al (2006) Pro-urokinase-type plasminogen activator is a substrate for hepsin. J Biol Chem 281:30439–30446
79. Kirchhofer D et al (2005) Hepsin activates pro-hepatocyte growth factor and is inhibited by hepatocyte growth factor activator inhibitor-1B (HAI-1B) and HAI-2. FEBS Lett 579:1945–1950
80. Kazama Y et al (1995) Hepsin, a putative membrane-associated serine protease, activates human factor VII and initiates a pathway of blood coagulation on the cell surface leading to thrombin formation. J Biol Chem 270:66–72

81. Chen Z et al (2003) Hepsin and maspin are inversely expressed in laser capture microdissectioned prostate cancer. J Urol 169:1316–1319
82. Magee JA et al (2001) Expression profiling reveals hepsin overexpression in prostate cancer. Cancer Res 61:5692–5696
83. Tanimoto H et al (1997) Hepsin, a cell surface serine protease identified in hepatoma cells, is overexpressed in ovarian cancer. Cancer Res 57:2884–2887
84. Lin B et al (1999) Prostate-localized and androgen-regulated expression of the membrane-bound serine protease TMPRSS2. Cancer Res 59:4180–4184
85. Thiery JP, Sleeman JP (2006) Complex networks orchestrate epithelial-mesenchymal transitions. Nat Rev Mol Cell Biol 7:131–142
86. Peinado H et al (2007) Snail, Zeb and bHLH factors in tumour progression: an alliance against the epithelial phenotype? Nat Rev Cancer 7:415–428
87. Kim S et al (2010) TMPRSS4 induces invasion and epithelial-mesenchymal transition through upregulation of integrin alpha5 and its signaling pathways. Carcinogenesis 31:597–606
88. Jung H et al (2008) TMPRSS4 promotes invasion, migration and metastasis of human tumor cells by facilitating an epithelial-mesenchymal transition. Oncogene 27:2635–2647
89. Cheng H et al (2009) Hepatocyte growth factor activator inhibitor type 1 regulates epithelial to mesenchymal transition through membrane-bound serine proteinases. Cancer Res 69:1828–1835
90. Min HJ et al (2014) TMPRSS4 upregulates uPA gene expression through JNK signaling activation to induce cancer cell invasion. Cell Signal 26:398–408
91. Wang CH et al (2015) TMPRSS4 facilitates epithelial-mesenchymal transition of hepatocellular carcinoma and is a predictive marker for poor prognosis of patients after curative resection. Sci Rep 5:12366
92. Min HJ et al (2014) TMPRSS4 induces cancer cell invasion through pro-uPA processing. Biochem Biophys Res Commun 446:1–7
93. Ramsay AJ et al (2008) The type II transmembrane serine protease matriptase-2--identification, structural features, enzymology, expression pattern and potential roles. Front Biosci 13:569–579
94. Velasco G et al (2002) Matriptase-2, a membrane-bound mosaic serine proteinase predominantly expressed in human liver and showing degrading activity against extracellular matrix proteins. J Biol Chem 277:37637–37646
95. Webb SL et al (2012) The influence of matriptase-2 on prostate cancer in vitro: a possible role for beta-catenin. Oncol Rep 28:1491–1497
96. Shi YE et al (1993) Identification and characterization of a novel matrix-degrading protease from hormone-dependent human breast cancer cells. Cancer Res 53:1409–1415
97. Ramsay AJ et al (2009) Matriptase-2 (TMPRSS6): a proteolytic regulator of iron homeostasis. Haematologica 94:840–849
98. Ramsay AJ et al (2009) Matriptase-2 mutations in iron-refractory iron deficiency anemia patients provide new insights into protease activation mechanisms. Hum Mol Genet 18:3673–3683
99. Sanders AJ et al (2010) The type II transmembrane serine protease, matriptase-2: possible links to cancer? Anti Cancer Agents Med Chem 10:64–69
100. Oberst M et al (2001) Matriptase and HAI-1 are expressed by normal and malignant epithelial cells in vitro and in vivo. Am J Pathol 158:1301–1311
101. Riddick AC et al (2005) Identification of degradome components associated with prostate cancer progression by expression analysis of human prostatic tissues. Br J Cancer 92:2171–2180
102. Tannock IF, Rotin D (1989) Acid pH in tumors and its potential for therapeutic exploitation. Cancer Res 49:4373–4384
103. McQueney MS et al (1997) Autocatalytic activation of human cathepsin K. J Biol Chem 272:13955–13960
104. Richter C et al (1998) Mechanism of activation of the gastric aspartic proteinases: pepsinogen, progastricsin and prochymosin. Biochem J 335(Pt 3):481–490

105. Tseng IC et al (2010) Matriptase activation, an early cellular response to acidosis. J Biol Chem 285:3261–3270
106. Lin CY et al (1997) Characterization of a novel, membrane-bound, 80-kDa matrix-degrading protease from human breast cancer cells. Monoclonal antibody production, isolation, and localization. J Biol Chem 272:9147–9152
107. Oberst MD et al (2003) The activation of matriptase requires its noncatalytic domains, serine protease domain, and its cognate inhibitor. J Biol Chem 278:26773–26779
108. Xu H et al (2012) Mechanisms for the control of matriptase activity in the absence of sufficient HAI-1. Am J Physiol Cell Physiol 302:C453–C462
109. Lee MS et al (2005) Simultaneous activation and hepatocyte growth factor activator inhibitor 1-mediated inhibition of matriptase induced at activation foci in human mammary epithelial cells. Am J Physiol Cell Physiol 288:C932–C941
110. Szabo R et al (2007) Matriptase inhibition by hepatocyte growth factor activator inhibitor-1 is essential for placental development. Oncogene 26:1546–1556
111. Kang JY et al (2003) Tissue microarray analysis of hepatocyte growth factor/met pathway components reveals a role for met, matriptase, and hepatocyte growth factor activator inhibitor 1 in the progression of node-negative breast cancer. Cancer Res 63:1101–1105
112. Saleem M et al (2006) A novel biomarker for staging human prostate adenocarcinoma: overexpression of matriptase with concomitant loss of its inhibitor, hepatocyte growth factor activator inhibitor-1. Cancer Epidemiol Biomark Prev 15:217–227
113. Bhatt AS et al (2005) Adhesion signaling by a novel mitotic substrate of src kinases. Oncogene 24:5333–5343
114. Engh RA et al (1995) Divining the serpin inhibition mechanism: a suicide substrate 'springe'? Trends Biotechnol 13:503–510
115. Irving JA et al (2000) Phylogeny of the serpin superfamily: implications of patterns of amino acid conservation for structure and function. Genome Res 10:1845–1864
116. Declerck PJ, Gils A (2013) Three decades of research on plasminogen activator inhibitor-1: a multifaceted serpin. Semin Thromb Hemost 39:356–364
117. Dass K et al (2008) Evolving role of uPA/uPAR system in human cancers. Cancer Treat Rev 34:122–136
118. Duffy MJ (2002) Urokinase plasminogen activator and its inhibitor, PAI-1, as prognostic markers in breast cancer: from pilot to level 1 evidence studies. Clin Chem 48:1194–1197
119. Bajou K et al (2008) Plasminogen activator inhibitor-1 protects endothelial cells from FasL-mediated apoptosis. Cancer Cell 14:324–334
120. Bajou K et al (1998) Absence of host plasminogen activator inhibitor 1 prevents cancer invasion and vascularization. Nat Med 4:923–928
121. Nishioka N et al (2012) Plasminogen activator inhibitor 1 RNAi suppresses gastric cancer metastasis in vivo. Cancer Sci 103:228–232
122. Zou Z et al (1994) Maspin, a serpin with tumor-suppressing activity in human mammary epithelial cells. Science 263:526–529
123. Sheng S et al (1994) Production, purification, and characterization of recombinant maspin proteins. J Biol Chem 269:30988–30993
124. Pemberton PA et al (1997) Maspin is an intracellular serpin that partitions into secretory vesicles and is present at the cell surface. J Histochem Cytochem 45:1697–1706
125. Bass R et al (2002) Maspin inhibits cell migration in the absence of protease inhibitory activity. J Biol Chem 277:46845–46848
126. McGowen R et al (2000) The surface of prostate carcinoma DU145 cells mediates the inhibition of urokinase-type plasminogen activator by maspin. Cancer Res 60:4771–4778
127. Biliran H Jr, Sheng S (2001) Pleiotrophic inhibition of pericellular urokinase-type plasminogen activator system by endogenous tumor suppressive maspin. Cancer Res 61:8676–8682
128. Sheng S et al (1998) Tissue-type plasminogen activator is a target of the tumor suppressor gene maspin. Proc Natl Acad Sci U S A 95:499–504
129. Zhang M et al (1997) Transactivation through Ets and Ap1 transcription sites determines the expression of the tumor-suppressing gene maspin. Cell Growth Differ 8:179–186

130. Domann FE et al (2000) Epigenetic silencing of maspin gene expression in human breast cancers. Int J Cancer 85:805–810
131. Qin L, Zhang M (2010) Maspin regulates endothelial cell adhesion and migration through an integrin signaling pathway. J Biol Chem 285:32360–32369
132. Zhang M et al (2000) Maspin is an angiogenesis inhibitor. Nat Med 6:196–199
133. Wang MC et al (2004) Maspin expression and its clinicopathological significance in tumorigenesis and progression of gastric cancer. World J Gastroenterol 10:634–637
134. Latha K et al (2005) Maspin mediates increased tumor cell apoptosis upon induction of the mitochondrial permeability transition. Mol Cell Biol 25:1737–1748
135. Chen EI et al (2005) Maspin alters the carcinoma proteome. FASEB J 19:1123–1124
136. Schwartz AL, Ciechanover A (1999) The ubiquitin-proteasome pathway and pathogenesis of human diseases. Annu Rev Med 50:57–74
137. Mercatali L et al (2006) RT-PCR determination of maspin and mammaglobin B in peripheral blood of healthy donors and breast cancer patients. Ann Oncol 17:424–428
138. Chim SS et al (2005) Detection of the placental epigenetic signature of the maspin gene in maternal plasma. Proc Natl Acad Sci U S A 102:14753–14758
139. Shi HY et al (2003) Modeling human breast cancer metastasis in mice: maspin as a paradigm. Histol Histopathol 18:201–206
140. Li Z et al (2005) Targeted expression of maspin in tumor vasculatures induces endothelial cell apoptosis. Oncogene 24:2008–2019
141. Ruegg MA et al (1989) Purification of axonin-1, a protein that is secreted from axons during neurogenesis. EMBO J 8:55–63
142. Hastings GA et al (1997) Neuroserpin, a brain-associated inhibitor of tissue plasminogen activator is localized primarily in neurons. Implications for the regulation of motor learning and neuronal survival J Biol Chem 272:33062–33067
143. Kaiserman D et al (2006) Mechanisms of serpin dysfunction in disease. Expert Rev Mol Med 8:1–19
144. Lloyd-Jones D et al (2009) Heart disease and stroke statistics--2009 update: a report from the American Heart Association statistics committee and stroke statistics subcommittee. Circulation 119:e21–181
145. Adibhatla RM, Hatcher JF (2008) Tissue plasminogen activator (tPA) and matrix metalloproteinases in the pathogenesis of stroke: therapeutic strategies. CNS Neurol Disord Drug Targets 7:243–253
146. Lebeurrier N et al (2005) The brain-specific tissue-type plasminogen activator inhibitor, neuroserpin, protects neurons against excitotoxicity both in vitro and in vivo. Mol Cell Neurosci 30:552–558
147. Yepes M et al (2002) Regulation of seizure spreading by neuroserpin and tissue-type plasminogen activator is plasminogen-independent. J Clin Invest 109:1571–1578
148. Chang WS et al (2000) Tissue-specific cancer-related serpin gene cluster at human chromosome band 3q26. Genes Chromosomes Cancer 29:240–255
149. Steele FR et al (1993) Pigment epithelium-derived factor: neurotrophic activity and identification as a member of the serine protease inhibitor gene family. Proc Natl Acad Sci U S A 90:1526–1530
150. He X et al (2015) PEDF and its roles in physiological and pathological conditions: implication in diabetic and hypoxia-induced angiogenic diseases. Clin Sci (Lond) 128:805–823
151. Ide H et al (2015) Circulating pigment epithelium-derived factor (PEDF) is associated with pathological grade of prostate cancer. Anticancer Res 35:1703–1708
152. Seruga B et al (2011) Drug resistance in metastatic castration-resistant prostate cancer. Nat Rev Clin Oncol 8:12–23
153. Dawson DW et al (1999) Pigment epithelium-derived factor: a potent inhibitor of angiogenesis. Science 285:245–248
154. Johnston EK et al (2015) Recombinant pigment epithelium-derived factor PEDF binds vascular endothelial growth factor receptors 1 and 2. In Vitro Cell Dev Biol Anim 51:730–738

155. Belkacemi L, Zhang SX (2016) Anti-tumor effects of pigment epithelium-derived factor (PEDF): implication for cancer therapy. A mini-review. J Exp Clin Cancer Res 35:4
156. Guan M et al (2007) Adenovirus-mediated PEDF expression inhibits prostate cancer cell growth and results in augmented expression of PAI-2. Cancer Biol Ther 6:419–425
157. Crawford SE et al (2001) Pigment epithelium-derived factor (PEDF) in neuroblastoma: a multifunctional mediator of Schwann cell antitumor activity. J Cell Sci 114:4421–4428
158. Filleur S et al (2005) Two functional epitopes of pigment epithelial-derived factor block angiogenesis and induce differentiation in prostate cancer. Cancer Res 65:5144–5152
159. Abramson LP et al (2003) Wilms' tumor growth is suppressed by antiangiogenic pigment epithelium-derived factor in a xenograft model. J Pediatr Surg 38:336–342
160. Mishur RJ et al (2008) Synthesis, X-ray crystallographic, and NMR characterizations of platinum(II) and platinum(IV) pyrophosphato complexes. Inorg Chem 47:7972–7982
161. Kazal LA et al (1948) Isolation of a crystalline trypsin inhibitor-anticoagulant protein from pancreas. J Am Chem Soc 70:3034–3040
162. Stenman UH et al (1982) Immunochemical demonstration of an ovarian cancer-associated urinary peptide. Int J Cancer 30:53–57
163. Kuwata K et al (2002) Functional analysis of recombinant pancreatic secretory trypsin inhibitor protein with amino-acid substitution. J Gastroenterol 37:928–934
164. Hirota M et al (2006) Genetic background of pancreatitis. Postgrad Med J 82:775–778
165. Gaber A et al (2009) High expression of tumour-associated trypsin inhibitor correlates with liver metastasis and poor prognosis in colorectal cancer. Br J Cancer 100:1540–1548
166. Soon WW et al (2011) Combined genomic and phenotype screening reveals secretory factor SPINK1 as an invasion and survival factor associated with patient prognosis in breast cancer. EMBO Mol Med 3:451–464
167. Ateeq B et al. (2011) Therapeutic targeting of SPINK1-positive prostate cancer. Sci Transl Med 3: 72ra17
168. McKeehan WL et al (1986) Two apparent human endothelial cell growth factors from human hepatoma cells are tumor-associated proteinase inhibitors. J Biol Chem 261:5378–5383
169. Ozaki N et al (2009) Serine protease inhibitor Kazal type 1 promotes proliferation of pancreatic cancer cells through the epidermal growth factor receptor. Mol Cancer Res 7:1572–1581
170. Stenman UH (2011) SPINK1: a new therapeutic target in cancer? Clin Chem 57:1474–1475

Role of Proteases in Diabetes and Diabetic Complications

13

P.V. Ravindra and T.K. Girish

Abstract

Proteases catalyze the breakdown of proteins by hydrolysis of peptide bonds. These enzymes are involved in a number of pathophysiological processes ranging from the cellular to organism level. These processes include cell growth, homeostasis, remodeling, renewal, division, metabolic pathways, tumor growth, metastasis, etc. A number of proteases are found to be involved in mediating the biochemical pathogenesis of metabolic syndrome such as diabetes and cardiovascular diseases. This chapter summarizes types of proteases, classification, and their proteolytic function in diabetes-associated complications in the kidney, eye, liver, heart, and lung. Understanding the role of proteases will provide insights into the development of preventive and therapeutic modalities for diabetes and diabetic complications.

Keywords

Proteases • Diabetes • Diabetic complications

13.1 Introduction

Proteases catalyze the breakdown of proteins by hydrolysis of peptide bonds. These are omnipresent in all forms of life. Proteases perform a plethora of complex physiological and pathological processes ranging from the cellular to organism level. These processes include cell growth, homeostasis, remodeling, renewal, division, tumor growth, metastasis, etc. These proteases also catalyze the proteolysis of pro-hormones and other precursor molecules to their active forms. They also promote

P.V. Ravindra (✉) • T.K. Girish
Department of Biochemistry, CSIR-CFTRI, Mysuru 570020, India
e-mail: raviravindra1@gmail.com

S. Chakraborti, N.S. Dhalla (eds.), *Proteases in Physiology and Pathology*,
DOI 10.1007/978-981-10-2513-6_13

the entry of infectious agents into target cells and tissues and release membrane-bound molecules facilitating the activation of signaling pathways in cells [1].

Extracellular proteases hydrolyze large proteins into amino acids for subsequent absorption by the cell, whereas intracellular proteases play a vital role in regulation of cell metabolism. Various proteases are shown to be involved in the pathogenesis of metabolic disorders including diabetes. However, the information about their mechanism of action is limited. Hence, this chapter summarizes types of proteases and their mechanism of action in the context of diabetes and diabetic complications.

13.2 Classification of Proteases (Types of Proteases)

Based on their site of action, proteases are subdivided into endopeptidases and exopeptidases. Endopeptidases cleave the peptide bonds that are away from the amino or carboxy termini of the substrate, while exopeptidases cleave the peptide bonds that are close to the termini of the substrate. Endopeptidases are classified into four major groups based on the presence of functional group at the active site—aspartic proteases, cysteine proteases, metalloproteases, and serine proteases [2].

13.2.1 Aspartic Acid Proteases

These proteases are commonly referred to as acidic proteases and have aspartic acid (Asp) in their active site situated in "Asp-Xaa-Gly" motif, where Xaa can be Thr or Ser. They have isoelectric points in the pH range of 3–4.5 and show optimal activity at acidic pH (pH 3 to 4). Their molecular mass ranges from 30 kDa to 45 kDa. The substrate-binding cleft can accommodate seven amino acid polypeptides [3, 6]. The mechanism of action involves removal of a proton by aspartate leading to the activation of the water molecule and attacking the scissile bond to produce a tetrahedral intermediate. Scissile amide's protonation following the rearrangement reaction of the intermediate results in the breakdown of the substrate into peptides [5].

13.2.2 Serine Proteases

Serine proteases are one of the most studied proteases with nearly 800 structures recorded in the protein data bank (PDB). One-third of these structures are of trypsin and thrombin [3]. Serine proteases have serine, histidine, and aspartate residues forming the catalytic triad in their active site. These are found in both endopeptidases and exopeptidases. Serine proteases show optimum activity ranging from the neutral pH (pH 7) to the alkaline pH (pH 11) and have isoelectric points ranging from pH 4 to pH 6. Their molecular mass ranges from 18 kDa to 35 kDa (except the protease found in the *Blakeslea trispora* with 126 kDa molecular mass) [4]. These proteases possess a broad substrate specificity including esterolytic and amidase activities. The catalytic mechanism involves concerted action by the catalytic triad. Histidine deprotonates the serine hydroxy group with aspartate and enables the nucleophilic attack on the substrate carbonyl carbon. One exception is that intramembrane serine proteases function with a catalytic diad due to lack of aspartate active site.

13.2.3 Cysteine Proteases

Cysteine proteases are found in both eukaryotic and prokaryotic organisms. They have cysteine and histidine residues forming the catalytic dyad in their active site. Cysteine proteases show maximum activity at the neutral pH, except lysosomal proteases that are maximally active at acidic pH. The proteolytic action of these proteases requires reducing agents, such as HCN or cysteine [8–10]. Similar to serine proteases, most cysteine proteases have a relatively shallow active site that can accommodate short substrate segments such as a strand or protein loops (e.g., endogenous inhibitors, cystatins). Sulfur in cysteine as opposed to the serine in serine proteases is used as the nucleophile for carrying out the proteolytic action on the substrates.

13.2.4 Metalloproteases

The most diverse among the proteases require a divalent metal ion (Zn^{2+}) for their activity. Approximately, 30 families are identified. Of them, 12 contained exopeptidases and 17 contained endopeptidases, while one of them contained both endo- and exopeptidases. Water molecule hydrogen bonded to a glutamate, and the three donor groups of the enzyme coordinate with the Zn^{2+} while carrying out the nucleophilic attack during the proteolysis [3].

13.3 Role of Proteases in Diabetes and Diabetic Complications

Proteases likely arose as simple destructive enzymes in protein evolution that are necessary for the generation of amino acids through protein catabolism in primitive organisms. For many years, studies on proteases focused only on their role in protein catabolism. However, recent research have shed light on other functions of the proteases; proteases act as site-specific scissors and catalyze specific proteolytic reactions to produce new protein products. Proteases can be detrimental or beneficial in the inflammatory process depending on the biological contexts, such as disease state, cell type, location, substrate availability, and inhibitors.

13.3.1 Kidney

Among all the diabetic complications, diabetic nephropathy (DN) is the leading cause of end-stage kidney disease worldwide. Moreover, the progressive decline in kidney function in diabetic patients is positively correlated with all-cause mortality and severe cardiovascular complications [11].

The kidney size is increased in the early phase of type I diabetes in human or in experimental diabetes in rats. This increase in size is mainly due deposition of protein as a result of decreased protein catabolism and/or increased protein synthesis. Both protein synthesis and degradation are energy dependent processes, which may be regulated by various proteases. The lysosomes play a pivotal role in the breakdown of proteins in mammalian cells which involves sequestration of these proteins in autophagic vacuoles, fusion of these vacuoles with primary lysosomes, and degradation of proteins within the newly formed secondary lysosomes by highly active proteases such as cathepsins B and cathepsin L.

The activities of these cathepsins were measured in kidneys from streptozotocin (STZ)-induced diabetic rats by Olbricht et al. (1992) [12]. Cathepsins' activities decreased with an increase in kidney weight, which indicates renal hypertrophy in STZ-injected rats. They also found decreased cathepsin activity in proximal tubule segments and kidney cortex. Liver was a positive control in this study, and it was found that liver weight and activities of these cathepsins were found elevated unlike in the kidney. This indicates that diabetes might be associated with decreased cathepsin activity independent of organ hypertrophy [12].

13.3.2 Proteases in Cataracted Eye

The clouding of the eye lens causes development of eye cataracts. Cataracted eye contributes majorly for the vision loss observed in people over 40 years of age. Diabetes is one of the primary contributing risk factor for the development of cataracts. The eye lens is made of specialized proteins, called crystallins. These include α-, β-, and γ-crystallins, which account for nearly 90% of the lens proteins. α-Crystallin is a predominant lens protein composed of αA and αB subunits with chaperone-like activity, and β- and γ-crystallins function as structural proteins. Diabetes induces the activation of polyol pathway and increases oxidative stress and nonenzymatic glycation of these lens proteins, which subsequently aggregate leading to the cataract development [13].

Various peptidases and proteases have been identified or isolated from lens and lens epithelial cells. These include acyl-peptide hydrolase, aminopeptidase III, calpains, dipeptidase, caspases [3, 6, 7], cathepsin (B, D), matrix metalloproteases, leucine aminopeptidase, serine-type protease, and trypsin-like protease. Lens proteins are degraded into amino acids by subsequent hydrolysis by proteases and aminopeptidases. Literature suggests both beneficial and detrimental effects of protease activity in the lens. Peptide chaperones such as αA and αB released following the cleavage of α-crystallin prevent aggregation and precipitation of unfolding proteins, similar to the full-length α-crystallin. On the other hand, αA-66-80 peptide generated following the proteolysis of αA-crystallin has been found to promote the formation of protein aggregates [13, 14]. Further, αA-66-80 peptides are resistant to downstream aminopeptidases and can suppress the degradation of other peptides. Incomplete hydrolysis of peptides leads to protein aggregation in lens leading to cataract formation [13–15].

13.3.3 Liver

Research findings on understanding the role of various proteases in the diabetic liver are limited. Lysosome proteases are the main digestive enzymes in autophagic vacuoles in hepatocytes. Recent works on liver autophagy focus on glycogen digestion and lipid digestion (lipophagy) [16–18]. Few studies have been done on the role of lysosomal proteases in the diabetic liver [19] and has shown decrease in cysteine proteases' activity, especially cathepsin B in STZ-induced rat liver [19]. This decrease in specific activity was attributed decreased expression levels of cathepsin B and in the diabetic rat liver [19]. In another study by Uchimura et al. (2014) found that serine protease prostasin ameliorates hepatic insulin insensitivity found in the type 2 diabetes by decreasing the activation of toll-like receptor (TLR4) signaling pathway [20].

13.3.4 Heart

Metabolic syndrome leads to cardiovascular diseases, and the underlying mechanisms are far from clear. Various proteases such as MMP, calpain, cathepsin, and caspase have been implicated in the pathogenesis of atherosclerosis, coronary heart disease, and heart disease associated with obesity, insulin resistance, and hypertension [21]. Cathepsins and MMPs influence cardiometabolic diseases by modifying the extracellular matrix. Additionally, MMPs and cathepsin also affect intracellular proteins leading to cardiometabolic diseases. On the other hand, activation of caspases and calpains influence NF-κB and apoptosis pathways. Clinically, proteases are used as biomarkers of cardiometabolic diseases. Moreover, the protease inhibitors have shown a beneficial cardiometabolic profile with unknown molecular mechanisms.

Even though the functional association between cardiometabolic diseases and proteases is well established, it is still unclear whether the increase in protease activity is the cause or the result of cardiometabolic disease. Inflammatory cytokines or reactive oxygen species could activate proteases that lyses substrates involved in cardiometabolic functions leading to diseased state. On the other hand, proteases can upregulate inflammatory mediators resulting in vicious cycle. Exploring the role of proteases in the pathophysiology of cardiometabolic disease may yield novel therapeutic targets [21].

13.3.5 Lung

Lung proteases were thought to be involved only in destruction of extracellular matrix. However, they were found to be involved in infections, local inflammation processes, and innate immunity too. Serine and metalloproteases modulate biological functions by activation of various specific cell surface receptors, promoting cytokine receptor shedding, cytokine and chemokine activation, and degradation and proteolysis of cytokine-binding proteins. The inflammatory process that is essential in host defense, if unregulated, leads to tissue injury, organ dysfunction, and lung diseases [22–24]. Due to larger pulmonary reserves, symptoms of lung damage due to diabetes appear at a later time than other organs. Hence, in spite of the first report three decades ago, less importance is given to pulmonary complications of diabetes. However, recent attention is on subclinical pulmonary complications in diabetes such as a reduced lung diffusing capacity, elastic recoil, and capillary volume [25, 26]. Another school of thought is that chronic high glucose level as seen in diabetes nonenzymatically glycates proteins and peptides of ECM leading to pathological changes in lungs [27]. Researchers have moved on from identification of diabetic-specific lung deficits to explore mechanisms by which diabetes affects the lungs. One hypothesis is that lung complications are due to elastin and collagen changes [28]. A dynamic balance between proteases and antiproteases maintains the lung connective tissue, and any imbalance leading to relative increase in proteases can cause lung injury [29]. Such an imbalance could be due to neutrophils that are the primary source of serine proteases, such as neutrophil elastase, cathepsin G, and neutrophil protease. They can digest collagen, elastin, laminin, and fibronectin in the lung ECM causing extensive lung damage [30]. However, the mechanism of imbalance between protease and antiprotease in diabetes remains obscure. Understanding the role of protease in lung inflammation will have significant implication in health and disease.

13.4 Conclusions

Significant advances have been made in understanding the pathophysiology of proteases, their regulators, and their receptors in diabetes; it is clear that proteases play critical roles in many metabolic syndromes including diabetes. Elucidating these roles is critical to our understanding of disease mechanisms. Targeting these proteases could present us an innovative approach to treat or control metabolic diseases. Animal studies show that the expression of these proteases is controlled by a variety of stimulants such as cytokines, hormones, and ROS. Activated proteases degrade intracellular proteins or ECM triggering metabolic diseases or tissue injury. These insights in the role of proteases in metabolic diseases would provide novel treatment avenues. Further studies are needed to understand the interaction between cellular signaling pathways and proteases.

Acknowledgement Authors are thankful to Prof. Ram Rajasekharan, Director, CSIR-CFTRI, for his interest and valuable suggestions. Dr. Ravindra P. V. thanks the Department of Biotechnology, New Delhi, for funding in the form of Ramalingaswami fellowship.

References

1. Dunn BM (2010) Introduction to the aspartic proteinase family. Aspartic Acid Proteases as Therapeutic Targets:1–21
2. Rao MB, Tanksale AM, Ghatge MS et al (1998) Molecular and biotechnological aspects of microbial proteases. Microbiol Mol Biol Rev 62:597–635
3. Tyndall JD, Nall T, Fairlie DP (2005) Proteases universally recognize beta strands in their active sites. Chem Rev 105:973–1000
4. Govind NS, Mehta B, Sharma M et al (1981) Protease and carotenogenesis in Blakeslea trispora. Phytochemistry 20:2483–2485
5. Erez E, Fass D, Bibi E (2009) How intramembrane proteases bury hydrolytic reactions in the membrane. Nature 459:371–378
6. Polgár L (1987) The mechanism of action of aspartic proteases involves 'push-pull'catalysis. FEBS Lett 219:1–4
7. Barrett AJ (1994) Proteolytic enzymes: serine and cysteine peptidases. Academic Press 244:1–765
8. Kotler MOSHE, Katz RA, Skalka AM (1988) Activity of avian retroviral protease expressed in Escherichia coli. J Virol 62:2696–2700
9. Kubo M, Imanaka T (1988) Cloning and nucleotide sequence of the highly thermostable neutral protease gene from Bacillus Stearothermophilus. Microbiology 134:1883–1892
10. Koszelak S, Ng JD, Day J et al (1997) The crystallographic structure of the subtilisin protease from Penicillium cyclopium. Biochemistry 36:6597–6604
11. Musante L, Tataruch D, Gu D et al (2015) Proteases and protease inhibitors of urinary extracellular vesicles in diabetic nephropathy. Journal of diabetes research 2015:1–15
12. Olbricht CJ, Geissinger B, Gutjahr E (1992) Renal hypertrophy in streptozotocin diabetic rats: role of proteolytic lysosomal enzymes. Kidney Int 41:966–972
13. Raju M, Santhoshkumar P, Sharma KK (2016) Alpha-crystallin-derived peptides as therapeutic chaperones. Biochimica et Biophysica Acta (BBA)-General Subjects 1860:246–251
14. Santhoshkumar P, Kannan R, Sharma KK (2015) Proteases in lens and cataract. In: Babizhayev MA, Li DW-C, Jacobi AK et al (eds) Studies on the cornea and lens. Springer, New York, pp 221–238
15. Hariharapura R, Santhoshkumar P, Sharma KK (2013) Profiling of lens protease involved in generation of αA-66-80 crystallin peptide using an internally quenched protease substrate. Exp Eye Res 109:51–59
16. Singh R, Kaushik S, Wang Y et al (2009) Autophagy regulates lipid metabolism. Nature 458:1131–1135
17. Papackova Z, Palenickova E, Dankova H et al (2012) Kupffer cells ameliorate hepatic insulin resistance induced by high-fat diet rich in monounsaturated fatty acids: the evidence for the involvement of alternatively activated macrophages. Nutrition & metabolism 9:1–15
18. Kalamidas SA, Kondomerkos DJ (2010) Autophagosomal glycogen-degrading activity and its relationship to the general autophagic activity in newborn rat hepatocytes: the effects of parenteral glucose administration. Microsc Res Tech 73:495–502
19. Peres GB, Juliano MA, Aguiar JAK et al (2014) Streptozotocin-induced diabetes mellitus affects lysosomal enzymes in rat liver. Braz J Med Biol Res 47:452–460
20. Uchimura K, Hayata M, Mizumoto T et al (2014) The serine protease prostasin regulates hepatic insulin sensitivity by modulating TLR4 signalling. Nat Commun 5:1–13

21. Hua Y, Nair S (2015) Proteases in cardiometabolic diseases: pathophysiology, molecular mechanisms and clinical applications. Biochimica et Biophysica Acta (BBA)-Molecular Basis of Disease 1852:195–208
22. Lungarella G, Cavarra E, Lucattelli M et al (2008) The dual role of neutrophil elastase in lung destruction and repair. Int J Biochem Cell Biol 40:1287–1296
23. Pitocco D, Fuso L, Conte EG et al (2012) The diabetic lung-a new target organ? The review of diabetic studies: Rev Diabet Stud 9:23–35
24. van den Borst B, Gosker HR, Zeegers MP et al (2010) Pulmonary function in diabetes: a meta-analysis. Chest J 138:393–406
25. Irfan M, Jabbar A, Haque AS et al (2011) Pulmonary functions in patients with diabetes mellitus. Lung India 28:89–92
26. Kuziemski K, Specjalski K, Jassem E (2011) Diabetic pulmonary microangiopathy—fact or fiction? Endokrynol Pol 62:171–176
27. Weynand B, Jonckheere A, Frans A et al (1999) Diabetes mellitus induces a thickening of the pulmonary basal lamina. Respiration 66:14–19
28. Ljubic S, Metelko Z, Car N et al (1998) Reduction of diffusion capacity for carbon monoxide in diabetic patients. Chest J 114:1033–1035
29. Tetley TD (1993) New perspectives on basic mechanisms in lung disease. 6. Proteinase imbalance: its role in lung disease. Thorax 48:560–565
30. Boxer LA, Smolen JE (1988) Neutrophil granule constituents and their release in health and disease. Hematol Oncol Clin North Am 2:101–134

Plant Latex Proteases: Natural Wound Healers

14

Amog P. Urs, V.N. Manjuprasanna, G.V. Rudresha, M. Yariswamy, and B.S. Vishwanath

Abstract

Proteases are ubiquitously present in several organisms including plants. In plants, one of the rich sources of protease is latex. Over 110 latices of different plant families are known to contain at least one proteolytic enzyme. The primary role of proteases in latices is defense against pests/insects. Apart from the defensive role in plants, latices are pharmacologically important and are integral components in herbal management of wounds, where it is extensively used in traditional medicines to stop bleeding and to promote healing of wounds. Plant latex proteases exhibit both clot-inducing and clot-hydrolyzing properties. Clot formation is vital for hemostasis, the initial phase of wound healing, whereas clot hydrolysis is a prerequisite for the events of regenerative phase. Overall, the plant latex proteases provide optimal conditions for physiological wound healing by complementing the endogenous proteases in hemostasis, wound debridement, microbial attenuation, cell proliferation, and angiogenesis. Further, complete functional characterization of purified proteases from latex along with physicochemical characterization is very crucial to strengthen the existing knowledge and will be pivotal in developing latex protease-based wound care supplements with minimal side effects.

Keywords

Ethnopharmacology • Hemostasis • Plasmin-like proteases • Thrombin-like proteases Wound healing

A.P. Urs • V.N. Manjuprasanna • G.V. Rudresha • B.S. Vishwanath (✉)
Department of Studies in Biochemistry, University of Mysore, Manasagangothri, Mysuru 570 006, Karnataka, India
e-mail: vishmy@yahoo.co.uk

M. Yariswamy
Department of Surgery, University of Missouri, Columbia, MO 65212, USA

S. Chakraborti, N.S. Dhalla (eds.), *Proteases in Physiology and Pathology*,
DOI 10.1007/978-981-10-2513-6_14

14.1 Introduction

Proteases are the major class of hydrolytic enzymes which cleave the peptide bond in the protein to give peptides and amino acids [1]. They are abundantly present in the human system evidenced by human genome sequencing which revealed that more than 2% of the genes encodes for proteases. Proteases mediate versatile and complex array of functions apart from the primary roles in food digestion and intracellular protein turnover [1–3]. Proteases are involved in the regulation of a large number of key physiological processes such as hemostasis (coagulation), tissue remodeling, wound healing, DNA replication, cell-cycle progression, cell proliferation, cell death, and immune response [2, 3]. In view of their ability to regulate key physiological processes, proteases have been employed for treating specified clinical conditions. Till date, the predominant use of proteases was restricted to cardiovascular disease [3, 4]. Of late, they are also emerging as useful therapeutic agents in the treatment and management of debilitating conditions including sepsis, chronic inflammatory disorders, cystic fibrosis, retinal disorders, and psoriasis [3, 5].

Although proteases are involved in the regulation of many key physiological processes, they have not been typically considered as a drug class despite their application in the clinic over the last several decades. In recent past, proteases have emerged as an expanding class of drugs that hold great promise and are used for treating a wide range of clinical conditions [3, 6]. In addition to the endogenous proteases, exogenous sources of proteases have been widely studied and employed for their therapeutic efficacies.

Microorganisms, insects, invertebrates, vertebrates, and plants constitute the major sources of proteases [7–12]. Among these sources, microbial, animal (venom), and plant latex proteases are widely employed for their therapeutic applications. Although there are few studies on therapeutic applications, still there is significant scope for exploring newer and potent proteases with therapeutic efficacy. Well-characterized plant latex proteases like papain, chymopapain, and ficin are used for treating wounds, cancer, digestive and viral disorders [13, 14]. In light of these evidences and the recent findings pertaining to latex proteases, the present chapter focuses on the role of plant latex proteases as wound healers and provides an overview of their application in wound healing.

14.2 Plant Latex Proteases

The presence of proteolytic enzymes in plant latex was first reported in 1940 [15]. Over 110 latices of different plant families are known to contain at least one proteolytic enzyme [16]. More than 70 proteases have been purified from latices of different plants belonging to various families, and the crystal structures of few proteases are known, which are listed in Tables 14.1 and 14.2, respectively. The majority of proteases found in latices belong to the cysteine and serine protease family; only one is a member of the aspartate and metalloprotease family, and none of the proteases reported till date belong to threonine, glutamic, and asparagine protease.

Table 14.1 List of purified plant latex proteases with reported characteristics and molecular weight (NR – no reports)

Protease	Plant	Functional characteristics		Molecular weight	References
		Purified protease	Crude latex		
1. Aspartate proteases					
A. Moraceae					
Ficins	*Ficus racemosa*	NR	Used for curing hemorrhoids, boils, edema, and chronic infected wounds	44.5 ± 05	[17, 18]
2. Cysteine proteases					
A. Apocynaceae					
Cg24-I	*Cryptostegia grandiflora*	Antifungal	Antioxidant	24,1	[19, 20]
Ervatamin A, B, C	*Ervatamia coronaria*	Papain-like	Gelatinolytic	27.6, 26, 23	[21–25]
Funastrain CII	*Funastrum clausum*	Papain-like	NR	23.636	[26]
Heynein	*Ervatamia heyneana*	NR	NR	23	[23]
Pergularain e I	*Pergularia extensa*	Thrombin-like	Plasmin- and thrombin-like	23.356	[27, 28]
Peruvianin-I	*Thevetia peruviana*	Germin-like	Antifungal and gelatinolytic	120	[25, 29, 30]
Philibertain g I and g II	*Philiberta gilliesii*	Papain-like	NR	23.530 23,9	[31, 32]
Plumerin-R	*Plumeria rubra*	Thrombin-like, plasmin-like, anti-inflammatory, and excision wound healer	Antioxidant	81.85	[19, 33]
B. Asclepiadaceae					
Araujiain H I, H II, H III	*Araujia hortorum*	NR	NR	24.03, 23.718, 23.546	[34, 35]
Asclepains A3, B5	*Asclepias syriaca*	NR	NR	23, 21	[39]

(continued)

Table 14.1 (continued)

Protease	Plant	Functional characteristics		Molecular weight	References
		Purified protease	Crude latex		
Asclepain C I, C II	*Asclepias curassavica*	Papain-like	Antifungal, plasmin-like, and thrombin-like	23.2	[27, 37, 38]
Asclepain F	*Asclepias fruticosa*	Papain-like	NR	23.652	[39]
Asclepain G (ten forms)	*Asclepias glaucescens*	NR	NR	Ag3 22.6, Ag6 23.5, Ag7 23, Ag8 23.5	[40]
Asclepain S	*Asclepias speciosa*	Milk coagulant	NR	NR	[15]
Calotropins DI, DII	*Calotropis gigantea*	NR	Gelatinolytic, plasmin- and thrombin-like, and wound healer	23.8, 24.2	[25, 27, 41–43]
Morrenain BI, BII	*Morrenia brachystephana*	Papain-like	NR	23.205, 25.5	[44, 45]
Morrenain OII	*Morrenia odorata*	NR	NR	25.8	[44]
Procerain, procerain B, CpCP-1, 2, and 3	*Calotropis procera*	Milk coagulant and thrombin-like	Anti-inflammatory and gelatinolytic	28.8	[25, 46–48]
C. Caricaceae					
Caricain, chymopapain, glycyl endopeptidases, papain	*Carica papaya*	Factor XIIIa-like, thrombin-like, and wound healer	Antifungal and gelatinolytic	23.429, 23.280, 23.650, 23.313	[16, 25, 49–51]
Endopeptidase, CCI, CCII, CCIII,CCIV, CC23, CC28, chymopapain isoform II, CMS1MS2, CMS2MS2	*Carica candamarcensis*	Mitogenic and papain-like	Angiogenic, cell proliferant, gastric ulcer, and wound healer	23–28.6	[16, 52–58]
Mexicain	*Jacaratia mexicana*	NR	NR	23.8	[59, 60]
Quercifoliain I, VQ-VII	*Vasconcellea quercifolia*	Milk coagulant and papain-like	NR	24.2, 23.98	[61, 62]

D. Euphorbiaceae					
Nivulian-I, Nivulian-II, and Nivulian-III	*Euphorbia nivulia*	Milk coagulant	Gelatinolytic and procoagulant	31.4, 43.6, and 52.8	[25, 63–65]
Pedilanthin	*Pedilanthus tithymaloides*	NR	Gelatinolytic and procoagulant	63.1	[25, 65, 66]
E. Moraceae					
Ficain	*Ficus glabrata*	NR	NR	NR	[16]
Ficins A, B, C, D, E, F, G, H, I, J, S	*Ficus carica*	NR	Activation of human coagulation factor X and antimicrobial	24–26	[67–71]
Ficain P I	*Ficus pumila*	NR	NR	28.6	[16]
Microcarpain	*Ficus microcarpa*	NR	NR	20	[72]
Protease	*Ficus hispida*	NR	Gelatinolytic	NR	[25, 73]
3. Metalloprotease					
A. Euphorbiaceae					
Cotinifolin	*Euphorbia cotinifolia*	NR	NR	79.76	[74]
4. Serine proteases					
A. Amaryllidaceae					
Crinumin	*Crinum asiaticum*	Antiplatelet, chymotrypsin-like, and thrombolytic	NR	67.7	[75, 76]
B. Apocynaceae					
Wrightin	*Wrightia tinctoria*	Trypsin-like	Collagenolytic, gelatinolytic and wound healer	57.9	[77, 78]
C. Asclepiadaceae					
Cryptolepain	*Cryptolepis buchananii*	NR	NR	79.5	[79]
D. Asteraceae					
Parthenain	*Parthenium argentatum*	NR	NR	63	[80]

(continued)

Table 14.1 (continued)

Protease	Plant	Functional characteristics		Molecular weight	References
		Purified protease	Crude latex		
Taraxalisin	*Taraxacum officinale*	Subtilisin-like	NR	65	[81]
E. Convolvulaceae					
Carnein	*Ipomoea carnea*	NR	Gelatinolytic	80.236	[25, 82]
F. Euphorbiaceae					
Euphorbains D1, D2	*Elaeophorbia drupifera*	NR	NR	117, 65	[16]
Euphorbain L	*Euphorbia lathyris*	NR	Gelatinolytic	43	[83]
Euphorbains la1, la2, la3	*Euphorbia lactea*	NR	NR	66, 44, 33	[84]
Euphorbain lc	*Euphorbia lactea cristata*	NR	NR	70	[84]
Euphorbain P	*Euphorbia pulcherrima*	NR	NR	74	[79]
Euphorbains T1, T2, T3,T4	*Euphorbia tirucalli*	Trypsin-like	Antioxidant	74, 74, 74, 74	[19, 85]
Euphorbains Y1, Y2, Y3	*Euphorbia cyparissias*	NR	NR	67, 33, 67	[86]
Hevains A, B, L	*Hevea brasiliensis*	Antifungal	Increases vascular permeability, angiogenesis, and wound healing	69, 58, 80	[87–91]
Hirtin	*Euphorbia hirta*	Fibrino(geno)lytic	Antimicrobial and gelatinolytic	34	[25, 92, 93]
Latex glycoprotein (LGP), proteases	*Synadenium grantii*	Fibrino(geno)lytic and procoagulant	Gelatinolytic and procoagulant	34.4, 76 ± 2	[25, 65, 94, 95]
Milin	*Euphorbia milii*	NR	Gelatinolytic	51.4	[25, 96]
Protease	*Euphorbia pseudochamaesyce*	NR	NR	82	[16]
Protease	*Euphorbia supine*	NR	NR	80	[97]
Protease	*Euphorbia heterophylla*	NR	NR	77.2	[98]
Prunifoline	*Euphorbia prunifolia*	Milk coagulant	Gelatinolytic	57.44	[25, 99]

G. Moraceae					
Amp48	*Artocarpus heterophyllus*	Antimicrobial and fibrino(geno)lytic	NR	48	[100, 101]
Artocarpin	*Artocarpus heterophyllus*	NR	NR	79.5	[102]
Benghalensin	*Ficus benghalensis*	NR	NR	47	[103]
Ficin E	*Ficus elastica*	NR	NR	50	[16]
Macluralisin	*Maclura pomifera*	Milk coagulant	NR	65	[104]
Protease	*Ficus carica*	Collagenolytic and gelatinolytic	Activation of human coagulation factor X, antimicrobial, and gelatinolytic	41	[25, 105]
Religiosin	*Ficus religiosa*	Milk coagulant	Gelatinolytic	43.4	[25, 106]
5. Type of protease not determined					
Curcain	*Jatropha curcas*	Wound healer	Antimicrobial	22	[107, 108]

Table 14.2 List of purified proteases with crystal structure

Protease	Plant	Molecular weight	References
1. Cysteine proteases			
A. Apocynaceae			
Ervatamin A	*Ervatamia coronaria*	27.6	[109]
Ervatamin B	*Ervatamia coronaria*	26	[110, 111]
Ervatamin C	*Ervatamia coronaria*	23	[110]
B. Asclepiadaceae			
Calotropins DI	*Calotropis gigantea*	23.8	[112, 113]
Calotropins DII	*Calotropis gigantea*	24.2	[112]
C. Caricaceae			
Caricain	*Carica papaya*	23.3	[114]
CMS1MS2	*Carica candamarcensis*	23	[115]
Glycyl endopeptidase	*Carica papaya*	23.3	[116]
Mexicain	*Jacaratia mexicana*	23.7	[60]
Papain	*Carica papaya*	27	[117]
2. Serine proteases			
A. Amaryllidaceae			
Crinumin	*Crinum asiaticum*	67.7	[76]
B. Asclepiadaceae			
Cryptolepain	*Cryptolepis buchananii*	79.5	[118]
C. Convolvulaceae			
Carnein	*Ipomea carnea*	80	[119]

The primary role of proteases in latices is plant defense against pests/insects. Many of these proteases are cysteine proteases, although few insecticidal metalloproteases and serine proteases have also been reported. The targets of protease toxicity range from the insect midgut to the hemocoel (body cavity) to the cuticle [9, 120]. Apart from the defensive role in plants, they are extensively used in industries as they are active over a range of temperature and pH (frequently used in food processing, tenderization of meat, brewing, cheese elaboration, bread manufacture, and in the leather and textile industries) [16, 121]. On the other hand, therapeutically, plant latex is one of the most preferred materials in traditional medicine for treating bleeding and wounds [16].

14.3 Need of Plant Latex Proteases as Wound Healing Agents

Wounds are the inevitable events encountered during the lifetime of an individual resulting from mechanical, chemical, or surgical damage, microbial infection, or an underlying pathological condition [122]. The host aptly responds to a wound through a series of events as a damage control response to restore the tissue integrity and function [123]. This complex physiological response, termed wound healing, is controlled and coordinated by immune cells, extracellular matrix (ECM), and components of hemostatic system [124]. In spite of the efficient host response, some pathophysiological conditions and/or due to infections result in altered healing response, in most cases ending up with delayed healing or non-healing chronic wounds [125]. To overcome the complications leading to altered healing, wound care and management is the remedy. Conventionally, the strategies of wound care involve the administration of synthetic antibiotics and anti-inflammatory agents, with or without debridement, with grafting as the last resort in extreme cases [126]. But these are associated with undesirable effects, including side effects, pain, discoloration of the skin, and complications of histocompatibility and tissue rejection [127]. These adverse effects need to be carefully addressed and overcome by complementary medicinal strategies. In this regard, herbal medicines would be suitable agents for wound care, since they provide optimal conditions to augment physiological healing process by offering minimal adverse effects [128]. Latex, an important plant-based component, is widely used for topical application on variety of wounds and is an integral part in herbal management of wounds [129, 130]. The list of latex-bearing plants used by tribal populations, folk medicinal practitioners, and researchers to promote wound healing is given in Table 14.3. In spite of vast information regarding the extensive usage of latex to promote wound healing, very few reports suggest the possible biochemical/molecular mechanism(s) of latex components in general and proteases in particular. The functional (biochemical) characteristics of only few purified proteases are reported. Complete functional characterization of purified proteases along with physicochemical characterization is very crucial as they can be developed into wound care supplements in the future with minimal side effects (Table 14.1).

14.4 Role of Plant Latex Proteases on Different Phases of Wound Healing

Wound healing is a complex response to tissue injury comprising of sequential and overlapping phases: (a) hemostasis, (b) inflammatory, (c) proliferative, and (d) remodeling phases [124] (Fig. 14.1). During physiological wound healing process, many endogenous proteases are involved in the different phases of wound healing; likewise, latex proteases used for treating wound will act on different phases of wound healing. For example, procoagulant and thrombin-like proteases act in the initial stages of wound healing and restore hemostasis (Fig. 14.2). Plasmin-like and other ECM-degrading proteases help in the later stages of wound healing for debridement, and some mitogenic proteases help in cell proliferation and angiogenesis (Table 14.1; Fig. 14.3a, b). Although, there is no conclusive evidence on plant latex proteases mediating remodeling of collagen and other extracellular matrix (ECM) components, the role of collagenolytic and other ECM-degrading proteases can be attributed with tissue remodeling ability (Fig. 14.3c).

Table 14.3 The list of latex-bearing plants used by tribal populations, folk medicinal practitioners, and researchers to promote wound healing and related disorders

Family	Plant names	Type of wounds	References
1. Altingiaceae	*Liquidambar orientalis*	Excision wounds	[131]
2. Amaranthaceae	*Achyranthes aspera*	Wounds	[131]
3. Anacardiaceae	*Mangifera indica*	Cracks, cuts, ulcer, and wounds	[132, 133]
	Pistacia atlantica	Wounds	[134]
	Semecarpus anacardium	Cuts and wounds	[135]
	Spondias pinnata	Cuts and wounds	[136]
4. Apiaceae	*Ferula assafoetida*	Ulcers	[137]
5. Apocynaceae	*Allamanda cathartica*	Excision and incision wounds	[138]
	Alstonia angustiloba	Abscesses, boils, and skin sores	[139]
	Alstonia scholaris	Boils, burns, and wounds	[131, 140]
	Alstonia venenata	Cuts and wounds	[141]
	Carissa carandas	Wounds	[135]
	Holarrhena pubescens	All types of wounds, aphthae and, ulcer in intestine	[133, 138, 142]
	Ichnocarpus frutescens	Bleeding wounds	[143, 144]
	Plumeria obtusa	Ulcers	[139, 145]
	Plumeria rubra	Excision wounds, gingival wounds, and ulcers	[33, 133]
	Rauvolfia serpentina	Cuts, otorrhea, and wounds	[133, 146]
	Strophanthus sarmentosus	Wounds	[131]
	Tabernaemontana divaricata	Cuts	[133]
	Tabernaemontana heyneana	Cuts and wounds	[133]
	Vallaris solanacea	Old wounds and sores	[147]
	Voacanga thouarsii	Wounds	[131]
	Wrightia tinctoria	Excision and fresh wounds	[77, 131]

(continued)

Table 14.3 (continued)

Family	Plant names	Type of wounds	References
6. Asclepiadaceae	*Araujia sericifera*	Warts	[148]
	Asclepias curassavica	Wounds	[143]
	Calotropis gigantea	Corns, cuts, excision wounds, gingival wounds, incision wounds, otorrhea, and thorn wounds	[43, 129, 133]
	Calotropis procera	Excision wounds	[138]
	Ceropegia juncea	Ulcers	[149]
	Cryptolepis buchananii	Cuts and wounds	[133]
	Cynanchum acutum	Ulcers	[149]
	Cynanchum callialatum	Ulcers	[149]
	Gymnema sylvestre	Cuts and wounds	[149]
	Hemidesmus indicus	Wounds	[143]
	Holostemma ada kodien Schultes	Blisters	[131]
	Hoya lanceolata	Boils	[150]
	Oxystelma esculentum	Ulcers	[149]
	Pergularia extensa	Wounds	[131, 138]
	Sarcostemma acidum	Burns, chronic ulcers, and wounds	[142, 149]
	Sarcostemma viminale	Bleeding wounds (hemorrhage)	[151]
	Tylophora indica	Ulcers and wounds	[149]
	Tylophora fasciculata	Wounds	[143]
7. Asteraceae	*Calendula officinalis*	Chronic ulcers and wounds	[152]
	Cichorium intybus	Ulcers and wounds	[153]
	Sonchus arvensis	Burns	[154]
	Taraxacum officinale	Corns, stomach ulcers, warts, and wounds	[152, 155]
	Tragopogon dubius	Heel wounds	[131]
8. Campanulaceae	*Asyneuma rigidum*	Burns	[134]
9. Caricaceae	*Carica candamarcensis*	Gastric ulcers and wounds	[58, 131]
	Carica papaya	Burns and wounds	[50]
10. Convolvulaceae	*Argyreia speciosa*	Boils, ulcers, and wounds	[156]
	Ipomoea pes-caprae	Wounds	[157]
	Ipomoea pes-tigridis	Cuts and wounds	[144]
11. Dipterocarpaceae	*Vateria indica*	Wounds	[131, 143]

(continued)

Table 14.3 (continued)

Family	Plant names	Type of wounds	References
12. Euphorbiaceae	*Croton bonplandianum*	Cuts and wounds	[158, 159]
	Croton macrostachyus	Sores, warts, and wounds	[160]
	Croton megalocarpus	Wounds	[160]
	Euphorbia agraria	Wounds	[134]
	Euphorbia antiquorum	Burns and wounds	[129]
	Euphorbia armena	Inflamed wounds	[134]
	Euphorbia caducifolia	Bleeding wounds and cutaneous eruptions	[131]
	Euphorbia candelabrum	Warts and wounds	[160]
	Euphorbia cuneata	Sores, warts, and wounds	[131]
	Euphorbia grantii	Bleeding open wounds and tissue healing	[131]
	Euphorbia helioscopia	Skin eruptions	[131]
	Euphorbia hirta	Warts and wounds	[129, 161]
	Euphorbia macroclada	Warts	[162]
	Euphorbia neriifolia	Wounds	[143]
	Euphorbia nivulia	Ulcers and wounds	[142, 163]
	Euphorbia pallens	Skin injuries and wounds	[157]
	Euphorbia pilosa	Wounds	[160]
	Euphorbia primulifolia	Syphilitic sores and warts	[131]
	Euphorbia prostrata	Abscesses and warts	[164]
	Euphorbia pseudograntii	Abscesses, warts, and wounds	[164]
	Euphorbia retusa	Eczema and wounds	[165]
	Euphorbia rothiana	Acne and boils	[166]
	Euphorbia seguieriana	Inflamed wounds	[134]
	Euphorbia tirucalli	Warts and wounds	[160, 164]
	Euphorbia thymifolia	Scabies, warts, and wounds	[143, 164]
	Euphorbia unispina	Leprosy sores	[164]
	Euphorbia virgata	Bleeding wounds	[183]
	Excoecaria benthamiana	Warts	[164]
	Excoecaria grahamii	Worm sores	[164]
	Hevea brasiliensis	Dermal and lip ulcers and wounds	[87, 168]
	Jatropha chevalieri	Abscesses, bleeding wounds, and boils	[169]
	Jatropha curcas	Cuts, ulcers, whitlow, and wounds	[107, 133]
	Jatropha glandulifera	Bleeding cuts and mouth ulcers	[142]
	Jatropha gossypifolia	Ulcers and wounds	[131]
	Jatropha heynei	Burns and cuts	[142]
	Jatropha mollissima	Wounds	[170]
	Jatropha multifida	Ulcers and wounds	[169]
	Pedilanthus tithymaloides	Wounds	[168]
	Phyllanthus emblica	Wounds with maggots in cattle	[133]
	Ricinus communis	Wounds	[168]

(continued)

Table 14.3 (continued)

Family	Plant names	Type of wounds	References
13. Moraceae	*Artocarpus heterophyllus*	Burns, dog bite, and wounds	[133, 172]
	Ficus auriculata	Cuts and wounds	[171]
	Ficus benghalensis	Cracks	[132]
	Ficus benjamina	Boils	[172]
	Ficus carica	Boils, inflamed wound	[134]
	Ficus hirta	Wounds	[172]
	Ficus hispida	Bleeding, dead space, and excision and incision wounds	[173]
	Ficus lacor	Blisters, boils, and ulcers	[172]
	Ficus neriifolia	Boils	[172]
	Ficus palmata	Thorn wounds	[136]
	Ficus racemosa	Aphthae, blisters, boils, cracks, cuts, and ulcers	[133, 172]
	Ficus religiosa	Cuts and wounds	[135, 172]
	Ficus sarmentosa	Boils	[172]
	Ficus semicordata	Boils and ulcers	[172]
	Ficus sycomorus	Boils and scabies	[160]
	Ficus virens	Boils	[146]
	Milicia excelsa	Excision wounds	[174]
	Morus nigra	Oral wounds	[162]
14. Papaveraceae	*Argemone Mexicana*	Blisters, dead space, and excision and incision wounds	[138, 168]
	Chelidonium majus	Warts	[175]
15. Phyllanthaceae	*Phyllanthus niruri*	Offensive sores and ulcers	[161]
16. Plumbaginaceae	*Plumbago zeylanica*	Scabies	[131]
17. Sapotaceae	*Bassia longifolia*	Wounds	[143]
	Madhuca indica	Aphthae, cuts, ulcers in intestine, and wounds	[133]
	Mimusops elengi	Wounds	[143]
18. Solanaceae	*Datura stramonium*	Wounds	[168]
19. Thymelaeceae	*Aquilaria agallocha*	Edema, ulcers, and wounds	[176, 177]
20. Urticaceae	*Cecropia peltata*	Warts and wounds	[178]

14.4.1 Hemostasis

Hemostasis is the physiological process which maintains the flowing blood in fluid state within the blood vessel, while it aims at providing thrombotic response following injury to limit the blood loss [179]. As an exogenous hemostatic agent, latex is an important plant-based component, which is widely employed in traditional system of medicine because of its ability to stop bleeding from fresh wounds [180]. The use of topical hemostatic agents from natural sources is gaining importance in wound care and management, owing to the efficacy and safety of naturally derived hemostatic agents [181]. The property of latex to stop bleeding is mostly attributed to proteases [182]. The majority of latex proteases exhibits procoagulant effect irrespective of the nature of proteases. *Calotropis gigantea* latex protease has been

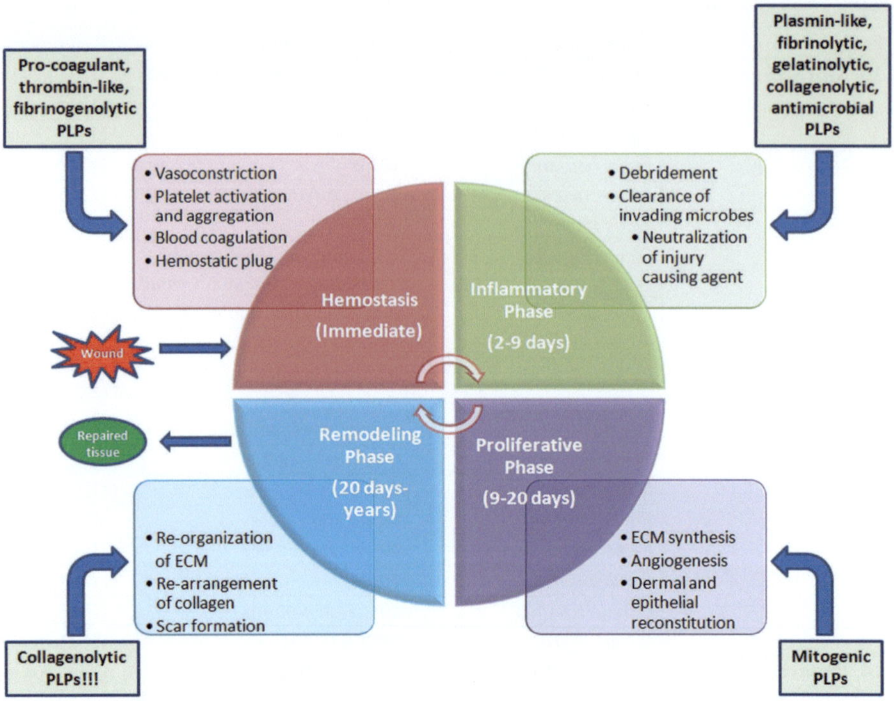

Fig. 14.1 Physiological events during the four phases of wound healing and the role of plant latex proteases (PLPs). PLPs exhibiting specific function are given in Table 14.1

shown with procoagulant, fibrin(ogen)olytic, and thrombin-like activities. In continuation, Shivaprasad et al. have reported that the cysteine proteases from Asclepiadaceae plant latices (*Asclepias curassavica*, *Cynanchum pauciflorum*, and *Pergularia extensa*) exhibit thrombin-like activity and facilitate the formation of clot even in the absence of Ca^{2+} ions [27]. Further, a cysteine protease "Pergularain e I" from *P. extensa* with thrombin-like activity was purified and characterized [28]. Procoagulant proteases may act in blood coagulation cascade or affect platelet function to aid the process of clot formation. The available reports suggest the interference of latex proteases only in common pathway, and their action on blood coagulation cascade is indicated in Fig. 14.2.

14.4.2 Inflammatory Phase

Clinically, inflammation is the second stage of wound healing that shows characteristic symptoms such as erythema, swelling, warmth, discomfort, and often associated with pain and functional disturbance [183]. The inflammatory phase ensures that the injury-causing agent is attenuated, diluted, and neutralized. Collectively,

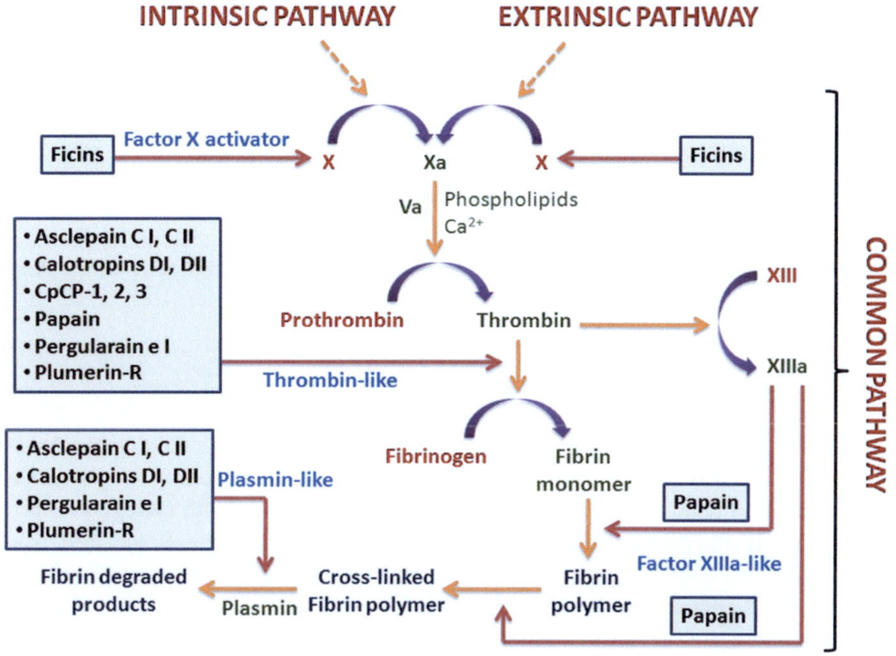

Fig. 14.2 Purified latex proteases interfere in the coagulation cascade mainly in the common pathway by activating factor X and exhibiting factor XIIIa-like, thrombin-like, and plasmin-like activities

during this phase, the injured region is prepared for healing [184]. Although there is fine balance between inflammatory mediators regulating inflammatory phase of wound healing, latex proteases have been reported to mediate the inflammatory phase and its smooth transition to repair phase. Few antimicrobial (Amp48 from *Artocarpus heterophyllus*, Cg24-I from *Cryptostegia grandiflora*, and hevains from *Hevea brasiliensis*) and anti-inflammatory (plumerin-R from *Plumeria rubra*) proteases have been isolated from latices which help in attenuating the infection-causing microbes. And mainly latex proteases have fibrinolytic (Amp48 from *Artocarpus heterophyllus*, crinumin from *Crinum asiaticum*, hirtin from *Euphorbia hirta*, LGP from *Synadenium grantii*, and plumerin-R from *Plumeria rubra*), gelatinolytic (serine protease from *Ficus carica*), and collagenolytic (serine protease from *Ficus carica*) activities which help in wound debridement (Fig. 14.3a and Table 14.1). Wound debridement is an important event during natural wound healing process and one of the vital aspects of successful wound management strategy. It is the removal of nonviable/dead, contaminated tissue and foreign material from the wound site, promoting the formation of granulation tissue and facilitating the progression of ordered wound healing [5, 185]. The presence of necrotic tissue in wound site mimics signs of infection and provides a suitable substrate for infecting microbes [186]. It also slows down the vital repair events including matrix formation, angiogenesis, granulation, tissue formation, and epidermal resurfacing [186].

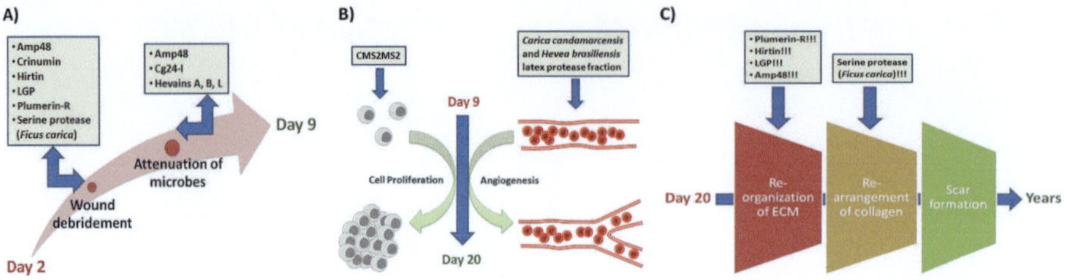

Fig. 14.3 Plant latex proteases aid (**A**) inflammatory, (**B**) proliferative, and (**C**) remodeling phases of wound healing

14.4.3 Proliferative Phase/Repair Phase

During the phase of repair, fresh tissue is layered in place of evacuated necrotic tissue [188]. The way for the formation of fresh tissue will be paved by the process of debridement of dead tissues by latex proteases in conjunction with endogenous proteases which are prerequisite for proliferative phase. Plumerin-R isolated from *Plumeria rubra* latex enhances wound healing by increasing the collagen formation (Fig. 14.3b). Collagen plays a central role in the healing of wounds as it is a principal component of connective tissue and provides a structural framework for regenerating tissue [33].

Angiogenesis, revascularization, and enhanced vascular permeability are crucial events to wound repair as they rescue tissues from ischemia. These events allow a variety of cytokines, growth factors, and nutrients to reach the damaged tissue and are also important for metabolite clearance. Further, proliferation of cells is important to replace the damaged tissues. Few latex proteases from *Carica candamarcensis* and *Hevea brasiliensis* have been shown with angiogenic and cell proliferative properties (mitogenic) along with the ability to increase vascular permeability and to activate the extracellular signal-regulated protein kinase (ERK) (Fig. 14.3b) [87, 189]. The ERK signaling cascade is a central MAPK pathway that plays a role in the regulation of cellular processes such as proliferation, differentiation, and development [190].

14.4.4 Remodeling Phase

The events of remodeling phase include the deposition of matrix and its subsequent changes including the alignment of ECM molecules along the tension lines [191]. Tissue remodeling occurs throughout the wound repair process and can overlap with the repair phase as it can begin as early as 1 week after injury and can last as long as 2 years, depending on the extent of wound [192]. Endogenous matrix metalloprotease expression is very important in the remodeling of collagen and non-collagen extracellular matrix (ECM) components [193]. In similar lines, collagenolytic and other ECM-degrading proteases of latex may also mediate the maturation and alignment of newly synthesized ECM proteins resulting in wound contraction apart from removal of dead tissue (Fig 14.3c). Interestingly, the latex from *Calotropis procera* has been reported with the ability to reduce the bundles of collagen fibers that are important for tissue remodeling [194]. *Wrightia tinctoria* latex serine protease has been shown with excision wound healing upon topical application with increase in collagen content. The same latex is also reported with collagenolytic and gelatinolytic activities that may be responsible for wound healing and tissue remodeling [77]. Even though there are no strong evidences regarding the role of latex proteases, it is worth looking for the role of latex proteases on the above lines.

Overall, the plant latex proteases provide optimal conditions for physiological wound healing by complementing the endogenous proteases in hemostasis, wound debridement, microbial attenuation, cell proliferation, and angiogenesis.

14.5 Experimental Validation for Plant Latex Proteases as Wound Healers

Wound healing potential of latex proteases is studied mostly in rodents (mice, rat, rabbit, and guinea pig) by excision, incision, and burn models. *In situ*, wound healing activity is assessed by cell proliferation assay in various cell lines. Purified proteases from latex of few plants have been shown to promote wound healing in experimental animals, and few are being used clinically as approved wound care supplements. Papain and chymopapain either singly or in combination with essential factors are being used in management of wound-related complications [13, 195]. Papain is widely used for clinical debridement in case of chronic wounds with overgranulated tissue. Few proteases from plant latices, namely, curcain (*Jatropha curcas*) and plumerin-R (*Plumeria rubra*), have been reported to promote wound healing in experimental animals. Although proteases are extensively used for debridement, some are hemostatic (ficins, Pergularain e I, LGP) or antimicrobial (Amp48 and Cg24-I) or promote the collagenization and fibrosis process (plumerin-R) [33, 107, 163]. Apart from the common mechanisms mentioned, few latex proteases exhibit mitogenic activity (CMS2MS2) evaluated in fibroblast cell culture and show increased neovascularization and epithelialization (P1G10) [52–54] (Table 14.1).

The reports from our lab also validate the use of latex for treating bleeding and wounds by traditional practitioners. The benefits of latex proteases in the first phase of wound healing, i.e., hemostasis, are due to procoagulant nature and thrombin-like activity of proteases. And they also have beneficial effects in second phase of wound healing, i.e., inflammatory phase, because of their fibrinolytic (plasmin-like activity), gelatinolytic, and collagenolytic activity which helps in wound debridement. These results were also evaluated using purified proteases in mice model giving substantial evidence [27, 28, 77, 94, 182, 196, 197]. Currently, we are engaged in the elucidation of molecular mechanisms of plant latex proteases in hemostasis and wound healing at the level of genus and species of latex-producing families.

14.6 Conclusion and Future Perspective

Irrespective of the underlying mechanisms, proteases exhibit positive effect toward wound healing. The scientific evaluation which explains the underlying molecular mechanisms, efficacy, and adverse effects, if any, would direct the findings toward development of potential wound care agents which can replace or supplement the conventional therapy for effective management of wounds.

Acknowledgements APU and RGV thank University Grants Commission (UGC), New Delhi and MVN thank IOE [funded by Ministry of Human Resource Development (MHRD), Government of India], UOM, Mysuru for fellowships. Authors also thank UGC-SAP, DST-PURSE and VGST for financial assistance to Department of Studies in Biochemistry, UOM and Vijnana Bhavan, UOM.

Conflict of interest The authors have no conflicts of interests to disclose.

References

1. Berg JM, Tymoczko JL, Stryer L (2002) Proteases: facilitating a difficult reaction.
2. Turk B (2006) Targeting proteases: successes, failures and future prospects. Nat Rev Drug Discov 5(9):785–799
3. Craik CS, Page MJ, Madison EL (2011) Proteases as therapeutics. Biochem J 435(1):1–16
4. Klingler D, Hardt M (2012) Targeting proteases in cardiovascular diseases by mass spectrometry-based proteomics. Circ Cardiovasc Genet 5(2):265–265
5. Ramundo J, Gray M (2008) Enzymatic wound debridement. J Wound Ostomy Cont Nurs 35(3):273–280
6. Fornbacke M, Clarsund M (2013) Cold-adapted proteases as an emerging class of therapeutics. Infect Dis Ther 2(1):15–26. doi:10.1007/s40121-013-0002-x
7. Matsubara K, Hori K, Matsuura Y, Miyazawa K (1999) A fibrinolytic enzyme from a marine green alga, Codium Latum. Phytochemistry 52(6):993–999
8. Hahn B-S, Cho SY, Ahn MY, Kim YS (2001) Purification and characterization of a plasmin-like protease from Tenodera Sinensis (Chinese mantis). Insect Biochem Mol Biol 31(6):573–581
9. Harrison RL, Bonning BC (2010) Proteases as insecticidal agents. Toxins 2(5):935–953
10. Amog PU, Manjuprasanna VN, Yariswamy M, Nanjaraj Urs AN, Joshi V, Suvilesh KN, Nataraju A, Vishwanath BS, Gowda TV (2016) Albizia lebbeck seed methanolic extract as a complementary therapy to manage local toxicity of Echis carinatus venom in a murine model. Pharm Biol 54(11):2568–2574
11. Hiremath V, Nanjaraj Urs AN, Joshi V, Suvilesh KN, Savitha MN, Amog PU, Rudresha GV, Yariswamy M, Vishwanath BS (2016) Differential action of medically important Indian BIG FOUR snake venoms on rodent blood coagulation. Toxicon 110:19–26
12. Amog PU, Yariswamy M, Joshi V, Nanjaraj Urs AN, Suvilesh KN, Manjuprasanna VN, Savitha MN, Rudresha GV, Nataraju A, Vishwanath BS, Gowda TV (2016) Local tissue damage induced by Echis carinatus venom: neutralization by Albizia Lebbeck seed aqueous extract in mice model. J Pharm Res 10(4):167–175
13. Udod V, Storozhuk V (1979) Treatment of suppurative diseases of soft tissues with proteolytic enzyme, papain. Klin Khir 1:37–38
14. González-Rábade N, Badillo-Corona JA, Aranda-Barradas JS, del Carmen O-SM (2011) Production of plant proteases in vivo and in vitro—a review. Biotechnol Adv 29(6):983–996

15. Winnick T, Davis AR, Greenberg DM (1940) Physicochemical properties of the proteolytic enzyme from the latex of the milkweed, Asclepilas speciosa torr. Some comparisons with other proteases: I. Chemical properties, activation-inhibition, pH activation-inhibition, pH-activity, and temperature-activity curves. J Gen Physiol 23:275–288
16. Domsalla A, Melzig MF (2008) Occurrence and properties of proteases in plant latices. Planta Med 74(7):699–711
17. Devaraj KB, Gowda LR, Prakash V (2008) An unusual thermostable aspartic protease from the latex of *Ficus Racemosa* (L.) Phytochemistry 69(3):647–655. doi:10.1016/j.phytochem.2007.09.003
18. Shiksharthi AR, Mittal S (2011) Ficus racemosa: Phytochemistry, traditional uses and pharmacological properties: a review. Int J Recent Adv Pharm Res 4:6–15
19. de Freitas CD, de Souza DP, Araújo ES, Cavalheiro MG, Oliveira LS, Ramos MV (2010) Anti-oxidative and proteolytic activities and protein profile of laticifer cells of *Cryptostegia grandiflora*, *Plumeria rubra* and *Euphorbia tirucalli*. Braz J Plant Physiol 22(1):11–22
20. Ramos M, Souza D, Gomes M, Freitas C, Carvalho C, Júnior P, Salas C (2014) A phytopathogenic cysteine peptidase from latex of wild rubber vine *Cryptostegia grandiflora*. Protein J 33(2):199–209
21. Nallamsetty S, Kundu S, Jagannadham MV (2003) Purification and biochemical characterization of a highly active cysteine protease ervatamin a from the latex of Ervatamia coronaria. J Protein Chem 22(1):1–13
22. Kundu S, Sundd M, Jagannadham MV (2000) Purification and characterization of a stable cysteine protease Ervatamin B, with two disulfide bridges, from the latex of Ervatamia coronaria. J Agric Food Chem 48(2):171–179
23. Patel BK, Jagannadham MV (2003) A high cysteine containing thiol proteinase from the latex of Ervatamia heyneana: purification and comparison with ervatamin B and C from Ervatamia coronaria. J Agric Food Chem 51(21):6326–6334. doi:10.1021/jf026184d
24. Thakurta PG, Biswas S, Chakrabarti C, Sundd M, Jagannadham MV, Dattagupta JK (2004) Structural basis of the unusual stability and substrate specificity of ervatamin C, a plant cysteine protease from Ervatamia coronaria. Biochemistry 43(6):1532–1540
25. Badgujar SB, Mahajan RT (2009) Proteolytic enzymes of some laticiferous plants belonging to Khandesh region of Maharashtra, India. J Pharm Res 2(9):1434–1437
26. Morcelle SR, Trejo SA, Canals F, Aviles FX, Priolo NS (2004) Funastrain c II: a cysteine endopeptidase purified from the latex of Funastrum clausum. Protein J 23(3):205–215
27. Shivaprasad H, Riyaz M, Kumar RV, Dharmappa K, Tarannum S, Siddesha J, Rajesh R, Vishwanath BS (2009) Cysteine proteases from the Asclepiadaceae plants latex exhibited thrombin and plasmin like activities. J Thromb Thrombolysis 28(3):304–308
28. Shivaprasad HV, Rajaiah R, Frey BM, Frey FJ, Vishwanath BS (2010) 'Pergularain e I'–a plant cysteine protease with thrombin-like activity from Pergularia Extensa latex. Thromb Res 125(3):e100–e105. doi:10.1016/j.thromres.2009.10.002
29. Sibi G, Wadhavan R, Singh S, Shukla A, Dhananjaya K, Ravikumar K, Mallesha H (2013) Plant latex: a promising antifungal agent for post harvest disease control. Pak J Biol Sci 16(23):1737
30. de Freitas CD, da Cruz WT, Silva MZ, Vasconcelos IM, Moreno FB, Moreira RA, Monteiro-Moreira AC, Alencar LM, Sousa JS, Rocha BA (2016) Proteomic analysis and purification of an unusual germin-like protein with proteolytic activity in the latex of Thevetia peruviana. Planta:1–14
31. Sequeiros C, Torres MJ, Trejo SA, Esteves JL, Natalucci CL, Lopez LM (2005) Philibertain g I, the most basic cysteine endopeptidase purified from the latex of Philibertia gilliesii hook. et Arn. (Apocynaceae). Protein J 24(7–8):445–453. doi:10.1007/s10930-005-7640-0
32. Sequeiros C, Torres MJ, Nievas ML, Caffini NO, Natalucci CL, López LM, Trejo SA (2016) The proteolytic activity of Philibertia gilliesii latex. Purification of Philibertain g II. Appl Biochem Biotechnol:1–15
33. Chanda I, Basu SK, Dutta SK, Das SRC (2011) A protease isolated from the latex of Plumeria Rubra Linn (Apocynaceae) 1: purification and characterization. Trop J Pharm Res 10(6):705–711

34. Priolo N, del Valle SM, Arribére MC, López L, Caffini N (2000) Isolation and characterization of a cysteine protease from the latex of Araujia Hortorum fruits. J Protein Chem 19(1):39–49
35. Obregón WD, Arribére MC, del Valle SM, Liggieri C, Caffini N, Priolo N (2001) Two new cysteine endopeptidases obtained from the latex of Araujia hortorum fruits. J Protein Chem 20(4):317–325
36. Brockbank WJ, Lynn KR (1979) Purification and preliminary characterization of two asclepains from the latex of Asclepias syriaca L. (milkweed). Biochim Biophys Acta 578(1):13–22
37. Moulin-Traffort J, Giordani R, Regli P (1989) Antifungal action of latex saps from Lactuca sativa L. and Asclepias curassavica L. Mycoses 33(7–8):383–392
38. Liggieri C, Arribere MC, Trejo SA, Canals F, Aviles FX, Priolo NS (2004) Purification and biochemical characterization of asclepain c I from the latex of Asclepias curassavica L. Protein J 23(6):403–411
39. Trejo SA, Lopez LM, Cimino CV, Caffini NO, Natalucci CL (2001) Purification and characterization of a new plant endopeptidase isolated from latex of Asclepias fruticosa L. (Asclepiadaceae). J Protein Chem 20(6):469–477
40. Tablero M, Arreguín R, Arreguín B, Soriano M, Sánchez RI, Rodríguez Romero A, Hernández-Arana A (1991) Purification and characterization of multiple forms of Asclepain g from Asclepias Glaucescens HBK. Plant Sci 74(1):7–15
41. Sengupta A, Bhattacharya D, Pal G, Sinha NK (1984) Comparative studies on calotropins DI and DII from the latex of Calotropis gigantea. Arch Biochem Biophys 232(1):17–25. doi:0003-9861(84)90517-4 [pii]
42. Pal G, Sinha NK (1980) Isolation, crystallization, and properties of calotropins DI and DII from Calotropis gigantea. Arch Biochem Biophys 202(2):321–329
43. Nalwaya N, Pokharna G, Deb L, Jain NK (2009) Wound healing activity of latex of Calotropis gigantea. Int J Pharm Pharm Sci 1(1):176–181
44. Cavalli SV, Cortadi A, Arribere MC, Conforti P, Caffini NO, Priolo N (2001) Comparison of two cysteine endopeptidases from latices of Morrenia Brachystephana Griseb. And Morrenia odorata (hook et Arn.) Lindley (Asclepiadaceae). Biol Chem 382(5):879–883. doi:10.1515/BC.2001.109
45. Cavalli SEV, Arribere MC, Cortadi A, Caffini NO, Priolo NS (2003) Morrenain b I, a papain-like endopeptidase from the latex of Morrenia Brachystephana Griseb. (Asclepiadaceae). J Protein Chem 22(1):15–22
46. Kumar V, Basu N (1994) Anti-inflammatory activity of the latex of Calotropis procera. J Ethnopharmacol 44(2):123–125
47. Dubey VK, Jagannadham MV (2003) Procerain, a stable cysteine protease from the latex of Calotropis procera. Phytochemistry 62(7):1057–1071
48. Ramos M, Araújo E, Jucá T, Monteiro-Moreira A, Vasconcelos I, Moreira R, Viana C, Beltramini LM, Pereira D, Moreno F (2013) New insights into the complex mixture of latex cysteine peptidases in Calotropis procera. Int J Biol Macromol 58:211–219
49. Giordani R, Siepaio M, Moulin-Traffort J, Regli P (1991) Antifungal action of Carica papaya latex: isolation of fungal cell wall hydrolysing enzymes. Mycoses 34(11–12):469–477
50. Gurung S, Škalko-Basnet N (2009) Wound healing properties of Carica papaya latex: in vivo evaluation in mice burn model. J Ethnopharmacol 121(2):338–341
51. Doolittle RF (2014) Clotting of mammalian fibrinogens by papain: a re-examination. Biochemistry 53(42):6687–6694
52. Gomes MT, Mello VJ, Rodrigues KC, Bemquerer MP, Lopes MT, Faça VM, Salas CE (2005) Isolation of two plant proteinases in latex from Carica Candamarcensis acting as mitogens for mammalian cells. Planta Med 71(3):244–248
53. Gomes MTR, Teixeira RD, Ribeiro HAL, Turchetti AP, Junqueira CF, Lopes MTP, Salas CE, Nagem RAP (2008) Purification, crystallization and preliminary X-ray analysis of CMS1MS2: a cysteine proteinase from *Carica Candamarcensis* latex. Acta Crystallogr Sect F: Struct Biol Cryst Commun 64(6):492–494

54. Gomes FS, Spínola CV, Ribeiro HA, Lopes MT, Cassali GD, Salas CE (2010) Wound-healing activity of a proteolytic fraction from *Carica Candamarcensis* on experimentally induced burn. Burns 36(2):277–283
55. Walraevens V, Vandermeers-Piret M-C, Vandermeers A, Gourlet P, Robberecht P (1999) Isolation and primary structure of the CCI papainlike cysteine proteinases from the latex of Carica candamarcensis hook. Biol Chem 380(4):485–488
56. Pereira MT, Lopes MT, Meira WO, Salas CE (2001) Purification of a cysteine proteinase from Carica candamarcensis L. and cloning of a genomic putative fragment coding for this enzyme. Protein Expr Purif 22(2):249–257
57. Silva C, Gomes M, Ferreira R, Rodrigues K, Val CG, Lopes M, Mello V, Salas C (2003) A mitogenic protein fraction in latex from Carica candamarcensis. Planta Med 69(10):926–932
58. Mello VJ, Gomes MTR, Lemos FO, Delfino JL, Andrade SP, Lopes MT, Salas CE (2008) The gastric ulcer protective and healing role of cysteine proteinases from Carica candamarcensis. Phytomedicine 15(4):237–244
59. Gavira JA, Gonzalez-Ramirez LA, Oliver-Salvador MC, Soriano-Garcia M, Garcia-Ruiz JM (2007) Structure of the mexicain-E-64 complex and comparison with other cysteine proteases of the papain family. Acta Crystallogr D Biol Crystallogr 63(Pt 5):555–563. doi:10.1107/S0907444907005616
60. Oliver-Salvador MC, Gonzalez-Ramirez LA, Gavira JA, Soriano-Garcia M, Garcia-Ruiz JM (2004) Purification, crystallization and preliminary X-ray analysis of mexicain. Acta Crystallogr D Biol Crystallogr 60(Pt 11):2058–2060. doi:10.1107/S0907444904021638
61. Torres MJ, Trejo SA, Martin MI, Natalucci CL, Avilés FX, López LM (2010) Purification and characterization of a cysteine endopeptidase from Vasconcellea quercifolia A. St-Hil latex displaying high substrate specificity. J Agric Food Chem 58(20):11027–11035
62. Torres MJ, Trejo SA, Natalucci CL, López LMI (2013) Biochemical characterization of VQ-VII, a cysteine peptidase with broad specificity, isolated from Vasconcellea quercifolia latex. Planta 237(6):1651–1659
63. Badgujar SB, Mahajan RT (2014) Identification and characterization of Euphorbia nivulia latex proteins. Int J Biol Macromol 64:193–201
64. Badgujar SB, Mahajan RT (2014) Nivulian-II a new milk clotting cysteine protease of Euphorbia nivulia latex. Int J Biol Macromol 70:391–398
65. Badgujar SB (2014) Evaluation of hemostatic activity of latex from three Euphorbiaceae species. J Ethnopharmacol 151(1):733–739
66. Bhowmick R, Prasanna Kumari N, Jagannadham M, Kayastha AM (2008) Purification and characterization of a novel protease from the latex of Pedilanthus tithymaloids. Protein Pept Lett 15(9):1009–1016
67. Kramer DE, Whitaker JR (1964) Ficus enzymes II. Properties of the proteolytic enzymes from the latex of Ficus Carica variety Kadota. J Biol Chem 239(7):2178–2183
68. Sugiura M, Sasaki M (1974) Studies on proteinases from Ficus Carica Var. Horaishi. V. Purification and properties of a sugar-containing proteinase (ficin S). Biochim Biophys Acta 350(1):38–47
69. Aref HL, Salah K, Chaumont JP, Fekih A, Aouni M, Said K (2010) In vitro antimicrobial activity of four Ficus carica latex fractions against resistant human pathogens (antimicrobial activity of Ficus carica latex). Pak J Pharm Sci 23(1):53–58
70. Baeyens-Volant D, Matagne A, El Mahyaoui R, Wattiez R, Azarkan M (2015) A novel form of ficin from Ficus carica latex: purification and characterization. Phytochemistry 117:154–167
71. Richter G, Schwarz HP, Dorner F, Turecek PL (2002) Activation and inactivation of human factor X by proteases derived from Ficus Carica. Br J Haematol 119(4):1042–1051
72. Mnif IH, Siala R, Nasri R, Mhamdi S, Nasri M, Kamoun AS (2015) A cysteine protease isolated from the latex of Ficus microcarpa: purification and biochemical characterization. Appl Biochem Biotechnol 175(3):1732–1744
73. Chetia D, Nath L, Dutta S (1999) Extraction, purification and physico-chemical properties of a proteolytic enzyme from the latex of Ficus Hispida Linn. Indian J Pharm Sci 61(1):29

74. Kumar R, Singh KA, Tomar R, Jagannadham MV (2011) Biochemical and spectroscopic characterization of a novel metalloprotease, cotinifolin from an antiviral plant shrub: Euphorbia cotinifolia. Plant Physiol Biochem 49(7):721–728
75. Singh KA, Kumar R, Rao G, Jagannadham MV (2010) Crinumin, a chymotrypsin-like but glycosylated serine protease from Crinum asiaticum: purification and physicochemical characterisation. Food Chem 119(4):1352–1358
76. Singh KA, Jagannadham M, Rao G, Celie PH (2011) Crystallization and preliminary X-ray analysis of crinumin, a chymotrypsin-like glycosylated serine protease with thrombolytic and antiplatelet activity. Acta Crystallogr Sect F: Struct Biol Cryst Commun 67(12):1545–1547
77. Yariswamy M, Shivaprasad H, Joshi V, Urs AN, Nataraju A, Vishwanath BS (2013) Topical application of serine proteases from Wrightia tinctoria R. Br.(Apocyanaceae) latex augments healing of experimentally induced excision wound in mice. J Ethnopharmacol 149(1):377–383
78. Tomar R, Kumar R, Jagannadham MV (2008) A stable serine protease, wrightin, from the latex of the plant Wrightia tinctoria (Roxb.) R. Br.: purification and biochemical properties. J Agric Food Chem 56(4):1479–1487. doi:10.1021/jf0726536
79. Lynn K, Clevette-Radford N (1984) Euphorbain p, a serine protease from Euphorbia pulcherrima. Phytochemistry 23(3):682–683
80. Lynn K (1988) Parthenain, a protease from Parthenium argentatum. Phytochemistry 27(7):1987–1991
81. Rudenskaya G, Bogacheva A, Preusser A, Kuznetsova A, Dunaevsky YE, Golovkin B, Stepanov V (1998) Taraxalisin–a serine proteinase from dandelion Taraxacum officinale Webb sl. FEBS Lett 437(3):237–240
82. Patel AK, Singh VK, Jagannadham MV (2007) Carnein, a serine protease from noxious plant weed Ipomoea carnea (morning glory). J Agric Food Chem 55(14):5809–5818
83. Lennox F, Ellis W (1945) Euphorbain, a protease occurring in the latex of the weed Euphorbia lathyris. Biochem J 39(5):465
84. Lynn K, Clevette-Radford N (1986) Isolation and characterization of proteases from Euphorbia lactea and Euphorbia lactea cristata. Phytochemistry 25(4):807–810
85. Lynn K, Clevette-Radford N (1985) Four serine proteases from the latex of Euphorbia tirucalli. Can J Biochem Cell Biol 63(10):1093–1096
86. Lynn K, Clevette-Radford N (1985) Three serine proteases from the latex of Euphorbia cyparissias. Phytochemistry 24(5):925–928
87. Mendonça RJ, Maurício VB, de Bortolli TL, Lachat JJ, Coutinho-Netto J (2010) Increased vascular permeability, angiogenesis and wound healing induced by the serum of natural latex of the rubber tree Hevea Brasiliensis. Phytother Res 24(5):764–768
88. Lynn K, Clevette-Radford N (1986) Hevains: serine-centred proteases from the latex of Hevea brasiliensis. Phytochemistry 25(10):2279–2282
89. Lynn K, Clevette-Radford N (1984) Purification and characterization of hevain, a serine protease from Hevea brasiliensis. Phytochemistry 23(5):963–964
90. Van Parijs J, Broekaert WF, Goldstein IJ, Peumans WJ (1991) Hevein: an antifungal protein from rubber-tree (Hevea brasiliensis) latex. Planta 183(2):258–264
91. Ferreira M, Mendonça RJ, Coutinho-Netto J, Mulato M (2009) Angiogenic properties of natural rubber latex biomembranes and the serum fraction of Hevea brasiliensis. Braz J Phys 39(3):564–569
92. Patel GK, Kawale AA, Sharma AK (2012) Purification and physicochemical characterization of a serine protease with fibrinolytic activity from latex of a medicinal herb Euphorbia hirta. Plant Physiol Biochem 52:104–111. doi:10.1016/j.plaphy.2011.12.004
93. Hussain M, Farooq U, Rashid M, Bakhsh H, Majeed A, Khan IA, Rana SL, Rehman M, Aziz A (2014) Antimicrobial activity of fresh latex juice and extract of Euphorbia hirta and Euphorbia Thymifolia: an in vitro comparative study. Int J Pharma Sci 4(3):546–553
94. Rajesh R, Nataraju A, Gowda C, Frey B, Frey F, Vishwanath BS (2006) Purification and characterization of a 34-kDa, heat stable glycoprotein from *Synadenium grantii* latex:

action on human fibrinogen and fibrin clot. Biochimie 88(10):1313–1322. doi:10.1016/j.biochi.2006.06.007
95. Menon M, Vithayathil P, Raju S, Ramadoss C (2002) Isolation and characterization of proteolytic enzymes from the latex of Synadenium grantii hook,'f'. Plant Sci 163(1):131–139
96. Yadav SC, Pande M, Jagannadham M (2006) Highly stable glycosylated serine protease from the medicinal plant Euphorbia milii. Phytochemistry 67(14):1414–1426
97. Arima K, Uchikoba T, Yonezawa H, Shimada M, Kaneda M (2000) Cucumisin-like protease from the latex of Euphorbia Supina. Phytochemistry 53(6):639–644
98. Singh SJ, Singh LR, Devi SK, Singh SS, Devi CB, Rully H (2015) Purification and characterization of a thermostable caseinolytic serine protease from the latex of Euphorbia heterophylla L. Protein Pept Lett 22(9):828–835
99. Mahajan RT, Adsul YD (2015) Isolation, purification and characterization of serine protease from latex of Euphorbia prunifolia Jacq. Int J Adv Res 3(1):388–395
100. Siritapetawee J, Thumanu K, Sojikul P, Thammasirirak S (2012) A novel serine protease with human fibrino(geno)lytic activities from Artocarpus heterophyllus latex. Biochim Biophys Acta 1824(7):907–912. doi:10.1016/j.bbapap.2012.05.002
101. Siritapetawee J, Thammasirirak S, Samosornsuk W (2012) Antimicrobial activity of a 48-kDa protease (AMP48) from Artocarpus heterophyllus latex. Eur Rev Med Pharmacol Sci 16(1):132–137
102. Renuka Prasad K, Virupaksha TK (1990) Purification and characterization of a protease from jackfruit latex. Phytochemistry 29(6):1763–1766
103. Sharma A, Kumari M, Jagannadham MV (2009) Benghalensin, a highly stable serine protease from the latex of medicinal plant Ficus benghalensis. J Agric Food Chem 57(23):11120–11126. doi:10.1021/jf902279u
104. Rudenskaya G, Bogdanova E, Revina L, Golovkin B, Stepanov V (1995) Macluralisin—a serine proteinase from fruits of Maclura pomifera (Raf.) Schneid. Planta 196(1):174–179
105. Raskovic B, Bozovic O, Prodanovic R, Niketic V, Polovic N (2014) Identification, purification and characterization of a novel collagenolytic serine protease from fig (Ficus carica Var. Brown Turkey) latex. J Biosci Bioeng 118(6):622–627
106. Kumari M, Sharma A, Jagannadham M (2010) Decolorization of crude latex by activated charcoal, purification and physico-chemical characterization of religiosin, a milk-clotting serine protease from the latex of Ficus religiosa. J Agric Food Chem 58(13):8027–8034
107. Nath LK, Dutta SK (1991) Extraction and purification of curcain, a protease from the latex of Jatropha Curcas Linn. J Pharm Pharmacol 43(2):111–114
108. Arekemase M, Kayode R, Ajiboye A (2011) Antimicrobial activity and phytochemical analysis of Jatropha curcas plant against some selected microorganisms. Int J Biol 3(3):52
109. Chakraborty S, Biswas S, Chakrabarti C, Dattagupta JK (2005) Crystallization and preliminary X-ray diffraction studies of the cysteine protease ervatamin A from Ervatamia coronaria. Acta Crystallogr Sect F: Struct Biol Cryst Commun 61(6):562–564
110. Chakrabarti C, Biswas S, Kundu S, Sundd M, Jagannadham MV, Dattagupta JK (1999) Crystallization and preliminary X-ray analysis of ervatamin B and C, two thiol proteases from Ervatamia coronaria. Acta Crystallogr D Biol Crystallogr 55(5):1074–1075
111. Biswas S, Chakrabarti C, Kundu S, Jagannadham MV, Dattagupta JK (2003) Proposed amino acid sequence and the 1.63 Å X-ray crystal structure of a plant cysteine protease, ervatamin B: some insights into the structural basis of its stability and substrate specificity. Proteins Struct Funct Bioinf 51(4):489–497
112. Pal GP, Sinha NK, Saenger W (1981) Crystallizations and preliminary X-ray studies of calotropins DI and DII. J Mol Biol 153(4):1157–1159
113. Heinemann U, Pal G, Hilgenfeld R, Saenger W (1982) Crystal and molecular structure of the sulfhydryl protease calotropin DI at 3.2 Å resolution. J Mol Biol 161(4):591–606
114. Katerelos NA, Taylor MA, Scott M, Goodenough PW, Pickersgill RW (1996) Crystal structure of a caricain D158E mutant in complex with E-64. FEBS Lett 392(1):35–39
115. Gomes MT, Teixeira RD, Lopes MT, Nagem RA, Salas CE (2012) X-ray crystal structure of CMS1MS2: a high proteolytic activity cysteine proteinase from Carica candamarcensis. Amino Acids 43(6):2381–2391

116. O'Hara BP, Hemmings AM, Buttle DJ, Pearl LH (1995) Crystal structure of glycyl endopeptidase from Carica papaya: a cysteine endopeptidase of unusual substrate specificity. Biochemist 34(40):13190–13195
117. Kozak M, Kozian E, Grzonka Z, Jaskólski M (1996) Crystallization and preliminary crystallographic studies of a new crystal form of papain from Carica papaya. Acta Biochim Pol 44(3):601–605
118. Pande M, Dubey VK, Jagannadham MV (2007) Crystallization and preliminary X-ray analysis of cryptolepain, a novel glycosylated serine protease from Cryptolepis buchanani. Acta Crystallogr Sect F: Struct Biol Cryst Commun 63(2):74–77
119. Patel AK, van Oosterwijk N, Singh VK, Rozeboom HJ, Kalk KH, Siezen RJ, Jagannadham MV, Dijkstra BW (2009) Crystallization and preliminary X-ray analysis of carnein, a serine protease from Ipomoea carnea. Acta Crystallogr Sect F: Struct Biol Cryst Commun 65(4):383–385
120. Konno K, Hirayama C, Nakamura M, Tateishi K, Tamura Y, Hattori M, Kohno K (2004) Papain protects papaya trees from herbivorous insects: role of cysteine proteases in latex. Plant J 37(3):370–378
121. Dubey VK, Pande M, Singh BK, Jagannadham MV (2007) Papain-like proteases: applications of their inhibitors. Afr J Biotechnol 6(9)
122. Ramzi SC, Vinay K, Stanley R (eds) (1994) Pathologic basis of diseases, vol 5, 5th edn. WB Saunders Company, Philadelphia
123. Crovetti G, Martinelli G, Issi M, Barone M, Guizzardi M, Campanati B, Moroni M, Carabelli A (2004) Platelet gel for healing cutaneous chronic wounds. Transfus Apher Sci 30(2):145–151
124. Stadelmann WK, Digenis AG, Tobin GR (1998) Physiology and healing dynamics of chronic cutaneous wounds. Am J Surg 176(2):26S–38S
125. Epstein FH, Singer AJ, Clark RA (1999) Cutaneous wound healing. N Engl J Med 341(10):738–746
126. Cowan L, Phillips P, Liesenfeld B, Mikhaylova A, Moore D, Stechmiller J, Schultz G (2011) Caution: when combining topical wound treatments, more is not always better. Wound Pract Res 19:60–64
127. Wysocki AB (2015) Dorsett-Martin WA (2008) enhance your knowledge of skin grafts. OR Nurse 2(9):30–38
128. Babu M (2000) Collagen based dressings—a review. Burns 26(1):54–62
129. Ayyanar M, Ignacimuthu S (2009) Some less known ethnomedicinal plants of Tirunelveli hills, Tamil Nadu. Journal of economic and taxonomic botany 33 (supplement):73–76
130. Suriya C, Kasatpibal N, Kunaviktikul W, Kayee T (2011) Diagnostic indicators for peptic ulcer perforation at a tertiary care hospital in Thailand. Clin Exp Gastroenterol 4:283
131. Uday P, Achar RR, Bhat PR, Rinimol V, Bindu J, Nafeesa Z, Swamy NS (2015) Laticiferous plant proteases in wound care. Int J Pharm Pharm Sci 7(13)
132. Muthu C, Ayyanar M, Raja N, Ignacimuthu S (2006) Medicinal plants used by traditional healers in Kancheepuram district of Tamil Nadu, India. J Ethnobiol Ethnomed 2(1):1
133. Bhat P, Hegde G, Hegde GR (2012) Ethnomedicinal practices in different communities of Uttara Kannada district of Karnataka for treatment of wounds. J Ethnopharmacol 143(2):501–514
134. Altundag E, Ozturk M (2011) Ethnomedicinal studies on the plant resources of east Anatolia, Turkey. Procedia Soc Behav Sci 19:756–777
135. Manandher N (1998) Ethnobotanical census on herbal medicines of Banke district, Nepal. Contribution to Nepalese Studies 25:57–63
136. Kunwar RM, Shrestha KP, Bussmann RW (2010) Traditional herbal medicine in far-west Nepal: a pharmacological appraisal. J Ethnobiol Ethnomed 6(1):1
137. Mahendra P, Bisht S (2012) Ferula asafoetida: traditional uses and pharmacological activity. Pharmacogn Rev 6(12):141–146. doi:10.4103/0973-7847.99948

138. Kumar B, Vijayakumar M, Govindarajan R, Pushpangadan P (2007) Ethnopharmacological approaches to wound healing—exploring medicinal plants of India. J Ethnopharmacol 114(2):103–113
139. Wong SK, Lim YY, Chan EW (2013) Botany, uses, phytochemistry and pharmacology of selected Apocynaceae species: a review. Phcog Commn 3(3):2
140. Arulmozhi S, Rasal V, Sathiyanarayanan L, Purnima A (2007) Screening of Alstonia scholaris Linn. R. Br., for wound healing activity. Orient Pharm Exp Med 7(3):254–260
141. Trivedi PC (2009) Indigenous Ethnomedicinal plants. Pointer Publishers
142. Mohan RK, Murthy PB (1992) Plants used in traditional medicine by tribals of Prakasam district, Andhra Pradesh. Anc Sci Life 11(3–4):176
143. Chopda M, Mahajan R (2009) Wound healing plants of Jalgaon district of Maharashtra state, India. Ethnobotanical leaflets 2009(1):1
144. Dey A, De JN (2010) A survey of ethnomedicinal plants used by the tribals of Ajoydha hill region, Purulia district, India. Am-Eurasian J Sustain Agric:280–291
145. Kadhirvel K, Ramya S, Sudha TPS, Ravi AV, Rajasekaran C, Selvi RV, Jayakumararaj R (2010) Ethnomedicinal survey on plants used by tribals in Chitteri Hills. Environ We Int J Sci Tech 5:35–46
146. Krishna NR, Varma Y, Saidulu C (2014) Ethnobotanical studies of Adilabad district, Andhra Pradesh, India. J Pharmacogn Phytochem 3(1)
147. Rajkumar M, Rajanna M (2011) Ex-situ conservation of climbing plants at University of Agricultural Sciences, Bangalore, Karnataka. Recent Res Sci Technol 3(4)
148. Watt J, Breyer-Brandwijk M (1962) The Medicinal and Poisonous Plants of southern and eastern Africa (2 edn.) Livingstone. Edinburgh, London 205–206
149. Boomibalagan P, S Ea, S R (2013) Traditional uses of medicinal plants of Asclepiadaceae by rural people in Madurai District, Tamil Nadu, India. Int J Bot 9 (3):133–139
150. Barukial J, Sarmah J (2011) Ethnomedicinal plants used by the people of Golaghat district, Assam, India. Int J Med Aromat Plants 1(3):203–211
151. Wondimu T, Asfaw Z, Kelbessa E (2007) Ethnobotanical study of medicinal plants around 'Dheeraa' town, Arsi zone, Ethiopia. J Ethnopharmacol 112(1):152–161
152. Rehecho S, Uriarte-Pueyo I, Calvo J, Vivas LA, Calvo MI (2011) Ethnopharmacological survey of medicinal plants in nor-Yauyos, a part of the landscape reserve nor-Yauyos-Cochas, Peru. J Ethnopharmacol 133(1):75–85
153. Koçyiğit M, Özhatay N (2006) Wild plants used as medicinal purpose in Yalova (Northwest Turkey). Turk J Pharm Sci 3(2):91–103
154. Manikandan PA (2008) Ethno-medico-botanical studies of Badaga population in the Nilgiri district of Tamilnadu, South India. Anc Sci Life 27(3):50
155. Manolova L (2003) Natural pharmacy. Dorrance Publishing
156. Joseph A, Mathew S, Skaria BP, Sheeja E (2011) Medicinal uses and biological activities of Argyreia Speciosa sweet (Hawaiian baby wood rose)-an overview. Indian J Nat Prod Resour 2(3):286–291
157. Chander MP, Kartick C, Gangadhar J, Vijayachari P (2014) Ethno medicine and healthcare practices among Nicobarese of car Nicobar–an indigenous tribe of Andaman and Nicobar Islands. J Ethnopharmacol 158:18–24
158. Rajendran K, Balaji P, Basu MJ (2008) Medicinal plants and their utilization by villagers in southern districts of Tamil Nadu. Indian J Tradit Knowl 7(3):417–420
159. Jain S, Jain R, Singh R (2009) Ethnobotanical survey of Sariska and Siliserh regions from Alwar district of Rajasthan, India. Ethnobotanical leaflets 2009(1):21
160. Njoroge GN, Bussmann RW (2007) Ethnotherapeautic management of skin diseases among the kikuyus of Central Kenya. J Ethnopharmacol 111(2):303–307
161. Malhotra S, Dutta B, Gupta RK, Gaur Y (1966) Medicinal plants of the Indian arid zone. Journal d'agriculture tropicale et de botanique appliquée 13(6–7):247–288
162. Çakılcıoğlu U, Şengün M, Türkoğlu İ (2010) An ethnobotanical survey of medicinal plants of Yazıkonak and Yurtbaşı districts of Elazığ province, Turkey. J Med Plants Res 4(7):567–572

163. Mahajan RT, Badgujar SB (2011) Bioprospecting of Euphorbia nivulia Buch.-ham. Int J Phytopharmacol 2(2):37–42
164. Plants M (2008) Plant resources of Tropical Africa, vol 1. PROTA Foundation, Wageningen
165. El-Ghazali GE, Al-Khalifa KS, Saleem GA, Abdallah EM (2010) Traditional medicinal plants indigenous to al-Rass province, Saudi Arabia. J Med Plants Res 4(24):2680–2683
166. Paulsamy S, Vijayakumar K, Murugesan M, Padmavathy S, Senthilkumar P (2007) Ecological status of medicinal and other economically important plants in the shola understories of Nilgiris, the Western Ghats. Nat Prod Rad 6(1):55–61
167. Özgen U, Kaya Y, Houghton P (2012) Folk medicines in the villages of Ilıca District (Erzurum, Turkey). Turk J Biol 36(1):93–106
168. Chaitanya S, Manthri S, Srilakshmi S, Ashajyothi V (2010) Wound healing herbs – a review. Int J Pharmacy Technol 2(4):603–624
169. Sabandar CW, Ahmat N, Jaafar FM, Sahidin I (2013) Medicinal property, phytochemistry and pharmacology of several Jatropha species (Euphorbiaceae): a review. Phytochemistry 85:7–29
170. de Oliveira RL, Neto EML, Araújo EL, Albuquerque UP (2007) Conservation priorities and population structure of woody medicinal plants in an area of caatinga vegetation (Pernambuco state, NE Brazil). Environ Monit Assess 132(1–3):189–206
171. Pant S, Samant S, Arya S (2009) Diversity and indigenous household remedies of the inhabitants surrounding Mornaula reserve forest in West Himalaya. Indian J Tradit Knowl 8(4):606–610
172. Kunwar RM, Bussmann RW (2006) Ficus (Fig) species in Nepal: a review of diversity and indigenous uses. Lyonia 11(1):85–97
173. Murti K, Lambole V, Panchal M (2011) Effect of Ficus hispida L. on normal and dexamethasone suppressed wound healing. Braz J Pharm Sci 47(4):855–860
174. Udegbunam S, Nnaji T, Udegbunam R, Okafor J, Agbo I (2013) Evaluation of herbal ointment formulation of Milicia excelsa (Welw) CC berg for wound healing. Afr J Biotechnol 12(21):3351
175. Watkins F, Pendry B, Corcoran O, Sanchez-Medina A (2011) Anglo-Saxon pharmacopoeia revisited: a potential treasure in drug discovery. Drug Discov Today 16(23):1069–1075
176. Islam MK, Saha S, Mahmud I, Mohamad K, Awang K, Uddin SJ, Rahman MM, Shilpi JA (2014) An ethnobotanical study of medicinal plants used by tribal and native people of Madhupur forest area, Bangladesh. J Ethnopharmacol 151(2):921–930
177. Sharma V, Hem K, Mishra A, Maurya SK (2016) Time tested remedies for wound care from ayurveda sciences. Innovare J Ayur Sci 6–10
178. Nayak BS (2006) Cecropia peltata L (Cecropiaceae) has wound-healing potential: a preclinical study in a Sprague Dawley rat model. Int J Low Extrem Wounds 5(1):20–26
179. Lundblad RL, Bradshaw RA, Gabriel D, Ortel TL, Lawson J, Mann KG (2004) A review of the therapeutic uses of thrombin. Thromb Haemost 91(5):851–860
180. Thankamma L (2003) Hevea latex as a wound healer and pain killer. Curr Sci 84(8):971–972
181. Samudrala S (2008) Topical hemostatic agents in surgery: a surgeon's perspective. AORN J 88(3):S2–S11
182. Rajesh R, Gowda CR, Nataraju A, Dhananjaya B, Kemparaju K, Vishwanath BS (2005) Procoagulant activity of Calotropis Gigantea latex associated with fibrin(ogen)olytic activity. Toxicon 46(1):84–92
183. Monaco JL, Lawrence WT (2003) Acute wound healing: an overview. Clin Plast Surg 30(1):1–12
184. Anderson JM (2001) Biological responses to materials. Annu Rev Mater Res 31(1):81–110
185. Sibbald RG, Williamson D, Orsted HL, Campbell K, Keast D, Krasner D, Sibbald D (2000) Preparing the wound bed--debridement, bacterial balance, and moisture balance. Ostomy Wound Manage 46(11.) 14–22, 24–8, 30–5; quiz 36–7
186. Leaper D (2002) Sharp technique for wound debridement. World Wide Wounds
187. Dreifke MB, Jayasuriya AA, Jayasuriya AC (2015) Current wound healing procedures and potential care. Mater Sci Eng C 48:651–662

188. Forrest L (1983) Current concepts in soft connective tissue wound healing. Br J Surg 70(3):133–140
189. Correa NC, Mendes IC, Gomes MT, Kalapothakis E, Chagas BC, Lopes MT, Salas CE (2011) Molecular cloning of a mitogenic proteinase from Carica Candamarcensis: its potential use in wound healing. Phytochemistry 72(16):1947–1954. doi:10.1016/j.phytochem.2011.06.010
190. Shaul YD, Seger R (2007) The MEK/ERK cascade: from signaling specificity to diverse functions. Biochim Biophys Acta 1773(8):1213–1226. doi:10.1016/j.bbamcr.2006.10.005
191. Hinz B, Mastrangelo D, Iselin CE, Chaponnier C, Gabbiani G (2001) Mechanical tension controls granulation tissue contractile activity and myofibroblast differentiation. Am J Pathol 159(3):1009–1020
192. Stroncek JD, Reichert WM (2008) Overview of wound healing in different tissue types. In: Indwelling neural implants: strategies for contending with the in vivo environment. CRC Press, Boca Raton
193. Kähäri VM, Saarialho-Kere U (1997) Matrix metalloproteinases in skin. Exp Dermatol 6(5):199–213
194. Aderounmua A, Omonisib A, Akingbasotec J, Makanjuolad M, Bejide R, Orafidiya L, Adelusolae K (2013) Wound-healing and potential antikeloidal properties of the latex of Calotropis Procera (Aiton) Asclepiadaceae in rabbits. Afr J Tradit Complement Altern Med 10(3):574–579
195. Maurer H (2001) Bromelain: biochemistry, pharmacology and medical use. Cellular and Molecular Life Sciences CMLS 58(9):1234–1245
196. Rajesh R, Shivaprasad HV, Gowda CD, Nataraju A, Dhananjaya BL, Vishwanath BS (2007) Comparative study on plant latex proteases and their involvement in hemostasis: a special emphasis on clot inducing and dissolving properties. Planta Med 73(10):1061–1067
197. Shivaprasad HV, Rajesh R, Nanda BL, Dharmappa KK, Vishwanath BS (2009) Thrombin like activity of *Asclepias Curassavica* L. latex: action of cysteine proteases. J Ethnopharmacol 123(1):106–109. doi:10.1016/j.jep.2009.02.016

Emerging Roles of Mitochondrial Serine Protease HtrA2 in Neurodegeneration

15

Ajay R. Wagh and Kakoli Bose

Abstract

High temperature requirement mitochondrial serine protease A2 (HtrA2), commonly known as Omi/PRSS25, is primarily known for its protein quality control function. Loss of this prime function of the protein results in neurodegenerative diseases such as Parkinson's, Alzheimer's, Huntington's and certain forms of epilepsy. HtrA2 belongs to the family of evolutionarily conserved proteases and is one of the core determinants of mitochondrial quality control. HtrA2 helps maintain normal mitochondrial functions during organelle biogenesis, metabolic remodelling and stress. It has been very well established that under stressful condition, HtrA2 is released from the mitochondria into the cytosol and facilitates apoptosis by binding to several members of the inhibitors of apoptosis protein (IAP) family. On the contrary, in vivo knockout studies showed a phenotype similar to Parkinson's suggesting its involvement in neurodegeneration along with maintenance of mitochondrial homeostasis. Therefore, presence of different cellular pathways and its unique multitasking ability makes HtrA2 a potential therapeutic target. This chapter discusses different facets of HtrA2 with main focus on its role as a quality control protease and its association with neurodegenerative disorders.

Keywords

Mitochondrial protein quality control • HtrA2 • Neurodegenerative disorders

A.R. Wagh • K. Bose (✉)
Integrated Biophysics and Structural Biology Lab, Advanced Centre for Treatment, Research and Education in Cancer (ACTREC), Tata Memorial Centre, Navi Mumbai 410210, Maharashtra, INDIA
e-mail: kbose@actrec.gov.in

S. Chakraborti, N.S. Dhalla (eds.), *Proteases in Physiology and Pathology*,
DOI 10.1007/978-981-10-2513-6_15

15.1 Introduction

15.1.1 Overview of Mitochondrial Protein Quality Control

Mitochondria are dynamic, semiautonomous organelles present in eukaryotic cells that play important role in energy metabolism and myriads of important physiological processes. The critical functions of mitochondria include generation of ATP through respiration, integration of several key metabolic and cofactor-generating pathways and regulation of ion homeostasis as well as apoptosis [1, 2]. Imbalance in mitochondrial homeostasis and integrity lead to severe pathophysiological consequences often with the onset of certain disease conditions (Fig. 15.1). Numerous studies implicate mitochondrial dysfunction as a fundamental factor for multiple pathologies in humans that include cardiovascular disorders, myopathies, certain cancers, type II diabetes and neurodegenerative diseases [1–12]. Neurodegenerative diseases such as Alzheimer's and Parkinson's (AD and PD) are particularly predominant in elderly population who have been linked with age-associated decline in mitochondrial health that leads to cellular damage [6, 8, 11, 13]. For example, several PD-associated genes interfere with pathways regulating mitochondrial function, morphology and dynamics and hence challenge mitochondrial integrity.

To circumvent these undesirable situations, our cells have efficient surveillance systems that ensure sustenance of mitochondrial functionality. This is achieved by

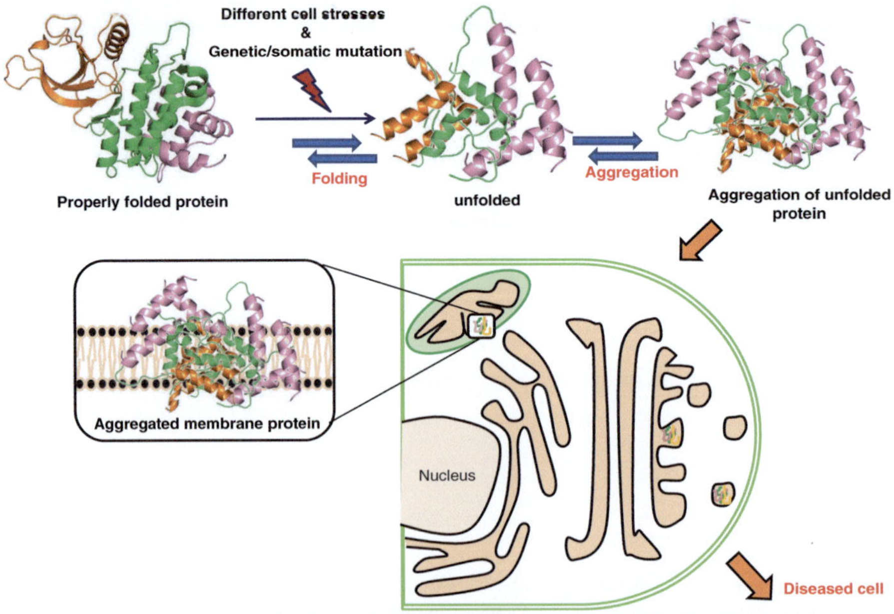

Fig. 15.1 Perturbations in protein quality control. Cellular stress and lack of protein quality control lead to aggregation of properly folded proteins in the mitochondrial membrane resulting in a diseased cell

detection and subsequent elimination of damage caused by harmful reactive oxygen species (ROS) as well as stress- and mutation-induced aggregation of damaged and/or denatured proteins.

Studies in the past have unravelled a system of interdependent quality control mechanisms that ensured cell survival, where mitochondrial chaperones and proteases are used as a first line of defence. Molecular chaperones promote proper protein folding and prevent aggregation, while proteases eliminate irreversibly damaged proteins [14].

Heat-shock proteins HSP70 and HSP60, which work efficiently in the matrix compartment of mitochondria, are responsible for the sorting, folding and disaggregation of proteins [15, 16]. Similarly, HSP70- and HSP90-type chaperones that work in the cytosol prevent aggregation and facilitate transport of unfolded newly synthesized or nascent polypeptides into mitochondria [17–19]. Several conserved mitochondrial proteases execute both protective and regulatory functions such as unfolded protein response (UPR) that requires proteolytic inactivation of negative regulators [20]. Among all the mitochondrial quality control factors for proteins, human HtrA2 belongs to a new class of oligomeric serine proteases [14], which are key components of a quality control system in the cell. HtrA2, the most well studied among the HtrA family members, acts as a sentinel at the mitochondrial intermembrane space and maintains homeostasis by degrading unfolded and damaged proteins.

In this chapter, we discuss the structural complexity, protease activity and allosteric property of serine protease HtrA2. Although its involvement in several different cellular processes has been mentioned in this chapter, here we attempt to highlight its role in protein quality control with implications in several neurodegenerative diseases.

15.1.2 The HtrA Family

The high temperature requirement A (HtrA) family of serine proteases belongs to the core set of proteases found in cells and is widely conserved from prokaryotes to humans [21]. They can be easily distinguished from other serine proteases based on their complex oligomeric structures, which comprise a conserved catalytic protease domain along with one or two carboxyterminal PDZ (*p*ostsynaptic density of 95 kDa, *d*isc large and *z*onula occludens 1) domains in each monomeric subunit [22]. The PDZ domains preferably bind to three to four C-terminal residues of target proteins that help in specific protein-protein interactions. The N-terminal domains of HtrAs exhibit significant sequence variability among the family members that include single transmembrane domain (prokaryotic DegS and human HtrA2), signal sequences, insulin-like growth factor-binding domains and serine protease inhibitor domains (human HtrA1, HtrA3 and HtrA4) as illustrated in Fig. 15.2. Another unique signature of this family is that their protease activities can be switched on and off by a distinct mechanism.

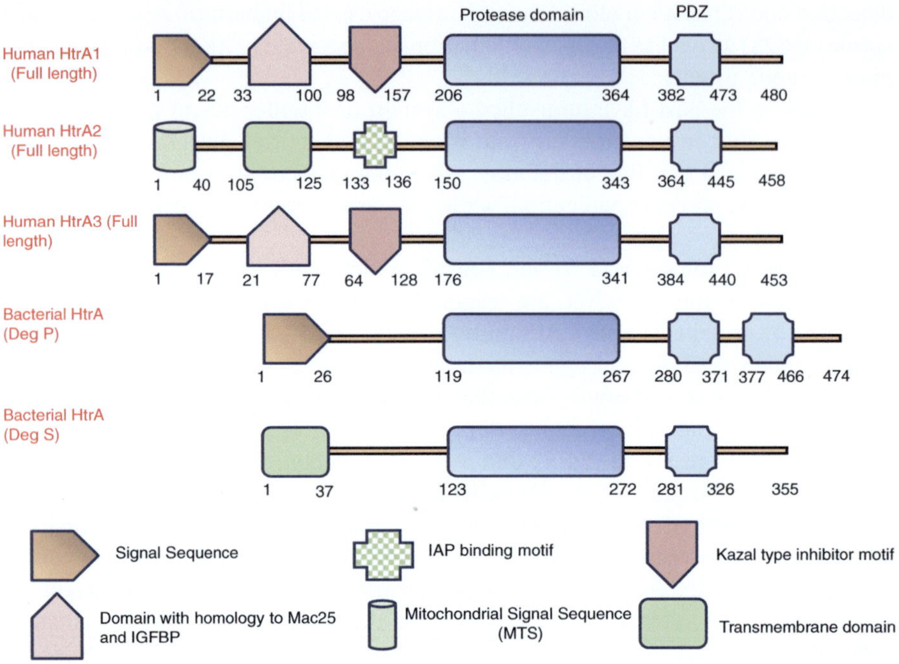

Fig. 15.2 Domain organization of HtrA family members. Domain organization of different HtrA protease family members. The numbers adjacent to each protein represent the size of each protease and their individual domains. The main structural characteristics of HtrA family member proteins are represented as boxes of different colours according to the key at the bottom of the figure

All HtrA family proteins are involved in a protein quality control [22]. One of the family members, DegP that is present in *Escherichia coli* (*E. coli*), has an additional role of a chaperone that stabilizes specific proteins present in the cell [23]. In addition, HtrA proteases can activate or regulate various signalling pathways. This is reflected in the multifaceted cellular functions with which they are associated such as bacterial virulence, maintenance of the photosynthetic apparatus, proliferation, cell migration and cell fate [22, 24, 25]. In mammals, several diseases are associated with loss protease activity of HtrAs that include arthritis, cancer, age-related macular degeneration as well as Parkinson's and Alzheimer's diseases [26–30].

The variety of cellular functions that HtrAs are associated with primarily depends on their cellular localization. Most HtrAs are localized extra-cytoplasmically; for example, while in Gram-negative bacteria they reside in the cell envelope, in case of Gram-positive bacteria, they are found in the extracellular space. Similarly, in eukaryotes, they reside in mitochondria, chloroplasts, nucleus and extracellular matrix.

Bacterial HtrAs are typically involved in various aspects of protein quality control, including the cellular response to protein folding stress and the degradation of misfolded and mislocalized cell envelope proteins [22]. Likewise, eukaryotic HtrAs eliminate damaged photosynthetic proteins in chloroplasts, remove misfolded proteins in diverse cellular organelles and extracellular space as well as regulate the availability of growth factors in mammalian cells [30–33].

Till date, four human members of this family have been identified, and it has been shown that they participate in protein quality control, regulation of cell proliferation and cell migration [34, 35]. They are HtrA1 (L56, PRSS11) [36, 37], HtrA2 (Omi) [38, 39], HtrA3 (PRSP) [40] and HtrA4 [22]. Among all of them, structures of HtrA1 and HtrA2 are well known. Recently, structure of unbound HtrA3 has been solved, however with several missing loops [41]. While HtrA2 is the most well-studied member of the human protease family, lesser information is available on HtrA1 that has been found to be associated with several diseases such as arthritis, cancer and Alzheimer's disease [26–28, 30, 42]. However, very little information on HtrA3 and no information on HtrA4 are currently available [43].

15.2 Mitochondrial Serine Protease HtrA2

Mitochondrion has an endosymbiotic origin; it contains several highly conserved quality control proteases, which are present in its different sub-compartments. For example, the mitochondrial matrix contains two soluble protease systems, Lon and Clp protease [44, 45], while inner mitochondrial membrane accommodates membrane-bound proteases, the i-AAA (*A*TPase *a*ssociated with a variety of cellular *a*ctivities) and m-AAA and rhomboid protease PARL/Pcp1 (PARL, presenilin-associated rhomboid like; Pcp1, *p*rocessed in *c*onjunction with m-AAA *p*rotease) [46–48]. Moreover, HtrA2 and a number of different oligopeptidases reside in the intermembrane space of mitochondria [29, 49, 50] (Fig. 15.3).

HtrA2 has been first identified as a proapoptotic molecule residing in mitochondria that contributes to apoptosis through both caspase-dependent and caspase-independent mechanisms [51–53]. The serine protease HtrA2 is a highly conserved enzyme, whose functions and mechanism of action are very similar to its bacterial counterparts DegP and DegS [35] that are stress-inducible quality control proteases. Similar to the bacterial proteases, HtrA2 is oligomeric in nature which, in its active form, functions as a pyramidal homo-trimer. It is the only mitochondrial protease bearing a PDZ domain reported till date that recognizes exposed hydrophobic stretches of misfolded proteins. Initial studies on HtrA2 have proposed its contribution to apoptosis, which is similar to proapoptotic Reaper family proteins found in *Drosophila melanogaster* [54]. However, the primary function was considered to be protein quality control based on the observation, where a transgenic mouse bearing an HtrA2 mutation led to motor neuron degeneration 2 (mnd2) that exhibited muscle wasting and premature death within 40 days. Interestingly, this mutation where a serine residue was replaced by a cysteine (S276C) led to complete disruption of HtrA2 enzymatic activity. Various studies from different groups implicate same mutation in its human counterpart to be associated with progressive neurodegenerative disorder especially Parkinson's disease [55, 56]. However, further studies are required to confirm its involvement in developing Parkinsonian phenotype.

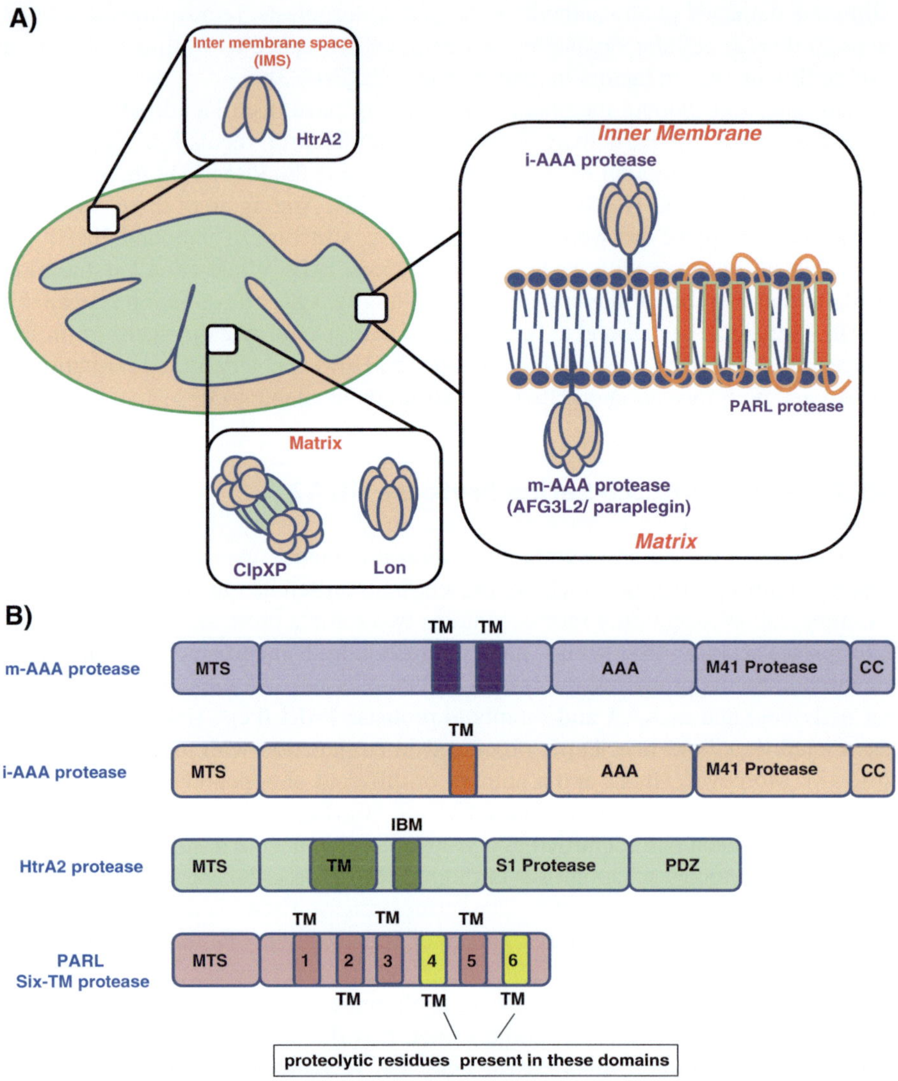

Fig. 15.3 Schematic representation of different mitochondrial quality control proteases. (**a**) Figure depicting localization of quality control proteases in mitochondria. (**b**) Cartoon representing domain organization of these mitochondrial proteases. Description of each one of these proteases has been provided in the text (Abbreviations: MTS, mitochondrial targeting sequence; TM, transmembrane domain; AAA, triple-A domain (ATPase associated with various cellular activities); M41 protease, metal binding proteolytic domain; CC, coiled coil; IBM, inhibitor of apoptosis (IAP)-binding motif; S1 protease, trypsin-like protease domain; PDZ, postsynaptic of 95 kDa, disc large, zonula occludens; IMS; intermembrane space; IM, inner membrane)

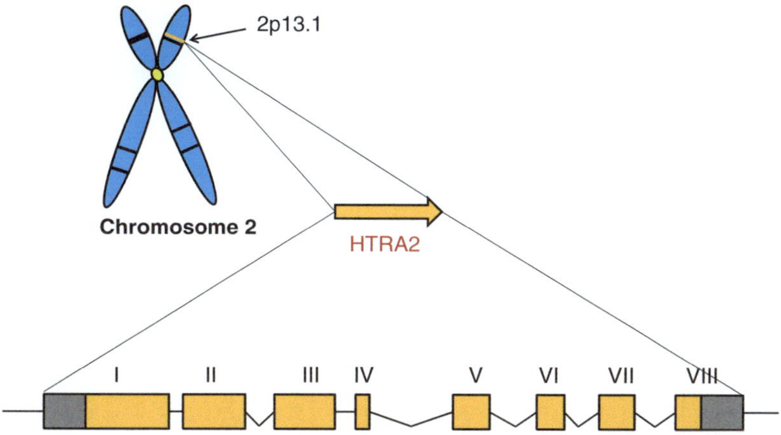

Fig. 15.4 Localization and schematic organization of the HTRA2 gene. Illustration of HTRA2 gene on chromosome 2p13.1. The chromosomal localization has been zoomed below to show the exons present in HTRA2 (*dark yellow*)

15.2.1 Chromosomal Localization and Maturation of HtrA2

The nuclear-encoded HtrA2 gene with eight exons has its chromosomal localization at 2p13.1 (Fig. 15.4). The gene codes for a 49 kDa polypeptide of 458 amino acid residues. After complete translation, the full-length protein is targeted primarily to the intermembrane space (IMS) [57], where it is attached through its N-terminal transmembrane anchor to the inner membrane of mitochondria. During maturation, the first 133 amino acids from the N-terminus get cleaved, and upon apoptotic stimulation, it is released from IMS into the cytosol as a 36 kDa mature protease [51–53, 57] (Fig. 15.5). This cleavage exposes an internal tetrapeptide motif (AVPS) that binds to inhibitors of apoptotic proteins (IAPs) such as XIAP, cIAP, etc. and relieves their inhibition on caspases, thus promoting apoptosis.

15.2.2 Structural Features of HtrA2 Protein

The structure of mature form of human HtrA2 in a substrate-unbound form has been solved by X-ray crystallography at 2.1 Å (Fig. 15.6) [58] that provides a wide overview of the global structural organization of the inactive protease. It has a trimeric pyramidal architecture with the short N-terminal region at the top and PDZ domains residing at the base of the pyramid. Each protease domain comprises *7α-helices* and *19β-strands*, which fold into a compact globular structure. Apart from the β-strands, the protease domain contains several loops, which are named according to the chymotrypsin nomenclature—LA (residues 170–174), L1 (302–306), L2 (323–329), L3 (275–295) and LD (259–266). These regulatory loops harbour active site pocket as well as accommodate the catalytic triad residues (Ser306, His198 and Asp228) in the hydrophobic core of the serine protease domain. Therefore these dynamic

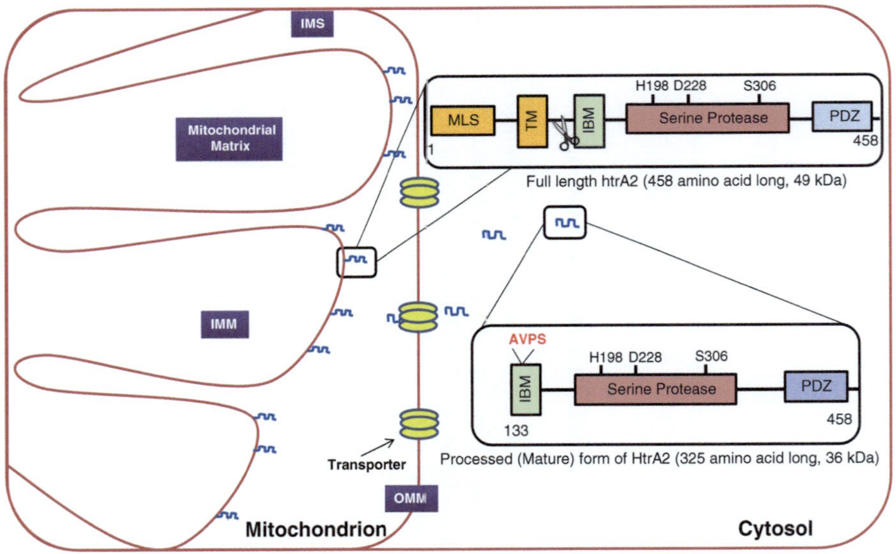

Fig. 15.5 Schematic representation of full-length HtrA2 and its maturation process. Full-length HtrA2 (1–458 aas), comprising a mitochondrial localization signal and a transmembrane domain (1–133 aas), is localized in the mitochondria. Maturation occurs as a result of cellular stress that removes first 133 amino acids and exposes a tetrapeptide IBM motif (AVPS) in its N-terminus. The catalytic triad comprising His 198, D228 and S306 has been shown in the figure. The full-length HtrA2 has been shown as light blue 'coil-like' structure attached to mitochondrial inner membrane. Here, amino acids are referred to as 'aas'. (Abbreviations: MLS, mitochondrial localization sequence; TM, transmembrane domain; IBM, IAP-binding motif; PDZ, postsynaptic of 95 kDa, disc large, zonula occludens)

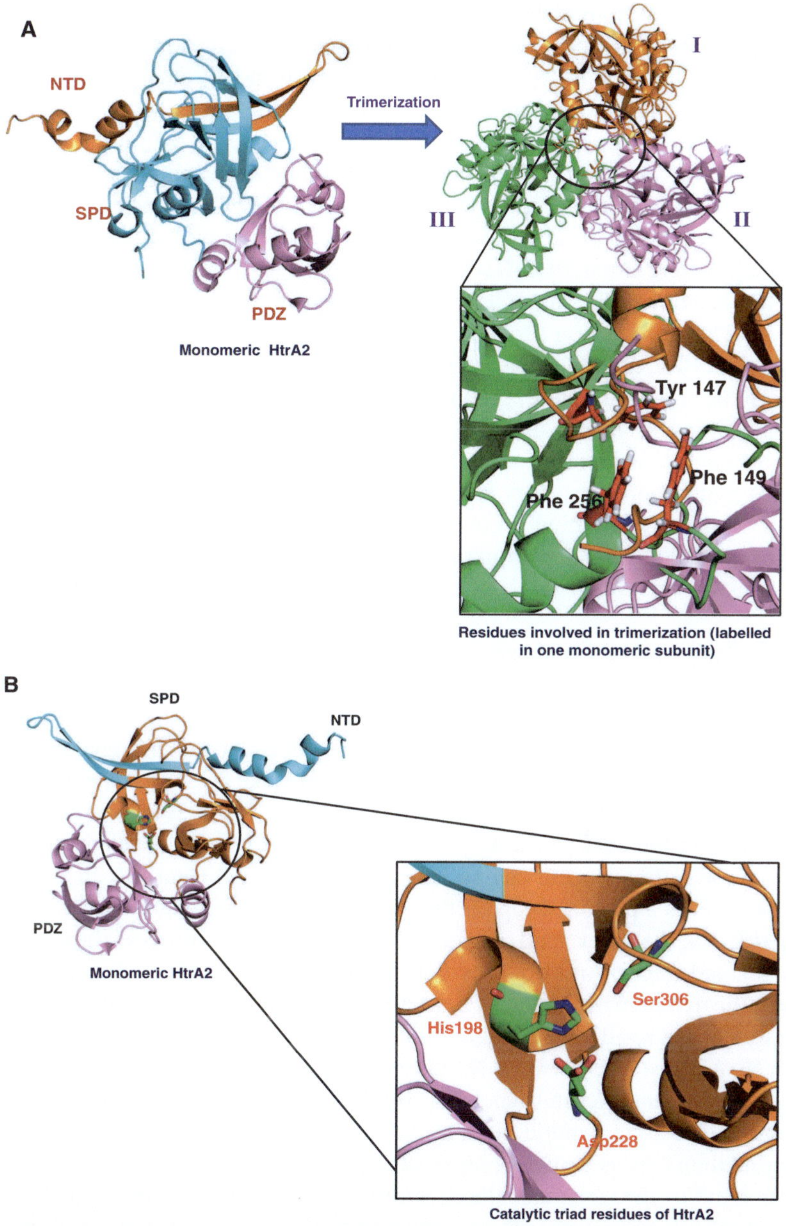

Fig. 15.6 Structure of mature HtrA2. (**a**) Crystal structure of mature inactive (S306A) HtrA2 (PDB: 1LCY) protein with 7α helix and 19β-strands [58]. The trimerization residues have been highlighted at the bottom of the figure. (**b**) Mature form of HtrA2 protein with catalytic triad (His 65, Asp 98, Ser 173). Different domains have been labelled and coloured differently. Images have been created using PyMol, Delano Scientific, USA

structural elements significantly define proteolytic activity, regulation and specificity of HtrA2 through a complex and concerted allosteric mechanism [35]. The protease has been found to be functionally active in its trimeric form, which is mediated primarily through its N-terminal region involving aromatic residues Tyr147, Phe149 and Phe256. The core serine protease domains that reside 25 Å above the base of the pyramid are surrounded by C-terminal PDZ domains on all three sides (Fig. 15.6). PDZ, the protein-protein interaction modules, is known to recognize and bind to specific hydrophobic residues usually in the C-termini of binding partners. The canonical binding site in PDZ (G-Φ-G-Φ motif, where Φ denotes hydrophobic residues) has been modified into YIGV in HtrA2, where the third residue (G) is invariant [59]. This recognition sequence is deeply embedded in the intimate interface between the PDZ and the protease domains. However, this linker region is mostly absent in the crystal structure, probably due to its high dynamic behaviour [58]. The PDZ domain packs against the protease domain through van der Waals contacts, and the hydrophobic residues on strands β11 and β12 of the protease domain interact with the hydrophobic residues from strand β14 and helix α5 of the PDZ domain. Therefore, the PDZ binding groove remains unavailable for interaction with other proteins in this 'closed' conformation.

15.2.3 Active Site Conformation

15.2.3.1 Catalytic Triad

A catalytic triad is a combination of three amino acids that are arranged in a specific conformation in three-dimensional spaces so as to make the active site environment conducive for substrate catalysis. In serine proteases, residues serine, histidine and aspartate form the catalytic triad. A series of concerted non-covalent interactions among these residues occur during substrate binding and catalysis. In general, the histidine acts as a proton acceptor, thus increasing the nucleophilicity of the active site serine. On the other hand, aspartate moiety aids in this process through several hydrogen and electrostatic bonds with the histidine residue. These observations hint towards the dynamic behaviour of the active site where rearrangement of side chains of catalytic triad residues is essential for substrate binding and subsequent hydrolysis. In HtrA2, distance between N^{ε} atom of His and O^{γ} atom of Ser for each molecule in the asymmetric unit has been found to increase from 4.1 Å to 5.5 Å with heptameric substrate peptide binding [60]. The reactive hydroxyl (−OH) group of the serine then acts as a nucleophile that attacks the carbonyl carbon of the scissile peptide bond of the substrate subsequently leading to its cleavage [61]. The active site pocket harbours several other residues that aid in the process of substrate recognition and binding. Their physicochemical properties and stereochemical arrangements are critical towards determining substrate specificity and affinity. However, further research is required to identify more natural substrates so as to unambiguously define HtrA2 substrate specificity.

15.2.3.2 Oxyanion Hole

Oxyanion hole that comprises backbone amides or positively charged residues neutralizes the negative charge on the tetrahedral transition state intermediate, thus promoting catalysis by reducing energy of activation [61]. Therefore, proper orientation of the oxyanion hole is extremely important for substrate catalysis. Activation of HtrA2 is tightly controlled at different levels in the cell. Its oligomeric structure and complex allosteric propagation provide a unique mechanism for proteolytic activation. Binding of proteins at the YIGV groove has been shown to relay a conformational change at the PDZ-protease interface, leading to flipping of a phenylalanine (Phe 303) towards histidine (His 198) of the catalytic traid, which is essential for proper oxyanion hole formation [62, 63].

15.2.4 Activation Mechanism of HtrA2

HtrA2 with a large trimeric structure (~110 kDa) undergoes a complex allosteric mechanism of activation [63]. Allostery, which is often a signature of multidomain proteins, helps enzymes to attain an active functional conformation through ligand binding at a site distal to the catalytic pocket.

Allosteric regulation occurs by receiving an allosteric signal at a distal site of a protein (e.g. binding of a modulator molecule). While binding of an allosteric activator causes shift in a protein's structure from inactive (or tensed 'T') state to an active (or relaxed 'R') state, the effect of an allosteric inhibitor is just the reverse. According to the classical theory of allostery, the two conformational states, 'T' and 'R', are always in a dynamic equilibrium [64].

The working model of HtrA2 activation was first proposed by Shi and co-workers way back in 2002. According to their model, 'YIGV' groove of PDZ when it interacts with substrate/modulator induces a significant conformational change at the PDZ-protease interface which unmasks the inhibitory effect of PDZ from the active site. However, this model fails to explain a few vital observations such as why the protease is active only in its trimeric form as well as what is the exact role of PDZ in enzyme catalysis and allostery. Recently, Bose and co-workers revisited the model and put forward their hypothesis on HtrA2 activation that highlights a complex allosteric mechanism involving a series of conformational changes leading to ligand binding and subsequent substrate cleavage [62, 63]. The new model emphasizes the role of N-terminal region in protease stabilization and intermolecular PDZ-protease movement in proper active site and oxyanion hole formation that subsequently lead to efficient substrate catalysis. The requirement of intermolecular PDZ-protease interaction demonstrates the necessity of trimeric mature HtrA2 in allosteric propagation and hence activation [60, 62]. HtrA2 activation is schematically presented in (Fig. 15.7).

Till date, a wide repertoire of proteins binding to the C-terminal PDZ domain has been found to stimulate the protease activity. Gupta et al. demonstrated that a peptide corresponding to the cytoplasmic C-terminal tail of presinilin-1 increased the proteolytic activity of HtrA2 towards generic serine protease substrate β-casein

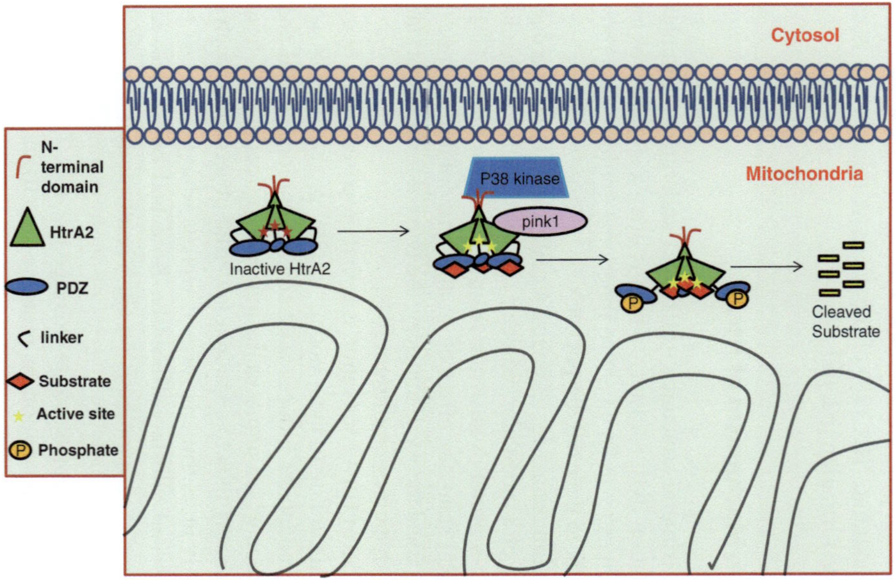

Fig. 15.7 Model for HtrA2 activation during stress. Binding of N-/C-terminal substrates result in intricate intermolecular PDZ-protease interactions and subtle conformational rearrangements in the regulatory loops of HtrA2 protease. This, along with phosphorylation by p38 and PINK1 kinases, leads to formation of the active enzyme

[65]. Furthermore, it has been shown that binding of certain peptides to the PDZ domain leads to enhanced HtrA2 activity. Ligand specificity at PDZ domain was determined to characterize its binding properties using peptide libraries fused to the C- or N-terminus of a phage coat protein. Series of peptides binding to the isolated PDZ domain were selected, and it was found that the peptide GQYYFV (termed PDZ_{opt}), which binds efficiently to PDZ, was able to stimulate HtrA2 activity that was quantitatively determined using synthetic substrate peptides [66]. Thus, engagement of binding partners with the PDZ domains results in opening up of access to the catalytic site.

However, recent studies suggest a dual regulatory switch in HtrA2, since apart from the classical mode of allosteric propagation, activation through N-terminal 'AVPS' tetrapeptide binding has also been observed, which adds complexity to its overall mode of action [60]. However, further studies are needed to follow the complex allosteric pathway at the molecular level and delineate the biological significance of this tight regulation on HtrA2 activity.

15.3 HtrA2 in Mitochondrial Protein Quality Control

HtrA2 in the mitochondria mainly functions as an ATP-independent serine protease. It is believed that the primary function of HtrA2 is the maintenance of mitochondrial homeostasis. Under normal physiological conditions, it acts as a quality control factor and promotes cell survival. Perturbations in its proteolytic activity lead to the accumulation of unfolded proteins in mitochondria, dysfunction of the mitochondrial respiration and generation of reactive oxygen species that result in overall loss of mitochondrial competency [33, 55, 56, 67]. Several studies such as the loss-of-function missense S276C mutation of HtrA2 in transgenic mice have led to motor neuron degeneration 2 or mnd2. Moreover, knockout mice carrying a homologous deletion of the HtrA2 gene exhibit phenotypes with features typical for the Parkinsonian syndrome. Both these studies showed accumulation of unfolded proteins in the mitochondria [33]. Ex vivo studies with both the mnd2 and HTRA2−/− cell lines exhibited an increased number of atypical mitochondria. In addition, they have been found to be more prone to death triggered by agents inducing intrinsic pathway of apoptosis (e.g. etoposide) as well as the ones affecting mitochondrial functions (e.g. rotenone) [55, 56]. Furthermore, two mutations (A141 and G399) in protease domain of HtrA2 in humans have been identified in patients with Parkinson's disease. Both mutations cause reduction in the proteolytic activity as well as influence morphology and function of mitochondria [68]. These observations indicate that the mitochondrial HtrA2 might act as a sentinel, which regulates the levels of misfolded proteins in the organelle in a manner very similar to its bacterial counterparts, DegP and DegS [35, 69].

15.4 Expression of HtrA2 in Brain Tissue

Normal brain tissue is represented by four different regions: *cerebellum*, *cerebral cortex*, *hippocampus* and *lateral ventricle* wall. Anatomically and histologically, the cerebral cortex is further subdivided into the outermost grey matter, the overlying white matter and the innermost deep grey matter components. The hippocampus, which is an important neuron-rich area in the brain, is closely associated with the cerebral cortex and is located in the temporal lobe. The cerebral cortex comprises neurons (nerve cells) and glial cells (supportive cells), whereas the white matter is made up of primarily glial cells and myelinated axons from neurons. The tissue distribution of HtrA2 protein varies from a ubiquitous to a highly specific expression. Northern blot analysis studies demonstrated that the expression of HtrA2 is highest in the cerebral cortex, while its expression in other parts of the brain is insignificant [39].

Defects in organelles such as mitochondria and the endoplasmic reticulum directly activate stress responses. It has been very much clear that mitochondrial dysfunction caused by loss of the serine protease HtrA2 results into a progressive movement disorder in mice and has been linked to Parkinsonian neurodegeneration in humans. Moisoi et al. found that loss of HtrA2 results in transcriptional upregulation of some of the important nuclear stress-response genes, including the transcription factor CHOP (C/EBP homologous protein) selectively in the cerebral cortex of the brain [33]. In another study, UCF-101, a novel small molecule HtrA2 inhibitor, has been successfully used to protect against cerebral ischemia/reperfusion injury in mice. The study clearly demonstrated that treatment with UCF-101 significantly reduced cerebral infarct size by about 16% with a concomitant improvement in neurological behaviour. UCF-101 has also been helpful in reducing apoptotic cell death in cerebral cortex [70].

Although HtrA2 has not been extensively studied in neurological disorders till date, these important observations would certainly lead towards devising therapeutic strategies to modulate HtrA2 functions with desired characteristics.

15.5 HtrA2 and Its Association with Neurological Disorders

Perturbation in the dynamic functions of HtrA2 protein, which include cell death and cellular protein quality control, leads to distinctive defects in neurons and is recognized as a key player in neurodegeneration. Two decades earlier, Gray and his collaborators demonstrated the interaction between HtrA2 and presenilin-1 using a yeast two-hybrid system, thus establishing its link with neurodegeneration [39]. Presenilin-1 is a catalytic component of γ-secretase enzyme that is implicated in the inherited forms of early-onset Alzheimer's disease. This interaction was later on validated in vitro along with characterization of presenilin-mediated regulation of HtrA2 protease activity [65]. Park et al. where in an in vitro system, demonstrated interaction between HtrA2 and AD-associated amyloid β [71]. Similarly, in a breakthrough in vivo study done by Jones et al. [55], a homozygous loss-of-function mutation (S276C) in the HtrA2 gene was identified in mnd2 mice leading to neurodegeneration, muscle wasting and death by 40 days of age. In another study, Martins and co-workers generated HtrA2 knockout mice to emphasize the physiological relevance of its protease activity. It was a clear observation that HtrA2 knockout mice displayed neurodegenerative phenotype with Parkinsonian features [56]. Taken together, the similarity with the activities of bacterial HtrA2 homologues and the fact that the loss of proteolytic activity in both mnd2 and HtrA2 knockout mice result in enhanced sensitivity to stress, it can be said that this protease might play an essential role in the mitochondria-related stress signals.

In the following sections, we summarize the available information pertaining to the involvement of HtrA2 in the onset and progression of neurodegenerative diseases, namely, Alzheimer's disease (AD), Parkinson's disease (PD) and Huntington's disease (HD).

15.5.1 Alzheimer's Disease

Alzheimer's disease (AD) is one of the major health problems in the world. It is a progressive and irreversible neurodegenerative disorder that is biochemically characterized by the occurrence of various pathological features such as formation of neurofibrillary tangles within neurons, progressive neuronal loss and accumulation of amyloid (Aβ) in the walls of the blood vessels and senile plaques. This represents the most common cause of dementia worldwide [72]. Mutations either in the precursor protein for Aβ (the β-amyloid precursor protein, APP) or in presenilin-1 (PS-1) or presenilin-2 (PS-2) are the exclusive reasons for causing familial early-onset forms of AD.

Amyloid beta (Aβ or Abeta) refers to a peptide comprising residues 36–43 (~4 kDa) of amyloid precursor protein (APP) that acts as the main component of the amyloid plaques, primarily found in the brains of Alzheimer's patients. APP protein is found in many tissues and organs, including the brain and spinal cord (central nervous system), with mostly unknown functions. Researchers hypothesize that it may be associated with other proteins on the surface of cells or help cells attach to one another. Recent studies suggest that it helps direct the movement of nerve cells during early brain development.

According to the amyloid cascade hypothesis, APP is cleaved by successive actions ofα, β- and γ-secretases, a family of proteolytic enzymes that process several transmembrane proteins including APP at the transmembrane region. While cleavage by α-secretase does not lead to Aβ formation, β- and γ-secretases cleave at N- and C-termini of APP, respectively, and are key players in Aβ deposition. γ-secretase attacks the C-terminal end of the Aβ peptide, cleaves within the transmembrane region of APP and generates a number of isoforms of 39–43 amino acid residues in length (Fig. 15.8) [73].

The most common isoforms are Aβ40 and Aβ42. The Aβ40 form is the more common of the two, but Aβ42 is the more fibrillogenic and toxic in nature due to its high aggregation property and is thus associated with disease conditions. Mutations

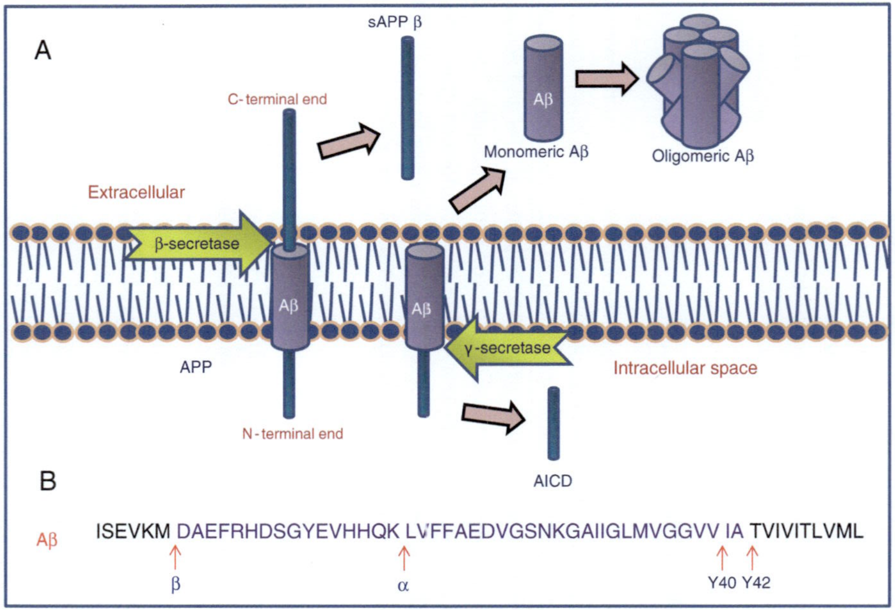

Fig. 15.8 **Molecular mechanism of Aβ peptide formation.** (**a**) Processing of the membrane-bound amyloid precursor protein (APP) by β- and γ-secretases leads to the formation of Aβ peptides. The cleavage products are finally secreted outside the cell. Cleavage by γ-secretase results in either Aβ 40 (soluble non-toxic) or Aβ 42 (fibrillar, toxic) peptides, the latter being responsible for production of amyloid plaques. (**b**) Amyloid beta sequence with α, β and γ-secretase cleavage sites (abbreviations: sAPPβ, soluble β-secretase producing APP; AICD, amyloid intracellular domain)

in APP associated with early-onset Alzheimer's have been noted to increase the relative production of Aβ42, and thus one suggested outcome of Alzheimer's therapy involves regulating the activity of β- and γ- secretases to produce mainly Aβ40 [74].

15.5.1.1 Formation of Amyloid Plaques

Considerable evidence has accumulated over the last 10 to 15 years that oligomers play a central role in AD pathogenesis. Accumulation of extracellular amyloid plaques in the brain is an essential feature of Alzheimer's disease (AD). These plaques principally consist of insoluble mass of oligomeric amyloid β-peptide (Aβ42), which comes from the sequential proteolytic processing of the amyloid precursor protein (APP) during its transport from the endoplasmic reticulum (ER) and Golgi to the plasma membrane. Aβ peptides spontaneously aggregate into soluble oligomers and mix to form insoluble fibrils specifically more in beta-sheet conformation and are finally deposited in diffused senile plaques [75]. Several studies have shown that oligomers are toxic components both in vitro [76] and in vivo [77] and that learning and memory deficits caused by oligomers in transgenic mouse models can be reduced when oligomer levels are decreased by accelerating fibril formation. So, it clearly seems at first sight that AD is linked up with an increased production and secretion of the Aβ42 to the extracellular space [78].

According to the previous reports, it has been observed that there is an accumulation of Aβ peptide within neurons and mitochondria from AD brains [79–81]. In addition to that, neurons from AD patient brain were found to contain abundance of mitochondrially targeted APP that interrupt the basic functions of mitochondria [82–84] as well as impair energy metabolism [82]. Based on these findings, it has been suggested that intracellular Aβ peptide accumulation and mitochondrial dysfunction play a central role in the pathogenesis of AD [12]. Therefore, turnover and degradation of APP and the Aβ peptide in the mitochondrial compartment appear to be important for neuronal survival. In neurons, proteolytic processing of APP constantly generates Aβ, and this intracellular Aβ is prone to form oligomers [85–88]. Since oligomeric Aβ is known to be the most neurotoxic form of the peptide, neurons need to adapt a mechanism for their detoxification under normal conditions [89–92].

15.5.1.2 HtrA2 and Its Association with Aβ Peptide

In the last few years, several studies reported some proteases that are involved in the process of mitochondrial quality control, which include presequence peptidase (PreP) [13] and serine protease HtrA2. Based on the structural similarities of HtrA2 with its bacterial homologue DegP, it is tempting to speculate that besides its proteolytic activity, HtrA2 may also have a chaperoning function in the intermembrane space of mitochondria. This property of HtrA2 assists in protein folding or in preventing amyloidogenic peptide aggregation. As mentioned earlier, HtrA2 interacts with C-terminal region of PS-1, which is a catalytic subunit of γ-secretase found to be involved in the processing of APP. Mutations in the gene encoding PS-1 selectively enhance the levels of highly amyloidogenic peptide Aβ42 and cause an increased death of neural cells by apoptosis and necrosis. This interaction has been hypothesized to regulate HtrA2 activity and determines its release from the mitochondria during apoptosis [39] [65]. Moreover, it has been demonstrated that HtrA2 selectively interacts with and disaggregates more neurotoxic oligomeric Aβ42 rather than its less toxic monomeric (Aβ42) form. Therefore, this interaction not only protects neurons from the neurotoxic Aβ accumulation but also aids in the decrease in proapoptotic activity of HtrA2, thus preventing death of neural cells. This phenomenon has been aptly termed as 'mutual detoxification' [93].

Thus, HtrA2 aids in reducing the toxic effects of oligomeric Aβ, which makes it a potential therapeutic target in neurodegenerative diseases.

15.5.1.3 Regulation of Aβ Levels by Proteolytic Degradation

In transgenic (Tg) worms, *C. elegans*, the heat-shock proteins (HSPs), prevents cellular stress through their interactions with intracellular Aβ, with subsequent attenuation in Aβ-induced toxicity. Presence of quality control proteins in mammalian mitochondria therefore suggests a similar mechanism, which might play an essential role in the neuronal detoxification of intracellular Aβ. In neurons, ER- and mitochondria-localized HtrA2 is a mammalian version of HSP that has been implicated in binding and detoxification of toxic Aβ for the following reasons. First, HtrA2 is a neuroprotective homologue of the bacterial survival factor HtrA, the major HSP that protects bacteria from heat stress [56, 96]. Second, HtrA2 expression is upregulated during neurogenesis, neuronal maturation as well as mouse brain development. This increase in expression has been found to be similar to the inducible HSPs, which protect Tg worms against intracellular Aβ accumulation, thus probably representing a protective response against accumulation of toxic metabolites [95]. Finally, similar to HSPs in Tg worms [94], HtrA2 protease does not directly degrade Aβ42 but clearly reduces the toxicity of Aβ42 in neuronal cells by disaggregating toxic oligomeric Aβ. As the major function of bacterial HSPs is to refold rather than to proteolytically degrade denatured proteins, it is therefore not surprising that HtrA2 reduces the neurotoxicity in a similar manner [93].

15.5.2 Parkinson's Disease

Parkinson's disease (PD) is the most common movement disorder and the second most common neurodegenerative disease after Alzheimer's that affects 1% of the population over the age of 65. It is pathologically characterized by an exclusive degeneration of dopamine releasing neurons of the *substantia nigra pars compacta* in the brain and the presence of characteristic proteinaceous intracytoplasmic inclusions, known as Lewy bodies in the affected brain areas [97]. Several reports proposed that environmental factors, genetic sensitivity and ageing are important components, which lead to the progression of this disorder [98–100].

Basically, the whole understanding of the molecular events in PD pathogenesis has been greatly advanced by the identification and analysis of PD-associated genes [101] that provided insights into the cellular mechanisms underlying PD

Table 15.1 Parkinson's disease-associated genes and their role in PD

Symbol	Gene locus	Name of gene	Role in PD
Park 1	4q21–22	SNCA	Confirmed
Park 2	6q25.2–q27	Parkin	Confirmed
Park 3	2p13	Unknown	Unconfirmed
Park 4	4q21–q23	SNCA triplication	----
Park 5	4p13	UCLH1	Unconfirmed but possible
Park 6	1p35–p36	Pink1	Confirmed
Park 7	1p36	DJ1	Confirmed
Park 8	12q12	LRRK2/dardarin	Confirmed
Park 9	1p36	ATP13A2	Confirmed
Park 10	1p32	Unknown	Confirmed
Park 11	2q36–27	Unknown	Unconfirmed
Park 12	Xq21–q25	Unknown	Confirmed
Park 13	2p12	HtrA2	Confirmed
Park 14	22q14.1	PLA2G6	Confirmed
Park 15	22q12–q13	FBX07	Confirmed
Park 16	1q32	Unknown	Confirmed
Park 17	16q11.2	VPS35	Confirmed
Park 18	3q27.1	EIF4G1	Unconfirmed

pathogenesis [102, 103]. Over the past 20 years, mutations in several genes have been definitively shown to mediate familial PD. Till date, 18 nuclear PD-related genes (*PARK*) and some of their mutants have been implicated in the pathogenesis of PD [104–111]. The *PARK* genes and their loci have been elaborated in Table 15.1. Likewise, mutations in SNCA that encodes *α-synuclein* protein (*PARK1)* [106] (a major component of Lewy bodies), *PARK4* [105] and *leucine-rich repeat kinase 2 (LRRK2)/dardarin* (*PARK8*) [108, 109] have been found to be associated with autosomal-dominant form of the disease. However, mutations in *PARKIN* (*PARK2*) [101, 104], *DJ-1* (*PARK7)* [107] and *PTEN-induced kinase 1* (*PINK1*, *PARK6*) [112] that are mitochondrially associated proteins are mainly involved in autosomal recessive forms of PD. Mutations in *ATP13A2* (*PARK9*), which encodes a lysosomal ATPase, have also been found in an atypical, autosomal recessive Parkinsonism [111]. More recently, mutations in two other PARK genes *VPS35* (*v*acuolar *p*rotein *s*orting-associated protein 35) *(PARK17)* [113, 114] and *EIF4G1* (*e*ukaryotic translation *i*nitiation *f*actor 4 *g*amma 1) *(PARK18)* [115] have been reported to cause autosomal dominant form of PD.

HTRA2 *(PARK13)* [68] has also been implicated in autosomal recessive PD post identification and characterization of two mutations that interfere with its protease activity [116]. The first connection between HtrA2 dysfunction and PD came from the characterization of mnd2 mutant mice as described in Sect. 3. The mnd2 mutation, leading to neurodegeneration, muscle wasting, involution of the spleen and thymus and death by 40 days of age, was identified in the HTRA2 gene [55]. Moreover, a neurodegenerative phenotype with Parkinsonian features has been described in the HTRA2 knockout mice [56]. A loss-of-HTRA2 study on the mouse model showed accumulation of unfolded proteins in mitochondria, a defective mitochondrial respiration and an enhanced production of reactive oxygen species in the brain tissue cells [33]. Implication of HtrA2 in Parkinson's has directed a lot of application-based research in that area. Several studies suggest that defects in mitochondrial respiratory chain, impaired mitochondrial dynamics and mitochondrial trafficking play a significant role in the mitochondrial dysfunction that takes place in neurodegenerative disorders. Thus, human HtrA2, which enhances mitochondrial bioenergetics, is an attractive potential therapeutics for betterment of mitochondrial dysfunction in Parkinson's disorder.

15.5.2.1 HtrA2 Variants and PD Pathogenesis

Most PD cases are due to sporadic mutations (i.e. they are of unknown cause). However, 15–20% of PD patients have a family history of the disease, suggesting that there is a strong genetic basis for development of Parkinson's in this subgroup. Since the molecular pathogenesis of sporadic PD and the basis for selective dopaminergic neuron loss remain unidentified, it is unclear whether gene mutations are involved in the development of this disease in sporadic PD patients.

PD occurs due to two important reasons: sporadic mutations (i.e. they are without known cause) and inherited (familial) mutations. According to the survey done on US population, 15–20% of PD patients carry inherited form of the disease, suggesting that how Parkinson's disease is associated to the genetic. As of the molecular pathogenesis of sporadic PD and unidentified basis for selective loss of dopamine-releasing neurons, it is still with a no known reason which shows association between genetic mutations and sporadic PD patients that leads to the development of disease.

Several single nucleotide polymorphisms (SNPs) of the HTRA2 gene have been identified, and their relevance in PD has been studied. Unlike mouse mnd2 mutation [55], genetic variation analyses in human have provided conflicting results regarding the involvement of the HTRA2 gene in PD. A mutation screen of the HTRA2 gene using candidate gene approach that has been performed in German PD patients resulted in the identification of a novel heterozygous G399S (rs72470545) mutation. Another interesting A141S (rs72470544) polymorphism has also been linked with PD (Fig. 15.9) [68]. Immunohistochemistry revealed that both these mutations induced mitochondrial dysfunction associated with the altered mitochondrial morphology. Moreover, studies on genetic variability of HtrA2 in Belgian population reconfirmed its association with PD, where mutational analysis of patient samples identified a new mutation (R404W) [117]. Similarly, from a study with a large group in Taiwan, one variant, c.427C > G (i.e. Pro143Ala) in exon 1, was found in early-onset PD patients. According to reports, the Pro143Ala variant is associated with a greater rate of mitochondrial dysfunction, mitochondrial morphology disruption and apoptosis under conditions of increased oxidative stress [118].

However, contrary to these observations, studies with North American PD population did not succeed in identifying HtrA2 as a PD risk gene [119]. Thus, genetic makeup and variability in different parts of the globe might determine the role of

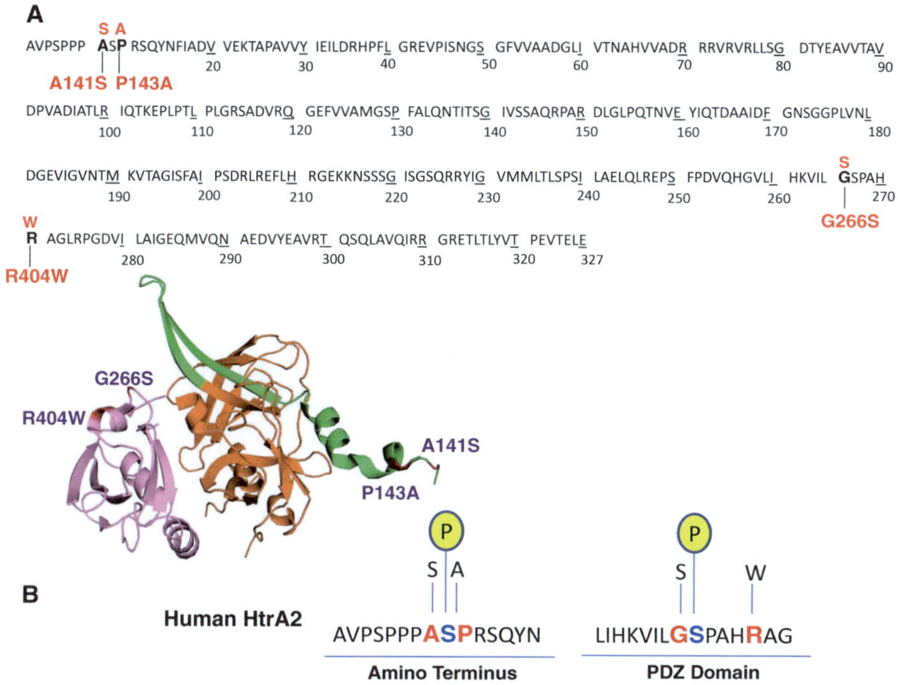

Fig. 15.9 Pathogenic mutations and putative phosphorylation sites in HtrA2. (**a**) Locations of PD-linked mutations A141S, G399S and R404W indicated in 'red' on the wild-type sequence as well as on the crystal structure of HtrA2 (PDB: 1LCY). A141, G399 and R404 are surface exposed residues. (**b**) The relative positions of phosphorylation sites in human HtrA2 are indicated by circles above the serine residues

HtrA2 in PD pathogenesis. Further studies in this area are required to unambiguously delineate the predisposition of HtrA2 mutations in PD pathogenesis.

Interestingly, the Scansite algorithm [120] that performs proteome-wide prediction of cell signalling interactions and post-translational modifications identified Ser142 and Ser400, in HtrA2 as putative phosphorylation sites for proline-directed serine/threonine kinases. The importance of this lies in the fact that both these residues reside in close proximity to the mutations obtained in PD populations. Out of the two sites, Ser400 is present in PDZ, a region well recognized as a protein-protein interaction domain [66].

The other HtrA2 variants A141S [68] and P143A [118] found in PD patients are close to the phosphorylation site S142, whereas G399S [68] and R404W [117] are in the proximity of the S400 phosphorylation position (Fig. 15.9). These mutations that are found in PD patients might be involved in blocking the important activity, i.e. phosphorylation on those residues, and therefore have a detrimental role in the activation of enzymes involved in important signalling pathways.

15.5.2.2 HtrA2 and PINK1

Recent experimental evidence supports interaction between PINK1 and HtrA2 in the mitochondria. It has been observed from an in vivo study that PINK1-dependent phosphorylation of HtrA2 increases its protease activity leading to increased survival against oxidative stress [121]. Study done on *Drosophila melanogaster* has also demonstrated strong interaction between PINK1 and HtrA2 proteins, implicating a common pro-survival pathway [122, 123]. Interestingly, both PINK1 and HtrA2 appear to be important regulators of mitochondrial protein quality control. While HtrA2 knockout mice showed an effective increase in ROS levels and an accumulation of misfolded proteins in brain mitochondria, post-mortem of brain tissue from PD patients with mutations in PINK1 revealed an increased level of misfolded mitochondrial respiratory complexes in the brain [124]. Interestingly, HtrA2 has been demonstrated to rescue PINK1 functions in *Drosophila* system but not vice versa [123, 125]. This observation reiterates the role of HtrA2 in maintenance of mitochondrial integrity.

15.5.3 Huntington's Disease (HD)

Huntington's disease is one of the neurodegenerative disorders manifested by unwanted choreatic movements, behavioural and psychiatric disturbances as well as dementia. HD is caused by the abnormal repetition of a triplet CAG (glutamine) repeat in exon-1 of the HD gene, resulting in elongated polyglutamine stretches in the ubiquitously expressed protein product known as mutant huntingtin (Htt) [126]. This disorder is characterized by selective degeneration of medium-sized spiny neurons in the striatum of brains in HD patients along with selective neuronal loss in striatum. Although the exact mechanism of toxicity development by this mutant Htt remains subtle, several processes including transcriptional dysregulation, abnormalities in mitochondrial energy metabolism, protein aggregation and oxidative damage [127, 128] might be involved.

15.5.3.1 HtrA2 and Its Implications in Huntington's Disease

The mutated gene that is responsible for causing HD has been identified more than two decades earlier [129]. Although the effects of this mutant Htt have been studied extensively, the mechanisms by which Htt causes neurodegeneration have not yet been fully determined. However, transcriptional deregulation and mitochondrial dysfunction [130] that contribute to the pathogenesis of HD have been implicated to be the two probable candidates for the same [131–133].

Research by Inagaki and his collaborators on rat primary neurons revealed a connection between the neuronal death and a selective downregulation of HtrA2 gene by mutant Htt in striatal neurons; this pattern was reflected at the protein level as well. These findings suggest a link between HtrA2 selective downregulation and striatal neuron-specific pathology in HD [134].

15.6 Concluding Remarks and Future Perspective

Recent progress in the structural and functional characterization of HtrA proteins from prokaryotic and eukaryotic species has greatly enhanced our understanding of this fascinating protein family. Human HtrA2 plays critical roles in protein quality control within mitochondrial intermembrane space. It eliminates damaged cellular components by protecting the cell from accumulation of toxic protein aggregates that are responsible for neurodegeneration in Alzheimer's and Parkinson's disease. Moreover, proapoptotic mammalian HtrA2 is implicated as a tumour repressor in certain cancers due to its ability to induce apoptosis both in caspase-dependent and caspase-independent manner. These unique features along with its ability to multitask make HtrA2 an important therapeutic target both in cancer and neurodegeneration. Synthetic HtrA2 inhibitor, UCF-101 [135, 136], has been found to reduce its catalytic activity and hence apoptosis, which might have important applications in several diseases including neurodegeneration. Further, understanding the molecular mechanism of processing of APP by HtrA2, dissection of its interaction with presenilin-1 and the inverse correlation between HtrA2 downregulation and Htt mutations might shed light into its way of involvement in PD, AD as well as HD. In addition, delineating the pathophysiological significance of newly identified phosphorylated residues in PD, and identification of HtrA2 modulators and substrates, would certainly provide an opportunity to challenge a variety of mitochondrial diseases associated with neurodegeneration.

References

1. Nunnari J, Suomalainen A (2012) Mitochondria: in sickness and in health. Cell 148(6):1145–1159
2. Vafai SB, Mootha VK (2012) Mitochondrial disorders as windows into an ancient organelle. Nature 491(7424):374–383
3. Ballinger SW (2005) Mitochondrial dysfunction in cardiovascular disease. Free Radic Biol Med 38(10):1278–1295
4. Brandon M, Baldi P, Wallace DC (2006) Mitochondrial mutations in cancer. Oncogene 25(34):4647–4662
5. DiMauro S, Garone C, Naini A (2010) Metabolic myopathies. Curr Rheumatol Rep 12(5):386–393
6. Karbowski M, Neutzner A (2012) Neurodegeneration as a consequence of failed mitochondrial maintenance. Acta Neuropathol 123(2):157–171
7. Koppenol WH, Bounds PL, Dang CV (2011) Otto Warburg's contributions to current concepts of cancer metabolism. Nat Rev Cancer 11(5):325–337
8. Lin MT, Beal MF (2006) Mitochondrial dysfunction and oxidative stress in neurodegenerative diseases. Nature 443(7113):787–795
9. Mootha VK et al (2003) PGC-1alpha-responsive genes involved in oxidative phosphorylation are coordinately downregulated in human diabetes. Nat Genet 34(3):267–273
10. Pedersen BL, Baekgaard N, Quistorff B (2009) Muscle mitochondrial function in patients with type 2 diabetes mellitus and peripheral arterial disease: implications in vascular surgery. Eur J Vasc Endovasc Surg 38(3):356–364

11. Rugarli EI, Langer T (2012) Mitochondrial quality control: a matter of life and death for neurons. EMBO J 31(6):1336–1349
12. Wallace DC (2005) A mitochondrial paradigm of metabolic and degenerative diseases, aging, and cancer: a dawn for evolutionary medicine. Annu Rev Genet 39:359–407
13. Baker BM, Haynes CM (2011) Mitochondrial protein quality control during biogenesis and aging. Trends Biochem Sci 36(5):254–261
14. Day CL, Hinds MG (2002) HtrA--a renaissance protein. Structure 10(6):737–739
15. Neupert W, Herrmann JM (2007) Translocation of proteins into mitochondria. Annu Rev Biochem 76:723–749
16. Voos W (2013) Chaperone-protease networks in mitochondrial protein homeostasis. Biochim Biophys Acta 1833(2):388–399
17. Fan AC, Bhangoo MK, Young JC (2006) Hsp90 functions in the targeting and outer membrane translocation steps of Tom70-mediated mitochondrial import. J Biol Chem 281(44):33313–33324
18. Hartl FU, Bracher A, Hayer-Hartl M (2011) Molecular chaperones in protein folding and proteostasis. Nature 475(7356):324–332
19. Young JC, Hoogenraad NJ, Hartl FU (2003) Molecular chaperones Hsp90 and Hsp70 deliver preproteins to the mitochondrial import receptor Tom70. Cell 112(1):41–50
20. Sidrauski C, Chapman R, Walter P (1998) The unfolded protein response: an intracellular signalling pathway with many surprising features. Trends Cell Biol 8(6):245–249
21. Page MJ, Di Cera E (2008) Evolution of peptidase diversity. J Biol Chem 283(44):30010–30014
22. Clausen T, Southan C, Ehrmann M (2002) The HtrA family of proteases: implications for protein composition and cell fate. Mol Cell 10(3):443–455
23. Krojer T et al (2008) Structural basis for the regulated protease and chaperone function of DegP. Nature 453(7197):885–890
24. Huesgen PF, Schuhmann H, Adamska I (2009) Deg/HtrA proteases as components of a network for photosystem II quality control in chloroplasts and cyanobacteria. Res Microbiol 160(9):726–732
25. Chien J et al (2009) Serine protease HtrA1 associates with microtubules and inhibits cell migration. Mol Cell Biol 29(15):4177–4187
26. Milner JM, Patel A, Rowan AD (2008) Emerging roles of serine proteinases in tissue turnover in arthritis. Arthritis Rheum 58(12):3644–3656
27. Chien J et al (2009) HtrA serine proteases as potential therapeutic targets in cancer. Curr Cancer Drug Targets 9(4):451–468
28. Coleman HR et al (2008) Age-related macular degeneration. Lancet 372(9652):1835–1845
29. Vande Walle L (2008) M. Lamkanfi, and P. Vandenabeele, *The mitochondrial serine protease HtrA2/Omi: an overview*. Cell Death Differ 15(3):453–460
30. Grau S et al (2005) Implications of the serine protease HtrA1 in amyloid precursor protein processing. Proc Natl Acad Sci U S A 102(17):6021–6026
31. Hou J, Clemmons DR, Smeekens S (2005) Expression and characterization of a serine protease that preferentially cleaves insulin-like growth factor binding protein-5. J Cell Biochem 94(3):470–484
32. Kapri-Pardes E, Naveh L, Adam Z (2007) The thylakoid lumen protease Deg1 is involved in the repair of photosystem II from photoinhibition in Arabidopsis. Plant Cell 19(3):1039–1047
33. Moisoi N et al (2009) Mitochondrial dysfunction triggered by loss of HtrA2 results in the activation of a brain-specific transcriptional stress response. Cell Death Differ 16(3):449–464
34. Singh N, Kuppili RR, Bose K (2011) The structural basis of mode of activation and functional diversity: a case study with HtrA family of serine proteases. Arch Biochem Biophys 516(2):85–96
35. Clausen T et al (2011) HTRA proteases: regulated proteolysis in protein quality control. Nat Rev Mol Cell Biol 12(3):152–162
36. Zumbrunn J, Trueb B (1996) Primary structure of a putative serine protease specific for IGF-binding proteins. FEBS Lett 398(2–3):187–192

37. Hu SI et al (1998) Human HtrA, an evolutionarily conserved serine protease identified as a differentially expressed gene product in osteoarthritic cartilage. J Biol Chem 273(51):34406–34412
38. Faccio L et al (2000) Characterization of a novel human serine protease that has extensive homology to bacterial heat shock endoprotease HtrA and is regulated by kidney ischemia. J Biol Chem 275(4):2581–2588
39. Gray CW et al (2000) Characterization of human HtrA2, a novel serine protease involved in the mammalian cellular stress response. Eur J Biochem 267(18):5699–5710
40. Nie GY et al (2003) Identification and cloning of two isoforms of human high-temperature requirement factor A3 (HtrA3), characterization of its genomic structure and comparison of its tissue distribution with HtrA1 and HtrA2. Biochem J 371(Pt 1):39–48
41. Glaza P et al (2015) Structural and functional analysis of human HtrA3 protease and its subdomains. PLoS One 10(6):e0131142
42. Hara K et al (2009) Association of HTRA1 mutations and familial ischemic cerebral small-vessel disease. N Engl J Med 360(17):1729–1739
43. Bhuiyan MS, Fukunaga K (2009) Mitochondrial serine protease HtrA2/Omi as a potential therapeutic target. Curr Drug Targets 10(4):372–383
44. Van Dyck L, Pearce DA, Sherman F (1994) PIM1 encodes a mitochondrial ATP-dependent protease that is required for mitochondrial function in the yeast Saccharomyces cerevisiae. J Biol Chem 269(1):238–242
45. Corydon TJ et al (1998) A human homologue of Escherichia coli ClpP caseinolytic protease: recombinant expression, intracellular processing and subcellular localization. Biochem J 331(Pt 1):309–316
46. Thorsness PE, White KH, Fox TD (1993) Inactivation of YME1, a member of the ftsH-SEC18-PAS1-CDC48 family of putative ATPase-encoding genes, causes increased escape of DNA from mitochondria in Saccharomyces cerevisiae. Mol Cell Biol 13(9):5418–5426
47. Lemberg MK et al (2005) Mechanism of intramembrane proteolysis investigated with purified rhomboid proteases. EMBO J 24(3):464–472
48. Arlt H et al (1996) The YTA10-12 complex, an AAA protease with chaperone-like activity in the inner membrane of mitochondria. Cell 85(6):875–885
49. Kambacheld M et al (2005) Role of the novel metallopeptidase Mop112 and saccharolysin for the complete degradation of proteins residing in different subcompartments of mitochondria. J Biol Chem 280(20):20132–20139
50. Buchler M, Tisljar U, Wolf DH (1994) Proteinase yscD (oligopeptidase yscD). Structure, function and relationship of the yeast enzyme with mammalian thimet oligopeptidase (metalloendopeptidase, EP 24.15). Eur J Biochem 219(1–2):627–639
51. Hegde R et al (2002) Identification of Omi/HtrA2 as a mitochondrial apoptotic serine protease that disrupts inhibitor of apoptosis protein-caspase interaction. J Biol Chem 277(1):432–438
52. Martins LM et al (2002) The serine protease Omi/HtrA2 regulates apoptosis by binding XIAP through a reaper-like motif. J Biol Chem 277(1):439–444
53. Verhagen AM et al (2002) HtrA2 promotes cell death through its serine protease activity and its ability to antagonize inhibitor of apoptosis proteins. J Biol Chem 277(1):445–454
54. Ekert PG, Vaux DL (2005) The mitochondrial death squad: hardened killers or innocent bystanders? Curr Opin Cell Biol 17(6):626–630
55. Jones JM et al (2003) Loss of Omi mitochondrial protease activity causes the neuromuscular disorder of mnd2 mutant mice. Nature 425(6959):721–727
56. Martins LM et al (2004) Neuroprotective role of the reaper-related serine protease HtrA2/Omi revealed by targeted deletion in mice. Mol Cell Biol 24(22):9848–9862
57. Suzuki Y et al (2001) A serine protease, HtrA2, is released from the mitochondria and interacts with XIAP, inducing cell death. Mol Cell 8(3):613–621
58. Li W et al (2002) Structural insights into the pro-apoptotic function of mitochondrial serine protease HtrA2/Omi. Nat Struct Biol 9(6):436–441
59. Cavallo L, Kleinjung J, Fraternali F (2003) POPS: a fast algorithm for solvent accessible surface areas at atomic and residue level. Nucleic Acids Res 31(13):3364–3366

60. Singh N et al (2014) Dual regulatory switch confers tighter control on HtrA2 proteolytic activity. FEBS J 281(10):2456–2470
61. Kraut J (1977) Serine proteases: structure and mechanism of catalysis. Annu Rev Biochem 46:331–358
62. Chaganti LK, Kuppili RR, Bose K (2013) Intricate structural coordination and domain plasticity regulate activity of serine protease HtrA2. FASEB J 27(8):3054–3066
63. Bejugam PR et al (2013) Allosteric regulation of serine protease HtrA2 through novel non-canonical substrate binding pocket. PLoS One 8(2):e55416
64. Laskowski RA, Gerick F, Thornton JM (2009) The structural basis of allosteric regulation in proteins. FEBS Lett 583(11):1692–1698
65. Gupta S et al (2004) The C-terminal tail of presenilin regulates Omi/HtrA2 protease activity. J Biol Chem 279(44):45844–45854
66. Martins LM et al (2003) Binding specificity and regulation of the serine protease and PDZ domains of HtrA2/Omi. J Biol Chem 278(49):49417–49427
67. Krick S et al (2008) Mpv17l protects against mitochondrial oxidative stress and apoptosis by activation of Omi/HtrA2 protease. Proc Natl Acad Sci U S A 105(37):14106–14111
68. Strauss KM et al (2005) Loss of function mutations in the gene encoding Omi/HtrA2 in Parkinson's disease. Hum Mol Genet 14(15):2099–2111
69. Ruiz N, Kahne D, Silhavy TJ (2006) Advances in understanding bacterial outer-membrane biogenesis. Nat Rev Microbiol 4(1):57–66
70. Su D et al (2009) UCF-101, a novel Omi/HtrA2 inhibitor, protects against cerebral ischemia/reperfusion injury in rats. Anat Rec (Hoboken) 292(6):854–861
71. Park HJ et al (2004) Alzheimer's disease-associated amyloid beta interacts with the human serine protease HtrA2/Omi. Neurosci Lett 357(1):63–67
72. Selkoe DJ (2001) Alzheimer's disease: genes, proteins, and therapy. Physiol Rev 81(2):741–766
73. Vassar R, Citron M (2000) Abeta-generating enzymes: recent advances in beta- and gamma-secretase research. Neuron 27(3):419–422
74. Salomone S et al (2012) New pharmacological strategies for treatment of Alzheimer's disease: focus on disease modifying drugs. Br J Clin Pharmacol 73(4):504–517
75. Hardy J (2009) The amyloid hypothesis for Alzheimer's disease: a critical reappraisal. J Neurochem 110(4):1129–1134
76. Lambert MP et al (1998) Diffusible, nonfibrillar ligands derived from Abeta1-42 are potent central nervous system neurotoxins. Proc Natl Acad Sci U S A 95(11):6448–6453
77. Walsh DM et al (2002) Naturally secreted oligomers of amyloid beta protein potently inhibit hippocampal long-term potentiation in vivo. Nature 416(6880):535–539
78. Borchelt DR et al (1996) Familial Alzheimer's disease-linked presenilin 1 variants elevate Abeta1-42/1-40 ratio in vitro and in vivo. Neuron 17(5):1005–1013
79. Lustbader JW et al (2004) ABAD directly links Abeta to mitochondrial toxicity in Alzheimer's disease. Science 304(5669):448–452
80. Manczak M et al (2006) Mitochondria are a direct site of a beta accumulation in Alzheimer's disease neurons: implications for free radical generation and oxidative damage in disease progression. Hum Mol Genet 15(9):1437–1449
81. Yamaguchi H et al (1992) Ultrastructural localization of Alzheimer amyloid beta/A4 protein precursor in the cytoplasm of neurons and senile plaque-associated astrocytes. Acta Neuropathol 85(1):15–22
82. Gouras GK, Almeida CG, Takahashi RH (2005) Intraneuronal Abeta accumulation and origin of plaques in Alzheimer's disease. Neurobiol Aging 26(9):1235–1244
83. Manfredi G, Beal MF (2000) The role of mitochondria in the pathogenesis of neurodegenerative diseases. Brain Pathol 10(3):462–472
84. Tabira T, Chui DH, Kuroda S (2002) Significance of intracellular Abeta42 accumulation in Alzheimer's disease. Front Biosci 7:a44–a49

85. Casas C et al (2004) Massive CA1/2 neuronal loss with intraneuronal and N-terminal truncated Abeta42 accumulation in a novel Alzheimer transgenic model. Am J Pathol 165(4):1289–1300
86. Knobloch M et al (2007) Intracellular Abeta and cognitive deficits precede beta-amyloid deposition in transgenic arcAbeta mice. Neurobiol Aging 28(9):1297–1306
87. Oakley H et al (2006) Intraneuronal beta-amyloid aggregates, neurodegeneration, and neuron loss in transgenic mice with five familial Alzheimer's disease mutations: potential factors in amyloid plaque formation. J Neurosci 26(40):10129–10140
88. Walsh DM et al (2000) The oligomerization of amyloid beta-protein begins intracellularly in cells derived from human brain. Biochemistry 39(35):10831–10839
89. Kim HJ et al (2003) Selective neuronal degeneration induced by soluble oligomeric amyloid beta protein. FASEB J 17(1):118–120
90. Kuo YM et al (1996) Water-soluble Abeta (N-40, N-42) oligomers in normal and Alzheimer disease brains. J Biol Chem 271(8):4077–4081
91. Lue LF et al (1999) Soluble amyloid beta peptide concentration as a predictor of synaptic change in Alzheimer's disease. Am J Pathol 155(3):853–862
92. McLean CA et al (1999) Soluble pool of Abeta amyloid as a determinant of severity of neurodegeneration in Alzheimer's disease. Ann Neurol 46(6):860–866
93. Liu ML et al (2009) Omi is a mammalian heat-shock protein that selectively binds and detoxifies oligomeric amyloid-beta. J Cell Sci 122(Pt 11):1917–1926
94. Cohen E et al (2006) Opposing activities protect against age-onset proteotoxicity. Science 313(5793):1604–1610
95. Fonte V et al (2008) Suppression of in vivo beta-amyloid peptide toxicity by overexpression of the HSP-16.2 small chaperone protein. J Biol Chem 283(2):784–791
96. Spiess C, Beil A, Ehrmann M (1999) A temperature-dependent switch from chaperone to protease in a widely conserved heat shock protein. Cell 97(3):339–347
97. Jellinger K (1990) New developments in the pathology of Parkinson's disease. Adv Neurol 53:1–16
98. Gandhi S, Wood NW (2005) Molecular pathogenesis of Parkinson's disease. Hum Mol Genet 14(18):2749–2755
99. Przedborski S (2005) Pathogenesis of nigral cell death in Parkinson's disease. Parkinsonism Relat Disord 11(Suppl 1):S3–S7
100. Sulzer D (2007) Multiple hit hypotheses for dopamine neuron loss in Parkinson's disease. Trends Neurosci 30(5):244–250
101. Farrer MJ (2006) Genetics of Parkinson disease: paradigm shifts and future prospects. Nat Rev Genet 7(4):306–318
102. Hardy J et al (2009) The genetics of Parkinson's syndromes: a critical review. Curr Opin Genet Dev 19(3):254–265
103. Martin I, Dawson VL, Dawson TM (2011) Recent advances in the genetics of Parkinson's disease. Annu Rev Genomics Hum Genet 12:301–325
104. Kitada T et al (1998) Mutations in the parkin gene cause autosomal recessive juvenile parkinsonism. Nature 392(6676):605–608
105. Singleton AB et al. (2003) alpha-Synuclein locus triplication causes Parkinson's disease. Science 302(5646):841
106. Polymeropoulos MH et al (1997) Mutation in the alpha-synuclein gene identified in families with Parkinson's disease. Science 276(5321):2045–2047
107. Bonifati V et al (2003) Mutations in the DJ-1 gene associated with autosomal recessive early-onset parkinsonism. Science 299(5604):256–259
108. Paisan-Ruiz C et al (2004) Cloning of the gene containing mutations that cause PARK8-linked Parkinson's disease. Neuron 44(4):595–600
109. Zimprich A et al (2004) Mutations in LRRK2 cause autosomal-dominant parkinsonism with pleomorphic pathology. Neuron 44(4):601–607

110. Greenamyre JT, Hastings TG (2004) Biomedicine. Parkinson's--divergent causes, convergent mechanisms. Science 304(5674):1120–1122
111. Ramirez A et al (2006) Hereditary parkinsonism with dementia is caused by mutations in ATP13A2, encoding a lysosomal type 5 P-type ATPase. Nat Genet 38(10):1184–1191
112. Valente EM et al (2004) Hereditary early-onset Parkinson's disease caused by mutations in PINK1. Science 304(5674):1158–1160
113. Vilarino-Guell C et al (2011) VPS35 mutations in Parkinson disease. Am J Hum Genet 89(1):162–167
114. Zimprich A et al (2011) A mutation in VPS35, encoding a subunit of the retromer complex, causes late-onset Parkinson disease. Am J Hum Genet 89(1):168–175
115. Chartier-Harlin MC et al (2011) Translation initiator EIF4G1 mutations in familial Parkinson disease. Am J Hum Genet 89(3):398–406
116. Dodson MW, Guo M (2007) Pink1, Parkin, DJ-1 and mitochondrial dysfunction in Parkinson's disease. Curr Opin Neurobiol 17(3):331–337
117. Bogaerts V et al (2008) Genetic variability in the mitochondrial serine protease HTRA2 contributes to risk for Parkinson disease. Hum Mutat 29(6):832–840
118. Lin CH et al (2011) Novel variant Pro143Ala in HTRA2 contributes to Parkinson's disease by inducing hyperphosphorylation of HTRA2 protein in mitochondria. Hum Genet 130(6):817–827
119. Simon-Sanchez J, Singleton AB (2008) Sequencing analysis of OMI/HTRA2 shows previously reported pathogenic mutations in neurologically normal controls. Hum Mol Genet 17(13):1988–1993
120. Obenauer JC, Cantley LC, Yaffe MB (2003) Scansite 2.0: proteome-wide prediction of cell signaling interactions using short sequence motifs. Nucleic Acids Res 31(13):3635–3641
121. Plun-Favreau H et al (2007) The mitochondrial protease HtrA2 is regulated by Parkinson's disease-associated kinase PINK1. Nat Cell Biol 9(11):1243–1252
122. Whitworth AJ et al (2008) Rhomboid-7 and HtrA2/Omi act in a common pathway with the Parkinson's disease factors Pink1 and Parkin. Dis Model Mech 1(2–3):168–174. discussion 173
123. Tain LS et al (2009) Drosophila HtrA2 is dispensable for apoptosis but acts downstream of PINK1 independently from Parkin. Cell Death Differ 16(8):1118–1125
124. de Castro P (2012) I., et al., *Genetic analysis of mitochondrial protein misfolding in Drosophila melanogaster*. Cell Death Differ 19(8):1308–1316
125. Yun J et al (2008) Loss-of-function analysis suggests that Omi/HtrA2 is not an essential component of the PINK1/PARKIN pathway in vivo. J Neurosci 28(53):14500–14510
126. Beal MF, Ferrante RJ (2004) Experimental therapeutics in transgenic mouse models of Huntington's disease. Nat Rev Neurosci 5(5):373–384
127. Browne SE, Beal MF (2004) The energetics of Huntington's disease. Neurochem Res 29(3):531–546
128. Chaturvedi RK, Beal MF (2008) Mitochondrial approaches for neuroprotection. Ann N Y Acad Sci 1147:395–412
129. A novel gene containing a trinucleotide repeat that is expanded and unstable on Huntington's disease chromosomes. The Huntington's Disease Collaborative Research Group. (1993) Cell 72(6): 971–83
130. Browne SE (2008) Mitochondria and Huntington's disease pathogenesis: insight from genetic and chemical models. Ann N Y Acad Sci 1147:358–382
131. La Spada AR, Morrison RS (2005) The power of the dark side: Huntington's disease protein and p53 form a deadly alliance. Neuron 47(1):1–3
132. Greenamyre JT (2007) Huntington's disease--making connections. N Engl J Med 356(5):518–520
133. Ross CA, Thompson LM (2006) Transcription meets metabolism in neurodegeneration. Nat Med 12(11):1239–1241

134. Inagaki R et al (2008) Omi / HtrA2 is relevant to the selective vulnerability of striatal neurons in Huntington's disease. Eur J Neurosci 28(1):30–40
135. Cilenti L et al (2003) Characterization of a novel and specific inhibitor for the pro-apoptotic protease Omi/HtrA2. J Biol Chem 278(13):11489–11494
136. Liu HR et al (2005) Role of Omi/HtrA2 in apoptotic cell death after myocardial ischemia and reperfusion. Circulation 111(1):90–96

Functional Relevance of Deubiquitinases in Life and Disease

16

Julia M. Fraile, Carlos López-Otín, and José M.P. Freije

Abstract

Deubiquitinases (DUBs) are critical regulators of ubiquitin-mediated signaling pathways through their ability to cleave the isopeptide bond that links ubiquitin to target proteins. The human genome encodes at least 100 DUBs, grouped into six families depending on sequence and structural properties. These proteolytic enzymes have pivotal roles in ubiquitin homeostasis and control of protein stability, and, consequently, their activities are tightly regulated by different mechanisms. Due to their wide diversity, DUBs are involved in multiple biological and pathological processes, including cancer. Accordingly, over the last years, many mutations in DUB genes or changes in their expression levels have been related to human malignancies. This chapter will focus on the description of the functional complexity of these enzymes in physiological and pathological conditions. We will review the involvement of DUBs in numerous biological processes and highlight their critical implications in cancer. Finally, we will discuss the growing relevance of DUBs for the development of novel therapeutic strategies against cancer.

Keywords

Ubiquitin • Deubiquitinases • DUB genes • Cancer

J.M. Fraile • C. López-Otín • J.M.P. Freije (✉)
Departamento de Bioquímica y Biología Molecular, Facultad de Medicina, Instituto Universitario de Oncología (IUOPA), Universidad de Oviedo, 33006 Oviedo, Spain
e-mail: clo@uniovi.es; jmpf@uniovi.es

S. Chakraborti, N.S. Dhalla (eds.), *Proteases in Physiology and Pathology*,
DOI 10.1007/978-981-10-2513-6_16

16.1 Introduction

Deubiquitinases, or DUBs, constitute a large group of proteases with the ability to hydrolyze the isopeptide bond that links the C-terminal group of ubiquitin to the ε-amino group of lysine side chains of target proteins [1, 2]. For this reason, DUBs have emerged as critical regulators of ubiquitin-mediated signaling pathways, affecting to the function and stability of multiple proteins. The human genome encodes at least 100 DUBs that are classified into six families according to structural features: ubiquitin-specific proteases (USPs), ubiquitin carboxy-terminal hydrolases (UCHs), ovarian tumor proteases (OTUs), Machado-Joseph disease protein domain proteases (MJDs), JAMM/MPN domain-associated metallopeptidases (JAMMs), and monocyte chemotactic protein-induced proteins (MCPIPs) (Fig. 16.1) [3]. The JAMM family is the only group with zinc metalloprotease activity, whereas all the rest are cysteine proteases and rely on a catalytic triad of cysteine, histidine, and aspartate residues located in their protease domain.

The USPs constitute the largest family of DUBs with more than 50 members [4]. Most of them share a conserved catalytic site within the USP domain, although the existence of "non-protease" USPs has also been described. Furthermore, the activity and specificity of many USPs are conferred by the presence of additional domains including the B-box domain that is present in CYLD; the zinc finger USP domain located in USP3, USP5, USP39, USP44, USP45, USP49, and USP51; the ubiquitin-interacting motif shared by USP25 and USP37; the ubiquitin-associated domain in USP5 and USP13; the DUSP domain found in USP4, USP11, USP15, USP20, USP33, and USP48; the exonuclease III domain present in USP52; and the ubiquitin-like domain of USP4, USP7, USP14, USP32, USP47, and USP48 [4].

The other five DUB families are composed of fewer members. There are four UCHs in humans, UCHL1, UCHL3, UCHL5/UCH37, and BAP1, all of which target small peptides from the C-terminus of ubiquitin. UCHL5 and BAP1 present an additional C-terminal extension that mediates the trimming of polyubiquitin chains from conjugated proteins and the interaction with the N-terminal ring finger of BRCA1, respectively [5].

Regarding OTUs, there are 18 protein-coding genes in the human genome that share an OTU domain and can be classified into four groups: otubains (OTUB1 and OTUB2), A20-like OTUs (A20/TNFAIP3, Cezanne, Cezanne 2, TRABID, and VCPIP1), OTUDs (OTUD1, OTUD2/YOD1, OTUD3, OTUD4, OTUD5, OTUD6A, OTUD6B, ALG13, and a pseudogene, HIN1L), and OTULIN-like OTUs (OTULIN and FAM105A) [6, 7]. The OTU core domain is composed of five β-strands situated between helical domains that vary in size among members of the OTU family [8]. Moreover, these enzymes exhibit additional domains, such as ubiquitin-binding domains (A20, TRABID, and OTUB1), ubiquitin-interacting motifs (OTUD1 and OTUD5), and ubiquitin-associated domains (Cezanne) [2, 9].

Ataxin-3 (ATXN3), ATXN3L, JOSD1, and JOSD2 conform the Josephin family. ATXN3 is the most studied member of this family due to the involvement of its mutated form in spinocerebellar ataxia type 3 or Machado-Joseph disease [10].

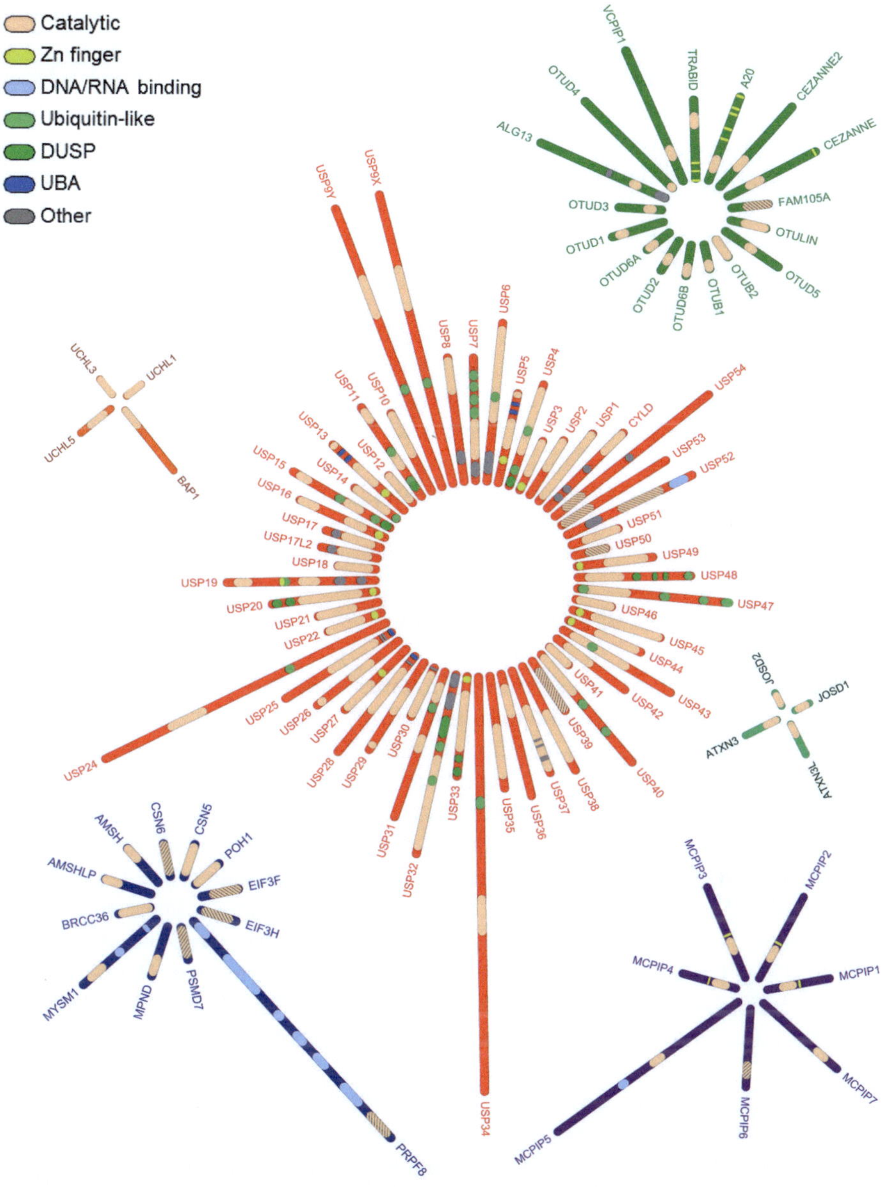

Fig. 16.1 Classification of human DUBs. Human DUBs are classified into six families represented separately and in different colors: USPs (red), UCHs (orange), OTUs (green), MJDs (light green), JAMMs (blue), and MCPIPs (purple). The catalytic core is indicated in plain light red if the DUB is active and stripped if inactive. Additional domains are shown in different colors. DUBs are represented with their N-termini oriented toward the center of the circle

Moreover, ATXN3 modulates the ubiquitination status of substrates involved in insulin-IGF-1 signaling, being indispensable for a normal life span [11].

The metalloproteinase family of JAMMs is composed of 12 proteins, wherein only 7 have catalytic activity: AMSH/STAMBP, AMSH-LP/STAMBPL1, BRCC36, POH1/PSMD14, MYSM1, MPND, and CSN5/JAB1. JAMM DUB domain has been studied in AMSH-LP and is composed of a JAMM core and two conserved insertions that were suggested to confer Lys63-linked polyubiquitin specificity. There is a high degree of sequence conservation between POH1, AMSH, and AMSH-LP, supporting the existence of a common mechanism for ubiquitin recognition and DUB catalysis in this family [12].

The last group of DUBs identified is the MCPIP1 family, which is composed of seven members. MCPIP1 contains a functional ubiquitin-associated domain at the N-terminus that is not required for DUB activity but mediates its interaction with ubiquitinated proteins. Moreover, there is an N-terminal conserved region and a CCCH-type zinc finger domain that is critical for MCPIP1 activity. Finally, there is a Pro-rich domain at the C-terminus. Interestingly, in this class the catalytic domain lacks the His box that is located outside the N-terminal conserved region [13].

16.2 Regulation of DUB Activities

DUBs have fundamental roles in ubiquitin homeostasis and protein stability. In this regard, DUBs generate free ubiquitin through the ability to process ubiquitin precursors and also promote the stability of ubiquitinated proteins by preventing both lysosomal and proteasomal degradation, depending on the nature of the ubiquitin linkage [14]. Consequently, an additional level of regulation of DUB function comes from the specificity for the ubiquitin chain linkages they processed (Fig. 16.2). USPs and UCHs cleave indiscriminately most ubiquitin chain types, being Lys48- or Lys63-linked polyubiquitin chains their most frequent targets. Interestingly, complex mechanisms of action have also been described for some members of the family. Thus, USP14 suppresses protein degradation by removing ubiquitin chains en bloc, independently of chain linkage type [15]. On the other hand, JAMMs and ATXN3 show a restricted specificity for Lys63-linked chains [12]. Similarly, most human OTUs show intrinsic linkage specificity, preferring one or a small defined subset of ubiquitin linkage types [7]. Although A20 only depolymerizes Lys63-linked polyubiquitin, its function in inflammation is critically regulated by linear ubiquitination levels [16]. Finally, OTULIN was described to exquisitely hydrolyze linear ubiquitin chains [17].

Due to their importance in protein homeostasis, DUB activities are regulated through a number of different mechanisms, such as transcriptional control of gene expression, posttranscriptional modifications, changes in subcellular localization, and activation mediated by interacting proteins [18]. Examples of transcriptional regulation affect *DUB-1*, *DUB-2*, and *DUB-3* [19–21], whose expression is induced by inflammatory cytokines. Moreover, CYLD is induced by the activation of the nuclear factor-κB (NF-κB) and MAPK kinase 3/6 (MKK3/6)-p38 pathways [22].

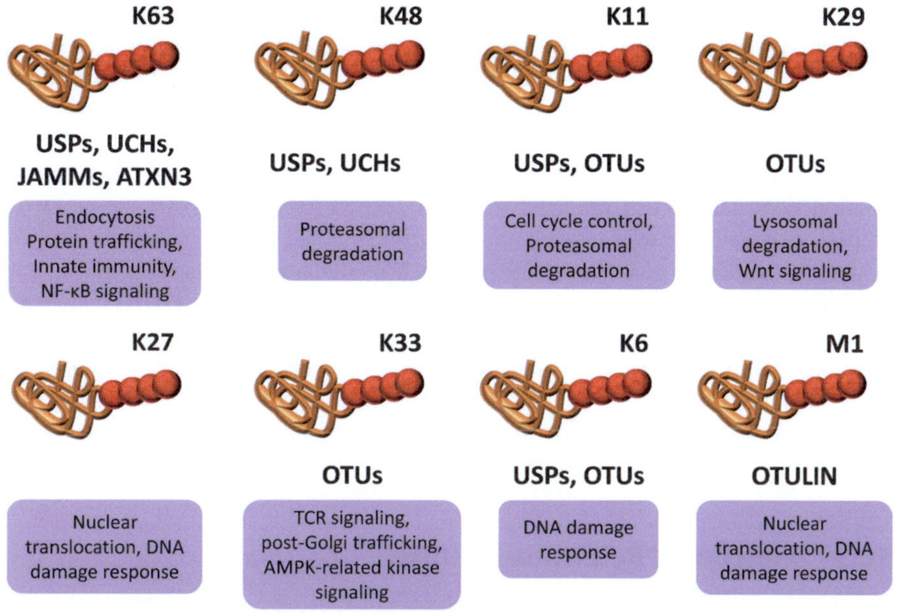

Fig. 16.2 Ubiquitin chain types processed by DUBs. Representation of the specificity of DUB families for different types of polyubiquitinated proteins and their cellular functions

Bcl10 promotes A20 expression through the activation of its promoter, which contains NF-κB-binding sites [23], and Snail1 mediates the transcriptional repression of Cezanne2 in hepatocellular carcinoma [24]. Furthermore, miR-26b and miR-4295 regulate USP9X and USP28, respectively, at posttranscriptional level [25, 26].

Many DUB functions are also regulated by posttranslational modifications. Phosphorylation inhibits CYLD and USP8, while it activates A20, OTUD5, USP1, USP4, USP7, USP15, USP16, USP19, USP28, USP34, and USP37 [27–29]. Furthermore, ubiquitin and ubiquitin-like modifications also modulate DUB function. Thus, ATXN3 and UCHL1 activities are modulated by ubiquitination [30, 31], whereas USP25, USP28, and CYLD are regulated by SUMO conjugation [32–34]. Reactive oxygen or nitrogen species are also involved in regulating DUB activities [35, 36]. Finally, USP1 and ATXN3 are inactivated by autoproteolytic cleavage, whereas CYLD and A20 are inhibited through interaction with other proteases [37–39].

As mentioned above, DUB activity can also be regulated by changes in subcellular localization, which can facilitate the interaction with specific substrates. USP36 localization in the nucleolus regulates its structure and function [40], whereas USP30 modulates the morphological properties of mitochondria [41]. Moreover, USP1 contains two nuclear localization signals that mediate nuclear import of the USP1/UAF1 complex [42]. The presence of ubiquitin-binding and ubiquitin-like domains also contributes to regulate the activity and specificity of several DUBs [2].

Finally, the regulation of many DUBs is mediated by the association with their interaction partners. Thus, USP1 plays a role in DNA damage repair by interacting with UAF1 [43]. Similarly, interaction of ASXL1 and ASXL2 with BAP1 is required for ubiquitin binding and H2A deubiquitination [44, 45], and valosin-containing protein (VCP/p97) interacts with ATXN3 to specify the cellular fate of its targets [46]. In addition, guanosine5'-monophosphate synthase and USP7 form a deubiquitination complex that is required to stabilize p53 [47]. Moreover, yeast DUB, Ubp10, possesses multiple binding modules that regulate protein interaction and are critical for ribosome biogenesis. Interestingly, its human homologue, USP36, contains the same regions flanking its catalytic domain [48]. Several DUBs have been reported to be integrated within large macromolecular complexes, such as proteasome or COP9 signalosome, to become active [49]. Finally, a global proteomics approach aimed at identifying proteins interacting with 75 DUBs uncovered a complex landscape of more than 770 putatively associated proteins, thus reflecting the complexity in the regulation of DUB functions [50].

16.3 Functional Relevance of DUBs

The wide functional diversity of DUBs has a profound impact on the regulation of multiple biological processes such as cell cycle regulation, DNA damage repair, chromatin remodeling, and several signaling pathways, which are frequently altered in human diseases, especially in cancer [3]. In this section, we will discuss the current knowledge of DUB functions within each of these biological processes.

16.3.1 Cell Cycle Regulation

Several members of the DUB superfamily are critical components of the core cell cycle machinery and cell cycle checkpoint. Functional analysis links USP28, USP36, and USP37 with the stability of c-Myc, a central modulator of cell growth, proliferation, and apoptosis [51–53]. By contrast, USP10 antagonizes the transcriptional activity of c-Myc and inhibits cell cycle proliferation through the stabilization of SIRT6 [54]. USP4 regulates the mono-ubiquitination of PDK1, a master growth factor signaling kinase that plays a critical role in cell proliferation and metabolism [55], whereas USP7 has an essential role in cell proliferation through the regulation of phosphatase and tensin homologue (PTEN) and FOXO localization [56, 57]. Furthermore, PTEN transcription is regulated by ATXN3 [58], and USP13 and OTUD3 regulate PTEN protein levels by deubiquitination [59, 60]. Similarly, UCHL5 deubiquitinating activity modulates the transcriptional activity of E2F1, an important transcription factor involved in cell cycle progression, DNA repair, and apoptosis response [61].

Many DUBs are crucial regulators of events occurring in mitosis, such as CYLD that modulates mitotic spindle orientation and regulates polo-like kinase 1 [62]. USP13 and USP37 antagonize anaphase-promoting complex (APC/C)-Cdh1 function by inhibiting the ubiquitination of Skp2 and Cyclin A, respectively [63, 64]. Moreover, USP37 interacts with chromatin-associated WAPL, contributing to sister chromatid resolution [65], and blocks cell proliferation by the deubiquitination and stabilization of p27 in medulloblastoma cells [66]. USP39 regulates Aurora kinase B levels and is essential for spindle checkpoint integrity [67], whereas USP44 prevents the premature activation of APC/C through the stabilization of its inhibitory Mad2-Cdc20 complex [68]. BRCC36 modulates functional bipolar spindle by deubiquitinating the essential spindle assembly factor nuclear mitotic apparatus [69]. In contrast, USP50 represses entry into mitosis through the regulation of HSP90-dependent Wee1 stability [70].USP2 and USP17L2 stabilize cyclin D1 and CDC25A, respectively [71, 72], and USP19 regulates cell proliferation in a cell context-dependent manner, through both E3 ligase KPC1-dependent and KPC1-independent mechanisms [73]. CSN5 prevents senescence and proper progression of the somatic cell cycle [74], and BAP1 influences cell cycle progression at G1/S by co-regulating transcription from host cell factor 1(HCF-1)/E2F-responsive promoters [75]. Moreover, BAP1 forms a transcriptional complex with KLF5 and HCF-1 that binds and partially represses *p27* gene expression, promoting cell proliferation [76]. Surprisingly, BAP1 deficiency in mice induces myeloid progenitor proliferation through elevated expression of enhancer of zeste 2 polycomb repressive complex 2 subunit (Ezh2) [77]. Finally, USP1 and USP33 are involved in the regulation of centrosome duplication [78, 79], whereas USP44 controls centrosome positioning, thus preventing aneuploidy [80].

16.3.2 DNA Damage Repair

Several DUBs are essential components of DNA repair mechanisms, which are frequently altered in human malignancies. USP1 deubiquitinates Fanconi's anemia protein (FANCD2), stabilizes CHK1, regulates the ubiquitination levels of proliferating cell nuclear antigen (PCNA), and interacts with UAF1 to promote double-strand break repair through homologous recombination [37, 43, 81, 82]. Moreover, USP3, USP16, USP26, USP37, USP44, OTUB1, BRCC36, and POH1 modulate the RNF8c/168 pathway of double-strand break (DSB) repair [83]. POH1 and BAP1 facilitate homologous recombination repair through loading RAD51 at DNA damage sites [84, 85], and USP11 and USP28 regulate the cellular response to mitomycin C-induced DNA damage within the BRCA2 pathway and CHK2-p53-PUMA pathway, respectively [86, 87]. USP9X regulates DNA damage responses through the stabilization of CLASPIN [88], whereas USP17L2 and USP51 control H2AX ubiquitination [89, 90]. Interestingly, Nishi and collaborators have demonstrated the role of UCHL5 in DNA damage repair by inhibiting NF-κB degradation [91]. Another important DUB with critical functions in DNA damage repair is USP7, which is a critical regulator of RAD18 protein levels [92] and suppresses oxidative stress-induced PCNA ubiquitination and mutagenesis [93]. Furthermore, ATM-dependent downregulation of USP7 by the phosphatase PPM1G leads to a p53-dependent DNA damage response [94] and similarly to USP1, USP7 also regulates CHK1 protein levels through deubiquitination [95]. Additionally, USP7 has a critical regulatory function in transcription-coupled nucleotide excision repair (TC-NER) by the stabilization of ERCC6 [96]. USP24 is also connected to NER by controlling the stability of damage-specific DNA-binding protein 2 [97]. Finally, USP47 has been identified as the enzyme responsible for the deubiquitination of the base excision repair DNA polymerase (Polβ), thus having an important role in DNA repair regulation and genome integrity maintenance [98].

16.3.3 Chromatin Remodeling

It is well known that posttranslational modifications of histones regulate chromatin structure dynamics and gene transcription. In this regard, several DUBs have been described to deubiquitinate both H2A and H2B, such as USP3, USP7, USP16, USP21, USP22, MYSM1, and BRCC36 [99, 100], although H2A is preferentially targeted [101]. MYSM1, USP7, USP22, and BRCC36 are part of the 2A-DUB, polycomb repressive complex 1, SAGA, and BRCA1-A multisubunit complexes, respectively [102–105]. Otherwise, USP3, USP16, and USP21 might exhibit different chromatin regulatory mechanisms since they have not been found in any of these complexes. Additionally, BAP1 associates with ASXL1 or ASXL2 in order to regulate H2A deubiquitination [44, 45]. By contrast, USP15 associates with SART3 and specifically deubiquitinates H2B in free histones [106]. Similarly, USP49 deubiquitinates H2B and regulates the cotranscriptional splicing of a large set of exons [107].

Besides histones, DUB can regulate gene expression by deubiquitinating other chromatin-related substrates. USP21 deubiquitinates and stabilizes TIP5, which is part of the NoRC, a chromatin-remodeling complex required for establishing repressive chromatin structure at rDNA promoters [108]. Furthermore, USP22 modulates the protein stability of telomeric repeat binding factor 1 [100], whereas USP7 and USP11 deubiquitinate MEL18 and BM1, two chromatin-bound polycomb repressive complex 1 components that affect the expression of $p16^{INK4a}$ [105]. In addition, USP7 has been recently described to deubiquitinate and stabilize the histone demethylase PHF8 [109] and UCHL5 interacts with Ino80 chromatin-remodeling complex [110].

16.3.4 Signaling Pathways

During the last decades, many DUBs have been described to regulate different signaling pathways, which are frequently altered in cancer. In this section, we will discuss the implication of DUBs in the regulation of several cancer-relevant pathways, such as those involving p53, NF-κB, receptor tyrosine kinases (RTKs), Wnt, and transforming growth factor-β (TGF-β) (Table 16.1).

p53 is a tumor suppressor gene whose function is essential for preventing cancer formation. So far, USP2, USP4, USP7, USP10, USP15, USP22, USP24, USP29, USP42, and OTUB1 are known to participate in p53 regulation. USP7 has a dual

Table 16.1 Functional relevance of USPs in signaling pathways

Signaling pathway	Function	USPs
p53	Activators	USP7, USP10, USP24, USP29, USP42
	Repressors	USP2, USP4, USP5, USP7, USP15, USP22
NF-κB	Activators	USP2, USP6, USP7, USP17, USP21, USP25
	Repressors	CYLD, USP4, USP7, USP10, USP11, USP15, USP18, USP21, USP25, USP34
RTK	Activators	USP2, USP8, USP18
	Repressors	USP8
Wnt	Activators	USP4, USP5, USP14, USP34, USP47
	Repressors	CYLD, USP4, USP15
TGF-β	Activators	USP4, USP9X, USP11, USP15
	Repressors	CYLD

role in regulating the stability of p53 since it can deubiquitinate both p53 and MDM2, an ubiquitin ligase that targets p53 for proteasomal degradation [111, 112]. USP2 and USP15 stabilize MDM2 but not p53 [113, 114], and OTUB1 directly suppresses MDM2-mediated p53 ubiquitination [115]. Moreover, USP10 stabilizes both mutated and wild-type p53, thus having a dual role in tumorigenesis depending on p53 status [116]. By contrast, USP4 promotes p53 degradation through the deubiquitination of ARF-BP1 ubiquitin ligase and the histone deacetylase HDAC2 [117, 118]. Furthermore, USP22 inhibits p53 transcriptional activation through the deubiquitination of SIRT1, leading to decreased levels of p53 acetylation and suppressing p53-mediated functions [119]. USP24 and USP29 deubiquitinate p53 and protect genomic stability by regulating UV damage and oxidative stress responses, respectively [120, 121]. Finally, USP42 forms a direct complex with and deubiquitinates p53 during the early phase of the response to a range of stress signals [122].

Nuclear factor-κB is a well-known modulator of innate and adaptive immune responses that is frequently deregulated in cancer [123]. A20 and CYLD act as tumor suppressors by inhibiting NF-κB signaling through the regulation of several components of the pathway [124]. Thus, both of them deubiquitinate TRAF6, whereas CYLD also controls the ubiquitin levels of TGF-β-activated kinase 1 (TAK1) [125], B-cell CLL/lymphoma 3 (Bcl3) [126], and mitogen-activated protein kinases [127]. On the other hand, A20 promotes the degradation of TRAF2 in lysosomes by means of its own E3 ligase activity [128] and is recruited into a TNF receptor signaling complex containing linear ubiquitin chain assembly complex (LUBAC) and IκB kinase (IKK) [129]. Additionally, A20 removes Lys63-linked ubiquitin of RIPK1 through its OTU domain and promotes its proteasomal degradation by Lys48 polyubiquitination [130].

RIPK1 ubiquitin levels can also be modulated by CYLD, USP4, USP7, and USP21 [131–134]. Furthermore, Cezanne deubiquitinates RIPK1 signaling intermediaries [135] and regulates noncanonical NF-κB signaling through inhibition of TRAF3 degradation [136]. Similarly, OTUD5 removes Lys63-linked ubiquitin of TRAF3, resulting in diminished type I interferon and IL-10 responses [137] and

USP25 inhibits IL-17-induced activation of NF-κB through the modulation of TRAF5 and TRAF6 ubiquitination [138]. By contrast, USP7 deubiquitinates NF-κB and promotes its transcriptional activity [139], whereas USP15 deubiquitinates and stabilizes IκBα [140]. USP2 modulates TNFα-induced NF-κB signaling through the regulation of IκB phosphorylation, nuclear translocation of NF-κB, and expression of NF-κB-dependent target genes [141]. Moreover, USP4 inhibits TNFα-induced NF-κB signaling by deubiquitinating TAK1 [142], TRAF2, and TRAF6 [143] and USP18 negatively regulates TAK1 activity during T helper 17 cell differentiation by deubiquitinating TAK1-TAB1 complex [144]. MCPIP1 deubiquitinates TRAF2, TRAF3, and TRAF6 and mediates USP10-dependent deubiquitination of IKKγ leading to the inhibition of NK-κB and the promotion of apoptosis [13, 145]. Furthermore, USP10 also prevents genotoxic NF-κB activation by inhibiting TRAF6 ubiquitination [146]. Finally, USP6 activates classical NF-κB in an atypical mechanism characterized by the absence of IκB degradation and the requirement of IKKα, IKKβ, and IKKγ [147].

There are at least five DUBs—USP2, USP8, USP18, AMSH, and POH1—that can interfere in the trafficking of RTKs such as epidermal growth factor receptor (EGFR), Met, and ErbB2. Thus, USP2 prevents EGFR degradation and, consequently, amplifies signaling activity from the receptor [148]. Regarding USP8, some studies support a role of this DUB in the stabilization of RTKs through deubiquitination [149], whereas other works suggest its implication in RTK degradation [150]. Furthermore, AMSH promotes EGFR recycling [151] and USP18 modulates EGFR translation [152]. Finally, POH1 regulates the ubiquitination levels of ErbB2, although it is not involved in its turnover [153].

The Wnt signaling is essential for embryonic development and is frequently activated in cancer [154]. Of note, CYLD, USP4, USP5, USP15, USP34, USP47, TRABID, and OTULIN are associated with this pathway. Thus, CYLD modulates Wnt signaling through the removal of Lys63-linked ubiquitin of the cytoplasmic effector Dishevelled (Dvl) [155]. Furthermore, USP4 and USP15 negatively modulate Wnt signaling by interacting with Nemo-like kinase and promoting β-catenin degradation, respectively [156, 157]. Conversely, USP4 has also been proposed as a positive regulator of Wnt pathway in colorectal carcinoma, through the regulation of β-catenin stabilization [158]. USP5 stabilizes FOXM1 by deubiquitination, which is essential for β-catenin recruitment to Wnt target gene promoters [159], and USP34 and USP47 positively regulate this pathway by influencing β-catenin-dependent transcription and preventing its degradation, respectively [160, 161]. Finally, TRABID deubiquitinates APC and is involved in T-cell factor-mediated transcription of Wnt genes [162], whereas OTULIN modulates Wnt signaling counteracting LUBAC through the cleavage of linear ubiquitin chains [163].

TGF-β is a multifunctional protein with dual role in oncogenesis, acting as a barrier to neoplastic transformation but promoting epithelial-to-mesenchymal transition at later stages [164]. So far, USP4, USP9X, USP11, USP15, CYLD, OTUB1, AMSH-LP, and UCHL5 are known to regulate this pathway. USP4 and USP11 strongly induce TGF-β signaling through deubiquitination of type I TGF-β receptor (TβR-I) [29, 165]. Similarly, USP15, identified as a DUB for type I bone morphogenetic protein (BMP) receptors and receptor-activated SMADs [166, 167], deubiquitinates and stabilizes TβR-I and SMURF2, the E3 ligase that targets the TβR complex for ubiquitin-mediated degradation [168, 169]. USP9X positively regulates this pathway by deubiquitinating SMAD4 and promoting its association with SMAD2 [170]. USP9X also regulates ubiquitin levels of the AMPK-related kinases NUAK1 and MARK4, modulating their LKB1-mediated phosphorylation and activation [171]. Moreover, CYLD regulates TGF-β signaling and the development of regulatory T cells through SMAD7 deubiquitination [172] and also decreases the stability of SMAD3 by deubiquitinating Lys63-polyubiquitinated AKT [173]. Finally, OTUB1 inhibits the ubiquitination and degradation of SMAD2/3 complex, independently of its catalytic activity [174], whereas AMSH-LP and UCHL5 potentiate TGF-β responses through their interaction with inhibitory I-SMADs [175, 176].

DUBs also play critical roles in other signaling pathways in addition to those described above. For instance, USP10 deubiquitinates and activates AMPK, a master regulator of metabolic homeostasis [177]. Moreover, USP9X negatively regulates the activity of mammalian target of rapamycin (mTOR) and muscle differentiation [178], and USP15 inhibits Nrf2, a key regulator of the antioxidant response, through deubiquitination of the E3 ligase Keap1 [179]. Notably, USP20 and USP33 are implicated in Von Hippel-Lindau (VHL) syndrome, a familiar cancer syndrome caused by germ line VHL mutations that predispose to various benign and malignant tumors. USP20 deubiquitinates and stabilizes HIF-1α, thereby inhibiting its degradation promoted by the E3 ubiquitin ligase VHL [180]. USP33 role is still unknown, but some of the disease-causing mutations in *VHL* block its interaction with this enzyme, suggesting a role of USP33 in the regulation of VHL. Moreover, Cezanne is essential for the stability of HIF-1α [181], and USP7 plays a role in hedgehog signaling by modulating GLI ubiquitination and stability [182]. Finally, an in vivo RNA interference screen in *Drosophila melanogaster* identified four DUBs (the *Drosophila* orthologs of vertebrate BAP1, USP10, and eIF3 complex subunits H and F) as modulators of Notch signaling activity [183].

16.3.5 Other Functional Roles of DUBs

Over the last few decades, several studies have demonstrated the involvement of DUBs in other physiological processes that play additional roles in cancer progression, such as epithelial-to-mesenchymal transition (EMT), cell migration, apoptosis, autophagy, and stem cell maintenance and differentiation. UCHL1 acts as a potent oncogene and regulates prostate cancer progression and metastasis by inducing EMT [184], whereas USP17 has a critical role in cell migration through the modulation of the subcellular localization of GTPases, which are essential for cell motility [185]. CYLD and USP9X also regulate cell migration through the GTPase Rac1 and the stabilization of the E3 ubiquitin ligase SMURF1, respectively [186, 187].

DUBs have dual and complex roles in the regulation of apoptotic processes, either promoting (USP2, USP4, USP7, USP8, USP9X, USP15, USP16, USP17, USP27, USP28, USP41, CYLD, A20, UCHL1, and ATXN3) or suppressing apoptosis (USP2, USP9X, USP18, UCHL3, and A20) [133, 188]. USP2 inhibits apoptosis in prostate cancer cells by stabilizing fatty acid synthase [189] and promotes cell death by deubiquitination and stabilization of the truncated form of the apoptosis-inducing factor, AIF [190]. As described above, USP4 and USP7 induce TNFα-mediated apoptosis by negatively regulating RIPK1 ubiquitination [133, 134], whereas USP7 deubiquitinates and stabilizes TIP60, an essential acetyl transferase required for p53-dependent induction of apoptosis [191]. USP9X leads to cell survival by deubiquitinating MCL1 [192] but also promotes apoptosis by stabilizing apoptosis signal-regulating kinase 1 [193]. Likewise, USP17 promotes apoptosis in cervical carcinoma through the regulation of the histone deacetylase activity of SDS3 [188]. Furthermore, USP27 deubiquitinates BH3-only protein and enhances its levels, counteracting the antiapoptotic effects of ERK activity [194]. Although A20 has been considered a tumor suppressor DUB-promoting cell death, it also acts as a potent prosurvival gene by inhibiting apoptosis in gliomas and breast carcinomas [195].

Several DUBs are also involved in autophagy, a critical intracellular catabolic mechanism that mediates the degradation of cytoplasmic proteins and organelles [196]. Thus, USP19 is considered a positive regulator of autophagy but a negative regulator of type I interferon signaling because it deubiquitinates Beclin-1 [197]. Ubiquitin levels of Beclin-1 are also modulated by A20, USP10, and USP13 [198, 199]. USP33 promotes autophagosome formation through the deubiquitination of the mono-ubiquitinated RAS-like GTPase RALB, providing the switch for the dual functions of RALB in autophagy and innate immune responses [200]. Moreover, USP36 modulates the selective autophagic degradation of protein aggregates [201]. Mitophagy, the degradation of mitochondria by autophagy, is also regulated by several DUBs. Thus, USP30 antagonizes mitophagy driven by the ubiquitin ligase parkin and protein kinase PINK1, which are encoded by two genes associated with Parkinson's disease [202]. Interestingly, USP8 and USP15, which are not located within mitochondria, have also been involved in promoting and inhibiting parkin-mediated mitophagy, respectively [203, 204].

Over the last years, the role of DUBs in stem cell maintenance and cellular reprogramming has been widely studied [205]. So far, several DUBs, such as USP7, USP16, USP22, USP28, USP44, POH1, and MYSM1, are involved in regulating stem cell biology. USP7 prevents neural stem/progenitor cell differentiation by the deubiquitination and stabilization of the repressor element 1-silencing transcription factor [206]. Furthermore, USP22 is required for c-Myc-induced transcription [103] and functions as a critical regulator of the transition from self-renewal to differentiation through the repression of the *Sox2* locus [207]. Moreover, USP28 stabilizes the chromatin modulator LSD1, which has a critical role in cellular pluripotency [208], whereas USP44 function regulates stem cell differentiation by inhibiting the mono-ubiquitination of H2B [209]. In addition, POH1 is required to maintain embryonic stem cell (ESC) self-renewal and pluripotency [210], and USP16 regulates ESC pluripotency gene expression and differentiation [211]. Interestingly, a mouse model that is trisomic for 132 genes homologous to genes on human chromosome 21 demonstrated that USP16 has an important role in antagonizing self-renewal in Down's syndrome [212]. Moreover, both USP16 and MYSM1 are essential for hematopoietic stem cell maintenance and differentiation [213, 214], and MYSM1 function is also critical for normal bone formation and mesenchymal stem cell differentiation [215].

16.4 Alterations of DUB Sequences and Functions in Cancer

There is a growing list of human malignancies in which several DUBs are mutated and behave as oncogenes or tumor suppressors [3]. Moreover, malignant tumors are frequently associated with profound changes in the expression levels of many DUBs. Furthermore, DUBs may function as tumor suppressors or oncogenes depending on the cellular context and the target affected by their regulation.

Mutations in *CYLD* have been described in familiar cylindromatosis, characterized by a predisposition to develop multiple head and neck skin tumors [216]. Chromosomal deletions and inactivating mutations in *A20* gene have been described in several lymphoma subtypes [217, 218]. Furthermore, point mutations and deletions in *BAP1* have been found in breast, lung, and clear cell renal cell carcinomas [219–221], as well as in metastasizing uveal melanomas, in malignant pleural mesotheliomas, and in myelodysplastic syndromes [220, 222, 223]. Finally, inactivating mutations of *USP9X* have been found in pancreatic ductal adenocarcinoma [224]. Regarding DUBs with pro-tumorigenic roles, somatic mutations in *USP6* and *USP28* are present in different human cancers. Thus, the fusion between the osteoblast cadherin 11 gene (*CDH11*) promoters and *USP6*, created by chromosomal rearrangement, leads to the overexpression of *USP6* in aneurismal bone cysts [225]. Furthermore, mutations in *USP6* and *USP28* have been reported in lobular breast cancer [226]. Finally, dominant mutations of *USP8* were found in four of ten corticotroph adenomas of the pituitary, thus causing Cushing's disease [227].

Many DUBs are linked to cancer through changes in their expression patterns. Thus, *UCHL1*, *JOSD1*, *CSN5*, and *USP9X* are overexpressed in non-small cell lung carcinomas, whereas *USP10*, *USP11*, *USP22*, *USP48*, and *CSN5* are overexpressed in malignant melanoma, correlating high levels of *USP10*, *USP11*, and *USP22* with a more aggressive and invasive phenotype [228]. Moreover, *CSN5* is overexpressed in early hepatocellular carcinoma [229], and *USP22* is upregulated in colorectal carcinomas and belongs to a set of marker genes that can predict metastatic potential and therapeutic outcome in human cancer [230]. Moreover, *USP22* and *OTUB1* overexpression correlates with poor prognosis in invasive breast [231] and ovarian cancer [232], respectively. *OTUB1* and *USP4* are overexpressed in colorectal cancer tissues and are associated with tumor size, differentiation, distant metastasis, and poor survival [233, 234]. Additionally, *USP4* is overexpressed in other human tumors [117, 235], but it is downregulated in small cell lung cancer cell lines [236]. Furthermore, *USP1* expression correlates with early steps of transformation in gastric cancer [228], whereas *USP2* is upregulated in ovarian and prostate carcinomas and is associated with lesions of poor prognosis [237]. In agreement with these results, overexpression of *USP2* protects prostate cancer cells from apoptosis and endows them with resistance to chemotherapeutic agents by promoting p53 degradation [238]. By contrast, *USP2* is downregulated in breast cancer, suggesting that this enzyme acts as an oncogene or tumor suppressor in a time- and tissue-specific manner [141]. *USP7* overexpression has also been associated with tumor aggressiveness in prostate cancer [56], whereas its expression and activity are repressed by STAT3 in colon cancer [239]. Additionally, overexpression of *USP17* has been found in primary lung, colon, esophagus, and cervix tumor biopsies [240], and aberrant expression of *USP14* has been found in a variety of cancers, such as multiple myeloma and colorectal, lung, and epithelial ovarian cancers [241, 242]. *USP39* overexpression correlates with poor prognosis in prostate cancer patients [243], whereas *USP44* expression is elevated in human T-cell leukemia and is associated with chromosomal instability [244]. By contrast, *USP44* is downregulated in lung adenocarcinomas and associates with poor prognosis [80]. Furthermore, *USP44* is epigenetically inactivated in colorectal adenomas, representing an early event in this pathology [245]. Likewise, *USP15* is downregulated in paclitaxel-resistant ovarian cancer [246] and *CYLD* in melanoma, hepatocellular carcinoma, and other malignant tumors [34, 247–249]. *USP18* expression is reduced in human leiomyosarcomas, and, accordingly, mutant mice deficient for this USP develop these tumors spontaneously [250]. Finally, *USP46* is downregulated in colorectal cancer samples [251] and *A20* in some types of lymphoma [252].

16.5 Targeting DUBs

Due to their wide functional diversity and critical implication in human pathologies, DUBs have emerged as appealing targets in the development of new specific therapies, especially against human malignancies [253]. Cyclopentenone prostaglandins were the first DUB inhibitors that targeted active sites [254]. So far, several small-molecule inhibitors against DUBs, mainly USPs and UCHLs, have been described. Thus, two different classes of inhibitors have been used to neutralized UCHL1: isatin *O*-acyl oximes and 3-amino-2-oxo-7*H*-thieno[2,3-*b*]pyridine-6-one derivatives [255]. Competitive inhibitors with similar dihydro-pyrrole skeletons and several isatin derivatives have been described to inhibit UCHL3 [3]. LS1 is another UCHL3 inhibitor identified in a FRET-based screen [253]. Regarding USPs, pimozide, GW7647, and ML323 are potent USP1-UAF1 inhibitors [13, 256], whereas 2-cyano-pyrimidine and triazine derivatives have been described to block USP2 function [257]. Moreover, HBX 41,108, HBX 19,818, HBX 28,258, P022077, P5091, and spongiacidin C have been described as selective USP7 inhibitors. In this sense, a more potent analog of P5091 inhibits USP7 and USP47 [253, 258]. HBX 90,397, HBX 90,659, and 9-oxo-9-H-indeno [1,2-b]pyrazine-2,3-dicarbonitrile and analogs selectively inhibit USP8 activity [257, 259]. Finally, mitoxantrone targets

USP11 and impacts pancreatic ductal adenocarcinoma cell survival [260], whereas IU1 inhibits USP14 and enhances proteasome activity [261].

There are also broad-spectrum inhibitors that target several DUBs. WP1130 blocks the activity of USP5, USP9X, USP14, and UCHL5 [253]. Furthermore, natural-occurring isothiocyanates exert anticancer effects by inhibiting DUBs, including USP9X and UCHL5 [262]. Similarly, small-molecule inhibitors against proteasome-associated DUBs have been described, such as RA-9, auranofin (Aur), 4-arylidene curcumin analog (AC17), and the chalcone derivative b-AP-15. RA-9 and Aur inhibit USP14 and UCHL5 and diminish tumor growth in vitro and in vivo [253, 263]. RA-9 displays more broad range activity and also inhibits UCHL1, UCHL3, USP2, and USP8 [263, 264]. AC17 also inhibits the DUB activity of 19S regulatory particles, resulting in inhibition of NF-κB signaling and reactivation of p53 [253], whereas b-AP15 blocks the activity of USP14 and UCHL5 and induces apoptosis that is insensitive to p53 status [265]. Furthermore, betulinic acid is a pan-DUB inhibitor that selectively kills prostate cancer but not normal cells [266]. Finally, gambogic acid inhibits cellular DUB activity, inducing the accumulation of polyubiquitinated proteins [253], and PR-619, another broad specificity DUB inhibitor, activates autophagic pathways [267].

Due to the increased cellular stresses produced by cancer, tumor cells might exhibit an exacerbated dependence on the normal function of certain DUBs [268]. In this scenario, DUBs will be essential for human malignancies, even without undergoing activating mutations. This concept is referred to as non-oncogene addiction and opens a new window of therapeutic opportunities for the treatment of human pathologies. Thus, the inhibition of USP2 has been described as an effective approach to suppress growth in cancer cells addicted to cyclin D1 expression [72]. Moreover, downregulation of *USP11* leads to spontaneous DNA damage repair activation and hypersensitivity to PARP inhibition, ionizing radiation and other sources of genotoxic stress [269]. Furthermore, gemcitabine disrupts the interaction of USP9X and MCL1 and sensitizes cells to ABT-737 treatment, inducing caspase-dependent apoptosis [270]. Additionally, depletion of USP4 diminishes TGF-β-induced EMT and metastasis [29], and *USP15* downregulation decreases TGF-β activity and oncogenesis [169]. Moreover, USP6 acts as an oncogene by positively regulating NF-κB, whose activity is essential for USP6-mediated tumorigenesis [147]. Finally, inhibition of USP37 blocks the proliferation of lung cancer cells by reducing c-Myc levels [52], and depletion of USP1-UAF1 overcomes the resistance of cancer cells to cisplatin [43] (Fig. 16.3).

As discussed above, multiple DUBs present tumor-protective properties, and, consequently, effective cancer therapies should spare their activity or even be designed to potentiate DUB activities lost as a consequence of malignant transformation. In this sense, demethylating drugs, in case of loss of expression by promoter hypermethylation, or exogenous compounds that compensate DUB activity could emerge as effective strategies to overcome DUB deficiency. As notable examples of this possibility, anti-inflammatory drugs, inhibition of tropomyosin kinase TRK, and downregulation of Snail have been proposed as strategies for *CYLD* inactivating mutations [271–273].

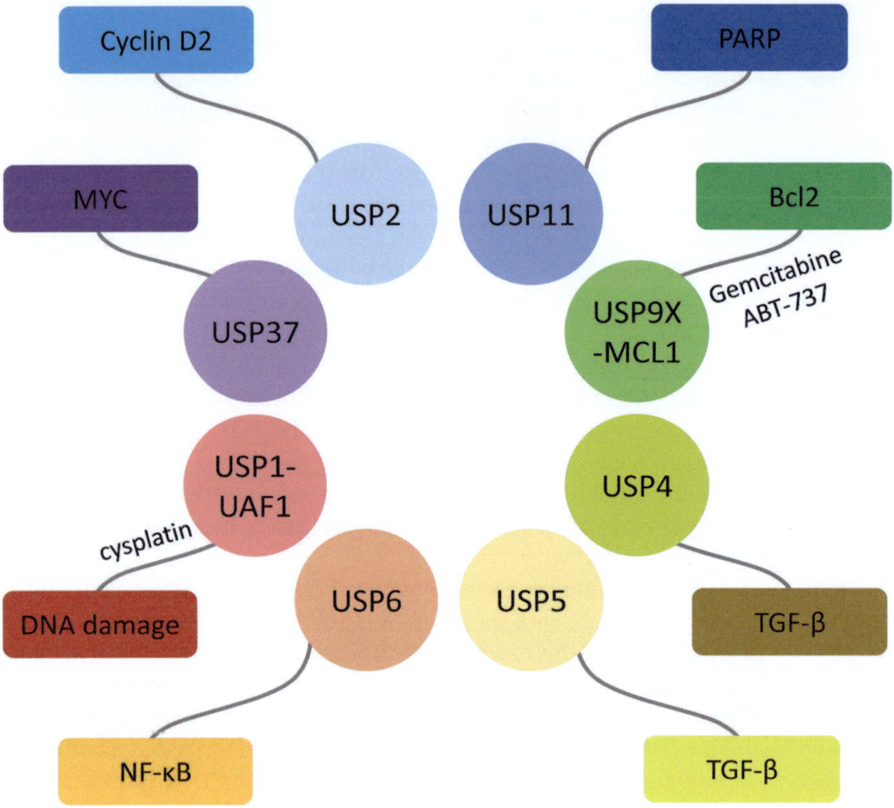

Fig. 16.3 Non-oncogene addiction of USPs. Several examples of non-oncogene addiction phenomena involving USPs and their functional interaction with different proteins in cancer

16.6 Conclusions and Perspectives

The human genome encodes at least 100 DUBs that, despite sharing a core domain, exhibit a wide structural and functional diversity. For this reason, DUBs are implicated in numerous physiological and pathological conditions, including cancer. In this sense, there is a broad spectrum of DUBs with critical functions at several stages of cancer development and progression. In fact, the number of DUBs known to be mutated that behave as oncogenes or tumor suppressor genes has kept growing over the last years. Moreover, transcriptional expression of several DUBs is also altered in many human malignancies. On this basis, DUBs emerge as appealing targets in the development of therapies against cancer, although further understanding of the activity and regulation of these proteolytic enzymes will be required for their functional and clinical validation as drugs targets. In this regard, the generation of gain- or loss-of-function mouse models for DUBs will help to understand their individual relevance in normal physiology, as well as their contribution to tumorigenic progression. Hopefully, these studies will contribute to clarify the functional complexity of this superfamily of proteases and pave the way for the development of new anticancer therapies based on DUB-targeting approaches.

Acknowledgments We thank Dr. Víctor Quesada for helpful comments and assistance. Our work is supported by grants from Ministerio de Economía y Competitividad, Instituto de Salud Carlos III (RTICC), and Principado de Asturias, Spain. The Instituto Universitario de Oncología is supported by Fundación Bancaria Caja de Ahorros de Asturias.

Conflict of Interest The authors declare no conflict of interest.

References

1. Clague MJ, Heride C, Urbe S (2015) The demographics of the ubiquitin system. Trends Cell Biol 25(7):417–426
2. Komander D, Clague MJ, Urbe S (2009) Breaking the chains: structure and function of the deubiquitinases. Nat Rev Mol Cell Biol 10(8):550–563
3. Fraile JM, Quesada V, Rodriguez D et al (2012) Deubiquitinases in cancer: new functions and therapeutic options. Oncogene 31(19):2373–2388
4. Quesada V, Diaz-Perales A, Gutierrez-Fernandez A et al (2004) Cloning and enzymatic analysis of 22 novel human ubiquitin-specific proteases. Biochem Biophys Res Commun 314(1):54–62
5. Koulich E, Li X, DeMartino GN (2008) Relative structural and functional roles of multiple deubiquitylating proteins associated with mammalian 26S proteasome. Mol Biol Cell 19(3):1072–1082
6. Quesada V, Ordonez GR, Sanchez LM et al (2009) The Degradome database: mammalian proteases and diseases of proteolysis. Nucleic Acids Res 37(Database issue):D239–D243
7. Mevissen TE, Hospenthal MK, Geurink PP et al (2013) OTU deubiquitinases reveal mechanisms of linkage specificity and enable ubiquitin chain restriction analysis. Cell 154(1):169–184
8. Reyes-Turcu FE, Ventii KH, Wilkinson KD (2009) Regulation and cellular roles of ubiquitin-specific deubiquitinating enzymes. Annu Rev Biochem 78:363–397
9. Wiener R, Zhang X, Wang T et al (2012) The mechanism of OTUB1-mediated inhibition of ubiquitination. Nature 483(7391):618–622
10. Nicastro G, Menon RP, Masino L et al (2005) The solution structure of the Josephin domain of ataxin-3: structural determinants for molecular recognition. Proc Natl Acad Sci U S A 102(30):10493–10498
11. Kuhlbrodt K, Janiesch PC, Kevei E et al (2011) The Machado-Joseph disease deubiquitylase ATX-3 couples longevity and proteostasis. Nat Cell Biol 13(3):273–281
12. Sato Y, Yoshikawa A, Yamagata A et al (2008) Structural basis for specific cleavage of Lys 63-linked polyubiquitin chains. Nature 455(7211):358–362
13. Liang J, Saad Y, Lei T et al (2010) MCP-induced protein 1 deubiquitinates TRAF proteins and negatively regulates JNK and NF-kappaB signaling. J Exp Med 207(13):2959–2973
14. Akutsu M, Dikic I, Bremm A (2016) Ubiquitin chain diversity at a glance. J Cell Sci 129(5):875–880
15. Lee BH, Lu Y, Prado MA et al (2016) USP14 deubiquitinates proteasome-bound substrates that are ubiquitinated at multiple sites. Nature 532(7599):398–401
16. Wertz IE, Newton K, Seshasayee D et al (2015) Phosphorylation and linear ubiquitin direct A20 inhibition of inflammation. Nature 528(7582):370–375
17. Keusekotten K, Elliott PR, Glockner L et al (2013) OTULIN antagonizes LUBAC signaling by specifically hydrolyzing Met1-linked polyubiquitin. Cell 153(6):1312–1326
18. Sahtoe DD, Sixma TK (2015) Layers of DUB regulation. Trends Biochem Sci 40(8):456–467

19. Jaster R, Baek KH, D'Andrea AD (1999) Analysis of cis-acting sequences and trans-acting factors regulating the interleukin-3 response element of the DUB-1 gene. Biochim Biophys Acta 1446(3):308–316
20. Burrows JF, McGrattan MJ, Rascle A et al (2004) DUB-3, a cytokine-inducible deubiquitinating enzyme that blocks proliferation. J Biol Chem 279(14):13993–14000
21. Baek KH, Kim MS, Kim YS et al (2004) DUB-1A, a novel deubiquitinating enzyme subfamily member, is polyubiquitinated and cytokine-inducible in B-lymphocytes. J Biol Chem 279(4):2368–2376
22. Yoshida H, Jono H, Kai H et al (2005) The tumor suppressor cylindromatosis (CYLD) acts as a negative regulator for toll-like receptor 2 signaling via negative cross-talk with TRAF6 AND TRAF7. J Biol Chem 280(49):41111–41121
23. Xu W, Xue L, Sun Y et al (2013) Bcl10 is an essential regulator for A20 gene expression. J Physiol Biochem 69(4):821–834
24. Xu Z, Pei L, Wang L et al (2014) Snail1-dependent transcriptional repression of Cezanne2 in hepatocellular carcinoma. Oncogene 33(22):2836–2845
25. Shen G, Lin Y, Yang X et al (2014) MicroRNA-26b inhibits epithelial-mesenchymal transition in hepatocellular carcinoma by targeting USP9X. BMC Cancer 14:393
26. Zhang L, Xu B, Qiang Y et al (2015) Overexpression of deubiquitinating enzyme USP28 promoted non-small cell lung cancer growth. J Cell Mol Med 19(4):799–805
27. Lopez-Otin C, Hunter T (2010) The regulatory crosstalk between kinases and proteases in cancer. Nat Rev Cancer 10(4):278–292
28. Huang OW, Ma X, Yin J et al (2012) Phosphorylation-dependent activity of the deubiquitinase DUBA. Nat Struct Mol Biol 19(2):171–175
29. Zhang L, Zhou F, Drabsch Y et al (2012) USP4 is regulated by AKT phosphorylation and directly deubiquitylates TGF-beta type I receptor. Nat Cell Biol 14(7):717–726
30. Todi SV, Winborn BJ, Scaglione KM et al (2009) Ubiquitination directly enhances activity of the deubiquitinating enzyme ataxin-3. EMBO J 28(4):372–382
31. Meray RK, Lansbury PT Jr (2007) Reversible monoubiquitination regulates the Parkinson disease-associated ubiquitin hydrolase UCH-L1. J Biol Chem 282(14):10567–10575
32. Zhen Y, Knobel PA, Stracker TH et al (2014) Regulation of USP28 deubiquitinating activity by SUMO conjugation. J Biol Chem 289(50):34838–34850
33. Meulmeester E, Kunze M, Hsiao HH et al (2008) Mechanism and consequences for paralog-specific sumoylation of ubiquitin-specific protease 25. Mol Cell 30(5):610–619
34. Kobayashi T, Masoumi KC, Massoumi R (2015) Deubiquitinating activity of CYLD is impaired by SUMOylation in neuroblastoma cells. Oncogene 34(17):2251–2260
35. Kulathu Y, Garcia FJ, Mevissen TE et al (2013) Regulation of A20 and other OTU deubiquitinases by reversible oxidation. Nat Commun 4:1569
36. Lee JG, Baek K, Soetandyo N et al (2013) Reversible inactivation of deubiquitinases by reactive oxygen species in vitro and in cells. Nat Commun 4:1568
37. Huang TT, Nijman SM, Mirchandani KD et al (2006) Regulation of monoubiquitinated PCNA by DUB autocleavage. Nat Cell Biol 8(4):339–347
38. Coornaert B, Baens M, Heyninck K et al (2008) T cell antigen receptor stimulation induces MALT1 paracaspase-mediated cleavage of the NF-kappaB inhibitor A20. Nat Immunol 9(3):263–271
39. Staal J, Driege Y, Bekaert T et al (2011) T-cell receptor-induced JNK activation requires proteolytic inactivation of CYLD by MALT1. EMBO J 30(9):1742–1752
40. Endo A, Matsumoto M, Inada T et al (2009) Nucleolar structure and function are regulated by the deubiquitylating enzyme USP36. J Cell Sci 122(Pt 5):678–686
41. Nakamura N, Hirose S (2008) Regulation of mitochondrial morphology by USP30, a deubiquitinating enzyme present in the mitochondrial outer membrane. Mol Biol Cell 19(5):1903–1911
42. Garcia-Santisteban I, Zorroza K, Rodriguez JA (2012) Two nuclear localization signals in USP1 mediate nuclear import of the USP1/UAF1 complex. PLoS One 7(6):e38570

43. Liang Q, Dexheimer TS, Zhang P et al (2014) A selective USP1-UAF1 inhibitor links deubiquitination to DNA damage responses. Nat Chem Biol 10(4):298–304
44. Daou S, Hammond-Martel I, Mashtalir N et al (2015) The BAP1/ASXL2 histone H2A deubiquitinase complex regulates cell proliferation and is disrupted in cancer. J Biol Chem 290(48):28643–28663
45. Sahtoe DD, van Dijk WJ, Ekkebus R et al (2016) BAP1/ASXL1 recruitment and activation for H2A deubiquitination. Nat Commun 7:10292
46. Laco MN, Cortes L, Travis SM et al (2012) Valosin-containing protein (VCP/p97) is an activator of wild-type ataxin-3. PLoS One 7(9):e43563
47. Reddy BA, van der Knaap JA, Bot AG et al (2014) Nucleotide biosynthetic enzyme GMP synthase is a TRIM21-controlled relay of p53 stabilization. Mol Cell 53(3):458–470
48. Reed BJ, Locke MN, Gardner RG (2015) A conserved deubiquitinating enzyme uses intrinsically disordered regions to scaffold multiple protein interaction sites. J Biol Chem 290(33):20601–20612
49. Adler AS, Littlepage LE, Lin M et al (2008) CSN5 isopeptidase activity links COP9 signalosome activation to breast cancer progression. Cancer Res 68(2):506–515
50. Sowa ME, Bennett EJ, Gygi SP et al (2009) Defining the human deubiquitinating enzyme interaction landscape. Cell 138(2):389–403
51. Popov N, Wanzel M, Madiredjo M et al (2007) The ubiquitin-specific protease USP28 is required for MYC stability. Nat Cell Biol 9(7):765–774
52. Pan J, Deng Q, Jiang C et al (2015) USP37 directly deubiquitinates and stabilizes c-Myc in lung cancer. Oncogene 34(30):3957–3967
53. Sun XX, He X, Yin L et al (2015) The nucleolar ubiquitin-specific protease USP36 deubiquitinates and stabilizes c-Myc. Proc Natl Acad Sci U S A 112(12):3734–3739
54. Lin Z, Yang H, Tan C et al (2013) USP10 antagonizes c-Myc transcriptional activation through SIRT6 stabilization to suppress tumor formation. Cell Rep 5(6):1639–1649
55. Uras IZ, List T, Nijman SM (2012) Ubiquitin-specific protease 4 inhibits mono-ubiquitination of the master growth factor signaling kinase PDK1. PLoS One 7(2):e31003
56. Song MS, Salmena L, Carracedo A et al (2008) The deubiquitinylation and localization of PTEN are regulated by a HAUSP-PML network. Nature 455(7214):813–817
57. van der Horst A, de Vries-Smits AM, Brenkman AB et al (2006) FOXO4 transcriptional activity is regulated by monoubiquitination and USP7/HAUSP. Nat Cell Biol 8(10):1064–1073
58. Sacco JJ, Yau TY, Darling S et al (2014) The deubiquitylase Ataxin-3 restricts PTEN transcription in lung cancer cells. Oncogene 33(33):4265–4272
59. Yuan L, Lv Y, Li H et al (2015) Deubiquitylase OTUD3 regulates PTEN stability and suppresses tumorigenesis. Nat Cell Biol 17(9):1169–1181
60. Zhang J, Zhang P, Wei Y et al (2013) Deubiquitylation and stabilization of PTEN by USP13. Nat Cell Biol 15(12):1486–1494
61. Mahanic CS, Budhavarapu V, Graves JD et al (2015) Regulation of E2 promoter binding factor 1 (E2F1) transcriptional activity through a deubiquitinating enzyme, UCH37. J Biol Chem 290(44):26508–26522
62. Yang Y, Liu M, Li D et al (2014) CYLD regulates spindle orientation by stabilizing astral microtubules and promoting dishevelled-NuMA-dynein/dynactin complex formation. Proc Natl Acad Sci U S A 111(6):2158–2163
63. Chen M, Gutierrez GJ, Ronai ZA (2011) Ubiquitin-recognition protein Ufd1 couples the endoplasmic reticulum (ER) stress response to cell cycle control. Proc Natl Acad Sci U S A 108(22):9119–9124
64. Huang X, Summers MK, Pham V et al (2011) Deubiquitinase USP37 is activated by CDK2 to antagonize APC(CDH1) and promote S phase entry. Mol Cell 42(4):511–523
65. Yeh C, Coyaud E, Bashkurov M et al (2015) The Deubiquitinase USP37 regulates chromosome cohesion and mitotic progression. Curr Biol 25(17):2290–2299
66. Das CM, Taylor P, Gireud M et al (2013) The deubiquitylase USP37 links REST to the control of p27 stability and cell proliferation. Oncogene 32(13):1691–1701

67. van Leuken RJ, Luna-Vargas MP, Sixma TK et al (2008) Usp39 is essential for mitotic spindle checkpoint integrity and controls mRNA-levels of aurora B. Cell Cycle 7(17):2710–2719
68. Stegmeier F, Rape M, Draviam VM et al (2007) Anaphase initiation is regulated by antagonistic ubiquitination and deubiquitination activities. Nature 446(7138):876–881
69. Yan K, Li L, Wang X et al (2015) The deubiquitinating enzyme complex BRISC is required for proper mitotic spindle assembly in mammalian cells. J Cell Biol 210(2):209–224
70. Aressy B, Jullien D, Cazales M et al (2010) A screen for deubiquitinating enzymes involved in the G(2)/M checkpoint identifies USP50 as a regulator of HSP90-dependent Wee1 stability. Cell Cycle 9(18):3815–3822
71. Pereg Y, Liu BY, O'Rourke KM et al (2010) Ubiquitin hydrolase Dub3 promotes oncogenic transformation by stabilizing Cdc25A. Nat Cell Biol 12(4):400–406
72. Shan J, Zhao W, Gu W (2009) Suppression of cancer cell growth by promoting cyclin D1 degradation. Mol Cell 36(3):469–476
73. Lu Y, Bedard N, Chevalier S et al (2011) Identification of distinctive patterns of USP19-mediated growth regulation in normal and malignant cells. PLoS One 6(1):e15936
74. Yoshida A, Yoneda-Kato N, Panattoni M et al (2010) CSN5/Jab1 controls multiple events in the mammalian cell cycle. FEBS Lett 584(22):4545–4552
75. Yu H, Mashtalir N, Daou S et al (2010) The ubiquitin carboxyl hydrolase BAP1 forms a ternary complex with YY1 and HCF-1 and is a critical regulator of gene expression. Mol Cell Biol 30(21):5071–5085
76. Qin J, Zhou Z, Chen W et al (2015) BAP1 promotes breast cancer cell proliferation and metastasis by deubiquitinating KLF5. Nat Commun 6:8471
77. LaFave LM, Beguelin W, Koche R et al (2015) Loss of BAP1 function leads to EZH2-dependent transformation. Nat Med 21(11):1344–1349
78. Jung JK, Jang SW, Kim JM (2016) A novel role for the deubiquitinase USP1 in the control of centrosome duplication. Cell Cycle 15(4):584–592
79. Li J, D'Angiolella V, Seeley ES et al (2013) USP33 regulates centrosome biogenesis via deubiquitination of the centriolar protein CP110. Nature 495(7440):255–259
80. Zhang Y, Foreman O, Wigle DA et al (2012) USP44 regulates centrosome positioning to prevent aneuploidy and suppress tumorigenesis. J Clin Invest 122(12):4362–4374
81. Nijman SM, Huang TT, Dirac AM et al (2005) The deubiquitinating enzyme USP1 regulates the Fanconi anemia pathway. Mol Cell 17(3):331–339
82. Murai J, Yang K, Dejsuphong D et al (2011) The USP1/UAF1 complex promotes double-strand break repair through homologous recombination. Mol Cell Biol 31(12):2462–2469
83. Kee Y, Huang TT (2015) Role of deubiquitinating enzymes in DNA repair. Mol Cell Biol 36(4):524–544
84. Morris JR (2012) Attenuation of the ubiquitin conjugate DNA damage signal by the proteasomal DUB POH1. Cell Cycle 11(22):4103–4104
85. Yu H, Pak H, Hammond-Martel I et al (2014) Tumor suppressor and deubiquitinase BAP1 promotes DNA double-strand break repair. Proc Natl Acad Sci U S A 111(1):285–290
86. Schoenfeld AR, Apgar S, Dolios G et al (2004) BRCA2 is ubiquitinated in vivo and interacts with USP11, a deubiquitinating enzyme that exhibits prosurvival function in the cellular response to DNA damage. Mol Cell Biol 24(17):7444–7455
87. Zhang D, Zaugg K, Mak TW et al (2006) A role for the deubiquitinating enzyme USP28 in control of the DNA-damage response. Cell 126(3):529–542
88. McGarry E, Gaboriau D, Rainey M et al. (2016) The deubiquitinase USP9X maintains DNA replication fork stability and DNA damage checkpoint responses by regulating CLASPIN during S-phase. Cancer Res 76(8):2384–2393
89. Delgado-Diaz MR, Martin Y, Berg A et al (2014) Dub3 controls DNA damage signalling by direct deubiquitination of H2AX. Mol Oncol 8(5):884–893
90. Wang Z, Zhang H, Liu J et al (2016) USP51 deubiquitylates H2AK13,15ub and regulates DNA damage response. Genes Dev 30(8):946–959

91. Nishi R, Wijnhoven P, le Sage C et al (2014) Systematic characterization of deubiquitylating enzymes for roles in maintaining genome integrity. Nat Cell Biol 16(10):1016–1026. 1011-1018
92. Zlatanou A, Sabbioneda S, Miller ES et al (2016) USP7 is essential for maintaining Rad18 stability and DNA damage tolerance. Oncogene 35(8):965–976
93. Kashiwaba S, Kanao R, Masuda Y et al (2015) USP7 is a suppressor of PCNA ubiquitination and oxidative-stress-induced mutagenesis in human cells. Cell Rep 13(10):2072–2080
94. Khoronenkova SV, Dianova II, Ternette N et al (2012) ATM-dependent downregulation of USP7/HAUSP by PPM1G activates p53 response to DNA damage. Mol Cell 45(6):801–813
95. Alonso-de Vega I, Martin Y, Smits VA (2014) USP7 controls Chk1 protein stability by direct deubiquitination. Cell Cycle 13(24):3921–3926
96. Schwertman P, Lagarou A, Dekkers DH et al (2012) UV-sensitive syndrome protein UVSSA recruits USP7 to regulate transcription-coupled repair. Nat Genet 44(5):598–602
97. Zhang L, Lubin A, Chen H et al (2012) The deubiquitinating protein USP24 interacts with DDB2 and regulates DDB2 stability. Cell Cycle 11(23):4378–4384
98. Parsons JL, Dianova II, Khoronenkova SV et al (2011) USP47 is a deubiquitylating enzyme that regulates base excision repair by controlling steady-state levels of DNA polymerase beta. Mol Cell 41(5):609–615
99. Joo HY, Zhai L, Yang C et al (2007) Regulation of cell cycle progression and gene expression by H2A deubiquitination. Nature 449(7165):1068–1072
100. Atanassov BS, Evrard YA, Multani AS et al (2009) Gcn5 and SAGA regulate shelterin protein turnover and telomere maintenance. Mol Cell 35(3):352–364
101. Zhang Y (2003) Transcriptional regulation by histone ubiquitination and deubiquitination. Genes Dev 17(22):2733–2740
102. Zhu P, Zhou W, Wang J et al (2007) A histone H2A deubiquitinase complex coordinating histone acetylation and H1 dissociation in transcriptional regulation. Mol Cell 27(4):609–621
103. Zhang XY, Varthi M, Sykes SM et al (2008) The putative cancer stem cell marker USP22 is a subunit of the human SAGA complex required for activated transcription and cell-cycle progression. Mol Cell 29(1):102–111
104. Feng L, Wang J, Chen J (2010) The Lys63-specific deubiquitinating enzyme BRCC36 is regulated by two scaffold proteins localizing in different subcellular compartments. J Biol Chem 285(40):30982–30988
105. Maertens GN, El Messaoudi-Aubert S, Elderkin S et al (2010) Ubiquitin-specific proteases 7 and 11 modulate Polycomb regulation of the INK4a tumour suppressor. EMBO J 29(15):2553–2565
106. Long L, Thelen JP, Furgason M et al (2014) The U4/U6 recycling factor SART3 has histone chaperone activity and associates with USP15 to regulate H2B deubiquitination. J Biol Chem 289(13):8916–8930
107. Zhang Z, Jones A, Joo HY et al (2013) USP49 deubiquitinates histone H2B and regulates cotranscriptional pre-mRNA splicing. Genes Dev 27(14):1581–1595
108. Khan A, Giri S, Wang Y et al (2015) BEND3 represses rDNA transcription by stabilizing a NoRC component via USP21 deubiquitinase. Proc Natl Acad Sci U S A 112(27):8338–8343
109. Wang Q, Ma S, Song N et al (2016) Stabilization of histone demethylase PHF8 by USP7 promotes breast carcinogenesis. J Clin Invest 126(6):2205–2220
110. Yao T, Song L, Jin J et al (2008) Distinct modes of regulation of the Uch37 deubiquitinating enzyme in the proteasome and in the Ino80 chromatin-remodeling complex. Mol Cell 31(6):909–917
111. Brooks CL, Li M, Hu M et al (2007) The p53--Mdm2--HAUSP complex is involved in p53 stabilization by HAUSP. Oncogene 26(51):7262–7266
112. Kon N, Kobayashi Y, Li M et al (2010) Inactivation of HAUSP in vivo modulates p53 function. Oncogene 29(9):1270–1279
113. Stevenson LF, Sparks A, Allende-Vega N et al (2007) The deubiquitinating enzyme USP2a regulates the p53 pathway by targeting Mdm2. EMBO J 26(4):976–986

114. Zou Q, Jin J, Hu H et al (2014) USP15 stabilizes MDM2 to mediate cancer-cell survival and inhibit antitumor T cell responses. Nat Immunol 15(6):562–570
115. Sun XX, Challagundla KB, Dai MS (2012) Positive regulation of p53 stability and activity by the deubiquitinating enzyme Otubain 1. EMBO J 31(3):576–592
116. Yuan J, Luo K, Zhang L et al (2010) USP10 regulates p53 localization and stability by deubiquitinating p53. Cell 140(3):384–396
117. Zhang X, Berger FG, Yang J et al (2011) USP4 inhibits p53 through deubiquitinating and stabilizing ARF-BP1. EMBO J 30(11):2177–2189
118. Li Z, Hao Q, Luo J et al (2016) USP4 inhibits p53 and NF-kappaB through deubiquitinating and stabilizing HDAC2. Oncogene 35(22):2902–2912
119. Lin Z, Yang H, Kong Q et al (2012) USP22 antagonizes p53 transcriptional activation by deubiquitinating Sirt1 to suppress cell apoptosis and is required for mouse embryonic development. Mol Cell 46(4):484–494
120. Liu J, Chung HJ, Vogt M et al (2011) JTV1 co-activates FBP to induce USP29 transcription and stabilize p53 in response to oxidative stress. EMBO J 30(5):846–858
121. Zhang L, Nemzow L, Chen H et al (2015) The deubiquitinating enzyme USP24 is a regulator of the UV damage response. Cell Rep 10(2):140–147
122. Hock AK, Vigneron AM, Carter S et al (2011) Regulation of p53 stability and function by the deubiquitinating enzyme USP42. EMBO J 30(24):4921–4930
123. Xia Y, Shen S, Verma IM (2014) NF-kappaB, an active player in human cancers. Cancer Immunol Res 2(9):823–830
124. Harhaj EW, Dixit VM (2011) Deubiquitinases in the regulation of NF-kappaB signaling. Cell Res 21(1):22–39
125. Reiley WW, Jin W, Lee AJ et al (2007) Deubiquitinating enzyme CYLD negatively regulates the ubiquitin-dependent kinase Tak1 and prevents abnormal T cell responses. J Exp Med 204(6):1475–1485
126. Massoumi R, Chmielarska K, Hennecke K et al (2006) Cyld inhibits tumor cell proliferation by blocking Bcl-3-dependent NF-kappaB signaling. Cell 125(4):665–677
127. Reiley W, Zhang M, Sun SC (2004) Negative regulation of JNK signaling by the tumor suppressor CYLD. J Biol Chem 279(53):55161–55167
128. Li L, Soetandyo N, Wang Q et al (2009) The zinc finger protein A20 targets TRAF2 to the lysosomes for degradation. Biochim Biophys Acta 1793(2):346–353
129. Tokunaga F, Nishimasu H, Ishitani R et al (2012) Specific recognition of linear polyubiquitin by A20 zinc finger 7 is involved in NF-kappaB regulation. EMBO J 31(19):3856–3870
130. Wertz IE, O'Rourke KM, Zhou H et al (2004) De-ubiquitination and ubiquitin ligase domains of A20 downregulate NF-kappaB signalling. Nature 430(7000):694–699
131. Moquin DM, McQuade T, Chan FK (2013) CYLD deubiquitinates RIP1 in the TNFalpha-induced necrosome to facilitate kinase activation and programmed necrosis. PLoS One 8(10):e76841
132. Xu G, Tan X, Wang H et al (2010) Ubiquitin-specific peptidase 21 inhibits tumor necrosis factor alpha-induced nuclear factor kappaB activation via binding to and deubiquitinating receptor-interacting protein 1. J Biol Chem 285(2):969–978
133. Hou X, Wang L, Zhang L et al (2013) Ubiquitin-specific protease 4 promotes TNF-alpha-induced apoptosis by deubiquitination of RIP1 in head and neck squamous cell carcinoma. FEBS Lett 587(4):311–316
134. Zaman MM, Nomura T, Takagi T et al (2013) Ubiquitination-deubiquitination by the TRIM27-USP7 complex regulates tumor necrosis factor alpha-induced apoptosis. Mol Cell Biol 33(24):4971–4984
135. McNally RS, Davis BK, Clements CM et al (2011) DJ-1 enhances cell survival through the binding of Cezanne, a negative regulator of NF-kappaB. J Biol Chem 286(6):4098–4106
136. Hu H, Brittain GC, Chang JH et al (2013) OTUD7B controls non-canonical NF-kappaB activation through deubiquitination of TRAF3. Nature 494(7437):371–374

137. Gonzalez-Navajas JM, Law J, Nguyen KP et al (2010) Interleukin 1 receptor signaling regulates DUBA expression and facilitates toll-like receptor 9-driven antiinflammatory cytokine production. J Exp Med 207(13):2799–2807
138. Zhong B, Liu X, Wang X et al (2012) Negative regulation of IL-17-mediated signaling and inflammation by the ubiquitin-specific protease USP25. Nat Immunol 13(11):1110–1117
139. Colleran A, Collins PE, O'Carroll C et al (2013) Deubiquitination of NF-kappaB by ubiquitin-specific protease-7 promotes transcription. Proc Natl Acad Sci U S A 110(2):618–623
140. Schweitzer K, Bozko PM, Dubiel W et al (2007) CSN controls NF-kappaB by deubiquitinylation of IkappaBalpha. EMBO J 26(6):1532–1541
141. Metzig M, Nickles D, Falschlehner C et al (2011) An RNAi screen identifies USP2 as a factor required for TNF-alpha-induced NF-kappaB signaling. Int J Cancer 129(3):607–618
142. Fan YH, Yu Y, Mao RF et al (2011) USP4 targets TAK1 to downregulate TNFalpha-induced NF-kappaB activation. Cell Death Differ 18(10):1547–1560
143. Xiao N, Li H, Luo J et al (2012) Ubiquitin-specific protease 4 (USP4) targets TRAF2 and TRAF6 for deubiquitination and inhibits TNFalpha-induced cancer cell migration. Biochem J 441(3):979–986
144. Liu X, Li H, Zhong B et al (2013) USP18 inhibits NF-kappaB and NFAT activation during Th17 differentiation by deubiquitinating the TAK1-TAB1 complex. J Exp Med 210(8):1575–1590
145. Niu J, Shi Y, Xue J et al (2013) USP10 inhibits genotoxic NF-kappaB activation by MCPIP1-facilitated deubiquitination of NEMO. EMBO J 32(24):3206–3219
146. Wang W, Huang X, Xin HB et al (2015) TRAF family member-associated NF-kappaB activator (TANK) inhibits genotoxic nuclear factor kappaB activation by facilitating Deubiquitinase USP10-dependent Deubiquitination of TRAF6 ligase. J Biol Chem 290(21):13372–13385
147. Pringle LM, Young R, Quick L et al (2012) Atypical mechanism of NF-kappaB activation by TRE17/ubiquitin-specific protease 6 (USP6) oncogene and its requirement in tumorigenesis. Oncogene 31(30):3525–3535
148. Liu Z, Zanata SM, Kim J et al (2013) The ubiquitin-specific protease USP2a prevents endocytosis-mediated EGFR degradation. Oncogene 32(13):1660–1669
149. Niendorf S, Oksche A, Kisser A et al (2007) Essential role of ubiquitin-specific protease 8 for receptor tyrosine kinase stability and endocytic trafficking in vivo. Mol Cell Biol 27(13):5029–5039
150. Alwan HA, van Leeuwen JE (2007) UBPY-mediated epidermal growth factor receptor (EGFR) de-ubiquitination promotes EGFR degradation. J Biol Chem 282(3):1658–1669
151. McCullough J, Clague MJ, Urbe S (2004) AMSH is an endosome-associated ubiquitin isopeptidase. J Cell Biol 166(4):487–492
152. Duex JE, Sorkin A (2009) RNA interference screen identifies Usp18 as a regulator of epidermal growth factor receptor synthesis. Mol Biol Cell 20(6):1833–1844
153. Liu H, Buus R, Clague MJ et al (2009) Regulation of ErbB2 receptor status by the proteasomal DUB POH1. PLoS One 4(5):e5544
154. Anastas JN, Moon RT (2013) WNT signalling pathways as therapeutic targets in cancer. Nat Rev Cancer 13(1):11–26
155. Tauriello DV, Haegebarth A, Kuper I et al (2010) Loss of the tumor suppressor CYLD enhances Wnt/beta-catenin signaling through K63-linked ubiquitination of Dvl. Mol Cell 37(5):607–619
156. Zhao B, Schlesiger C, Masucci MG et al (2009) The ubiquitin specific protease 4 (USP4) is a new player in the Wnt signalling pathway. J Cell Mol Med 13(8B):1886–1895
157. Huang X, Langelotz C, Hetfeld-Pechoc BK et al (2009) The COP9 signalosome mediates beta-catenin degradation by deneddylation and blocks adenomatous polyposis coli destruction via USP15. J Mol Biol 391(4):691–702
158. Yun SI, Kim HH, Yoon JH et al (2015) Ubiquitin specific protease 4 positively regulates the WNT/beta-catenin signaling in colorectal cancer. Mol Oncol 9(9):1834–1851
159. Chen Y, Li Y, Xue J et al (2016) Wnt-induced deubiquitination FoxM1 ensures nucleus beta-catenin transactivation. EMBO J 35(6):668–684

160. Lui TT, Lacroix C, Ahmed SM et al (2011) The ubiquitin-specific protease USP34 regulates axin stability and Wnt/beta-catenin signaling. Mol Cell Biol 31(10):2053–2065
161. Shi J, Liu Y, Xu X et al (2015) Deubiquitinase USP47/UBP64E regulates beta-catenin ubiquitination and degradation and plays a positive role in Wnt signaling. Mol Cell Biol 35(19):3301–3311
162. Tran H, Hamada F, Schwarz-Romond T et al (2008) Trabid, a new positive regulator of Wnt-induced transcription with preference for binding and cleaving K63-linked ubiquitin chains. Genes Dev 22(4):528–542
163. Rivkin E, Almeida SM, Ceccarelli DF et al (2013) The linear ubiquitin-specific deubiquitinase gumby regulates angiogenesis. Nature 498(7454):318–324
164. Akhurst RJ, Padgett RW (2015) Matters of context guide future research in TGFbeta superfamily signaling. Sci Signal 8 (399):re10
165. Al-Salihi MA, Herhaus L, Macartney T et al (2012) USP11 augments TGFbeta signalling by deubiquitylating ALK5. Open Biol 2(6):120063
166. Inui M, Manfrin A, Mamidi A et al (2011) USP15 is a deubiquitylating enzyme for receptor-activated SMADs. Nat Cell Biol 13(11):1368–1375
167. Herhaus L, Al-Salihi MA, Dingwell KS et al (2014) USP15 targets ALK3/BMPR1A for deubiquitylation to enhance bone morphogenetic protein signalling. Open Biol 4(5):140065
168. Iyengar PV, Jaynes P, Rodon L et al (2015) USP15 regulates SMURF2 kinetics through C-lobe mediated deubiquitination. Sci Rep 5:14733
169. Eichhorn PJ, Rodon L, Gonzalez-Junca A et al (2012) USP15 stabilizes TGF-beta receptor I and promotes oncogenesis through the activation of TGF-beta signaling in glioblastoma. Nat Med 18(3):429–435
170. Dupont S, Mamidi A, Cordenonsi M et al (2009) FAM/USP9x, a deubiquitinating enzyme essential for TGFbeta signaling, controls Smad4 monoubiquitination. Cell 136(1):123–135
171. Al-Hakim AK, Zagorska A, Chapman L et al (2008) Control of AMPK-related kinases by USP9X and atypical Lys(29)/Lys(33)-linked polyubiquitin chains. Biochem J 411(2):249–260
172. Zhao Y, Thornton AM, Kinney MC et al (2011) The deubiquitinase CYLD targets Smad7 protein to regulate transforming growth factor beta (TGF-beta) signaling and the development of regulatory T cells. J Biol Chem 286(47):40520–40530
173. Lim JH, Jono H, Komatsu K et al (2012) CYLD negatively regulates transforming growth factor-beta-signalling via deubiquitinating Akt. Nat Commun 3:771
174. Herhaus L, Al-Salihi M, Macartney T et al (2013) OTUB1 enhances TGFbeta signalling by inhibiting the ubiquitylation and degradation of active SMAD2/3. Nat Commun 4:2519
175. Ibarrola N, Kratchmarova I, Nakajima D et al (2004) Cloning of a novel signaling molecule, AMSH-2, that potentiates transforming growth factor beta signaling. BMC Cell Biol 5:2
176. Wicks SJ, Haros K, Maillard M et al (2005) The deubiquitinating enzyme UCH37 interacts with Smads and regulates TGF-beta signalling. Oncogene 24(54):8080–8084
177. Deng M, Yang X, Qin B et al (2016) Deubiquitination and activation of AMPK by USP10. Mol Cell 61(4):614–624
178. Agrawal P, Chen YT, Schilling B et al (2012) Ubiquitin-specific peptidase 9, X-linked (USP9X) modulates activity of mammalian target of rapamycin (mTOR). J Biol Chem 287(25):21164–21175
179. Villeneuve NF, Tian W, Wu T et al (2013) USP15 negatively regulates Nrf2 through deubiquitination of Keap1. Mol Cell 51(1):68–79
180. Li Z, Wang D, Messing EM et al (2005) VHL protein-interacting deubiquitinating enzyme 2 deubiquitinates and stabilizes HIF-1alpha. EMBO Rep 6(4):373–378
181. Bremm A, Moniz S, Mader J et al (2014) Cezanne (OTUD7B) regulates HIF-1alpha homeostasis in a proteasome-independent manner. EMBO Rep 15(12):1268–1277
182. Zhou Z, Yao X, Li S et al (2015) Deubiquitination of ci/Gli by Usp7/HAUSP regulates hedgehog signaling. Dev Cell 34(1):58–72
183. Zhang J, Liu M, Su Y et al (2012) A targeted in vivo RNAi screen reveals deubiquitinases as new regulators of notch signaling. G3 (Bethesda) 2(12):1563–1575

184. Jang MJ, Baek SH, Kim JH (2011) UCH-L1 promotes cancer metastasis in prostate cancer cells through EMT induction. Cancer Lett 302(2):128–135
185. de la Vega M, Kelvin AA, Dunican DJ et al (2011) The deubiquitinating enzyme USP17 is essential for GTPase subcellular localization and cell motility. Nat Commun 2:259
186. Gao J, Sun L, Huo L et al (2010) CYLD regulates angiogenesis by mediating vascular endothelial cell migration. Blood 115(20):4130–4137
187. Xie Y, Avello M, Schirle M et al (2013) Deubiquitinase FAM/USP9X interacts with the E3 ubiquitin ligase SMURF1 protein and protects it from ligase activity-dependent self-degradation. J Biol Chem 288(5):2976–2985
188. Ramakrishna S, Suresh B, Baek KH (2011) The role of deubiquitinating enzymes in apoptosis. Cell Mol Life Sci 68(1):15–26
189. Graner E, Tang D, Rossi S et al (2004) The isopeptidase USP2a regulates the stability of fatty acid synthase in prostate cancer. Cancer Cell 5(3):253–261
190. Oh KH, Yang SW, Park JM et al (2011) Control of AIF-mediated cell death by antagonistic functions of CHIP ubiquitin E3 ligase and USP2 deubiquitinating enzyme. Cell Death Differ 18(8):1326–1336
191. Dar A, Shibata E, Dutta A (2013) Deubiquitination of Tip60 by USP7 determines the activity of the p53-dependent apoptotic pathway. Mol Cell Biol 33(16):3309–3320
192. Schwickart M, Huang X, Lill JR et al (2010) Deubiquitinase USP9X stabilizes MCL1 and promotes tumour cell survival. Nature 463(7277):103–107
193. Nagai H, Noguchi T, Homma K et al (2009) Ubiquitin-like sequence in ASK1 plays critical roles in the recognition and stabilization by USP9X and oxidative stress-induced cell death. Mol Cell 36(5):805–818
194. Weber A, Heinlein M, Dengjel J et al. (2016) The deubiquitinase Usp27x stabilizes the BH3-only protein Bim and enhances apoptosis. EMBO Rep 17(5):724–738
195. Vendrell JA, Ghayad S, Ben-Larbi S et al (2007) A20/TNFAIP3, a new estrogen-regulated gene that confers tamoxifen resistance in breast cancer cells. Oncogene 26(32):4656–4667
196. Magraoui FE, Reidick C, Meyer HE et al (2015) Autophagy-related deubiquitinating enzymes involved in health and disease. Cell 4(4):596–621
197. Jin S, Tian S, Chen Y et al (2016) USP19 modulates autophagy and antiviral immune responses by deubiquitinating Beclin-1. EMBO J 35(8):866–880
198. Shi CS, Kehrl JH (2010) TRAF6 and A20 regulate lysine 63-linked ubiquitination of Beclin-1 to control TLR4-induced autophagy. Sci Signal 3 (123):ra42
199. Liu J, Xia H, Kim M et al (2011) Beclin1 controls the levels of p53 by regulating the deubiquitination activity of USP10 and USP13. Cell 147(1):223–234
200. Simicek M, Lievens S, Laga M et al (2013) The deubiquitylase USP33 discriminates between RALB functions in autophagy and innate immune response. Nat Cell Biol 15(10):1220–1230
201. Taillebourg E, Gregoire I, Viargues P et al (2012) The deubiquitinating enzyme USP36 controls selective autophagy activation by ubiquitinated proteins. Autophagy 8(5):767–779
202. Bingol B, Tea JS, Phu L et al (2014) The mitochondrial deubiquitinase USP30 opposes parkin-mediated mitophagy. Nature 510(7505):370–375
203. Cornelissen T, Haddad D, Wauters F et al (2014) The deubiquitinase USP15 antagonizes Parkin-mediated mitochondrial ubiquitination and mitophagy. Hum Mol Genet 23(19):5227–5242
204. Durcan TM, Tang MY, Perusse JR et al (2014) USP8 regulates mitophagy by removing K6-linked ubiquitin conjugates from parkin. EMBO J 33(21):2473–2491
205. Suresh B, Lee J, Kim KS et al (2016) The importance of ubiquitination and Deubiquitination in cellular reprogramming. Stem Cells Int 2016:6705927
206. Huang Z, Wu Q, Guryanova OA et al (2011) Deubiquitylase HAUSP stabilizes REST and promotes maintenance of neural progenitor cells. Nat Cell Biol 13(2):142–152
207. Sussman RT, Stanek TJ, Esteso P et al (2013) The epigenetic modifier ubiquitin-specific protease 22 (USP22) regulates embryonic stem cell differentiation via transcriptional repression of sex-determining region Y-box 2 (SOX2). J Biol Chem 288(33):24234–24246

208. Wu Y, Wang Y, Yang XH et al (2013) The deubiquitinase USP28 stabilizes LSD1 and confers stem-cell-like traits to breast cancer cells. Cell Rep 5(1):224–236
209. Fuchs G, Shema E, Vesterman R et al (2012) RNF20 and USP44 regulate stem cell differentiation by modulating H2B monoubiquitylation. Mol Cell 46(5):662–673
210. Buckley SM, Aranda-Orgilles B, Strikoudis A et al (2012) Regulation of pluripotency and cellular reprogramming by the ubiquitin-proteasome system. Cell Stem Cell 11(6):783–798
211. Yang W, Lee YH, Jones AE et al (2014) The histone H2A deubiquitinase Usp16 regulates embryonic stem cell gene expression and lineage commitment. Nat Commun 5:3818
212. Adorno M, Sikandar S, Mitra SS et al (2013) Usp16 contributes to somatic stem-cell defects in Down's syndrome. Nature 501(7467):380–384
213. Nijnik A, Clare S, Hale C et al (2012) The critical role of histone H2A-deubiquitinase Mysm1 in hematopoiesis and lymphocyte differentiation. Blood 119(6):1370–1379
214. Gu Y, Jones AE, Yang W et al (2016) The histone H2A deubiquitinase Usp16 regulates hematopoiesis and hematopoietic stem cell function. Proc Natl Acad Sci U S A 113(1):E51–E60
215. Li P, Yang YM, Sanchez S et al (2016) Deubiquitinase MYSM1 is essential for normal bone formation and mesenchymal stem cell differentiation. Sci Rep 6:22211
216. Saggar S, Chernoff KA, Lodha S et al (2008) CYLD mutations in familial skin appendage tumours. J Med Genet 45(5):298–302
217. Honma K, Tsuzuki S, Nakagawa M et al (2009) TNFAIP3/A20 functions as a novel tumor suppressor gene in several subtypes of non-Hodgkin lymphomas. Blood 114(12):2467–2475
218. Novak U, Rinaldi A, Kwee I et al (2009) The NF-{kappa}B negative regulator TNFAIP3 (A20) is inactivated by somatic mutations and genomic deletions in marginal zone lymphomas. Blood 113(20):4918–4921
219. Jensen DE, Proctor M, Marquis ST et al (1998) BAP1: a novel ubiquitin hydrolase which binds to the BRCA1 RING finger and enhances BRCA1-mediated cell growth suppression. Oncogene 16(9):1097–1112
220. Harbour JW, Onken MD, Roberson ED et al (2010) Frequent mutation of BAP1 in metastasizing uveal melanomas. Science 330(6009):1410–1413
221. Pena-Llopis S, Vega-Rubin-de-Celis S, Liao A et al (2012) BAP1 loss defines a new class of renal cell carcinoma. Nat Genet 44(7):751–759
222. Dey A, Seshasayee D, Noubade R et al (2012) Loss of the tumor suppressor BAP1 causes myeloid transformation. Science 337(6101):1541–1546
223. Bott M, Brevet M, Taylor BS et al (2011) The nuclear deubiquitinase BAP1 is commonly inactivated by somatic mutations and 3p21.1 losses in malignant pleural mesothelioma. Nat Genet 43(7):668–672
224. Perez-Mancera PA, Rust AG, van der Weyden L et al (2012) The deubiquitinase USP9X suppresses pancreatic ductal adenocarcinoma. Nature 486(7402):266–270
225. Oliveira AM, Hsi BL, Weremowicz S et al (2004) USP6 (Tre2) fusion oncogenes in aneurysmal bone cyst. Cancer Res 64(6):1920–1923
226. Shah SP, Morin RD, Khattra J et al (2009) Mutational evolution in a lobular breast tumour profiled at single nucleotide resolution. Nature 461(7265):809–813
227. Reincke M, Sbiera S, Hayakawa A et al (2015) Mutations in the deubiquitinase gene USP8 cause Cushing's disease. Nat Genet 47(1):31–38
228. Luise C, Capra M, Donzelli M et al (2011) An atlas of altered expression of deubiquitinating enzymes in human cancer. PLoS One 6(1):e15891
229. Kaposi-Novak P, Libbrecht L, Woo HG et al (2009) Central role of c-Myc during malignant conversion in human hepatocarcinogenesis. Cancer Res 69(7):2775–2782
230. Glinsky GV (2005) Death-from-cancer signatures and stem cell contribution to metastatic cancer. Cell Cycle 4(9):1171–1175
231. Zhang Y, Yao L, Zhang X et al (2011) Elevated expression of USP22 in correlation with poor prognosis in patients with invasive breast cancer. J Cancer Res Clin Oncol 137(8):1245–1253
232. Wang Y, Zhou X, Xu M et al. (2016) OTUB1-catalyzed deubiquitination of FOXM1 facilitates tumor progression and predicts a poor prognosis in ovarian cancer. Oncotarget 7(24):36681–36697

233. Xing C, Lu XX, Guo PD et al (2016) Ubiquitin-specific protease 4-mediated Deubiquitination and stabilization of PRL-3 is required for potentiating colorectal oncogenesis. Cancer Res 76(1):83–95
234. Zhou Y, Wu J, Fu X et al (2014) OTUB1 promotes metastasis and serves as a marker of poor prognosis in colorectal cancer. Mol Cancer 13:258
235. Gray DA, Inazawa J, Gupta K et al (1995) Elevated expression of Unph, a proto-oncogene at 3p21.3, in human lung tumors. Oncogene 10(11):2179–2183
236. Frederick A, Rolfe M, Chiu MI (1998) The human UNP locus at 3p21.31 encodes two tissue-selective, cytoplasmic isoforms with deubiquitinating activity that have reduced expression in small cell lung carcinoma cell lines. Oncogene 16(2):153–165
237. Yang Y, Hou JQ, Qu LY et al (2007) Differential expression of USP2, USP14 and UBE4A between ovarian serous cystadenocarcinoma and adjacent normal tissues. Xi Bao Yu Fen Zi Mian Yi Xue Za Zhi 23(6):504–506
238. Priolo C, Tang D, Brahamandan M et al (2006) The isopeptidase USP2a protects human prostate cancer from apoptosis. Cancer Res 66(17):8625–8632
239. Yang Z, Huo S, Shan Y et al (2012) STAT3 repressed USP7 expression is crucial for colon cancer development. FEBS Lett 586(19):3013–3017
240. McFarlane C, Kelvin AA, de la Vega M et al (2010) The deubiquitinating enzyme USP17 is highly expressed in tumor biopsies, is cell cycle regulated, and is required for G1-S progression. Cancer Res 70(8):3329–3339
241. Wang Y, Wang J, Zhong J et al (2015) Ubiquitin-specific protease 14 (USP14) regulates cellular proliferation and apoptosis in epithelial ovarian cancer. Med Oncol 32(1):379
242. Shinji S, Naito Z, Ishiwata S et al (2006) Ubiquitin-specific protease 14 expression in colorectal cancer is associated with liver and lymph node metastases. Oncol Rep 15(3):539–543
243. Huang Y, Pan XW, Li L et al. (2016) Overexpression of USP39 predicts poor prognosis and promotes tumorigenesis of prostate cancer via promoting EGFR mRNA maturation and transcription elongation. Oncotarget 7(16):22016–22030
244. Zhang Y, van Deursen J, Galardy PJ (2011) Overexpression of ubiquitin specific protease 44 (USP44) induces chromosomal instability and is frequently observed in human T-cell leukemia. PLoS One 6(8):e23389
245. Sloane MA, Wong JW, Perera D et al (2014) Epigenetic inactivation of the candidate tumor suppressor USP44 is a frequent and early event in colorectal neoplasia. Epigenetics 9(8):1092–1100
246. Xu M, Takanashi M, Oikawa K et al (2009) USP15 plays an essential role for caspase-3 activation during paclitaxel-induced apoptosis. Biochem Biophys Res Commun 388(2):366–371
247. Jenner MW, Leone PE, Walker BA et al (2007) Gene mapping and expression analysis of 16q loss of heterozygosity identifies WWOX and CYLD as being important in determining clinical outcome in multiple myeloma. Blood 110(9):3291–3300
248. Massoumi R, Kuphal S, Hellerbrand C et al (2009) Down-regulation of CYLD expression by snail promotes tumor progression in malignant melanoma. J Exp Med 206(1):221–232
249. Nikolaou K, Tsagaratou A, Eftychi C et al (2012) Inactivation of the deubiquitinase CYLD in hepatocytes causes apoptosis, inflammation, fibrosis, and cancer. Cancer Cell 21(6):738–750
250. Chinyengetere F, Sekula DJ, Lu Y et al (2015) Mice null for the deubiquitinase USP18 spontaneously develop leiomyosarcomas. BMC Cancer 15:886
251. Li X, Stevens PD, Yang H et al (2013) The deubiquitination enzyme USP46 functions as a tumor suppressor by controlling PHLPP-dependent attenuation of Akt signaling in colon cancer. Oncogene 32(4):471–478
252. Durkop H, Hirsch B, Hahn C et al (2003) Differential expression and function of A20 and TRAF1 in Hodgkin lymphoma and anaplastic large cell lymphoma and their induction by CD30 stimulation. J Pathol 200(2):229–239
253. D'Arcy P, Wang X, Linder S (2015) Deubiquitinase inhibition as a cancer therapeutic strategy. Pharmacol Ther 147:32–54
254. Mullally JE, Fitzpatrick FA (2002) Pharmacophore model for novel inhibitors of ubiquitin isopeptidases that induce p53-independent cell death. Mol Pharmacol 62(2):351–358

255. Liu Y, Lashuel HA, Choi S et al (2003) Discovery of inhibitors that elucidate the role of UCH-L1 activity in the H1299 lung cancer cell line. Chem Biol 10(9):837–846
256. Chen J, Dexheimer TS, Ai Y et al (2011) Selective and cell-active inhibitors of the USP1/UAF1 deubiquitinase complex reverse cisplatin resistance in non-small cell lung cancer cells. Chem Biol 18(11):1390–1400
257. Guedat P, Colland F (2007) Patented small molecule inhibitors in the ubiquitin proteasome system. BMC Biochem 8(Suppl) 1:S14
258. Yamaguchi M, Miyazaki M, Kodrasov MP et al (2013) Spongiacidin C, a pyrrole alkaloid from the marine sponge Stylissa Massa, functions as a USP7 inhibitor. Bioorg Med Chem Lett 23(13):3884–3886
259. Daviet L, Colland F (2008) Targeting ubiquitin specific proteases for drug discovery. Biochimie 90(2):270–283
260. Burkhart RA, Peng Y, Norris ZA et al (2013) Mitoxantrone targets human ubiquitin-specific peptidase 11 (USP11) and is a potent inhibitor of pancreatic cancer cell survival. Mol Cancer Res 11(8):901–911
261. Lee BH, Lee MJ, Park S et al (2010) Enhancement of proteasome activity by a small-molecule inhibitor of USP14. Nature 467(7312):179–184
262. Lawson AP, Long MJ, Coffey RT et al (2015) Naturally occurring Isothiocyanates exert anti-cancer effects by inhibiting deubiquitinating enzymes. Cancer Res 75(23):5130–5142
263. Coughlin K, Anchoori R, Iizuka Y et al (2014) Small-molecule RA-9 inhibits proteasome-associated DUBs and ovarian cancer in vitro and in vivo via exacerbating unfolded protein responses. Clin Cancer Res 20(12):3174–3186
264. Anchoori RK, Khan SR, Sueblinvong T et al (2011) Stressing the ubiquitin-proteasome system without 20S proteolytic inhibition selectively kills cervical cancer cells. PLoS One 6(8):e23888
265. D'Arcy P, Brnjic S, Olofsson MH et al (2011) Inhibition of proteasome deubiquitinating activity as a new cancer therapy. Nat Med 17(12):1636–1640
266. Mullauer FB, Kessler JH, Medema JP (2010) Betulinic acid, a natural compound with potent anticancer effects. Anti-Cancer Drugs 21(3):215–227
267. Seiberlich V, Borchert J, Zhukareva V et al (2013) Inhibition of protein deubiquitination by PR-619 activates the autophagic pathway in OLN-t40 oligodendroglial cells. Cell Biochem Biophys 67(1):149–160
268. Freije JM, Fraile JM, Lopez-Otin C (2011) Protease addiction and synthetic lethality in cancer. Front Oncol 1:25
269. Wiltshire TD, Lovejoy CA, Wang T et al (2010) Sensitivity to n(ADP-ribose) polymerase (PARP) inhibition identifies ubiquitin-specific peptidase 11 (USP11) as a regulator of DNA double-strand break repair. J Biol Chem 285(19):14565–14571
270. Zhang C, Cai TY, Zhu H et al (2011) Synergistic antitumor activity of gemcitabine and ABT-737 in vitro and in vivo through disrupting the interaction of USP9X and mcl-1. Mol Cancer Ther 10(7):1264–1275
271. Brummelkamp TR, Nijman SM, Dirac AM et al (2003) Loss of the cylindromatosis tumour suppressor inhibits apoptosis by activating NF-kappaB. Nature 424(6950):797–801
272. Kuphal S, Shaw-Hallgren G, Eberl M et al (2011) GLI1-dependent transcriptional repression of CYLD in basal cell carcinoma. Oncogene 30(44):4523–4530
273. Rajan N, Elliott R, Clewes O et al (2011) Dysregulated TRK signalling is a therapeutic target in CYLD defective tumours. Oncogene 30(41):4243–4260

Part II

General Aspects of Proteases

17 Submitochondrial Calpains in Pathophysiological Consequences

Pulak Kar, Krishna Samanta, Tapati Chakraborti, Md Nur Alam, and Sajal Chakraborti

Abstract

It has now been well established that different mitochondrial compartments contain varieties of calpains. The expression levels of these calpains are tissue and cell type specific. Although, mitochondrial compartments contain different types of calpains, the precise location within mitochondria and their functions remain imprecise. The aim of the present review is to confer information concerning the localization of calpains in different mitochondrial compartments affiliated with their function, particularly in pathophysiological conditions. For instance, mitochondrial μ-calpain is located within the inner membrane, intermembrane space, and mitochondrial matrix depending on cell types. μ-Calpain activity facilitates cleavage of apoptosis-inducing factor (AIF) within inner membrane and intermembrane space, while the activated μ-calpain within matrix is associated with cleavage of complex I subunits and metabolic enzymes. Understandably, inhibition of the μ-calpain could be a potential strategy to ameliorate ischemia–reperfusion-associated injuries.

Keywords

Submitochondria • Calcium • Calpain • Apoptosis • Ischemia–reperfusion • Disease

---

P. Kar • K. Samanta
Department of Physiology, Anatomy & Genetics, University of Oxford, Oxford, UK
e-mail: pulak_2006@yahoo.com; pulak_kar@rediffmail.com

T. Chakraborti • M.N. Alam • S. Chakraborti (✉)
Department of Biochemistry and Biophysics, University of Kalyani, Kalyani 741235, West Bengal, India
e-mail: saj_chakra@rediffmail.com

S. Chakraborti, N.S. Dhalla (eds.), *Proteases in Physiology and Pathology*,
DOI 10.1007/978-981-10-2513-6_17

17.1 Introduction

Calpains constitute a superfamily of intracellular calcium-dependent cysteine proteases, which is divided into two major types: classical and nonclassical [1]. Alternative splice variants in human calpain, 9 genes out of the 15, have been shown to code for the classical calpains; out of these, six genes have been demonstrated to cleave proteolytically a number of substrates in a tissue-specific manner, thereby modulating their functions that may implicate in a variety of pathological conditions [2].

Calpains are classically considered as cytosolic proteins, and activation of which plays a pivotal role in ischemia–reperfusion injury of the myocardium [3]. In isolated rabbit hearts, ischemia–reperfusion increases both the cytosolic μ-calpain and m-calpain activity [4]. During ischemia–reperfusion injury in the myocardium, cytosolic calpains are activated leading to cleavage of several proteins such as Na^+/K^+-ATPase, bid, Ca^{2+}-ATPase, α-fodrin, and troponin T [4]. Cardiac injury is critically associated with mitochondrial dysfunction following ischemia–reperfusion or during heart failure [4]. It is corroborated by the fact that calpains have been localized within mitochondria [4–7] and μ-calpain activity is increased in the mouse heart following ischemia–reperfusion [4]. Expression of ubiquitous μ-and m-calpains is activated in apoptosis models involving elevated $[Ca^{2+}]_i$. It is further substantiated by the proteolysis of Na^+/Ca^{2+} exchanger (NCX) by μ-calpain within inner mitochondrial membrane, which in turn facilitates Ca^{2+} elevation and cell demise, via release of apoptosis-inducing factor (AIF) [5, 8]. This review will focus on mitochondrial localized calpains and their involvement in some pathophysiological conditions.

17.2 Submitochondrial Distribution of Calpains

Calpains are known to be cytoplasmic enzymes, but we and other investigators have shown that calpains also localized within mitochondria [4–7]. Mitochondrial compartmentalization of calpains is tissue specific and cell type dependent. Mitochondrial μ-calpain was initially localized within the liver mitochondrial intermembrane space [6]. Our previous study showed that μ-calpain is also present in the inner membrane of the isolated pulmonary artery smooth muscle mitochondria [9]. We also reported that the pulmonary smooth muscle mitochondria contain calpastatin [9]. Kar et al. [9] also demonstrated that μ-calpain–calpastatin association exists in the inner mitochondrial membrane of pulmonary artery smooth muscle under physiological Ca^{2+} level. Chen et al. [10] reported that μ-calpain immunoactivity was detected in a component including inner membrane and matrix in cardiac mitochondria. The large subunit of μ-calpain contains the leader sequence in its N-terminus, which plays a critical role in importing its corresponding small regulatory subunit [7]. The biochemical characteristics of mitochondrial μ-calpain are similar to cytosolic μ-calpain with an 80 kDa large catalytic subunit as well as a 28 kDa regulatory small subunit [5].

Activation of mitochondrial m-calpain, localized in IMS, increases the permeability of the OMM by augmented interactions with VDAC in liver mitochondria [11]. Not only in liver mitochondria, but m-calpain activity has also been identified in the brain [12]. On the contrary, m-calpain is scarcely detected in the trypsin-purified heart mitochondria [10] pointing toward the assumption that m-calpain is less likely to be located in the heart mitochondria. However Chen et al. [4] suggested that identification of calpain in mitochondrial sub-compartment depends on the purification processes and detergent used. For example, m-calpain has been identified and localized in the mitochondrial matrix when purified without protease from rat heart mitochondria [4].

Calpain 10 is a different calpain found in renal cortex mitochondria; they are found in outer membrane, intermembrane space, inner membrane, and matrix of the mitochondria [13]. Calpain 10 activity is involved in calcium-induced mitochondrial dysfunction and associated with impairment of the mitochondrial respiratory chain and cell injury [13]. Recently, Bround et al. [14] reported the presence of calpain 10 in cardiac mitochondria and also its involvement in the disruption of ryanodine receptor-mediated apoptosis [14]. The abovementioned findings lead us to a conviction that both mitochondrial μ-calpain and calpain 10 contribute to cell injury during ischemia–reperfusion.

17.3 Regulation of Submitochondrial Calpain Activity

$[Ca^{2+}]_i$ regulate calpain activity; however, influx of calcium into mitochondria is facilitated by calcium uniporter (H^+/Ca^{2+}) in the IMM [4], whereas calcium is extruded from mitochondria into cytosol through Na^+/Ca^{2+} exchange [4, 8] and MPTP opening [15] (Fig. 17.1). μ-Calpain and m-calpain are activated by 3–50 μM Ca^{2+} and 400–800 μM Ca^{2+} concentrations, respectively [6]. In the resting cardiac myocyte, calcium concentration in the mitochondrial matrix is similar to that of cytosol which has found to be about 100 nM [4, 16]. Increase in ROS generation due to mitochondrial respiratory chain impairment by ischemia–reperfusion is also substantiated by several studies [17, 18]. Therefore, combined effect of calcium overload and oxidative stress during ischemia–reperfusion leads to mitochondrial μ- and m-calpain activation [4].

Calpastatin, an endogenous inhibitor of calpain, has already been identified within mitochondria [6, 19]. Kar et al. [5] demonstrated the existence of μ-calpain–calpastatin association in the inner mitochondrial membrane under physiological Ca^{2+} concentration, suggesting omnipotent role of calpastatin in regulation of mitochondrial μ-calpain. Not only by calpastatin, but some reports demonstrated that the μ-calpain and m-calpain activities can be regulated by calpain 4 gene knockout as well [20, 21].

Ozaki et al. [22] came up with an intriguing observation that ERp57, a cytosolic protein involved in correcting the folding and rearrangement of the disulfide bond of misfolded glycoproteins, also regulates μ-calpain [4]. ERp57 is localized within ER and functions as molecular chaperone. However, ERp57 that exists mainly in the

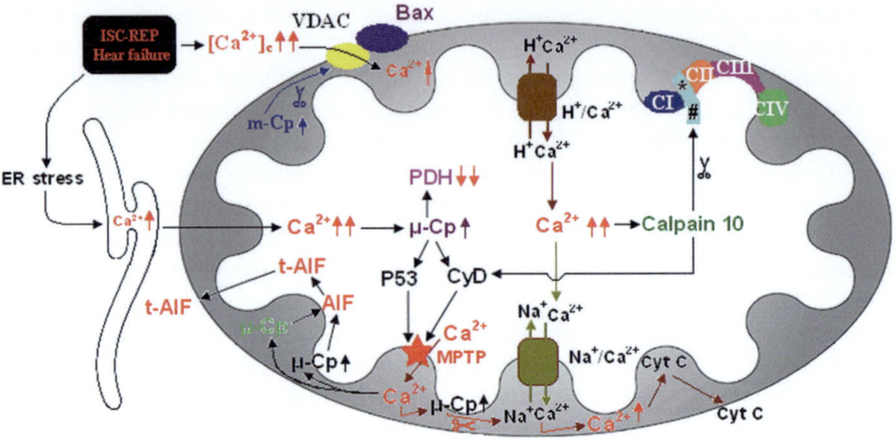

Fig. 17.1 Schematic representation of sub-mitochondrial calpains and their functions. Mitochondrial Ca^{2+} overload is occurred by cytosolic Ca^{2+} overload as well as by ER stress in several pathophysiological consequences, such as, ischemia–reperfusion or heart failure. Mitochondrial Ca^{2+} dynamics depends upon different ion channels or exchangers located in different mitochondria compartments, which transport Ca^{2+} into or out of the sub-mitochondrial compartments. Ca^{2+} entrance into the mitochondrial intermembrane space (IMS) is occurred by outer membrane voltage dependent anion channel (VDAC). H^+/Ca^{2+} of the inner-mitochondrial membrane (IMM) influxes Ca^{2+} into the mitochondrial matrix from IMS. Mitochondrial matrix Ca^{2+} is exported by Na^+/Ca^{2+} exchanger. Mitochondrial Ca^{2+} overload activates µ-calpain in the IMM or IMS and m-calpain in the IMS [4]. Activated µ-calpain involves in AIF processing as well as cleaves the mitochondrial $Na^+/^{2+}Ca$ exchanger, which leads to mitochondrial Ca^{2+} overload and causes mitochondrial permeability transition pore (MPTP) opening and cytochrome C release. Activated m-calpain truncates VDAC, which associates with Bax in the mitochondrial outer membrane. After releasing into the cytosol through VDAC/Baxmediated pores or Bax/Bax-mediated pores tAIF translocates to the nucleus and fragmentes DNA [5]. Activated mitochondrial matrix µ-calpain and calpain 10 cleave the complex I of mitochondrial respiratory chain [4, 13]. Activated µ-calpian interacts with the cyclophilin D (CyD) and p53 and facilitates the MPTP opening (see text for detail)

mitochondrial OM is different from cytosolic one [22]. Ozaki et al. [22] have demonstrated that mitochondrial µ-calpain associates with ERp57, but cytosolic µ-calpain and m-calpain do not. Their studies also suggested that ERp57 is involved in µ-calpain large subunit refolding, by forming disulfide bonds to shape up functional conformations. Not only ERp57, but IMS proteins also are implicated for refolding and formation of disulfide bonds by protein oxidation in the IMS [23]. Therefore, the association between mitochondrial µ-calpain with ERp57 may provide important clues in understanding mitochondrial µ-calpain functions [22]. In addition to µ-calpain, studies have indicated m-calpain's association with ERp75 as well. Importantly, ERp75 is an important molecular chaperone belonging to the heat shock protein 70 family, which releases tAIF (truncated apoptosis-inducing factor) from mitochondria [24, 25]. It has multiple functions including the import of proteins into mitochondria and the proper folding of newly synthesized chromosomal (nuclear) and extrachromosomal (mitochondrial) encoded proteins. ERp75 response is associated with many forms of stress including depleted glucose levels, oxidative injury, and ultraviolet irradiation [26].

Calcium concentration is increased to approximately 1 μM and 2.6 μM at the end of 25 min and 30 min reperfusion, respectively [4]. This calcium overload is believed to be associated with the activation of calpain 10; however, some isoforms of calpain 10 are not exclusively calcium dependent [13]. Although phosphorylation and post-translational modifications may affect calpain 10 activity, the actual mechanisms associated with the regulation of calpain 10 are still not clear [27]. Calpain activity is not only dependent on endogenous regulatory mechanisms; it is also affected by other different physiological conditions. Under acidic conditions the activities of calpains are decreased [4, 28], albeit different calpains show varied sensitivity toward acidification. A number of synthetic and nature-derived calpain inhibitors are available to study and manipulate calpain activity under pathophysiological conditions and in experimental models. MDL 28170, for example, a typical synthetic inhibitor of calpain, is used to inhibit both μ- and m-calpains [4].

17.4 Submitochondrial Calpains in Pathophysiology

Calpains play a key role in a variety of physiological processes such as embryonic development, intracellular signal transduction, and cell cycle regulation [4]. During ischemia–reperfusion, $[Ca^{2+}]_i$ overload activates cytosolic μ- and m-calpains . The activated cytosolic μ- and m-calpains in turn cleave calcium regulator proteins including Na^+–K^+-ATPases, Ca^{2+}-ATPases, H^+-ATPases, Na^+/H^+ exchanger, and Na^+/Ca^{2+} exchanger and further augment calcium overload [4, 28–30] (Fig. 17.1). Thus, calpain activation complements with the failure of intracellular calcium control mechanisms that result in a disproportionate $[Ca^{2+}]_i$ accumulation [3, 31]. Intracellular calcium overload is implicated in apoptosis and necrosis by induction of mitochondrial calcium overload with subsequent activation of mitochondrial μ-calpain and, conceivably, m-calpain (Fig. 17.1).

17.4.1 Mitochondrial Inner, Outer, and Intermembrane Space μ- and m-Calpains and Apoptosis

Proapoptotic proteins such as cytochrome C, AIF, Smac/DIABLO, and Omi/HtrA2 are known to be released from the intermembrane space (IMS) of mitochondria, and that has been demonstrated to be a major phenomenon in apoptosis [5]. The active form of AIF (truncated AIF) translocates from mitochondria to the nucleus and induces DNA damage setting off caspase-independent cell death process [4]. Pre-cleaved form of AIF (62 kDa) is anchored at the inner mitochondrial membrane within the intermembrane space [32]. The release of AIF from the mitochondria into cytosol is facilitated by its cleavage. Addition of calcium in the isolated liver or heart mitochondria has been shown to cleave AIF (62 kDa) to a truncated form (57 kDa), which is further substantiated by the use of calpain inhibitor, as it prevents the calcium-mediated AIF cleavage [6]. These aforesaid results support the fact that activation of mitochondrial calpain leads to AIF cleavage. Even though μ-calpain is identified in the intermembrane space of cardiac mitochondria [10] and implicated in the cleavage and activation of AIF in heart mitochondria, reports show that μ-calpain is not involved in AIF cleavage in brain mitochondria [33]. Therefore, complementation of AIF cleavage with mitochondrial μ-calpain activation may be tissue dependent. The translocation of the cleaved AIF from the intermembrane space to the nucleus is eased by the permeation of the outer mitochondrial membrane [6], which is accomplished by the activation of mitochondrial m-calpain by cleaving VDAC and induction of Bax on the outer membrane in isolated liver mitochondria [6] (Fig. 17.1).The aforementioned phenomenon induced by m-calpain is not universal; rather it seems to be species specific [10, 34]. Ischemia–reperfusion is known to increase the outer membrane permeability by opening of the mitochondrial permeability transition pore (MPTP) [35]. Not only that, but in ischemic condition, induction of imbalance between anti-apoptotic and proapoptotic bcl-2 family proteins also increases the permeability of the outer membrane [4]. The aforesaid observations assure that AIF release is controlled by both mitochondrial μ-calpain and mitochondrial m-calpain. Conceivably, inhibition of mitochondrial calpains could prove useful as a therapeutic measure in preventing many disorders such as retinitis pigmentosa/retinal degeneration and cerebral ischemia [24].

17.4.2 Mitochondrial Matrix Calpains in Pathophysiology

Cytosolic calpain activation can cleave bid to t-bid, which in turn can elevate outer mitochondrial membrane permeability [36]. μ-Calpain is also localized on the endoplasmic reticulum (ER) [37, 38]. So its activation can be associated with calcium overload through ER stress [37, 38], and induction of intracellular calcium overload and subsequent mitochondrial calcium overload can be associated with MPTP sensitivity and opening via activation of cytosolic calpains. In ischemic cardiac mitochondria, μ-calpain activity is increased; conversely calpain inhibition decreased the MPTP opening [4]. In isolated perfused heart, calpain inhibitors have been shown to ameliorate ischemia–reperfusion injury by improving oxidative phosphorylation [4]. Chen et al. [4] proposed that μ-calpain within the mitochondrial matrix damages complex I. Not only μ-calpain, but m-calpain is also implicated in complex I damage and MPTP permeation (Fig. 17.1). Thus, multiple studies converge to support the pivotal role of mitochondrial calpain in disruption of mitochondrial metabolism and cell death. Not only ETC, but ischemia–reperfusion impairs metabolic enzymes in the tricarboxylic acid (TCA) cycle as well [4, 39, 40]. More precise proteomic studies show that ischemia–reperfusion leads to degradation of several metabolic enzymes including pyruvate dehydrogenase (PDH), malate dehydrogenase (MDH), and succinate dehydrogenase (SDH) in rat heart mitochondria [4, 39, 40].

Along with μ- and m-calpain, calpain 10 is also localized within mitochondrial matrix, which has recently gained importance because of its role in type 2 diabetes [5]. In renal mitochondria, calpain 10 activity is associated with ETC impairment by proteolytic digestion of complex I subunits [13, 41]. Genetic ablation of complex I triggers MPTP opening in mouse heart mitochondria [42]. Therefore, the impaired "complex I" is not only implicated in reducing the rate of respiration but also in sensitizing mitochondria to undergo MPTP opening which is a key mechanism to induce cell death during ischemia–reperfusion [4] (Fig. 17.1).

17.5 Mitochondrial Oxidative Stress and Submitochondrial Calpain in Pathophysiology

In aerobic cells mitochondria represent the key site of molecular oxygen consumption and ROS production [43]. Reactive oxygen species (ROS) production is associated with stress and can lead to cell death, if not detoxified. ROS produced by the mitochondria can oxidize proteins and induce lipid peroxidation, which in turn compromise the versatility of membrane properties of the mitochondria. Not only proteins and lipids, but mitochondrial DNA (mtDNA) is also susceptible to ROS-mediated damage. ROS-mediated damage of mtDNA is highly plausible as they are devoid of protective histones and localize in close proximity to the respiratory chain. It is further substantiated by the studies that tell oxidative modification in mtDNA bases is 10- to 20-fold higher than nuclear DNA [44]. Some mitochondrial proteins are encoded by its DNA, which are essential mainly for the respiratory chain and ATP synthesis by oxidative phosphorylation. ROS production is not only associated with direct oxidation and damages. It also augments cellular deterioration via induction of Ca^{2+}-mediated pathways. ROS-mediated sustained Ca^{2+} activates mitochondrial intermembrane space, and inner mitochondrial membrane localized μ-calpain which then cleaves the apoptosis-inducing factor (AIF) [5, 45, 46] (Fig. 17.1). Mitochondrial ROS mediates oxidative modification of AIF which further trigger proteolysis and also carbonylation of the protein, thus increasing susceptibility to calpain cleavage [47]. Apoptosis-inducing factor (AIF) is a 62 kDa flavoprotein and is anchored to the inner mitochondrial membrane (IMM) in the close proximity of complex I. AIF needs to be cleaved from the anchor peptide so that the 57 kDa pro-apoptotic fragment can be released into the cytosol for further translocation into the nucleus, where it promotes large-scale DNA damage. Therefore, oxidative stress, ROS generation, Ca^{2+} elevation, calpain activation, release and translocation of AIF, and DNA fragmentation are associated with cell death [46, 48].

Ischemia–reperfusion and associated injury have often been implicated with mitochondrial permeability transition, and pretreatment with cyclosporine A (CsA) has been found to exert a protective effect. Therefore the paradigm suggests that PTP formation is the cause of the damage [45]. The precise molecular mechanism by which activation of the mitochondrial calpains sensitizes the MPTP opening in cardiac mitochondria is not clear. p53 is a tumor suppressor protein, and its content is normally maintained at a low level by interaction with Mdm2 (mouse double minute 2 homolog). During oxidative stress Mdm2 dissociates from the p53-Mdm2 complex by phosphorylation, thus rapidly increasing its level in cytosol and nucleus. The phosphorylated p53 translocates to and accumulates within mitochondria to form a complex with cyclophilin D, which in turn increases MPTP opening [4]. The translocation of p53 from cytosol to brain mitochondria is aided by the activation of cytosolic calpain [4], implicating cytosolic calpain's possible role in induction of MPTP opening (Fig. 17.1). Complex I and cyclophilin D are associated with MPTP opening via activation of mitochondria-localized μ-calpain or m-calpains [4, 34]. Cyclophilin D, which is located within the matrix of the mitochondria, can be accessed by the mitochondrial μ-calpain and plays a decisive role in the regulation of MPTP opening. Genetic inhibition of complex I leads to increased protein acetylation of MPTP [4]. During ischemia–reperfusion, calpain-mediated MPTP opening eventually damages complex I, and that can be attenuated by calpain inhibitors [4, 34]. Therefore, activation of mitochondrial μ-calpain corroborates with the increase in MPTP opening, albeit by impairing complex I in the respiratory chain or by facilitating translocation of cyclophilin D from the matrix to the inner mitochondrial membrane [4] (Fig. 17.1).

Mitochondrial μ-calpain also plays a critical role in cardiomyopathic changes in type 1 diabetes (T1D). In cardiomyocytes of transgenic mice, capn4 deletion or injection of streptozotocin in the wild types induces T1D [49].

An increase in the expression of μ-calpain in mitochondria has been observed in diabetic mouse hearts and that augments ROS levels leading to a reduction in ATP synthase (ATP5A1) activity. The μ-calpain has been shown to be induced in both type 1 and type 2 diabetic heart mitochondria. In some [49], but not all [50, 51], system genetic inhibition of μ-calpain increases the ATP synthase activity due to a marked attenuation of mitochondrial ROS generation with concomitant decrease in damage of diabetic heart.

Upon activation, calpain contributes to myocardial dysfunction in diabetes by triggering proteolysis of some cytosolic proteins such as protein kinase C and nuclear factor-kB [52, 53], myofibrils, and intracellular Ca^{2+} regulatory proteins [49, 54, 55].

17.6 Concluding Remarks and Future Perspective

This review summarizes some pathophysiological aspects of mitochondrial calpains. Research in the recent past suggested that μ-calpain, m-calpain, and also calpain 10 are compartmentalized within the mitochondria and their expression depends on the cell type and tissue specificity. Activation of μ-calpain induces specific proteolytic cleavage of AIF to a truncated form within the intermembrane space, thus facilitating the release of AIF from mitochondria into cytosol (Fig. 17.1). On the other hand, activation of μ-calpain and m-calpain within mitochondrial matrix inhibits mitochondrial metabolism by degrading metabolic enzymes, for example, the complex I subunits. Added to that, hyperactivation of the μ-calpain or m-calpain following ischemia–reperfusion may sensitize MPTP opening in cardiac mitochondria, while cytosolic calpain inhibition during ischemia–reperfusion ameliorates cardiac injury [4]. The pathophysiological outcome associated with deteriorated cell death pathways is evident, albeit due to alteration of Ca^{2+} fluxes and activation of mitochondrial calpains. A comprehensive research on the above could prove useful to develop novel strategies of therapeutic interventions in the near future.

Acknowledgment Financial assistance from the DST-PURSE program of the University of Kalyani is acknowledged.

References

1. Ono Y, Sorimachi H (2012) Calpains: an elaborate proteolytic system. Biochim Biophys Acta 1824:224–236
2. Sorimachi H, Hata S, Ono Y (2011) Impact of genetic insights into calpain biology. J Biochem 150:23–37
3. Neuhof C, Neuhof H (2014) Calpain system and its involvement in myocardial ischemia and reperfusion injury. World J Cardiol 6:638–652
4. Chen Q, Lesnefsky EJ (2015) Heart mitochondria and calpain 1: location, function, and targets. Biochim Biophys Acta 1852:2372–2378
5. Kar P, Samanta K, Shaikh S, Chowdhury A, Chakraborti T, Chakraborti S (2010) Mitochondrial calpain system: an overview. Arch Biochem Biophys 495:1–7
6. Ozaki T, Tomita H, Tamai M, Ishiguro S (2007) Characteristics of mitochondrial calpains. J Biochem 142:365–376
7. Badugu R, Garcia M, Bondada V, Joshi A, Geddes JW (2008) N terminus of calpain 1 is a mitochondrial targeting sequence. J Biol Chem 283:3409–3417
8. Kar P, Chakraborti T, Samanta K, Chakraborti S (2009) μ-Calpain mediated cleavage of the Na^+/Ca^{2+} exchanger in isolated mitochondria under A23187 induced Ca^{2+} stimulation. Arch Biochem Biophys 482:67–76
9. Kar P, Chakraborti T, Samanta K, Chakraborti S (2008) Submitochondrial localization of associated mu-calpain and calpastatin. Arch Biochem Biophys 470(2):176–186
10. Chen Q, Paillard M, Gomez L, Ross T, Hu Y, Xu A, Lesnefsky EJ (2011) Activation of mitochondrial mu-calpain increases AIF cleavage in cardiac mitochondria during ischemia–reperfusion. Biochem Biophys Res Commun 415:533–538
11. Ozaki T, Yamashita T, Ishiguro S (2009) Mitochondrial m-calpain plays a role in the release of truncated apoptosis-inducing factor from the mitochondria. Biochim Biophys Acta 1793:1848–1859
12. Kosenko E, Poghosyan A, Kaminsky Y (2011) Sub cellular compartmentalization of proteolytic enzymes in brain regions and the effects of chronic beta-amyloid treatment. Brain Res 1369:184–193
13. Arrington DD, Van Vleet TR, Schnellmann RG (2006) Calpain 10: a mitochondrial calpain and its role in calcium-induced mitochondrial dysfunction. Am J Physiol Cell Physiol 291:C1159–C1171
14. Bround MJ, Wambolt R, Luciani DS, Kulpa JE, Rodrigues B, Brownsey RW, Allard MF, Johnson JD (2013) Cardiomyocyte ATP production, metabolic flexibility, and survival require calcium flux through cardiac ryanodine receptors in vivo. J Biol Chem 288:18975–18986
15. Di Lisa F, Carpi A, Giorgio V, Bernardi P (2011) The mitochondrial permeability transition pore and cyclophilin D in cardioprotection. Biochim Biophys Acta 1813:1316–1322
16. Gunter TE, Yule DI, Gunter KK, Eliseev RA, Salter JD (2004) Calcium and mitochondria. FEBS Lett 567:96–102
17. Chen Q, Camara AK, Stowe DF, Hoppel CL, Lesnefsky EJ (2007) Modulation of electron transport protects cardiac mitochondria and decreases myocardial injury during ischemia and reperfusion. Am J Physiol Cell Physiol 292:C137–C147

18. Turrens JF, Beconi M, Barilla J, Chavez UB, McCord JM (1991) Mitochondrial generation of oxygen radicals during reoxygenation of ischemic tissues. Free Radic Res Commun 2:681–689
19. Kar P, Chakraborti T, Roy S, Choudhury R, Chakraborti S (2007) Identification of calpastatin and μ-calpain and studies of their association in pulmonary smooth muscle mitochondria. Arch Biochem Biophys 466:166–176
20. Tan Y, Dourdin N, Wu C, De Veyra T, Elce JS, Greer PA (2006) Conditional disruption of ubiquitous calpains in the mouse. Genesis 44:297–303
21. Rao MV, McBrayer MK, Campbell J, Kumar A, Hashim A, Sershen H, Stavrides PH, Ohno M, Hutton M, Nixon RA (2014) Specific calpain inhibition by calpastatin prevents tauopathy and neurodegeneration and restores normal lifespan in tau P301L mice. J Neurosci 34:9222–9234
22. Ozaki T, Yamashita T, Ishiguro S (2008) ERp57-associated mitochondrial μ-calpain truncates apoptosis-inducing factor. Biochim Biophys Acta 1783:1955–1963
23. Herrmann JM, Kohl R (2007) Catch me if you can! Oxidative protein trapping in the inter membrane space of mitochondria. J Cell Biol 176:559–563
24. Ozaki T, Yamashita T, Ishiguro S (2009) Mitochondrial m-calpain plays a role in the release of truncated apoptosis-inducing factor from the mitochondria. Biochim Biophys Acta 1793(12):1848–1859
25. Mizzen LA, Chang C, Garrels JI, Welch WJ (1989) Identification, characterization, and purification of two mammalian stress proteins present in mitochondria, grp 75, a member of the hsp 70 family and hsp 58, a homolog of the bacterial groEL protein. J Biol Chem 264:20664–20675
26. Liu Y, Liu W, Song XD, Zuo J (2005) Effect of GRP75/mthsp70/PBP74/mortalin overexpression on intracellular ATP level, mitochondrial membrane potential and ROS accumulation following glucose deprivation in PC12 cells. Mol Cell Biochem 268(1–2):45–51
27. Turner MD, Cassell PG, Hitman GA (2005) Calpain-10: from genome search to function. Diabetes Metab Res 21:505–514
28. Inserte J, Hernando V, Garcia-Dorado D (2012) Contribution of calpains to myocardial ischaemia/reperfusion injury. Cardiovasc Res 96:23–31
29. Inserte J, Barba I, Hernando V, Garcia-Dorado D (2009) Delayed recovery of intracellular acidosis during reperfusion prevents calpain activation and determines protection in postconditioned myocardium. Cardiovasc Res 81:116–122
30. Neuhof C, Fabiunk V, Speth M, Moller A, Fritz F, Tillmanns H, Neuhof H, Erdogan A (2008) Reduction of myocardial infarction by postischemic administration of the calpain inhibitor A-705253 in comparison to the Na^+/H^+ exchange inhibitor Cariporide in isolated perfused rabbit hearts. Biol Chem 389:1505–1512
31. Smith MA, Schnellmann RG (2012) Calpains, mitochondria, and apoptosis. Cardiovasc Res 96:32–37
32. Otera H, Ohsakaya S, Nagaura Z, Ishihara N, Mihara K (2005) Export of mitochondrial AIF in response to proapoptotic stimuli depends on processing at the inter membrane space. EMBO J 24:1375–1386
33. Wang Y, Kim NS, Li X, Greer PA, Koehler RC, Dawson VL, Dawson TM (2009) Calpain activation is not required for AIF translocation in PARP-1-dependent cell death (parthanatos). J Neurochem 110:687–696
34. Shintani-Ishida K, Yoshida KI (2015) Mitochondrial m-calpain opens the mitochondrial permeability transition pore in ischemia–reperfusion. Int J Cardiol 197:26–32
35. Weiss JN, Korge P, Honda HM, Ping P (2003) Role of the mitochondrial permeability transition in myocardial disease. Circ Res 93:292–301
36. Chen M, He H, Zhan S, Krajewski S, Reed JC, Gottlieb RA (2001) Bid is cleaved by calpain to an active fragment in vitro and during myocardial ischemia/reperfusion. J Biol Chem 276:30724–30728
37. Zheng D, Wang G, Li S, Fan GC, Peng T (2015) Calpain-1 induces endoplasmic reticulum stress in promoting cardiomyocyte apoptosis following hypoxia/reoxygenation. Biochim Biophys Acta 1852:882–892

38. Samanta K, Kar P, Chakraborti T, Chakraborti S (2013) An overview of endoplasmic reticulum calpain system. In: Chakraborti S, Dhalla NS (eds) Proteases in health and disease, advances in biochemistry in health and disease, vol 7. Springer, New York, pp 3–19
39. Ussher JR, Wang W, Gandhi M, Keung W, Samokhvalov V, Oka T, Wagg CS, Jaswal JS, Harris RA, Clanachan AS, Dyck JR, Lopaschuk GD (2012) Stimulation of glucose oxidation protects against acute myocardial infarction and reperfusion injury. Cardiovasc Res 94:359–369
40. Deng N, Zhang J, Zong C, Wang Y, Lu H, Yang P, Wang W, Young GW, Wang Y, Korge P, Lotz C, Doran P, Liem DA, Apweiler R, Weiss JN, Duan H, Ping P (2011) Phosphoproteome analysis reveals regulatory sites in major pathways of cardiac mitochondria. Mol Cell Proteomics 10:M110 000117
41. Giguere CJ, Covington MD, Schnellmann RG (2008) Mitochondrial calpain 10 activity and expression in the kidney of multiple species. Biochem Biophys Res Commun 366:258–262
42. Karamanlidis G, Lee CF, Garcia-Menendez L, Kolwicz SC Jr, Suthammarak W, Gong G, Sedensky MM, Morgan PG, Wang W, Tian R (2013) Mitochondrial complex I deficiency increases protein acetylation and accelerates heart failure. Cell Metab 18:239–250
43. Boveris A, Chance B (1973) The mitochondrial generation of hydrogen peroxide. General properties and effect of hyperbaric oxygen. Biochem J 134:707–716
44. Richter C (1995) Oxidative damage to mitochondrial DNA and its relationship to ageing. Int J Biochem Cell Biol 27:647–653
45. Orrenius S, Gogvadze V, Zhivotovsky B (2015) Calcium and mitochondria in the regulation of cell death. Biochem Biophys Res Commun 460:72–81
46. Norberg E, Gogvadze V, Ott M, Horn M, Uhlen P, Orrenius S, Zhivotovsky B (2008) An increase in intracellular Ca^{2+} is required for the activation of mitochondrial calpain to release AIF during cell death. Cell Death Differ 15:1857–1864
47. Norberg E, Gogvadze V, Vakifahmetoglu H, Orrenius S, Zhivotovsky B (2010) Oxidative modification sensitizes mitochondrial apoptosis-inducing factor to calpain-mediated processing. Free Radic Biol Med 48:791–797
48. Susin SA, Zamzami N, Castedo M, Hirsch T, Marchetti P, Macho A, Daugas E, Geuskens M, Kroemer G (1996) Bcl-2 inhibits the mitochondrial release of an apoptogenic protease. J Exp Med 184:1331–1341
49. Ni R, Zheng D, Xiong S, Hill DJ, Sun T, Gardiner RB, Fan GC, Lu Y, Abel ED, Greer PA, Peng T (2016) Mitochondrial Calpain-1 disrupts ATP synthase and induces superoxide generation in type 1 diabetic hearts: a novel mechanism contributing to diabetic cardiomyopathy. Diabetes 65:255–268
50. Bugger H, Boudina S, Hu XX et al (2008) Type 1 diabetic Akita mouse hearts are insulin sensitive but manifest structurally abnormal mitochondria that remain coupled despite increased uncoupling protein 3. Diabetes 57:2924–2932
51. Herlein JA, Fink BD, O'Malley Y, Sivitz WI (2009) Superoxide and respiratory coupling in mitochondria of insulin-deficient diabetic rats. Endocrinology 150:46–55
52. Zhang Y, Matkovich SJ, Duan X, Diwan A, Kang MY, Dorn GW 2nd (2011) Receptor-independent protein kinase C alpha (PKCalpha) signaling by calpain generated free catalytic domains induces HDAC5 nuclear export and regulates cardiac transcription. J Biol Chem 286: 26943–26951
53. Ma J, Wei M, Wang Q et al (2012) Deficiency of Capn4 gene inhibits nuclear factor-kB (NF-kB) protein signaling/inflammation and reduces remodeling after myocardial infarction. J Biol Chem 287:27480–27489
54. French JP, Quindry JC, Falk DJ et al (2006) Ischemia-reperfusion-induced calpain activation and SERCA2a degradation are attenuated by exercise training and calpain inhibition. Am J Physiol Heart Circ Physiol 290:H128–H136
55. Pedrozo Z, Sánchez G, Torrealba N et al (2010) Calpains and proteasomes mediate degradation of ryanodine receptors in a model of cardiac ischemic reperfusion. Biochim Biophys Acta 1802:356–362

Serine Proteases in the Lectin Pathway of the Complement System

18

Fabiana A. Andrade, Kárita C.F. Lidani, Sandra J. Catarino, and Iara J. Messias-Reason

Abstract

The complement system plays a crucial role in host defense against pathogen infections and in the recognition and removal of damaged or altered *self*-components. Complement system activation can be initiated by three different pathways—classical, alternative, and lectin pathways—resulting in a proteolytic cascade, which culminates in multiple biological processes including opsonization and phagocytosis of intruders, inflammation, cell lysis, and removal of immune complexes and apoptotic cells. Furthermore, it also functions as a link between the innate and adaptive immune responses. The lectin pathway (LP) activation is mediated by serine proteases, termed mannan-binding lectin (MBL)-associated serine proteases (MASPs), which are associated with the pattern recognition molecules (PRMs) that recognize carbohydrates or acetylated compounds on surfaces of pathogens or apoptotic cells. These result in the proteolysis of complement C2 and C4 generating C3 convertase (C4b2a), which carries forward the activation cascade of complements, culminating in the elimination of foreign molecules. This chapter presents an overview of the complement system focusing on the characterization of MASPs and its genes, as well as its functions in the immune response.

Keywords

Serine proteases • Complement system • Lectin pathway

F.A. Andrade • K.C.F. Lidani • S.J. Catarino • I.J. Messias-Reason (✉)
Department of Clinical Pathology, Hospital de Clínicas, Universidade Federal do Paraná (UFPR), General Carneiro, 181 Curitiba, PR, Brazil
e-mail: iarareason@hc.ufpr.br

S. Chakraborti, N.S. Dhalla (eds.), *Proteases in Physiology and Pathology*,
DOI 10.1007/978-981-10-2513-6_18

18.1 The Complement System

The human immune system is an extraordinary complex of biochemical mechanisms that provides effective defense against a large number of pathogens while also protecting against improper responses to *self*-components. The immune system exhibits innate and adaptive responses that cooperate together to facilitate appropriate host defense. Innate immunity provides the first line of defense by recognizing specific patterns present on the surface of microbes (PAMPs, pathogen-associated molecular patterns) or damaged cells (DAMPs, damage-associated molecular patterns) through innate pattern recognition molecules and receptors (PRMs and PRRs, respectively). The effectors of innate immunity include epithelial barriers, phagocytes and natural killer cells, cytokines, and a whole complex of proteins known as the complement system [1, 2].

The complement system is comprised of more than 35 plasma proteins and cell surface receptors/regulators, which enables the recognition, tagging, and elimination of various microbial intruders and foreign cells. Most of the soluble proteins circulate in functionally inactive forms called proenzymes or zymogens, which share identical domain organization and overall structure, but differ in enzymatic properties and physiological significance in health and disease [1]. Upon proteolytic cleavage, inactive proteins become activated, resulting in a proteolytic cascade that culminates in multiple biological processes such as opsonization and phagocytosis of intruders, inflammation, cell lysis, and removal of immune complexes in addition to being a link between the innate and adaptive immune responses [3]. Furthermore, the complement system plays an important role in the removal of apoptotic cells by recognizing damaged or altered *self*-components, thereby contributing to tissue homeostasis and preventing autoimmunity [4, 5]. However, excessive complement activation may be deleterious and is associated with tissue damage in certain diseases. Conversely, insufficient activity has also been associated with susceptibility to infection and autoimmune diseases [6]. Complement system is also involved in noninflammatory functions in the brain, such as basal and ischemia-induced neurogenesis [7] and synapse remodeling and pruning [8]. Further, the complement system also interacts with the coagulation system, although the precise molecular mechanism underlying the interaction has not been elucidated [9].

Complement activation involves a remarkably powerful degree of amplification and thus requires an appropriate and efficient checking system of regulatory molecules to maintain homeostatic balance to ensure efficient destruction of pathogens and recognition of *self*-components. The regulation predominantly occurs at the level of the convertases and during assembly of the membrane attack complex (MAC) [4]. The regulatory proteins, both, soluble proteins (such as Factor H and Factor I) and proteins on host cell membranes (such as CR1, CD46, CD55, and CD59) are necessary to ensure that complement activation is not exacerbated or deficient to prevent tissue damage or physiological disorders, respectively [6].

18.2 Pathways of Complement Activation

Complement activation can be initiated by three different pathways: classical, alternative, and lectin pathways [9]. Each pathway is activated by different components that converge in the formation of active enzyme complexes (C3 and C5 convertases), followed by the assembly of the terminal pathway and MAC (C5b-9), which is inserted into to the target cell membrane to lyse the cell. Complement activation also results in the release of chemoattractants (C4a, C3a, and C5a), which are potent inflammatory molecules, and opsonins (C3b and C4b), which mediate phagocytosis (Fig. 18.1). Serine proteases play an important role in human physiology and pathology, activating each other to promote initiation and amplification of the complement cascade [10]. They present a common domain containing the catalytic triad of histidine, aspartic acid and serine residues [11]. The serine proteases of the complement system include Clr (85 KDa) and Cls (85 KDa) of the classical pathway (CP),

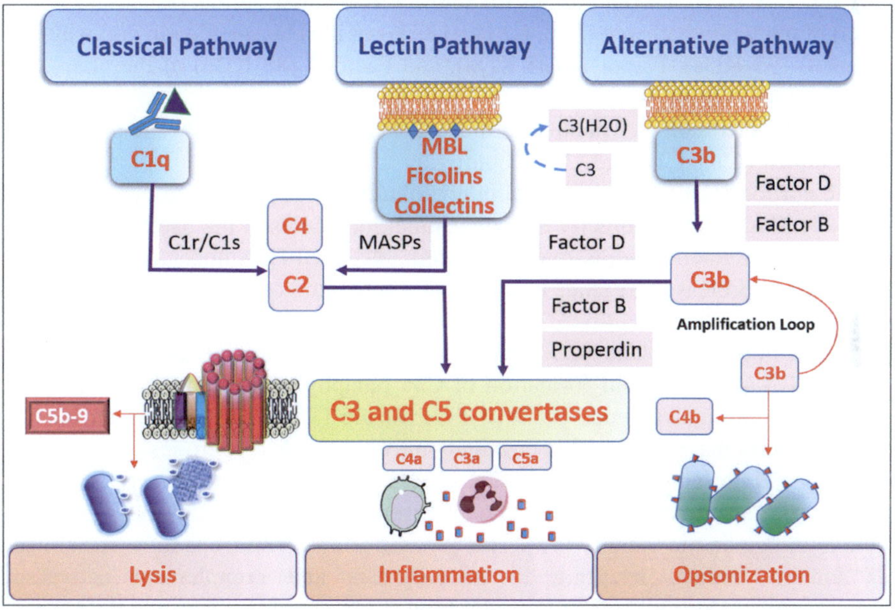

Fig. 18.1 Complement activation by the classical pathway (CP), lectin pathway (LP), and alternative pathway (AP). CP typically requires an antigen-antibody complex on pathogen surface and binding to C1 complex (C1q, C1r, and C1s) for its activation. LP recognizes mannose-terminating glycan or acetylated residues on pathogens leading to MBL/ficolins/collectins-MASP complex activation. Both pathways induce the formation of C3 convertase, C4b2a. AP is permanently activated at a low level by spontaneous hydrolysis of C3 into C3(H2O). Lack of complement inhibitors on pathogens induces AP activation by the C3bBb assembly. Complement activation leads to opsonization and phagocytosis of pathogens owing to C3b and C4b deposition, bacterial lysis by C5b-9 complex formation, and inflammation by C4a, C3a, and C5a, leading to recruitment of immune cells, endothelial and epithelial cell activation, and platelet activation

Table 18.1 Serine proteases of the complement system [9, 32]

Protease	Complement pathway	Active form	Function
C1r	Classical	C1 complex (C1q, C1r, C1s)	Clr autoactivation and Cls cleavage
C1s	Classical	C1 complex (C1q, C1r, C1s)	C2 and C4 cleavage
Factor I	Alternative	Factor I complex with C3b or C4b	C3b and C4b cleavage
Factor B	Alternative	C3bBb	C3 and C5 cleavage
Factor D	Alternative	C3bBD complex	Cleaves factor B bound to C3b
MASP-1	Lectin	MBL/MASPs complex	C2 (but not C4), C3 and MASP-2 cleavage and MASP-1 autoactivation
MASP-2	Lectin	MBL/MASPs complex	C2 and C4 cleavage
MASP-3	Lectin	MBL/MASPs complex	Remains unclear
C2	Classical/ lectin	C4b2a	C3 and C5 cleavage

MASPs 1–3 (mannan-binding lectin (MBL)-associated serine proteases; 80–90 KDa) of the lectin pathway (LP), C2 (110 KDa) of the classical/lectin pathway, and Factor B (93 KDa), Factor D (25 KDa), and Factor I (88 KDa) of the alternative pathway (AP) (Table 18.1) [12].

The CP activation is typically antibody dependent and requires the presence of C1 complex (C1q, C1r, and C1s subunits) by the binding of subcomponent C1q to the Fc portion (CH2 domain) of immunoglobulins M or G [13]. In the absence of antibody, C1q can also directly recognize other molecules of the bacterial cell wall, viral envelope membrane, C-reactive protein, etc. [1, 14]. Autocatalytic activation of the serine protease C1r leads to subsequent activation of C1s, that in turn cleaves C4 and C2 into larger (C4b, C2a) and smaller (C4a, C2b) fragments to form the enzyme complex C4bC2a (C3 convertase) [4]. The formation of C3 convertase leads to C3 activation and formation of C3a (anaphylatoxin) and C3b (opsonin), with C3 as the convergence point of the cascade [15]. C3b exposes an internal thioester bond that allows stable covalent binding to hydroxyl groups of any carbohydrates and proteins on the target surface. C3 convertase activity is very efficient, leading to the formation of approximately 1000 molecules of C3b that are able to bind to targets in the vicinity [16]. This process allows pathogens to be recognized as foreign bodies, resulting in phagocytosis and complement activation. Subsequently, additional C3b molecules bind to C3 convertase forming the C5 convertase (C4bC2aC3b) that cleaves C5 in to C5a and C5b, initiating the terminal pathway and assembly of MAC (Fig. 18.1) [4, 17].

The AP occurs on microbial surfaces in the absence of specific antibody. The AP activation occurs on the surface of foreign bodies at a low level by the spontaneous hydrolysis of the internal thioester bond in C3, leading to the formation of C3b analog, C3(H2O). Factor B binds the C3(H2O) and is then cleaved by Factor D, generating a distinct C3 convertase (C3bBb) that further cleaves C3 molecules. In the presence of an activating surface (e.g., a bacterial cell wall), C3b is protected from inactivation by regulatory proteins such as Factors I and H. However, in the

AP, a more active C3 convertase (C3bBb) is formed instead, which is further stabilized by properdin. In contrast to other pathways, AP functions as an amplification loop providing a strong positive feedback activation of C3, thereby increasing the production of pro-inflammatory mediators [18]. In fact, 80–90% of pathological complement activation in disease is driven by the AP [19]. Furthermore, the alternative convertase assembly may also be initiated by non-covalent attachment of properdin to some target surfaces (Fig. 18.1) [20, 21].

18.3 The Lectin Pathway

The existence of the LP was first discovered in the 1970s when the plant lectin mannose-binding protein (concanavalin A) was found to activate the complement system [22]. This pathway was further characterized by using proteins isolated from rabbit liver and serum; however, its function remained unclear initially [23, 24]. In 1992, Matsushita and Fujita reported that MBL and MASPs activated the LP, which was a landmark study on the mechanism of LP activation [25]. Thus far, 6 different PRMs that initiate the activation of the LP have been identified: 3 ficolins (M-ficolin, L-ficolin, and H-ficolin, also known as ficolin-1, ficolin-2, and ficolin-3, respectively), and 3 collectins (MBL, collectin 11 or collectin kidney-1 or CL-K1, and collectin 10 or colletin-L1or CL-L1). Similar to AP, the LP may be activated in the absence of immune complexes by the binding of PRMs to carbohydrates or acetylated compounds on the surfaces of pathogens (PAMPs) or apoptotic cells (DAMPs) (Fig. 18.2). The PRMs form complexes with the serine proteases, MASPs (MASPs

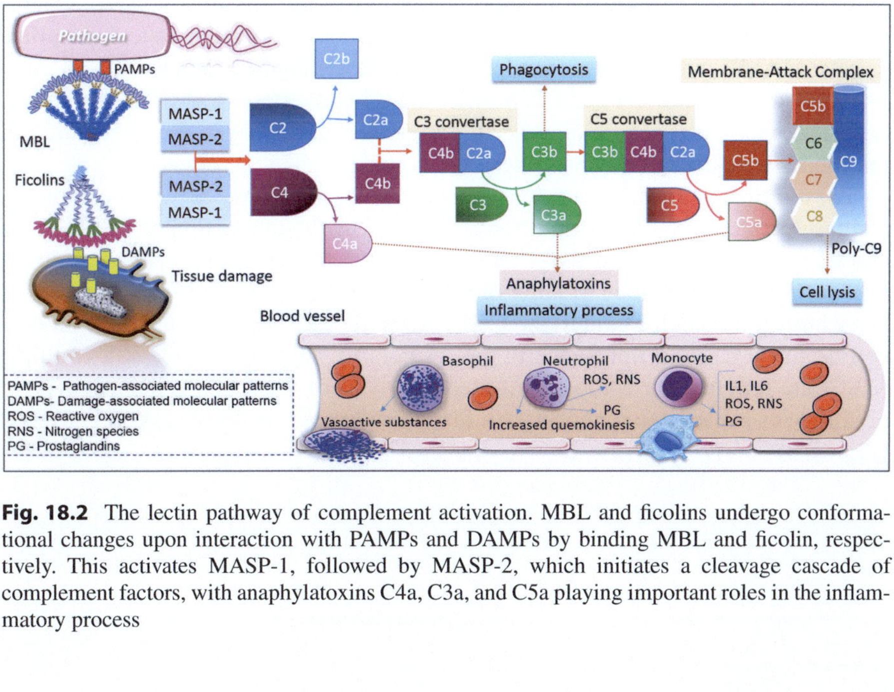

Fig. 18.2 The lectin pathway of complement activation. MBL and ficolins undergo conformational changes upon interaction with PAMPs and DAMPs by binding MBL and ficolin, respectively. This activates MASP-1, followed by MASP-2, which initiates a cleavage cascade of complement factors, with anaphylatoxins C4a, C3a, and C5a playing important roles in the inflammatory process

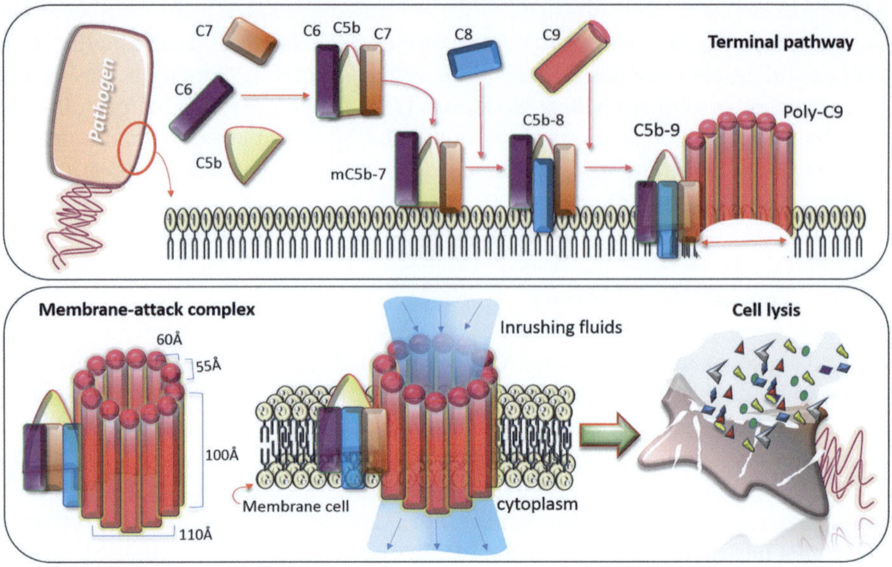

Fig. 18.3 Membrane attack complex (MAC) formation and the resultant consequences in target cell. Newly formed C5b reacts with C6 to form the stable C5b6 complex. Binding of C7 results in a hydrophobic complex that targets the membrane (mC5b-7). Membrane insertion is initiated upon binding of C8 (C5b-8) after which 12–18 copies of C9 polymerize to form the pore-forming ring structure to induce lysis of microbial membranes

1, 2, and 3), and two nonenzymatic splice products MBL-associated proteins (MAps19 and 44) [26–28]. Upon binding of PRM/MASP complexes to appropriate targets, MASPs get activated from pro-enzymes (zymogens) to active forms catalyzing the cleavage of C4 and C2, to generate C3 convertase (C4bC2a), which carries the complement activation cascade forward, culminating in the elimination of microbial intruders by phagocytosis or cells lysis [29].

The terminal pathway occurs in a similar manner in all three activation pathways and results in the assembly of the MAC, initiated by the interaction of C5b with C6 and C7 molecules, yielding the C5bC6C7 (C5b-7) complex. The membrane insertion event is initiated upon binding of C8 to C5b-7 complex. Subsequently, 12–18 copies of C9 molecules bind to the C5b-7, forming the lytic pore (C5b-9) inducing cell death by causing imbalance in cell osmolarity (Fig. 18.2) [14, 30]. Multiple MACs are required for complement-mediated lysis of nucleated cells; however, in erythrocytes it has been demonstrated that a single pore could cause cell lysis [31] (Fig. 18.3).

18.4 MBL-Associated Serine Proteases (MASPs) of the Lectin Pathway

There are three serine proteases associated with the LP PRMs, MASP-1, MASP-2, and MASP-3, in addition to two nonenzymatic proteins MAp19 and MAp44 [1]. The three MASP enzymes have an identical domain organization (Fig. 18.4), which is also similar to that of the two classical pathway serine proteases, C1r and C1s. The regulatory domain (A-chain) is composed of C1r/C1s, Uegf, and bone morphogenetic protein 1 (CUB1), followed by the epidermal growth factor (EGF), a second CUB domain (CUB2), and two contiguous complement control proteins (CCPs) 1 and 2 [33, 34]. The regulatory domain is responsible for dimerization of MASP polypeptides and binding to PRMs [35–37]. The regulatory domain is followed by the module with the catalytic activity (B-chain), the serine protease (SP) domain [33, 34]. The CCP2 and SP domains are connected through a linker peptide (also termed the activation peptide), where an Arg-Ile bond is cleaved through autolysis when MASP/PRM complexes bind to pathogens, linking the A- and B-chain connected via a disulfide bond [33].

All MASPs are generated from two genes. MASP-1, MASP-3, and MAp44 are encoded by the *MASP1* gene through an alternative splicing process [38, 39], while MASP-2 and MAp19 are alternatively spliced products of *MASP2* gene [40].

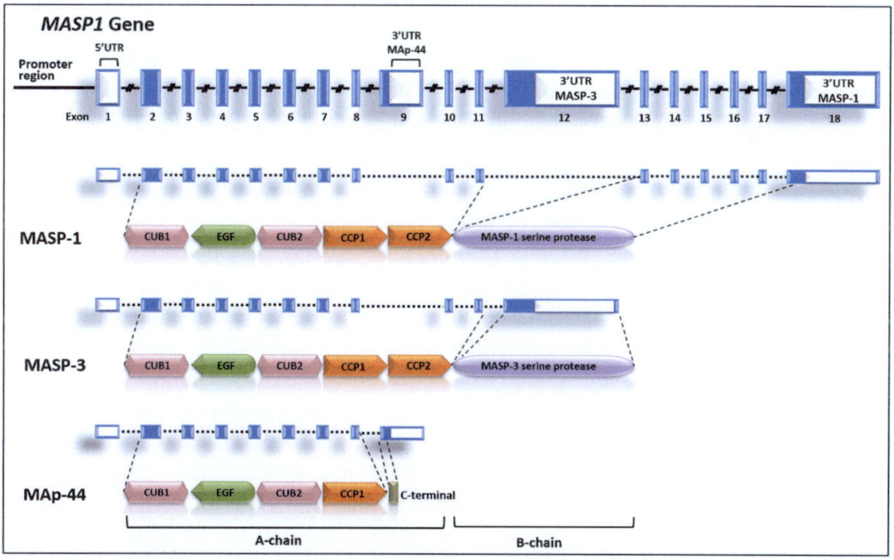

Fig. 18.4 *MASP1* gene and transcripts. The primary transcript can be spliced into three different mRNAs encoding the MASP-1, MASP-3, and MAp44 proteins. *Blue boxes* indicate the translated part of the exons. MASP: mannose-binding lectin associated serine protease. MAp44: mannose-binding lectin-associated protein of 44 kDa. CUB, C1r/C1s, Uegf, and bone morphogenetic protein; EGF, epidermal growth factor; CCP, complement control protein. Exons are drawn to scale and introns are truncated

18.4.1 *MASP1* Gene

The *MASP1* gene contains 18 exons and is located on chromosome 3q27–q28 spanning 76 kb (Fig. 18.4) [41, 42]. The gene encodes a primary pre-mRNA transcript, which is spliced differentially to yield three distinct mRNAs encoding the MASP-1, MASP-3, and MAp44 (also termed MAp 1) proteins [38, 39]. MASP-1 and MASP-3 regulatory domains (CUB1-EGF-CUB2-CCP1-CCP2) are encoded by exons 2–8 and exons 10 and 11, while the SP domain is encoded by exons 13–18 and exon 12 in MASP-1 and MASP-3, respectively. MAp44 lacks the SP domain but shares the first four domains (CUB1-EGF-CUB2-CCP) with MASP-1 and MASP-3 that are encoded by exons 2–8. Exon 9 is unique to MAp44 [39, 43]. The mRNA encoding MASP-1 is largely observed in the liver, while mRNA for MASP-3 is primarily observed in the liver and cervix, followed by bladder, brain, colon prostate, and placenta [39]. The highest expression of MAp44 is observed in the heart; it was weakly expressed in cervix, colon, and liver [39].

Some *MASP1* gene polymorphisms are associated with the serum levels of MASP-1, MASP-3, and MAp44 (Table 18.2); most associations were observed in healthy individuals. In Danish blood donors, heterozygotes of rs190590338 (G > A) lead to increase in MASP-1 median concentration, while the minor allele of rs7625133 (A > C) decreased MAp44 concentration. The minor alleles of SNPs rs3774275 (A > G), rs698090 (T > C), and rs67143992 (G > A) result in an increase in MASP-1 and MAp44 and a decrease in MASP-3 serum concentrations; SNPs rs72549154 (G > T) and rs35089177 (T > A) showed the opposite effect—the minor alleles result in an increase of MASP-3 and a decrease of MASP-1 and MAp44 [44]. The additive effect of some *MASP1* SNPs in haplotypes on MASP-1, MASP-3, and MAp44 serum concentrations has also been described. The *MASP1 TGAG* haplotype (rs35089177 (T > A), rs62292785 (G > A), rs7625133 (A > C), and rs72549254 (G > A)), for example, leads to an increase in MASP-1 and MAp44 and decrease in MASP-3 concentration in healthy blood donors [44].

In patients with cystic fibrosis homozygous (A/A) and heterozygous (G/A) alleles, SNP rs850312 (G > A) was associated with the earlier onset of *Pseudomonas aeruginosa* colonization [45]. These same genotypes were associated with higher on-admission MASP-3 levels in critically ill children, exhibiting a protective effect, as higher MASP-3 levels are related to a better outcome [46]. The T/T genotype of rs710469 (C > T) was also considered a protective genotype in critically ill children by increasing on-admission MASP-3 levels, although the genotype was equally distributed among controls and patients [46]. A non-synonymous polymorphism (rs38343199) in exon 10 (G > A) located in the MASP-1 and MASP-3 CCP2 domain was evaluated in systemic lupus erythematosus (SLE), systemic inflammatory response syndrome (SIRS), and/or sepsis patients. However, no association was found between this amino acid substitution and the diseases [47]. Some mutations in *MASP1* gene are also related to the autosomal-recessive 3MC syndrome (Carnevale, Mingarelli, Malpuech, and Michels) [48–50].

Table 18.2 *MASP1* gene polymorphisms associated with MASP-1, MASP-3, and MAp44 concentration and diseases

dbSNP	Allele	MAF	Gene region	Gene position	Amino acid position	Protein region	Serum levels[a]	Disease association
rs190590338	G > A	<1%	Promoter	−2464	n.a.	n.a.	G/A: Increase MASP-1 levels	–
rs7625133	A > C	3%	Promoter	−961	n.a.	n.a.	A/C, C/C: Decrease MAp44 levels	–
rs35089177	T > A	28%	Promoter	−1418	n.a.	n.a.	T/A, AA: Decrease MASP-1 and MAp44 levels	–
rs75284004	A > G	1%	Promoter	−1479	n.a.	n.a.	A/G decrease MASP-3 levels	–
rs62292785	G > A	10%	Promoter	−1251	n.a.	n.a.	G/A: Decrease MASP-1 levels	–
rs72549254	G > A	17%	Intron 1	9	n.a.	n.a.	A/G: Increase MASP-3 levels	–
							AG, AA: Decrease MAp44 levels	
rs710469	C > T	49%	Intron 2	24,903	n.a.	n.a.	T/T: Higher on-admission MASP-3 levels in critically ill children	Protective effect on critically ill children [46]
rs3774275	A > G	24%	Intron 8	44,153	n.a.	n.a.	A/G, G/G: Increase MASP-1 and MAp44 and decrease MASP-3 levels	–

(continued)

Table 18.2 (continued)

dbSNP	Allele	MAF	Gene region	Gene position	Amino acid position	Protein region	Serum levels[a]	Disease association
rs113938200	C > T	<1%	Exon 9	44,259	p.Asn368Asp	C-terminal MAp44	C/T: Decrease MAp44 levels	–
rs698090	C > T	46%	Exon 9	45,121	n.a.	3'UTR MAp44	C/T: Increase MAp44 levels	–
rs850312	G > A	21%	Exon 12	55,613	p.Leu617Leu	CCP2 MASP-3	A/A, A/G: Higher on-admission MASP-3 levels in critically ill children	Earlier *Pseudomonas aeruginosa* colonization [45], protective effect on critically ill children [46]
rs72549154	G > T	7%	Exon 12	55,489	p.Arg576Met	SP MASP-3	G/T: Decrease MASP-1 levels	–
rs67143992	G > A	9%	Exon12	56,100	n.a.	3′ UTR MASP-3	G/A: Increase MASP-1, MAp44 and decrease of MASP-3 levels	–
							A/A: Increase MAp44 and decrease MASP-3 levels	

dbSNP, Single Nucleotide Polymorphism Database; n.a., not applicable; MAF, minor allele frequency of 1000 genomes project (all populations); CCP, complement control protein; SP, serine protease; UTR, untranslated

[a]Compared to the homozygote state of the major allele in [44, 46]

18.4.2 MASP-1

MASP-1 was characterized by Matsushita and Fujita (1992) as the first serine protease C1s-like and was designated as mannose-binding protein (MBP)-associated serine protease (MASP). This serine protease plays a central role in the initiation of the LP, by carrying out the activation of MASP-2. It is considered a promiscuous protease since its substrate binding groove is wide and resembles that of trypsin rather than early complement proteases [51].

Recent findings supported MASP-1 as an essential component of the LP, whose concentration is 20-fold higher than MASP-2 in the plasma. MASP-1 undergoes autoactivation to subsequently activate MASP-2 efficiently—acting in a manner analogous to that of C1r and C1s in the CP, being responsible for 60% of the C2 cleaved and C3 convertase formation [52, 53]. MASP-1 autoactivation seems to control the initiation of the LP [54], but does not cleave C4, being not capable of generating C3 convertase by itself, although direct activation of C3 by MASP-1 can occur at a relatively low efficiency [55, 56].

MASP-1 is primarily expressed in the liver, with mean plasma levels of 11 μg/ml (range 4–30 μg/ml) [57], and significantly contributes to the development of the inflammatory reaction by proteolytic activity. MASP-1 induces Ca^{2+} signaling, NF-κB and p38 MAPK pathways in endothelial cells through protease-activated receptor 4 (PAR4) [58]. This activity leads to the release of IL-6 and IL-8, activating the chemotaxis of neutrophil granulocytes [59]. MASP-1 is also able to modulate the immune response by the release of pro-inflammatory bradykinin from high-molecular-weight kininogen [60].

MASP-1 is immediately activated after microbial infection by the binding of PRM complexes to targets leading to opsonization, cell lysis, release of anaphylatoxins, chemotaxis of neutrophils, and inflammation. In fact, MASP-1 plasma levels have been associated with some inflammatory disorders, and the activity of MBL/MASP-1 complex has been associated to disease severity in post-streptococcal acute glomerulonephritis and hepatitis C virus (HCV) infection, leading to glomerular fibrinogen deposits and sustained hematuria [61], and liver fibrosis [62], respectively. In addition, MASP-1 plasma levels were also higher in patients who suffered myocardial infarction and lower in patients with acute ischemic stroke [63]. High levels of MASP-1 were also observed in patients with type 1 diabetes mellitus [64].

In autoimmune diseases, high plasma levels of MASP-1 were associated with SLE [65]. In contrast, MASP-1 levels were reduced in patients with hereditary angioedema in response to the degree of complement C4 consumption, which was expected to contribute to the pathophysiology and severity of the disease [66].

Furthermore, MASP-1 was shown to play a role in coagulation, cleaving factor XIII and fibrinogen and mediating the formation of cross-linked fibrin, although with lower catalytic efficiency compared to thrombin [67]. In fact, antithrombin in the presence of heparin is a more potent inhibitor of MASP-1 then C1 inhibitor. The ancient origin of MASP-1 and its thrombin-like activity suggests its involvement in a coagulation-based defense mechanism in the early evolution of innate immunity [68]. Interestingly, components of the coagulation cascade amplify complement activation in such a manner that both complement and coagulation cascade are interconnected through an important crosstalk [9]. In addition, MASP-1 was associated with thrombus formation in a mouse model of arterial injury [69], and in patients with diabetes, contributing to an enhanced thrombotic environment and consequent vascular complications [64].

18.4.3 MASP-3

MASP-3 is an alternative spliced product of *MASP1* gene, which contains an identical A-chain, but an entirely different B chain and is highly conserved [70]. MASP-3 is mainly expressed in the pancreas, skeletal muscle, spleen, thymus, prostate, and ovary [56]. The mean serum concentration is 5.2 μg/ml (range 1.8–10.6 μg/ml) [71], mainly occurring in association with ficolin-3 and in lower amounts with ficolin-2 and MBL [38].

MASP-3 does not cleave any complement components and it is not inhibited by C1-inhibitor [56, 72]. MASP-3 may reduce the LP activity as it has to compete for MASP binding sites on the LP recognition molecules [39]. Similar to C1s, MASP-3 cleaves insulin-like growth factor-binding protein-5 (IGFBP-5), an important regulator of physiological processes in the bone, kidney, and mammary glands [73]. MASP-3 has also been implicated in the activation of the AP in mice [74]; however, in humans MASP-3 is not required for activation of AP [52].

Along with CL-K1, MASP-1, and MAp44, MASP-3 seems to have an important role in early embryonic development, as shown by the effect of five rare MASP-3 exon 12 mutations in four independent families with autosomal recessive 3MC syndrome, characterized by several development disorders. All the implicated mutations are predicted to damage the SP domain, eliminating the enzymatic activity [49, 75]. According to Venkatraman et al., this disorder is probably a result of structural defects caused by disruption of $Ca^{(2+)}$ binding during biosynthesis of CL-K1, causing structural changes in the protein and in the consequent CL-K1/MASP-3 complexes [76]. In this context, MASP-3 also cleaves IGFBP-5 [73], regulating physiological processes in kidney, bone, among others, and interestingly, is expressed in the craniofacial region during mouse embryonic development [49].

In addition, MASP-3 levels were associated to infections in children admitted to the intensive care unit (ICU). Low MASP-3 levels on-admission were associated with an increased risk of acquiring new infection in critically ill children [46].

18.4.4 MAp44

MAp44 is an alternative splice product of the *MASP1* gene, which lacks the SP domain and consequently, its functional activity. The polypeptide was named MAp44 due to its molecular mass of 44 kDa. MAp44 is mainly expressed in the heart and skeletal muscle, with a mean serum concentration of 1.7 μg/ml (range 0.8–3.2 μg/ml) [39, 43].

Although MAp44 does not contain the SP domain, the other domains interact with MBL or ficolins, thereby competing with MASP-1, MASP-2, and MASP-3 and resulting in the inhibition of C4 deposition and consequently the inhibition of downstream complement activation [39, 43, 77]. In addition to inhibiting the incorporation of MASPs into MBL/ficolin complexes, MAp44 was shown to prevent MBL deposition on MBL ligands and restricting complement activation and C3 deposition [78].

MAp44 has been associated with cardioprotective effects, preserving cardiac function, decreasing infarct area, and preventing thrombogenesis in murine models of ischemia/reperfusion injury by inhibiting MBL and C3 deposition [69, 78]. Due to its protective effects on cardiovascular system, MAp44 has been suggested to be used in a therapeutic approach for the treatment of myocardial ischemia/reperfusion injury and thrombogenesis [78]. In contrast, Frauenknecht et al. demonstrated that MAp44 levels were not directly related to the pathophysiology of cardio- and cerebrovascular diseases, but instead was associated with cardiovascular risk factors such as dyslipidemia, obesity, and hypertension [63].

18.4.5 *MASP2* Gene

The *MASP2* gene comprises 12 exons and is located on chromosome 1p36.23–31 spanning about 20 kb [79, 80]. The primary gene transcript gives rise to two different mRNAs generated by alternative splicing/polyadenylation, encoding the MASP-2 serine protease and a truncated MASP-2-related plasma protein, termed MAp19 or sMAP (Fig. 18.5) [79]. For MASP-2, the regulatory domains, CUB1-EGF-CUB2-CCP1-CCP2, are encoded by exons 2–4 and exons 6–11, while the serine protease domains are encoded by exon 12. MAp19 is encoded by 4 exons, of which 3 (exons 2–4) are shared with MASP-2 and encode the CUB1-EGF regulatory domains, whereas exon 5 encodes four specific C-terminal amino acids. MAp19 does not have a serine protease domain [40, 81].

Some *MASP2* polymorphisms are associated with modulation of MASP-2 and MAp19 serum levels (Table 18.3). The rs72550870 (T > C) responsible for the Asp > Gly substitution in residue 120 (p.D120G) occurs in the CUB1 domain [82] and affects both MASP-2 and MAp19 leading to a reduced serum concentration by eliminating the binding to MBL and ficolins and affecting complement activation [83]. The MASP-2 levels in heterozygous p.D120G healthy subjects is about half of those in subjects with the wild-type allele [82]. The rs12085877 (G > A) leads to an amino acid substitution (p.R439H) in the MASP-2 serine protease domain leading to a reduction in MASP-2 concentration in heterozygotes. MASP-2 with the p.R439H polymorphism is able to bind to MBL, however, showing reduced enzymatic activity in the MBL-MASP2 complexes [84, 85]. Several other *MASP2* polymorphisms, including rs7548659 (G > T) in the promoter region, rs61735600 (C > T) and rs56392418 (G > A) in exon 3, rs2273344(C > T) in intron 4, rs9430347 (T > C) in intron 5, rs17409276 (G > A) in intron 9, rs12711521(C > A) and rs2273346 (A > G) in exon 10, and rs12085877 (C > T) and rs1782455 (G > A) in

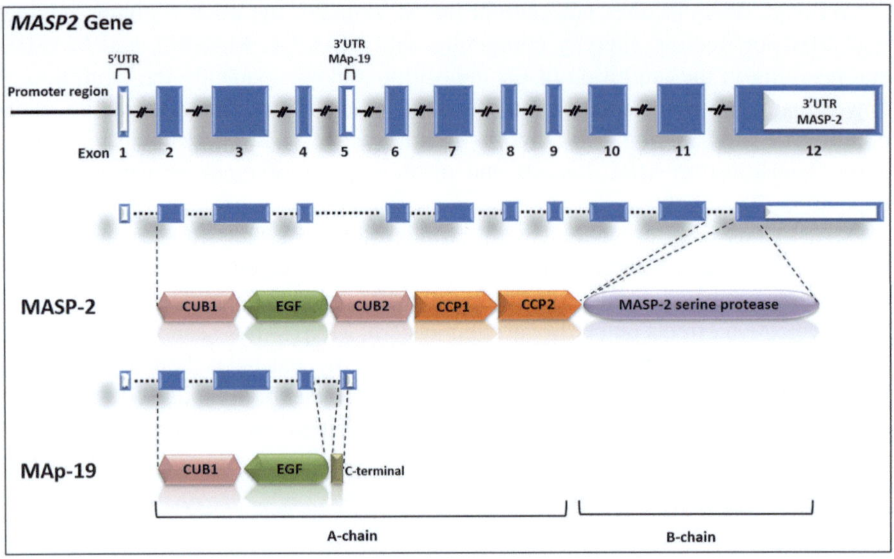

Fig. 18.5 *MASP2* gene and transcripts. Alternative splicing of primary transcript gives rise to two different mRNAs encoding MASP-2 and MAp19 proteins. *Blue boxes* indicate translated part of the exons. MASP: mannose-binding lectin-associated serine protease. MAp19: mannose-binding lectin-associated protein of 19 kDa. CUB, C1r/C1s, Uegf, and bone morphogenetic protein; EGF, epidermal growth factor; CCP, complement control protein. Exons are numbered and drawn to scale; introns are truncated

exon 12, were found to be associated with the modulation of serum levels [84–87]. Some of them were associated with the susceptibility to leprosy [87], hepatitis C [88], malaria [89], bacterial infections after orthotopic liver transplantation [90], Chagas disease [91], rheumatoid arthritis [92], tuberculosis [93], rheumatic fever [94], and endemic pemphigus foliaceus [95].

18.4.6 MASP-2

The second MASP was identified in 1997 by Thiel et al., which showed notable homology with the first reported MASP (MASP-1) and the serine proteases, C1s and C1r, of the CP. Nevertheless, subsequent analysis demonstrated that despite the homology, MASP-2 was entirely different in assembly and function to C1s [56, 68, 96] with a 1000-fold higher catalytic activity and could be inhibited by C1-inhibitor 50-fold more rapidly [97]. The almost identical substrate specificity of MASP-2 and C1s is mediated through different group of enzyme-substrate interactions, and it is very probable that the major functional difference between them is reflected in the different loop structures of the two enzymes [96].

MASP-2 is synthesized as single-chain proenzyme, and its activation proceeds through the cleavage of a single Arg-Ile bond, generating the two disulfide-linked

Table 18.3 *MASP2* gene polymorphisms associated with MASP-2 and MAp19 concentration and diseases

dbSNP	Allele	MAF	Gene region	Gene position	Amino acid position	Protein region	Serum levels[a]	Disease associations
rs7548659	G > T	43%	Promoter	−175	n.a.	n.a.	High MASP-2 and low MAp19 concentration	Susceptibility to leprosy [87]
rs72550870	T > C	1%	Exon 3	599	p.D120G	CUB1	Low MASP-2 and MAp19 concentration	Endemic pemphigus foliaceus [95], rheumatic fever [94], persistent inflammatory disease, and severe pneumococcal pneumonia [82]
rs61735600	C > T	2%	Exon 3	537	p.R99Q	CUB1	High MASP-2 concentration	Susceptibility to leprosy [87]
rs56392418	C > T	4%	Exon 3	620	p.P126L	CUB1	Low MASP-2 concentration	–
rs2273343	T > C	1%	Exon 4	1689	p.H155R	EGF	Low MASP-2 concentration	–
rs2273344	C > T	16%	Intron 4	2143	n.a.	n.a.	High MASP-2 and low MAp19 concentration	–
rs9430347	G > A	15%	Intron 5	2420	n.a.	n.a.	High MASP-2 and low MAp19 concentration	–
rs17409276	G > A	16%	Intron 9	16,060	n.a.	n.a.	High MASP-2 and low MAp19 concentration	–
rs2273346	A > G	12%	Exon 10	16,368	p.V377A	CCP2	Low MASP-2 concentration	Susceptibility to tuberculosis [93], rheumatic fever [94]

(continued)

Table 18.3 (continued)

dbSNP	Allele	MAF	Gene region	Gene position	Amino acid position	Protein region	Serum levels[a]	Disease associations
rs12711521	C > A	42%	Exon 10	16,349	p.D371Y	CCP2	High MASP-2 and low MAp19 concentration	Susceptibility to leprosy [87], HCV [88], complications after orthotopic liver transplantation [90]
rs12085877	G > A	3%	Exon 12	19,578	p.R439H	SP	Low MASP-2 concentration	Susceptibility to leprosy [87] and rheumatic fever [94], protective to malaria [89]
rs1782455	G > A	31%	Exon 12	19,741	p.S493=	SP	High MASP-2 and low MAp19 concentration	Susceptibility to leprosy [87]

dbSNP Single Nucleotide Polymorphism Database, *n.a.* not applicable, *MAF* minor allele frequency of 1000 genomes project (all populations), *CCP* complement control protein, *SP* serine protease

[a]Effect of the homozygote of the minor allele in [83–85, 87, 91]

chains, A and B [3]. The MASP-2 protease is comprised of 3 N-terminal non-catalytic domains (CUB1-EGF-CUB2) and 3 catalytic domains (CCP1-CCP2-SP). The non-catalytic domain is responsible for the binding of the protease to the recognition molecules, such as MBL. The catalytic domains are responsible for protein conformation and help to ensure the narrow selectivity for protein substrates by restricting access to the substrate binding [36, 56, 96]. The binding interface of the protease is located on all the fragments of CCP1-CCP2-SP, binding C4 with similar affinity [98].

In contrast to MASP-1, MASP-2 is a very specific protease, which very efficiently cleaves C4 and proconvertase C2, thus having the ability to generate the C3 convertase on its own [29, 34, 99]. MASP-2 can autoactivate, but under physiological conditions, MASP-1 is the essential MASP-2 activator [34]. MASP-1 is 20-fold more abundant than MASP-2 [57], having a much higher propensity for autoactivation, thus causing a dramatic increase in the rate of activation of MASP-2 [52].

MASP-2 is mainly expressed in the liver [80, 100] and is stable over time in healthy individuals, with concentration around 400–500 ng/mL in serum/plasma (range 70–1200 ng/mL) [101, 102].

The first clinical effect of MASP-2 deficiency was reported in 2003 when a patient with an inherited deficiency of MASP-2 showed several and recurrent infectious and autoimmune disease manifestations. Sequence analysis of DNA revealed a point mutation in exon 3, causing substitution of glycine for aspartic acid at position 105 (D105G) [82]. In 2005 another report with the same mutation and similar clinical condition confirmed the importance of MASP-2 deficiency in human health [83].

Further investigations showed that MASP-2 levels may be associated with several other diseases, with levels lower than 100 ng/mL being considered deficient [3]. Low MASP-2 levels were reported in acute stroke when compared with normal coronary vessel individuals [63]. This finding is in line with the observation that myocardial infarction induces complement activation with MASP-2 consumption [63, 103]. In contrast, MASP-2 deficiency appears to protect mice from gastrointestinal post-ischemic reperfusion injury [104].

Furthermore, low MASP-2 levels were associated with malignancy among critically ill children [46] and with rheumatic fever [94]. The authors suggested that low MASP-2 levels may reflect protein consumption due to complement activation, which may be involved in the establishment of rheumatic heart disease [94].

On the other hand, high MASP-2 levels appear to protect against rheumatoid arthritis and articular symptoms suggesting that MASP-2 levels might be used as a biomarker in the follow-up of individuals with familial predisposition to the disease [92]. High MASP-2 levels were also associated with the development of severe infections in adult patients with hematological cancer undergoing chemotherapy [105], type 1 diabetes mellitus [64], and juvenile idiopathic arthritis [106]. Similar to the complement system as a whole, MASP-2 represents a dual role in diseases. In general, low MASP-2 can lead to a compromised immune response against pathogens, thereby facilitating infection and disease progression, but on the other hand, high MASP-2 level can lead to exacerbated inflammatory response and tissue injury.

Additionally, MASP-2 levels have been related to a number of other diseases, including schizophrenia [107], septic shock [108], acute lymphoblastic leukemia, non-Hodgkin lymphoma, central nervous system tumors [109], and colorectal cancer [110, 111].

Finally, MASP-2 is also known to trigger the coagulation cascade by cleaving prothrombin to thrombin in a similar manner as factor Xa, generating cross-linked fibrin covalently bound on bacterial surfaces. This MASP-2 function may be protective by limiting the dissemination of infection [67, 112].

18.4.7 MAp19

MAp19 is a truncated 19 kDa product of alternative splicing and polyadenylation of the primary RNA transcript of the *MASP2* gene [56]. It contains the same CUB1 and EGF domain as MASP-2, but has an additional four unique amino acids at the C-terminal end of the protein, with no serine protease activity [29, 56, 113]. MAp19 forms homodimers via the CUB1 and EGF domains, like MASP-2, and associates with MBL and ficolins in a calcium-dependent manner [37]. It is secreted by the liver in to the plasma and expressed by Kupffer cells with a similar median level as MASP-2 (217 ng/ml, 26–675 ng/ml) [114].

The function of MAp19 is not entirely understood, but because of its ability to bind to MBL and ficolins, it was speculated that MAp19 competes with MASPs, thus acting as a downregulator to the LP. In fact, MAp19 was shown to reduce the activation of C4, by being an attenuator of the activation of LP [115]. Nevertheless, only a minor fraction of MAp19 is associated with MBL and ficolins, and binding to MBL/ficolins occurs with about ten times lower affinity compared with MASP-2 [116].

In a recent study, MAp19 was not related to inflammatory markers in patients with systemic and oligoarticular juvenile idiopathic arthritis differently as observed for the others MASPs [106].

Finally, MAp19 is excreted in human urine and may play a role in the inhibition of calcium oxalate renal stone formation [114, 117]. The nucleocapsid N protein of severe acute respiratory syndrome coronavirus interacts with MAp19 in vitro, but the functional significance of this remains unknown [118].

18.5 Conclusions

This chapter discussed several aspects and research findings that point out the importance of serine proteases of the LP and its gene polymorphisms in the human physiology and pathology. The activation of complement by complex PRMs/MASP-1/2 has been associated not only with immune response but also with other biological processes, such as coagulation and embryonic development. However, future studies are required in order to clarify the role of MASP-3, MAp19, and MAp44 proteins in the activation of the LP.

In conclusion, serine proteases of the LP have an essential role in maintaining physiological homeostasis. The activation of complement requires an effective regulatory system that is able to perform a complex checking mechanism in order to prevent pathological disorders. The impact of plasma MASP levels and its genetic polymorphisms in health and diseases processes should be encouraged in order to improve the knowledge about its real role in the maintenance of homeostasis and development of diseases. This may disclose new therapeutic and/or preventive strategies.

References

1. Degn SE, Thiel S (2013) Humoral pattern recognition and the complement system. Scand J Immunol 78:181–193
2. Hoffmann JA, Kafatos FC, Janeway CA, Ezekowitz RA (1999) Phylogenetic perspectives in innate immunity. Science 284:1313–1318
3. Thiel S (2007) Complement activating soluble pattern recognition molecules with collagen-like regions, mannan-binding lectin, ficolins and associated proteins. Mol Immunol 44:3875–3888
4. Dunkelberger JR, Song W-C (2010) Complement and its role in innate and adaptive immune responses. Cell Res 20:34–50
5. Trouw LA, Blom AM, Gasque P (2008) Role of complement and complement regulators in the removal of apoptotic cells. Mol Immunol 45:1199–1207
6. Carroll MV, Sim RB (2011) Complement in health and disease. Adv Drug Deliv Rev 63:965–975
7. Rahpeymai Y, Hietala MA, Wilhelmsson U et al (2006) Complement: a novel factor in basal and ischemia-induced neurogenesis. EMBO J 25:1364–1374
8. Stevens B, Allen N, Vazquez L et al (2007) The classical complement cascade mediates CNS synapse elimination. Cell 131:1164–1178
9. Ricklin D, Hajishengallis G, Yang K, Lambris JD (2010) Complement: a key system for immune surveillance and homeostasis. Nat Immunol 11:785–797
10. Gál P, Dobó J, Beinrohr L et al (2013) Inhibition of the serine proteases of the complement system. Adv Exp Med Biol 735:23–40
11. Wetsel R (2000) C5. In: Morley B, Walport M (eds) Complement Factsb. London, pp 104–109
12. Sim RB, Laich A (2000) Serine proteases of the complement system. Biochem Soc Trans 28:545–550
13. Duncan AR, Winter G (1988) The binding site for C1q on IgG. Nature 332:738–740
14. Tegla CA, Cudrici C, Patel S et al (2011) Membrane attack by complement: the assembly and biology of terminal complement complexes. Immunol Res 51:45–60
15. Janeway CA, Travers P, Walport M, Shlomchk M (2005) Immunobiology: the immune system in health and disease, 6th edn. Garland Science Taylor & Francis Group, New York
16. Khan FH (2009) The elements of immunology. Pearson Education, Delhi
17. Walport MJ (2001) Complement. First of two parts. N Engl J Med 344:1058–1066
18. Emlen W, Li W, Kirschfink M (2010) Therapeutic complement inhibition: new developments. Semin Thromb Hemost 36:660–668
19. Harboe M, Mollnes TE (2008) The alternative complement pathway revisited. J Cell Mol Med 12:1074–1084
20. Spitzer D, Mitchell LM, Atkinson JP, Hourcade DE (2007) Properdin can initiate complement activation by binding specific target surfaces and providing a platform for de novo convertase assembly. J Immunol 179:2600–2608

21. Xu W, Berger SP, Trouw LA et al (2008) Properdin binds to late apoptotic and necrotic cells independently of C3b and regulates alternative pathway complement activation. J Immunol 180:7613–7621
22. Milthorp P, Forsdyke DR (1970) Inhibition of lymphocyte activation at high ratios of concanavalin A to serum depends on complement. Nature 227:1351–1352
23. Kawasaki T, Etoh RYI (1978) Isolation and characterization of a mannan-binding protein from rabbit liver. Biochem Biophys Res Commun 81:1018–1024
24. Kozutsumi Y, Kawasaki T, Yamashina I (1980) Isolation and characterization of a mannan-binding protein from rabbit serum. Biochem Biophys Res Commun 95:658–664
25. Matsushita M, Fujita T (1992) Activation of the classical complement pathway by mannose-binding protein in association with a novel C1s-like serine protease. J Exp Med 176:1497–1502
26. Matsushita M, Endo Y, Taira S et al (1996) A novel human serum lectin with collagen- and fibrinogen-like domains that functions as an opsonin. J Biol Chem 271:2448–2454
27. Honoré C, Hummelshoj T, Hansen BE et al (2007) The innate immune component ficolin 3 (Hakata antigen) mediates the clearance of late apoptotic cells. Arthritis Rheum 56:1598–1607
28. Drickamer K (1988) Two distinct classes of carbohydrate-recognition domains in animal lectins. J Biol Chem 263:9557–9560
29. Thiel S, Vorup-Jensen T, Stover CM et al (1997) A second serine protease associated with mannan-binding lectin that activates complement. Nature 386:506–510
30. Hadders MA, Bubeck D, Roversi P et al (2012) Assembly and regulation of the membrane attack complex based on structures of C5b6 and sC5b9. Cell Rep 1:200–207
31. Koski CL, Ramm LE, Hammer CH et al (1983) Cytolysis of nucleated cells by complement: cell death displays multi-hit characteristics. Proc Natl Acad Sci U S A 80:3816–3820
32. Sim RB, Tsiftsoglou SA (2004) Proteases of the complement system. Biochem Soc Trans 32:21–27
33. Sato T, Endo Y, Matsushita M, Fujita T (1994) Molecular characterization of a novel serine protease involved in activation of the complement system by mannose-binding protein. Int Immunol 6:665–669
34. Kjaer TR, Thiel S, Andersen GR (2013) Toward a structure-based comprehension of the lectin pathway of complement. Mol Immunol 56:413–422
35. Thielens NM, Enrie K, Lacroix M et al (1999) The N-terminal CUB-epidermal growth factor module pair of human complement protease C1r binds Ca^{2+} with high affinity and mediates Ca^{2+}–dependent interaction with C1s. J Biol Chem 274:9149–9159
36. Wallis R, Dodd RB (2000) Interaction of mannose-binding protein with associated serine proteases: effects of naturally occurring mutations. J Biol Chem 275:30962–30969
37. Gregory LA, Thielens NM, Matsushita M et al (2004) The X-ray structure of human mannan-binding lectin-associated protein 19 (MAp 19) and its interaction site with mannan-binding lectin and L-ficolin. J Biol Chem 279:29391–29397
38. Skjoedt M-O, Palarasah Y, Munthe-Fog L, et al (2010) MBL-associated serine protease-3 circulates in high serum concentrations predominantly in complex with Ficolin-3 and regulates Ficolin-3 mediated complement activation. Immunobiology 215:921–931
39. Degn SE, Hansen AG, Steffensen R et al (2009) MAp 44, a human protein associated with pattern recognition molecules of the complement system and regulating the lectin pathway of complement activation. J Immunol 183:7371–7378
40. Stover CM, Thiel S, Thelen M et al (1999) Two constituents of the initiation complex of the mannan-binding lectin activation pathway of complement are encoded by a single structural gene. J Immunol 162:3481–3490
41. Takada F, Seki N, Matsuda Y et al (1995) Localization of the genes for the 100-kDa complement-activating components of Ra-reactive factor (CRARF and Crarf) to human 3q27-q28 and mouse 16B2-B3. Genomics 25:757–759
42. Skjoedt M, Hummelshoj T, Palarasah Y et al (2011) Serum concentration and interaction properties of MBL/ficolin associated protein-1. Immunobiology 216:625–632

43. Skjoedt M-O, Hummelshoj T, Palarasah Y et al (2010) A novel mannose-binding lectin/ficolin-associated protein is highly expressed in heart and skeletal muscle tissues and inhibits complement activation. J Biol Chem 285:8234–8243
44. Ammitzbøll CG, Steffensen R, Jørgen Nielsen H et al (2013) Polymorphisms in the MASP1 gene are associated with serum levels of MASP-1, MASP-3, and MAp 44. PLoS One 8:e73317
45. Haerynck F, Van Steen K, Cattaert T et al (2012) Polymorphisms in the lectin pathway genes as a possible cause of early chronic Pseudomonas Aeruginosa colonization in cystic fibrosis patients. Hum Immunol 73:1175–1183
46. Ingels C, Vanhorebeek I, Steffensen R et al (2014) Lectin pathway of complement activation and relation with clinical complications in critically ill children. Pediatr Res 75:99–108
47. Weiss G, Madsen HO, Garred P (2007) A novel mannose-binding lectin-associated serine protease 1/3 gene variant. Scand J Immunol 65:430–434
48. Degn SE, Jensenius JC, Thiel S (2011) Disease-causing mutations in genes of the complement system. Am J Hum Genet 88:689–705
49. Sirmaci A, Walsh T, Akay H et al (2010) MASP1 mutations in patients with facial, umbilical, coccygeal, and auditory findings of Carnevale, Malpuech, OSA, and Michels syndromes. Am J Hum Genet 87:679–686
50. Atik T, Koparir A, Bademci G et al (2015) Novel MASP1 mutations are associated with an expanded phenotype in 3MC1 syndrome. Orphanet J Rare Dis 10:128
51. Dobó J, Harmat V, Beinrohr L et al (2009) MASP-1, a promiscuous complement protease: structure of its catalytic region reveals the basis of its broad specificity. J Immunol 183:1207–1214
52. Degn SE, Jensen L, Hansen AG et al (2012) Mannan-binding lectin-associated serine protease (MASP)-1 is crucial for lectin pathway activation in human serum, whereas neither MASP-1 nor MASP-3 is required for alternative pathway function. J Immunol 189:3957–3969
53. Héja D, Kocsis A, Dobó J et al (2012) Revised mechanism of complement lectin-pathway activation revealing the role of serine protease MASP-1 as the exclusive activator of MASP-2. Proc Natl Acad Sci U S A 109:10498–10503
54. Megyeri M, Harmat V, Major B et al (2013) Quantitative characterization of the activation steps of mannan-binding lectin (MBL)-associated serine proteases (MASPs) points to the central role of MASP-1 in the initiation of the complement lectin pathway. J Biol Chem 288:8922–8934
55. Ambrus G, Gál P, Kojima M et al (2003) Natural substrates and inhibitors of mannan-binding lectin-associated serine protease-1 and -2: a study on recombinant catalytic fragments. J Immunol 170:1374–1382
56. Yongqing T, Drentin N, Duncan RC et al (2012) Mannose-binding lectin serine proteases and associated proteins of the lectin pathway of complement: two genes, five proteins and many functions? Biochim Biophys Acta 1824:253–262
57. Thiel S, Jensen L, Degn SE et al (2012) Mannan-binding lectin (MBL)-associated serine protease-1 (MASP-1), a serine protease associated with humoral pattern-recognition molecules: normal and acute-phase levels in serum and stoichiometry of lectin pathway components. Clin Exp Immunol 169:38–48
58. Megyeri M, Makó V, Beinrohr L et al (2009) Complement protease MASP-1 activates human endothelial cells: PAR4 activation is a link between complement and endothelial function. J Immunol 183:3409–3416
59. Jani PK, Kajdacsi E, Megyeri M et al (2014) MASP-1 induces a unique cytokine pattern in endothelial cells: a novel link between complement system and neutrophil granulocytes. PLoS One 9:10–13
60. Dobó J, Major B, Kékesi KA et al (2011) Cleavage of Kininogen and subsequent Bradykinin release by the complement component: mannose-binding lectin-associated serine protease (MASP)-1. PLoS One 6:1–8
61. Hisano S, Matsushita M, Fujita T et al (2007) Activation of the lectin complement pathway in post-streptococcal acute glomerulonephritis. Pathol Int 57:351–357

62. Brown KS, Ryder SD, Irving WL et al (2007) Mannan binding lectin and viral hepatitis. Immunol Lett 108:34–44
63. Frauenknecht V, Thiel S, Storm L et al (2013) Plasma levels of mannan-binding lectin (MBL)-associated serine proteases (MASPs) and MBL-associated protein in cardio- and cerebrovascular diseases. Clin Exp Immunol 173:112–120
64. Jenny L, Ajjan R, King R et al (2015) Plasma levels of mannan-binding lectin-associated serine proteases MASP-1 and MASP-2 are elevated in type 1 diabetes and correlate with glycaemic control. Clin Exp Immunol 180:227–232
65. Troldborg A, Thiel S, Laska MJ et al (2015) Levels in plasma of the serine proteases and associated proteins of the lectin pathway are altered in patients with systemic lupus erythematosus. J Rheumatol 42:948–951
66. Hansen CB, Csuka D, Munthe-Fog L et al (2015) The levels of the lectin pathway serine protease MASP-1 and its complex formation with C1 inhibitor are linked to the severity of hereditary angioedema. J Immunol 195:3596–3604
67. Krarup A, Wallis R, Presanis JS et al (2007) Simultaneous activation of complement and coagulation by MBL-associated serine protease 2. PLoS One 2:e623
68. Presanis JS, Hajela K, Ambrus G et al (2004) Differential substrate and inhibitor profiles for human MASP-1 and MASP-2. Mol Immunol 40:921–929
69. La Bonte LR, Pavlov VI, Tan YS et al (2012) MBL-associated serine protease -1 (MASP-1) is a significant contributor to coagulation in a murine model of Occlusive thrombosis. J Immunol 188:885–891
70. Stover C, Lynch NJ, Dahl M et al (2003) Murine serine proteases MASP-1 and MASP-3, components of the lectin pathway activation complex of complement, are encoded by a single structural gene. Genes Immun 4:374–384
71. Degn SE, Jensenius JC, Bjerre M (2011) The lectin pathway and its implications in coagulation, infections and auto-immunity. Curr Opin Organ Transpl 16:21–27
72. Zundel S, Cseh S, Lacroix M et al (2004) Characterization of recombinant Mannan-binding lectin-associated serine protease (MASP) 3 suggests an activation mechanism different from that of MASP-1 and MASP-2. J Immunol 172:4342–4350
73. Cortesio CL, Jiang W (2006) Mannan-binding lectin-associated serine protease 3 cleaves synthetic peptides and insulin-like growth factor-binding protein 5. Arch Biochem Biophys 449:164–170
74. Iwaki D, Kanno K, Takahashi M et al (2011) The role of mannose-binding lectin-associated serine protease-3 in activation of the alternative complement pathway. J Immunol 187:3751–3758
75. Rooryck C, Diaz-Font A, Osborn DPS et al (2011) Mutations in the lectin complement pathway genes COLEC11 and MASP1 cause 3MC syndrome. Nat Genet 43:197–203
76. Venkatraman Girija U, Furze CM, Gingras AR et al (2015) Molecular basis of sugar recognition by collectin-K1 and the effects of mutations associated with 3MC syndrome. BMC Biol 13:27
77. Degn SE, Jensen L, Olszowski T et al (2013) Co-complexes of MASP-1 and MASP-2 associated with the soluble pattern-recognition molecules drive lectin pathway activation in a manner inhibitable by MAp 44. J Immunol 191:1334–1345
78. Pavlov VI, Skjoedt M-O, Siow Tan Y et al (2012) Endogenous and natural complement inhibitor attenuates myocardial injury and arterial thrombogenesis. Circulation 126:2227–2235
79. Stover CM, Schwaeble WJ, Lynch NJ et al (1999) Assignment of the gene encoding mannan-binding lectin-associated serine protease 2 (MASP2) to human chromosome 1p36.3-->p36.2 by in situ hybridization and somatic cell hybrid analysis. Cytogenet Cell Genet 84:148–149
80. Stover C, Endo Y, Takahashi M et al (2001) The human gene for mannan-binding lectin-associated serine protease-2 (MASP-2), the effector component of the lectin route of complement activation, is part of a tightly linked gene cluster on chromosome 1p36.2-3. Genes Immun 2:119–127

81. Schwaeble W, Dahl MR, Thiel S et al (2002) The Mannan-binding lectin-associated serine proteases (MASPs) and MAp 19: four components of the lectin pathway activation complex encoded by two genes. Immunobiology 205:455–466
82. Stengaard-Pedersen K, Thiel S, Gadjeva M et al (2003) Inherited deficiency of mannan-binding lectin-associated serine protease 2. N Engl J Med 349:554–560
83. Sørensen R, Thiel S, Jensenius JC (2005) Mannan-binding-lectin-associated serine proteases, characteristics and disease associations. Springer Semin Immunopathol 27:299–319
84. Thiel S, Steffensen R, Christensen IJ et al (2007) Deficiency of mannan-binding lectin associated serine protease-2 due to missense polymorphisms. Genes Immun 8:154–163
85. Thiel S, Kolev M, Degn S et al (2009) Polymorphisms in mannan-binding lectin (MBL)-associated serine protease 2 affect stability, binding to MBL, and enzymatic activity. J Immunol 182:2939–2947
86. Boldt AB, Grisbach C, Steffensen R et al (2011) Multiplex sequence-specific polymerase chain reaction reveals new MASP2 haplotypes associated with MASP-2 and MAp 19 serum levels. Hum Immunol 72:753–760
87. Boldt ABW, Goeldner I, Stahlke ERS et al (2013) Leprosy association with low MASP-2 levels generated by MASP2 haplotypes and polymorphisms flanking MAp 19 exon 5. PLoS One 8:e69054
88. Tulio S, Faucz FR, Werneck RI et al (2011) MASP2 gene polymorphism is associated with susceptibility to hepatitis C virus infection. Hum Immunol 72:912–915
89. Holmberg V, Onkamo P, Lahtela E et al (2012) Mutations of complement lectin pathway genes MBL2 and MASP2 associated with placental malaria. Malar J 11:61
90. de Rooij B-JF, van Hoek B, ten Hove WR et al (2010) Lectin complement pathway gene profile of donor and recipient determine the risk of bacterial infections after orthotopic liver transplantation. Hepatology 52:1100–1110
91. Boldt A, Luz PR, Messias-Reason IJ (2011) MASP2 haplotypes are associated with high risk of cardiomyopathy in chronic Chagas disease. Clin Immunol 140:63–70
92. Goeldner I, Skare T, Boldt ABW et al (2014) Association of MASP-2 levels and MASP2 Gene polymorphisms with rheumatoid arthritis in patients and their relatives. PLoS One 9:e90979
93. Chen M, Liang Y, Li W et al (2015) Impact of MBL and MASP-2 gene polymorphism and its interaction on susceptibility to tuberculosis. BMC Infect Dis 15:151
94. Catarino SJ, Boldt ABW, Beltrame MH, et al (2014) Association of MASP2 polymorphisms and protein levels with rheumatic fever and rheumatic heart disease. Hum Immunol.
95. Messias-Reason I, Bosco DG, Nisihara RM et al (2008) Circulating levels of mannan-binding lectin (MBL) and MBL-associated serine protease 2 in endemic pemphigus foliaceus. Clin Exp Dermatol 33:495–497
96. Gál P, Dobó J, Závodszky P, Sim RBM (2009) Early complement proteases: C1r, C1s and MASPs. A structural insight into activation and functions. Mol Immunol 46:2745–2752
97. Kerr FK, Thomas AR, Wijeyewickrema LC et al (2008) Elucidation of the substrate specificity of the MASP-2 protease of the lectin complement pathway and identification of the enzyme as a major physiological target of the serpin, C1-inhibitor. Mol Immunol 45:670–677
98. Duncan RC, Bergström F, Coetzer TH et al (2012) Multiple domains of MASP-2, an initiating complement protease, are required for interaction with its substrate C4. Mol Immunol 49:593–600
99. Rossi V, Cseh S, Bally I et al (2001) Substrate specificities of recombinant mannan-binding lectin-associated serine proteases-1 and -2. J Biol Chem 276:40880–40887
100. Stover CM, Lynch NJ, Hanson SJ et al (2004) Organization of the MASP2 locus and its expression profile in mouse and rat. Mamm Genome 15:887–900
101. Møller-Kristensen M, Jensenius JC, Jensen L et al (2003) Levels of mannan-binding lectin-associated serine protease-2 in healthy individuals. J Immunol Methods 282:159–167
102. Ytting H, Christensen IJ, Thiel S et al (2007) Biological variation in circulating levels of mannan-binding lectin (MBL) and MBL-associated serine protease-2 and the influence of age, gender and physical exercise. Scand J Immunol 66:458–464

103. Zhang M, Hou YJ, Cavusoglu E et al (2011) MASP-2 activation is involved in ischemia-related necrotic myocardial injury in humans. Int J Cardiol 166:499–504
104. Schwaeble WJ, Lynch NJ, Clark JE et al (2011) Targeting of mannan-binding lectin-associated serine protease-2 confers protection from myocardial and gastrointestinal ischemia/reperfusion injury. Proc Natl Acad Sci U S A 108:7523–7528
105. Ameye L, Paesmans M, Thiel S et al (2012) M-ficolin levels are associated with the occurrence of severe infections in patients with haematological cancer undergoing chemotherapy. Clin Exp Immunol 167:303–308
106. Petri C, Thiel S, Jensenius JC, Herlin T (2015) Investigation of complement-activating pattern recognition molecules and associated enzymes as possible inflammatory markers in oligoarticular and systemic juvenile idiopathic arthritis. J Rheumatol 42:1252–1258
107. Mayilyan KR, Arnold JN, Presanis JS et al (2006) Increased complement classical and mannan-binding lectin pathway activities in schizophrenia. Neurosci Lett 404:336–341
108. Charchaflieh J, Wei J, Labaze G et al (2012) The role of complement system in septic shock. Clin Dev Immunol 2012:407324
109. Fisch UP, Zehnder A, Hirt A et al (2011) Mannan-binding lectin (MBL) and MBL-associated serine protease-2 in children with cancer. Swiss Med Wkly 141:w13191
110. Ytting H, Christensen IJ, Thiel S et al (2005) Serum mannan-binding lectin-associated serine protease 2 levels in colorectal cancer: relation to recurrence and mortality. Clin Cancer Res 11:1441–1446
111. Ytting H, Christensen IJ, Thiel S et al (2008) Pre- and postoperative levels in serum of mannan-binding lectin associated serine protease-2 -a prognostic marker in colorectal cancer. Hum Immunol 69:414–420
112. Gulla KC, Gupta K, Krarup A et al (2010) Activation of mannan-binding lectin-associated serine proteases leads to generation of a fibrin clot. Immunology 129:482–495
113. Takahashi M, Endo Y, Fujita T, Matsushita M (1999) A truncated form of mannose-binding lectin-associated serine protease (MASP)-2 expressed by alternative polyadenylation is a component of the lectin complement pathway. Int Immunol 11:859–863
114. Degn S, Thiel S, Nielsen O (2011) MAp 19, the alternative splice product of the MASP2 gene. J Immunol Methods:89–101
115. Iwaki D, Kanno K, Takahashi M et al (2006) Small mannose-binding lectin-associated protein plays a regulatory role in the lectin complement pathway. J Immunol 177:8626–8632
116. Degn SE, Thiel S, Nielsen O et al (2011) MAp 19, the alternative splice product of the MASP2 gene. J Immunol Methods 373:89–101
117. Kang I, Kim J, Chang S et al (1999) Mannan-binding lectin (MBL)-associated plasma protein present in human urine inhibits calcium oxalate crystal growth. FEBS Lett 462:89–93
118. Liu J, Ali MAM, Shi Y et al (2009) Specifically binding of L-ficolin to N-glycans of HCV envelope glycoproteins E1 and E2 leads to complement activation. Cell Mol Immunol 6:235–244

Pups, SAMPs, and Prokaryotic Proteasomes

19

Subrata Ganguli and C. Ratna Prabha

Abstract

In eukaryotes, barrel-shaped proteases known as 26S proteasomes are responsible for removing misfolded proteins and regulatory proteins after they serve their function. 26S proteasomes are ATP-dependent proteases with three different proteolytic activities. The proteolytic active sites are segregated into the inner compartment of proteasome to prevent nonspecific degradation of cytosolic proteins. Eukaryotic cells tag the proteins with ubiquitin, in order to selectively target them for degradation by proteasomes. The presence of proteasomes in some species of actinobacteria and archaea is known for more than two decades. However, the details of the molecules used as tags and the mechanism of tagging are coming to light only in the recent times. In actinobacteria prokaryotic ubiquitin-like molecules (Pups) and in archaea small archaeal modifier proteins (SAMPs) are used as tags. Though prokaryotic proteasomes show homology to their eukaryotic counterparts, the prokaryotic tagging mechanism is vastly different suggesting convergent evolution. The structure of prokaryotic proteasomes, Pups, and SAMPs and the tagging mechanisms are presented here in detail, and the similarities and differences with eukaryotic system are highlighted. The possible applications of the knowledge generated in this area to the treatment of tuberculosis are underscored at the end of the chapter.

S. Ganguli
R&D, Consortium for Training, Research, and Development, AE 641, Salt Lake, Kolkata 700064, India

C. Ratna Prabha (✉)
Department of Biochemistry, Faculty of Science, The Maharaja Sayajirao University of Baroda, Vadodara 390002, India
e-mail: chivukula_r@yahoo.com

S. Chakraborti, N.S. Dhalla (eds.), *Proteases in Physiology and Pathology*,
DOI 10.1007/978-981-10-2513-6_19

Keywords
Prokaryotic ubiquitin-like protein • Pup • Small archaeal modifier proteins • SAMP1 • SAMP2 • Prokaryotic proteasomes

19.1 Introduction

The phenomena and the routes leading to in vivo degradation of biomolecules in general, and proteolysis in particular, had drawn the attention of molecular biologists, biochemists, physiologists, and pathologists alike. The natural routes of degrading undesirable proteins include proteins which have served their function and proteins which have lost their structure during the course of action, and improperly folded proteins are as important as ridding the cell of invading foreign molecules. Protein degradation is crucial to maintenance of homeostasis under normal physiological condition and in the performance of cell-mediated defense mechanism during an abnormal pathological situation. The introductory section in this book chapter narrates the account of both lysosome-mediated and proteasome-dependent degradation, with emphasis on the ubiquitin-proteasome system (UPS) found in eukaryotes, and how this knowledge helped in understanding proteasomes and proteasome-mediated degradation in archaea and prokaryotes, with later sections being devoted to the details of structure, function, and pathogenesis related to the prokaryotic ubiquitin-like proteins (Pups) of actinobacteria and structure and function of small archaeal modifier proteins (SAMPs) of archaea.

Historically lysosomal degradation was the only well-studied pathway for scavenging proteins, certain organelles, and some biomolecules [1, 2]. Lysosomes, described in 1955 by the Belgian biologist Christian de Duve, which contain hydrolytic enzymes are meant for degrading proteins and other biomolecules. Mutations in the genes encoding lysosomal enzymes cause genetic diseases, collectively called lysosomal storage diseases [3–5]. Christian de Duve received Nobel Prize in Physiology or Medicine in the year 1974 for the discovery of lysosomes and degradative pathways of biomolecules associated with them. Monopoly of lysosomes as the only natural degradative machinery for cellular proteins in vivo ended with the discovery of proteasomes and the ubiquitin-proteasome system (UPS) for the degradation of proteins [6–8].

19.2 Proteasomes of Eukaryotic Cells

Ion eukaryotic cell self-compartmentalized proteases known as the 26S proteasomes are present in the cytosol [6]. They are responsible for degrading proteins, which have served their function, besides removing damaged and truncated proteins. 26S proteasome is a protein complex of about 2.5 MDa. It is made up of 20S proteasome or the core particle and 19S particle or PA700 particle formed by 19 regulatory subunits present at one or both ends of the particle. The 20S proteasome is a hollow cylindrical structure formed by four stacked rings comprising of seven protein subunits each [9]. Each of the inner two rings of the complex is made up of seven -subunits, whereas the outer two rings are made up of seven α-subunits each. There are three protease active sites in the β-ring, located on the interior surface of the central chamber. The protease activities associated with proteasome are described as trypsin-like, chymotrypsin-like, and caspase-like based on the cleavage pattern. The outer two rings control the gate through which proteins must enter the inner chamber for proteolysis. The entrance to the complex could be as narrow as 13 Å with the interior chamber at most 53 Å wide and is 150 Å by 115 Å in case of mammals. In eukaryotes the 20S core proteasome is essential for survival.

19.3 Ubiquitin: The Protein Tag Used for Posttranslational Modification of Eukaryotic Proteins

In the late 1970s, studies with reticulocytes lacking lysosomes led to identification of a novel ATP-dependent intracellular proteolytic mechanism [10, 11]. Further studies identified several protein chains making up the protease system. It was also observed that modification of lysine residues in substrate proteins is a prerequisite for proteolytic degradation by this novel mechanism. Finally the entire system was described as the ubiquitin-mediated protein degradation pathway, wherein proteins are degraded by the 26S proteasome after being tagged by ubiquitin. This arrangement is referred to as the ubiquitin-proteasome system (UPS) in the eukaryotes. The Nobel Prize in Chemistry for the year 2004 was awarded to Aaron Ciechanover, Avram Hershko, and Irwin Rose for the discovery of this novel proteolytic pathway.

Ubiquitin is a small protein modifier, about 8.5 KDa in molecular size, and is ubiquitously found in eukaryotic cells ranging from single-celled microorganisms to plants and mammals [12]. Early work on ubiquitin identified it as ATP-dependent proteolysis factor or APF1 [10, 11]. Eukaryotic cells conjugate the protein ubiquitin to modify other proteins in a process called ubiquitination [7, 8]. Conjugation of ubiquitin to substrate proteins is achieved by formation of a covalent isopeptide bond between the carboxy terminus of ubiquitin and lysine side chain of the target protein. Ubiquitination is catalyzed by the serial action of three enzymes: the ubiquitin-activating enzyme (E1), ubiquitin-conjugating enzyme (E2), and ubiquitin ligase (E3). A molecule of ubiquitin is conjugated to substrate protein in monoubiquitination, or ubiquitin molecules can form chains on the protein by polyubiquitination. Ubiquitination serves many different functions inside the cell; however one of the most important functions is guiding substrate proteins to proteasomes for degradation [10]. Ubiquitination of substrate protein is a prerequisite for degradation by ubiquitin-proteasome system (UPS). The deregulation of ubiquitin pathway has been implicated in Alzheimer's disease, Parkinson's disease, cancers, and several other genetic disorders [13].

19.3.1 Proteasomes in Prokaryotes

Among prokaryotes, archaea and gram-positive bacteria belonging to the order Actinomycetales are known to possess 20S proteasomes [14]. The prokaryotic 20S proteasome like its eukaryotic counterpart has four seven-membered rings arranged one above the other in the order of αββα forming a hollow cylinder [15, 16]. However, in prokaryotic proteasome, the α-rings and β-rings are homoheptameric in their composition. In actinobacteria they are encoded by two genes, *PrcA* and *PrcB*. Proteolytic active sites are present on the inner surface of the two inner β-rings, while the two outer α-rings interact with the ATPases. All seven subunits in the β-rings are associated with proteolytic activity. The bacterial α- and β-subunits are homologous to eukaryotic α- and β-subunits. The structure of 20S proteasome has been studied in several bacteria including *Frankia* [17], *Rhodococcus* [18], *Streptomyces* [19], and *Mycobacterium* [16]. Proteasomes of all these bacteria except those of *Mycobacterium* exhibit chymotryptic activity. The proteolytic activity of proteasomes of *Mycobacterium* resembles that of eukaryotic proteasomes with three different cleavage patterns, though it is not clear how a single type of β-subunit exhibits three different proteolytic activities [16]. Proteasomes from archaea also display three different proteolytic activities. Further, with *Mycobacterium tuberculosis* (*Mtb*) proteasomes, it was observed that the N-terminal region of α-subunits forms protuberances into the mouth of the proteasomal tunnel, preventing indiscriminate entry and degradation of cytosolic proteins [16]. The α-rings of 20S proteasomes from eukaryotes, archaea, and other members of actinobacteria display obstructions similar to those of *Mtb* to prevent nonspecific degradation of proteins.

Like the eukaryotic 26S proteasome, prokaryotic proteasome has two subcomplexes, namely, the core particle and the regulatory particle [20]. In *Haloferax volcanii*, a halophilic archaeon, the core particle of proteasome is made up of two types of α-subunits, either α1 or α2, and a single type of β-subunit [21, 22]. In archaea the regulatory particle is called proteasome-activating nucleotidase (PAN). PAN is of 650 KDa and comprises six homologous subunits with AAA ATPase activity [23]. Once again in *Haloferax volcanii*, two different types of ATPase rings were identified, namely, PAN-A and PAN-B [24]. The opening at the α-ring is of 1.3 nm in diameter, which makes prior unfolding of substrate proteins obligatory for entering the proteasomal tunnel. PAN of *Methanococcus jannaschii*, heterologously expressed in *E. coli*, could associate with 20S proteasome of *Thermoplasma acidophilum* [23, 25, 26], establishing the fact that the structure of PAN is well conserved among the species of archaea. The ATPase activity of PAN is responsible for unfolding and translocating the proteins into proteasomes.

The subcomplex exhibiting ATPase-dependent unfoldase activity is a homohexamer known as Mpa in mycobacteria [27] and ARC in actinobacteria [28]. In the ATPase ring of eukaryotes and archaea on top of the AAA domain responsible for the ATPase activity, a single ring formed by oligonucleotide-binding (OB) domain is present, whereas in bacteria there are two OB domains arranged in two tiers [29]. The C-terminus of ATPase ring interacts with the α-ring of 20S proteasome in archaea as well as in eukaryotes. It was observed that the second residue from the C-terminus of the ATPases is aromatic, and it sits in a groove in the α-ring, bringing about a conformational change in the latter leading to opening of the obstructive gate formed by N-termini of the α-chains [30, 31]. The assembly of purified ATPase ring and 20S proteasome could not be demonstrated with bacteria, in spite of the aromatic residue as the penultimate residue in the C-terminus of ATPase chains and the similarity in the architecture of proteasome with that of archaea [32, 33]. The possible reasons for this failure are speculated to be either conformational changes introduced into the subunits or the loss of accessory factors which may be required for the assembly, during the process of purification.

19.4 Prokaryotic Ubiquitin-Like Protein (Pup)

Substrate proteins are recruited to bacterial proteasome using the mechanism analogous to ubiquitination found in eukaryotes. Here the protein substrate is conjugated to a protein known as prokaryotic ubiquitin-like protein (Pup) [34, 35]. A screen setup to find proteins associated with Mpa-proteasome using bacterial two-hybrid system identified Pup as a protein bound to Mpa [35]. The process of tagging the substrate with Pup protein is called pupylation. Pupylation was initially observed in *Mtb* and *Mycobacterium smegmatis* (*Msm*), and later it was concluded that all the actinobacterial species which show proteasome-mediated degradation use the same mechanism for marking proteins. Pup is a small protein of 60 to 70 residues in length [36]. However, its sequence and structure do not bear any homology to ubiquitin [37–39]. The C-terminal residue in Pup is glutamate or glutamine, which is preceded by diglycine. Pup displays disordered structure in unbound state. The C-terminal region takes up coiled coil conformation, showing a weak propensity for attaining helical conformation.

19.5 Pupylation in Bacteria

During pupylation an isopeptide linkage is formed between the C-terminus of Pup and lysine side chain of the substrate protein [34, 35]. The C-terminal glutamine of PupQ is deamidated and converted to PupE, which has glutamate at its carboxyl end (Fig. 19.1). The carboxyl of resultant glutamate residue at the end of PupE is conjugated to protein substrate. Efforts to find proteins involved in the conjugation of Pup

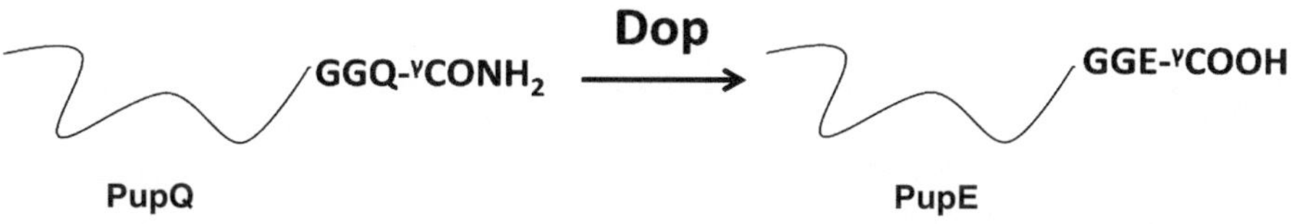

Fig. 19.1 Deamidation of PupQ to PupE by Dop

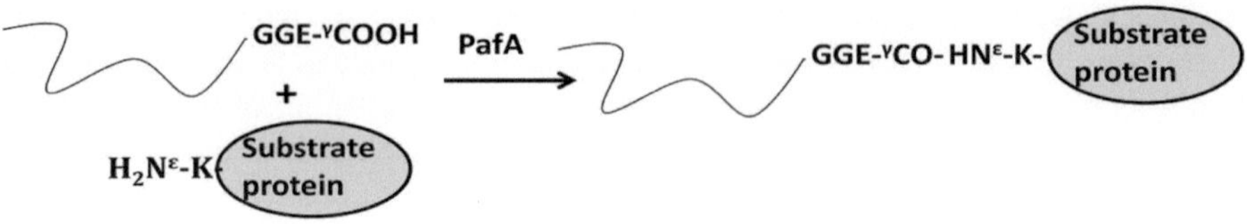

Fig. 19.2 Pupylation of substrate protein catalyzed by PafA

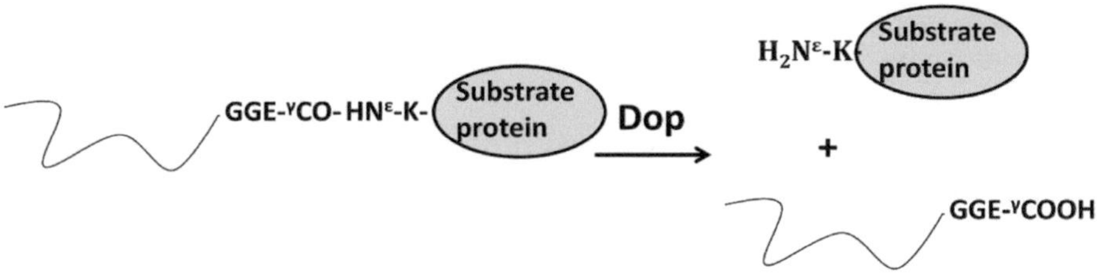

Fig. 19.3 Depupylation of pupylated protein by Dop

using pull-down assays identified two proteins, PafA and Dop [40]. Later on it was found with Dop knockouts that Dop catalyzes the deamidation reaction on Pup in mycobacteria, and in its absence the conjugation between PupQ and substrate proteins fails to occur leading to accumulation of proteasomal substrates [41, 42]. Several other actinobacteria, which synthesize PupE with glutamate as the last residue, could carry out pupylation even in the absence of Dop.

On the other hand, PafA acts as Pup ligase catalyzing conjugation of PupE to substrate proteins (Fig. 19.2). In *Mtb* strains carrying *pafA* deletion, pupylated proteins were not found and proteasomal substrates were accumulated, establishing PafA's role as Pup ligase [35, 43]. PafA catalyzes the reaction in two steps. PafA initially binds ATP and activates PupE to phosphorylated PupE, by transferring the phosphate group from ATP [44]. While ADP remains bound to PafA, PafA catalyzes the second step in which activated Pup protein is transferred to substrate releasing phosphate.

Only a single Pup ligase and/or monopupylation has been identified so far [40]. There is no experimental evidence suggesting the possibility for existence of more than one Pup ligase or polypupylation.

19.6 Dop Catalyzes Depupylation to Maintain Protein Homeostasis

However, the presence of Dop even in those bacteria which produce PupE instead of PupQ suggested another, not so far identified function [40]. Studies on these organisms revealed that Dop acts as depupylase (Fig. 19.3), removing Pup from substrate proteins [45, 46]. Dop uses ATP as a cofactor for both deamidase and depupylase reactions. In Dop-catalyzed reactions, ATP can be replaced by ADP successfully as there is no energy requirement [40].

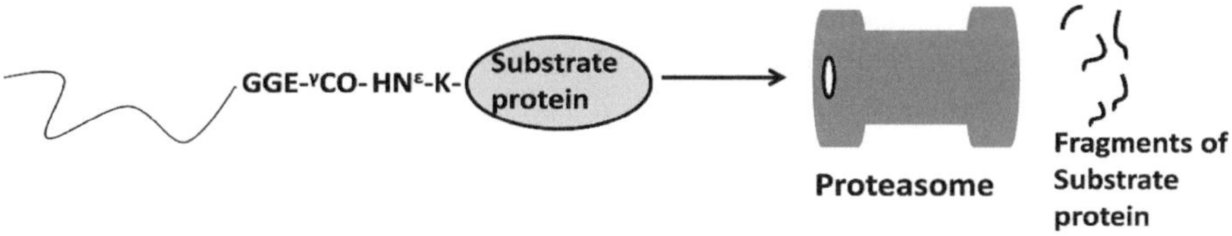

Fig. 19.4 Degradation of pupylated substrate protein by prokaryotic proteasome

19.7 Binding of Pupylated Substrate to Mpa and Its Subsequent Degradation by Proteasomes

The pupylated substrate is recognized and bound by Mpa-proteasome complex [35]. Residues 20–58 of Pup serve as docking site and are bound by Mpa [37]. In the bound state, region corresponding to 21–64 residues of Pup adopts helical conformation [47]. The ATPase activity of Mpa unfolds the substrate protein [33]. The N-terminus of Pup is oriented into proteasomal pore [33, 48]. First 8–9 residues of Pup initiate translocation into proteasome, and slowly the substrate protein is threaded into proteolytic chamber leading to its degradation (Fig. 19.4). Hence, the N-terminal 8–9 residues and the stretch of residues 21–64 in Pup are described as a two-part degron, for their deterministic role in protein degradation. Here, Pup protein undergoes degradation along with the substrate.

19.8 Pupylation as a Regulator of Protein Degradation

Mpa ring undergoes pupylation at its C-terminal on Lys591 residue [49–51]. After pupylation the Mpa ring is capable of binding to pupylated proteins and unfolding them. However, it will not be able to interact with 20S proteasome, which makes degradation of proteins impossible [52]. Depupylation by Dop can nullify the effect of pupylation of Mpa [52]. Hence, while pupylation of proteins directs them for degradation, pupylation of Mpa prevents protein degradation.

19.9 Pup-Proteasome System in Relation to Pathogenesis of Mycobacteria

In most of the actinobacteria, proteasomes are not essential under normal conditions of growth as there are other compartmentalized proteases such as Clp proteases, Lon protease, and FtsH [53–55]. *Mycobacterium tuberculosis* (Mtb) does not require proteasomes for its in vitro growth, while it cannot survive without Clp protease genes. However, Mtb requires proteasomes for its persistence in the macrophages inside the host [55–59], which may be attributed to the observed absence of Lon protease [60]. Infected macrophages produce NO and reactive nitrogen intermediates (RNI) with the help of inducible nitric oxide synthase (iNOS) induced by IFN-γ [61]. Proteasomes are required to remove proteins damaged by the onslaught of NO and RNI for the survival of the pathogen. Conversely, even iNOS-deficient mice mutations of *Mpa* and *PafA* led to attenuation of pathogen [27, 43, 62], suggesting role for the proteasomes in the regulated degradation of other stress-related proteins required for survival and persistence of the pathogen [36]. Interestingly, several pathogenesis-associated proteins such as Icl1 [62] and Ino1 [63] have been shown to undergo degradation by PPS.

With the emergence of multidrug-resistant Mtb, there is a continuous search for novel drug targets in the organism. Mycobacterial proteasome is increasingly perceived as a promising target. However, most of the inhibitors [16, 43, 55, 57, 64] including bortezomib [65] have been found to affect both human and mycobacterial proteasomes alike. More extensive searches of chemical libraries for inhibitors with high selectivity identified two oxathiazol-2-one compounds, namely, GL5 and HT1171 [66]. These compounds can kill mycobacteria, which were already debilitated by NO treatment. The two compounds modify the active site threonine of mycobacterial proteasome. These compounds initially induce a conformational change in the β-subunits of mycobacterial proteasome, which leads to chemical modification of the active site threonine by the compounds subsequently. In human proteasome the residues responsible for conformational change of β-subunit are not conserved, leaving the proteasome unharmed by these compounds.

19.10 Evolutionary Aspects of PPS

The 20S proteasomes of actinobacteria, archaea, and eukaryotes are clearly homologous. 20S proteasomes in all these organisms are threonine proteases. There is an ongoing debate on the origin of 20S proteasomes. Many evolutionary biologists are of the view that proteasomes originated in actinobacteria and were passed on to archaea as actinobacterial ancestors gave rise to archaea [67], while others believe that actinobacteria came into possession of proteasomal genes through horizontal gene transfer from either archaea or eukaryotes [68, 69]. Interestingly, the hexameric ATPase rings present in all three lineages also show strong homology. The basic structural plan of heptameric barrel-shaped protease associating with hexameric ATPase ring is also uniformly found in all these organisms.

Even though tagging substrate proteins is common to both PPS and UPS, Pup and ubiquitin do not bear any homology either in their sequence or structure. While ubiquitin is a globular protein with β-grasp fold, Pup is an unstructured protein.

The process of ubiquitination and pupylation also is analogous. A cascade of three enzymes transfers ubiquitin to substrate protein. Activation of ubiquitin involves ubiquitin adenylate. There is a single enzyme PafA which catalyzes conjugation of Pup to substrate protein. Though the reaction is ATP driven, the intermediate formed in the reaction is phosphorylated Pup. Pup has glutamate or glutamine as the C-terminal residue, whereas diglycine is found at the C-terminal of ubiquitin. The isopeptide bond formed by Pup is between the γ-carboxyl of glutamate and lysine of substrate, and in the case of ubiquitin, the isopeptide bond is formed between α-carboxyl of glycine and lysine of substrate. There is a single Pup ligase, and only monopupylation occurs on substrate in contrast to ubiquitination, where more than six hundred ubiquitin ligases and polyubiquitination with different branching patterns are present. Another interesting difference is the sequences of PafA and DopA that are homologous to glutamine synthetase [70, 71].

19.11 Small Archaeal Modifier Proteins: SAMP1 and SAMP2

In eukaryotes proteins are marked for proteasomal degradation by tagging with ubiquitin posttranslationally. Ubiquitin is a small globular protein with a characteristic β-grasp fold in its structure. Another hallmark of ubiquitin is a diglycine motif at its C-terminal through which it is conjugated to other proteins. Though existence of proteasomes was established in archaca [14], the mechanism of targeting protein substrates for degradation remained elusive for very long. The search for proteins with the characteristic features of β-grasp fold and diglycine motif identified several proteins [72], chiefly involved in sulfur transfer such as MoaD and ThiS [73]. One of them is Urm1, a sulfur carrier in tRNA modification pathway in yeast [74, 75]. It can also conjugate with alkyl hydroperoxide reductase [76], establishing the possibility for existence of ubiquitin-like protein modifiers in archaea closely related to sulfur-carrying proteins. Following these observations, Humbard's group looked for potential protein modifiers in *Haloferax volcanii* with smaller size and with a diglycine motif in C-terminus and selected five proteins [77]. Further, narrowing down the search to β-grasp fold structure, they suggested three proteins as possible protein modifiers. The three proteins were N-terminally FLAG tagged, and the possibility of their conjugation to other proteins was investigated using immunoprecipitation. The experiments finally helped in identifying two small archaeal modifier proteins (SAMPs), namely, SAMP1 and SAMP2. They are conjugated to Lys residues in substrates. Proteasomal mutants of *panA* (encoding PAN-A) and *psmA* (encoding α1-subunit) accumulated SAMP1ylated proteins [77]. However, the above double mutations and single mutations of *panA*, *panB*, and *psmA* did not lead to elevation of SAMP2ylated products, suggesting feedback regulation [78]. SAMP2 with Lys in 58th position can undergo conjugation to form a chain of poly-SAMP2 [78].

19.12 SAMPylation in Archaea

In archaea SAMPylation occurs between the α-carboxyl of C-terminal glycine and the ε-amino group of lysine residue of the substrate, resembling ubiquitination of eukaryotic cells. The process is catalyzed by the enzyme MoeB, which is homologous to Uba4p, the E1 enzyme of Urm1 pathway [74, 79, 80]. The role of MoeB in SAMPylation could be established easily as it copurified with SAMP1 and SAMP2. Based on the properties of MoeB, the process of activation of SAMP1 or SAMP2 is proposed to be through adenylation [78]. There are no known equivalents of E2 and E3 enzymes in this system. Hence, it is possible that the process of SAMPylation is accomplished by just one enzyme, MoeB. The existence of de-SAMPylases with JAMM motif was predicted using bioinformatics. Though, there are many gaps to be bridged in our understanding of SAMPylation. SAMPylation is a possible ancestor of ubiquitination.

19.13 Conclusions

Proteaomes are present in all three lineages of life. Proteasome is a barrel-shaped protease with its active sites sequestered into its hollow inner compartment. Proteins are targeted for proteasomal degradation by ear marking them with proteinaceous tag. The most interesting fact learned from the discovery of PPS (Pup-proteasome system) is posttranslational modification of proteins by a protein tag that is not limited to eukaryotes. Prokaryotes have evolved an analogous system to target specific proteins for degradation. Unlike UPS of eukaryotes, PPS is not essential for survival of bacteria, the only exceptions being survival of *Mtb* in macrophages and *Msm* under starvation, which are dependent on functional PPS.

There are several unanswered questions regarding PPS. How does the bacterium exercise selectivity with respect to substrates with a single Pup ligase? Are there any accessory factors involved in substrate selection? How are pupylation and depupylation balanced? Can we interfere with the process of pupylation as a therapeutic route instead of blocking the proteasomes? The answers to these questions may find ways for the treatment of tuberculosis.

SAMP proteins of archaea and SAMPylation suggest that posttranslational modification of proteins with protein tags for degradation by proteasomes evolved much before the origin of eukaryotic lineage. However, there are many gaps to be filled in our understanding of SAMPylation. There is a need for establishing the connection between SAMPylation and proteasomes, as the accumulation of SAMPylated proteins in strains with proteasome subunit deletion does not say whether the accumulation is a direct effect of failure of degradation. Similarly, poly-SAMPylation that has any roles intracellularly other than protein degradation needs to be unraveled.

References

1. de Duve C, Pressman BC, Gianetto R, Wattiaux R, Appelmans F (1955) Tissue fractionation studies. 6. Intracellular distribution patterns of enzymes in rat-liver tissue. Biochem J 60:604–617
2. de Duve C (2005) The lysosome turns fifty. Nat Cell Biol 7:847–849
3. Schultz ML, Tecedor L, Chang M, Davidson BL (2011) Clarifying lysosomal storage diseases. Trends Neurosci 34:401–410
4. Lieberman AP, Puertollano R, Raben N, Slaugenhaupt S, Walkley SU, Ballabio A (2012) Autophagy in lysosomal storage disorders. Autophagy 8:719–730
5. Parenti G, Pignata C, Vajro P, Salerno M (2013) New strategies for the treatment of lysosomal storage diseases. Int J Mol Med 31:11–20
6. Xie Y (2010) Structure, assembly and homeostatic regulation of the 26S proteasome. J Mol Cell Biol 2:308–317

7. Varshavsky A (2012) The ubiquitin system, an immense realm. Annu Rev Biochem 81:167–176
8. Hershko A, Ciechanover A (1998) The ubiquitin system. Annu Rev Biochem 67:425–479
9. Glickman MH (2000) Getting in and out of the proteasome. Semin Cell Dev Biol 11:149–158
10. Ciechanover A, Hod Y, Hershko A (2012) A heat-stable polypeptide component of an ATP-dependent proteolytic system from reticulocytes. 1978. Biochem Biophys Res Commun 425:565–570
11. Goldstein G, Scheid M, Hammerling U, Schlesinger DH, Niall HD, Boyse EA (1975) Isolation of a polypeptide that has lymphocyte-differentiating properties and is probably represented universally in living cells. Proc Nat Acad Sci U S A 72:11–15
12. Wilkinson KD (2005) The discovery of ubiquitin-dependent proteolysis. Proc Nat Acad Sci U S A 102:15280–15282
13. Ciechanover A (2005) Intracellular protein degradation: from a vague idea, through the lysosome and the ubiquitin-proteasome system, and onto human diseases and drug targeting (Nobel lecture). Angew Chem Int Ed Engl 44:5944–5967
14. De Mot R, Nagy I, Walz J, Baumeister W (1999) Proteasomes and other self-compartmentalizing proteases in prokaryotes. Trends Microbiol 7:88–92
15. Kwon YD, Nagy I, Adams PD, Baumeister W, Jap BK (2004) Crystal structures of the *Rhodococcus* proteasome with and without its pro-peptides: implications for the role of the pro-peptide in proteasome assembly. J Mol Biol 335:233–245
16. Lin G, Hu G, Tsu C, Kunes YZ, Li H, Dick L, Parsons T, Li P, Chen Z, Zwickl P, Weich N, Nathan C (2006) Mycobacterium tuberculosis prcBA genes encode a gated proteasome with broad oligopeptide specificity. Mol Microbiol 59:1405–1416
17. Pouch M-N, Cournoyer B, Baumeister W (2000) Characterization of the 20S proteasome from the actinomycete *Frankia*. Mol Microbiol 35:368–377
18. Tamura T, Nagy I, Lupas A, Lottspeich F, Cejka Z, Schoofs G, Tanaka K, De Mot R, Baumeister W (1995) The first characterization of a eubacterial proteasome: the 20S complex of Rhodococcus. Curr Biol 5:766–774
19. Nagy I, Tamura T, Vanderleyden J, Baumeister W, De Mot R (1998) The 20S proteasome of *Streptomyces coelicolor*. J Bacteriol 180:5448–5453
20. Akopian TN, Kisselev AF, Goldberg AL (1997) Processive degradation of proteins and other catalytic properties of the proteasome from *Thermoplasma acidophilum*. J Biol Chem 272:1791–1798
21. Kaczowka SJ, Maupin-Furlow JA (2003) Subunit topology of two 20S proteasomes from *Haloferax volcanii*. J Bacteriol 185:165–174
22. Humbard MA, Zhou G, Maupin-Furlow JA (2009) The N-terminal penultimate residue of 20S proteasome alpha1 influences its N(alpha) acetylation and protein levels as well as growth rate and stress responses of *Haloferax volcanii*. J Bacteriol 191:3794–3803
23. Benaroudj N, Goldberg AL (2000) PAN, the proteasome-activating nucleotidase from archaebacteria, is a protein-unfolding molecular chaperone. Nat Cell Biol 2:833–839
24. Reuter CJ, Kaczowka SJ, Maupin-Furlow JA (2004) Differential regulation of the PanA and PanB proteasome-activating nucleotidase and 20S proteasomal proteins of the haloarchaeon *Haloferax volcanii*. J Bacteriol 186:7763–7772
25. Benaroudj N, Zwickl P, Seemuller E, Baumeister W, Goldberg AL (2003) ATP hydrolysis by the proteasome regulatory complex PAN serves multiple functions in protein degradation. Mol Cell 11:69–78
26. Smith DM, Kafri G, Cheng Y, Ng D, Walz T, Goldberg AL (2005) ATP binding to PAN or the 26S ATPases causes association with the 20S proteasome, gate opening, and translocation of unfolded proteins. Mol Cell 20:687–698
27. Darwin KH, Lin G, Chen Z, Li H, Nathan C (2005) Characterization of a Mycobacterium tuberculosis proteasomal ATPase homologue. Mol Microbiol 55:561–571
28. Wolf S, Nagy I, Lupas A, Pfeifer G, Cejka Z, Muller SA, Engel A, De Mot R, Baumeister W (1998) Characterization of ARC, a divergent member of the AAA ATPase family from *Rhodococcus erythropolis*. J Mol Biol 277:13–25

29. Djuranovic S, Hartmann MD, Habeck M, Ursinus A, Zwickl P, Martin J, Lupas AN, Zeth K (2009) Structure and activity of the N-terminal substrate recognition domains in proteasomal ATPases. Mol Cell 34:580–590
30. Rabl J, Smith DM, Yu Y, Chang SC, Goldberg AL, Cheng Y (2008) Mechanism of gate opening in the 20S proteasome by the proteasomal ATPases. Mol Cell 30:360–368
31. Smith DM, Chang SC, Park S, Finley D, Cheng Y, Goldberg AL (2007) Docking of the proteasomal ATPases' carboxyl termini in the 20S proteasome's alpha ring opens the gate for substrate entry. Mol Cell 27:731–744
32. Wang T, Li H, Lin G, Tang C, Li D, Nathan C, Darwin KH (2009) Structural insights on the *Mycobacterium tuberculosis* proteasomal ATPase Mpa. Structure 17:1377–1385
33. Striebel F, Hunkeler M, Summer H, Weber-Ban E (2010) The mycobacterial Mpa–proteasome unfolds and degrades pupylated substrates by engaging Pup's N-terminus. EMBO J 29:1262–1271
34. Burns KE, Liu WT, Boshoff HI, Dorrestein PC, Barry CE III (2009) Proteasomal protein degradation in mycobacteria is dependent upon a prokaryotic ubiquitin-like protein. J Biol Chem 284:3069–3075
35. Pearce MJ, Mintseris J, Ferreyra J, Gygi SP, Darwin KH (2008) Ubiquitin-like protein involved in the proteasome pathway of *Mycobacterium tuberculosis*. Science 322:1104–1107
36. Striebel F, Imkamp F, Özcelik D, Weber-Ban E (2014) Pupylation as a signal for proteasomal degradation in bacteria. Biochim Biophys Acta 1843:103–113
37. Sutter M, Striebel F, Damberger FF, Allain FH, Weber-Ban E (2009) A distinct structural region of the prokaryotic ubiquitin-like protein (pup) is recognized by the N-terminal domain of the proteasomal ATPase Mpa. FEBS Lett 583:3151–3157
38. Chen X, Solomon WC, Kang Y, Cerda-Maira F, Darwin KH, Walters KJ (2009) Prokaryotic ubiquitin-like protein pup is intrinsically disordered. J Mol Biol 392:208–217
39. Liao S, Shang Q, Zhang X, Zhang J, Xu C, Tu X (2009) Pup, a prokaryotic ubiquitin like protein, is an intrinsically disordered protein. Biochem J 422:207–215
40. Striebel F, Imkamp F, Sutter M, Steiner M, Mamedov A, Weber-Ban E (2009) Bacterial ubiquitin-like modifier pup is deamidated and conjugated to substrates by distinct but homologous enzymes. Nat Struct Mol Biol 16:647–651
41. Imkamp F, Rosenberger T, Striebel F, Keller PM, Amstutz B, Sander P, Weber-Ban E (2010) Deletion of *dop* in *Mycobacterium smegmatis* abolishes pupylation of protein substrates in vivo. Mol Microbiol 75:744–754
42. Cerda-Maira FA, Pearce MJ, Fuortes M, Bishai WR, Hubbard SR, Darwin KH (2010) Molecular analysis of the prokaryotic ubiquitin-like protein (pup) conjugation pathway in Mycobacterium tuberculosis. Mol Microbiol 77:1123–1135
43. Pearce MJ, Arora P, Festa RA, Butler-Wu SM, Gokhale RS, Darwin KH (2006) Identification of substrates of the *Mycobacterium tuberculosis* proteasome. EMBO J 25: 5423–5432
44. Guth E, Thommen M, Weber-Ban E (2011) Mycobacterial ubiquitin-like protein ligase PafA follows a two-step reaction pathway with a phosphorylated pup intermediate. J Biol Chem 286:4412–4419
45. Burns KE, Cerda-Maira FA, Wang T, Li H, Bishai WR, Darwin KH (2010) "Depupylation" of prokaryotic ubiquitin-like protein from mycobacterial proteasome substrates. Mol Cell 39:821–827
46. Imkamp F, Striebel F, Sutter M, Özcelik D, Zimmermann N, Sander P, Weber-Ban E (2010) Dop functions as a depupylase in the prokaryotic ubiquitin-like modification pathway. EMBO Rep 11:791–797
47. Wang T, Darwin KH, Li H (2010) Binding-induced folding of prokaryotic ubiquitin-like protein on the *Mycobacterium* proteasomal ATPase targets substrates for degradation. Nat Struct Mol Biol 17:1352–1357
48. Burns KE, Pearce MJ, Darwin KH (2010) Prokaryotic ubiquitin-like protein provides a two-part degron to Mycobacterium proteasome substrates. J Bacteriol 192:2933–2935
49. Festa RA, McAllister F, Pearce MJ, Mintseris J, Burns KE, Gygi SP, Darwin KH (2010) Prokaryotic ubiquitin-like protein (pup) proteome of Mycobacterium tuberculosis [corrected]. PLoS One 5:e8589

50. Watrous J, Burns K, Liu WT, Patel A, Hook V, Bafna V, Barry CE III, Bark S, Dorrestein PC (2010) Expansion of the mycobacterial "PUPylome". Mol BioSyst 6:376–385
51. Poulsen C, Akhter Y, Jeon AH, Schmitt-Ulms G, Meyer HE, Stefanski A, Stuhler K, Wilmanns M, Song YH (2010) Proteome-wide identification of mycobacterial pupylation targets. Mol Syst Biol 6:386
52. Delley CL, Striebel F, Heydenreich FM, Ozcelik D, Weber-Ban E (2012) Activity of the mycobacterial proteasomal ATPase Mpa is reversibly regulated by pupylation. J Biol Chem 287:7907–7914
53. Hong B, Wang L, Lammertyn E, Geukens N, Van Mellaert L, Li Y, Anne J (2005) Inactivation of the 20S proteasome in Streptomyces lividans and its influence on the production of heterologous proteins. Microbiology 151:3137–3145
54. Knipfer N, Shrader TE (1997) Inactivation of the 20S proteasome in *Mycobacterium smegmatis*. Mol Microbiol 25:375–383
55. Gandotra S, Lebron MB, Ehrt S (2010) The Mycobacterium tuberculosis proteasome active site threonine is essential for persistence yet dispensable for replication and resistance to nitric oxide. PLoS Pathog 6
56. Gandotra S, Schnappinger D, Monteleone M, Hillen W, Ehrt S (2007) In vivo gene silencing identifies *the Mycobacterium tuberculosis* proteasome as essential for the bacteria to persist in mice. Nat Med 13:1515–1520
57. Darwin KH, Ehrt S, Gutierrez-Ramos JC, Weich N, Nathan CF (2003) The proteasome of *Mycobacterium tuberculosis* is required for resistance to nitric oxide. Science 302:1963–1966
58. Blumenthal A, Trujillo C, Ehrt S, Schnappinger D (2010) Simultaneous analysis of multiple *Mycobacterium tuberculosis* knockdown mutants in vitro and in vivo. PLoS One 5:e15667
59. Sassetti CM, Rubin EJ (2003) Genetic requirements for mycobacterial survival during infection. Proc Natl Acad Sci U S A 100:12989–12994
60. Cerda-Maira F, Darwin KH (2009) The *Mycobacterium tuberculosis* proteasome: more than just a barrel-shaped protease. Microbes Infect 11:1150–1155
61. MacMicking JD, North RJ, LaCourse R, Mudgett JS, Shah SK, Nathan CF (1997) Identification of nitric oxide synthase as a protective locus against tuberculosis. Proc Natl Acad Sci U S A 94:5243–5248
62. McKinney JD, Honer zu Bentrup K, Munoz-Elias EJ, Miczak A, Chen B, Chan WT, Swenson D, Sacchettini JC, Jacobs Jr. WR, Russell DG (2000) Persistence of Mycobacterium tuberculosis in macrophages and mice requires the glyoxylate shunt enzyme isocitrate lyase. Nature 406: 735–738
63. Movahedzadeh F, Smith DA, Norman RA, Dinadayala P, Murray-Rust J, Russell DG, Kendall SL, Rison SC, McAlister MS, Bancroft GJ, McDonald NQ, Daffe M, Av-Gay Y, Stoker NG (2004) The *Mycobacterium tuberculosis* ino1 gene is essential for growth and virulence. Mol Microbiol 51:1003–1014
64. Lin G, Tsu C, Dick L, Zhou XK, Nathan C (2008) Distinct specificities of *Mycobacterium tuberculosis* and mammalian proteasomes for N-acetyl tripeptide substrates. J Biol Chem 283: 34423–34431
65. Kropff MH, Bisping G, Wenning D, Volpert S, Tchinda J, Berdel WE, Kienast J (2005) Bortezomib in combination with dexamethasone for relapsed multiple myeloma. Leuk Res 29:587–590
66. Lin G, Li D, de Carvalho LP, Deng H, Tao H, Vogt G, Wu K, Schneider J, Chidawanyika T, Warren JD, Li H, Nathan C (2009) Inhibitors selective for mycobacterial versus human proteasomes. Nature 461:621–626
67. Valas RE, Bourne PE (2011) The origin of a derived superkingdom: how a gram-positive bacterium crossed the desert to become an archaeon. Biol Direct 6:16
68. Volker C, Lupas AN (2002) Molecular evolution of proteasomes. Curr Top Microbiol Immunol 268:1–22
69. Gille C, Goede A, Schloetelburg C, Preissner R, Kloetzel PM, Gobel UB, Frommel C (2003) A comprehensive view on proteasomal sequences: implications for the evolution of the proteasome. J Mol Biol 326:1437–1448

70. Eisenberg D, Gill HS, Pfluegl GM, Rotstein SH (2000) Structure–function relationships of glutamine synthetases. Biochim Biophys Acta 1477:122–145
71. Gill HS, Eisenberg D (2001) The crystal structure of phosphinothricin in the active site of glutamine synthetase illuminates the mechanism of enzymatic inhibition. Biochemistry 40:1903–1912
72. Iyer LM, Burroughs AM, Aravind L (2006) The prokaryotic antecedents of the ubiquitin signaling system and the early evolution of ubiquitin-like beta-grasp domains. Genome Biol 7:R60
73. Kessler D (2006) Enzymatic activation of sulfur for incorporation into biomolecules in prokaryotes. FEMS Microbiol Rev 30:825–840
74. Leidel S, Pedrioli PG, Bucher T, Brost R, Costanzo M, Schmidt A, Aebersold R, Boone C, Hofmann K, Peter M (2009) Ubiquitin-related modifier Urm1 acts as a sulphur carrier in thiolation of eukaryotic transfer RNA. Nature 458:228–232
75. Schlieker CD, Van der Veen AG, Damon JR, Spooner E, Ploegh HL (2008) A functional proteomics approach links the ubiquitin-related modifier Urm1 to a tRNA modification pathway. Proc Natl Acad Sci U S A 105:18255–18260
76. Goehring, A.S. Rivers DM, Sprague GF Jr (2003) Urmylation: a ubiquitin-like pathway that functions during invasive growth and budding in yeast. Mol Biol Cell 14: 4329–4341
77. Humbard MA, Miranda HV, Lim JM, Krause DJ, Pritz JR, Zhou G, Chen S, Wells L, Maupin-Furlow JA (2010) Ubiquitin-like small archaeal modifier proteins (SAMPs) in *Haloferax volcanii*. Nature 463:54–60
78. Darwin KH, Hofmann K (2010) SAMPyling proteins in archaea. Trends Biochem Sci 35:348–351
79. Noma A, Sakaguchi Y, Suzuki T (2009) Mechanistic characterization of the sulfur-relay system for eukaryotic 2-thiouridine biogenesis at tRNA wobble positions. Nucleic Acids Res 37:1335–1352
80. Goehring AS, Rivers DM, Sprague GF Jr. (2003) Attachment of the ubiquitin-related protein Urm1p to the antioxidant protein Ahp1p. Eukaryot Cell 2: 930–936

Role of Proteases in Photo-aging of the Skin

20

Rita Ghosh

Abstract

Aging is an inevitable process in living organisms that results from molecular damage over time. The skin being the most exposed part of the body, time and environmental aggressor leave their indelible mark on the skin. So, skin aging consists of two clinically and biologically independent processes – the intrinsic chronological aging and the aging through extrinsic factors. While intrinsic aging process proceeds at a genetically determined pace due to buildup of damaging products from cellular metabolism, exposure to solar radiations produces biological damages to the cells, known as photo-aging. It adds up to the effects of chronological aging, and it is the most prominent and important among the extrinsic factors. The normal architecture of the skin is disrupted due to degradation of skin components like collagens, fibers, etc. Photo-aged skin presents fine and coarse wrinkles with blotchy pigmentation, increased fragility, and rough texture. It results from complex biological phenomena that lead to activation of several proteases; the most crucial among them are the matrix metalloproteases (MMPs). UV irradiation generates reactive oxygen species and activates a number of transcription factors like AP1, NF-κB, p53, and growth factor like TGFβ. These, in turn, stimulate the MMPs and other proteases. UV radiation also inhibits the expression of natural inhibitors of MMP (TIMP), thereby enhancing the activity of the MMPs. Understanding of the molecular basis of photo-aging is important for its prevention and effective recovery. Antioxidants and other com-

R. Ghosh (✉)
Department of Biochemistry & Biophysics, University of Kalyani,
Kalyani 741235, West Bengal, India
e-mail: ritadg2001@yahoo.co.in

S. Chakraborti, N.S. Dhalla (eds.), *Proteases in Physiology and Pathology*,
DOI 10.1007/978-981-10-2513-6_20

pounds that inhibit the molecular pathways that result in expression of the proteases have proved to be useful in prevention/reversal of skin aging.

Keywords
Photo-aging • Ultraviolet radiation • Skin • Proteases • Prevention

20.1 Introduction

Aging is an inevitable process common to all living organisms that result from molecular damage over time. The skin is the largest of all organs of the human body that constitutes about 16% of the total body weight. It maintains body temperature and hydration; it also has roles in sensory perception and immune surveillance. It acts as a barrier between the inner body parts and the environment [1]. Being the most exposed part of the body, different environmental aggressors leave their mark on the skin. The skin is thus a good model for studying the consequences of aging. Skin aging consists of two clinically and biologically independent processes – the intrinsic chronological aging and the other is due to aging through extrinsic factors. The intrinsic aging process proceeds at a genetically determined pace due to buildup of damaging products from cellular metabolism and is similar in all parts of the body [2]. Among the extrinsic causative agents, the most significant is the exposure to solar radiations that produce biological damages to the cells. UV exposure has been associated with photo-carcinogenesis, photo-aging, and photo-immunosuppression [3, 4]. Photo-aging adds up to the effects of chronological aging.

20.2 Action Spectra for Photo-aging

The ultraviolet (UV) radiations from the solar spectra are primarily responsible for the photo-aging of the skin. The UV spectrum has been divided into three ranges – UVA (400–320 nm), UVB (320–280 nm), and UVC (280–200 nm). The UVC is readily absorbed by the most important biomolecules like cellular and mitochondrial DNA and proteins, but the atmosphere blocks out this range of radiation, and therefore, it does not contribute toward photoaging. UVB is also damaging to different cellular components, but the ozone layer is capable of blocking this range of radiation from reaching the earth's surface. However, there is a serious concern about exposure to solar UVB due to the thinning of the ozone layer. UVB rays can only penetrate up to the basal layer of the epidermis of the skin. Ninety six percent of the solar radiation that reaches the earth surface is UVA. It penetrates the skin more deeply than UVB through the epidermis and dermis and contributes most significantly toward photo-aging [5].

20.3 Structure of the Skin

The process of photo-aging of the skin is multifactorial and complex; it affects various layers of the skin with the major damage being observed in the connective tissues of the dermis. The skin is composed of three layers – the outer epidermis, the thin basement membrane (BM), and the thicker inner layer, called the dermis. The epidermis is about a tenth the thickness of the dermis. The epidermis is made up of three layers. The stratum corneum is the outermost layer that is made up of a few layers of nonnucleated dead cells. It is followed by the granular layer that consists of granulated interconnected cells. They have a granular appearance due to deposition of keratin. The next Malpighian layer contains the squamous epithelial cells and the dendritic melanocytes. The melanocytes produce the melanin that is transported outward through these dendrites [6].

The basement membrane is important for maintaining a healthy skin. The BM is present at the junction of the dermis and epidermis to provide mechanical support to the outer protective epidermis. It also prevents contact between the two layers and determines the polarity of the epidermis. The proliferating cells of the epidermis remain attached to it, while the daughter cells migrate outward. It also consists of three different layers – the lamina lucida, the lamina densa, and the lamina fibroreticularis [7].

The dermis consists of the connective tissues. Both the dermis and the BM contain various types of collagens, elastic fibers, and glycosaminoglycans in the extracellular matrix (ECM). The ECM is most important for the skin structure and elasticity. The ECM contains different collagens of which about 85% is type I procollagen [8]. The elastic fiber network constitutes about 2–4% of the ECM and is a key component of the dermal connective tissue that provides elasticity to the skin. Morphologically, it consists of two distinct components – the elastin and the microfibrils [9]. The elastic fiber network accounts for 2–4% of the ECM in sun-protected skin and provides elasticity to the skin [10]. The glycosaminoglycan and the proteoglycans comprise only 0.1–0.35 of the dry weight of the skin. Their role is in hydrating the skin and biological signaling [11].

20.4 Characteristics of Photo-aging

Both chronological and photoaged skin show alteration in skin microrelief. Deep furrows appear in the skin leading to wrinkles. Intrinsically aged skin is thin, and there is laxity of the skin due to reduced elasticity [12]. There are no pigmentary changes or deep wrinkles. In contrast, the photoaged skin appears with fine and deep coarse wrinkles, shallowness, uneven blotchy pigmentation, and rough and leathery skin texture with increased fragility. There is loss of mature collagen with a distinctive basophilic appearance called basophilic degeneration [13]. No vascular damage is observed in intrinsically aged skin. Both acute and chronic exposures to UV radiation increase skin vascularization and angiogenesis [14]. Emerging information has revealed that though phenotypically it differs extensively from the intrinsic chronological aging [15], they share some fundamental molecular pathways [16].

Collagen is a major structural protein found in the dermal ECM that provides stability and tensile strength to the dermis [17]. Procollagens are synthesized in the dermal fibroblasts; these are secreted into the extracellular space where these are enzymatically processed to mature collagen. The mature collagen spontaneously forms fibrils that are responsible for the strength and resilience of the skin. These fibrils are stabilized by cross-linkages. These collagen fibrils have a half-life of 17 years. The accumulation of fragmented collagen has a lasting consequence on the structure of the skin that contributes to the chronological aging. UV radiation also alters the dermal collagen. It promotes the breakdown of collagen and also inhibits the biosynthesis of procollagen. It has been observed that exposure to UV radiation can result in complete loss of procollagen synthesis for 24 hours [18]. Both collagens I and III decreased on UV exposure, but between types I and III collagen present in the ECM, the relative amount of collagen I was more [19]. Collagen VII contains the anchoring fibrils, and they contribute toward stabilization of the dermis-epidermis junction (DEJ). Collagen VII is also severely reduced in photoaged skin. Loss of collagen VII and elastic fibers was found in fibroblasts. Wrinkling is due to the loss of collagen fibers and deposition of the abnormal degenerative elastic material [13]. *In vitro* UVA induces elafin in fibroblasts. It forms the elafin-elastin complex that inhibits the binding of elastase to elastin. Formation of this complex prevents elastic fibers from elastolytic degradation resulting in accumulation of elastic fibers. Elafin is thus believed to be integral to actinic elastosis [5].

Elastic fibers constitute the structural elements of the connective tissues that have a central core of amorphous, cross-linked hydrophobic elastin surrounded by fibrillin-rich microfibril. The elastic fiber network extends from the DEJ into the dermis. Disruption and reduplication of the BM at the DEJ are observed in sun-exposed skin [7]. Loss of elasticity is an important clinical feature of photo-aging. A leathery weather-beaten appearance known as solar elastosis is the result of elastic fiber degradation and is a hallmark of photo-aging.

Tenascin is a large glycoprotein that is found just below the DEJ. In sun-exposed skin, this protein is increased both in the dermis and epidermis; while its increase is small in the dermis, there is an appreciable increase of this protein in the epidermis of the skin [20]. The increase in deposition of glycosaminoglycans and dystrophic elastosis was revealed through immune-positive staining of the severely disorganized tropoelastin and its associated microfibrillar component, fibrillin. Fibrillin is truncated and depleted in the upper dermis and DEJ of photoaged skin.

Plasminogen activators (PAs) are produced by different cell types including the epidermal keratinocytes. They degrade the zymogen, plasminogen, to plasmin directly and also by activating other proteases. Activity of urokinase-type plasminogen activator (uPA) was found in the stratum corneum and in the basal layer after disruption of the barrier, which is responsible for the delay in recovery from photoaging [7].

Skin color or skin phototype (SPT) is a good indicator of a person's susceptibility to damaging effects of UV. White skin, having low SPT (SPT I–III), is more susceptible to damage than darker skin that have higher SPT. Photo-aging requires several years of exposure. However, initial pigmentary changes of photo-aging (large-sized lentigines) can be seen within weeks after exposure to blistering sunburn. Chronic high degree of exposure to sunlight results in altered melanin distribution. There is a great variability in the degree of melanosome distribution within keratinocytes. Some keratinocytes are sparsely melanized, while there is abundance of melanosomes organized in the perinuclear space.

Apart from alterations in the structural organization of the connective tissue, the fibroblast cells in the dermis also reveal characteristic changes in photoaged skin. The fibroblasts adopt a stellate phenotype. At the ultrastructural level, the presence of highly activated rough endoplasmic reticulum indicated their enhanced biosynthetic activity [4].

20.5 Mechanism of Photo-aging

Multiple mechanisms are believed to be involved in the process of photo-aging although only a few aspects have been studied. The most intensively investigated is the involvement of proteases. In photo-aging, the degradation of the ECM proteins in the skin is through the matrix metalloproteases (MMPs), and it contributes greatly to the dermal remodeling [21]. Exposure to UV light also decreases the expression of tissue inhibitors of metalloproteases (TIMPs) [22]. TIMP-1 has a protective role; it is involved in recovery from cutaneous photodamage through suppression of ECM degradation and inflammation [23]. Normally, MMPs that are released in the extracellular space are prevented from exerting their effect by the endogenous TIMP. Four different forms of TIMPs are known: TIMP-1, TIMP-2, TIMP-3, and TIMP-4. They inhibit the MMPs with low selectivity by forming tight 1:1 complexes. They also participate in morphology modulation, suppression of cell growth promotion, tumor growth, induction of apoptosis, and inhibitions of angiogenesis, invasion, and metastasis [24]. Overexpression of TIMP-1 resulted in significant inhibition of UVB-induced ECM degradation; it also led to suppression of roughness of the skin and the decrease of skin elasticity. Inhibition of TIMP-1 had the opposite effects [23]. There are also a few reports related to other proteases, such as cathepsins. Cathepsins are a class of secretory proteolytic proteases that exert diverse effects. Cathepsin K is needed for processing of cutaneous elastin; its activity is lost in UVA-treated cells [25]. There is dual inactivation of cathepsins B and L by UVA; recent studies have indicated the impairment of cathepsin B in the UV-induced photo-aging process [26]. The production of cathepsin G and other elastases – like enzymes by dermal fibroblasts – is stimulated by UVA, which possibly contributes to elastosis areas in sun-damaged skin [27]. Using inhibitors of cathepsin G, it has been shown that UVB-induced photo-aging could be prevented through inhibition of fibronectin fragmentation [28].

UVA-induced photo-oxidative stress modulates the different structural proteins – actin, collagen, elastin, and keratin – through different enzymes like p38, mitogen-activated protein kinase (MAPK), MMPs, and other proteases as well as different transcription factors like activator protein (AP)-1/AP-2 and nuclear factor-κB (NF-κB). Apart from skin aging, MMPs are known for their roles of tissue destruction through proteolytic events for a wide range of physiological and pathophysiological conditions that includes embryogenesis, wound healing, inflammation, arthritis, tumor invasion, angiogenesis, cancer, and metastasis. Various extracellular stimuli like UV or infrared radiation, growth factors, cytokines, and tumor promoters can activate the MMPs that are involved in the degradation of the epidermal basement membrane [29, 30]. MMP-1 breaks the triple helix of collagen and the collagen cross-link in the ECM. As 70–80% of the skin is composed of collagen, the damage caused by MMP-1 leads to deeper wrinkles in the skin.

Some studies indicate that epidermal keratinocytes are the major source of MMPs. Dermal cells also contribute MMPs by indirect paracrine mechanism through release of growth factors and cytokines [21].The mast cells are widely distributed in the connective tissues of the skin, and their occurrence increases in UV-irradiated skin. The tryptase derived from mast cells degrades the ECM by digesting collagen I. These tryptases also activate MMP-9. This was established by demonstrating its gel-lytic activity in collagen IV [30]. The collagen IV also degrades the BM. In UVB-damaged skin, the MMP inhibitor CGS27023A enhanced the assembly of BM at the DEJ, indicating the role of MMP in disruption of lamina densa in sun-damaged skin [M]. The damage to the BM is also related to the plasmin and the MMPs [7]. The conditioned medium from UV-irradiated cells in culture contains uPA secreted from the irradiated cells [31]. The plasminogen increased degradation of the BM and impaired its assembly at the DEJ. Using inhibitors of both MMP and plasmin, it was shown that both are necessary to prevent the damage to BM of irradiated cells [7]. Lysozymes and α-antitrypsin are also associated with damaged elastic fiber and correlated with photodamage [20].

In non-dermatological conditions, neutrophils are responsible for ECM degradation. In mice model also, it has been shown that the neutrophil elastase is responsible for solar elastosis [32]. Exposing human skin to a combination of UVA and UVB resulted in a rapid infiltration of neutrophils. They are the major source of MMP-1 and MMP-9. The keratinocytes and fibroblasts are also capable of producing these enzymes in small amounts [33]. MMP-8 was found to increase in human skin [34]. MMP-12 too was found in human skin 24 hours after UV exposure. Both MMP-8 and MMP-12 are found in human skin as a result of influx of neutrophils and macrophages [35]. Some investigators have shown that skin neutrophils and macrophages are terminally differentiated cells; therefore, they no longer transcribe new mRNA, and the residual mRNAs of these MMPs are below the level of detection. They have shown that the MMP-8 proteins found in the skin remain in an inactive precursor form [34]; it did not have collagenolytic activity in the skin [36].

Other investigators have shown that MMPs are induced even at low doses of UV radiation, but they have argued that only high doses of exposure are necessary for photo-aging; therefore, they have proposed that at sub-erythemogenic doses of

UV, MMPs may be involved in processes other than ECM degradation; higher doses of irradiation can only lead to immediate changes associated with ECM degradation by neutrophil [29, 32]. The cumulative damage to ECM by the MMPs through chronic exposure was proposed by Fisher et al. [37, 38]. They have proposed that skin damage is only repaired partially and the cumulative effects of such damages are involved in photo-aging by MMP-1, MMP-3, and MMP-9. Neutrophil-derived elastase is responsible for photoaging [39]. At erythemogenic doses of UV, the elastase induced is linked to neutrophils, while at sub-erythemogenic doses, MMPs may be derived from the keratinocytes and fibroblasts. Therefore, in the pathogenesis of photo-aging, the contribution from direct damage induced to ECM by the generated reactive oxygen species (ROS), neutrophil-derived proteases, as well as the MMPs secreted by keratinocytes and fibroblasts may be involved.

20.6 Reactive Oxygen Species

Various physiological processes generate a low level of ROS that contribute toward chronological aging. Both UVB and UVA are known to generate ROS [40]. UVB is directly absorbed by DNA, while other chromophores present in the skin, like trans-urocanic acid present in the epidermis, absorb UVA [41]. These excited photosensitizers subsequently react with oxygen and produce ROS like superoxide anions ($\dot{O}_2^-$) and singlet oxygen ($^1O_2$). UV light also increases the NADPH oxidase and xanthine oxidase activity, which is also involved in the generation of ROS [42]. Neutrophils also produce $\dot{O}_2^-$ and $^1O_2$. The $\dot{O}_2^-$ is converted to hydrogen peroxide (H_2O_2) by superoxide dismutase. The H_2O_2 produced crosses the cell membrane and generates the highly toxic hydroxyl radical ($O\dot{H}$) through Fenton reaction with Fe (II). $^1O_2$ and $O\dot{H}$ initiate lipid peroxidation of cellular membranes. $O\dot{H}$ and intermediates of lipid peroxidation are also involved in induction of MMP-1 and MMP-3 by UVB [22]. UVA-generated $^1O_2$ initiates the c-Jun N-terminal kinase (JNK) and p38 MAPK leading to expression of MMPs [13]. UVB-induced $O\dot{H}$ and lipid peroxidation products also stimulated ERK, p38 MAPK, and subsequently induced JNK2. UV light upregulates c-Jun and c-Fos that increases the activation of transcription factor AP-1, required for the expression of the MMPs [43].

ROS plays an important role in collagen metabolism both directly and indirectly. ROS not only destroys the interstitial collagen but also inactivates tissue inhibitors of matrix metalloproteases (TIMPs) and induces the synthesis and activation of the matrix metalloproteases (MMPs) [44]. It has been demonstrated through different approaches that $^1O_2$ and H_2O_2 are involved in the induction of MMP-1, MMP-2, and MMP-3 by UVA.

ROS have been implicated as a causative agent for intrinsic aging. Mutations in mitochondrial DNA impair the function of mitochondria leading to defects in electron transport and oxidative phosphorylation, which in turn enhances the generation of ROS. It has also been demonstrated that $^1O_2$ generated by UVA causes the same mutations in the mitochondrial DNA that are involved in aging [45]. ROS also results in shortening of the telomere [46]. Shortening of telomere length provides signal for replicative senescence. Thus, there exists some overlap in the causes of photo-aging with that of intrinsic aging through ROS.

20.7 Proteases in Photo-aging

MMPs are a family of zinc proteinases that consist of 28 members of which 24 are expressed in mammals. They are classified in relation to their substrate specificities and whether they are bound to the cell membranes or secreted as soluble proteins. The different classes of MMPs are collagenases (MMP-1, MMP-8, and MMP-13), gelatinases (MMP-2 and MMP-9), stromelysins (MMP-3, MMP-10, and MMP-11), matrilysin (MMP-7 and MMP-26), membrane-type MMPs (MMP-15, MMP-16, MMP-17, MMP-24, and MMP-25), and other types (MMP-12,MMP-19, MMP-20, MMP-21, MMP-22, MMP-23, and MMP28) [47]. In normal healthy, sun-protected, adult human skin, the basal mRNA expression levels of the MMP family members were found to be extremely low, MMP-14 being the only exception. While the transcripts for MMP-8, MMP-10, MMP-12, MMP-20, and MMP-26 were not detected at all, the transcripts for the remaining MMPs were almost near the level of detection, being almost thousand-fold lower than the housekeeping gene, 36B4, that was used as internal control. Basal expression of MMP-14 mRNA was approximately 35-fold higher than that of other detectable MMPs. It has been demonstrated that there is induction of MMP-1, MMP-2, MMP-3, MMP-7, MMP-9, MMP-10, and MMP-12 in response to UVA and UVB irradiation. In an *in vivo* study where expression of 19 MMPs was observed in normal human skin, only three MMPs – MMP-1, MMP-3, and MMP-9 – were significantly induced in response to UV irradiation. There was several thousand-fold increase in the induction of MMP-1 and MMP-3 mRNA, whereas MMP-9 was only modestly induced by six-fold 24 hours after irradiation. MMP-1 initiates the cleavage at a single site within its central triple helix of type I and III fibrillar collagen in the skin [21]. Once cleaved by MMP-1, collagen can be further degraded by elevated levels of MMP-3 and MMP-9. In contrast, following irradiation, MMP-14 was reduced nearly 80% by 8 hours and remained so till 24 hours [23, 47]. The physiological function of MMP-14 in human skin, however, remains to be determined. MMP-8 (neutrophil collagenase) and MMP-12 (macrophage elastase) proteins were also present in human skin 24 hours after UV irradiation.

UV irradiation of organ culture resulted in collagen fragmentation, together with alteration in the structure and the organization of collagen fibrils in the dermis through MMP-1 that resembled changes observed in photo-aged skin. Different studies suggested that MMP-1, MMP-3, and MMP-9 are the primary UV-inducible collagenolytic enzymes, MMP-1 being the major protease that is capable of initiating the degradation of native fibrillar collagens in human skin in vivo [21, 36]. MMP-2 degrades the elastin as well as the basement membrane molecules like collagen IV and VII [48]. MMP-3 has broader substrate specificity for collagen IV, proteoglycan, fibronectin, and laminin. Fibrillin is attacked by MMP and MMP-independent serine proteases like neutrophil elastase [13]. While the degradation is regulated by the MMPs, their natural inhibitors, the TIMPs, also have a considerable role in the process.

20.8 Molecular Mechanisms That Stimulate Proteases

Multiple mechanisms are involved in photo-aging that includes different proteases. The MMPs are the most extensively studied, and they are recognized as the major proteases involved in the process of photo-aging. Upregulation of AP-1 by UV leads to increase in expression of the MMPs responsible for collagen degradation of the ECM. Jun and Fos can associate either as homo- or heterodimers to form the transcription factor AP-1 that binds to the AP-1-binding site in DNA to regulate the expression of several proteins that includes the MMPs like MMP-1, MMP-3, and MMP-9 [21].

MMPs are also regulated by different members of the MAPK family – the extracellular regulated protein kinase (ERK), Jun N-terminal kinase (JNK), and p38 MAPK [49]. The roles of the MAPKs are to control the activation of different transcription factors and to transfer extracellular signals to the nucleus to stimulate expression of target genes. The MAPKs on phosphorylation activate the transcription factors AP-1, NF-κB, and p53 [50]. These transcription factors in turn activate the MMPs.

Phosphorylation of MAPK 1/2, JNK 1/2, and p38 is mediated through UVB-induced ROS. UV light activates MAPKs which in turn stimulates AP-1 that is also responsible for expression of MMPs [22]. Apart from AP1, MAPK also regulates the activation of NF-κB and p53. JNK and p53 also stimulate AP-1 [51]. UV irradiation can activate various intracellular signal pathways known as UV response. UV response includes activation of NF-κB that is also implicated in the transcriptional regulation of UV-induced MMPs [52]. NF-κB has pro-apoptotic function and is the key molecule in the p53-mediated apoptosis [50].

UVA induces cytokines involved in inflammation, such as tumor necrosis factor (TNF)-α, interleukin-1, and also ROS. Both UVA and UVB generate H_2O_2 within 15 min that in turn induces AP-1, which remains elevated for 24 hours [53]. It leads to breakdown of collagen and also downregulates procollagen synthesis [54]. In addition, UVA induces the phosphorylation MAPK in dermal fibroblasts. The activation of the p38 and c-Jun N-terminal kinase (JNK) signals stimulates the level of the c-Fos and c-Jun transcription factor AP-1. After upregulation, AP-1 binds to the MMP genes to increase their mRNA level. Although the NF-κB of activated B cell does not bind directly to the promoter site of the MMP-1 and the basic fibroblast growth factor (bFGF), it upregulates both MMP-1 and bFGF [55]. Exposure to UVA causes secretion of TNF-α inducing MMP-1 in keratinocytes. The secreted TNF-α is transported to the fibroblasts and induces MMP-1 in normal human dermal fibroblast.

Transforming growth factor β (TGFβ) is often used to stimulate the synthesis of collagen. ERK pathways mediate TGFβ-induced ECM production and fibrosis. UV irradiation simultaneously induces TGFβ along with ERK signaling [56]. Activated TGFβ increases collagenase production to cause photo-aging. Production of type I collagen is mainly reduced by TGFβ. Induction of AP-1 by UV also signals the reduction of procollagen synthesis through TGFβ [57].

20.9 Recovery and Prevention of Photo-aging

As the pathogenesis of photo-aging involves damage to ECM through ROS, different antioxidants have been used to prove their efficacy in reversing/preventing photoinduced damages. A number of compounds having antioxidant properties have demonstrated anti-photo-aging function. Some of them are N-acetyl cysteine (NAC), pyrrolidine dithiocarbamate (PDTC), epigallocatechin 3-O-gallate, and lipoteichoic acid. NAC and PDTC inhibit NF-κB activation in a variety of cell lines [58]. Vitamin C has been shown to increase collagen production [53]. Topical applications of retinoids not only repair photoaged skin but also prevent skin aging [40, 59]. The retinoic acid precursor, β-carotene, inhibited the expression of MMPS [60]. Astaxanthin prevents photo-aging through inhibition of MMP-1 [20]. Effectiveness of several phyto-products has been evaluated for their antiaging/photo-aging roles [47, 53, 61]. Green tea has been found to have antiaging properties. A green tea polyphenol, EGCG, acts as an antioxidant and has anti-inflammatory and immune-modulatory role. A concentration-dependent EGCG treatment effectively decreased MMP expression in human fibroblast by decreasing ROS level and through decrease of the transcription factor NF-κB [62]. Glycyrrhizic acid from licorice root prevents MMP-1 activation and NF-κB signaling and therefore has a role in prevention of photo-aging [52]. Bog blueberry anthocyanins protected UVB-induced skin photo-aging by blocking collagen destruction and inflammatory responses through NF-κB and MAPK signaling [63]. Pomegranate extract, which is rich in anthocyanidins and ellagitannins, was found to protect cells from oxidative damage; it prevented the upregulation of the MMP-1, MMP-3, MMP-7, and MMP-9 and activated the TIMP-1 [22]. Astragaloside IV from *Astragalus membranaceus*, a Chinese medicinal herb, also has a role in inhibiting photo-aging [64]. Forskolin, a pharmacologically active compound derived from plant root, was found to protect the skin against UV light-induced damages and prevented symptoms of aging by upregulating melanin production [65]. Luteolin-rich extract of Reseda seed has a protective role from UV-induced damages and, therefore, may have a role in prevention of photo-aging by protecting the collagen and other skin components from solar radiation-induced damage [66]. Fucoxanthin, a carotenoid found in different edible algae, exhibited antiaging property through its antioxidant effects [67]. CoQ10 protects skin from photo-aging by inhibiting the production of IL-6 which stimulates fibroblasts in the dermis to upregulate MMP production by paracrine manner [68]. Non-heme iron is found to increase in the skin of hairless mice upon UV irradiation. Topical application of iron chelators could delay photo-aging (13).

Some bacterial and insect products too have a role in preventing photoaging. Lipoteichoic acid from *Lactobacillus sakei* inhibits the pro-inflammatory cytokine, TNFα, and blocks the phosphorylation of MAPK family in THP-1 cells. It also decreased the expression of MMP-1. Lipoteichoic acid has the potential to suppress UVA-induced damage and act as anti-photo-aging agent [63]. A 9-mer peptide CopA3 that was synthesized from a natural peptide, coprisin, isolated from dung beetle also has proven activity in prevention and treatment of skin aging [69].

20.10 Future Perspective and Conclusion

The loss of structural and functional stability of the skin is an inevitable consequence of aging that is caused by both intrinsic and extrinsic factors. Although it is almost impossible to prevent aging completely, premature aging and damage from photo-aging can be delayed. As there is considerable overlap between the features of chronological and photo-induced aging as well as in the general mechanism of both processes, understanding photo-aging is important as it would help to elucidate aging in general. Such knowledge would help in development of strategies for prevention as well as treatment protocols for reversal of the aging symptoms. Considering the increase in the aged population worldwide, it is of prime significance.

Most intervention methods developed so far have been from in vitro studies that have to be proved in vivo. The depletion of stratospheric ozone layer and its implications in photo-aging needs to be highlighted. Emphasis on preventive strategies is imperative; this can help to enhance the rationale for development of photo-protective agents. Oxidative damages through generation of ROS have a prime role in photoinduced aging process. For effective protection, therefore, many antioxidant compounds have been found to be useful. While some have proven benefits when applied topically or taken through diet, in some other cases, their role has been uncertain [70]. Sunscreens are known to provide protection from sun-induced damages, but their mode of action is different from antioxidants. In principle they might act synergistically with antioxidants to prevent photo-aging, but research in this area is needed as this has not been tested. Treatments with antioxidants are often associated with side effects [71, 72]; this aspect therefore needs critical evaluation.

UV exposure is involved in the etiology of skin cancer. In photoaged skin various signal pathways are triggered that turn the cells resistant to apoptosis [73]. Thus, deterrence of photo-aging may also help to avoid cancer. Thus, an in-depth understanding of the physiological basis of skin aging is needed particularly in relation to photo-induced carcinogenesis. It can help for the advancement in treatment of the unwelcome consequences of both cosmetic and pathogenic skin aging. Development of novel preventive as well as therapeutic approaches would not only be beneficial for cosmetic science, but it would also enhance the overall quality of life.

Acknowledgments The author acknowledges assistance from the University of Kalyani, Kalyani, and DST-PURSE, Government of India, for supporting the work.

References

1. Farage MA, Miller KW, Elsner P et al (2013) Characteristics of the aging skin. Adv Wound Care 2:5–10
2. Farage MA, Miller KW, Elsner P et al (2008) Functional and physiological characteristics of the aging skin. Aging Clin Exp Res 20:195–200
3. L N, Altieri A, Imberti GL et al (2005) Sun exposure, phenotypic characteristics, and cutaneous malignant melanoma. An analysis according to different clinic - pathological variants and anatomic locations (Italy). Cancer Causes Control 16(8):893–899
4. Herrmann MWG, Ma W, Kuh L et al (2000) Photoaging of the skin from phenotype to mechanisms. Exptal. Geront 35:307–316
5. Ichihashi M (2009) Photoaging of the skin. Anti-Aging Medi 6(6):46–59
6. Kulka M (2013) Mechanisms and treatment of photoaging and photodamage. In: Using old solutions to new problems – natural drug discovery in the 21st century pp 255–276.
7. Amano S (2009) Possible involvement of basement membrane damage in skin Photoaging. J Invest Dermatol 14:2–7
8. Seo JE, Kim S, Shin MH (2010) Ultraviolet irradiation induces thrombospondin-1 which attenuates type-I pro-collagen down regulation in human dermal fibroblasts. J Derm Sci 59:16–24
9. Ohnishi Y, Tajima S, Akiyama M et al (2000) Expression of elastin-related proteins and matrix metalloproteinases in actinic elastosis of sun-damaged skin. Arch Dermatol Res 292:27–31
10. Uitto J (1979) Biochemistry of the elastic fibres in normal connective tissues and its alterations in disease. J Invest Dermatol 72:1–10
11. Davidson EA (1965) Polysaccharide structure and metabolism. In: Montagna W (ed) Ageing: biology of skin 6:255–270
12. Sjerobabski-Masnec I, Šitum M (2010) Skin Aging. Acta Clin Croat 49:515–519
13. Scharffetter–Kochanek K, Brenneisen P, Wenk J et al (2000) Photoaging of the skin from phenotype to mechanisms. Exptal Geront 35:307–316
14. Yano K, Kajuya K, Ishiwata M et al (2004) Ultraviolet B-induced skin angiogenesis is associated with a switch in the balance of vascular endothelial growth factor and thrombospondin-1 expression. J Invest Dermatol 122:201–208
15. Sakura M, Chiba Y, Kamiya E et al (2014) Differences in the histopathology and cytokine expression pattern between chronological aging and Photoaging of hairless mice skin. Mod Res Inflamm 3:82–89
16. Fisher GJ, Kang S, Varani J et al (2002) Mechanisms of Photoaging and chronological skin aging. Arch Dermatol 138(11):1462–1470
17. Trautinger F, Mazzucco K, Knobler RM et al (1994) UVA- and UVB-induced changes in hairless mouse skin collagen. Arch Dermatol Res 286:490–494
18. Fisher GJ, Datta S, Wang Z et al (2000) C-Jun – dependent inhibition of cutaneous procollagen transcription following ultraviolet irradiation is reversed by all-*trans* retinoic acid. J Clin Invest 106(5):663–670
19. Talwar HS, Griffiths CEM, Fisher G et al (1995) Reduced type I and type III procollagens in Photodamaged adult human skin. J Invest Dermatol 105:285–290
20. Seite S, Tison-Regnier S, Chistiaens F et al (1998) Effect of repeated low doses of solar simulated UVR in human: comparison with severe photodamaged skin. Protection of skin from ultraviolet radiation Eds: Rougier A, Schaefer H:59–71
21. Quan T, Qin Z, Xia W et al (2009) Matrix-degrading metalloproteinases in Photoaging. J Invest Dermatol 14:20–24
22. Zaid MA, Afaq F, Syed DN et al (2007) Inhibition of UVB-mediated oxidative stress and markers of Photoaging in immortalized HaCaT keratinocytes by pomegranate polyphenol extract POMx. Photochem Photobiol 83:882–888
23. Yokose U, Hachiya A, Sriwiriyanont P et al (2012) The endogenous protease inhibitor TIMP-1 mediates protection and recovery from cutaneous photodamage. J Invest Dermatol 132(12):2800–2809

24. Khokha R, Denhardt DT (1989) Matrix metalloproteinases and tissue inhibitor of metalloproteinases: a review of their role in tumorigenesis and tissue invasion. Invasion Metastasis 9(6):391–405
25. Codriansky KA, Quintanilla-Dieck MJ, Gan S et al (2009) Intracellular degradation of elastin by cathepsin K in skin fibroblasts – a possible role in photoaging. Photochem Photobiol 85:1356–1363
26. Lai W, Zheng Y, Ye ZZ et al (2010) Changes of cathepsin B in human photoaging skin both in vivo and in vitro. Chin Med J 123:527–531
27. Cavarra E, Fimiani M, Lungarella G et al (2002) UVA light stimulates the production of cathepsin G and elastase-like enzymes by dermal fibroblasts: a possible contribution to the remodeling of elastotic areas in sun-damaged skin. Biol Chem 383:199–206
28. Son ED, Shim JH, Choi H et al (2012) Cathepsin G inhibitor prevents ultraviolet B-induced photoaging in hairless mice via inhibition of fibronectin fragmentation. Dermatology 224:352–360
29. Ito K, Tanaka K, Kojima H et al (2008) Mast cell tryptase and photoaging: possible involvement in the degradation of extra cellular matrix and basement membrane proteins, Dermatol. Res 300(Suppl 1):S69–S76
30. Pillai S, Oresajo C, Hayward J (2005) Ultraviolet radiation and skin aging: roles of reactive oxygen species, in inflammation and protease activation, and strategies for prevention of inflammation induced matrix degradation - a review. Int J Cosmet Sci 27:17–34
31. Chao SC, Hu DN, Yang PY et al (2013) Ultraviolet-A irradiation upregulated urokinase-type plasminogen activator in pterygium fibroblasts through ERK and JNK pathways. Invest Opthal Vis Sci 54(2):999–1007
32. Starcher B, Conrad M (1995) A role for neutrophil elastase in the progression of solar elastosis. Connect Tissue Res 31(2):133–140
33. Rijken F, Rebecca C, Kiekens M et al (2006) Pathophysiology of photoaging of human skin: focus on neutrophils. Photochem Photobiol Sci 5:184–189
34. Fisher GJ, Choi HC, Bata-Csorgo Z et al (2001) Ultraviolet irradiation increases matrix metalloproteinase-8 protein inhuman skin in vivo. J Invest Dermatol 117:219–226
35. Chung JH, Seo JY, Lee MK et al (2002) Ultraviolet modulation of human macrophage metalloelastase in human skin in vivo. J Invest Ther Dermatol 119:507–512
36. Brennan M, Bhatti H, Nerusu K et al (2003) Matrix mettalloproteinase-1 is the collagenolytic enzyme responsible for collagen damage in UV-irradiated human skin. Photochem Photobiol 78:43–48
37. Fisher GJ, Datta SC, Talwar HS et al (1996) Molecular basis of sun-induced premature skin ageing and retinoid antagonism. Nature 379:335–339
38. Fisher GJ, Wang ZQ, Datta SC et al (1997) Pathophysiology of premature skin aging induced by ultraviolet light. N Engl J Med 337:1419–1428
39. Rijken F, Bruijnzeel PLB (2009) The pathogenesis of Photoaging: the role of neutrophils and neutrophil-derived enzymes. J Invest Dermatol 14:67–72
40. Wlaschek M, Tantcheva-Poor I, Naderi L et al (2001) Solar UV irradiation and dermal photoaging. J. Photochem Photobiol B Biol 63:41–51
41. Hanson KM, Simon JD (1998) Epidermal trans-urocanic acid and the UV-A-induced photoaging of the skin. Proc Natl Acad Sci (USA) 95(18):10576–10578
42. Valencia A, Kochevar IE (2008) Nox1-based NADPH oxidase is the major source of UVA-induced reactive oxygen species in human keratinocytes. J Invest Dermatol 128:214–222
43. Quan T (2016) Molecular mechanism of human skin connective tissue aging. In: Quan T (ed) Molecular mechanisms of skin aging and age-related diseases. CRC Press, Boca Raton, pp 1–40
44. Löffek S, Schilling O, Franzke CW (2011) Biological role of matrix metalloproteinases: a critical balance. Euro Resp J 38:191–208
45. Birch-Machin MA, Swalwell H (2010) How mitochondria record the effects of UV exposure and oxidative stress using human skin as a model tissue. Mutagenesis 25(2):101–107

46. Richter T, Zglinicki TV (2007) A continuous correlation between oxidative stress and telomere shortening in fibroblasts. Exptal Gerontol 42(11):1039–1122
47. Pittayapruek P, Meephansan J, Prapapan O et al (2016) Role of matrix metalloproteinases in Photoaging and Photocarcinogenesis. Int J Mol Sci 17:868–887
48. Philips N, Auler S, Hugo R et al (2011) Beneficial Regulation of Matrix Metalloproteinases for Skin Health Enzyme Research Article ID:427285. 4 pageshttp://dx.doi.org/10.4061/2011/427285
49. Shin MH, Moon YJ, Seo JE (2008) Reactive oxygen species produced by NADPH oxidase, xanthine oxidase, and mitochondrial electron transport system mediate heat shock-induced MMP −1 and MMP −9 expression. Free Rad Biol Medi 44:635–645
50. O'Prey J, Crighton D, Martin AG et al (2010) p53-mediated induction of Noxa and p53AIP1 requires NFκB. Cell Cycle 9:947–952
51. Shaulian E, Karin M (2001) AP1 in cell proliferation and survival. Nature 20(19):2390–2400
52. Afnan Q, Adil MD, Nissar-Ul A et al (2012) Glycyrrhizic acid (GA), a triterpenoid saponin glycoside alleviates ultraviolet-B irradiation-induced photoaging in human dermal fibroblasts. Phytomedicine 19:658–664
53. Pandel R, Poljšak B, Godic A et al (2013) Skin Photoaging and the role of antioxidants in its prevention. ISRN Dermatology Article ID 930164. 11 pages http://dx.doi.org/10.1155/2013/930164
54. Kim HH, Kim DH, MH O et al (2015) Inhibition of MMP1 and type I procollagen expression by phenolic compounds isolated from the leaves of *Quercus mongolica* in ultraviolet – irradiated human fibroblast cells. Arch Pharma Res 38:11–17
55. Tanaka K (2005) Prevention of UVB mediated skin photoaging in a NFκB inhibitor. Parthnolide J Pharmacol Exptal Therap 315:624–630
56. Jian J, Pelle E, Yang Q et al (2011) Iron sensitizes keratinocytes and fibroblasts to UVA-mediate matrix metalloproteinase-1 through TNF-α and ERK activation. Exp Dermatol 20:249–254
57. Quan T, He T, Kang S (2004) Solar ultraviolet irradiation reduces collagen in photoaged human skin by blocking transforming growth factor-beta type II receptor/Smad signaling. Am J Pathol 165(3):741–751
58. Schreck R, Rieber P, Baeuerle PA (1991) Reactive oxygen intermediates as apparently widely used messengers in the activation of the NF-kappa B transcription factor and HIV-1. EMBO J 10:2247–2258
59. Griffiths CEM, Russman AN, Majmudar G et al (1993) Restoration of collagen formation in photodamaged human skin by tretinoin (retinoic acid). New Engl J Med 329:530–535
60. Wertz K, Seifert N, Hunziker PB et al (2004) Beta-carotene inhibits UVA-induced matrix metalloprotease 1 and 10 expression in keratinocytes by a singlet oxygen-dependent mechanism. Free Rad Biol Med 37(5):654–670
61. Nichols JA, Katiyar SK (2010) Skin photoprotection by natural polyphenols: anti-inflammatory anti-oxidant and DNA repair mechanisms. Arch Dermatol Res 302 (2):71–83
62. Bae JY, Cho JS, Choi YJ (2008) Epigallocatechin gallate hampers collagen destruction and collagenase activation in ultraviolet-B-irradiated human dermal fibroblasts: involvement of mitogen-activated protein kinase. Food Chem Toxicol 46:1298–1307
63. Bae JY, Lim SS, Kim SJ et al (2009) Bog blueberry anthocyanins alleviate photoaging in ultraviolet-B irradiation-induced human dermal fibroblasts. Mol Nutr Food Res 53:726–738
64. Yang B, Ji C, Chen X et al (2011) Protective effect of Astragaloside IV against matrix metalloproteinase-1 expression in ultraviolet-irradiated human dermal fibroblasts. Arch Pharm Res 34(9):1553–1560
65. Amaro-Ortiz A, Yan B, D'Orazio JA (2014) Ultraviolet radiation, aging and the skin: prevention of damage by topical cAMP manipulation. Molecules 19 (5): 6202 – 6219
66. Casetti F, Jung W, Wölfle U et al (2009) Topical application of solubilized Reseda luteola extract reduces ultraviolet B-induced inflammation in vivo. J Photochem Photobiol B: Biology 96:260–265
67. Urikura I, Sugawara T, Hirata T (2011) Protective effect of Fucoxanthin against UVB- induced skin Photoaging in hairless mice. Biosci Biotechnol Biochem 75(4):757–760

68. Inui M, Ooe M, Fujii K et al (2008) Mechanisms of inhibitory effects of CoQ10 on UVB-induced wrinkle formation in vitro and in vivo. Biofactors 32(1–4):237–243
69. Dong-Hee K, Han-Hyuk K, Hyeon-Jeong K et al (2014) CopA3 peptide prevents ultraviolet-induced inhibition of type-I procollagen and induction of matrix metalloproteinase-1 in human skin fibroblasts. Molecules 19:6407–6414
70. Greul AK, Grundmann JU, Heinrich F et al (2002) Photoprotection of UV-irradiated human skin: an antioxidative combination of vitamins E and C, carotenoids, selenium and proanthocyanidins. Skin Pharmacol Appl Ski Physiol 15:307–315
71. Weiss JS, Ellis CN,.Headington JT et al (1988) Topical tretinoin in the treatment of aging skin. J Am Acad Dermatol. 19 (1):169–175
72. Rittie L, Fisher GJ, Voorhees JJ (2006) Retinoid therapy for photoaging, In: Gilchrest B, Krutmann J (eds) Skin aging, vol 13. Springer, Berlin, pp 143–156
73. Salminen A, Ojala J, Kaarniranta K (2011) Apoptosis and aging: increased resistance to apoptosis enhances the aging process. Cell Mol Life Sci 68(6):1021–1031

Insect Proteases: Structural-Functional Outlook

21

Shounak Jagdale, Sneha Bansode, and Rakesh Joshi

Abstract

Proteases perform various activation and effector functions in development, growth, and survival in insects. In this chapter, we have focused on the catalogue of insect proteases and their structural and functional aspects. We have reviewed the proteases involved in insect's vital life processes like reproduction, development, immunity, and defence. The indispensability of seminal fluid and egg protease during fertilization highlights their evolutionary primitiveness. Furthermore, various cellular proteases like cathepsins and caspases take over the earlier one's functions. The role of cellular proteases is well documented in the developmental process like embryogenesis, metamorphosis, moulting, and eclosion. Cellular proteases are further supported by haemolymph and digestive proteases to facilitate the growth and survival of the insect. Apart from developmental cathepsin and caspases, haemolymph contains a diverse pool of proteases that serves a pivotal role in immunity against various pathogens. Amongst various insect proteases, digestive proteases show highest structural and functional variability according to developmental stage, food content and stress level. This chapter provides an insight of structural-functional aspects of insect proteases and their role in insect physiology.

Keywords

Insect • Development • Digestion • Immunity • Protease

S. Jagdale • R. Joshi (✉)
Institute of Bioinformatics and Biotechnology, Savitribai Phule Pune University (Formerly University of Pune), Pune 411007, Maharashtra, India
e-mail: rakesh.joshi@unipune.ac.in; rakeshjoshi687@gmail.com

S. Bansode
Department of Chemistry, University of Cambridge, Lensfield Road, Cambridge CB2 1EW, UK

S. Chakraborti, N.S. Dhalla (eds.), *Proteases in Physiology and Pathology*, DOI 10.1007/978-981-10-2513-6_21

21.1 Introduction

An insect's life cycle is driven by a plethora of metabolic enzymes. Amongst these are the proteases, which are vital to the tissue-specific functioning in various processes like development, immune response, and digestion. The developmental processes of metamorphosis and eclosion are performed by acidic pH favouring enzymes, like cathepsins and caspases, which are typically stored in the lysosomes [1]. The digestive process, on the other hand, is dominated by alkaline pH favouring serine proteases like trypsin, chymotrypsin, and elastase, which function in the insects' midgut [2]. Most of these proteases often function in various signalling and activation cascade pathways. These pathways are also involved in insect embryonic development, apoptosis and insect immune responses [3–5].

There are limited reports that catalogue insect proteases and their functionality [6–9]. In this chapter, we discuss the structural-functional aspects of insect cellular and digestive proteases (Table 21.1).

Table 21.1 Various insect proteases and their functions

Sr. No.	Name	Type	Function	References
Seminal proteases				
01	Seminase	Trypsin-like serine protease	Enhances level of oogenesis and of egg-laying capacity	[10]
			Binding of sex peptide to the sperm	
			Proteolytic activation of metalloproteases	
02	Initiatorin	Trypsin-like serine protease	Sperm activation	[11]
03	CG11864	Metalloprotease	Downstream processing of Sfps	[12]
04	CG6168	Metalloprotease	Protection of females from systemic bacterial infection.	[13]
Egg and embryonic proteases				
05	30kP protease A	Serine protease	Degradation of vitellin	[14]
			Degradation of yolk proteins during late embryogenesis	
06	Cathepsin L-like protein	Cysteine protease	Digestion of vitellin	[15]
07	Cathepsin B-like protein			[16]
08	Nudel	Serine protease	Eggshell biosynthesis	[1718]
09	Gastrulation defective		Dorsal-ventral patterning	
10	Snake			
11	Ester			

(continued)

Table 21.1 (continued)

Sr. No.	Name	Type	Function	References
Developmental and cellular proteases				
12	Cathepsin L	Cysteine protease	Degradation of cuticle, epidermis and labial gland	[16, 21–22]
			Differentiation of imaginal discs	
			Granulocyte to macrogranulocyte transformation	
			Histolysis of fat bodies and silk gland	
			Normal expression of apoptosis effector gene caspase 1	
			Autophagy of degraded fat body cells	
13	Cathepsin F	Cysteine protease	Required for cuticle and epidermis layer degradation during moulting	[16]
14	Cathepsin O			
15	Cathepsin B	Cysteine protease	Metamorphosis from pupa to adult	[16, 23–27]
			Histolysis of fat bodies and silk gland	
			Degradation of labial gland	
			DNA fragmentation in fat bodies	
			Immune response against parasitic infection	
16	Cathepsin D	Cysteine protease	Degradation of larval fat bodies	[16, 25, 26]
			DNA fragmentation in larval midgut cells	
			Immune response against parasitic infection	
17	Cocoonase	Trypsin protease	Softening insect cocoons to permit the escape of adult moths	[6]
18	Dronc	Caspases	Stress-induced apoptosis	[28–44>]
19	Dredd			
20	Strica			
21	Crice			
22	DCP-1			
23	Decay			
24	Damm			

(continued)

Table 21.1 (continued)

Sr. No.	Name	Type	Function	References
25	DM1-MMP	Matrix metalloprotease	Remodelling of extracellular matrix	[45–47]
26	DM2-MMP	Matrix metalloprotease	Development of the central nervous system and tracheae	[48–50
27	ADAMs	Metalloprotease	Dendrite reshaping in optic lobes of the brain and eye imaginal discs	[5152–56]
28	Neprilysin-like	Metalloprotease	Axon extension in nervous system development	[57]
			Activation of signalling pathway resulting in patterning of imaginal discs	
			Proteolytic cleavage of active signalling peptides in cardiovascular, nervous, inflammatory and immune functions	
Haemolymph proteases				
29	Modular serine protease	Serine protease	Toll signalling pathway	[58]
30	Grass		Toll signalling pathway	[58]
31	PAPs		Melanization, degradation of tracheal epithelium and larval hypodermis during pupation	[59–63]
32	Sp22D			
Digestive proteases				
33	Trypsin	Serine proteases	Digestion of proteins to oligopeptides	[2, 64]
34	Chymotrypsin		Digestion of elastin	
35	Elastase			
36	Cathepsin L	Cysteine proteases	Digestion of oligopeptides with hydrophobic amino acids	[65]
37	Cathepsin D	Aspartic proteases	Killing the ingested bacteria	[66]
38	Aminopeptidase N	Aminopeptidases	Hydrolysis of oligopeptides into dipeptides	[64]
39	Aminopeptidase A			
40	Carboxypeptidase A	Carboxypeptidase	Hydrolysis of single amino acids from the C-terminus of the peptide chain	[66]
41	Carboxypeptidase B			
42	Dipeptidases		Hydrolysis of dipeptides	[67–69]

21.2 Proteases Involved in Insect Reproduction

Reproduction is the most important activity for the survival of any species. In insects, eggs produced in the female insect body get fertilized by sperm, and eventually the fertilized eggs are delivered outside the body, where they hatch. Recent molecular understanding has suggested that proteases have a major role to play in the process of insect reproduction [70–73]. Interestingly, the male and female reproductive tract proteases and protease homologs show adaptive evolution, indicating their intersexual co-evolution [71, 74–78].

21.2.1 Seminal Proteases

The seminal fluid of insects contain various proteases that get transferred to the eggs during fertilization. These proteases are expressed in almost all the tissues of the male reproductive system. In *Drosophila* sp., the ejaculate has about 21 seminal fluid proteases (SFPs), 17 protease inhibitors (PIs) and 10 putative serine protease homologs [79, 80, 10, 81]. Most insect SFPs are conserved at structural and functional levels across different taxa. SFPs have been identified in insects like *Anopheles gambiae*, *Ceratitis capitata*, *Heliconius* sp., *Bombyx mori*, *Tribolium castaneum*, *Gryllus* sp. and *Amblyomma hebraeum* as well [82–85, 11, 86–88]. 'Seminase', a trypsin-like serine protease from the male gets transferred to the female body via the seminal fluid after mating. It enhances the levels of oogenesis and the egg-laying capacity of female insects (Fig. 21.1). Additionally, seminase interacts with the other SFPs which leads to the binding of a sex peptide to the sperm [10, 89, 90]. This interaction is required for the efficient release of sperm from the spermathecae and also for post-mating maintenance of egg-laying [91]. Females mated with seminase knockdown mutants were observed to have a significant reduction in their egg-laying capacity over time. In such a case, it has been illustrated that the males can store the sperm normally, but in females, it remains stored in the seminal receptacle and spermathecae even 10 days post fertilization [10]. Seminase has been observed to induce the proteolytic activation of a metalloprotease CG11864, which is required for the downstream processing of other SFPs like Acp36De and an ovulation prohormone 'ovulin' [12]. This shared activity of SFPs in male and female insects suggests their crosstalk during the evolutionary course of insect sex.

In males, trypsin-like serine proteases are essential for the activation of sperm motility [92] and sperm activation; for example, in *B. mori*, a serine protease known as 'initiatorin' has been identified to be necessary for sperm activation and has been found to be conserved in various classes of insects from phylogenetic and evolutionary analysis (Fig. 21.1) [11]. In female insects, metallo-SFPs like CG6168 have been found to protect females from systemic bacterial infections transmitted through the seminal fluid [13]. The function of SFPs has been further complemented with eggs and embryonic proteases. Various other studies have identified numerous SFPs, but their role still remains to be investigated.

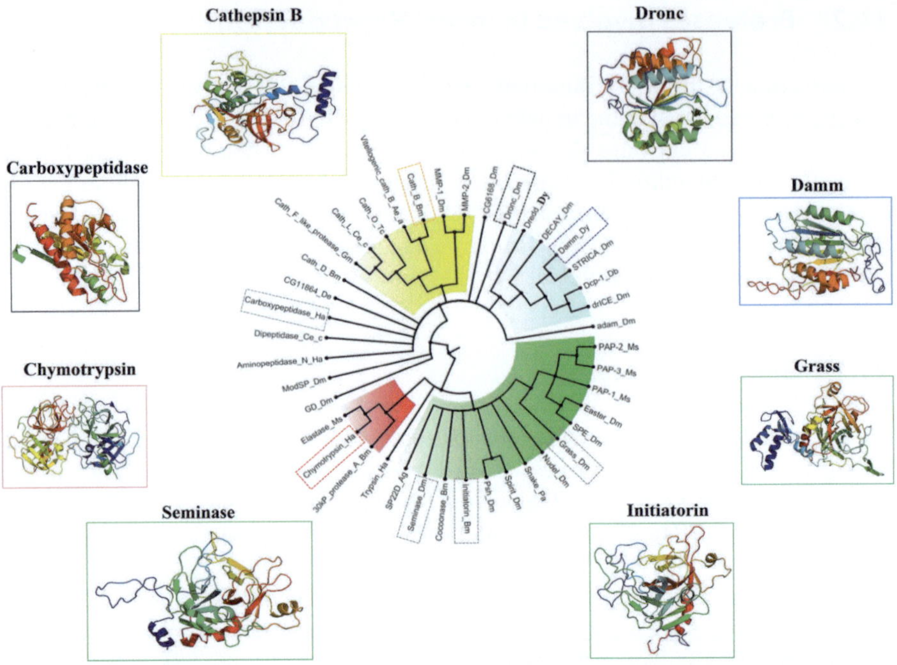

Fig. 21.1 Diversity of insect proteases: Cladogram of insect proteases was generated by MEGA7: Molecular Evolutionary Genetics Analysis Version 7.0 using a neighbour-joining algorithm with 1000 bootstrap iterations and p-distance amino acid substitution method. Structures of proteases were retrieved from PDB. PDB IDs are 2FP3 (Dronc), 1JQG (carboxypeptidase), 1EQ9 (chymotrypsin) and 2XXL (Grass). Other proteases were modelled by using Modeller 9.15. All the sequences were retrieved from NCBI protein database

21.2.2 Egg and Embryonic Proteases

Insect eggs contain a macromolecular glycoprotein called 'vitellogenin' which is synthesized in fat bodies. It is then proteolytically processed and taken up by oocytes, where it is stored in yolk granules in a crystalline form called 'vitellin'. Vitellin serves as the primary source of amino acid supplements in eggs and is digested by proteases to release the amino acids required for embryonic development [93]. These vitellin-degrading enzymes are primarily serine proteases which were initially isolated and characterized from *B. mori* eggs [94]. Mostly these proteases are expressed in their zymogen form in the ovaries and are then proteolytically activated during embryogenesis. They are expressed temporally with differential expression in early, mid and late embryogenesis. In *B. mori* eggs, yolk proteins are degraded specifically by a 30 kP protease A whose expression is higher during late embryogenesis [14]. In *Aedes aegypti*, the carboxypeptidase expressed in fat bodies during embryogenesis is transported by the haemolymph and is eventually taken up by oocytes [95]. Along with serine proteases, there are multiple cysteine proteases that participate in the degradation of yolk protein during embryogenesis such as, a papain-like C1 family of cysteine proteases which assists the major serine proteases in degradation pathways in eggs. In *B. mori* eggs, a 47 kD cysteine protease similar to cathepsin L is expressed as a zymogen in the ovary and the fat bodies, and it helps in vitellin degradation [15], while in insects like *Helicoverpa armigera* and *Antheraea pernyi*, vitellin degradation is carried out by an acid protease similar to cathepsin B. The expression and activity of the cathepsin B-like protease gradually decrease as the embryo development progresses [16]. A cysteine protease similar to cathepsin B has been recognized as an enzyme that digests egg yolk proteins in *D. melanogaster* and *Musca domestica* and is activated by a serine protease, after proteolytic processing (Fig. 21.1) [96, 97]. Several other serine proteases play a crucial role in embryogenesis and embryonic patterning as follows.

21.2.2.1 Proteases in Dorsal-Ventral Patterning of Embryo

An extracellular serine protease cascade activates a signal transduction system that regulates dorsal-ventral patterning in *D. melanogaster* embryo [98, 17, 99]. The proteins involved in this cascade, namely, Nudel, gastrulation defective (GD), Snake and Easter, are of maternal origin and are stored in the perivitelline space. All these proteases contain amino-terminal clip domains and carboxyl-terminal serine protease domains [100]. Nudel is a 350 kD serine protease secreted by the ovarian follicle cells. It activates the signalling cascade associated with the ventral signalling cascade [101, 102] and is also involved in the eggshell biosynthesis. Nudel is responsible for the proteolytic cleavage of peroxidases and for the release of H_2O_2 from follicle cells, which induces the cross-linking of the vitelline membrane, during embryogenesis. Cross-linking of the vitelline membrane results in a specific matrix structure which may be necessary for the function of one or more dorsoventral proteases [18]. In silico studies have suggested that Nudel interacts with several proteins of the ventral signalling pathway and activates GD. GD is expressed in germline-derived nurse cells of germarium, and it activates another protease called Snake in downstream pathways (Fig. 21.2A) [103, 104]. Snake is assisted by Easter and activates 'Spätzle', a Toll receptor ligand. Spätzle binds to Toll receptors and initiates the dorsal signalling pathway, one of the key pathways in embryonic development.

21.3 Developmental and Cellular Proteases

21.3.1 Cathepsins

Cathepsins (Cath) are cellular proteases active at acidic pH and play a crucial role in metamorphosis, moulting, eclosion and other related developmental processes [16]. They display differential expression in various tissues and developmental stages. Insects undergo metamorphosis through four life cycle stages: eggs, larvae, pupae, and moths. During the process of moulting, insects transit from neonates to

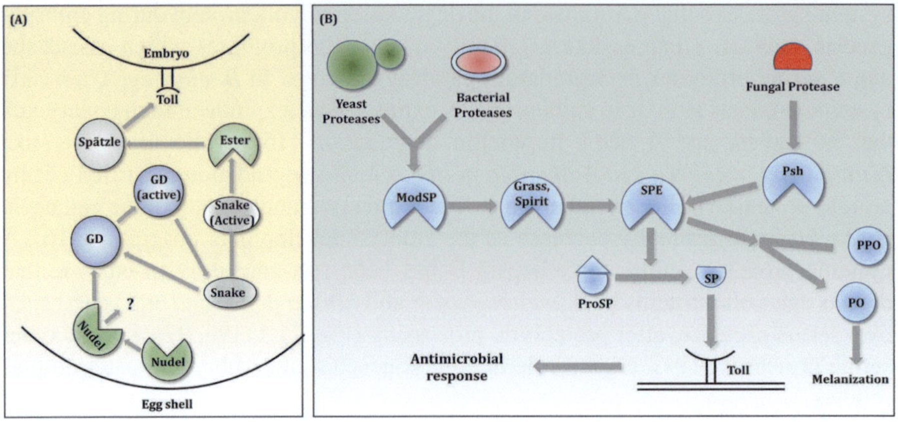

Fig. 21.2 (**A**) Protease cascade involved in the dorsoventral patterning of *Drosophila* embryo: protease cascade which is activated by Nudel which leads to initiation of dorsal signalling pathway. (**B**) Protease cascade involved in the insect's immune response: antigens activate the protease cascade which leads to activation of AMPs by Toll signalling pathway. Melanization pathway is activated by prophenoloxidase-activating proteins which are clip serine proteases

six larval instars and resynthesize the exoskeleton by replacing the older one. Insect metamorphosis and moulting, and thus the tissue remodelling of fat bodies and haemolymph, are dominated by Cath degradation.

Lepidopterans show maximum expression of Cath L during moulting and prepupation which suggest its role in the degradation of cuticle and epidermis during moulting. During larva-pupa metamorphosis, Cath L transforms granulocytes to macrogranulocytes which travel back to the fat bodies and release Cath L at an early pupal stage for further utilization. Cath L proteolytically degrades and releases fat body cells into the haemolymph. In a *D. melanogaster* haemocyte expressed cell line, Cath L-like protease helps degrade phagocytosed material [19]. In *Sarcophaga peregrina*, Cath L is involved in the differentiation of imaginal discs [20, 21]. In dipteran insects like *Delia radicum*, Cath L functions in the metamorphosis of the midgut, as is indicated by its higher expression in the midgut, at the beginning of the late third instar [105].

Cath F and Cath O which are expressed in the midgut and the haemolymph, respectively, are known to supplement the activity of Cath L in lepidopterans [16]. Another cathepsin, Cath B, is expressed in granulocytes and plasmatocytes at the sixth instar stage in most insects, implying its role in larval to pupal transition. In *B. mori*, Cath B and Cath L play an important role in the histolysis of the silk gland and fat bodies, during pupation. Cath B is reported to be involved in the metamorphosis from pupa to adult and is found active in the fat bodies of both adult males and females, indicating its involvement in the ageing process. Cath B and Cath D are highly expressed in the fifth instar larval fat bodies and the pupal gut and work together to degrade fat bodies. Cath D is overexpressed in the larval midgut and remains so until eclosion, aiding the formation of adult fat bodies [16].

The differential role of proteases in development and metamorphosis also suggests their role in apoptosis. For example, in *Manduca sexta*, the cathepsins, Cath B and Cath L, take part in the degradation of cytoskeletal proteins in labial glands during metamorphosis [23]. Cath L-dependent programmed cell death (PCD) is observed in the fat body cells of *H. armigera*. Further, its mandate includes the expression of the apoptosis effector gene caspase-1 and the regulation of its function. In the absence of Cath L, apoptosis inhibitor gene 'survivin' is upregulated. In *H. armigera,* phagocytes show high expression of Cath L, which is involved in the remodeling of the fat body cells and wing development, whereas Cath B is required for degradation of adult fat bodies [22, 24]. It has been noted that silencing Cath B and Cath D results in the inhibition of DNA fragmentation and apoptosis in fat body cells [21, 105].

Cathepsin expression also regulates an insect's defence system against various pathogens. In *B. mori*, larvae infected with baculoviruses overexpress Cath B and Cath D [25]. Similarly, in *M. sexta*, parasitic infection by *Cotesia congregata* bracovirus results in the elevated expression of MsCath1, MsCath2, Cath B and Cath L [106]. Parasitic infection of *Plutella xylostella* by *Cotesia plutellae* increases the Cath B expression [27] . In a bean bug *Riptortus pedestris*, Cath L-like protease is responsible for maintaining the population of *Burkholderia* sp. which is an important gut symbiont [107]. Apart from biotic stress, abiotic factors such as increased temperature and oxidative stress also lead to the overexpression of Cath B and Cath D [25, 108]. Starvation also leads to the upregulation of Cath B, Cath O and Cath L [109].

21.3.2 Serine Protease

With respect to tissue remodelling, some trypsin-like serine proteases are involved in the softening of insect cocoons to permit the escape of adult moths, with the help of the enzyme 'cocoonase'. This enzyme, which had initially been identified and characterized from *B. mori*, is secreted from a specialized mouth part called Galea in its zymogen form – 'prococoonase' – and is then activated by proteolytic cleavage [6]. Homologues of cocoonase such as trypsin proteases are found in many other lepidopterans and dipterans.

21.3.3 Caspases

Caspases – cysteinyl aspartate-specific proteases – are responsible for metamorphosis in holometabolous insects, as their activation induces massive PCD. They have a canonical structure containing a prodomain and a peptidase C14 domain. The peptidase C14 domain is a catalytic domain with conserved binding (L/S-T/S-H-G) and active (Q-A/R-C-R/Q-G) sites. Few comparative studies have shown that caspases exhibit conservation at structural and functional levels in various classes of insects [28].

The role of caspases has been widely studied in *D. melanogaster* and *A. aegypti*. Caspases are usually present in the cytoplasm in an inactive state until the initiation of the cell death signalling pathway. Activation of the initiator caspases triggers the effector caspases, resulting in morphological and biochemical changes associated with apoptosis, such as DNA fragmentation and plasma membrane blebbing [29]. Caspases are also involved in the autophagic activity and immune response [30]. It is reported that the inhibitor of apoptosis (IAP) regulates the pro-apoptotic regulation of caspases and its removal is responsible for the activation of caspases. In *D. melanogaster*, seven caspases have been reported, namely, death regulator Nedd2-like caspase (Dronc), death-related ced-3/Nedd2-like caspase (Dredd), Ser-/Thr-rich caspase (Strica), death-related ICE-like caspase (Drice), decapping protein 1(DCP-1), death executioner caspase-related to Apopain/Yama (Decay) and death-associated molecule related to MCH-2 (Damm). Initiation of apoptosis depends on the death signals that converge on the initiator caspases, namely, Dredd, Dronc and Strica (Fig. 21.1). Dredd has a long prodomain with two death-inducing domains (DID) which interact with Fas-associated death domain-containing protein (dFADD) [31]. Dredd is also involved in the innate immune response and is activated upon interacting with Toll receptors. Once it binds to Toll, Dredd proteolytically cleaves Relish, a transcription factor that regulates expression of antimicrobial peptides and is homologous to mammalian NF-κB [32]. It has been reported that Dredd mutants show reduced immune response against Gram-negative bacteria [33].

Another apoptosis caspase is 'Dronc'. It has an N-terminal CARD domain that interacts with death-associated APAF1-related killer (DARK), an adaptor protein which activates Dronc [34]. Dronc is expressed in all body tissues and is required for apoptosis in larval tissues in response to toxic agents and X- and γ-radiations [35–37]. Dronc mutants show inhibition of IAP-induced apoptosis [38]. Zygotic Dronc mutants showed reduced levels of apoptosis and resulting in delayed histolysis of salivary glands [39]. The next caspase involved in apoptosis is Strica, which has a long prodomain, rich in serine and threonine. It plays a crucial role in oogenesis and is also required for the removal of interommatidial cells of pupal retina and salivary glands [40, 41]. In vitro studies show that Strica is essential for Hid-mediated apoptosis, but its mechanism of action is still unknown [41]. Strica and Dronc double mutants show significant defects in egg chamber during mid- and late oogenesis indicating their redundancy in PCD during oogenesis [42].

Initiator caspases activate effector caspases like Drice, Dcp-1, Decay and Damm which lead to the transmission of cell death signals. Drice is the primary target of Dronc and acts downstream of cell death activators like Reaper and Hid which are required for apoptosis [32, 35, 38–40]. It has been observed that Drice mutants show reduced apoptosis in the pupal retina, embryonic nervous system, and adult wings. Another effector caspase 'Dcp-1' – the first caspase to be discovered in insects – acts similar to Drice, downstream of Reaper and Hid [43]. In Dcp-1 mutants, ecdysone-induced apoptosis is reduced while in some cases, Dcp-1 can substitute for Drice [44]. The structural homologue of Drice and Dcp-1 is Decay, and it is also involved in Hid-mediated apoptosis [35, 44]. Decay mutant shows activation of cell death pathways which suggest that its function is redundant during development. Although it is known to induce cell death in *Drosophila* eye [110], there is comparatively limited data available for Damm.

In lepidopteran, putative caspases have been consolidated into six clades (caspases 1–6). Caspases 1, 2 and 3 are considered to be effector caspases and show similarity to *Drosophila* DCP-1, Drice and Decay [111]. Furthermore, caspases of clade 4 show similarity to Strica and Damm. In case of caspases 5 and 6, they exhibit high structural similarity with Dredd and Dronc, respectively. Few candidates of initiator caspases such as Drice, Dronc and Strica have been identified in *A. aegypti* and *A. gambiae* [112]. These effector caspases are further divided into two clades, in which clade 1 caspases are similar to Decay of *Drosophila* and include two caspases from *Aedes* and eight caspases from *Anopheles*, while clade 2 caspases show homology to Dcp-1 and Drice of *Drosophila* and include two caspases from *Aedes* and *Anopheles* each.

21.3.4 Metalloproteases

During tissue remodelling the extracellular matrix is degraded by matrix metalloproteases (MMPs) which are the integral membrane proteins present on the outer surface of cells. Three families of metalloproteases have been reported in insects that have a major role in histolysis and degradation of peptide hormones. In insects like *Tribolium castaneum*, knockdown of MMP results in defect in both embryonic development and metamorphosis [45].

D. melanogaster has two MMP genes, Dm1-MMP and Dm2-MMP, which play a crucial role in metamorphosis. Dm1-MMP shows higher expression in embryos and remodels the extracellular matrix leading to the development of the central nervous system and tracheae [46, 47]. Dm2-MMP is highly expressed in the optic lobes of the brain and the eye imaginal discs and is involved in dendrite reshaping after adult eclosion [48–50]. While the indispensability of Dm1-MMP in the larval tracheal development and pupal head eversion is highlighted by mutation analysis, Dm2-MMP mutants do not show any defect in tissue remodelling during metamorphosis [113, 51]. Another class of metalloproteases known as ADAMs (also known as kuzbanian), containing a disintegrin and metalloprotease domain, has been reported to function in axon extension in nervous system development [52, 53]. These metalloproteases interact with Notch, a transmembrane protein receptor, resulting in the activation of the downstream signalling pathway which is linked to the patterning of imaginal discs [114, 54–56].

A third family of zinc metalloproteases similar to mammalian neprilysins has been identified in insects and it is involved in the degradation of peptide hormones. These metalloproteases are widely distributed in various tissues and are involved in the proteolytic cleavage of active signalling peptides in cardiovascular, nervous, inflammatory, immune and reproductive functions. In insects of different classes, these enzymes can degrade tachykinin-related peptides as well. Information about substrates and the physiological roles of these proteases is limited, but it seems likely that they act as negative regulators of peptide signalling molecular hormones [57].

21.4 Haemolymph Proteases

Haemolymph is a circulatory fluid in arthropods comprising water, inorganic salts, and organic molecules. The presence and concentration of proteins in the haemolymph vary during the course of development. These proteins are classified as per their functions and categorized as proteolysis regulators, lipid transporters, and enzymes. Haemolymph contains haemocytes which play an important role in the immune response.

21.4.1 Proteases Involved in Immune Response

Amongst the many diverse extracellular proteases, those involved in insect immune response are the serine proteases. A cascade of serine proteases invokes a rapid response to infection and wounding. When faced with a microbial challenge, *Drosophila* activates a series of antimicrobial peptides (AMP) in the fat bodies. AMP activation is brought about by the Toll signalling pathway, where effector molecules are expressed upon the activation of several serine proteases and the processing of Spätzle by Easter (Fig. 21.2B). Several clip domains (SPE) and modular serine proteases (ModSPs) are involved upstream of the *Drosophila* Toll cascade [58, 115]. Serine proteases such as Grass (Fig. 21.1) and Spirit have been identified to play a role in connecting ModSP and SPE in terms of their activities and functions albeit their limited biochemical functional analysis. Persephone (Psh) which is a clip serine protease identifies virulence factors that can trigger the Toll pathway, such as proteases from Gram-positive bacteria and fungi. It has been hypothesized that the Psh zymogen plays a sensory role in the detection of these proteolytic virulent factors.

Usually, a response to wounds or infection leads to activation of the melanization pathway. In this pathway, prophenoloxidase (proPO) is proteolytically processed by haemolymph serine proteases called prophenoloxidase-activating proteinases or PAPs, to phenoloxidase, which oxidizes phenol to produce quinine. Quinines besides being lethal for most pathogens can also polymerize into melanin and seal wounds. In *M. sexta*, PAP1 has a single regulatory clip domain at its amino-terminal, whereas PAP2 and PAP3 have two clip domains. PAP1 is a 44 kD protein with two peptide chains of 31 kD and 13 kD. Apart from the haemolymph, PAP1 is also present in the fat bodies and the cuticle of the insect and is highly expressed in haemocytes and fat bodies during bacterial infection [59]. PAP2 and PAP3 are found in fat bodies and haemolymph [60]. PAPs require serine protease homologs (SPH) 1 and 2 in order to cleave proPO. Several other PAPs have been identified, characterized and purified from other insects like *Holotrichia diomphalia* and *B. mori* [61, 62]. *M. sexta* PAP1 shows structural homology to *H. diomphalia* PPAF-I, whereas *M. sexta* PAP2 and PAP3 show structural homology to *B. mori* PPAE.

In *M. sexta* several haemolymph serine proteases (HP1–HP22) have been identified, and most of these HPs contain regulatory clip domain(s) at their amino-terminal. HP1, HP2, HP6, HP8, HP13, HP17, HP18, HP21, and HP22 have a single clip domain, while HP12 and HP15 have two clip domains. Increased levels of HP2, HP7, HP9, HP10, and HP12–HP22 are seen in case of bacterial infection. This indicates the presence of a complex serine protease network in *M. sexta* haemolymph which responds to bacterial infection and wounding [116]. Other than immune response proteases, some chitin-binding proteases are also present in the haemolymph. In mosquitoes, a modular serine protease Sp22D (serine protease from chromosomal division 22D) is expressed in various tissues with the highest expression in haemocytes and is terminally secreted in the haemolymph [63], during early and late embryonic development. It has a trypsin-like domain linked to a low-density lipoprotein receptor-like domain (LDLr), scavenger receptor cysteine-rich (SRCR) domain, mucin-like repeats and two chitin-binding domains (CBDs). Sp22D is expressed in early and late embryonic development. It exhibits slow expression in larval development, high expression during pupal development and moderate expression in adults. This protein is spatially expressed, showing abundance in the head, thorax, and abdomen. It is expressed in haemocytes and shows high chitin-binding activity. It has been noted that exposure of chitin to haemolymph results in mechanical disruption causing tissue remodelling during development. In pupae, chitin is in direct contact with haemolymph, which causes the degradation of tracheal epithelium and larval hypodermis. Chitin-binding activity of Sp22D and its high expression during pupal development suggests that the primary role of Sp22D is to detect the exposed chitin and then trigger the appropriate physiological response.

After the completion of developmental cascade, there are a series of growth cycles, which are primarily driven by physiological processes like feeding, digestion, etc. Proteases involved in digestion are well known and characterized and will be discussed in the next section in greater detail.

21.5 Digestive Proteases

The insect midgut harbours a complex digestive environment consisting of various proteolytic enzymes with a variety of specificities. Proteases present in insect gut are mostly trypsins, chymotrypsins, elastases, Cath B-like proteases, aminopeptidases, and carboxypeptidases, all of which are responsible for protein digestion. Serine proteases dominate the larval gut environment and contribute to about 95% of the total digestive activity.

21.5.1 Trypsins

Trypsins are ubiquitous digestive proteases that cleave peptide bonds at the carboxyl-terminal of basic amino acids, displaying a preference for Arg over Lys. In insects, both anionic and cationic trypsins are found. Anionic trypsins were first reported in the larvae of *Tineola bisselliella* [117]. From midgut extracts of *T. molitor* and *Locusta migratoria*, three anionic and one cationic trypsinogens have been identified [118]. In highly evolved insects, trypsin subsites are more hydrophobic [119], the binding strength of which varies in different insects.

On the basis of sequence conservation, trypsin-like enzymes are characterized by conserved N-terminal residues – IVGG, a conserved catalytic triad of serine proteinase (His57, Asp102, and Ser195) and Asp189. The active form of trypsin contains single polypeptide chain which forms two β-barrels connected by a disordered loop which contains the catalytic residues [120]. The catalytic mechanism of trypsin has been reported and is well characterized.

Variation in sequences and the number of trypsin genes harbour the key molecular mechanisms of differential digestive activity in insects. Various insects have pool serine protease, which usually exists as an array of diverse protease isoforms. In many lepidopterans and hemipterans, digestive flexibility has been recorded to be related to development, food content, and growth [2]. While adapting to specific diets, multiple mutations in structural genes, regulatory genes and adaptive expression patterns contribute to the organism's fitness. Variation and positive selection of trypsin genes from multigene families lead to new gene functions along with new tissue, development and environment-specific regulatory regimes [2, 64]. This results in enhanced physiological flexibility and an improved matching of digestion and heterogeneous food environment with the developmental changes in an insect's nutritional needs [2, 64].

21.5.2 Chymotrypsins

The digestive role of trypsin is supported by chymotrypsins, which cleave peptide chains at the carboxyl-terminal of aromatic amino acids (Fig. 21.1). These proteases are present in the insect's gut and are active at alkaline pH 8–11. In the case of heteropterans, chymotrypsins are also found in salivary glands. These enzymes have conserved signature sequences, namely, the signal peptide and an activation peptide that terminate with a catalytic triad (His57, Asp102 and Ser195). Additionally, they have three conserved pairs of cysteine residues, an IVGG sequence at the N-terminal and Ser/Gly/Tyr at the 189th position. In Lepidoptera, chymotrypsin activity is usually low [64]. In *Vespa orientalis*, chymotrypsin is shown to be similar to vertebrate chymotrypsin as it acts on glucagon and β-chain of oxidized insulin. Although there are fewer reports on insect chymotrypsin as compared to insect trypsin, the distribution of chymotrypsin amongst different insect taxa is similar to that of trypsin [7].

21.5.3 Elastases

An elastase-like enzyme in insects was first described in the cricket *Teleogryllus commodus* [121]. True elastases were first isolated from gypsy moth midguts [122] and from the larvae of *Solenopsis invicta* [123]. This enzyme has since been described in many other insects. Many of the heteropteran insects are zoophytophagous. The bug, *Lygus hesperus* which is a notorious crop pest and a capable predator for a variety of insects, shows the presence of elastase in its salivary gland and gut. The average elastase activity of gut elastase, however, is only 18–25% of the activity of salivary gland elastase. Elastase production is induced in response to the presence of specific food components, i.e. elastin which is present in specific plants and is also a key component of the extracellular matrix of various insects on which the *L. hesperus* feeds [124]. More research needs to be carried out in order to ascertain the exact role elastases play in other insect species.

21.5.4 Cysteine Proteases

Cysteine proteases are found in the insect midgut in heteropterans and coleopterans. Their widespread occurrence in the heteropteran midgut is probably attributed to the loss of serine proteases after adaptation to sap feeding. In coleopterans, insects belonging to the infra-order Cucujiformia show the presence of cysteine proteases, as they are adapted to feed on seeds rich in naturally occurring trypsin inhibitors.

Intestinal cysteine proteases are endopeptidases. Cath L is a true endopeptidase that specifically cleaves peptide bonds that have hydrophobic amino acid residues. Most of the insect midgut cysteine proteases are Cath L-like enzymes. These proteases have been isolated from *Diabrotica virgifera*, *Acyrthosiphon pisum*, and *T. molitor*. They have pH optima of 5–6 and a mass of 20–40 kD. Their zymogens possess an N-terminal peptide that is cleaved to activate the enzyme, along with the catalytic triad (Cys25, His169, and Asn175) and the ERFNIN motif [120, 65].

Though these enzymes are very important in insect digestion processes, relatively less information is available about them.

21.5.5 Aspartic Proteases

Aspartic proteases in insects were first reported in 1955 by Greenberg and Paretsky in *Musca domestica*. Aspartic proteases are Cath D-like enzymes that are present in the midgut of insects with acidic pH optima in the range 2.5–3.5 and cleave internal peptide bonds in proteins. In *M. domestica,* three Cath D-like (CAD) enzymes are expressed. CAD1 is expressed throughout the body, whereas CAD2 and CAD3 are expressed only in the midgut of the insect. They all show the presence of catalytic Asp33 and Asp229 [125]. The significance of acidic proteases in Diptera stems from the larvae ingesting food rich in bacteria, such as decaying organic matter. Bacteria once ingested are killed in the middle midgut by the combined action of low pH and acidic proteases. Aspartic proteases similar to Cath D are found in many coleopterans and hemipterans [66].

21.5.6 Aminopeptidases

Aminopeptidases specifically remove amino acids from the N-terminal of proteins. They are classified on the basis of the cofactor they require which is usually Zn^{2+} or Mn^{2+}. Aminopeptidase N has a broad specificity as compared to aminopeptidase A. In less evolved insects such as orthopterans and coleopterans, less amount of soluble aminopeptidase is seen, whereas more amount is seen in highly evolved insects such as dipterans and lepidopterans. Aminopeptidases are mainly found bound to the microvilli of insect midgut cells. Insect intestinal aminopeptidases have pH optima of 7.2–9 and a molecular mass of 90–130 kD. Sequences of aminopeptidase N from different insects such as *Trichoplusia ni*, *A. pisum*, *A. aegypti*, *H. armigera*, *Ostrinia nubilalis* and *B. mori* show the presence of a signal peptide, a conserved RLP motif near the N-terminal, a zinc-binding motif, a GAMEN motif and a long hydrophobic C-terminal with a glycosylphosphatidylinositol anchor [120].

Soluble aminopeptidases act on oligopeptides formed by the action of trypsin on proteins. These oligopeptides diffuse through the peritrophic membrane into the ectoperitrophic space. The aminopeptidases hydrolyze them into smaller peptides which are then further hydrolyzed by microvillar aminopeptidases into dipeptides [126, 127].

21.5.7 Carboxypeptidases

Carboxypeptidases hydrolyse a single amino acid residue from the C-terminal of a protein. Insect intestinal carboxypeptidases have been categorized as carboxypeptidase A and carboxypeptidase B. Both these have similar structures and kinetic properties, suggesting that they are derived from a common ancestral enzyme [128]. Carboxypeptidase A is a well-studied enzyme, with pH optima of 7.5–9 and mass of 20–50 kD. It shows the presence of signal and activation peptides, a Zn-binding catalytic triad (His69, Glu72, and His196), a substrate binding site (Arg71, Asn144, Arg145, and Tyr248), Arg127 and Glu270, all of which are responsible for catalysis [66]. Carboxypeptidase B has pH optima of 7.5–8, and it is partially characterized based on its isolates from dipterans like *Stomoxys calcitrans*, *Glossina morsitans and Rhynchosciara americana* [66, 129, 130]. Other than these two carboxypeptidases, cysteine carboxypeptidases are present in some hemipterans like *Rhodnius prolixus* which have acidic pH optima [131]. These enzymes have a lysosomal origin similar to other cysteine proteases.

21.5.8 Dipeptidases

Dipeptidases hydrolyze those dipeptides that are produced due to the action of aminopeptidases. The first true dipeptidase was isolated from the midgut of a hemipteran *R. prolixus* [67]. In *M. domestica*, dipeptidases are only found in the microvillar membrane and show substrate specificity to Gly-Leu [68]. In *S. frugiperda*, two dipeptidases are found in the midgut cells which are specific to Gly-Leu [69]. In *R. americana*, three dipeptidases are present [132]. Two of them are soluble while one is membrane bound. Soluble dipeptidases hydrolyse Gly-Leu and Pro-Gly but show different efficiencies for the two dipeptides. The substrate specificity of membrane-bound dipeptidase in still unknown, leaving dipeptidases to be the least studied insect proteases, providing scope for further research.

21.6 Conclusion and Future Perspectives

It is evident that the insect proteases are involved in most of the cellular, physiological and biochemical processes. The role of seminal and egg proteases in the reproductive processes is very crucial; however, detailed structural studies need to be done for functional understanding of these proteases. Seminal and egg proteases show interdependent functionality; thus, high-throughput analysis at gene and protein level can provide comprehensive understanding of the co-evolution of these proteases. The physiological and biochemical role of cathepsins and caspases in the development processes is well studied. Although, due to limited structural information on insect cathepsins and caspases, it is difficult to understand their mechanistic action as activator or effectors. Insect metalloproteases show similarity across the invertebrates as well as the vertebrates. The role of metalloproteases in the CNS development can provide a framework for the study of these proteins across different model systems. The digestive proteases show high variability in different insects on basis of diet content, development and stress level. These enzymes are very well studied at biochemical level; still significant structural and molecular information are unavailable. Thus, detailed understanding of insect proteases at molecular level can be utilized to provide new arena for basic research and also to be applicable in the insect control field. The insect proteases can be used as new targets for developing effective and sustainable insect control strategies.

Acknowledgements Funding under the Divisional Research and Development Program (DRDP) to the Institute of Bioinformatics and Biotechnology, Savitribai Phule Pune University, Pune, India, is greatly acknowledged. Dr. Sneha acknowledges the Royal Society, UK, for its financial support under the Newton International Fellowship scheme. Editorial assistance by Mr. Aniruddha Agnihotri and Ms. Shriya Lele is highly acknowledged. The authors declare no conflict of interest.

References

1. Turk B, Turk D, Turk V (2000) Lysosomal cysteine proteases: more than scavengers. Biochimica et Biophysica Acta (BBA)-Protein Structure and Molecular Enzymology 1477(1):98–111
2. Srinivasan A, Giri AP, Gupta VS (2006) Structural and functional diversities in lepidopteran serine proteases. Cell Mol Biol Lett 11(1):132–154
3. Krem MM, Di Cera E (2002) Evolution of enzyme cascades from embryonic development to blood coagulation. Trends Biochem Sci 27(2):67–74

4. Feinstein-Rotkopf Y, Arama E (2009) Can't live without them, can live with them: roles of caspases during vital cellular processes. Apoptosis 14(8):980–995
5. Cerenius L, Kawabata SI, Lee BL, Nonaka M, Söderhäll K (2010) Proteolytic cascades and their involvement in invertebrate immunity. Trends Biochem Sci 35(10):575–583
6. Law JH, Dunn PE, Kramer KJ (1977) Insect proteases and peptidases. Adv Enzymol Relat Areas Mol Biol 45:389–425
7. Applebaum S. (1985) Biochemistry of digestion. In: Comprehensive insect physiology biochemistry and pharmacology. Pergamon Press, Oxford, pp 279–311
8. Terra WR, Ferreira C, Jordao BP, Dillon RJ (1996) Digestive enzymes. In: Biology of the insect midgut. Springer, Dordrecht, pp 153–194
9. Reeck G, Oppert B, Denton M, Kanost M, Baker J, Kramer K (1999) Insect proteinases. In: Proteases: new perspective. Birkhäuser Basel, pp 125–148
10. LaFlamme BA, Ram KR, Wolfner MF (2012) The Drosophila melanogaster seminal fluid protease "seminase" regulates proteolytic and postmating reproductive processes. PLoS Genet 8(1):e1002435
11. Nagaoka S, Kato K, Takata Y, Kamei K (2012) Identification of the sperm-activating factor initiatorin, a prostatic endopeptidase of the silkworm, Bombyx mori. Insect Biochem Mol Biol 42(8):571–582
12. Ram KR, Sirot LK, Wolfner MF (2006) Predicted seminal astacin-like protease is required for processing of reproductive proteins in Drosophila melanogaster. Proc Natl Acad Sci 103(49):18674–18679
13. Mueller JL, Page JL, Wolfner MF (2007) An ectopic expression screen reveals the protective and toxic effects of Drosophila seminal fluid proteins. Genetics 175(2):777–783
14. Maki N, Yamashita O (2001) The 30kP protease a responsible for 30-kDa yolk protein degradation of the silkworm, Bombyx mori: cDNA structure, developmental change and regulation by feeding. Insect Biochem Mol Biol 31(4):407–413
15. Kageyama T, Takahashi SY (1990) Purification and characterization of a cysteine proteinase from silkworm eggs. Eur J Biochem 193(1):203–210
16. Saikhedkar N, Summanwar A, Joshi R, Giri A (2015) Cathepsins of lepidopteran insects: aspects and prospects. Insect Biochem Mol Biol 64:51–59
17. Morisalo D, Anderson KV (1995) Signaling pathways that establish the dorsal-ventral pattern of the Drosophila embryo. Annu Rev Genet 29(1):371–399
18. LeMosy EK, Hashimoto C (2000) The nudel protease of Drosophila is required for eggshell biogenesis in addition to embryonic patterning. Dev Biol 217(2):352–361
19. Tryselius Y, Hultmark D (1997) Cysteine proteinase 1 (CP1), a cathepsin L-like enzyme expressed in the Drosophila melanogaster haemocyte cell line mbn-2. Insect Mol Biol 6(2):173–181
20. Homma KI, Natori S (1996) Identification of substrate proteins for cathepsin L that are selectively hydrolyzed during the differentiation of imaginal discs of Sarcophaga peregrina. Eur J Biochem 240(2):443–447
21. Homma KI, Matsushita T, Natori S (1996) Purification, characterization, and cDNA cloning of a novel growth factor from the conditioned medium of NIH-Sape-4, an embryonic cell line of Sarcophaga peregrina (flesh fly). J Biol Chem 271(23):13770–13775
22. Wang LF, Chai LQ, He HJ, Wang Q, Wang JX, Zhao XF (2010) A cathepsin L-like proteinase is involved in moulting and metamorphosis in Helicoverpa armigera. Insect Mol Biol 19(1):99–111
23. Facey CO, Lockshin RA (2010) The execution phase of autophagy associated PCD during insect metamorphosis. Apoptosis 15(6):639–652
24. Yang XM, Hou LJ, Dong DJ, Shao HL, Wang JX, Zhao XF (2006) Cathepsin B-like proteinase is involved in the decomposition of the adult fat body of Helicoverpa armigera. Arch Insect Biochem Physiol 62(1):1–10
25. Wu FY, Zou FM, Jia JQ, Wang SP, Zhang GZ, Guo XJ, Gui ZZ (2011) The influence of challenge on Cathepsin B and D expression patterns in the silkworm Bombyx mori L. Int J Ind Entomol 23(1):129–135

26. Gui Z, Lee K, Kim B, Choi Y, Wei Y, Choo Y et al (2006) Functional role of aspartic proteinase cathepsin D in insect metamorphosis. BMC Dev Biol 6(1):1
27. Song KH, Jung MK, Eum JH, Hwang IC, Han SS (2008) Proteomic analysis of parasitized Plutella xylostella larvae plasma. J Insect Physiol 54(8):1271–1280
28. Accorsi A, Zibaee A, Malagoli D (2015) The multifaceted activity of insect caspases. J Insect Physiol 76:17–23
29. Nicholson DW (1999) Caspase structure, proteolytic substrates, and function during apoptotic cell death. Cell Death Differ 6(11):1028–1042
30. Mariño G, Niso-Santano M, Baehrecke EH, Kroemer G (2014) Self-consumption: the interplay of autophagy and apoptosis. Nat Rev Mol Cell Biol 15(2):81–94
31. Hu S, Yang X (2000) dFADD, a novel death domain-containing adapter protein for the Drosophilacaspase DREDD. J Biol Chem 275(40):30761–30764
32. Stöven S, Silverman N, Junell A, Hedengren-Olcott M, Erturk D, Engström Y et al (2003) Caspase-mediated processing of the Drosophila NF-κB factor relish. Proc Natl Acad Sci 100(10):5991–5996
33. Leulier F, Rodriguez A, Khush RS, Abrams JM, Lemaitre B (2000) The Drosophila caspase Dredd is required to resist gram-negative bacterial infection. EMBO Rep 1(4):353–358
34. Dorstyn L, Read S, Cakouros D, Huh JR, Hay BA, Kumar S (2002) The role of cytochrome c in caspase activation in Drosophila melanogaster cells. J Cell Biol 156(6):1089–1098
35. Waldhuber M, Emoto K, Petritsch C (2005) The Drosophila caspase DRONC is required for metamorphosis and cell death in response to irradiation and developmental signals. Mech Dev 122(7):914–927
36. Quinn LM, Dorstyn L, Mills K, Colussi PA, Chen P, Coombe M et al (2000) An essential role for the caspase dronc in developmentally programmed cell death in Drosophila. J Biol Chem 275(51):40416–40424
37. Chew SK, Akdemir F, Chen P, Lu WJ, Mills K, Daish T et al (2004) The apical caspase dronc governs programmed and unprogrammed cell death in Drosophila. Dev Cell 7(6):897–907
38. Muro I, Hay BA, Clem RJ (2002) The Drosophila DIAP1 protein is required to prevent accumulation of a continuously generated, processed form of the apical caspase DRONC. J Biol Chem 277(51):49644–49650
39. Cooper DM, Granville DJ, Lowenberger C (2009) The insect caspases. Apoptosis 14(3):247–256
40. Lee CY, Clough EA, Yellon P, Teslovich TM, Stephan DA, Baehrecke EH (2003) Genome-wide analyses of steroid- and radiation-triggered programmed cell death in Drosophila. Curr Biol 13(4):350–357
41. Leulier F, Ribeiro PS, Palmer E, Tenev T, Takahashi K, Robertson D et al (2006) Systematic in vivo RNAi analysis of putative components of the Drosophila cell death machinery. Cell Death Differ 13(10):1663–1674
42. Baum JS, Arama E, Steller H, McCall K (2007) The Drosophila caspases Strica and Dronc function redundantly in programmed cell death during oogenesis. Cell Death Differ 14(8):1508–1517
43. Song Z, McCall K, Steller H (1997) DCP-1, a Drosophila cell death protease essential for development. Science 275(5299):536–540
44. Xu D, Wang Y, Willecke R, Chen Z, Ding T, Bergmann A (2006) The effector caspases drICE and dcp-1 have partially overlapping functions in the apoptotic pathway in Drosophila. Cell Death Differ 13(10):1697–1706
45. Knorr E, Schmidtberg H, Vilcinskas A, Altincicek B (2009) MMPs regulate both development and immunity in the Tribolium model insect. PLoS One 4(3):e4751
46. Llano E, Pendás AM, Aza-Blanc P, Kornberg TB, López-Otín C (2000) Dm1-MMP, a matrix metalloproteinase from Drosophila with a potential role in extracellular matrix remodeling during neural development. J Biol Chem 275(46):35978–35985
47. Glasheen BM, Robbins RM, Piette C, Beitel GJ, Page-McCaw A (2010) A matrix metalloproteinase mediates airway remodeling in Drosophila. Dev Biol 344(2):772–783

48. Guha A, Lin L, Kornberg TB (2009) Regulation of Drosophila matrix metalloprotease Mmp2 is essential for wing imaginal disc: trachea association and air sac tubulogenesis. Dev Biol 335(2):317–326
49. Yasunaga KI, Kanamori T, Morikawa R, Suzuki E, Emoto K (2010) Dendrite reshaping of adult Drosophila sensory neurons requires matrix metalloproteinase-mediated modification of the basement membranes. Dev Cell 18(4):621–632
50. Llano E, Adam G, Pendás AM, Quesada V, Sánchez LM, Santamaría I, … López-Otín C (2002) Structural and enzymatic characterization of Drosophila Dm2-MMP, a membrane-bound matrix metalloproteinase with tissue-specific expression. J Biol Chem 277(26):23321–23329
51. Page-McCaw A, Ewald AJ, Werb Z (2007) Matrix metalloproteinases and the regulation of tissue remodelling. Nat Rev Mol Cell Biol 8(3):221–233
52. Fambrough D, Pan D, Rubin GM, Goodman CS (1996) The cell surface metalloprotease/disintegrin Kuzbanian is required for axonal extension in Drosophila. Proc Natl Acad Sci 93(23):13233–13238
53. Rooke J, Pan D, Xu T, Rubin GM (1996) KUZ, a conserved metalloprotease-disintegrin protein with two roles in Drosophila neurogenesis. Science 273(5279):1227–1231
54. Lieber T, Kidd S, Young MW (2002) Kuzbanian-mediated cleavage of Drosophila Notch. Genes Dev 16(2):209–221
55. Bland CE, Kimberly P, Rand MD (2003) Notch-induced proteolysis and nuclear localization of the Delta ligand. J Biol Chem 278(16):13607–13610
56. Delwig A, Rand MD (2008) Kuz and TACE can activate Notch independent of ligand. Cell Mol Life Sci 65(14):2232–2243
57. Bland ND, Pinney JW, Thomas JE, Turner AJ, Isaac RE (2008) Bioinformatic analysis of the neprilysin (M13) family of peptidases reveals complex evolutionary and functional relationships. BMC Evol Biol 8(1):1
58. Buchon N, Poidevin M, Kwon HM, Guillou A, Sottas V, Lee BL, Lemaitre B (2009) A single modular serine protease integrates signals from pattern-recognition receptors upstream of the Drosophila toll pathway. Proc Natl Acad Sci 106(30):12442–12447
59. Jiang H, Wang Y, Kanost MR (1998) Pro-phenol oxidase activating proteinase from an insect, Manduca sexta: a bacteria-inducible protein similar to Drosophila easter. Proc Natl Acad Sci 95(21):12220–12225
60. Jiang H, Wang Y, Yu XQ, Zhu Y, Kanost M (2003) Prophenoloxidase-activating proteinase-3 (PAP-3) from Manduca sexta hemolymph: a clip-domain serine proteinase regulated by serpin-1J and serine proteinase homologs. Insect Biochem Mol Biol 33(10):1049–1060
61. Satoh D, Horii A, Ochiai M, Ashida M (1999) Prophenoloxidase-activating enzyme of the silkworm, Bombyx mori. Purification, characterization, and cDNA cloning. J Biol Chem 274(11):7441–7453
62. Lee SY, Kwon TH, Hyun JH, Choi JS, Kawabata SI, Iwanaga S, Lee BL (1998) In vitro activation of pro-phenol-oxidase by two kinds of prophenol-oxidase-activating factors isolated from hemolymph of coleopteran, Holotrichia diomphalia larvae. Eur J Biochem 254(1):50–57
63. Danielli A, Loukeris TG, Lagueux M, Müller HM, Richman A, Kafatos FC (2000) A modular chitin-binding protease associated with hemocytes and hemolymph in the mosquito Anopheles gambiae. Proc Natl Acad Sci 97(13):7136–7141
64. Christeller JT, Laing WA, Markwick NP, Burgess EPJ (1992) Midgut protease activities in 12 phytophagous lepidopteran larvae: dietary and protease inhibitor interactions. Insect Biochem Mol Biol 22(7):735–746
65. Barrett AJ, Rawlings ND, Woessner JF (1998) Handbook of proteolytic enzymes. Academic Press, San Diego
66. Terra WR, Ferreira C (1994) Insect digestive enzymes: properties, compartmentalization and function. Comp Biochem Physiol B Comp Biochem 109(1):1–62
67. Garcia ES, Guimarães JA (1979) Proteolytic enzymes in the Rhodnius prolixus midgut. Experientia 35(3):305–306
68. Jordão BP, Terra WR (1989) Distribution, properties, and functions of midgut carboxypeptidases and dipeptidases from Musca domestica larvae. Arch Insect Biochem Physiol 11(4):231–244

69. Ferreira C, Capella AN, Sitnik R, Terra WR (1994) Properties of the digestive enzymes and the permeability of the peritrophic membrane of Spodoptera frugiperda (Lepidoptera) larvae. Comp Biochem Physiol A Physiol 107(4):631–640
70. Swanson WJ, Clark AG, Waldrip-Dail HM, Wolfner MF, Aquadro CF (2001) Evolutionary EST analysis identifies rapidly evolving male reproductive proteins in Drosophila. Proc Natl Acad Sci 98(13):7375–7379
71. Swanson WJ, Wong A, Wolfner MF, Aquadro CF (2004) Evolutionary expressed sequence tag analysis of Drosophila female reproductive tracts identifies genes subjected to positive selection. Genetics 168(3):1457–1465
72. Braswell WE, Andrés JA, Maroja LS, Harrison RG, Howard DJ, Swanson WJ (2006) Identification and comparative analysis of accessory gland proteins in Orthoptera. Genome 49(9):1069–1080
73. Sirot LK, Poulson RL, McKenna MC, Girnary H, Wolfner MF, Harrington LC (2008) Identity and transfer of male reproductive gland proteins of the dengue vector mosquito, Aedes aegypti: potential tools for control of female feeding and reproduction. Insect Biochem Mol Biol 38(2):176–189
74. Panhuis TM, Swanson WJ (2006) Molecular evolution and population genetic analysis of candidate female reproductive genes in Drosophila. Genetics 173(4):2039–2047
75. Haerty W, Jagadeeshan S, Kulathinal RJ, Wong A, Ram KR, Sirot LK et al (2007) Evolution in the fast lane: rapidly evolving sex-related genes in Drosophila. Genetics 177(3):1321–1335
76. Lawniczak MK, Begun DJ (2007) Molecular population genetics of female-expressed mating-induced serine proteases in Drosophila melanogaster. Mol Biol Evol 24(9):1944–1951
77. Findlay GD, Yi X, MacCoss MJ, Swanson WJ (2008) Proteomics reveals novel Drosophila seminal fluid proteins transferred at mating. PLoS Biol 6(7):1417–1426
78. Prokupek A, Hoffmann F, Eyun SI, Moriyama E, Zhou M, Harshman L (2008) An evolutionary expressed sequence tag analysis of Drosophila spermatheca genes. Evolution 62(11):2936–2947
79. Mueller JL, Ripoll DR, Aquadro CF, Wolfner MF (2004) Comparative structural modeling and inference of conserved protein classes in Drosophila seminal fluid. Proc Natl Acad Sci U S A 101(37):13542–13547
80. Kelleher ES, Watts TD, LaFlamme BA, Haynes PA, Markow TA (2009) Proteomic analysis of Drosophila mojavensis male accessory glands suggests novel classes of seminal fluid proteins. Insect Biochem Mol Biol 39(5):366–371
81. LaFlamme BA, Wolfner MF (2013) Identification and function of proteolysis regulators in seminal fluid. Mol Reprod Dev 80(2):80–101
82. Dottorini T, Nicolaides L, Ranson H, Rogers DW, Crisanti A, Catteruccia F (2007) A genome-wide analysis in Anopheles gambiae mosquitoes reveals 46 male accessory gland genes, possible modulators of female behavior. Proc Natl Acad Sci 104(41):16215–16220
83. Davies SJ, Chapman T (2006) Identification of genes expressed in the accessory glands of male Mediterranean fruit flies (Ceratitis capitata). Insect Biochem Mol Biol 36(11):846–856
84. Walters JR, Harrison RG (2008) EST analysis of male accessory glands from Heliconius butterflies with divergent mating systems. BMC Genomics 9(1):592
85. Walters JR, Harrison RG (2010) Combined EST and proteomic analysis identifies rapidly evolving seminal fluid proteins in Heliconius butterflies. Mol Biol Evol 27(9):2000–2013
86. South A, Sirot LK, Lewis SM (2011) Identification of predicted seminal fluid proteins in Tribolium castaneum. Insect Mol Biol 20(4):447–456
87. Andrés JA, Maroja LS, Bogdanowicz SM, Swanson WJ, Harrison RG (2006) Molecular evolution of seminal proteins in field crickets. Mol Biol Evol 23(8):1574–1584
88. Weiss BL, Stepczynski JM, Wong P, Kaufman WR (2002) Identification and characterization of genes differentially expressed in the testis/vas deferens of the fed male tick, Amblyomma hebraeum. Insect Biochem Mol Biol 32(7):785–793
89. Ram KR, Wolfner MF (2007) Seminal influences: Drosophila Acps and the molecular interplay between males and females during reproduction. Integr Comp Biol 47(3):427–445
90. Ram KR, Wolfner MF (2009) A network of interactions among seminal proteins underlies the long-term postmating response in Drosophila. Proc Natl Acad Sci 106(36):15384–15389

91. Chapman T, Bangham J, Vinti G, Seifried B, Lung O, Wolfner MF et al (2003) The sex peptide of Drosophila melanogaster: female post-mating responses analyzed by using RNA interference. Proc Natl Acad Sci 100(17):9923–9928
92. Friedländer M, Jeshtadi A, Reynolds SE (2001) The structural mechanism of trypsin-induced intrinsic motility in Manduca sexta spermatozoa in vitro. J Insect Physiol 47(3):245–255
93. Raikhel AS, Dhadialla TS (1992) Accumulation of yolk proteins in insect oocytes. Annu Rev Entomol 37(1):217–251
94. Indrasith LS, Sasaki T, Yamashita O (1988) A unique protease responsible for selective degradation of a yolk protein in Bombyx mori. Purification, characterization, and cleavage profile. J Biol Chem 263(2):1045–1051
95. Cho WL, Deitsch KW, Raikhel AS (1991) An extraovarian protein accumulated in mosquito oocytes is a carboxypeptidase activated in embryos. Proc Natl Acad Sci 88(23):10821–10824
96. Medina M, Vallejo CG (1989) A serine proteinase in Drosophila embryos: yolk localization and developmental activation. Insect Biochem 19(7):687–691
97. Ribolla PE, De Bianchi AG (1995) Processing of procathepsin from Musca domestica eggs. Insect Biochem Mol Biol 25(9):1011–1017
98. LeMosy EK, Hong CC, Hashimoto C (1999) Signal transduction by a protease cascade. Trends Cell Biol 9(3):102–107
99. Moussian B, Roth S (2005) Dorsoventral axis formation in the Drosophila embryo—shaping and transducing a morphogen gradient. Curr Biol 15(21):R887–R899
100. Smith CL, Delotto R (1992) A common domain within the proenzyme regions of the Drosophila snake and easter proteins and Tachypleus proclotting enzyme defines a new subfamily of serine proteases. Protein Sci 1(9):1225–1226
101. Turcotte CL, Hashimoto C (2002) Evidence for a glycosaminoglycan on the nudel protein important for dorsoventral patterning of the Drosophila embryo. Dev Dyn 224(1):51–57
102. Hong CC, Hashimoto C (1995) An unusual mosaic protein with a protease domain, encoded by the nudel gene, is involved in defining embryonic dorsoventral polarity in Drosophila. Cell 82(5):785–794
103. LeMosy EK, Tan YQ, Hashimoto C (2001) Activation of a protease cascade involved in patterning the Drosophila embryo. Proc Natl Acad Sci 98(9):5055–5060
104. Dissing M, Giordano H, DeLotto R (2001) Autoproteolysis and feedback in a protease cascade directing Drosophila dorsal–ventral cell fate. EMBO J 20(10):2387–2393
105. Hegedus D, O'grady M, Chamankhah M, Baldwin D, Gleddie S, Braun L, Erlandson M (2002) Changes in cysteine protease activity and localization during midgut metamorphosis in the crucifer root maggot (Delia radicum). Insect Biochem Mol Biol 32(11):1585–1596
106. Serbielle C, Moreau S, Veillard F, Voldoire E, Bézier A, Mannucci MA et al (2009) Identification of parasite-responsive cysteine proteases in Manduca sexta. Biol Chem 390(5/6):493–502
107. Byeon JH, Seo ES, Lee JB, Lee MJ, Kim JK, Yoo JW et al (2015) A specific cathepsin-L-like protease purified from an insect midgut shows antibacterial activity against gut symbiotic bacteria. Dev Comp Immunol 53(1):79–84
108. Kim BY, Lee KS, Sohn MR, Kim KY, Choi KH, Kang PD, Jin BR (2011) Bombyx mori cathepsin D expression is induced by high temperature and H 2 O 2 exposure. J Asia Pac Entomol 14(3):285–288
109. Ge ZY, Wan PJ, Li GQ, Xia YG, Han ZJ (2014) Characterization of cysteine protease-like genes in the striped rice stem borer, Chilo suppressalis. Genome 57(2):79–88
110. Harvey NL, Daish T, Mills K, Dorstyn L, Quinn LM, Read SH et al (2001) Characterization of the DrosophilaCaspase, DAMM. J Biol Chem 276(27):25342–25350
111. Courtiade J, Pauchet Y, Vogel H, Heckel DG (2011) A comprehensive characterization of the caspase gene family in insects from the order Lepidoptera. BMC Genomics 12(1):357
112. Bryant B, Blair CD, Olson KE, Clem RJ (2008) Annotation and expression profiling of apoptosis-related genes in the yellow fever mosquito, Aedes aegypti. Insect Biochem Mol Biol 38(3):331–345

113. Page-McCaw A, Serano J, Santé JM, Rubin GM (2003) Drosophila matrix metalloproteinases are required for tissue remodeling, but not embryonic development. Dev Cell 4(1):95–106
114. Sotillos S, Roch F, Campuzano S (1997) The metalloprotease-disintegrin Kuzbanian participates in Notch activation during growth and patterning of Drosophila imaginal discs. Development 124(23):4769–4779
115. Jang IH, Chosa N, Kim SH, Nam HJ, Lemaitre B, Ochiai M et al (2006) A Spätzle-processing enzyme required for toll signaling activation in Drosophila innate immunity. Dev Cell 10(1):45–55
116. Jiang H, Wang Y, Gu Y, Guo X, Zou Z, Scholz F et al (2005) Molecular identification of a bevy of serine proteinases in Manduca sexta hemolymph. Insect Biochem Mol Biol 35(8):931–943
117. Ward CW (1975) Properties and specificity of the major anionic trypsin-like enzyme in the keratinolytic larvae of the webbing clothes moth. Biochimica et Biophysica Acta (BBA)-Enzymology 391(1):201–211
118. Lam W, Coast GM, Rayne RC (2000) Characterisation of multiple trypsins from the midgut of Locusta migratoria. Insect Biochem Mol Biol 30(1):85–94
119. Lopes AR, Juliano MA, Marana SR, Juliano L, Terra WR (2006) Substrate specificity of insect trypsins and the role of their subsites in catalysis. Insect Biochem Mol Biol 36(2):130–140
120. Gilbert LI (2012) Insect molecular biology and biochemistry. Academic Press
121. Christeller JT, Laing WA, Shaw BD, Burgess EPJ (1990) Characterization and partial purification of the digestive proteases of the black field cricket, Teleogryllus commodus (Walker): elastase is a major component. Insect Biochemistry 20(2):157–164
122. Valaitis AP (1995) Gypsy moth midgut proteinases: purification and characterization of luminal trypsin, elastase and the brush border membrane leucine aminopeptidase. Insect Biochem Mol Biol 25(1):139–149
123. Whitworth ST, Kordula, T.,&Travis J. (1999) Molecular cloning of Soli E2: an esterase-like serine proteinase from the imported fire ant (Solenopsis invicta). Insect Biochem Mol Biol 29:249–254
124. Zeng F, Cohen AC (2001) Induction of elastase in a zoophytophagous heteropteran, Lygus hesperus (Hemiptera: Miridae). Ann Entomol Soc Am 94(1):146–151
125. Padilha MH, Pimentel AC, Ribeiro AF, Terra WR (2009) Sequence and function of lysosomal and digestive cathepsin D-like proteinases of Musca domestica midgut. Insect Biochem Mol Biol 39(11):782–791
126. Terra WR, Ferreira C, De Bianchi AG (1979) Distribution of digestive enzymes among the endo-and ectoperitrophic spaces and midgut cells of Rhynchosciara and its physiological significance. J Insect Physiol 25(6):487–494
127. Ferreira C, Terra WR (1985) Minor aminopeptidases purified from the plasma membrane of midgut caeca cells of an insect (Rhynchosciara americana) larva. Insect Biochem 15(5):619–625
128. Vonk HJ, Western JRH (1984) Comparative biochemistry and physiology of enzymatic digestion. Academic Press
129. Houseman JG, Campbell FC, Morrison PE (1987) A preliminary characterization of digestive proteases in the posterior midgut of the stable fly Stomoxys calcitrans (L.)(Diptera: Muscidae). Insect Biochem 17(1):213–218
130. Gooding RH, Rolseth BM (1976) Digestive processes of haematophagous insects. XI. Partial purification and some properties of six proteolytic enzymes from the tsetse fly Glossina morsitans morsitans Westwood (Diptera: Glossinidae). Can J Zool 54(11):1950–1959
131. Houseman JG, Downe AER (1981) Exoproteinase activity in the posterior midgut of Rhodnius prolixus Stål (Hemiptera: Reduviidae). Insect Biochem 11(5):579–582
132. Klinkowstrom AM, Terra WR, Ferreira C (1995) Midgut dipeptidases from 36 Rhynchosciara americana (diptera) larvae. Properties of soluble and membrane- bound forms. Insect Biochem Mol Biol 25(3):303–310

Protease-Antiprotease Interactions: An Overview of the Process from an "In Silico" Perspective

22

Angshuman Bagchi

Abstract

Most if not all of the cellular processes involve protein-protein interactions (PPIs). The detailed information of the amino acid residues involved in PPIs may, therefore, be used in many important aspects like drug development, elucidation of molecular pathways, generation of protein mimetic, understanding of disease mechanisms, and development of docking methodologies to build structural models of protein complexes. Among the different physiological PPIs, protease-antiprotease interactions play a significant role. An imbalance between proteases and antiproteases is involved in many pathogenic reactions. This special class of PPI, therefore, needs a thorough scrutiny. There are different PPIs determining experimental tools. However, these tools are time-consuming and expensive. In response to these difficulties, a number of bioinformatic software tool have been developed. The algorithms are meant for prediction of three-dimensional structures of proteins as well as protein complexes. The structure prediction methods involve homology modeling, threading, and ab initio modeling. These methods have nearly 75%–80% overall accuracies. The other method is molecular docking which is meant to generate the three-dimensional conformations of protein complexes. The docking methods can broadly be classified as rigid body docking and flexible docking. In this chapter, the different aspects of experimental and computational modeling and docking strategies will be discussed. The basic terminologies will be revisited. This chapter is aimed at providing a firsthand knowledge on protein interaction methods using protease-antiprotease interactions as an example.

A. Bagchi (✉)
Department of Biochemistry & Biophysics, University of Kalyani,
Kalyani 741235, West Bengal, India
e-mail: angshu@klyuniv.ac.in; angshuman_bagchi@yahoo.com

S. Chakraborti, N.S. Dhalla (eds.), *Proteases in Physiology and Pathology*,
DOI 10.1007/978-981-10-2513-6_22

Keywords
Protease • Antiprotease • PPI • Bioinformatics

22.1 Introduction

In the biological systems, almost all the biochemical reactions are the outcomes of different forms of protein-protein interactions (PPIs). Proteins not only react among themselves but also with other biomolecules like nucleic acids, organic or inorganic cofactors, etc. [1–10]. However, one of such PPIs having a huge physiological significance is protease-antiprotease interactions. Such interactions are at the heart of many different diseases like chronic neutrophilic lung disease [11–20]. It has therefore become essential to physiologists to have a good understanding of PPIs. Thus, various experimental and computational approaches have constantly being invented by different scientific groups to predict the PPIs. The PPI identification technologies have various degrees of accuracies. The experimental tools to study PPIs include X-ray crystallography, nuclear magnetic resonance imaging, electron microscopy, microarray analysis, co-immunoprecipitation techniques, etc. These techniques are fairly accurate and produce good results. But the major drawbacks of these techniques are cost and time. The failure rates are quite high too for these techniques [20–30]. In order to have an alternative strategy to identify PPIs, different computational algorithms have constantly been being proposed. These computational techniques have varying degrees and accuracies. These computational tools though not as accurate as the experimental tools come up with fairly good predictive models of PPIs. The computational approaches may, therefore, be considered to be the start point of PPI prediction methodologies [31–40]. The present chapter is aimed to give a firsthand knowledge of different computational PPI prediction methodologies taking protease-antiprotease interactions as a reference. However, before going into the technical details of the PPI prediction methods, the basic definitions need to be revisited.

22.2 Basic Definitions

22.2.1 Protease

A protease is an enzyme which hydrolyzes the peptide linkages in proteins. There are different classes of proteases present in our cellular systems. However, the basic common functionality of a protease is splitting the peptide linkages in proteins; the mechanisms of actions of the proteases vary between the different classes of proteases. The proteases are classified into the following seven broad categories depending upon the active site of amino acid residues present in them [41–43]:

- Serine proteases – serine proteases have a functional serine residue and cleave a peptide linkage with the side chain of the serine.
- Cysteine proteases – cysteine proteases have a functional cysteine residue and cleave a peptide linkage with the side chain of the histidine.
- Threonine proteases – threonine proteases have a functional threonine residue and cleave a peptide linkage with the side chain of the threonine.
- Aspartic proteases – aspartic proteases have a functional aspartate residue and cleave a peptide linkage with the side chain of the aspartate.
- Glutamic proteases – glutamic proteases have a functional glutamate residue and cleave a peptide linkage with the side chain of the glutamate
- Metalloproteases – generally these types of proteases function using a metal ion-like zinc.
- Asparagine peptide lyases – such proteases use an asparagine to perform and do not require water for the hydrolysis of the peptide linkages [44, 45].

22.2.2 Antiprotease

These are inhibitors of proteases. The protease inhibitors can be another protein or any other molecule that can inhibit the function of a protease. In physiology, a protease inhibitor is often used interchangeably with the enzyme alpha-1 antitrypsin (A1AT). This protease inhibitor is often found to be associated with the disease alpha-1 antitrypsin deficiency [41–45].

22.2.3 PPI Interface

PPI interface refers to the area between the two protein chains. If the two protein chains have the same amino acid compositions, the interface is called homomeric interface otherwise heteromeric interface. The PPI interface has the following characteristics:

- Surface area of interface: For heterodimer, the surface area is generally around 600Å^2. For a homodimer, it is even larger.
- Shape of the PPI interface: In general, the PPI interface is nearly flat. A PPI interface has two separate zones, viz., the core which is buried in the interface and the rim which is solvent accessible.
- Composition of amino acids at the PPI interface: In general, it is observed that PPI interface has an abundance of aromatic amino acid residues and Arg. However, Cys is not generally found at the PPI interfaces.
- Secondary structural distribution at the PPI interface: Generally a PPI interface is made up of beta sheet regions [46–55].

22.2.4 Classification of PPI Interface

The PPIs can broadly be classified into several different classes based on the nature of the interacting partners, stability of the PPI complexes, the life-span of the interactions between the protein partners, and the nature of the PPI interface between the proteins.

- Nature of interacting protein partners: If the interacting protein partners have the same amino acid compositions, they form homo-oligomers, with structural symmetry. On the other hand, nonidentical protein partners form hetero-oligomers. Hemoglobin is a homo-tetramer and a protease-antiprotease complex is a heteromer [46–50].
- Stability of interacting protein complexes: If the individual protein partners forming the PPI complex cannot exist in free state and are stable only in multimeric association, they are called obligate oligomers (homo-obligomers and/or hetero-obligomers), like the Arc repressor dimer where dimerization is essential for DNA binding. On the other hand, when the protein partners can exist in free states on their own, they are called non-obligate partners like antigen-antibody complex [46–52].
- Lifetime of PPI: When an association between the protein partners is highly stable and needs external agencies to break them, they are called permanent complexes. Hetero-trimeric G protein (Gα, Gβγ, and GDP) forms this type of PPI. In contrast, the interacting partners of sperm lysin, a homodimer, exist in a dynamic equilibrium consisting of association and dissociation of oligomeric forms. This type of PPI is named as transient complex [46–55].
- Nature of the interaction interface: When the individual protein partners in a PPI use the same interacting interface to join each other, they are called isologous complexes. On the other hand, in heterologous assembly, the individual protein partners in a PPI complex use different interfaces to form PPI without any closed symmetry [46–55].

22.3 Mechanism of PPI

In order to interact, two protein molecules must be in close proximity. There are mainly non-covalent forces responsible for the PPIs. However, the only covalent binding interaction is disulfide linkage. The most important non-covalent interaction found in PPIs is the hydrogen bonding between the polar atoms in proteins. The hydrogen bonding involves both the main and side chain atoms of the amino acids in proteins. The second type of non-covalent interaction leading to the formation of PPIs is the ion pair formation. This occurs mainly between the side chains of an acidic amino acid with that of the basic amino acid. There are other interactions like stacking interactions, between the nonpolar hydrophobic side chains of the amino acids in proteins, and cation-pi interactions, between the aromatic side chains of Phe, Tyr, and Trp with the positively charged side chains of Lys and Arg. However, the binding interactions are specific for a protein complex. The biomolecular mechanisms of protease-antiprotease interactions follow some specific rules. The protease inhibitors are produced by different organisms to counteract the exogenous proteases. As mentioned earlier, there are several classes of protease inhibitors. Among the different protease inhibitors, the serine proteases are the best studied ones. They are mostly 379–390 amino acid residue long proteins. These proteases generally bind adjacent to the S-H-D complex present in the active site regions of the proteases [46–56].

However, there is another class of protease inhibitor called the high molecular weight macroglobulins. The characteristic of such inhibitors is that these inhibitors can exhibit their functionalities irrespective of the catalytic mechanisms of the proteases. These protease inhibitors exhibit their effects by forming covalent bonds with the proteases. This covalent interaction would lead to some kind of conformational changes in the inhibitors, and the inhibitors can entrap the proteases. This type of binding interactions between the protease inhibitors with proteases would leave active site of the protease free to interact with the low molecular weight substrates but not with high molecular weight substrates [46–56].

22.4 Detection of PPI

It is an already established fact that more than 80% of proteins function in association rather in isolation. Therefore, studying and understanding of PPI are becoming more and more relevant for fields like system biology, molecular medicine, etc. The study of PPI is required to understand the molecular mechanisms behind mutations which lead to disease onset. PPI detection methodologies can broadly be classified as experimental and computational. As mentioned earlier in this chapter, experimental tools give authentic results, but they are often very time-consuming and expensive. The failure rates are also very high for experimental PPI prediction methods. On the other hand, in order to have a firsthand knowledge about PPIs for which experimental findings are not yet available, computational tools are developed. Thus, it has become a common practice to perform computational analyses of PPIs before going into the experiments. In this chapter, the various aspects of PPI predictions using computational tools are discussed [57–71].

The various computational algorithms for prediction of PPI can broadly be classified mainly into the following categories:

- Numerical value-based methods
- Probabilistic methods
- Interference of interactions from homologous structures
- Association methods

22.4.1 Numerical Value-Based Methods

These methods are dependent on a function derived from the amino acid residues of the protein chain under investigation. The different kinds of information collected involve the details of secondary structure, sequence conservation, solvent accessibility, physicochemical characteristics of the amino acids, etc. These different sets of information are used to build a training model which has information about the PPI and non-PPI amino acid residues. The training model may then be used to detect the nature of a user-defined input. The possible sources of the information are the Protein Data Bank (PDB) and literature mining. This method is used in support vector machines (SVMs), random forests, etc. The method is heavily dependent on the accuracy of the training dataset. And depending on the nature of the training dataset, the method can produce a result with around 80% accuracy. However, the most difficult problem is to find a suitable negative dataset [57–71].

22.4.2 Probabilistic Methods

Such methods are based on conditional probability of whether a particular amino acid residue is at protein-protein interaction interface or not with the probability of getting a particular value of a parameter of that residue. The method relies on the hypothesis that potentially interacting protein pairs should coevolve. They should have orthologs in closely related species. In other words, proteins that form complexes or are part of a biochemical pathway should be present simultaneously together in order to exhibit their functions. A phylogenetic profile is then constructed for each protein under consideration. A phylogenetic profile is nothing but a record of whether the protein is present in certain genomes. If two proteins are found to be present or absent in the same genomes, those proteins are deemed likely to be functionally related and elicit their functions by protein-protein interactions [57–71].

22.4.3 Interference of Interactions from Homologous Structures

The method uses the information of known protein complexes and then builds a model from such information. The method uses sequence conservation data from the protein complexes to first build the model. Then structural information like secondary structures, solvent accessibility, etc. are incorporated into the model. This method is by far the most accurate method. However, the severe drawback of the method is the presence of less number of protein complexes [57–71].

22.4.4 Association Method

This method is somewhat similar to the previous method that the method uses information of characteristic sequence motifs in proteins. Each motif has a special affinity toward binding another specific motif. The method uses this information to search for interacting protein partners. The method uses a log odd score which is calculated as $\log_2(P_{ij}/P_iP_j)$, where P_{ij} is the observed frequency of motifs i and j occurring together in one protein pair and P_i and P_j are the background frequencies of motifs i and j in the data. Those interactions are considered positive for which the log odd score is positive [57–71].

22.5 Conclusion and Future Direction

Prediction of PPIs is a very daunting task. As previously mentioned, PPIs remain at the heart of most if not all of the biochemical reactions. The most important among them is the protease-antiprotease interactions. The abundance of protein sequence information instigated the scientists to come up with protein interaction prediction methodologies that use the protein sequence information only. It is also a well-established fact that sequence is more conserved than structure. So, similar sequences mean similar structures. However, this assumption fails below a sequence similarity level of 30%. So, the sequence-based PPI prediction methods have very low accuracy levels. On the other hand, methods based on protein structures are fairly accurate, but the drawback is there are very less number of good protein-protein complex structures that are available. Therefore, it is very much essential to come up with a good sequence-based predictor. Nonetheless, the bioinformatic tools may come up with a firsthand knowledge of PPIs for which experimentation is not yet possible.

Acknowledgment The author would like to acknowledge the help rendered by the DBT-sponsored Bioinformatics Infrastructural Facility of the University of Kalyani. The author would also like to thank the Department of Biotechnology (DBT, India) for the financial support (SAN No. 102/IFD/SAN/1824/2015-2016). The author is grateful to the Virologie et Immunologie Moléculaires, INRA. UR892, Domaine de Vilvert 78352 Jouy-en-Josas, France, for the infrastructural support. The infrastructural help from the Department of Biochemistry and Biophysics, University of Kalyani, is duly acknowledged.

References

1. Lesk AM (2010) Introduction to protein science: artchitechture, function, and Genetics, 2nd Edition, Pg. no.: 17–38. Oxford University Press, New York
2. Branden C, Tooze A., (1998) Introduction to protein structure, 2nd edn. Garland Publishing Inc., New York, pp 373–392
3. Kessel A, Ben-Tal N (2010) Introduction to proteins: structure, function, and motion, 1st edn. Chapman & Hall CRC, Florida, pp 36–65
4. Whiteford D (2005) Proteins: structure and function, 1st edn. Wiley, Chichester, pp 189–244

5. Kurian J, Conforti B, Wemmer D (2012) The molecules of life: physical and chemical principles, 1st edn. Garland Science, New York, pp 530–787
6. Nelson DL, Cox MM (2012) Principles of biochemistry, 5th edn. W.H. Freemann & Company, New York, pp 157–237
7. Walsh G (2002) Proteins: biotechnology and biochemistry, 1st edn. Wiley, Chichester, pp 251–278
8. Creighton TE (1992) Proteins: structures and molecular properties, 2nd edn. W.H. Freemann & Company, New York
9. Park JS, Cochran JR (2009) Protein engineering and design, 1st edn. CRC Press, Florida, 131–150
10. Tropp BE (2011) Molecular biology: genes to proteins, 4th edn. Jones & Bartlett Learning, UK, pp 27–75
11. Greene CM, McEvanely NG (2009) Proteases and antiproteases in chronic neutrophilic lung disease – relevance to drug discovery. Br J Pharmacol 158:1048–1058
12. Meyer M, Jaspers I (2015) Respiratory protease/antiprotease balance determines susceptibility to viral infection and can be modified by nutritional antioxidants. Am J Physiol Lung Cell Mol Physiol 308:L1189–L2010
13. Hutchison DC (1987) The rôle of proteases and antiproteases in bronchial secretions. Eur J Respir Dis Suppl 153:78–85
14. Testa V, Capasso G, Maffulli N et al (1994) Proteases and antiproteases in cartilage homeostasis. A brief review Clin Orthop Relat Res (308):79-84
15. Twigg MS, Brockbank S, Lowry P et al (2015) The role of serine proteases and Antiproteases in the cystic fibrosis lung. Mediat Inflamm 293053
16. Sandholm L (1986) Proteases and their inhibitors in chronic inflammatory periodontal disease. J Clin Periodontol 13:19–26
17. Hubbard RC, Crystal RG (1986) Antiproteases and antioxidants: strategies for the pharmacologic prevention of lung destruction. Respiration 50(Suppl 1):56–73
18. Bourin M, Gautron J, Berges M et al (2012) Transcriptomic profiling of proteases and antiproteases in the liver of sexually mature hens in relation to vitellogenesis. BMC Genomics 13:457
19. Kuhn, C 3rd (1986) The biochemical pathogenesis of chronic obstructive pulmonary diseases: protease-antiprotease imbalance in emphysema and diseases of the airways. J Thorac Imaging 1:1–6
20. Erickson S (1978) Proteases and protease inhibitors in chronic obstructive lung disease. Acta Med Scand 203:449–455
21. Golemis E (2002) Protein-protein interactions: a molecular cloning manual, 2nd edn. Cold Spring Harbor Laboratory Press, Cold Spring Harbor/New York, pp 1–50
22. Kerppola, T.K., (2008) Bimolecular fluorescence complementation: visualization of Molecular Interactions in Living Cells. Methods in Cell Biol 9:789–798
23. Bollag DM, Rozycki MD, Edelstein SJ (1996) Protein methods, 2nd edn. Wiley Publishers, New York, pp 1–83
24. Ausubel FM (1987) Current protocols in molecular biology. Wiley, New York/Boston, pp 15.1.1–15.1.14
25. Piehler J (2005) New methodologies for measuring protein interactions in vivo and in vitro. Current opinions in. Struct Biol 15:4–14
26. Puig O, Caspari F, Riquat G et al (2001) The tandem affinity purification (TAP) method: a general procedure of protein complex purification. Methods 24:218–229
27. Rao VS, Srinivas K, Sujini GN et al (2014) Protein-protein interaction detection: methods and analysis. 147648
28. Braun, P. & Gingras., A.C. (2012) History of protein-protein interactions: from egg-white to complex networks. Proteomics 12: 1478–1498
29. Phizicky EM, Fields S (1995) Protein-protein interactions: methods for detection and analysis. Microbiol Rev 59:94–123
30. Rigaut G, Shevchenko A, Rutz B et al (1999) A generic protein purification method for protein complex characterization and proteome exploration. Nat Biotechnol 17:1030–1032

31. Shoemaker BA, Panchenko AR (2007) Deciphering protein-protein interactions.Part II. Computational methods to predict protein and domain interaction partners. PLoS Comput Biol 3:e43
32. Theofilatos K, Dimitrakopoulos C, Tsakalidis A et al (2011) Computational approaches for the prediction of protein-protein interactions-a survey. Curr Bioinforma 6:398–414
33. Tuncbag N, Kar G, Keskin O et al (2009) A survey of available tools and web servers for analysis of protein-protein interactions and interfaces. Brief Bioinform 10:217–232
34. Xiaoli L., Wu M, Chee-Kewong K et al (2010) Computational approaches for detecting protein complexes from protein interaction networks: a survey. BMC Genomics;11.
35. Skrabanek L, Saini HK, Bader GD et al (2008) Computational prediction of protein-protein interactions. Mol Biotechnol 38:1–17
36. Bader G et al (2003) BIND the biomolecular interaction network database. Nucleic Acids Res 31:248–250
37. Chatr-aryamontri A et al (2007) MINT the molecular INTeraction database. Nucleic Acids Res 35(Database):D572–D574
38. Peri S et al (2004) Human protein reference database as a discovery resource for proteomics. Nucleic Acids Res 32(Database issue):D497–D501
39. Hermjakob L et al (2004) IntAct an open source molecular interaction database. Nucleic Acids Res 32:D452–D455
40. Breitkreutz BJ et al (2008) The BioGRID interaction database: 2008 update. Nucleic Acids Res 36(Database issue):D637–D640
41. Puente XS, López-Otín C (2004) A genomic analysis of rat proteases and protease inhibitors. Genome Res 14:609–622
42. Sims GK, Wander MM (2002) Proteolytic activity under nitrogen or sulfur limitation. Appl Soil Ecol 568:1–5
43. van der Hoorn RA (2008) Plant proteases: from phenotypes to molecular mechanisms. Annu Rev Plant Biol 59:191–223
44. Woessner, edited by Barrett AJ, Rawlings ND, Fred J (2004) Handbook of proteolytic enzymes, 3rd edn . London: Academic, Elsevier, pp 1–16
45. Oda K (2012) New families of carboxyl peptidases: serine-carboxyl peptidases and glutamic peptidases. J Biochem 151:13–25
46. Ofran Y, Rost B (2003a) Analysing six types of protein-protein interfaces. J Mol Biol 325:377–387
47. Bahadur RP, Chakrabarti P, Rodier F, Janin J (2004) A dissection of specific and non-specific protein-protein interfaces. J Mol Biol 336:943–955
48. Bogan AA, Thorn KS (1998) Anatomy of hot spots in protein interfaces. J Mol Biol 280:1–9
49. Keskin O, Ma B, Nussinov R (2005) Hot regions in protein-protein interactions: the organization and contribution of structurally conserved hot spot residues. J Mol Biol 345:1281–1294
50. Ofran Y, Rost B (2003b) Predicted protein-protein interaction sites from local sequence information. FEBS Lett 544:236–239
51. Sheinerman FB, Norel R, Honig B (2000) Electrostatic aspects of protein – protein interactions. Curr Opin Struct Biol:153–159
52. Schreiber G. (2002) Kinetic studies of protein – protein interactions.Curr. Opin. Struc. Bio.:41–7
53. Shenoy SR, Jayaram B (2010) Proteins: sequence to structure and function - Current status. Curr Protein Pept Sci 11:498–514
54. Nooren IM, Thornton JM (2003) Structural characterization and functional significance of transient protein-protein interactions. J Mol Biol 325:991–1018
55. Nooren IM (2003) New EMBO member's review : diversity of protein-protein interactions. EMBO J 22:3486–3492
56. Faisal M, Oliver JL, Kaattari SL (1999) Potential role of protease-anti-protease interactions in *Perkinsus Marinus* infection in *Crassostrea sp*. Bull Eur Ass Fish Pathol 19:269–276
57. Bradford JR, Needham CJ, Bulpitt AJ, Westhead DR (2006) Insights into protein-protein interfaces using a Bayesian network prediction method. J Mol Biol 362:365–386

58. Choong YS, Tye GJ, Lim TS (2013) Minireview: applied structural bioinformatics in proteomics. Protein J 32:505–511
59. Gallet X, Charloteaux B, Thomas a BR (2000) A fast method to predict protein interaction sites from sequences. J Mol Biol 302:917–926
60. Li JJ, Huang DS, Wang B, Chen P (2006) Identifying protein-protein interfacial residues in heterocomplexes using residue conservation scores. Int J Biol Macromol 38:241–247
61. Murakami Y, Mizuguchi K (2014) Homology-based prediction of interactions between proteins using averaged one-dependence estimators. BMC Bioinformatics 15:213
62. Neuvirth H, Raz R, Schreiber G (2004) ProMate: a structure based prediction program to identify the location of protein-protein binding sites. J Mol Biol 338:181–199
63. Wang B, Chen P, Huang DS, Li JJ, Lok TM, Lyu MR (2006) Predicting protein interaction sites from residue spatial sequence profile and evolution rate. FEBS Lett 580:380–384
64. Lua RC, Marciano DC, Katsonis P, Adikesavan AK, Wilkins AD, Lichtarge O (2014) Prediction and redesign of protein–protein interactions. Prog Biophys Mol Biol 116:194–202
65. Lage K (2014) Protein-protein interactions and genetic diseases: the interactome. Biochim Biophys Acta Mol Basis Dis Elsevier BV 1842:1971–1980
66. Cukuroglu E, Engin HB, Gursoy A, Keskin O (2014) Hot spots in protein-protein interfaces: towards drug discovery. Prog. Biophys Mol Biol Elsevier Ltd 1–9
67. Kobzar OL, Trush VV, Tanchuk VY, Zhilenkov AV, Troshin PA, Vovk AL (2014) Fullerene derivatives as a new class of inhibitors of protein tyrosine phosphatases. Bioorg Med Chem Lett Elsevier Ltd 24:3175–3179
68. Kushwaha SK, Shakya M (2010) Protein interaction network analysis-approach for potential drug target identification in mycobacterium tuberculosis. J Theor Biol [internet]. Elsevier 262:284–294
69. You Z-H, Lei Y-K, Zhu L, Xia J, Wang B (2013) Prediction of protein-protein interactions from amino acid sequences with ensemble extreme learning machines and principal component analysis. BMC Bioinformatics. BioMed Central ltd 14. Suppl 8
70. Zahiri J, Yaghoubi O, Mohammad-Noori M, Ebrahimpour R, Masoudi-Nejad A (2013) PPIevo: protein-protein interaction prediction from PSSM based evolutionary information. Genomics Elsevier Inc 102:237–242
71. Pawson T, Nash P (2000) Protein-protein interactions define specificity in signal transduction. Genes Dev 14:1027–1047

Snake Venom Proteinases as Toxins and Tools

23

K.N. Suvilesh, A.N. Nanjaraj Urs, M.N. Savitha, M.D. Milan Gowda, and B.S. Vishwanath

Abstract

Proteinases from snake venoms have long been fascinating targets due to their structural, functional, and domain architectural diversities. Depending on these differences, snake venom proteinases are broadly classified as snake venom serine proteinases (SVSPs) and snake venom metalloproteinases (SVMPs). Unlike SVSPs, additional domains along with catalytic domain present in SVMPs are responsible for the subclassification. Non-catalytic domains of SVMPs direct catalytic domain to site-specific target and thereby assist to amplify the toxicities associated. The presence of additional domains along with catalytic domain renders SVMPs more toxic than SVSPs. Though non-catalytic domains function to facilitate the site-specific action of SVMPs, catalytic domain with the metal ion zinc in the active site is critical in eliciting the toxic action. Despite having a lot of reports regarding the toxic action of SVMPs and SVSPs, they prove to be promising tools when studied individually. In many cases their isolation and characterization have led to pharmacologically active drugs or research/diagnostic tool. This chapter initially describes the SVMP-induced local tissue damage such as hemorrhage and its neutralization by employing a novel strategy; zinc specific chelation therapy. Secondly, venom proteinase-induced systemic alterations such as perturbations in the complement and hemostatic system along with their applications as tools in the similar area are discussed. Finally, diagnostic applications of both SVSPs and SVMPs in coagulation laboratories and also their use in the identification of the snake species responsible for bite are discussed.

K.N. Suvilesh • A.N. Nanjaraj Urs • M.N. Savitha • M.D. Milan Gowda • B.S. Vishwanath (✉)
Department of Studies in Biochemistry, University of Mysore,
Manasagangotri, Mysore 570 006, Karnataka, India
e-mail: vishmy@yahoo.co.uk

S. Chakraborti, N.S. Dhalla (eds.), *Proteases in Physiology and Pathology*,
DOI 10.1007/978-981-10-2513-6_23

Keywords
Chelators • Coagulation cascade • Complement system • Diagnostic tools • Local tissue damage • Proteinases

Abbreviations

BM	Basement membrane
CS	Complement system
EC	Endothelial cells
GPIb	Glycoprotein Ib
SVMPs	Snake venom metalloproteinases
SVSPs	Snake venom serine proteinases
SVTL	snake venom thrombin-like enzymes
vWF	von Willebrand factor
α2β1	alpha2beta1

23.1 Introduction

Snake venoms have motivated interests and curiosity in many aspects of research including pharmacy and medicine since ancient times. The endeavor for better understanding of snake venom components and their effects extrapolated to application has drawn commercial attention too [1]. Fundamentally snake venom is produced by the venom gland and secreted to capture, kill, and digest the prey, which is executed by the complex pool of proteins (90% of the dry weight) present [2, 3]. Further, the protein composition of the venom can be broadly classified as enzymatic and non-enzymatic (Fig. 23.1); among these, proteins with enzymatic activity are thought to be the principal component in killing and digesting the prey [4].

In spite of the existing structural and functional differences among the myriad blends of enzymes present in snake venom, they all are secretory in nature and follow typical eukaryotic secretory protein synthesis mechanism involving mRNA, ribosomes, endoplasmic reticulum, and Golgi apparatus [10]. Before secretion, individual enzymes of snake venom undergo variable posttranslational modifications (PTMs), and proteinases being the high molecular weight toxins show the higher variability in terms of structure, function, and PTMs [11]. This variability makes snake venom proteinase a fascinating target, and to date more than 85 sequences of venom proteinases are available at UniProt database.

Further, proteinases present in snake venoms by themselves are diverse collection of proteins with a broad array of molecular weights ranging between 15 and 380 kDa and are generally classified as serine and metalloproteinases [12, 13]. Snake venom serine proteinases (SVSPs), like other venom enzymes, are synthesized and secreted

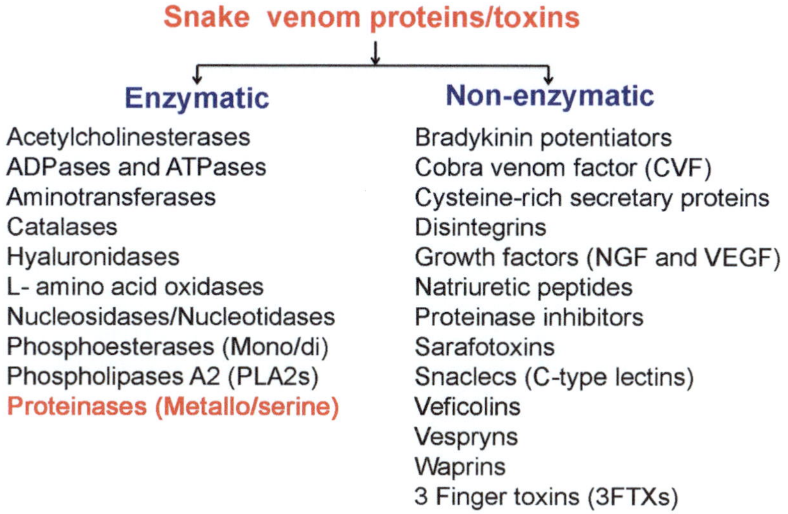

Fig. 23.1 General classification of snake venom proteins/toxins. NGF-nerve growth factor and VEGF-vascular endothelial growth factor [5–9]

with simple PTMs [14], whereas snake venom metalloproteinases (SVMPs) are synthesized as large precursors with different domains and undergo extensive PTMs.

SVMPs are synthesized in the secretory cells of the venom gland as a nascent structure comprising constantly pre, pro, M (Metalloproteinase) with or without D/DL (Disintegrin/Disintegrin-Like), and C (Cysteine-rich) domains in which pre is a signal peptide through which the protein gets transported to the endoplasmic reticulum where it undergoes sequential posttranslational processing like removal of signal peptide followed by disulfide bond formation, glycosylation, and in some specific circumstances dimerization. Final proteolytic processing of pro-forms of SVMPs occurs in the Golgi apparatus to convert them as active proteins and thereby get secreted in vesicles to the lumen. This proteolytic processing along with PTMs results in different final products which leads to the classification of SVMPs (Fig. 23.2).

Based on the presence or absence of D/DL, C domains as observed through mRNA transcripts, or cDNA and proteins isolated from the venom, SVMPs are categorized into three classes: P-I to P-III, where P-II and P-III are further divided into subclasses. P-I SVMPs being the simplest class are composed of only M domain; P-II SVMPs in continuation to M domain contain a D domain to the carboxyl terminal which is in most cases posttranslationally and proteolytically processed; based on which P-II class is further divided into P-IIa to P-IIe subclasses. Further, P-III class along with M domain, instead of D, contains DL domain followed by C domain. Based on posttranslational processing such as dimerization, proteolysis, and attachment of C-type lectin-like (CTL) domain, P-III class is further subdivided into P-IIIa to PIII-d. Presence of additional CTL domain along with M, DL, and C domains is evident in PIII-d class of SVMPs. Earlier, even lectin-like

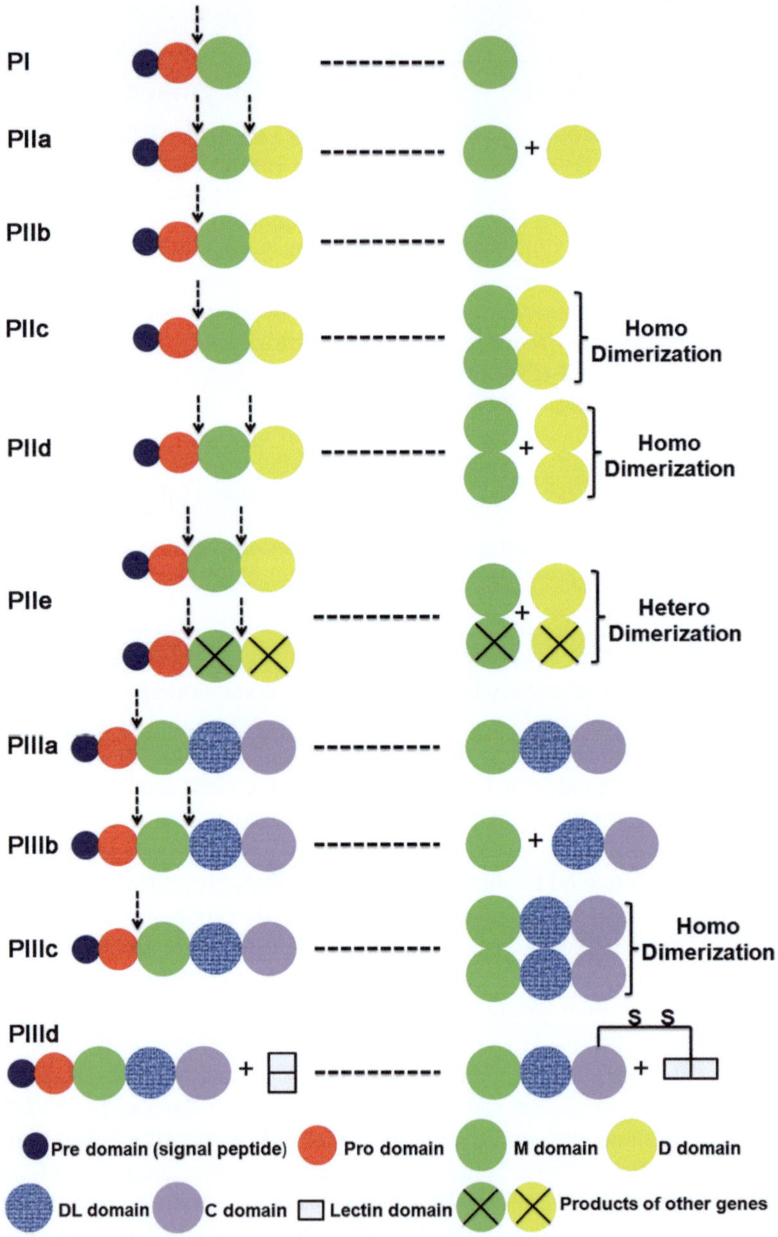

Fig. 23.2 Schematic of SVMPs classification. SVMPs are classified into three major classes and subclasses depending on their domain architecture. A *dashed arrowhead* on the *left-hand side* points the cleavage site of nascent protein. The final processed products are depicted on the *right-hand side*

domains were thought to be coded along with the other three domains, and SVMPs with this domain were considered under separate class P-IV. To date, no P-IV mRNA transcript has been observed and probably represents another PTM of the P-III structure thereby merged into P-III class as P-IIId [3, 11, 15, 16]. Table 23.1 lists the examples of unprocessed proteinase(s) and their processed domains in each different class of SVMPs.

Further, SVMPs as symbolized by their name are metal-/zinc-dependent proteolytic enzymes and contain a consensus zinc-binding sequence HEXXHXXGXXH (single-letter code of amino acids; X, any residue) in M domain [30].

23.2 Snake Venom Proteinase(s) as Toxins

Commonly associated systemic toxicities with SVMPs and SVSPs are fibrin(ogen) olytic, platelet aggregation or inhibition, and blood coagulation factor activation/depletion leading to consumption coagulopathy and intravascular bleeding. Few SVSPs and SVMPs attack the complement system leading to rapid generation of excess anaphylatoxins [31]. However, these systemic toxicities can be effectively managed by the administration of commercially available anti-snake venom (ASV) [32].

Additional to the above said complications, majority of SVMPs induces local toxicities such as hemorrhage by degrading the components of the basement membrane (BM) underlying the endothelial cells (EC) of the capillaries allowing the circulating blood to enter the extracellular space which in turn leads to hypoxia, eventually resulting in tissue necrosis [33], and administration of ASV generally fails to neutralize this pathological effect [32].

In general, additional domains (D, DL, C, and CTL) of SVMPs have been shown to direct the M domain (catalytic subunit) to specific substrates at relevant targets by facilitating their backbone flexibility in specific surface regions. For instance, P-II and P-III SVMPs have been reported to preferentially bind to the microvasculature leading to co-localization with type IV collagen and perlecan thereby causing intense local tissue damage [34–36]. In contrast, P-I SVMPs, devoid of these additional domains, show a widespread localization in the extracellular matrix and are considered to be less toxic than P-II or P-III SVMPs [37]. In continuation, due to the presence of D, DL, C, and CTL domains in P-II and P-III SVMPs, they escape the trap mechanism adopted by α2 macroglobulin (a plasma proteinase inhibitor) to inhibit endogenous proteinases, whereas P-I SVMPs are readily inhibited by α2 macroglobulin [38, 39].

In the past, different approaches such as immunohistochemistry of tissue sections, immunoblot from tissue homogenates, and proteomic analysis of exudates collected in the vicinity of affected tissue have provided novel clues for understanding the pathogenesis of SVMP-induced hemorrhage [40–42]. Some observations demonstrate the disruption and formation of gaps between the EC at the site of envenomation suggesting direct catalytic activity of SVMPs in hydrolyzing the key substrates (laminin, nidogen, perlecan, type IV collagen) at the BM surrounding EC in capillaries/microvessel thereby damaging the subendothelial structural integrity

Table 23.1 Examples for proteinases and their processed domains

Class	Snake species	Example for proteinase	Example for disintegrin	Remarks	References
P-I	*Crotalus atrox*	Atrolysin B	NA	Contains only M domain	[17]
P-IIa	*Crotalus atrox*	Atrolysin E	MVD	Despite of having MVD sequence, the processed disintegrin is a potent platelet aggregation inhibitor	[18, 19]
P-IIb	*Protobothrops jerdonii*	Jerdonitin	RGD	This protein does not undergo proteolytic processing to release the disintegrin domain	[20]
P-IIc	*Agkistrodon bilineatus*	Bilitoxin-1	MGD	Undergoes homodimerization; due to the presence of MGD in D domain no platelet aggregation inhibitory effect is seen	[21, 22]
P-IId	*Agkistrodon contortrix*	ND	Contortrostatin-RGD	Inhibits platelet aggregation, blocks cancer cell adhesion to fibronectin and vitronectin. The M domain and proteinase are not detected in venom so far	[23]
P-IIe	*Agkistrodon contortrix*	ND	Acostatin-RGD	The protein undergoes heterodimerization with the product of other gene. The M domain is not detected in venom so far	[24]
P-IIIa	*Crotalus atrox*	Atrolysin A	NA	Contains DL domain instead of D domain	[25]
P-IIIb	*Crotalus atrox*	VAP 2	NA	Catrocollastatin is the part of VAP 2 and has DL and C domains through which it inhibits platelet aggregation	[26, 27]
P-IIIc	*Crotalus atrox*	VAP 1	NA	Induces apoptosis in vascular endothelial cells and hence the name VAP	[28]
P-IIId	*Daboia russelii*	RVV-X	NA	Contains lectin-like domain in addition to DL and C domains	[29]

Notes: *ND* not detected in venom so far; *NA* not applicable to that class

eventually leading to extravasation of blood components through gaps formed between intercellular junctions which is otherwise known as hemorrhage [43, 44] (Fig. 23.3a).

Some other findings suggested that hemorrhage occurs by extravasation of erythrocytes through damaged EC and not through widened intercellular junctions. Additional domains of SVMPs specifically interact with integrins present on the membrane surface of EC leading to disruption and formation of gaps within the cells through which erythrocytes escapes. This process also leads to infiltration and activation of leukocytes which further amplifies the ongoing event by releasing the chemotactic agents and increasing the production of pro-inflammatory cytokines such as TNF-α, IL-1β, and IL-6 [45, 46]. On the other hand, extravasation of RBCs at the site of inflammation/vascular damage is followed by massive hemolysis by auto-oxidation. This event leads to the accumulation of methemoglobin and ferric heme which mediates oxidative stress and inflammatory reactions. These events taken together may operate in combinational fashion and lead to the progression and aggravation of local tissue damage [47, 48] (Fig. 23.3b).

In addition to the conventional concepts of SVMP-induced tissue damage, a recent report introduces the concept of NETosis to explain the mechanism of progressive tissue destruction induced by *Echis carinatus* venom. The finding highlights the fact that the neutrophils (first line of defense in immune system) infiltrate to the bite site, where they release DNA and entrap venom toxins. These neutrophil extracellular traps (NETs) further block the blood vessels and prevent the venom from entering into the circulation. This causes the accumulation of venom toxins, especially SVMPs at the vicinity of BM, thereby causing continued tissue degradation at the bite site [49] (Fig. 23.3c).

23.3 A New Approach Toward the Neutralization of Local Tissue Damage Induced by SVMPs: Chelating Agents

The inability of ASV therapy to neutralize the local tissue damage induced by SVMPs has urged scientists and medical fraternities to develop an alternate therapy. Despite the involvement of non-catalytic domains in the above mentioned mechanisms underlying extensive hemorrhage caused by SVMPs, the role of catalytic domain is a must, and any SVMP with truncated catalytic domain will not cause hemorrhage or severe local tissue damage. Thus, inhibition of catalytically active M domain is a sensible and vital strategy for limiting the progression of local tissue damage induced by SVMPs. As M domain requires zinc (Zn^{2+}) for its catalytic activity, use of chelation therapy (employing chelating agents) will be a better approach to counteract the debilitating action of SVMPs.

In view of these structural and functional aspects of SVMPs, our research group used pharmacologically approved chelating agents such as ethylenediaminetetraacetic acid (EDTA) and ethylene glycol-bis(β-aminoethyl ether)-N,N,N′,N′-tetraacetic acid (EGTA) to study the inhibitory effects of these compounds against *Trimeresurus malabaricus* venom-induced local tissue damage [50]. Later on, other

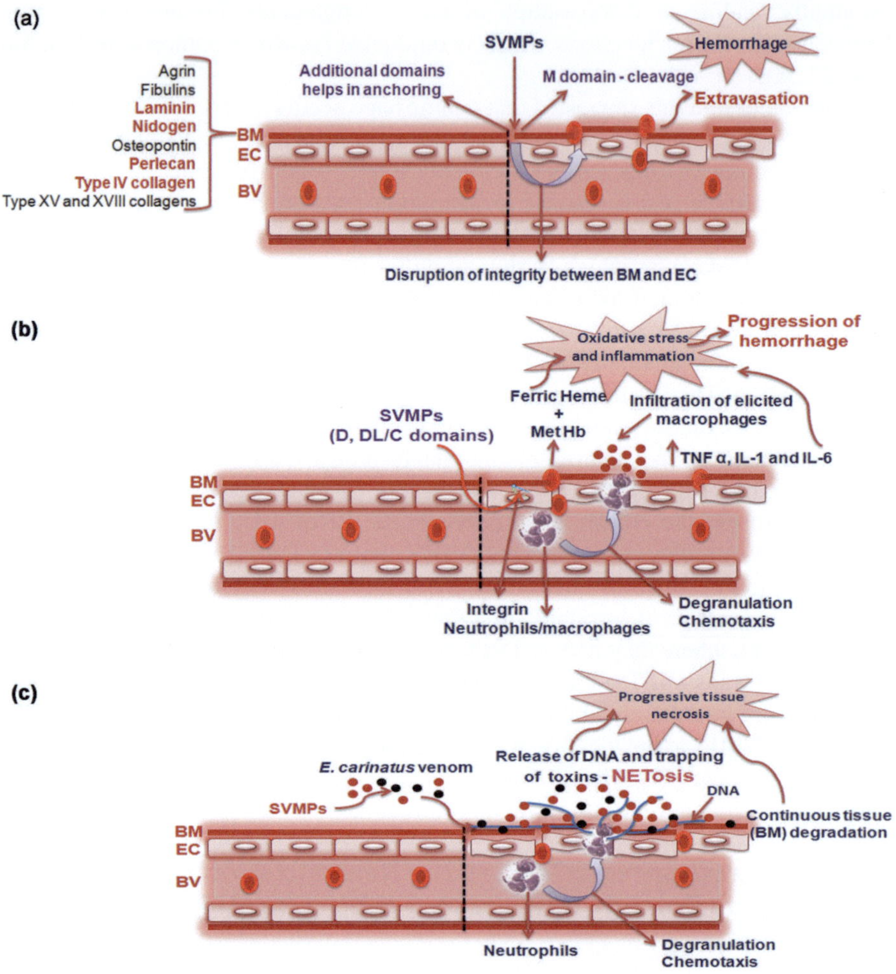

Fig. 23.3 Proposed mechanisms underlying SVMP-induced hemorrhage: (**a**) SVMPs through non-catalytic domains anchor to the basement membrane (BM) underlying endothelial cells (EC) of blood vessel (BV) and degrade BM proteins such as laminin, collagen, and others resulting in the formation of intercellular gaps between the EC through which blood components escape to the extracellular space. (**b**) Additional domains of SVMPs interact with integrins present on membrane surface of EC and lead to disruption and formation of gaps within the cells through which erythrocytes escape. Subsequently leukocyte infiltration at the bite site occurs with an increase in the level of pro-inflammatory cytokines TNF-α, IL-1β, and IL-6. Hemolysis of the infiltrated RBCs leads to accumulation of methemoglobin (Met Hb) and ferric heme which in turn results in oxidative stress and inflammation. (**c**) Infiltrated neutrophils at the bite site release filamentous DNA which traps the venom toxins leading to the formation of neutrophil extracellular traps (NETs). Trapped SVMPs continuously degrade extracellular matrix components resulting in progressive tissue damage

report also came up with respect to chelation therapy against snake venom-induced hemorrhage and myotoxicity [33]. The aforesaid divalent metal ion chelators were shown to be highly effective in neutralizing the activity of SVMPs in vitro. However, their nonspecific binding with physiologically vital divalent metal ions, particularly calcium (Ca^{2+}), poses an obstacle for their in vivo applications or pharmacological use [51]. Thus, pharmacological use of specific Zn^{2+} chelators rather than nonspecific divalent metal chelators might be effective in the management of local toxicity as they can serve as adjunctive therapeutic molecules to aid ASV therapy by limiting local tissue destruction in treatment mode [52].

Keeping the above mentioned aspects as background, recently our research group designed a study using three Zn^{2+}-specific chelators TPEN (N,N,N′,N′-tetrakis (2-pyridylmethyl) ethane-1,2-diamine), DTPA (diethylene triamine pentaacetic acid), and TTD (tetraethyl thiuram disulfide) in which their inhibitory potential was evaluated against *Echis carinatus* venom metalloproteinase(s) in vitro. Further, the results obtained were extended to the animal model to evaluate the inhibitory potential, where upon challenging experiments, the selected chelators were very efficient in inhibiting the *Echis carinatus* venom-induced local tissue damage, and unlike EDTA and EGTA, they were nontoxic at minimal dose required for inhibition studies [52].

All the abovementioned toxic symptoms associated with snake venom proteinases are cumulative effects, observable when they act together. However, individually they can be exploited to study many different unknown aspects of Biochemistry and Pharmacology owing to their stringent specific action.

23.4 Snake Venom Proteinases as Tools

23.4.1 Snake Venom Proteinases and Complement System

Different biochemical processes occurring in the body under apt regulation to maintain homeostasis are reported to be disturbed upon envenomation by snake venom proteinases, and one such susceptible area is the complement system (CS) [53]. Both SVMPs and SVSPs have been shown to affect complement pathway leading to inappropriate complement inhibition or activation which in most cases leads to complement consumption [54, 55]. Further, CS is a central part of both innate and adaptive immune responses and mediates a series of functions ranging from the modulation of local inflammatory responses to the promotion of phagocytosis and the lysis of pathogens [56]. The human CS is composed of more than 40 plasma and cell surface proteins that participate in balanced activation and regulation of it [57]. The complement activation is triggered by a series of proteolytic activities arbitrated by serine proteinases similar to blood coagulation cascade. The CS consists of and can be activated by any of the three activation pathways, i.e., classical, alternative, and mannose-binding lectin (MBL) pathways, which merge at the proteolytic activation step of C3 resulting in the formation of C3 convertases, a central component of the system. C3 convertase will cleave C3 to yield C3a and C3b; thus, generated C3b in turn forms complex with C3 convertases to activate C5 which leads to the formation of membrane attack complex (MAC) by assembling C6, C7, C8, and multiples of C9 which promotes the lysis of pathogen and is referred as lytic pathway [58].

So far, more than 15 snake venom proteinases including both SVMPs and SVSPs which directly or indirectly activate CS have been isolated and characterized (Table 23.2). In finding reasons to the question as to why snake venom proteinases should target the complement, reports suggest that generation of a large amount of anaphylatoxins such as C5a and C3a upon envenomation may play an important role in the inflammatory process and they may also assist, due to their vasodilatory effects, to enhance the spreading of other venom components [53].

In addition to some case reports stating fatal bites, victims died due to anaphylactic shock, and the mechanism behind this uncontrolled anaphylaxis may be the generation of huge quantities of anaphylatoxins [55]. These grounds clearly suggest the inappropriate activation of the CS by snake venom proteinases which will lead to uncontrolled inflammation and anaphylaxis. Apart from undesired activation of the CS by SVMPs and SVSPs, they also prove to be promising tools in studying complement pathway in detail and create a platform to design drugs in complement deficiency disorders and in transplantation to minimize the graft rejection [70]. Figure 23.4 summarizes the reported snake venom proteinase action on different proteins and activation pathways of CS.

Among the reported SVMPs and SVSPs affecting CS, few are best characterized with respect to their site of action on one or more complement protein(s) in comparison to endogenous complement activating proteinase complexes. Cobrin is a metalloproteinase isolated and characterized from the *Naja siamensis* venom, belongs to the P-III class which specifically cleaves C3 of the CS, and is the first SVMP acting on complement which was studied in detail with respect to its site of action along with the role of C3 fragments generated thereby [64, 65]. C3 (180 kDa) being a multidomain protein with having a central role in the CS is made up of two polypeptide chains (β and α) [74]. α chain of C3 is specifically cleaved by cobrin, but at a different site compared to endogenous C3 convertases and factor I, yielding two novel cleavage products termed C3o and C3p (Fig. 23.5a, b) [64, 75].

The venom proteinase removes the C3p fragment from the C3dg region of the C3 α chain. The major cleavage fragment C3o contains the unaltered β chain of C3 and two α chain-derived polypeptides. When tested for homology, C3o and C3p resembled physiologically generated C3c and C3d fragments, respectively (see Fig. 23.5b). C3o being structurally similar but in contrast to C3c is capable of supporting factor B activation in the presence of factor D, and it is believed that the presence of extra carboxyl terminal residues of the C3g region is the only difference which is responsible for this effect (Fig. 23.5a, b).

Cobra venom factor (CVF) is a nonenzymatic protein present in cobra venom which can form structural subunit in the complex CVFBb, making it more stable C3 convertase than C3bBb. CVF though binds to factor B structurally resembles more to C3c and C3o than C3b [76]. Looking at this structural similarity, one can clearly speculate that the CVF is made in the venom gland of the cobra by posttranslational

Table 23.2 Snake venom proteinases affecting complement system with their targets

Name of proteinase	Snake species	Type of proteinase	MW (kDa)	Target	Class	Remarks	References
Atrase B	*Naja atra*	Metallo	49.4	Factor B, C6, C7, and C8	P-III	Factor B and C6 are the major targets	[59]
Bap-1	*Bothrops asper*	Metallo	22.7	C5	P-I	Generation of anaphylatoxin C5a and neutrophil chemotaxis	[60]
B1	*Crotalus basiliscus*	Metallo	27	ND	ND	Activates complement-mediated cell lysis at lower concentration and inhibits at higher concentration	[61]
B2	*Crotalus basiliscus*	Metallo	27.5	ND	ND	Inhibits complement-mediated cell lysis	[61]
BpirSP27	*Bothrops pirajai*	Serine	27.12	ND	NA	Inhibits complement-mediated cell lysis and induces leukocyte chemotaxis	[62, 63]
BpirSP41	*Bothrops pirajai*	Serine	40.63	ND	NA	Same effect as that of BpirSP27 but percentage of inhibition is more	[62, 63]
Cobrin	*Naja siamensis*	Metallo	67.66	C3	P-III	Cleaves human C3 to form C3o and C3p which are different from physiologically derived fragments C3b and C3a	[64, 65]
C-SVMP	*Bothrops pirajai*	Metallo	23.14	C3, C4, and C5	P-I	Activates complement pathway by cleaving the α chain of C3, C4, and C5	[54]
Flavoxobin	*Trimeresurus flavoviridis*	Serine	26.7	C3	NA	Cleaves human C3 at the same site where C3 convertase cleaves and acts as a novel heterologous C3 convertase	[55]
F5	*Crotalus scutulatus*	Metallo	27	C3	ND	C3 product generated was able to mediate cell lysis	[66, 67]
M5	*Crotalus molossus*	Metallo	25	C2, C3, and C4		M5 completely hydrolyzed C2, C3, and C4. Hydrolysis of C3 by M5 generated a fragment of ≈ 130 kDa, indicating a different cleavage site from that of C3 convertases	[68]
Oxiagin	*Naja oxiana*	Metallo	49.8	C4b2a complex	P-III	Only reprolysin-lacking proteinase activity inhibits complement pathway through its lectin-like domain	[69]
rFII	*Agkistrodon acutus*	Metallo	26.71	C5, C6, and C9	P-III	Inhibition of MAC formation	[70, 71]

Notes: *NA* not applicable and *ND* not determined. Molecular weight of proteinases are retrieved from UniProt database

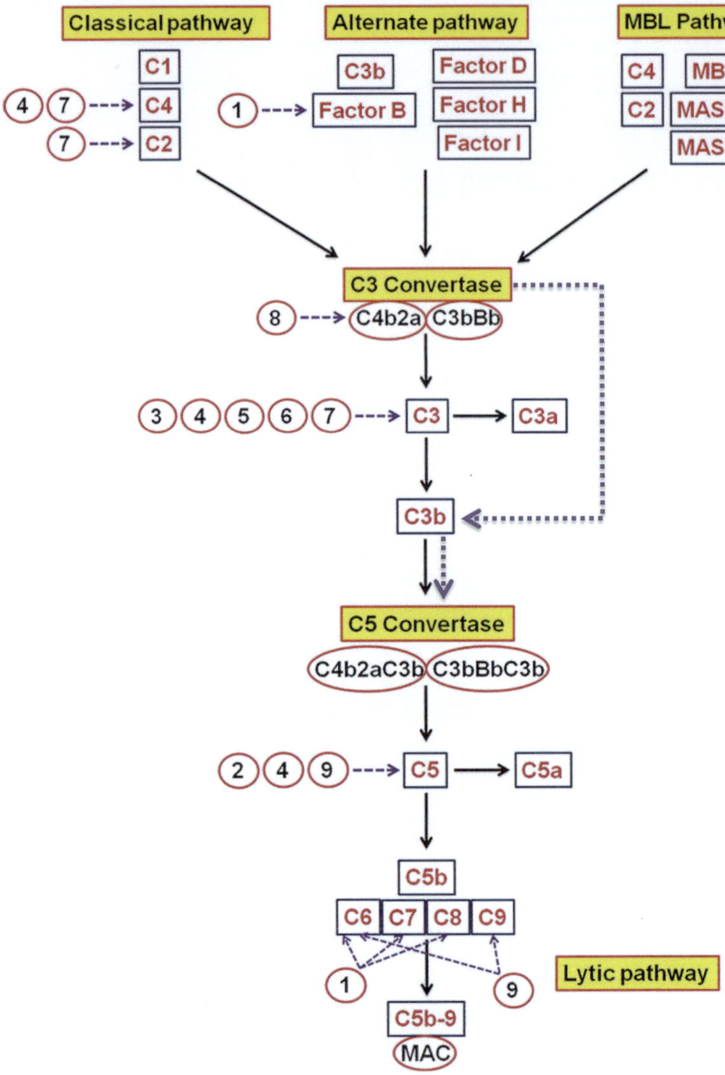

Fig. 23.4 Schematic of complement pathway and snake venom proteinases acting on different components of complement system: Complement system (CS) can be activated by any of the three pathways, viz., classical, alternative, and MBL pathway. Classical pathway involves the participation of complement components C1, C2, and C4 and is triggered by the activation of C1. Activated C1 in turn activates C4 and C2 resulting in the formation of C4b, C4a, C2b, and C2a of which C4b and C2a bind to form the classical pathway C3 convertase (C4b2a). MBL pathway is activated by MASPs (MBL-associated serine proteinases) the MASP-1 and MASP-2, which then activate C4 and C2, to form the C3 convertase, C4b2a complex. Alternative pathway is just an amplification of classical and MBL pathways in which traces of C3b present will bind to factor B to form C3B. Later, factor D will activate B in C3B complex resulting in the removal of Ba and leading to the formation of C3bBb, alternative pathway C3 convertase. C3 convertases formed will activate C3 to yield C3b plus C3a, of which C3b binds to C3 convertases C4b2a and C3bBb to form C5 convertases C4b2aC3b and C3bBbC3b, respectively. C5 convertases will act on C5 to release C5b and C5a, of which C5b gets attached to the target surface and recruits C6, C7, C8, and multiples of C9 to form MAC which mediates the lytic pathway [72, 73]. The dashed arrows with numbers pointing toward the complement protein(s) scattered all along the pathway indicate the targets of snake venom proteinases, and the numbers are in alphabetical order: (*1*) atrase B, (*2*) Bap-1, (*3*) cobrin, (*4*) C-SVMP, (*5*) flavoxobin, (*6*) F5, (*7*) M5, (*8*) oxiagin, and (*9*) rFII

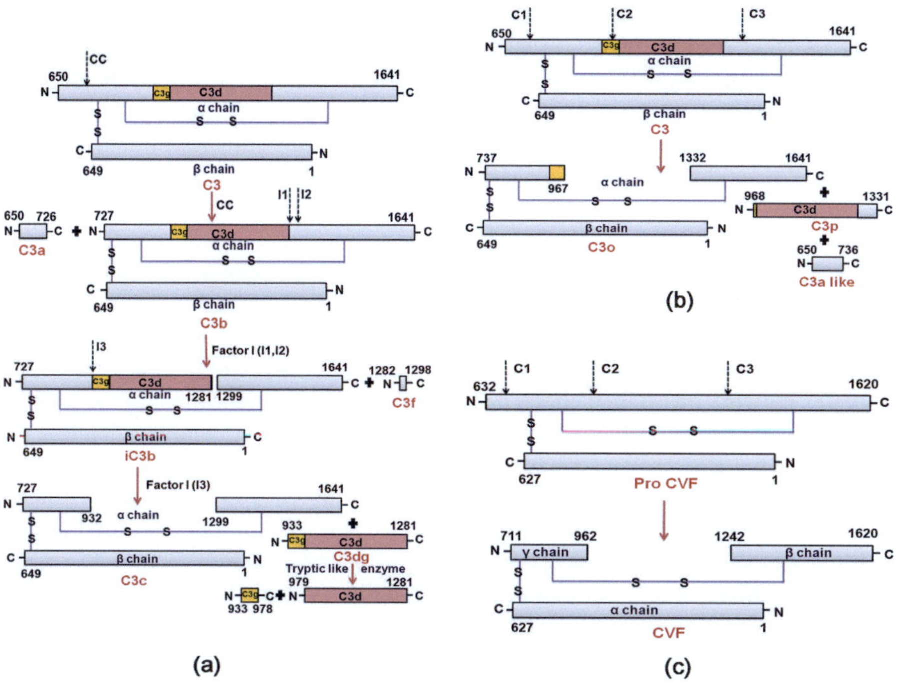

Fig. 23.5 Action of endogenous C3 convertase and cobrin on human complement C3 in comparison with cobra venom factor (CVF). (**a**) C3 convertase (CC) cleaves C3 α chain between residues 726 and 727 to acquiesce C3a and C3b. After the targeted function of C3b, excess C3b produced will be degraded by factor I as a regulatory step. First by using two cleavage sites (designated here as I1 and I2), factor I cleaves between residues 1281 and 1282 and 1298 and 1299 to remove C3f and yield iC3b. Later, using the third cleavage site (designated here as I3), factor I cleaves between residues 932 and 933 to remove C3dg fragment and yield C3c. The released C3dg fragment is further hydrolyzed by tryptic-like enzymes to produce C3g (*yellow*) and C3d (*magenta*) [75]. (**b**) Cobrin cleaves complement C3 through its three cleavage sites (designated here as C1, C2, and C3) between the residues 736 and 737, 967 and 968, and 1331 and 1332 to yield C3o, C3p, and C3a-like fragments. (**c**) Speculated action of cobrin on pro-CVF is synthesized in the venom gland before secretion to yield active CVF. N and C represent the amino terminus and carboxyl terminus of the polypeptide chains, respectively. The numbers used represent the terminal amino acid residues in the whole protein and cleavage products

processing of a pro-CVF with an intact C3-like α chain which may be proteolytically processed by cobrin (Fig. 23.5c).

A serine proteinase flavoxobin isolated from *Trimeresurus flavoviridis* venom cleaves C3 at the Arg726-Ser727 site in the α chain which is also the cleaving target for natural C3 convertase of the human CS. Hence, flavoxobin acts as a heterologous C3 convertase. This selective specificity of flavoxobin may make it a promising tool to activate C3 in laboratory animal models to keep complement activation continuously rolling [55].

There are only two metalloproteinases known to degrade downstream complement proteins C5-C9; these are atrase B and rfII (recombinant fibrinogenase II) isolated from the venom of *Naja atra* and *Agkistrodon acutus*, respectively [70, 71]. Selective cleavage of downstream complement proteins responsible for MAC formation can be exploited to design drugs for hyperacute graft rejection syndrome wherein formation of MAC is very critical to destroy the graft [70].

23.4.2 Snake Venom Proteinases and Blood Coagulation System

Another major target for snake venom proteinases is the blood coagulation process which includes both thrombosis and hemostasis [77, 78]. Blood coagulation is a complex network involving cellular and protein compartments which can be broadly divided as primary and secondary hemostasis [79]. When circulating platelets are exposed to the BM proteins such as von Willebrand factor (vWF) and collagen at the site of injury, interaction takes place between the receptor(s) on platelet membrane and the exposed proteins to form platelet plug; the process is called as primary hemostasis or thrombosis [80].

Snake venom proteinases can activate or inhibit primary hemostasis which can be accomplished by catalytic and non-catalytic mechanisms [81]. SVMPs using their non-catalytic D, DL, C, and CTL domains inhibit platelet aggregation. The non-catalytic domains bind to the receptors on the platelet membranes, thereby blocking the interaction of platelets with the physiological ligands such as fibrinogen, vWF, thrombin, and collagen eventually resulting in platelet aggregation inhibition [82–84]. Observations on platelet activation/inhibition mediated by snake venom proteinases through catalytic mechanism rather than non-catalytic provide insight into various aspects of thrombosis.

It is a well-known fact that binding of membrane glycoproteins GPIbα, GPVI, and α2β1 to their ligands vWF and collagen, respectively, leads to platelet activation and chain of downstream reactions such as phosphorylation of tyrosine residues, secretion of aggregation inducers, and so on, which culminate at platelet aggregation [85]. Still there are certain unknown aspects underlying thrombosis, and these gaps in fact can be filled by exploiting differential catalytic mechanisms of snake venom proteinases on platelet aggregation.

Some SVMPs mediate platelet aggregation inhibition by catalytically destructing the platelet membrane glycoprotein(s) (e.g., GPIbα and GPVI) receptors or otherwise hydrolyzing the ligands required for thrombosis (Table 23.3). To take a few examples, Barnettlysin-I, a metalloproteinase isolated from the venom of *Bothrops barnetti*, selectively inhibits collagen and vWF-induced platelet aggregation, while thrombin-induced aggregation is not affected. Barnettlysin-I degrades high molecular weight vWF into low molecular weight fragments which lose the function to aggregate platelets, and at the same time vWF A1 domain which contains the stretch of amino acid to bind GPIb is also degraded. On the other hand, Barnettlysin-I degrades the receptor GPIb into two fragments making it further unable to bind vWF. Conversely, Barnettlysin-I inhibits collagen-induced platelet aggregation by

Table 23.3 SVMPs acting on platelet membrane receptors and ligands

Name of SVMP	Snake species	Class	Molecular weight in kDa	Target receptors and ligands	Remarks	References
Acurhagin	*Deinagkistrodon acutus*	P-III	68.54	Collagen and vWF	Impairs collagen and ristocetin-induced platelet aggregation by cleaving collagen and vWF	[93]
Alborhagin	*Trimeresurus albolabris*	P-III	ND	GP VI	Induces platelet aggregation by ectodomain shedding of GP VI	[92]
Barnettlysin-I	*Bothrops barnetti*	P-I	23.38	vWF, GPIb, and α2β1	Cleaves peptide bond between basic residues Lys 413-Leu 414 and Arg 696-Leu 697	[86]
Jararhagin	*Bothrops jararaca*	P-III	63.98	α2β1 and vWF	Binds to α2 subunit to degrade β1 chain	[87, 88]
Kistomin	*Calloselasma rhodostoma*	P-I	25	GPVI, GPIb, and vWF	Differentially cleaves GPVI and GPIb	[89, 94]
Mocarhagin	*Naja mossambica*	P-III	68.17	GPIb	Removes a peptide with highly sulfated tyrosine residues from GPIb α which is functionally important for binding to ligands	[90]
Natrahagin	*Naja atra*	ND	ND	GPIb	Significantly inhibits thrombin-induced platelet aggregation	[95]
Kaouthiagin	*Naja kaouthia*	P-III	44.49	vWF	Cleaves vWF by binding to N-terminal domain and inhibits platelet aggregation	[91]
Triflamp	*Trimeresurus flavoviridis*	ND	28	GPIbα	More studies need to be carried out to explore the actual mechanism	[96]

Notes: *ND* not determined. Molecular weights of proteinases are retrieved from UniProt database

degrading specifically α2 subunit in the α2β1 integrin receptor creating truncated protein which is unable to bind collagen; unlike vWF, collagen is not degraded [86].

Another metalloproteinase named jararhagin, isolated from the venom of *Bothrops jararaca*, binds to α2 subunit and degrades β1 subunit of the integrin α2β1 thereby rendering it unable to bind its ligand collagen which results in platelet aggregation inhibition [87]. However, there is a report which also showed the degradation of vWF by jararhagin indicating that platelet aggregation mediated by vWF may also get inhibited [88]. Along with α2β1 integrin, GPVI serves another receptor for the ligand collagen and is selectively cleaved by a metalloproteinase kistomin (isolated from *Calloselasma rhodostoma* venom) to yield a 35 kDa fragment, and the truncated receptor is not capable of inducing downstream signal transduction events upon ligand binding.

Further, the data presented using synthetic peptides to determine the cleavage site of kistomin confirms two sites between Glu-Ala and Val-Phe, respectively [88]. Report from a study suggested that kistomin inhibits vWF-induced platelet aggregation by digesting vWF and its receptor GPIb. Cleavage of platelet α chain of GPIb by kistomin resulted in the release of 45 and 130 kDa soluble fragments, indicating that kistomin cleaves GPIbα at two distinct sites. Simultaneously, kistomin-mediated degradation of high molecular weight vWF to low molecular weight fragments is also seen [89]. In a study carried out by Ward and colleagues [90], showed that mocarhagin, a novel metalloproteinase isolated from the venom of *Naja mossambica*, cleaves the platelet vWF receptor GPIbα and also identified sulfated tyrosine/anionic sequence Tyr 276-Glu 282 of GPIbα as a binding site for vWF and thrombin.

A vWF binding and cleaving metalloproteinase termed kaouthiagin from *Naja kaouthia* venom was shown to bind to the N-terminal region of vWF between residues 1 and 190 and cleaved exactly between Pro-708 and Asp-709 [91]. In contrast to the typical platelet aggregation inhibitory effect of SVMPs, alborhagin a metalloproteinase isolated from the venom of *Trimeresurus albolabris* induces platelet aggregation by ectodomain shedding of GPVI, like endogenous ADAMs [92].

Taking these reports together, one can clearly say that these proteinases will be useful as tools to carry out functional dissection studies of ligands and receptors responsible for platelet aggregation. In particular, narrow specificity of these proteinases to act on any one or two receptor on platelets can be used to study the signal transduction events mediated by other receptors than the vulnerable ones. Figure 23.6 is a pictorial presentation of SVMPs acting on specific ligand or receptor which are critical for platelet aggregation. SVSPs activate platelet aggregation in a similar manner to thrombin. A few SVSPs which act as thrombin-like enzymes are called as snake venom thrombin-like enzymes (SVTLEs) which affect thrombosis by cleavage of PARs (protease-activated receptors; Fig. 23.6) thereby activating platelets which finally aggregates [83]. The serine proteinase thrombin activates platelet aggregation by cleaving and releasing a tethered N-terminus peptide from the G-protein-coupled receptors PARs [97].

Thromocytin, cerastocytin, PA-BJ, TLBm, and gyroxin are a few examples among the SVSPs reported to induce platelet aggregation similar to thrombin [98–103].

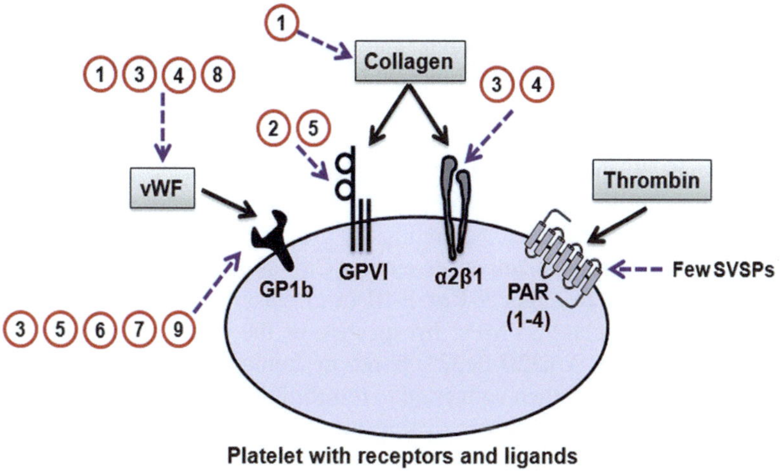

Fig. 23.6 Action of SVMPs and SVSPS on different ligands and platelet membrane receptors: *Numbers* with the *dashed arrows* pointing either ligand or receptor indicate action of particular SVMP toward it, and SVSPs with *dashed arrow* pointing protease-activated receptor (PAR) symbolize the only target. *Numbers* used to represent the name of SVMPs in alphabetical order are (*1*) acurhagin, (*2*) alborhagin, (*3*) Barnettlysin-I, (*4*) jararhagin, (*5*) kistomin, (*6*) mocarhagin, (*7*) natrahagin, (*8*) kaouthiagin, and (*9*) triflamp. Abbreviations used: glycoprotein1b (GP1b), glycoprotein VI (GPVI), α2β1 (integrin α2β1 or glycoprotein Ia/IIa complex)

PA-BJ, serine proteinase isolated from the venom of *Bothrops jararaca,* induces platelet aggregation through activation of PAR1 and PAR4. The cleavage occurs at Arg41-Ser42 (like thrombin cleavage), and Arg46-Asn47 a site different and extra compared to thrombin [104]. Determination of site-specific cleavage of these SVSPs in comparison to thrombin will potentially help to fill the gap in the existing knowledge on PAR-directed platelet aggregation.

In continuation, secondary hemostasis is a process which occurs by the combined action of intrinsic, extrinsic, and common pathways of blood coagulation involving series of serine proteinases activating the successive one culminating in conversion of prothrombin to thrombin which converts fibrinogen to fibrin to hold and stabilize the platelet plug formed as a result of primary hemostasis [105]. Both SVMPs and SVSPs disturb blood coagulation pathways by functioning as thrombin-like enzymes; prothrombin activators; fibrino(geno)lytic enzymes; blood coagulation factor V, VII, and X activators; protein C activators; and plasminogen activators [106].

Briefly, SVTLEs are SVSPs and, like thrombin, able to clot fibrinogen by cleaving from the N-terminal end. Based on catalytic action, they are divided into three groups: SVTLE-A cleaves off fibrinopeptide A from fibrinogen Aα chain, SVTLE-B cleaves off fibrinopeptide B from Bβ, and SVTLE-AB cleaves off both fibrinopeptides A and B. Commonly, majority of SVTLEs cleave off either fibrinopeptide A or B, but rarely both fibrinopeptides as thrombin [107].

As a result of acting at only one position, SVTLEs lead to the formation of loose clots but not stable ones like thrombin. These exceptional functional features allow their clinical use as defibrinogenating agents, for example, ancrod (Arvin®; from *Calloselasma rhodostoma*) and batroxobin (Defibrase®; from *Bothrops moojeni*) [31]. Further, both SVMPs and SVSPs are reported to be prothrombin activators and are divided into four groups A, B, C and D based on the mechanism of activation and cofactor requirement. Prothrombin activators belonging to group A and B are SVMPs in which proteinases belonging to group A function without cofactor requirement, while group B activators require Ca^{2+} as cofactor. Group A prothrombin activators belong to either P-I or P-III (a-c) class SVMPs, whereas Group B activators are P-IIId class SVMPs. Irrespective of the class, both group A and B activators cleave the Arg320-Ile321 bond in human prothrombin to produce meizothrombin, which is then converted to α thrombin (active thrombin) by autolysis. Group C and D prothrombin activators are SVSPs in which group C activators require Ca^{2+} and negatively charged phospholipids for activity, whereas group D activators requires Ca^{2+}, phospholipids, and activated blood coagulation factor V (FVa) [108].

Apart from having thrombin-like and prothrombin-activating properties, both SVMPs and SVSPs possess fibrino(geno)lytic activities, but in contrast to SVTLEs and thrombin, these fibrinogenases cleave fibrinogen at abrupt sites preferentially from C-terminal end producing truncated form which is no longer a functional fibrinogen [109]. Venom proteinases which degrade fibrinogen may be classified as either α or β chain fibrinogenases depending on the chain they degrade. Only one or few venom proteinase(s) which activate blood coagulation factor V, VII, and X, protein C, and plasminogen have been isolated and studied [110–115].

As discussed, upon envenomation SVSPs or SVMPs exert action individually or in combination resulting in pathological conditions such as disseminated intravascular coagulopathy (DIC), consumption coagulopathy, and excess bleeding eventually resulting in hypoxia and death of the victim. Nevertheless, on individual consideration, these proteinases have application in coagulation laboratories and in diagnosis of the snake species responsible for bite.

23.4.3 Snake Venom Proteinases as Diagnostic Tools in Coagulation Laboratories

Evident from the above discussion, for almost every factor involved in blood coagulation, there is a venom proteinase that can activate or inhibit it. Many of these venom proteinase(s) are resistant to endogenous inhibitors, and this property extends their application to diagnose problems pertaining to coagulation abnormalities in hematology laboratories and also aid in identification of the snake species responsible for bite. Batroxobin, a SVTLE, belongs to group A and is used in diagnosing the problems related to dysfibrinogenemia or hypofibrinogenemia under the name

Reptilase® time (Pentapharm, Basel, Switzerland). Principally, batroxobin cleaves the Aα chain of fibrinogen leading to the formation of clot. Conditions like dysfibrinogenemia or hypofibrinogenemia will lead to a prolongation of Reptilase time [116].

Prothrombin is converted to α thrombin in the final step of coagulation by prothrombinase complex comprising of activated factor X (FXa), activated factor V (FVa), phospholipids, and Ca^{2+} ions which involve the generation of meizothrombin as an intermediate [117]. Ecarin clotting time (ECT) involves the same mechanism in which a metalloproteinase named ecarin from the venom of *Echis carinatus* directly generates meizothrombin, and hence, the test may be termed as meizothrombin generation test [118, 119]. Normally prothrombin before secretion undergoes posttranslational gamma carboxylation and is vitamin K dependent. On the other hand, des-carboxy prothrombin is an abnormal protein lacking gamma carboxyl group and can be observed in patients undergoing oral anticoagulant therapy such as warfarin (a vitamin K antagonist) and in certain liver disorders [120]. Ecarin also converts des-carboxy prothrombin to meizothrombin which makes ECT useful for anticoagulant management to provide point-of-care test in comparison with normal prothrombin time or thrombin time. Meizothrombin is readily inhibited by hirudin (an anticoagulant from leech saliva), and prolonged ECT is observed in blood plasma of those patients who are undergoing hirudin therapy. Clinicians manage hirudin therapy by observing the changes with respect to ECT [121]. Ecarin® is also manufactured and marketed by Pentapharm, Basel, Switzerland.

Textarin is a serine proteinase isolated from the venom of *Pseudonaja textilis*, capable of activating prothrombin and belongs to group D [122]. The Textarin®/Ecarin® ratio (Pentapharm, Basel, Switzerland) is a simple test for a lupus anticoagulant (LA). Presence of high concentrations of antilipid antibody in the biological fluids is the characteristic feature of LA. These antibodies appear to act by interfering with the binding of phospholipid to form prothrombin activator, thereby affecting both the intrinsic and extrinsic pathways of blood coagulation [123]. Textarin/Ecarin ratio is based on the differential dependence of these two snake venoms on phospholipid to activate prothrombin. Textarin, requires FVa, Ca^{2+}, and phospholipid to activate prothrombin, whereas Ecarin activates prothrombin to form meizothrombin in the absence of phospholipid. In a condition like LA, the Textarin time is prolonged due to its phospholipid dependence, but the Ecarin time is not [124].

RVV-V® (Pentapharm, Basel, Switzerland) is a serine proteinase FV activator from *Daboia siamensis* venom which specifically activates FV [110]. Though the use of RVV-V in diagnosing FV deficiency is limited, it is extensively used to prepare FV-free plasma to study FV deficiencies by correction methods [125].

Among the SVSPs, only protein C activators exhibit direct anticoagulant effects. Physiologically, the zymogen of protein C circulating in the blood is activated by thrombin. This activated protein C degrades FVa and FVIIIa thereby controls undesirable blood coagulation [126]. Protac® (Pentapharm, Basel, Switzerland) is a direct activator of protein C isolated from *Agkistrodon contortrix* venom [127]. Resistance to activated protein C (APC-R) is a genetic abnormality leading to debilitating conditions such as hypercoagulability and pulmonary embolism. The most common APC-R is FV Leiden which is a variant or otherwise mutated form of human FV and resistant to protein C degradation which eventually leads to hypercoagulability [128].

RVV-V and Protac together with Noscarin, a serine proteinase and group D prothrombin activator isolated from *Notechis scutatus* venom, are used to diagnose FV Leiden mutation or APC-R under the product name Pefakit® APC-R Factor V Leiden (Pentapharm, Basel, Switzerland) [129]. The test involves addition of RVV-V and Protac followed by the addition of Noscarin. The prothrombin activator converts prothrombin to thrombin and induces coagulation of the sample. If the FVa molecules in the sample are digested by Protac-activated protein C, the velocity of prothrombin activation by the FV-dependent Noscarin is slow, and therefore, the clotting time is long. If the FVa elimination is curtailed due APC-R or FV Leiden, the velocity of prothrombin activation is high, and the clotting time is short [130].

RVV-X, a high molecular weight metalloproteinase from *Daboia siamensis* venom, is an activator of FX and employed for the detection of deficiencies of FVII and FX and also in LA to measure change in clotting time due to the presence of antiphospholipid antibodies [131–133]. Measurement of clotting time using RVV-X® (Pentapharm, Basel, Switzerland) involves the activation of FX by RVV-X in the presence of FVa, phospholipids, and Ca^{2+} ions leading to the formation of prothrombinase complex. Further, the prothrombinase complex thus formed activates prothrombin to thrombin. Normal RVV-X-induced clotting time and prolonged prothrombin time indicates FVII deficiency [extrinsic pathway of blood coagulation [134]], whereas prolonged RVV-X-induced clotting time signifies FX deficiency [135]. Proteinases which are in use as tools in coagulation laboratories are listed in Table 23.4.

23.4.4 Snake Venom Proteinases as Tools to Identify Snake Species Responsible for Bite

Recent updates show the existence of more than 600 venomous snake species around the world, and all of them belong to any of the Colubridae, Viperidae, or Elapidae families [136]. In India more than 200 snakes have been identified, and only 52 of them are reported to be venomous. Among the 52 venomous snakes, *Naja naja*, *Bungarus caeruleus*, *Daboia russelii*, and *Echis carinatus* are accountable for the majority of envenomations reported and are collectively termed as "BIG FOUR" venomous snakes of India [137, 138]. *Naja naja* and *Bungarus caeruleus* belong to Elapidae family, whereas *Daboia russelii* and *Echis carinatus* belong to Viperidae family.

Snake bite is considered as an occupational health hazard, and survey reports show an annual estimated death of 1300 to 50,000 in India [139, 140]. The only available treatment strategy against snake bite in India is the administration of

Table 23.4 Snake venom proteinases as tools in diagnostics

Name	Type of proteinase	Snake species	Trade name	Manufacturer	Remarks	References
Batroxobin	Serine	*Bothrops atrox*	Pefakit® Reptilase® time	Pentapharm, Basel, Switzerland	SVTLE-A, used for the determination of fibrinogen polymerization disorders and other clinical conditions connected with the last phase of coagulation	[143]
Ecarin	Metallo	*Echis carinatus*	Ecarin®	Pentapharm, Basel, Switzerland	Group A prothrombin activator, used in determination of prothrombin levels in patients undergoing anticoagulant therapy	[119]
Carinactivase (CA-1)	Metallo	*Echis carinatus*	Not yet available in market	NA	Group B prothrombin activator, measures only normal thrombin but not des-carboxy thrombin as evident with patients undergoing warfarin therapy	[144]
Textarin	Serine	*Pseudonaja textilis*	Textarin®	Pentapharm, Basel, Switzerland	Group D prothrombin activator, used in combination with Ecarin to detect delay in clotting time due to the presence of lupus anticoagulants (LA)	[124]
RVV-V	Serine	*Daboia siamensis*	RVV-V®	Pentapharm, Basel, Switzerland	RVV-V is used to destabilize and selectively inactivate FV in plasma and thus to prepare a routine reagent for the FV determination	[145]
RVV-X	Metallo	*Daboia siamensis*	RVV-X®	Pentapharm, Basel, Switzerland	Used for the detection of FVII and FX deficiencies. Also used to detect prolonged clotting time in LA	[146, 147]
ACC-C	Serine	*Agkistrodon contortrix*	Protac®	Pentapharm, Basel, Switzerland	Determination of Leiden mutation in FV	[148]
Noscarin	Serine	*Notechis scutatus*	Pefakit® APC-R Factor V Leiden	Pentapharm, Basel, Switzerland	Group D prothrombin activator, used in combination with RVV-X and ACC-C to detect Leiden mutation in FV	[130]

Note: *NA* not applicable

polyvalent ASV raised against the BIG FOUR venoms in equine. Administration of ASV poses some challenges and risks such as anaphylactic shock and allergic symptoms as secondary complications which may also lead to death [141]. Hospital records also evidence the administrations of prophylactic and antihistamine drugs to majority of inpatients to combat the secondary complications mounted by ASV [142]. These complications can be minimized by administering species-specific monovalent ASV and for which identification of snake species responsible for bite is very critical.

As of now, Australia and Papua New Guinea are the only two countries which are using monovalent ASV through the means of snake venom detection kits. These kits aid in the detection of the antigen(s)/protein(s) of the species-specific venom present in the blood or plasma of the victims based on the ELISA method, and depending on the test outcome, a clinician may decide to administer species-specific monovalent ASV. However, recent reports have shown the cross-reactivity existing between these monovalent ASV against different species which may also reflect with snake venom detection kits. These cross-reactivities are attributed to the antigenic similarity shared by snake venom proteins [38, 74, 79, 149–151]. Hence, there is a need for an alternative detection method other than ELISA or perhaps a non-antigen-based method to identify the snake species responsible for bite.

Evident from the aforementioned aspects, hemostatic system comprising plenty of proteins is a preferred target for snake venom enzymes. Immediately after envenomation, hemostatic system gets affected and undergoes plenty of observable changes, and these changes show marked differences depending on the family of snake which has offended [152, 153]. On these grounds, our research group designed a study in which different groups of rats were injected with BIG FOUR venoms and the plasma obtained thereafter was used for the assessment of common routine coagulation parameters such as recalcification time, prothrombin time, activated partial thromboplastin time, and coagulation factor assays including fibrinogen levels. In all the tests, striking differences existed between the plasma obtained from viperid and elapid venom-injected groups, and results were promising enough to differentiate between viperid and elapid species. If these result are consistent with the human subjects, bivalent ASV specific against two elapids or viperids can be administered rather than polyvalent ASV which in turn brings down the load of ASV thereby reducing the secondary complications [154].

23.5 Concluding Remarks

Among the diverse classes of toxins present in snake venom, proteinases being high molecular weight components exhibit structural and functional diversity. In addition, it has given a wide platform for toxinologists not only to explore their toxic properties but also to extrapolate the same to their pharmacological applications. Although different domains contribute to structure and functions of SVMPs, catalytic domain is vital in inducing pathological burden. Targeting the catalytic domain will prove invaluable in SVMP-induced pathologies. Thus, chelation therapy might be a vital strategy in mitigating the extensive tissue damage induced by these proteinases. Further, both SVSPs and SVMPs are capable of inducing systemic alteration by specifically cleaving/inhibiting/activating the components of complement system and blood coagulation pathway. Despite the observed toxicities, differential selectivity and specificity of these toxins toward their substrates can be used as tools to explore and expand the existing knowledge. In addition, the ability of SVMPs and SVSPs to act similarly as blood coagulation factors has extended their application to coagulation laboratories. Further, potential of these proteinases to differentially alter the coagulation cascade upon bite by different snake species may be harvested to identify the snake responsible for bite.

Acknowledgments The authors thank University Grants Commission (UGC) for providing research fellowship to carry out the work and thank Dr. Yariswamy, M. and Mr. Vikram Joshi for the critical reading of the manuscript.

Conflict of Interest The authors have no conflicts of interests to disclose.

References

1. Okuda J, Kiyokawa R (2000) Snake as a symbol in medicine and pharmacy–a historical study. Yakushigaku Zasshi 35(1):25–40
2. Aird SD, Watanabe Y, Villar-Briones A, Roy MC, Terada K, Mikheyev AS (2013) Quantitative high-throughput profiling of snake venom gland transcriptomes and proteomes (Ovophis okinavensis and Protobothrops flavoviridis). BMC Genomics 14:790. doi:10.1186/1471-2164-14-790
3. Ramos OH, Selistre-de-Araujo HS (2006) Snake venom metalloproteases–structure and function of catalytic and disintegrin domains. Comp Biochem Physiol C Toxicol Pharmacol 142(3–4):328–346. doi:S1532-0456(05)00240-1 [pii] 10.1016/j.cbpc.2005.11.005
4. Jorge da Silva N, Jr, & Aird SD (2001) Prey specificity, comparative lethality and compositional differences of coral snake venoms. Comp Biochem Physiol C Toxicol Pharmacol 128 (3):425–456. doi:S1532045600002155 [pii]
5. McCleary RJ, Kini RM (2013) Non-enzymatic proteins from snake venoms: a gold mine of pharmacological tools and drug leads. Toxicon 62:56–74. doi:S0041-0101(12)00751-9 [pii]10.1016/j.toxicon.2012.09.008
6. Kang TS, Georgieva D, Genov N, Murakami MT, Sinha M, Kumar RP, Kaur P, Kumar S, Dey S, Sharma S, Vrielink A, Betzel C, Takeda S, Arni RK, Singh TP, Kini RM (2011) Enzymatic toxins from snake venom: structural characterization and mechanism of catalysis. FEBS J 278(23):4544–4576. doi:10.1111/j.1742-4658.2011.08115.x
7. Tu AT, Passey RB (1966) Snake venoms and catalase and peroxidase activities. Toxicon 3(4):307–308
8. Dhananjaya BL, Nataraju A, Rajesh R, Raghavendra Gowda CD, Sharath BK, Vishwanath BS, D'Souza CJ (2006) Anticoagulant effect of Naja naja venom 5'nucleotidase: demonstration through the use of novel specific inhibitor, vanillic acid. Toxicon 48(4):411–421. doi:S0041-0101(06)00210-8 [pii] 10.1016/j.toxicon.2006.06.017
9. Kemparaju K, Girish KS (2006) Snake venom hyaluronidase: a therapeutic target. Cell Biochem Funct 24(1):7–12. doi:10.1002/cbf.1261
10. Oron U, Bdolah A (1973) Regulation of protein synthesis in the venom gland of viperid snakes. J Cell Biol 56(1):177–190

11. Fox JW, Serrano SM (2008) Insights into and speculations about snake venom metalloproteinase (SVMP) synthesis, folding and disulfide bond formation and their contribution to venom complexity. FEBS J 275 (12):3016–3030. doi:EJB6466 [pii] 10.1111/j.1742-4658.2008.06466.x
12. Kini RM, Evans HJ (1992) Structural domains in venom proteins: evidence that metalloproteinases and nonenzymatic platelet aggregation inhibitors (disintegrins) from snake venoms are derived by proteolysis from a common precursor. Toxicon 30(3):265–293
13. Jagadeesha DK, Shashidhara murthy R, Girish KS, Kemparaju K (2002) A non-toxic anticoagulant metalloprotease: purification and characterization from Indian cobra (Naja naja naja) venom. Toxicon 40 (6):667–675. doi:S0041010101002161 [pii]
14. Serrano SM, Maroun RC (2005) Snake venom serine proteinases: sequence homology vs. substrate specificity, a paradox to be solved. Toxicon 45(8):1115–1132. doi:S0041-0101(05)00075-9 [pii] 10.1016/j.toxicon.2005.02.020
15. Fox JW, Serrano SM (2005) Structural considerations of the snake venom metalloproteinases, key members of the M12 reprolysin family of metalloproteinases. Toxicon 45(8):969–985. doi:S0041-0101(05)00064-4 [pii] 10.1016/j.toxicon.2005.02.012
16. Fox JW, Serrano SM (2009) Timeline of key events in snake venom metalloproteinase research. J Proteome 72(2):200–209. doi:S1874-3919(09)00041-4 [pii] 10.1016/j.jprot.2009.01.015
17. Hite LA, Jia LG, Bjarnason JB, Fox JW (1994) cDNA sequences for four snake venom metalloproteinases: structure, classification, and their relationship to mammalian. Arch Biochem Biophys 308(1):182–191. doi:S0003-9861(84)71026-5 [pii] 10.1006/abbi.1994.1026
18. Hite LA, Shannon JD, Bjarnason JB, Fox JW (1992) Sequence of a cDNA clone encoding the zinc metalloproteinase hemorrhagic toxin e from Crotalus atrox: evidence for signal, zymogen, and disintegrin-like structures. Biochemistry 31(27):6203–6211
19. Shimokawa K, Jia LG, Shannon JD, Fox JW (1998) Isolation, sequence analysis, and biological activity of atrolysin E/D, the non-RGD disintegrin domain from Crotalus atrox venom. Arch Biochem Biophys 354(2):239–246. doi:S0003986198906981 [pii]
20. Zhu L, Yuan C, Chen Z, Wang W, Huang M (2010) Expression, purification and characterization of recombinant Jerdonitin, a P-II class snake venom metalloproteinase comprising metalloproteinase and disintegrin domains. Toxicon 55(2–3):375–380. doi:S0041-0101(09)00439-5 [pii] 10.1016/j.toxicon.2009.08.016
21. Nikai T, Taniguchi K, Komori Y, Masuda K, Fox JW, Sugihara H (2000) Primary structure and functional characterization of bilitoxin-1, a novel dimeric P-II snake venom metalloproteinase from Agkistrodon bilineatus venom. Arch Biochem Biophys 378(1):6–15. doi:S0003-9861(00)91795-8 [pii] 10.1006/abbi.2000.1795
22. Markland FS Jr, Swenson S (2013) Snake venom metalloproteinases. Toxicon 62:3–18. doi:S0041-0101(12)00747-7 [pii] 10.1016/j.toxicon.2012.09.004
23. Zhou Q, Hu P, Ritter MR, Swenson SD, Argounova S, Epstein AL, Markland FS (2000) Molecular cloning and functional expression of contortrostatin, a homodimeric disintegrin from southern copperhead snake venom. Arch Biochem Biophys 375(2):278–288. doi:10.1006/abbi.1999.1682 S0003-9861(99)91682-X [pii]
24. Moiseeva N, Bau R, Swenson SD, Markland FS Jr, Choe JY, Liu ZJ, Allaire M (2008) Structure of acostatin, a dimeric disintegrin from Southern copperhead (Agkistrodon contortrix contortrix), at 1.7 A resolution. Acta Crystallogr D Biol Crystallogr 64(Pt 4):466–470. doi:S0907444908002370 [pii] 10.1107/S0907444908002370
25. Kamiguti AS, Gallagher P, Marcinkiewicz C, Theakston RD, Zuzel M, Fox JW (2003) Identification of sites in the cysteine-rich domain of the class P-III snake venom metalloproteinases responsible for inhibition of platelet function. FEBS Lett 549(1–3):129–134. doi:S0014579303007993 [pii]
26. Masuda S, Maeda H, Miao JY, Hayashi H, Araki S (2007) cDNA cloning and some additional peptide characterization of a single-chain vascular apoptosis-inducing protein, VAP2. Endothelium 14(2):89–96. doi:778554271 [pii] 10.1080/10623320701346882
27. Takeda S (2016) ADAM and ADAMTS family proteins and snake venom metalloproteinases: a structural overview. Toxins (Basel) 8(5.) doi:toxins8050155 [pii] 10.3390/toxins8050155

28. Takeda S, Igarashi T, Mori H, Araki S (2006) Crystal structures of VAP1 reveal ADAMs' MDC domain architecture and its unique C-shaped scaffold. EMBO J 25(11):2388–2396. doi:7601131 [pii] 10.1038/sj.emboj.7601131
29. Takeda S, Igarashi T, Mori H (2007) Crystal structure of RVV-X: an example of evolutionary gain of specificity by ADAM proteinases. FEBS Lett 581(30):5859–5864. doi:S0014-5793(07)01216-1 [pii] 10.1016/j.febslet.2007.11.062
30. Takeda S, Takeya H, Iwanaga S (2012) Snake venom metalloproteinases: structure, function and relevance to the mammalian ADAM/ADAMTS family proteins. Biochim Biophys Acta 1824(1):164–176. doi:S1570-9639(11)00095-1 [pii] 10.1016/j.bbapap.2011.04.009
31. Kini RM (2006) Anticoagulant proteins from snake venoms: structure, function and mechanism. Biochem J 397(3):377–387. doi:BJ20060302 [pii] 10.1042/BJ20060302
32. Nanjaraj Urs AN, Yariswamy M, Joshi V, Suvilesh KN, Sumanth MS, Das D, Nataraju A, Vishwanath BS (2015) Local and systemic toxicity of Echis carinatus venom: neutralization by Cassia auriculata L. leaf methanol extract. J Nat Med 69(1):111–122. doi:10.1007/s11418-014-0875-3
33. Howes JM, Theakston RD, Laing GD (2007) Neutralization of the haemorrhagic activities of viperine snake venoms and venom metalloproteinases using synthetic peptide inhibitors and chelators. Toxicon 49(5):734–739. doi:S0041-0101(06)00443-0 [pii] 10.1016/j.toxicon.2006.11.020
34. Bjarnason JB, Fox JW (1994) Hemorrhagic metalloproteinases from snake venoms. Pharmacol Ther 62(3):325–372. doi:0163-7258(94)90049-3 [pii]
35. Kamiguti AS, Theakston RD, Desmond H, Hutton RA (1991) Systemic haemorrhage in rats induced by a haemorrhagic fraction from Bothrops jararaca venom. Toxicon 29(9):1097–1105. doi:0041-0101(91)90207-8 [pii]
36. Escalante T, Rucavado A, Fox JW, Gutierrez JM (2011) Key events in microvascular damage induced by snake venom hemorrhagic metalloproteinases. J Proteome 74(9):1781–1794. doi:S1874-3919(11)00128-X [pii] 10.1016/j.jprot.2011.03.026
37. Gutierrez JM, Rucavado A, Escalante T, Diaz C (2005) Hemorrhage induced by snake venom metalloproteinases: biochemical and biophysical mechanisms involved in microvessel damage. Toxicon 45(8):997–1011. doi:S0041-0101(05)00066-8 [pii] 10.1016/j.toxicon.2005.02.029
38. Baramova EN, Shannon JD, Bjarnason JB, Gonias SL, Fox JW (1990) Interaction of hemorrhagic metalloproteinases with human alpha 2-macroglobulin. Biochemistry 29(4):1069–1074
39. Arakawa H, Nishigai M, Ikai A (1989) Alpha 2-macroglobulin traps a proteinase in the midregion of its arms. An immunoelectron microscopic study. J Biol Chem 264(4):2350–2356
40. Homma M, Tu AT (1971) Morphology of local tissue damage in experimental snake envenomation. Br J Exp Pathol 52(5):538–542
41. Escalante T, Ortiz N, Rucavado A, Sanchez EF, Richardson M, Fox JW, Gutierrez JM (2011) Role of collagens and perlecan in microvascular stability: exploring the mechanism of capillary vessel damage by snake venom metalloproteinases. PLoS One 6(12):e28017. 10.1371/journal.pone.0028017 PONE-D-11-17100 [pii]
42. Herrera C, Escalante T, Voisin MB, Rucavado A, Morazan D, Macedo JK, Calvete JJ, Sanz L, Nourshargh S, Gutierrez JM, Fox JW (2015) Tissue localization and extracellular matrix degradation by PI, PII and PIII snake venom metalloproteinases: clues on the mechanisms of venom-induced hemorrhage. PLoS Negl Trop Dis 9(4):e0003731. doi:10.1371/journal.pntd.0003731 PNTD-D-15-00037 [pii]
43. Ohsaka A (1979) Hemorrhagic, necrotizing and edema-forming effects of snake venoms. In: Lee C-Y (ed) Snake venoms. Springer, Berlin/Heidelberg, pp 480–546. doi:10.1007/978-3-642-66913-2_14
44. Gutierrez JM, Escalante T, Rucavado A, Herrera C (2016) Hemorrhage caused by snake venom metalloproteinases: a journey of discovery and understanding. Toxins (Basel) 8(4.) doi:toxins8040093 [pii] 10.3390/toxins8040093
45. Ownby CL, Kainer RA, Tu AT (1974) Pathogenesis of hemorrhage induced by rattlesnake venom. An electron microscopic study. Am J Pathol 76(2):401–414

46. Ownby CL, Bjarnason J, Tu AT (1978) Hemorrhagic toxins from rattlesnake (Crotalus atrox) venom. Pathogenesis of hemorrhage induced by three purified toxins. Am J Pathol 93(1):201–218
47. Sharma RD, Katkar GD, Sundaram MS, Paul M, NaveenKumar SK, Swethakumar B, Hemshekhar M, Girish KS, Kemparaju K (2015) Oxidative stress-induced methemoglobinemia is the silent killer during snakebite: a novel and strategic neutralization by melatonin. J Pineal Res 59(2):240–254. doi:10.1111/jpi.12256
48. Katkar GD, Sundaram MS, Hemshekhar M, Sharma DR, Santhosh MS, Sunitha K, Rangappa KS, Girish KS, Kemparaju K (2014) Melatonin alleviates Echis carinatus venom-induced toxicities by modulating inflammatory mediators and oxidative stress. J Pineal Res 56(3):295–312. doi:10.1111/jpi.12123
49. Katkar GD, Sundaram MS, NaveenKumar SK, Swethakumar B, Sharma RD, Paul M, Vishalakshi GJ, Devaraja S, Girish KS, Kemparaju K (2016) NETosis and lack of DNase activity are key factors in Echis carinatus venom-induced tissue destruction. Nat Commun 7:11361. doi:ncomms11361 [pii] 10.1038/ncomms11361
50. Gowda CD, Rajesh R, Nataraju A, Dhananjaya BL, Raghupathi AR, Gowda TV, Sharath BK, Vishwanath BS (2006) Strong myotoxic activity of Trimeresurus malabaricus venom: role of metalloproteases. Mol Cell Biochem 282(1–2):147–155. doi:10.1007/s11010-006-1738-3
51. Banfi G, Salvagno GL, Lippi G (2007) The role of ethylenediamine tetraacetic acid (EDTA) as in vitro anticoagulant for diagnostic purposes. Clin Chem Lab Med 45(5):565–576. doi:10.1515/CCLM.2007.110
52. Nanjaraj Urs AN, Ramakrishnan C, Joshi V, Suvilesh KN, Veerabasappa Gowda T, Velmurugan D, Vishwanath BS (2015) Progressive hemorrhage and myotoxicity induced by echis carinatus venom in murine model: neutralization by inhibitor cocktail of N,N,N',N'-Tetrakis (2-Pyridylmethyl) Ethane-1,2-diamine and silymarin. PLoS One 10(8):e0135843. doi:10.1371/journal.pone.0135843 PONE-D-15-18679 [pii]
53. Pidde-Queiroz G, Furtado Mde F, Filgueiras CF, Pessoa LA, Spadafora-Ferreira M, van den Berg CW, Tambourgi DV (2010) Human complement activation and anaphylatoxins generation induced by snake venom toxins from Bothrops genus. Mol Immunol 47(16):2537–2544. doi:S0161-5890(10)00504-3 [pii] 10.1016/j.molimm.2010.07.003
54. Pidde-Queiroz G, Magnoli FC, Portaro FC, Serrano SM, Lopes AS, Paes Leme AF, van den Berg CW, Tambourgi DV (2013) P-I snake venom metalloproteinase is able to activate the complement system by direct cleavage of central components of the cascade. PLoS Negl Trop Dis 7(10):e2519. doi:10.1371/journal.pntd.0002519 PNTD-D-13-01135 [pii]
55. Yamamoto C, Tsuru D, Oda-Ueda N, Ohno M, Hattori S, Kim ST (2002) Flavoxobin, a serine protease from Trimeresurus flavoviridis (habu snake) venom, independently cleaves Arg726-Ser727 of human C3 and acts as a novel, heterologous C3 convertase. Immunology 107(1):111–117. doi:1490 [pii]
56. Sim RB, Laich A (2000) Serine proteases of the complement system. Biochem Soc Trans 28(5):545–550
57. Merle NS, Noe R, Halbwachs-Mecarelli L, Fremeaux-Bacchi V, Roumenina LT (2015) Complement system part II: role in immunity. Front Immunol 6:257. doi:10.3389/fimmu.2015.00257
58. Forneris F, Wu J, Gros P (2012) The modular serine proteases of the complement cascade. Curr Opin Struct Biol 22(3):333–341. doi:S0959-440X(12)00067-X [pii] 10.1016/j.sbi.2012.04.001
59. Sun QY, Bao J (2010) Purification, cloning and characterization of a metalloproteinase from Naja atra venom. Toxicon 56(8):1459–1469. doi:S0041-0101(10)00332-6 [pii] 10.1016/j.toxicon.2010.08.013
60. Farsky SH, Goncalves LR, Gutierrez JM, Correa AP, Rucavado A, Gasque P, Tambourgi DV (2000) Bothrops asper snake venom and its metalloproteinase BaP-1 activate the complement system. Role in leucocyte recruitment. Mediat Inflamm 9(5):213–221. doi:10.1080/09629350020025728
61. Molina O, Seriel RK, Martinez M, Sierra ML, Varela-Ramirez A, Rael ED (1990) Isolation of two hemorrhagic toxins from Crotalus basiliscus basiliscus (Mexican west coast rattlesnake) venom and their effect on blood clotting and complement. Int J Biochem 22(3):253–261

62. Menaldo DL, Bernardes CP, Santos-Filho NA, Moura Lde A, Fuly AL, Arantes EC, Sampaio SV (2012) Biochemical characterization and comparative analysis of two distinct serine proteases from bothrops pirajai snake venom. Biochimie 94(12):2545–2558. doi:S0300-9084(12)00289-1 [pii] 10.1016/j.biochi.2012.07.007
63. Menaldo DL, Bernardes CP, Pereira JC, Silveira DS, Mamede CC, Stanziola L, Oliveira F, Pereira-Crott LS, Faccioli LH, Sampaio SV (2013) Effects of two serine proteases from Bothrops pirajai snake venom on the complement system and the inflammatory response. Int Immunopharmacol 15(4):764–771. doi:S1567-5769(13)00082-9 [pii] 10.1016/j.intimp.2013.02.023
64. O'Keefe MC, Caporale LH, Vogel CW (1988) A novel cleavage product of human complement component C3 with structural and functional properties of cobra venom factor. J Biol Chem 263(25):12690–12697
65. Guan HH, Goh KS, Davamani F, Wu PL, Huang YW, Jeyakanthan J, Wu WG, Chen CJ (2010) Structures of two elapid snake venom metalloproteases with distinct activities highlight the disulfide patterns in the D domain of ADAMalysin family proteins. J Struct Biol 169(3):294–303. doi:S1047-8477(09)00313-X [pii] 10.1016/j.jsb.2009.11.009
66. Rael ED, Jones LP (1983) Isolation of an anticomplement factor from the venom of the Mojave rattlesnake (Crotalus scutulatus scutulatus). Toxicon 21(1):57–65. doi:0041-0101(83)90049-1 [pii]
67. Martinez M, Rael ED, Maddux NL (1990) Isolation of a hemorrhagic toxin from Mojave rattlesnake (Crotalus scutulatus scutulatus) venom. Toxicon 28(6):685–694
68. Chen T, Rael ED (1997) Purification of M5, a fibrinolytic proteinase from Crotalus molossus molossus venom that attacks complement. Int J Biochem Cell Biol 29(5):789–799. doi:S1357-2725(96)00139-2 [pii]
69. Shoibonov BB, Osipov AV, Kryukova EV, Zinchenko AA, Lakhtin VM, Tsetlin VI, Utkin YN (2005) Oxiagin from the Naja oxiana cobra venom is the first reprolysin inhibiting the classical pathway of complement. Mol Immunol 42(10):1141–1153. doi:S0161-5890(04)00470-5 [pii] 10.1016/j.molimm.2004.11.009
70. Lin X, Qi JZ, Chen MH, Qiu BT, Huang ZH, Qiu PX, Chen JS, Yan GM (2013) A novel recombinant fibrinogenase of Agkistrodon acutus venom protects against hyperacute rejection via degradation of complements. Biochem Pharmacol 85(6):772–779. doi:S0006-2952(12)00753-8 [pii] 10.1016/j.bcp.2012.11.012
71. Wang R, Qiu P, Jiang W, Cai X, Ou Y, Su X, Cai J, Chen J, Yin W, Yan G (2008) Recombinant fibrinogenase from Agkistrodon acutus venom protects against sepsis via direct degradation of fibrin and TNF-alpha. Biochem Pharmacol 76(5):620–630. doi:S0006-2952(08)00381-X [pii] 10.1016/j.bcp.2008.05.033
72. Sim RB, Tsiftsoglou SA (2004) Proteases of the complement system. Biochem Soc Trans 32(Pt 1):21–27. doi:10.1042/
73. Sarma JV, Ward PA (2011) The complement system. Cell Tissue Res 343(1):227–235. doi:10.1007/s00441-010-1034-0
74. Mathern DR, Heeger PS (2015) Molecules great and small: the complement system. Clin J Am Soc Nephrol 10(9):1636–1650. doi:CJN.06230614 [pii] 10.2215/CJN.06230614
75. Sahu A, Lambris JD (2001) Structure and biology of complement protein C3, a connecting link between innate and acquired immunity. Immunol Rev 180:35–48
76. Vogel CW, Fritzinger DC, Gorsuch WB, Stahl GL (2015) Complement depletion with humanised cobra venom factor: efficacy in preclinical models of vascular diseases. Thromb Haemost 113(3):548–552. doi:14-04-0300 [pii] 10.1160/TH14-04-0300
77. Kini RM, Evans HJ (1990) Effects of snake venom proteins on blood platelets. Toxicon 28(12):1387–1422
78. Sajevic T, Leonardi A, Krizaj I (2011) Haemostatically active proteins in snake venoms. Toxicon 57(5):627–645. doi:S0041-0101(11)00011-0 [pii] 10.1016/j.toxicon.2011.01.006
79. Lippi G, Franchini M, Montagnana M, Favaloro EJ (2012) Inherited disorders of blood coagulation. Ann Med 44(5):405–418

80. Broos K, Feys HB, De Meyer SF, Vanhoorelbeke K, Deckmyn H (2011) Platelets at work in primary hemostasis. Blood Rev 25(4):155–167. doi:S0268-960X(11)00022-1 [pii] 10.1016/j.blre.2011.03.002
81. Kamiguti AS (2005) Platelets as targets of snake venom metalloproteinases. Toxicon 45(8):1041–1049. doi:S0041-0101(05)00069-3 [pii] 10.1016/j.toxicon.2005.02.026
82. Serrano SM, Wang D, Shannon JD, Pinto AF, Polanowska-Grabowska RK, Fox JW (2007) Interaction of the cysteine-rich domain of snake venom metalloproteinases with the A1 domain of von Willebrand factor promotes site-specific proteolysis of von Willebrand factor and inhibition of von Willebrand factor-mediated platelet aggregation. FEBS J 274(14):3611–3621. doi:EJB5895 [pii] 10.1111/j.1742-4658.2007.05895.x
83. Lu Q, Clemetson JM, Clemetson KJ (2005) Snake venoms and hemostasis. J Thromb Haemost 3(8):1791–1799. doi:JTH1358 [pii] 10.1111/j.1538-7836.2005.01358.x
84. Clemetson KJ (2010) Snaclecs (snake C-type lectins) that inhibit or activate platelets by binding to receptors. Toxicon 56(7):1236–1246. doi:S0041-0101(10)00111-X [pii] 10.1016/j.toxicon.2010.03.011
85. Rumbaut RE, Thiagarajan P (2010) doi:NBK53450 [bookaccession]
86. Sanchez EF, Richardson M, Gremski LH, Veiga SS, Yarleque A, Niland S, Lima AM, Estevao-Costa MI, Eble JA (2016) A novel fibrinolytic metalloproteinase, barnettlysin-I from Bothrops barnetti (Barnett s pitviper) snake venom with anti-platelet properties. Biochim Biophys Acta 1860(3):542–556. doi:S0304-4165(15)00360-8 [pii] 10.1016/j.bbagen.2015.12.021
87. Kamiguti AS, Markland FS, Zhou Q, Laing GD, Theakston RD, Zuzel M (1997) Proteolytic cleavage of the beta1 subunit of platelet alpha2beta1 integrin by the metalloproteinase jararhagin compromises collagen-stimulated phosphorylation of pp72. J Biol Chem 272(51):32599–32605
88. Kamiguti AS, Hay CR, Theakston RD, Zuzel M (1996) Insights into the mechanism of haemorrhage caused by snake venom metalloproteinases. Toxicon 34(6):627–642. doi:0041010196000177 [pii]
89. Hsu CC, Wu WB, Chang YH, Kuo HL, Huang TF (2007) Antithrombotic effect of a protein-type I class snake venom metalloproteinase, kistomin, is mediated by affecting glycoprotein Ib-von Willebrand factor interaction. Mol Pharmacol 72(4):984–992. doi:mol.107.038018 [pii] 10.1124/mol.107.038018
90. Ward CM, Andrews RK, Smith AI, Berndt MC (1996) Mocarhagin, a novel cobra venom metalloproteinase, cleaves the platelet von Willebrand factor receptor glycoprotein Ibalpha. Identification of the sulfated tyrosine/anionic sequence Tyr-276-Glu-282 of glycoprotein Ibalpha as a binding site for von Willebrand factor and alpha-thrombin. Biochemistry 35(15):4929–4938. doi:10.1021/bi952456c bi952456c [pii]
91. Hamako J, Matsui T, Nishida S, Nomura S, Fujimura Y, Ito M, Ozeki Y, Titani K (1998) Purification and characterization of kaouthiagin, a von Willebrand factor-binding and -cleaving metalloproteinase from Naha kaouthia cobra venom. Thromb Haemost 80(3):499–505. doi:98090499 [pii]
92. Wijeyewickrema LC, Gardiner EE, Moroi M, Berndt MC, Andrews RK (2007) Snake venom metalloproteinases, crotarhagin and alborhagin, induce ectodomain shedding of the platelet collagen receptor, glycoprotein VI. Thromb Haemost 98(6):1285–1290. doi:07121285 [pii]
93. Wang WJ, Huang TF (2002) Purification and characterization of a novel metalloproteinase, acurhagin, from Agkistrodon acutus venom. Thromb Haemost 87(4):641–650
94. Hsu CC, Wu WB, Huang TF (2008) A snake venom metalloproteinase, kistomin, cleaves platelet glycoprotein VI and impairs platelet functions. J Thromb Haemost 6(9):1578–1585. doi:JTH3071 [pii] 10.1111/j.1538-7836.2008.03071.x
95. Guo D, Hou L, Chen Z, Wu S, Chen N (2000) Effect of natrahagin, a Chinese cobra venom proteinase on hemorrheology in rat. Zhong Yao Cai 23(6):339–341
96. Tseng YL, Lee CJ, Huang TF (2004) Effects of a snake venom metalloproteinase, triflamp, on platelet aggregation, platelet-neutrophil and neutrophil-neutrophil interactions: involvement of platelet GPIbalpha and neutrophil PSGL-1. Thromb Haemost 91(2):315–324. doi:10.1160/TH03-07-0426 04020315 [pii]

97. De Candia E (2012) Mechanisms of platelet activation by thrombin: a short history. Thromb Res 129(3):250–256. doi:S0049-3848(11)00577-9 [pii] 10.1016/j.thromres.2011.11.001
98. Kirby EP, Niewiarowski S, Stocker K, Kettner C, Shaw E, Brudzynski TM (1979) Thrombocytin, a serine protease from Bothrops atrox venom. 1. Purification and characterization of the enzyme. Biochemistry 18(16):3564–3570
99. Kubisz P, Arabi A, Seghier F, Cronberg S (1984) Investigations on the effect of thrombocytin on platelets. Thromb Res 33(2):225–227. doi:0049-3848(84)90183-X [pii]
100. Marrakchi N, Zingali RB, Karoui H, Bon C, el Ayeb M (1995) Cerastocytin, a new thrombin-like platelet activator from the venom of the Tunisian viper Cerastes cerastes. Biochim Biophys Acta 1244(1):147–156
101. Vilca-Quispe A, Ponce-Soto LA, Winck FV, Marangoni S (2010) Isolation and characterization of a new serine protease with thrombin-like activity (TLBm) from the venom of the snake Bothrops marajoensis. Toxicon 55(4):745–753. doi:S0041-0101(09)00524-8 [pii] 10.1016/j.toxicon.2009.11.006
102. Yamashiro ET, Oliveira AK, Kitano ES, Menezes MC, Junqueira-de-Azevedo IL, Paes Leme AF, Serrano SM (2014) Proteoforms of the platelet-aggregating enzyme PA-BJ, a serine proteinase from Bothrops jararaca venom. Biochim Biophys Acta 1844(12):2068–2076. doi:S1570-9639(14)00235-0 [pii] 10.1016/j.bbapap.2014.09.012
103. Yonamine CM, Kondo MY, Nering MB, Gouvea IE, Okamoto D, Andrade D, da Silva JA, Prieto da Silva AR, Yamane T, Juliano MA, Juliano L, Lapa AJ, Hayashi MA, Lima-Landman MT (2014) Enzyme specificity and effects of gyroxin, a serine protease from the venom of the South American rattlesnake Crotalus durissus terrificus, on protease-activated receptors. Toxicon 79:64–71. doi:S0041-0101(13)00458-3 [pii] 10.1016/j.toxicon.2013.12.002
104. Santos BF, Serrano SM, Kuliopulos A, Niewiarowski S (2000) Interaction of viper venom serine peptidases with thrombin receptors on human platelets. FEBS Lett 477(3):199–202. doi:S0014-5793(00)01803-2 [pii]
105. Gale AJ (2011) Continuing education course #2: current understanding of hemostasis. Toxicol Pathol 39(1):273–280. doi:0192623310389474 [pii] 10.1177/0192623310389474
106. Matsui T, Fujimura Y, Titani K (2000) Snake venom proteases affecting hemostasis and thrombosis. Biochim Biophys Acta 1477(1–2):146–156. doi:S0167-4838(99)00268-X [pii]
107. Castro HC, Zingali RB, Albuquerque MG, Pujol-Luz M, Rodrigues CR (2004) Snake venom thrombin-like enzymes: from reptilase to now. Cell Mol Life Sci 61(7–8):843–856. doi:10.1007/s00018-003-3325-z
108. Kini RM (2005) The intriguing world of prothrombin activators from snake venom. Toxicon 45(8):1133–1145. doi:S0041-0101(05)00076-0 [pii] 10.1016/j.toxicon.2005.02.019
109. Cortelazzo A, Guerranti R, Bini L, Hope-Onyekwere N, Muzzi C, Leoncini R, Pagani R (2010) Effects of snake venom proteases on human fibrinogen chains. Blood Transfus 8(Suppl 3):s120–s125. doi:10.2450/2010.019S
110. Tokunaga F, Nagasawa K, Tamura S, Miyata T, Iwanaga S, Kisiel W (1988) The factor V-activating enzyme (RVV-V) from Russell's viper venom. Identification of isoproteins RVV-V alpha, -V beta, and -V gamma and their complete amino acid sequences. J Biol Chem 263(33):17471–17481
111. Parry MA, Jacob U, Huber R, Wisner A, Bon C, Bode W (1998) The crystal structure of the novel snake venom plasminogen activator TSVPA: a prototype structure for snake venom serine proteinases. Structure 6(9):1195–1206. doi:S0969-2126(98)00119-1 [pii]
112. Zhang Y, Wisner A, Xiong Y, Bon C (1995) A novel plasminogen activator from snake venom. Purification, characterization, and molecular cloning. J Biol Chem 270(17):10246–10255
113. Chen HS, Chen JM, Lin CW, Khoo KH, Tsai IH (2008) New insights into the functions and N-glycan structures of factor X activator from Russell's viper venom. FEBS J 275(15):3944–3958. doi:EJB6540 [pii] 10.1111/j.1742-4658.2008.06540.x
114. Nakagaki T, Lin P, Kisiel W (1992) Activation of human factor VII by the prothrombin activator from the venom of Oxyuranus scutellatus (Taipan snake). Thromb Res 65(1):105–116. doi:0049-3848(92)90230-8 [pii]

115. Orthner CL, Bhattacharya P, Strickland DK (1988) Characterization of a protein C activator from the venom of Agkistrodon contortrix contortrix. Biochemistry 27(7):2558–2564
116. Karapetian H (2013) Reptilase time (RT). Methods Mol Biol 992:273–277. doi:10.1007/978-1-62703-339-8_20
117. Rosing J, Zwaal RF, Tans G (1986) Formation of meizothrombin as intermediate in factor Xa-catalyzed prothrombin activation. J Biol Chem 261(9):4224–4228
118. Kornalik F, Blomback B (1975) Prothrombin activation induced by Ecarin – a prothrombin converting enzyme from Echis carinatus venom. Thromb Res 6(1):57–63
119. Nowak G (2003) The ecarin clotting time, a universal method to quantify direct thrombin inhibitors. Pathophysiol Haemost Thromb 33(4):173–183. doi:81505 [pii] 81505
120. Pei G, Laue TM, Aulabaugh A, Fowlkes DM, Lentz BR (1992) Structural comparisons of meizothrombin and its precursor prothrombin in the presence or absence of procoagulant membranes. Biochemistry 31(30):6990–6996
121. Casserly IP, Kereiakes DJ, Gray WA, Gibson PH, Lauer MA, Reginelli JP, Moliterno DJ (2004) Point-of-care ecarin clotting time versus activated clotting time in correlation with bivalirudin concentration. Thromb Res 113(2):115–121. doi:10.1016/j.thromres.2004.02.012 S0049384804001082 [pii]
122. Stocker K, Hauer H, Muller C, Triplett DA (1994) Isolation and characterization of Textarin, a prothrombin activator from eastern brown snake (Pseudonaja textilis) venom. Toxicon 32(10):1227–1236
123. Hughes GR (1983) Thrombosis, abortion, cerebral disease, and the lupus anticoagulant. BMJ 287(6399):1088–1089
124. Triplett DA, Stocker KF, Unger GA, Barna LK (1993) The Textarin/Ecarin ratio: a confirmatory test for lupus anticoagulants. Thromb Haemost 70(6):925–931
125. Perchuc AM, Wilmer M (2011) Diagnostic use of snake venom components in the coagulation laboratory. In: Kini MR, Clemetson JK, Markland SF, McLane AM, Morita T (eds) Toxins and hemostasis: from bench to bedside. Springer Netherlands, Dordrecht, pp 747–766. doi:10.1007/978-90-481-9295-3_43
126. Fay WP, Owen WG (1989) Platelet plasminogen activator inhibitor: purification and characterization of interaction with plasminogen activators and activated protein C. Biochemistry 28(14):5773–5778
127. Stocker K, Fischer H, Meier J (1988) Practical application of the protein C activator Protac from Agkistrodon contortrix venom. Folia Haematol Int Mag Klin Morphol Blutforsch 115(3):260–264
128. Kujovich JL (2010) Factor V Leiden thrombophilia. Genet Med 13(1):1–16
129. Gable PS, Le DT, McGehee W, Rapaport SI A ProtacÂ®-based screening test for activated protein C-resistant factor Va and other defects activator (CA-1) during warfarin anticoagulation. J Heart Valve Dis 10(3):388–392
130. Schoni R (2005) The use of snake venom-derived compounds for new functional diagnostic test kits in the field of haemostasis. Pathophysiol Haemost Thromb 34(4-5):234–240. doi:92430 [pii] 10.1159/000092430
131. Bachmann F, Duckert F, Koller F (1958) The Stuart-Prower factor assay and its clinical significance. Thromb Diath Haemorrh 2(1–2):24–38
132. Quick AJ (1971) Thromboplastin generation: effect of the bell-alton reagent and russell viper venom on prothrombin consumption. Am J Clin Pathol 55(5):555–560
133. Segal JB, Lehmann HP, Petri M, Mueller L, Kickler TS (2002) Testing strategies for diagnosing lupus anticoagulant: decision analysis. Am J Hematol 70(3):195–205
134. Palta S, Saroa R, Palta A (2014) Overview of the coagulation system. Indian J Anaesth 58(5):515
135. Marsh NA (2001) Diagnostic uses of snake venom. Haemostasis 31(3–6):211–217. doi:48065 [pii] 48065
136. O'Shea M (2008) Venomous snakes of the world. New Holland Publishers

137. Sharma LR, Lal V, Simpson ID (2008) Snakes of medical significance in india: the first reported case of envenoming by the levantine viper (Macrovipera lebetina). Wilderness Environ Med 19(3):195
138. Ahmed SM, Ahmed M, Nadeem A, Mahajan J, Choudhary A, Pal J (2008) Emergency treatment of a snake bite: pearls from literature. J Emerg Trauma Shock 1(2):97
139. Swaroop S, Grab B (1954) Snakebite mortality in the world. Bull World Health Organ 10(1):35–76
140. Mohapatra B, Warrell DA, Suraweera W, Bhatia P, Dhingra N, Jotkar RM, Rodriguez PS, Mishra K, Whitaker R, Jha P, Gyapong JO (2011) Snakebite mortality in India: a nationally representative mortality survey. PLoS Negl Trop Dis 5(4):e1018
141. Amin MR, Mamun SMH, Rashid R, Rahman M, Ghose A, Sharmin S, Rahman MR, Faiz MA (2008) Anti-snake venom: use and adverse reaction in a snake bite study clinic in Bangladesh. J Venom Anim Toxins Includ Trop Dis 14(4):660–672
142. Gaidhankar SL, Ghanghas RR, Pore SM, Ramanand SJ, Patil PT, Gore AD, Pawar MP (2015) A retrospective study of use of polyvalent anti-snake venom and risk factors for mortality from snake bite in a tertiary care setting. Indian J Pharm 47(3):270
143. Cunningham MT, Brandt JT, Laposata M, Olson JD (2002) Laboratory diagnosis of dysfibrinogenemia. Arch Pathol Lab Med 126(4):499–505. doi:10.1043/0003-9985(2002)126<0499:LDOD>2.0.CO;2
144. Iwahashi H, Kimura M, Nakajima K, Yamada D, Morita T (2001) Determination of plasma prothrombin level by Ca2+-dependent prothrombin activator (CA-1) during warfarin anticoagulation. J Heart Valve Dis 10(3):388–392
145. Kisiel W, Canfield WM (1981) Snake venom proteases that activate blood-coagulation factor V. Methods Enzymol 80 Pt C:275–285
146. Denson KW (2004) The specific assay of prower-stuart factor and factor VII. Acta Haematol 25(2):105–120
147. Pengo V, Biasiolo A, Rampazzo P, Brocco T (1999) dRVVT is more sensitive than KCT or TTI for detecting lupus anticoagulant activity of antibeta2-glycoprotein I autoantibodies. Thromb Haemost 81(2):256–258. doi:99020256 [pii]
148. Gempeler-Messina PM, Volz K, Buhler B, Muller C (2001) Protein C activators from snake venoms and their diagnostic use. Haemostasis 31(3–6):266–272. doi:48072 [pii] 48072
149. Judy O, Haiart S, Galluccio S, White J, Weinstein SA (2015) An instructive case of presumed brown snake (spp.) envenoming. Clin Toxicol 53(8):834–839
150. O'Leary MA, Isbister GK (2009) Commercial monovalent antivenoms in Australia are polyvalent. Toxicon 54(2):192–195
151. Isbister GK, O'Leary MA, Hagan J, Nichols K, Jacoby T, Davern K, Hodgson WC, Schneider JJ (2010) Cross-neutralisation of Australian brown snake, taipan and death adder venoms by monovalent antibodies. Vaccine 28(3):798–802
152. Moriarity R, Dryer S, Replogle W, Summers R (2012) The role for coagulation markers in mild snakebite envenomations. West J Emerg Med 13(1):68–74
153. Harshavardhana H, Pasha I, Prabhu NS, Ravi P (2014) Snake bite induced coagulopathy: a study of clinical profile and predictors of poor outcome. Int J Scient Stud 2(1):2–5
154. Vilas H, Nanjaraj Urs AN, Vikram J, Suvilesh KN, Savitha MN, Urs P, Amog GVR, Yariswamy M, Vishwanath BS (2016) Differential action of medically important Indian BIG FOUR snake venoms on rodent blood coagulation. Toxicon 110:19–26

The World of Proteases Across Microbes, Insects, and Medicinal Trees

24

Ratnakar Chitte and Sushma Chaphalkar

Abstract

Various types of proteases are present in all living organisms, and they play important roles in physiology. There is interest in protease types of enzymes because of their vital applications in industries. Proteases have specific inhibitors are used for control of larval insect targets. Microbes produce proteolytic–fibrinolytic enzymes, which play important roles in pathological clot lysis. Proteases from microbes, insects, medicinal plants, and endophytes have been studied. Antimicrobial peptides (AMPs) produced by medicinal plants, as well endophytic bacteria, play important roles in control of plant pathogens. Medicinal trees in the Nakshatra Garden produce proteases and peptides, each of which has its own physiological role and warrants investigation for various applications.

Keywords

Microbial • Insect & medicinal proteases • Antimicrobial peptides (AMPs) • Endophytes • Medicinal trees

24.1 Introduction

Microorganisms are found in inhospitable environments, such as those with extremes of pH, temperature, nutrient concentrations, and pressure, providing clues about the diversity and origins of microbial life. Archaea is one of the three domains of life. Archaea differ from Eukaryotes and Eubacteria in terms of genetic, biochemical, and structural features. Many archaeal genomes have been sequenced for

R. Chitte (✉) • S. Chaphalkar
Vidya Pratishthan's School of Biotechnology,
Vidyanagari, Baramati Dist, Pune 413 133, India
e-mail: rrc10@rediffmail.com; rrchitte10@rediffmail.com

S. Chakraborti, N.S. Dhalla (eds.), *Proteases in Physiology and Pathology*,
DOI 10.1007/978-981-10-2513-6_24

understanding of the unique cellular processes of Archaea and mechanisms for their adaptation to extreme environments. Over half of all archaeal genes encode unique proteins with unknown functions [1].

Methanogens have been identified in the hindguts of many terrestrial arthropods, including millipedes, cockroaches, termites, and scarab beetles [2]. The strict anaerobic conditions required for methanogen survival in the human colon and adaptation to specific local nutrient environments must also apply to other members of the microflora.

It is assumed that Eukaryotes originated from Archaea. This has been supported by phylogenetic analyses in which Eukaryotes were found to be linked with Archaea. Though Archaea and Eukaryotes share many basic features at the molecular level; the archaeal mobilome resembles the bacterial mobilome more closely than the eukaryotic mobilome.

Archaea and bacteria are also quite similar at the genomic level. Both have small circular genomes. Otherwise, the gene organization of Archaea is not similar to that of bacteria and exhibits a lot of "eukaryotic features" at the molecular level [3–6]. Many archaeal operational systems—such as ATP production, protein secretion, cell division, and protein modification process—use proteins homologous to those of Eukaryotes rather than the bacterial system [7–9].

Archaea and Eukaryotes share many features in all aspects of their cellular physiology. Moreover, Archaea and Eukaryotes share a more complex evolutionary relationship than has previously been understood.

To date, several scenarios have been proposed to explain the origins of Archaea and Eukaryotes [10–12].

These evolutionary domains also reflect changes in the nature of biomolecules for adaptation to physiological conditions.

Thermophilic bacteria produce multiple proteases, such as keratinase, metalloproteases, and serine endopeptidases. Fibrinolytic proteases play important roles in disease and thrombosis. Likewise, many peptidases play important roles in strategies against disease targets (24.1).

24.2 Physiological Significance of Proteases

Proteases play important roles in diseases in terms of physiology. Fibrinolytic proteases are important mediators of inflammation in diseases such as atherosclerosis, rheumatoid arthritis, and cancer. The zymogens of these proteases circulating in the blood lead to tissue damage as a consequence of vascular leaks, contributing to coagulation or fibrinolysis. A large amount of extravascular fibrin is a specific hallmark of lung injury and disease, including acute lung injury (ALI) [13] and Idiopathic Pulmonary fibrosis (IPF) [14]. In rheumatoid arthritis, fibrin is deposited into inflamed hyperplastic synovial tissue and the fluid of arthritic joints [15]. Deposits of insoluble fibrin on various bone parts lead to progression of arthritis [16].

24.2.1 Coagulation in Disease

Extracellular deposition of fibrin by the coagulation cascade is found in a number of diseases [13, 17].

In cases of cancer, fibrin is also detected surrounding carcinoma cells, particularly at the interface of stromal cells and blood vessels [18]. In disease conditions, fibrin in the matrix acts as a scaffold to support proliferation, migration, and growth of mesenchymal or tumor cells.

In rheumatoid arthritis, fibrin becomes autoantigenic by posttranslational modification, possibly leading to inflammation [19].

24.2.2 Fibrinolysis in Disease

Fibrinolysis is the counterpart of coagulation. Plasmin plays a key role in fibrinolysis, generated by proteolytic activation of plasminogen by tPA and uPA [20, 21].

Fibrinolysis is associated with tPA which, unlike uPA, shows fibrin-enhanced proteolytic activity [22]. Plasmin plays an important role in wound repair processes in damaged tissue by degrading fibrin [23]. In other ways, excessive formation of plasmin is harmful.

In rheumatoid arthritis, synovial levels of fibrin D-dimer are a measure of fibrinolysis and are used to assess the progress of the disease and the effects of therapy [24].

tPA-triggered plasmin generation is a critical component of extravascular proteolytic damage in immune brains and leads to hypoxic ischemia [25].

In cardiovascular disease conditions, generation of fibrin degradation products (FDPs) and D-dimers is semiquantitatively measured to assess the disease condition [26].

24.2.3 Regulation of Coagulation and Fibrinolysis

Coagulation and fibrinolysis in physiological wound repair are interlinked processes.

An important negative regulator of coagulation, thrombomodulin, binds thrombin to restrict it from cleavage of fibrinogen or activation of PAR-1.

Protein C is activated by the thrombomodulin–thrombin complex. A hereditary deficiency of protein C is an established risk factor for venous thrombosis [27]. Plasminogen and plasminogen activator, which accelerate and localize plasmin formation to the cell surface, are important regulators of fibrinolysis. Fibrinolysis is negatively regulated by the serpin plasminogen activator inhibitor 1 (PAI-1), which covalently binds to and inactivates plasminogen activators. It has been noted that PAI-1 levels are higher in many respiratory diseases [28]. The direct effects of PAI-1 on cells may also lead to disease pathology conditions [29].

Hyperfibrinolytic bleeding occurs in a number of diseases, including chronic liver disease, as a result of decreased concentrations of alpha antiplasmin [30]. Plasmin leads to proteolysis of FXa, revealing a cryptic binding site for tPA in the cleaved product (denoted as beta FXa) [31]. Binding of beta FXa to tPA accelerates fibrinolysis [32].

Carboxypeptidase removes C-terminal lysine and arginine residues on fibrin, with the consequence of reducing plasminogen binding, triggering activation and subsequent breakdown of fibrin.

Increased levels of activated TAFI are associated with diseases such as cardiovascular [33] and inflammatory bowel diseases [34].

24.3 Microbial Proteases

Purification and biochemical characterization of the fibrinolytic enzyme nattokinase from shrimp shell with *Bacillus subtilis TKU007* has been carried out [35]. A novel nattokinase has been reported from *Pseudomonas* sp. *TKU015*, using shrimp shells as a substrate [36]. A fibrinolytic serine protease has been physicochemically characterized from latex of the medicinal herb *Euphorbia hirta* [37]. Proteases affecting coagulation and fibrinolysis have been isolated and characterized from plant latex from the Euphorbiaceae family [38, 39]. These proteases produced from latex have physiological significance and may help to prevent pathological infections in plants. Lattices from different plants have traditionally been used to stop bleeding, and it is now known that they contain proteases that affect blood coagulation.

Microbes produce proteases that have their own significance in biochemical pathways. Some insects, such as honey bees and larvae, produce proteases in their guts. These proteases play important roles in digestion of food material. Some medicinal plants secreate fibrinolytic proteases may be for defense against plant pathogen and insects. Some fruits and seeds have also been reported to contain fibrinolytic proteases.

The fibrinolytic protease actinokinase is produced by thermophilic *Streptomyces* sp. Such enzymes produced by microbes may protect them against infectious agents.

Production of enzymes with fermentation has been reported to significantly increase the yields and sources of fibrinolytic proteases [40]. Cloning and expression of the ackS gene in various host strains and enzyme secretion have been described [41]. Spore preservation techniques using a filter disk technique for proteases with multiple enzyme strains have also been described [42]. Such proteases not only have physiological importance but also are being explored for commercial interest in various sectors for food, pharmaceutical, diagnostic, and agricultural applications.

24.4 Insect Proteases

24.4.1 Expression of Protease Inhibitors in Insect-Resistant Transgenic Plants

Different proteases are classified according to their mechanism of catalysis: (1) serine proteinases; (2) cysteine proteinases; (3) aspartic proteinases; and (4) metalloproteinases.

Two major proteinase classes present in the digestive systems of phytophagous insects are serine and cysteine proteinases [43], dominating the larval gut environment. These proteases are generally targeted for inhibitors for control of insect pests. Proteinase inhibitors are one of the most abundant classes of proteins in plants.

A number of transgenic plants have been developed with proteinase inhibitors for resistance to different families of insects. Serine proteinase inhibitors have been the subject of more research than any other class of inhibitors in the guts of many insect families. One of the recent developments in the field of plant genetic engineering is manipulation of plants for disease and insect resistance. A single defensive gene can be transferred from one plant to another either with own promoter or with a constitutive promoter. In such a way, insect resistance of transgenic plants can be developed for different families of insects.

There is commercial interest in plants to integrate and express foreign genes and produce recombinant protein molecules. New inhibitors for use against predatory insects, with the potential for use in recombinant DNA technology to develop transgenic resistance plants, have been studied [44]. Plant protease inhibitors are the main defensive system used against plant pathogens.

Medicinal plants produce inhibitors as small molecules. These small molecules need to be screened for inhibitory action of enzymes such as serine and cysteine protease inhibitors. These small molecules have great potential to bind the active site residues of key enzymes involved in disease conditions. Thus, there is scope to search for novel inhibitors for their potential uses. These inhibitors could be cloned and expressed in heterologous expression systems and could be used for commercial applications in direct or indirect ways.

Aspartic proteases are one of the major catalytic classes of proteases, found in microbes, insects, and medicinal plants. Insects have digestive aspartic peptidases in their guts [45]. The effects of peptidase inhibitors on the activity of aspartic proteases play significant roles in plant protection. Different classes of peptidases had been reported in relation to various disease condition (Table 24.1).

Table 24.1 Different classes of peptidases used in structure-based drug design

Peptidase	Biological function	Disease
Serine peptidases		
Thrombin	Proteolysis of fibrinogen	Thrombosis
Factor Xa	Conversion of prothrombin to thrombin	Thrombosis
Factor VIIa	Activation of factors IX and X	Thrombosis
Urokinase	Activation of plasminogen	Cancer
Cysteine peptidases		
FP-2	Hemoglobin degradation	Malaria
Calpains 1 and 2	Degradation of cytoskeletal proteins	Stroke, neural injuries
Picomain cysteine peptidases	Processing of viral pro-protein	Virus infection
Cathepsin B	Antigen processing	Acute pancreatitis, cancer
Aspartic peptidases		
Plasmepsin	Hemoglobin degradation	Malaria
Renin	Processing of angiotensinogen	Blood pressure
Memapsin 2	Secretase activity	Alzheimer disease
Metallopeptidases		
Matrix metallopeptidase 1	Degradation of connective tissue	Tissue damage in tumor invasion
Carboxypeptidases B and U	Cleavage of tissue plasminogen activator	Blood coagulation

24.5 Medicinal Plant Proteases

Some medicinal plants produce antimicrobial peptides (AMPs) as part of their own defensive mechanisms. These peptides also have potential as therapeutic peptides.

The Nakshatra Garden consists of medicinal trees. These trees produce varieties of biologically active compounds such as alkaloids, flavonoids, terpenoids, peptides, and enzymes. The production of these bioactive molecules by plants has physiological significance. Processes to extract these bioactive molecules have been attempted. Significant activities of proteases have been detected. These medicinal trees are associated with endophytic consortia. These endophytes produce important metabolites, such as plant growth hormones and promoters, which are symbiotically helpful to the plants and vice versa. Characterization of endophytic bacteria and plasmids has been carried out [46].

The proteases produced by the Nakshatra Garden's medicinal trees need to be screened for various applications, and these plants may produce enzymes suitable for food and pharmaceutical uses.

Proteases have been extracted from medicinal plants such as *Azadirachta indica* and *Terminalia arjuna* to be studied for their potential as antiviral targets, and analytical studies of proteases extracted from *Azadirachta indica* have been carried out [47, 48].

There are many bioactive compounds to be isolated and identified using mass spectroscopy and NMR techniques, and further work is needed to identify as yet unknown compounds. From our laboratory, various in vitro activities of compounds—such as antidiabetic, anti-inflammatory, anticancer, and antimicrobial activities—have been reported, and some data are still unpublished.

There is a lot of diversity and variation in the specificity of proteases from microbes, insects, and medicinal trees, and they could be useful as effective therapeutic agents. Depending on the specificity of their hydrolytic actions, proteases could be used for various applications in the food, detergent, leather, and pharmaceutical industries.

Gene cloning technology is being widely used to design molecules for better understanding of structures and functional relationships.

24.6 Applications for Proteases

Proteases are physiologically essential for all living organisms. They have wide diversity, depending on the sources from which they are obtained, such as microbes, insects, and medicinal plants [49]. Proteases such as papain, bromelain, and keratinases are proteases of plant origin.

Proteases have a variety of functions according to their applications. Proteases at cellular levels contribute greatly to metabolic pathways. They produce cascade systems for hemostasis and inflammation for normal physiological regulation of cells. Proteases are involved in the life cycles of disease-causing organisms and thus are potential targets for development of therapeutic agents for deadly diseases such as cancer, AIDS, and cardiovascular diseases [17]. Microbial proteases are used in the treatment of many disorders, such as cancer, inflammation, cardiovascular disorders, necrotic wounds, etc. [50].

Proteases are used in the pharmaceutical industry for preparation of medicines, such as fibrinolytic proteases for use as clot lysis agents.

Proteases also have wide applications in the detergent and food industries, where they are prepared in bulk quantities and used as crude preparations. Keratinolytic enzymes are used for dehairing and degradation of feathers to generate amino acids, and they could be used for animal feed preparation.

Proteases used in medicines are produced in small amounts and are required in absolutely pure forms. Proteases also have widespread applications in laundry detergents [51, 52].

Alkaline proteases are used in different brands of detergents for day-to-day uses. They are also used for removal of stains such as milk and blood. Protease enzymes remove body secretions and foods such as milk, egg, fish, and meat. The stability of the enzymes is most important, and they need to be stable and active in the detergent solution, with temperature stability at different washing temperatures.

Proteases are used in the food industry, such as for cheese making, baking, preparation of soya hydrolysates, and meat tenderization [53].

The main application of proteases in the dairy industry is for manufacturing of cheese. Coagulating proteases are classified into three main categories based on their source: (1) animal rennet; (2) microbial milk coagulant; and (3) genetically engineered chymosin.

24.7 Conclusion

Microbes are well known for production of therapeutic enzymes, and medicinal plants have been in use from ancient times for treatment of diseases. These medicinal plants are well known to show a broad spectrum of activities due to the presence of important metabolites. The active molecules are being specifically targeted toward diseases. Microbes, insects, and medicinal plants produce proteases for their own physiological significance, and their isolation and characterization have been carried out and reported. Further study of the diversity of these proteases is needed for understanding of their mechanisms of action for substrate catalysis to implement applications in the food, pharmaceutical, and agricultural industries.

References

1. Graham et al (2000) An archael genomic signature. PNAS 97(7):3304–3308
2. Sustr V (2014) Methane production and methanogenic Archaea in the digestive tracts of millipedes (Diplopoda). PLoS One 9(7):e102659
3. Zillig W (1991) Comparative biochemistry of Archaea and bacteria. Curr Opin Genet Dev 1(4):544–551
4. Olsen G, Woese CR (1997) Archaeal genomics: an overview. Cell 89(7):991–994
5. Garrett R, Klenk HP (2007) Archaea. Blackwell, Oxford
6. Brochier-Armanet C, Forterre P, Gribaldo S (2011) Phylogeny and evolution of the Archaea: one hundred genomes later. Curr Opin Microbiol 14(3):274–281
7. Grüber G, Marshansky V (2008) New insights into structure–function relationships between archeal ATP synthase (A1A0) and vacuolar type ATPase (V1V0). BioEssays (11–12):1096–1109
8. Van den Berg BWM, Clemons Jr WM, Collinson I et al (2004) X-ray structure of a protein-conducting channel. Nature 427(6969):36–44
9. Nunoura T, Takaki Y, Kakuta J et al (2011) Insights into the evolution of Archaea and eukaryotic protein modifier systems revealed by the genome of a novel archaeal group. Nucleic Acids Res 39(8):3204–3223
10. Koonin EV (2010) The origin and early evolution of eukaryotes in the light of phylogenomics. Genome Biol 11(5):209
11. Martijn J, Ettema TJ (2013) From archaeon to eukaryote: the evolutionary dark ages of the eukaryotic cell. Biochem Soc Trans 41(1):451–457
12. Lombard J, López-García P, Moreira D (2012) The early evolution of lipid membranes and the three domains of life. Nat Rev Microbiol 10(7):507–515
13. Tucker T, Idell S (2013) Plasminogen–plasmin system in the pathogenesis and treatment of lung and pleural injury. Semin Thromb Hemost 39(4):373–381
14. Noble PW, Barkauskas CE, Jiang D (2012) Pulmonary fibrosis: patterns and perpetrators. J Clin Invest 122 (8):2756–2762

15. Raghu H, Flick MJ (2011) Targeting the coagulation factor fibrinogen for arthritis therapy. Curr Pharm Biotechnol 12(9):1497–1506
16. Carmassi F, De Negri F, Morale M, Song KY, Chung SI (1996) Fibrin degradation in the synovial fluid of rheumatoid arthritis patients: a model for extravascular fibrinolysis. Semin Thromb Hemost 22(6):489–496
17. Wagers SS, Norton RJ, Rinaldi LM, Bates JHT, Sobel BE, Irvin CG, (2004) Extravascular fibrin, plasminogen activator, plasminogen activator inhibitors, and airway hyperresponsiveness. J Clin Invest 114(1):104–111
18. Schardt FW, Schmausser B, Bachmann E (2013) Monoclonal antibodies for immunodetection of fibrin deposits on cancer cells. Histol Histopathol 28(8):993–998
19. Sanchez-Pernaute O, Filkova M, Gabucio A et al (2013) Citrullination enhances the pro-inflammatory response to fibrin in rheumatoid arthritis synovial fibroblasts. Ann Rheum Dis 72(8):1400–1406
20. Kim J, Hajjar KA (2002) Annexin II: a plasminogen–plasminogen activator co-receptor. Front Biosci 7:d341–d348
21. Madureira PA, Surette AP, Phipps KD, Taboski MAS, Miller VA, Waisman DM (2011) The role of the annexin A2 heterotetramer in vascular fibrinolysis. Blood 118(18):4789–4797
22. Wagers S, Lundblad LKA, Ekman M, Irvin CG, Bates JHT (2004) The allergic mouse model of asthma: normal smooth muscle in an abnormal lung? J Appl Physiol 96(6):2019–2027
23. Swaisgood CM, French EL, Noga C, Simon RH, Ploplis VA (2000) The development of bleomycin-induced pulmonary fibrosis in mice deficient for components of the fibrinolytic system. Am J Pathol 157(1):177–187
24. Bloom BJ, Alario AJ, Miller LC (2009) Persistent elevation of fibrin D-dimer predicts long-term outcome in systemic juvenile idiopathic arthritis. J Rheumatol 36(2):422–426
25. Yang D, Kuan CY (2014) Anti-tissue plasminogen activator (tPA) as an effective therapy of neonatal hypoxia–ischemia with and without inflammation. CNS Neurosci Ther 21(4):367–373
26. Chitte RR, Deshmukh SV, Kanekar PP (2011) Production, purification and biochemical characterization of a fibrinolytic enzyme from thermophilic *Streptomyces* sp. MCMB−379. Appl Biochem Biotechnol 165:1406–1413
27. Tjeldhorn L, Sandset PM, Haugbro K, Skretting G (2010) Hereditary protein C deficiency caused by the Ala267Thr mutation in the protein C gene is associated with symptomatic and asymptomatic venous thrombosis. Thromb Res 125(3):230–234
28. Brims FJH, Chauhan AJ, Higgins B, Shute JK (2010) Up-regulation of the extrinsic coagulation pathway in acute asthma—a case study. J Asthma 47(6):695–698
29. Iwaki T, Urano T, Umemura K (2012) PAI-1, progress in understanding the clinical problem and its aetiology. Br J Haematol 157(3):291–298
30. Tripodi A, Mannucci PM (2011) The coagulopathy of chronic liver disease. N Engl J Med 365(2):147–156
31. Pryzdial ELG, Kessler GE (1996) Autoproteolysis or plasmin mediated cleavage of factor Xa exposes a plasminogen binding site and inhibits coagulation. J Biol Chem 271(28):16614–16620
32. Talbot K, Meixner SC, Pryzdial ELG (2010) Enhanced fibrinolysis by proteolysed coagulation factor Xa. Biochim Biophys Acta, Proteins Proteomics 1804(4):723–730
33. Tregouet DA, Schnabel R, Alessi MC et al (2009) Activated thrombin activatable fibrinolysis inhibitor levels are associated with the risk of cardiovascular death in patients with coronary artery disease: the AtheroGene study. J Thromb Haemost 7(1):49–57
34. Saibeni S, Bottasso B, Spina L et al (2004) Assessment of thrombin-activatable fibrinolysis inhibitor (TAFI) plasma levels in inflammatory bowel diseases. Am J Gastroenterol 99(10):1966–1970
35. Wang S, Wu Y, Liang T (2011) Purification and biochemical characterization of a nattokinase by conversion of shrimp shell with *Bacillus subtilis* TKU 007. New Biotechnol 28(2):196–202
36. Wang S, Chen H, Liang T, Lin Y (2009) A novel nattokinase produced by *Pseudomonas* sp. TKU015 using shrimp shell as substrate. Process Biochem 44:70–76

37. Patel G, Kawale A, Sharma A (2012) Purification and physiological characterization of a serine protease with fibrinolytic activity from latex of a medicinal herb *Euphorbia hirta*. Plant Physiol Biochem 52:104–111
38. Rajesh R, Natarajua A, Gowda CDR, Frey FJ, Vishwanath BS (2006) Purification and characterization of a 34 kDa, heat stable glycoprotein from *Synadenium grantii* latex: action on human fibrinogen and fibrin clot. Biochimie 88:1313–1322
39. Fonseca KC, Morais NCG, Queiroz MR et al (2010) Purification and biochemical characterization of Eumillin from *Euphorbia milli* var. *hislopii* latex. Phytochemistry 71:708–715
40. Chitte R (2013) Microbial source of fibrinolytic protease and its biotechnological potential. In Prof Chakraborti and Prof Dhalla (ed) Proteases in health and diseases. Springer, New York, 7:79–88
41. Chitte RR, Dey S (2001) Cloning and expression of a thermophilic *Streptomyces* actinokinase gene in *E. coli*. Indian J Exp Biol 39:414–415
42. Kulkarni GA, Chitte RR (2015) Preservation of thermophilic bacterial spores using filter disc techniques. J Bioprocess Biotech 5:223
43. Haq SK, Atif SM, Khan RH (2004) Protein proteinase inhibitor genes in combat against insects pests and pathogens: natural and engineered phytoprotection. Arch Biochem Biophys 431:145–159
44. Ussuf KK, Laxmi NH, Mitra R (2001) Proteinase inhibitors plant derived genes of insecticidal protein for developing insect-resistant transgenic plants. Curr Sci 80:847–853
45. Wright MK, Brandt SL et al (2006) Characterization of digestive proteolytic activity in *Lygus hesperus Knight* (Hemiptera: Miridae). J Insect Physiol 52:717–728
46. Berde CV, Bhosale PP, Chaphalkar SR (2010) Plasmids of endophytic bacteria as vectors for transformation in plants. Int J Integr Biol 9(3):113–118
47. Gupta A, Shah AP, Chaphalkar SR (2016) Extraction of proteases from medicinal plants and their potential as anti-viral targets. J Biotechnol Biomater 6: 228
48. Gupta A, Chaphalkar SR (2015) Analytical studies of protease extracted from *Azadirachta indica*. World J Pharm Res 4:1391–1398
49. Rao MB, Tanskale AM, Ghatger MS, Deshpande VV (1998) Molecular and biotechnological aspects of microbial proteases. Microbiol Mol Biol Rev 63(3):596–635
50. Chanalia P, Gandhi D, Singh JD (2011) Applications of microbial proteases in pharmaceutical industry: an over view. Rev Medical Microbial 22(4):96–101
51. Kalpana Devi M, Rasheedha Banu A, Gnanaprabhal GR, Pradeep BV, Palaniswamy M (2008) Purification, characterization of alkaline protease enzyme from native isolate *Aspergillu niger* and its compatibility with commercial detergents. Indian J Sci Technol 1:1–6
52. Bholay AD, Niranjan P (2012) Bacterial extracellular alkaline proteases and its industrial applications. Int Res J Biol Sci 1(7):1–5
53. Cheong C, Chun SS, Kim YH (1993) Production and properties of an alkaline protease from *Pseudomonas* sp. *SJ-320*. Korean Biochemical J 26:479–484

A Review on the Mode of the Interactions of Bacterial Proteases with Their Substrates

25

Sanchari Bhattacharjee, Rakhi Dasgupta, and Angshuman Bagchi

Abstract

Maintaining the quality of cellular proteome is a huge challenge to living cell as the proper functioning of protein directly affects proper functioning as well as survival of the cell. Due to different cellular stresses, the cellular proteins face recurring threats which has a negative impact on cellular proteins. As a result of these threats, proteins become damaged as well as misfolded, and these defective proteins impose a load to the cellular machinery by elevating the level of cytotoxicity. Protease enzymes are a type of cellular machinery that is specifically used to eliminate the damaged and short-lived regulatory proteins by their proteolytic mechanisms. Bacteria evolves different types of bacterial proteases that are highly diverse which correspond to their localization, sequence, structure, active sites, proteolytic mechanism, substrate specificity as well as function. Bacterial proteases not only eliminate the damaged protein but also act as chaperones in some special situation and thus act as charonin which makes a promising effect in cellular protein quality control. This study illustrates the detailed comparison of bacterial FtsH, Lon and Clp protease and their modes of interaction with substrates.

Keywords

FtsH protease • Lon protease • Clp protease

S. Bhattacharjee • R. Dasgupta (✉) • A. Bagchi (✉)
Department of Biochemistry and Biophysics, University of Kalyani,
Kalyani 741235, West Bengal, India
e-mail: rdgadg@gmail.com; angshumanb@gmail.com

S. Chakraborti, N.S. Dhalla (eds.), *Proteases in Physiology and Pathology*,
DOI 10.1007/978-981-10-2513-6_25

25.1 Introduction

Maintaining the integrity of proteome is vital for cell viability. Cells however face with a continuous stream of misfolded proteins that hamper the survival of cells [1]. Proteins are highly susceptible to misfolding not only during their native folding mechanism but also due to cellular stress. Protein misfolding results in loss of function of proteins and, if remain unchecked, can potentially form toxic protein inclusions leading to cytotoxicity of the cells [2]. So, these consequences threaten the protein homeostasis of the cell. To counteract these problems, cells emerge with a mechanism, called stress response mechanism that protects a cell in unfavourable environmental conditions, by employing mechanisms that reduce acute damage to cellular machinery as well as provide the cell a recovery measure against adverse conditions.

Cellular stress responses are primarily mediated through stress proteins. Stress proteins are of two categories: those that operate only under stress and those that are triggered on both in stress responses as well as in normal cellular functioning. These stress proteins, because of their essential role in cell survival during normal condition as well as in stressed condition, are highly conserved throughout the phyla, from simplest eukaryotic to the most complex prokaryotic cells. There are different pathways to encounter stress. A particular anti-stress response is however dependent on the type of stressor. Heat shock response, for example, is initiated by the presence of denatured proteins [3]. The fate of the stressed cells is ultimately determined by whether they adopt a protective or a destructive pathway depending on the nature and duration of the stress as well as the cell type.

With this cellular stress response, the cells also co-evolve the protein quality control network, a network that maintains the quality of the cellular proteins by setting different checkpoints at different levels and is mainly composed of molecular chaperones and proteases. Due to cellular stresses, the proteins become misfolded, aggregated as well as damaged. The molecular chaperones with their foldase activity not only refold the aggregated proteins to their native folding state but also prevent further aggregation of misfolded proteins via their holdase activity, whereas the protease eliminates the damaged proteins to reduce the cytotoxicity which will compromise with the viability of the cell [4]. Thus this quality control network functions in a concerted way to ensure the quality of the cellular proteins which is essential for the proper functioning and survival of the cell.

Proteases play an important role in the elimination of stressed proteins which accumulate during stress, thereby activating different types of stress response pathways [5]. In *Escherichia coli* there are several different proteases to maintain the cellular protein quality [6].

25.2 Protease: The Cellular Degradation Machinery

The misfolded proteins which are formed due to cellular stress impose a burden to normal cellular functioning as well as cell survival. To combat these conditions, post-translational cellular control is being carried out by degrading the short-lived regulatory proteins and protein quality control by eliminating defective and potentially damaged proteins from the cell. Proteases, a special kind of enzymes, facilitate the above phenomena. Usually these enzymes are ATP-dependent proteases, having both ATPase and proteolytic activities in domains within a single polypeptide chain or in individual subunits forming a complex assembly.

The misfolded proteins which are formed due to cellular stress impose a burden to normal cellular functioning as well as cell survival. Protease, a special kind of enzymes, facilitates degradation and elimination of misfolded proteins. Rapid proteolysis eliminates defective and potentially damaged proteins from the cell [7] exhibiting a promising role in post-translational cellular control by degrading the short-lived regulatory proteins and thus contributes in protein quality control. In all cells, protein degradation is mediated by ATP-dependent proteases, which contains both ATPase and protease as separate domains within a single polypeptide chain or as individual subunits in complex assemblies.

Protease breaks the proteins into smaller polypeptide fragments by accelerating the hydrolysis of peptide bonds. A unique feature of proteases is their preference to cleave the peptide bond associated with a particular amino acid. This particular amino acid belongs to the active site of the protease. The active site is actually in a cavity of the protein, called catalytic pocket. The type of amino acid residues within the pocket will determine the particular type of the proteases. Proteases hydrolyse the peptide bonds by activating a nucleophile; this activated nucleophile in turn attacks the carbon of the peptide bond [8]. As the nucleophile attaches itself, the electrons in the carbon-oxygen double bond migrate onto the oxygen forming a high-energy tetrahedral intermediate. So to stabilize this, the intermediate will then decompose, usually releasing the two peptide fragments.

25.3 Protease Families and Catalytic Mechanisms

There are usually six main classes of protease such as serine protease, cysteine protease, aspartate protease, threonine protease, glutamic acid protease and metalloprotease. These different classes of protease utilize either a different nucleophile or a different mechanism to activate it.

25.3.1 Serine Protease

Serine protease is found in both prokaryotes and eukaryotes. In this protease, the serine present in the active site of the protein acts as nucleophilic amino acid and cleaves peptide bond in proteins. The active site of the enzyme is a catalytic triad consisting of three amino acids: His 57, Ser 195 and Asp 102. This triad is preserved in all serine protease enzymes [9].

These three catalytic triad residues make a charge relay system and serve a pivotal role in proteolysis. Though the amino acid members of the triad are located far from one another in the primary sequence, due to folding these residues come very close to one another in the core of the native enzyme forming the triad.

The catalytic mechanism generates several intermediates. The catalysis of the peptide cleavage follows a ping-pong mechanism, in which a substrate binds, a product is released, another substrate binds and another product is released. In the case of serine protease, the –OH group of serine serves as nucleophile [10] attacking the carbonyl carbon of the scissile peptide bond of the substrate. In this case, as a result of substrate binding to the polypeptide, the N-terminus 'half' of the peptide is being cleaved. Again water binds as another substrate and the C-terminus 'half' of the peptide is released as product.

25.3.2 Cysteine Protease

Cysteine protease, also known as thiol protease, involves a catalytic mechanism where cysteine residue serves as nucleophile in a catalytic triad. In the first step, in the enzyme's active site, the cysteine residue is deprotonated by a neighbouring amino acid which have a basic side chain, generally a histidine residue. This follows a step, where the substrate carbonyl carbon confronts a nucleophilic attack by the anionic sulphur of deprotonated cysteine. In the following step, a fragment of the substrate is released with an amine terminus; a thioester intermediate joins the new carboxy-terminus of the substrate as well [11].

The thioester bond is hydrolysed to generate a carboxylic acid moiety on the remaining substrate fragment to produce the free enzyme. And so they are considered as thiol proteases.

25.3.3 Aspartate Protease

In case of aspartic proteases, to catalyze their peptide substrates; an activated water molecule interacts to one or more aspartate residues. In their active site, there are two highly conserved aspartates which are optimally active at acidic pH. Aspartyl group not only specifically cleave dipeptide bonds with hydrophobic residues but also a beta-methylene group as well. They mediate the proteolysis in a single step [12] and do not form covalent intermediates like serine or cysteine protease.

It follows the general acid-base mechanism in which a water molecule is coordinated between two highly conserved aspartate residues. One aspartate attracts a proton from water and activates it which exhibits a nucleophilic attack on the carbonyl carbon of the substrate scissile bond and forms an oxyanion intermediate. Then rearrangement of these intermediates results in the formation of two peptide fragments by cleaving the substrate peptide [13].

25.3.4 Threonine Protease

Threonine proteases involve a threonine (Thr) residue in their active site. The secondary alcohol of N-terminal threonine of threonine protease is used for nucleophilic attack to exhibit catalysis. The N-terminal amide of N-terminal threonine acts as a general base which polarizes ordered water and deprotonates the alcohol to increase its reactivity as a nucleophile [14].

Catalysis takes place in two steps:

- Firstly the nucleophilic attack to the substrate forming a covalent acyl-enzyme intermediate
- Secondly the hydrolysis of the intermediate restoring the free enzyme and release of product

25.3.5 Glutamic Acid Protease

Glutamic proteases contain a glutamic acid residue in the active. These residues perform the nucleophilic attack, where the glutamic acid serves as a general acid that donates a proton to the carbonyl oxygen of the peptide bond of the substrate. In this reaction, one or two water molecules supply a hydroxyl group, and the glutamic acid again donates a proton to the amide nitrogen, which results in the breakage of the peptide bond. The glutamine and the glutamic acid restore to its initial state [15].

25.3.6 Metalloprotease

In metalloproteinase, or metalloprotease, the catalytic mechanism comprises of a metal, mostly zinc that activates the water molecule. The metal ion is positioned in its place by three amino acids acting as ligands which are usually His, Glu, Asp or Lys, and at least one other residue is required for catalysis to employ the electrophilic role [16]. Most of the metalloproteases possess an HEXXH motif [17] which contributes in the formation of the metal-binding site.

More precisely, the HEXXH motif is as 'abXHEbbHbc' for metalloproteases, where 'a' is usually valine or threonine, 'b' is an uncharged residue and 'c' is a hydrophobic residue, but proline is never found in this site, because it would break the helical structure adopted by this motif in metalloproteases [18].

25.3.7 Proteases in *Escherichia coli*

In *Escherichia coli*, damaged proteins are degraded and eliminated by proteases. Some ATP-dependent proteases also have intrinsic chaperone activity and are termed as charonin [19]. Both these activities require either two polypeptide chains or a single polypeptide chain. Clp protease have two different chains, one having a proteolytic component and another an intrinsic chaperone ATPase component. Lon protease and FtsH protein are however examples of a single polypeptide chain charonin.

25.3.8 FtsH Protease

Filamentation temperature sensitive (FtsH) is a Zn^{2+}-dependent metalloprotease which belongs to the AAA class of proteins. It is first discovered in *Escherichia coli*, but it is distributed throughout the evolutionary phyla.

25.3.9 Genomic Organization of Ftsh Genes

In *E. coli*, ftsH is the second gene of a bicistronic operon. The first gene, ftsJ, affects cell division. These two genes have potential promoters, ftsJ is recognized by σ^{70} which is the vegetative sigma factor and ftsH is recognized by σ^{32} which is the heat shock sigma factor. Though the function of ftsH gene is highly conserved in all bacterial species, in different genera, the position of ftsh gene in their operon with respect to the other neighbouring genes is not conserved. So, ftsH may belong to either monocistronic, bicistronic or polycistronic operon [20].

25.3.10 Structure of FtsH Protein

AAA+ proteases are self-compartmentalizing proteolytic complexes. Monomers outline a cavity, more precisely a narrow pore. The residues of the narrow pore or the adaptor molecule recognize the proteins destined for proteolytic degradation. ATP hydrolysis causes conformational changes in the AAA+ ring, thereby pulling the substrate towards the degradation pore [21].

25.3.10.1 Primary Structure

In FtsH both the ATPase and the proteolytic activity are present on a 650 amino acid long single polypeptide chain. The N-terminal consists of two transmembrane helices with a small periplasmic region in between and is connected to the AAA module by a 15–20 amino acid long glycine-rich linker. The AAA module has Walker A motif as the pore residues and Walker B motif as the second region of homology fingerprint [22]. The HEXXH motif (frequently HEAGH) also known as characteristic 'zincin' motif identifies the protease active centre (Fig. 25.1). The leucine-rich motif at the C-terminus region has been involved in corecognition of certain substrates [23]. *E. coli* FtsH has been reported to autocatalytically prune its C-terminus, predicted to be mostly unstructured with a variable length. However the significance of this modification remains unknown.

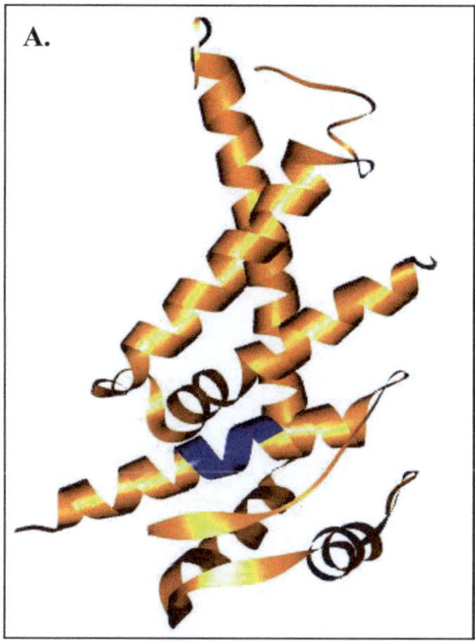

B.

>FtsH
TISPKEKEKIAI**HEAGH**ALMGLVSDD
DDKVHKISIIKHIYDKKDLYNKILVLL
GGRAAEEVFFGKDGITTGAENDLQRA
TDLAYRMVSMWGMSDKVGPIAIRRT
AVDTSPDLLREIDEEVKRIITEQYEKA
KAIVEEYKEPLKAVVKKLLEKETITCE
EFVEVFKLYGIELKDKCK

Fig. 25.1 (**a**) Crystal structure of protease domain of FtsH protease monomer (in *flat ribbon view*). The region marked in *blue* indicates the location of the HEXXH motif. (**b**) FASTA sequence of FtsH protease domain (retrieved from UniProt). The HEAGH motif required for proteolysis is marked as *bold* in *red colour*

25.3.10.2 Tertiary Structure

The FtsH is a homohexamer molecule with the monomeric AAA domain forming the sixfold symmetric rings [24]. Previously, all the intermediates of the ATPase cycle were believed to have sixfold symmetry, but biochemical evidence have shown that in at least some nucleotide states, nonequivalent nucleotide binding sites are present. Earlier, AAA domain interactions together with the involvement of the TM helices were thought to be responsible for hexamerization. This concept proved to be wrong as the TM-helix interactions confer increased stability to the hexamer [25]. FtsH is an assembly of two hexameric rings, one formed by the AAA modules and the other by the protease moieties. The hexameric ring formed by the protease moieties exhibits perfect sixfold symmetry and encircles a proteolytic chamber within it (Fig. 25.2) [26].

In the ADP-bound state, the aromatic amino acid residues move inwards, and as a result the pore residues rearrange themselves changing the conformation from the APO to the ADP state. The mechanical and the kinetic force generated by these large rearrangements promote substrate unfolding and its translocation into the proteolytic chamber [27].

25.3.10.3 Function of ftsH Protease

Domain structure analysis revealed that FtsH acts as a Zn^{+2} and ATP-dependent proteases, and this protease activity is measured by the lysogenization frequency of

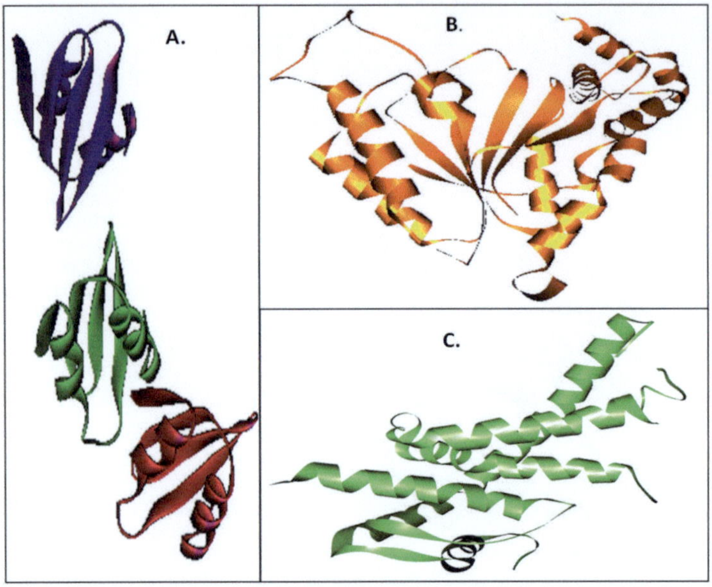

Fig. 25.2 (**a**) Different orientation of periplasmic N-terminal domain of FtsH protease crystal structure (*flat ribbon view*) (PDB ID: 4VOB). (**b**) Crystal structure of AAA domain of FtsH protease (*flat ribbon view*) (PDB ID: 1LV7). (**c**) Crystal structure of cytoplasmic protease domain of FtsH protease (flat ribbon view) (PDB ID: 2DI4)

λ-phage. It was observed that ftsh allele produces a temperature-sensitive protein which shows a loss in its activity at extreme heat shock temperature. At this heat shock temperature, the proteolytic entity of ftsH is reduced, and the lysogenization frequency of this allele is increased [28].

Another finding suggests that the transcription activator protein CII which is highly unstable is a potential target for FtsH. When the cell carrying ftsh allele is in physiological temperature, the half-life of the CII protein is only 2 min, i.e. it becomes highly unstable. But the same cell faces a thermal upshift, the proteolytic activity of the ftsh is reduced, and then the half-life of CII protein increases up to 30 mins. Actually FtsH employs the ATP hydrolysis to degrade the CII protein [29].

Another phage λ-repressor protein shows instability in the presence of active FtsH. λ-repressor protein CI, which possesses an 11 amino acid to form a non-polar as well as destabilizing tail, faces degradation by purified FtsH, but on the other hand, short non-polar carboxy-termini shows stabilization in *E. coli* cells when active FtsH is absent. When cells have diminished FtsH, they accumulated σ^{32} which have a minute half-life in physiological condition, but when the cell faces thermal upshift, it turns on different heat shock proteins including FtsH. When this heat-shock protein production reaches an optimal level, FtsH promotes degradation of the σ factor which is carried by ATP-dependent pathway and accelerated by Zn^{+2}. σ^{32} is degraded in part by FtsH-mediated degradation pathway [30].

FtsH is unique among the ATP-dependent proteases because it is a cytoplasmic membrane protein with two N-terminal membrane-spanning regions [31]. The cytoplasmic domain is of 200 amino acid residues and homologous to the AAA family members. The active site of the protease along with the Zn^{2+} metal is positioned at the C-terminal of the AAA domain, which promotes the proteolysis [32]. The FtsH protease, being a protease, not only engaged in the selective degradation of damaged proteins but also performs as a molecular chaperone. FtsH can bind to denatured proteins, and this binding does not necessarily ensure the proteolysis of the bound protein [33], which is a unique characteristic feature of molecular chaperones. Some molecular chaperones such as the DnaK and GroE help in folding of denatured proteins and considered as foldase protein, whereas the small heat-shock proteins bind with unfolded proteins, prevent further denaturation, but are not involved in their folding, and so they are called holdase protein. FtsH actively participates in folding of non-native proteins by binding to these non-native proteins and then sends them either to its own protease pathway or to the folding pathway. By satisfying both the protease as well chaperone activity, FtsH can be considered as a charonin.

FtsH recognizes at least three classes of substrate proteins with different proteolytic pathways. Under in vivo conditions, both the DnaK chaperone activity and ATPase activity are necessary to degrade σ32, whereas under in vitro condition, the DnaK system is expendable. This finding points the view of different conformational stabilities of these proteins within the cell. As FtsH cannot directly attack σ32, firstly it has to interact with the DnaK system which brings about partial unfolding, whereas FtsH with its innate chaperone activity promotes the unfolding of CII. On the other hand, in the case of CIII, it might already be present in partially unfolded form and so directly interact with FtsH protease to exhibit proteolytic response [34].

25.3.11 Lon Protease

Lon protease was the principle protease to be identified in *Escherichia coli* which is actually an ATP-dependent protease. It is conserved in all living organisms and catalyses the damaged as well as degraded proteins. Prokaryotic Lons are the prime enzymes to execute proteolysis and eliminate mutant, abnormal, degraded proteins as well as the catalysis of short-lived regulatory proteins which in turn maintains protein quality control and cellular homeostasis [35]. Apart from the proteolytic activity, AAA + proteins are contributed in many cellular functions such as membrane fusion, protein and organelle translocation, DNA and RNA unwinding, assembly and disassembly of multi-protein complexes and microtubule severing. The cellular activity of AAA + protease is largely defined by the interacting functional partners. In the case of Lon, the AAA + domain is interacted with protease domain to execute protease activity [36].

It contains homohexamers having multidomains. The homohexamers of Lon protease in *E. coli* form an encircled degradation cavity within which the misfolded proteins are degraded, but in yeast, the Lon protease forms a seven-membered ring [37].

25.3.11.1 Domain Organization of Lon Protease

On the basis of differences in the domain numbers and domain sequences, Lons are categorized into two subfamilies, Lon A and Lon B. Each monomer comprises of a single polypeptide chain which is 784 amino acids long. Each subunit is made up of three functional domains: (1) a central domain or A domain, an ATPase domain belonging to AAA+ superfamily, (2) a short N-terminal domain or N domain that interacts with target substrate proteins and (3) a C-terminal domain or P domain, i.e. a proteolytic domain involves in proteolysis [38].

25.3.11.2 Structure and Function of Lon Protease

The N Domain

The N domain is present in only Lon A subfamily members. The N-terminal domain is composed of approximately 300 amino acid open reading frame but with unknown biological functions. The N domain is divided into two sub-domains which are generated by limited proteolysis. Limited proteolysis of *E. coli* Lon by different proteases generates several transiently stable N-terminal fragments of varying length, and the termination residue ranges from 223 to 240 [39]. Extended incubation with protease results in the reduction of N-terminal domain to Lon N-209 which is highly stable and exists as a monomer. Another possible suggested boundary is residue 119. Lon N119 has a unique fold with three twisted β-sheets which are folded to form a shallow U shape. The depression of the shallow U shape occurred due to the presence of a single α-helix [40]. Most of the regions around α-helix of the N domain are contributed by hydrophilic residues, whereas the exposed surfaces have hydrophobic residues located across the β-sheet which are opposite in position to the alpha helix. It was observed that the two sub-domains of N domain of Lon protease are interacted to each other to form a single bimodal structural unit [41] (Fig. 25.3).

The N domain of Lon protease has protein-binding ability, thus helping in substrate recognition. In vitro experiments revealed that *E. coli* Lon protease which lacks 107 N-terminal residues shows a drastic reduction in protein degradation activity. In *Mycobacterium smegmatis* Lon protease deletion in 90, 225 or 277 N-terminal residues contributes in complete removal of proteolytic activity, reduction in protein binding activity as well as an alteration in oligomeric state.

These deletions make some structural distortions that affect the different protein activity as well as oligomerization state [42]. Experiments revealed that mutated Lon N domain changes its substrate specificity. From the experimental evidences, it can be assumed that N domain takes part during the interaction with specific substrate protein which involves the disordered regions in substrates with some recognizable specific motifs, domains or sequences that are being identified by Lon protease.

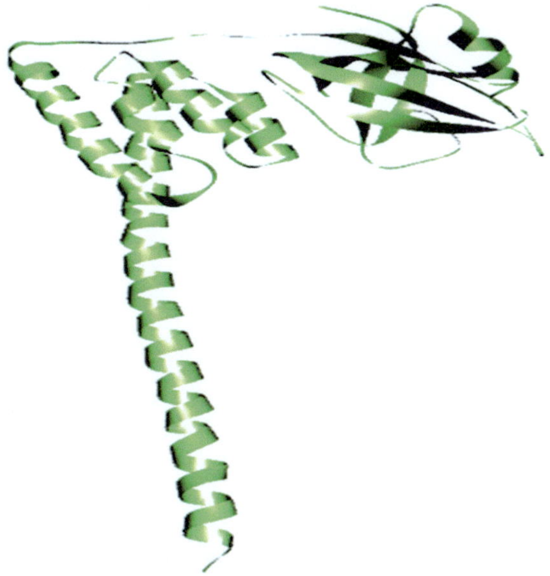

Fig. 25.3 Crystal structure of N-terminal domain of Lon protease (*flat ribbon view*) (PDB ID: 2ANE)

The A Domain or AAA+ Module

The AAA+ module or A domain, composed of 220–250 amino acids, has two structural domain such as an α-/β-domain and an α-domain with Walker A and B motif with high sequence conservation [43]. These two structural domains interact with substrate proteins. ATP hydrolysis confers alteration in conformation and orientation within domain. Binding to substrate proteins with AAA+ module causes unfolding of target proteins, transfer of damaged protein to protease domain as well as activation of functional domain in a concerted fashion [44]. The conserved sequence within this domain has a conserved helix-strand-helix-helix-strand-helix topology. In the middle of helix 1, there is a slight bend which contributes for the space and angle between α- and β-domains. Whether the helix 1 is straight or have a bend in the middle, the AAA+ proteins differ. Helix 3 is a long helix, followed by β-strand 1 and helix 2. The following β-strand loops with β-strand 1 form a parallel β-sheet. The C-terminal domain being open at the end remains connected at the beginning of P domain [45] (Fig. 25.4).

The AAA functions as a typical AAA module of ATPase protein. The α-domain attaches to the α-/β-domain as a rigid body with a loop which is susceptible to the nucleotide state of module. The α-domain of one subunit interacts with the α-/β--domain of adjacent subunit which gives strength to the subunit interactions and leads to the formation of assembled ring [46].

A relative rotation and separation between these α- and β-domains is generated due to simultaneous binding and release of nucleotides to substrates. A force is charged on the load by the relative movement of state of assembly and the bound load present which again generates a force on a polypeptide which remains connected to one of these domains. In this way, nucleotide binding or release causes

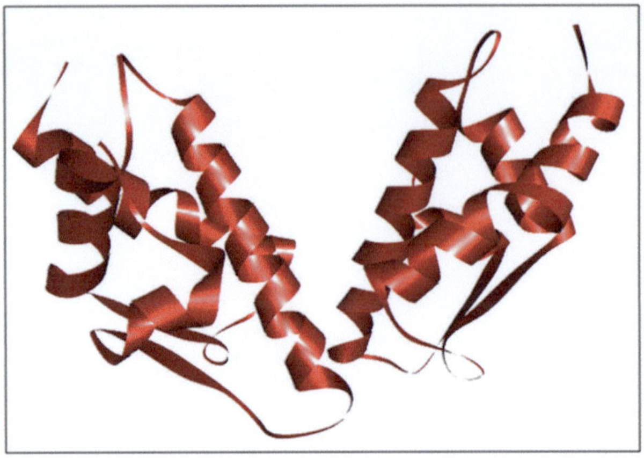

Fig. 25.4 Crystal structure of AAA domain of Lon protease (*flat ribbon view*) (PDB ID: 4GIT)

allosteric changes that affect the configuration of the domain involving the catalytic residues that make an alteration in the interactions, and thus a decision is being made whether or not the bound substrate protein within AAA module will transfer to protease domain. In oligomeric structure of Lon protease, this effect arises in different subunits with multiple active sites in a divergent way but in ordered or sequential manner [47].

P Domain or Protease Domain

P domain of Lon protease which possesses amino acid residues ranges from 585 to 784, shows proteolytic activity and reveals a unique fold in catalytic pocket. The P domain subunit has 6 α-helices and 10 β-strands and consists of two sub-domains in which the residues range from 585 to 697 and 698 to 784. Residues from 585 to 589, i.e. the first 9 residues of P domain, are disordered. A long β-hairpin loop is formed of β-strand 1 and antiparallel β-strand 2. Helix 1 makes a separation between the loop and parallel β-strand 3 and 4 from that of the first large β-sheet. To strengthen the sub-domain, the disulphide bridge is formed between Cys617 and Cys691 which connects the end of helix 2 with the end of β-strand 2. At the base of the sub-domain of strand 5, again a small β-sheet is formed which generates a shallow groove towards the centre of the ring. This strand 5 is connected to the helix 2 by the formation of a loop with the catalytic Ser679 [48].

Following helix 2, a bridge to the second sub-domain is formed by the random coils. A short β-strand 6 forms another β-loop formed by antiparallel strands 7 and 8, followed by helix 3 which contains the second catalytic Lys722 [49]. Strands 6, 9 and 10 form a third small β-sheet are sandwiched by helix 3 and C-terminal helix 6 (Fig. 25.5).

Though the proteolytic domain of Lon contains the residues of classical catalytic triad (His665, His667 and Asp767), it does not form the conventional catalytic triad.

It is observed that, within the proteolytic domain, Lon exhibits Ser679-Lys722 dyad in active site to promote catalysis [50]. The classical catalytic triad residues take part in intra- and intermolecular interactions. In the case of proteases with catalytic triad mechanism, substrates are attacked from the re-face of the amide bond, whereas the protease with the catalytic dyad mechanism performs the attack from si-face in the active site. The side chain of Asp676 forms hydrogen bonds with Val633 and Met634 amino acid residues. His665 and His667 are placed on the surface of the molecule and interact with Leu709 and Thr643; His667 also produces an ion pair with Glu614 [51]. If the catalytic triad residues are mutated, these do not affect the oligomerization of Lon protease which is a prerequisite for proteolytic activity. It illustrates that these triad residues are not utilized to form an active site. These residues all belong to the 15 amino acid fragment HVHVPEGATPKDGPS (665–679), preceded and included the catalytic Ser679. Mutation in Ser679 with alanine causes the loss of proteolytic activity, but there is no change in ATPase activity (Fig. 25.6).

Proteolytic domain of Lon protease contains a single conserved lysine at amino acid position 722 which is located 43 residues beyond the catalytic serine 679 and if this lysine722 is mutated to glutamine by site directed mutagenesis, the Lon K722Q mutant loses its hydrolytic activity and generates the similar properties of S679A mutant, i.e. to be proteolytically inactive, but it does not affect the property of solubility [52].

Experimental data explains that Lys722 forms a hydrogen bond with Gly717, and this hydrogen bond remains strictly conserved in Lon superfamilies. ε-amino group of Lys722 is in its position to form the hydrogen bond with the catalytic Ser679 residue. In addition to maintain the position of the general base, the conserved Thr704 residue is involved in hydrogen bonding with the Lys722 side chain. So, the experimental analysis clearly portrays that P domain of Lon protease utilizes a unique Ser^{679}-Lys^{722} catalytic dyad [53].

Lon-mediated degradation is mediated by peptide bond hydrolysis but without the dissociation of substrate. Both the prokaryotic and eukaryotic Lon proteases exhibit proteolysis of substrates by producing short peptide products consisting of ~5–30 amino acids. Hydrolysis of peptide bond proceeds via sequential linear manner from the amino to the carboxyl-termini or vice versa.

It is possible that protein substrates are proteolysed from the bulk solvent by the Lon complex and degraded by repetitive rounds of substrate binding, cleavage and release and rebinding to the proteolytic site in a sequential way, thereby resulting in small hydrolysed peptide products [54]. Results also show that Lon does not necessarily cleave substrates at a specific peptide consensus sequence; rather it shows a preference for hydrophobic residues. In addition to have cleavage specificity, peptide sequences within an exposed or unstructured region of a substrate also may serve to trigger substrate recognition and interaction, which facilitates the initiation of degradation mechanism [55].

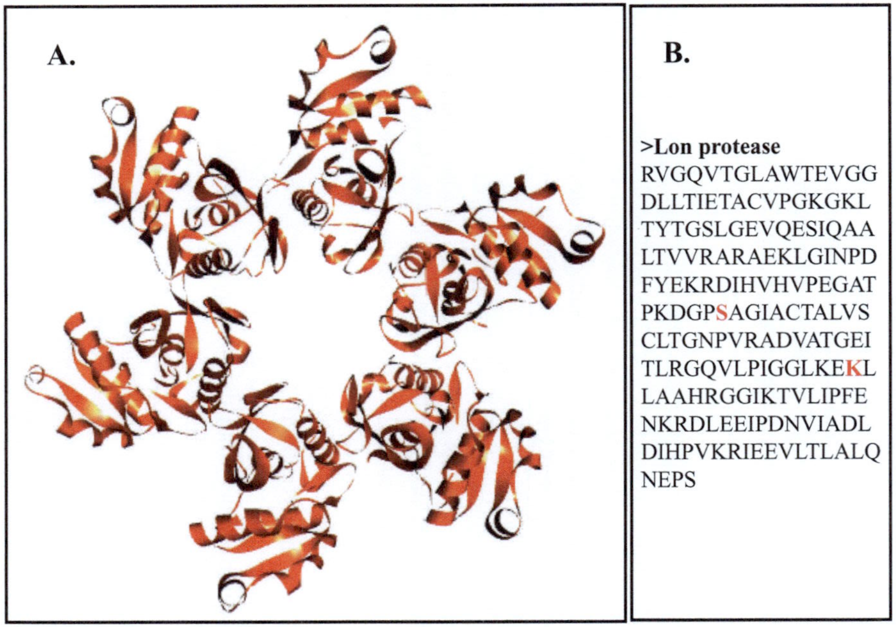

Fig. 25.5 (**a**) Crystal structure of homohexamers of protease domain of Lon protease (flat ribbon view) (PDB ID: 1RRE). (**b**) FASTA sequence of Lon protease proteolytic domain (retrieved from UniProt). Catalytic dyad residue serine 679 and lysine 722 are marked as *bold* in *red colour*

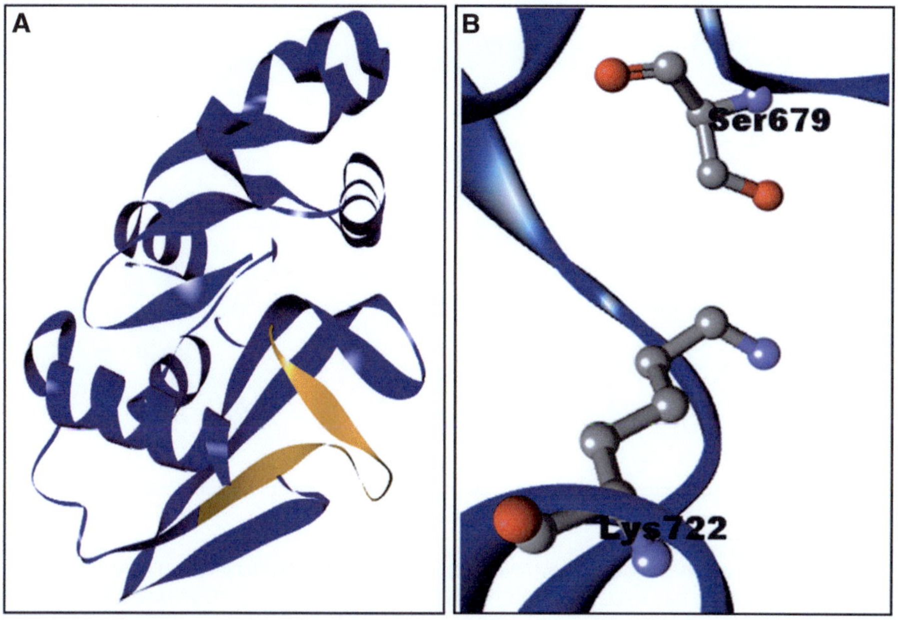

Fig. 25.6 (**a**) Crystal structure of protease domain of Lon protease (flat ribbon view). The region marked in yellow indicates the location of HVHVPEGATPKDGPS fragment. (**b**) Catalytic Ser^{679}-Lys^{722} dyad of Lon protease

25.3.12 Clp Protease

The functional Clp protease consists of two components: a proteolytic component, ClpP, and several regulatory ATPase components. The two components interact in an ATP-dependent manner with the ATPase subunits that confer substrate specificity and activate ClpP for proteolysis. In *E. coli*, ClpA and ClpX act as regulatory components of Clp, while in other organisms, ClpC shows the evolutionary equivalent with ClpA. Homologues of ClpP and the ATPase components of Clp have been identified [56]. In *E. coli*, the in vivo substrates of Clp include abnormal proteins and several short-lived regulatory proteins. The three known ATPase components of Clp are members of the Clp/Hsp100 family of molecular chaperones with ClpA and ClpC composed of two ATP-binding motifs which is a Walker consensus motif, whereas ClpX contains a single motif only. Electron microscopic studies of *E. coli* ClpA depict that it is hexameric and point to a symmetry mismatch in the ClpAP complex. The ATPase subunits of Clp participate both in proteolysis and as molecular chaperones in several cellular reactions, and this participation of the Clp ATPase subunits in both of these functions suggests that Clp protease serves a role in cellular protein quality control [57] (Fig. 25.7).

ClpP is synthesized as a 207-residue proprotein that is processed to generate the final mature protein of 193 residue through autocatalysis during assembly or folding. Native ClpP forms a stable tetradecamer composed of two rings with seven

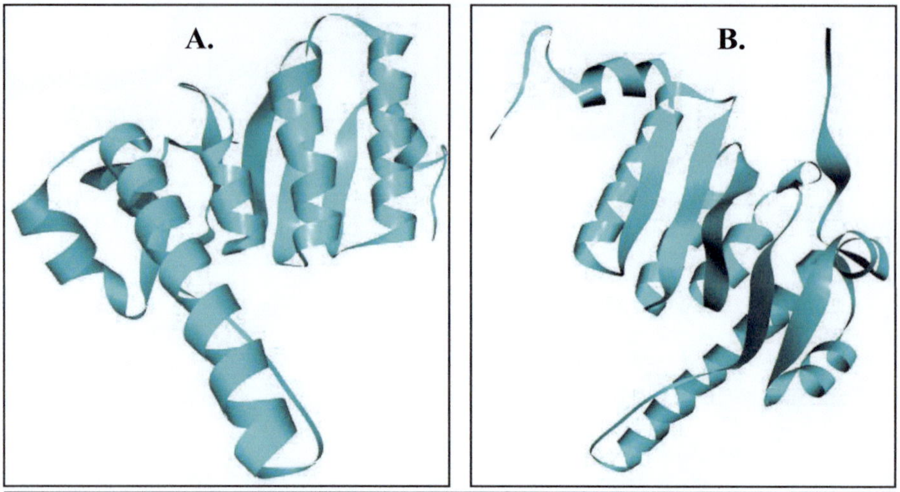

C.

>Clp protease
SRGERSFDIYSRLLKERVIFLTGQVEDHMANLIVAQMLFLEAENPEKDIYLYINSPGGVITAGMSIYDTMQ
FIKPDVSTICMGQAA**S**MGAFLLTAGAKGKRFCLPNSRVMI**H**QPLGGYQGQATDIEIHAREILKVKGRMN
ELMALHTGQSLEQIERDTER**D**RFLSAPEAVEYGLVDSILTHRN

Fig. 25.7 Crystal structure of Clp protease in (**a**). Front view and (**b**) back view (flat ribbon view) (PDB ID 1TYF) (**c**) FASTA sequence of protease domain of Clp protease (retrieved from UniProt). Catalytic triad residues (serine 97, histidine 122, aspartate 171) are marked as *bold* in *red colour*

subunits each. When the peptides are fewer than 6 residues, it exhibits limited serine-peptidase activity and preferentially cleaves them after hydrophobic residues. Degradation of peptides longer than approximately 6 residues needs the involvement of an ATPase subunit such as ClpA to form the active protease (ClpAP), but not ATP hydrolysis; by contrast, proteolysis of protein substrates demands ATP hydrolysis as well as ClpAP. Protein substrates are catalysed in a highly sequential manner, producing 7- to 10-residue peptides; the pattern of cleavage does not show any clear sequence specificity [58].

Each ClpP subunit composes an α-/β-fold buildup of six repeats of the α-/β-unit along with an additional protruding α-/β-unit. The ten β-strands form two layers of β-sheets that are packed against a layer of α-helices. The layers of sheets remain perpendicular to each other to create one side of the substrate-binding cleft. In projection, the ClpP monomer shows a similarity with a hatchet having wedge-shaped head (head domain) and a short handle. Residues 28–120 and160–188 comprises the head domain, and the 'handle' is formed by residues 125–130 and 132–157. Residues 11–27 belong to the amino termini, and residues 189–193 belong to the carboxyl-termini which lie at opposite ends of the head domain, extending away from its central core [59]. The catalytic triad, Ser-97, His-122 and Asp-171, is situated in a cleft at the connection point of the head domain and handle. The structure of ClpP defines a fifth family of serine proteases which is different from the other four families of serine protease, i.e. trypsin-like family, chymotrypsin-like family, cytomegalovirus proteases and subtilisin-like family [60]. The catalytic triad consists of Ser-97, His-122 and Asp-171. In the triad, a hydrogen bond between Asp-171 and His-122 is evident; additionally, the side chain of Asp-171 produces two hydrogen bonds with the side chains of the nonconserved amino acids His-138 and Tyr-128, both of which are contributed by a neighbouring intraring subunit [61]. Mutation of Asp-171 with alanine results in loss of proteolytic activity. ClpP's catalytic triad is situated in a cleft and is composed of β-strand 4 and strand 9 from the head domain and the handle, respectively [62].

Tetradecameric ClpP with an overall cylindrical shape, approximately 90 A° in both height and diameter, has a central channel that helps in penetration of denatured protein. Access to the chamber from outside is controlled by two axial entrance pores. The interior surface of the pore is hydrophilic and formed by residues 11–17, whose main-chain atoms lie parallel to the sevenfold axis of the oligomers [63]. Two conserved aromatic residues (Phe-17 in helix A and Phe-49 in helix B) from each subunit are positioned at the outer perimeter of the pores. These residues may involve in aiding the passage of substrates into and products out of the proteolytic chamber. It is proposed that the entrance pores in ClpP will allow a particular hydrodynamic radii of polypeptides that can enter freely into and out of the proteolytic chamber during proteolysis and thus strongly constrict the conformation and size of substrates that enters into the proteolytic chamber and on the length of proteolytic products leaving it [64].

According to the size of the entrance pores hypothesis, ClpP alone effciently degrades peptides shorter than 6 residues and more slowly degrades oligopeptides of up to approximately 30 residues but does not degrade larger oligopeptides. Mechanistically, the translocation of substrates into the proteolytic chamber may resemble the energy-dependent transport of newly synthesized polypeptides across membranes [65].

Each of the three major solvent-exposed surfaces (exterior sides, interior proteolytic chamber and exterior endon) of the ClpP oligomer has a distinct character. The exterior side surfaces of ClpP are hydrophilic and composed of nonconserved residues. The deep notches and grooves in this surface facilitate the tight packing of the oligomers in the crystal. On the other hand, the residues committing to the exposed surface of the chamber are partially conserved, presenting of hydrophobic surface area to the solvent. The hydrophobic surfaces in the chamber having a key role to maintain the substrates in an unfolded as well as easily degradable state.

Large hydrophobic surfaces are also found in the interior chambers of two other ring-like oligomers. The ATPase components of Clp are thought to bind to the exterior end-on surface of ClpP. This surface, viewed looking down the central pore of the oligomer, has a sprocket-like appearance due to seven ridges extending radially outwards from the pore's edges. Their surface is hydrophobic, comprised of conserved aromatic residues including Phe-82, Tyr-60 and Tyr-62. The conserved residues of these grooves or edges may fix the complexes of ClpP with its ATPase components [66].

The ClpP protease remains as a tetradecamer, where two heptameric rings stacked back-to-back one upon another. The catalytic triad of each subunit is situated at the interface of three monomers, and so the oligomerization occurred to structure the functional form of the protease. The subunits remain in association with each other to provide a mechanism for establishing a contact between active sites that may exhibit a role to activate the allosteric form of ClpP. The extensive interactions among the head domains of ClpP contribute to the stabilization of individual rings of the protease [67].

In the proteolytic chamber of ClpP, two continuous and parallel grooves are present which are perpendicular to the entrance pores, engraving the circumference of its inner walls. The grooves are hydrophobic in nature, and each connects the active sites in one heptamer ring to form a substrate-binding surface in a continuous fashion. The diameter of the entrance pores to the proteolytic chamber will also permit the short peptide products to diffuse freely out of the chamber, whereas the larger, partially degraded substrates will be retained. Combination of these two factors may account for the observed distribution of product sizes [68].

25.4 Discussion

Even though the bacterial proteases perform the same proteolytic activity, there are many differences regarding different aspects. These proteases vary in their subcellular localization. Both Lon and Clp are cytoplasmic protease, whereas Ftsh has a periplasmic domain, membrane-bound domain as well as cytoplasmic domain. Sequence analysis of these proteins shows huge diversities in respect of sequence length, amino acid composition and even the structures of catalytic pockets. FtsH and Clp both utilize conventional catalytic triad consisting of serine, histidine and aspartate, whereas Lon employs a unique catalytic dyad consisting of serine and lysine to promote catalysis. Hydrophobicity analysis of catalytic pockets (Fig. 25.8) of FtsH protease, Lon protease and Clp protease reveals that even if they perform the same proteolytic activity, there are some differences regarding their hydrophobicity. The catalytic pocket of FtsH indicates less hydrophobicity, Lon protease has neutral hydrophobicity, whereas Clp protease reveals increased hydrophobicity.

Though these three proteases belong to *Escherichia coli*, they also possess a substrate diversity as well as functional diversity. Lon and Clp both promote the proteolysis and degradation of misfolded proteins, but FtsH, being a protease, not only engaged in the selective degradation of damaged proteins but also performs as a molecular chaperone. FtsH can bind to denatured proteins, and this binding does not necessarily ensure the proteolysis of the bound protein which is a unique characteristic feature of molecular chaperones. FtsH is actively participated in folding activity of non-native proteins, i.e. acts as foldase and then sends them either to its own protease pathway or to the folding pathway. By satisfying both the protease and the chaperone activity, FtsH can be considered as a charonin.

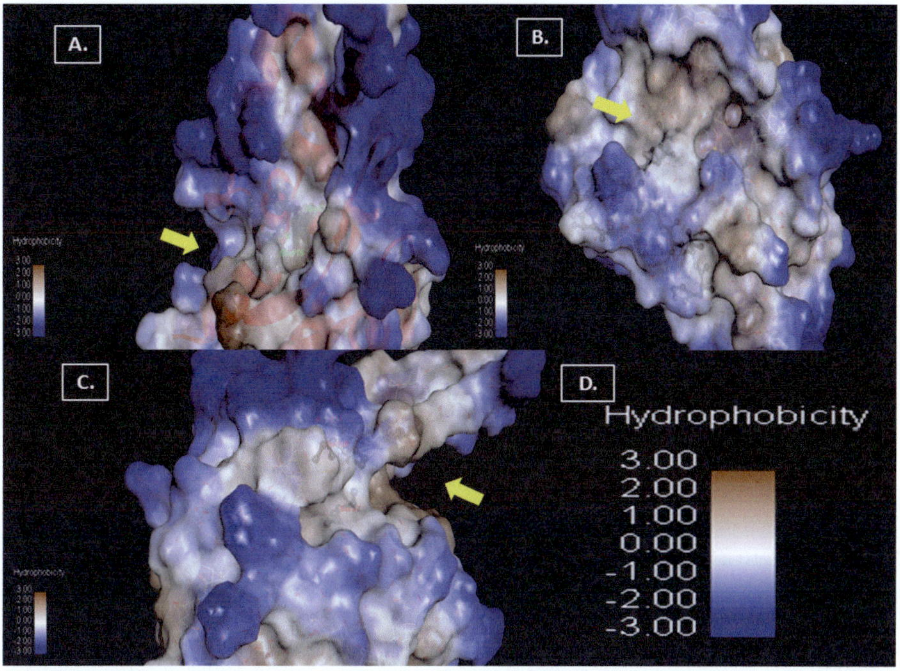

Fig. 25.8 (**a**) Catalytic pocket of FtsH protease. (**b**) Catalytic pocket of Lon protease. (**c**) Catalytic pocket of Clp protease. (**d**) Hydrophobicity scale where the blue region denotes less hydrophobicity, white region denotes neutral hydrophobicity and brown region denotes increased hydrophobicity. *Yellow arrow* denotes the exact location of catalytic pocket

25.5 Conclusion and Future Direction

Bacterial proteases having differences in various aspects exhibit the proteolytic functions not only by eliminating the damaged proteins, but some of them also promote chaperone activity to initiate proper folding of misfolded proteins to maintain cellular protein quality control. In the future, detailed analyses of the effect of mutation of these bacterial proteases in their structure, catalytic pocket, chemical properties as well as proteolytic mechanism with their substrates are going to be observed. These three bacterial proteases interact with each other to maintain the cellular protein quality. The functioning of interacting protease network of these bacterial proteases in *Escherichia coli* is yet to be deciphered to elucidate the entire proteolytic machinery for cellular proteome quality control.

Acknowledgement The authors are thankful to the Department of Biochemistry and Biophysics and Bioinformatics Infrastructure Facility Centre, University of Kalyani, for their continuous support and for providing the necessary instruments to carry out the experiments. The authors would like to acknowledge the DBT (project no.BT/PR6869/BID/7/417/2012) and the ongoing DST-PURSE programme (2012-2015) for their support.

References

1. Welch WJ (1993) How cells respond to stress. Sci Am 268(5):56–64
2. Fulda S, Gorman AM, Hori O et al. (2010) Cellular stress responses: cell survival and cell death. Int J Cell Biol Article ID 214074, 23 pages
3. Richter K, Haslbeck M, Buchner J (2010) The heat shock response: life on the verge of death. Mol Cell:253–266
4. Burston SG, Clarke AR (1995) Molecular chaperones: physical and mechanistic properties. Essays Biochem 29:125–136
5. Spiess C, Beil A, Ehrmann M (1999) A temperature-dependent switch from chaperone to protease in a widely conserved heat shock protein. Cell 97(3):339–347
6. Micevski D, Dougan DA (2013) Proteolytic regulation of stress response pathways in Escherichia coli. Subcell Biochem 66:105–128
7. Tompa P, Prilusky J, Silman I et al (2008) Structural disorder serves as a weak signal for intracellular protein degradation. Proteins 71(2):903–909
8. Kahne D, Clark Still W (1988) Hydrolysis of a peptide bond in neutral water. J Am Chem Soc 110(22):7529–7534
9. Hedstrom L (2002) Serine protease mechanism and specificity. Chem Rev 102(12):4501–4524
10. Radisky ES, Lee JM, Lu CK et al (2006) Insights into the serine protease mechanism from atomic resolution structures of trypsin reaction intermediates. Proceedings of the. Nat Acad Sci U S A 103(18):6835–6840
11. Chapman HA, Riese RJ, Shi GP (1997) Emerging roles for cysteine proteases in human biology. Annu Rev Physiol 59:63–88
12. Szecsi PB (1992) The aspartic proteases. Scand J Clin Lab In vest Suppl 210:5–22
13. Rawlings ND, Barrett AJ, Bateman A (2011) Asparagine peptide lyases: a seventh catalytic type of proteolytic enzymes. J Biol Chem 286(44):38321–38328
14. Dodson G, Wlodawer A (1998) Catalytic triads and their relatives. Trends Biochem Sci 23(9):347–352
15. Oda K (2012) New families of carboxyl peptidases: serine-carboxyl peptidases and glutamic peptidases. J Biochem 151(1):13–25
16. Wu S, Zhang C, Xu D et al (2010) Catalysis of carboxypeptidase A: promoted-water versus nucleophilic pathways. J Phys Chem B 114:9259–9267
17. Kilshtain AV, Warshel A (2009) On the origin of the catalytic power of carboxypeptidase A and other metalloenzymes. Proteins 77:536–550
18. Pelmenschikov V, Bloomberg M, Siegbahn EM (2002) A theoretical study of the mechanism for peptide hydrolysis by thermolysin. J Biol Inorg Chem 7:284–298
19. Schumann W (1999) FtsH-a single-chain charonin? FEMS Microbiol Rev 23:1–11
20. Tomoyasu T, Yamanaka K, Murata K et al (1993) Topology and subcellular localization of FtsH protein in Escherichia coli. J Bacteriol 175:1352–1357
21. Akiyama Y, Ehrmann M, Kihara A et al (1998) Polypeptide binding of Escherichia coli FtsH (HflB). Mol Microbiol 28:803–812
22. Krzywda S, Brzozowski AM, Verma C et al (2002) The crystal Structure of the AAA domain of the ATP-dependent protease FtsH of Escherichia coli at 1.5A°resolution. Structure 10:1073–1083
23. Suno R, Niwa H, Tsuchiya D et al (2006) Structure of the whole cytosolic region of ATP-dependent protease FtsH. Mol Cell 22:575–585
24. Neuwald AF, Aravind L, Spouge JL et al (1999) AAA+: a class of chaperone-like ATPases associated with the assembly, operation, and disassembly of protein complexes. Genome Res 9:27–43
25. Krzywda S, Brzozowski AM, Karata K et al (2002) Crystallization of the AAA domain of the ATP dependent protease FtsH of Escherichia coli. Acta Crystallogr. D Biol Crystallogr 58:1066–1067

26. Akiyama Y, Ito K (2000) Roles of multimerization and membrane association in the proteolytic functions of FtsH (HflB). EMBO J 19:3888–3895
27. Yamada-Inagawa T, Okuno T, Karata K et al (2003) Conserved pore residues in the AAA protease FtsH are important for proteolysis and its coupling to ATP hydrolysis. J Biol Chem 278:50182–50187
28. Kihara A, Akiyama Y, Ito K (1997) Host regulation of lysogenic decision in bacteriophage lambda: Transmembrane modulation of FtsH (HflB), the cII degrading protease, by HfKC (HflA). Proc Natl Acad Sci U S A 94:5544–5549
29. Shotland Y, Koby S, Teff D et al (1997) Proteolysis of the phage lambda CII regulatory protein by FtsH (HflB) of Escherichia coli. Mol Microbiol 24:1303–1310
30. Tomoyasu T, Gamer J, Bukau B et al (1995) Escherichia coli FtsH is a membrane-bound, ATP-dependent protease which degrades the heat-shock transcription factor σ32. EMBO J 14:2551–2560
31. Makino S, Makino T, Abe K et al (1999) Second transmembrane segment of FtsH plays a role in its proteolytic activity and homo-oligomerization. FEBS Lett 460:554–558
32. Confalonieri F, Duguet M (1995) A 200-amino acid ATPase module in search of a basic function. BioEssays 17:639–650
33. Tomoyasu T, Yuki T, Morimura S et al (1993) The Escherichia coli FtsH protein is a prokaryotic member of a protein family of putative ATPases involved in membrane functions, cell cycle control, and gene expression. J Bacteriol 175:1344–1351
34. Kihara A, Akiyama Y, Ito K et al (1998) Different pathways for protein degradation by the FtsH/HflKC membrane-embedded protease complex: an implication from the interference by a mutant form of a new substrate protein. YccA J Mol Biol 279:175–188
35. Goldberg AL, Moerschell RP, Chung CH et al (1994) ATP-dependent protease La (lon)from Escherichia coli. Methods Enzymo 244:350–375
36. Vieux EF, Wohlever ML, Chen JZ et al (2013) Distinct quaternary structures of the AAA+ Lon protease control substrate degradation. Proc Natl Acad Sci U S A 110(22):1–7
37. Van Dijl JM, Kutejová E, Suda K et al (1998) The ATPase and protease domains of yeast mitochondrial Lon: roles in proteolysis and respiration-dependent growth. Proc. Natl Acad Sci U S A 95(18):10584–10589
38. Mel'nikov EE, Tsirul'nikov KV, Rotanova TV (2000) Coupling of proteolysis with ATP hydrolysis by Escherichia coli Lon proteinase. I. Kinetic aspects of ATP hydrolysis. Bioorg Khim 26(7):530–538
39. Li M, Gustchina A, Rasulova FS et al (2010) Crystal structure of the N-terminal domain of E. coli Lon protease. Acta Cryst. D 66:865–873
40. Rotanova TV, Rasulova FS, Dergousova NI et al (1998) Synthesis and characterisation of ATP-dependent forms of Lon-proteinase with modified N-terminal domain from Escherichia coli. Bioorg Khim 24:370–375
41. Li M, Rasulova F, Melnikov EE et al (2005) Crystal structure of the N-terminal domain of E. coli Lon protease. Protein Sci 14:2895–2900
42. Roudiak SG, Shrader TE (1998) Functional role of the N-terminal region of the Lon protease from Mycobacterium smegmatis. Biochemistry 37:11255–11263
43. Lee AY, Chen YD, Chang YY et al. (2014) Structural basis for DNA-mediated allosteric regulation facilitated by the AAA(+) module of Lon protease. Acta Crystallogr Sect D 70:218–230
44. Lee I, Berdis AJ, Suzuki CK et al (2006) Recent developments in the mechanistic enzymology of the ATP-dependent Lon protease from Escherichia coli: highlights from kinetic studies. Mol BioSyst 2:477–483
45. Botos I, Melnikov EE, Cherry S et al (2004) Crystal structure of the AAA+ alpha domain of E. coli Lon protease at 1.9A resolution. J Struct Biol 146(1–2):113–122
46. Rotanova TV, Botos I, Melnikov EE et al (2006) Slicing a protease: structural features of the ATP-dependent Lon proteases gleaned from investigations of isolated domains. Protein Sci 15(8):1815–1828
47. Lin CC, Su SC, Su MY et al (2016) Structural insights into the allosteric operation of the Lon AAA+ protease. Structure 24(5):667–675

48. Van Melderen L, Aertsen A (2009) Regulation and quality control by Lon-dependent proteolysis. Res Microbiol 160(9):645–651
49. Cha SS, An YJ, Lee CR et al (2010) Crystal structure of Lon protease: molecular architecture of gated entry to a sequestered degradation chamber. EMBO J 29(20):3520–3530
50. Gur E, Sauer RT (2009) Degrons in protein substrates program the speed and operating efficiency of the AAA+ Lon proteolytic machine. Proc Natl Acad Sci U S A 106(44):18503–18508
51. Van Melderen L, Gottesman S (1999) Substrate sequestration by a proteolytically inactive Lon mutant. Proc Natl Acad Sci U S A 96(11):6064–6071
52. Rotanova TV, Starkova NN, Koroleva EP et al (1998) Mutations in the proteolytic domain of Escherichia coli protease Lon impair the ATPase activity of the enzyme. FEBS Lett 422(2):218–220
53. Botos I, Melnikov EE, Cherry S et al (2004) The catalytic domain of Escherichia coli Lon protease has a unique fold and a Ser-Lys dyad in the active site. J Biol Chem 279(9):8140–8148
54. Sauer RT, Baker TA (2011) AAA+ proteases: ATP-fueled machines of protein destruction. Annu Rev Biochem 80:587–612
55. Tsilibaris V, Maenhaut-Michel G, Van Melderen L (2006) Biological roles of the Lon ATP-dependent protease. Res Microbiol 157:701–713
56. Gottesman S, Roche E, Zhou Y et al (1998) The ClpXP and ClpAP proteases degrade proteins with carboxy-terminal peptide tails added by the SsrA-tagging system. Genes Dev 12(9):1338–1347
57. Shin DH, Lee CS, Chung CH et al (1996) Molecular symmetry of the ClpP component of the ATP-dependent Clp protease, an Escherichia coli homolog of 20S proteasome. J Mol Biol 262:71–76
58. Thompson MW, Maurizi MR (1994) Activity and specificity of Escherichia coli ClpAP protease in cleaving model peptide substrates. J Biol Chem 269:18201–18208
59. Wang J, Hartling JA, Flanagan JM et al (1997) The structure of ClpP at 2.3 Å resolution suggests a model for ATP-dependent proteolysis. Cell 91:447–456
60. Szyk A, Maurizi MR (2006) Crystal structure at 1.9 Å of E. coli ClpP with a peptide covalently bound at the active site. J Struct Biol 156:165–174
61. Kim DY, Kim KK (2008) The structural basis for the activation and peptide recognition of bacterial ClpP. J Mol Biol 379:760–771
62. Kang SG, Maurizi MR, Thompson M et al (2004) Crystallography and mutagenesis point to an essential role for the N-terminus of human mitochondrial ClpP. J Struct Biol 148:338–352
63. Maurizi MR, Singh SK, Thompson MW et al (1998) Molecular properties of ClpAP protease of Escherichia coli. ATP-dependent association of ClpA and clpP Biochemistry 37:7778–7786
64. Kimber MS, Yu AY, Borg M et al (2010) Structural and theoretical studies indicate that the cylindrical protease ClpP samples extended and compact conformations. Structure 18:798–808
65. Maglica Z, Kolygo K, Weber-Ban E et al (2009) Optimal efficiency of ClpAP and ClpXP chaperone-proteases is achieved by architectural symmetry. Structure 17:508–516
66. Gribun A, Kimber MS, Ching R et al (2005) The ClpP double ring tetradecameric protease exhibits plastic ring-ring interactions, and the N termini of its subunits form flexible loops that are essential for ClpXP and ClpAP complex formation. J Biol Chem 280:16185–16196
67. Geiger SR, Böttcher T, Sieber SA et al (2011) A conformational switch underlies ClpP protease function. Angew Chem Int Ed Engl 50:5749–5752
68. Porankiewicz J, Wang J, Clarke AK et al (1999) New insights into the ATP-dependent Clp protease. Escherichia coli and beyond. Mol Microbiol 32:449–458

The Ubiquitin Proteasome System with Its Checks and Balances

26

Prranshu Yadav, Ankita Doshi, Yong Joon Yoo, and C. Ratna Prabha

Abstract

Cells need to quickly change according to changing environment to survive, and for that, they must not just make new proteins but also degrade others equally promptly. For this purpose, cells have evolved the ubiquitin system, which consists of ubiquitin molecules which are used to tag proteins in a process called ubiquitination; E1, E2 and E3 enzymes which carry out the process of ubiquitination; and deubiquitinating enzymes (DUBs) that remove the ubiquitin from the substrate proteins in a process called deubiquitination. Ubiquitination involves various lysine residues on ubiquitin; among them K48 and K63 are the most significant and well understood. Ubiquitination with K48 linkage leads to degradation of substrate proteins by a multi-protein complex called proteasome. Proteasome-mediated degradation is involved in numerous different processes in cells, due to which defects in it are responsible for several diseases. But due to the high diversity of E3 enzymes and ubiquitin target proteins, there are many drug targets that can be utilized to treat diseases. This makes it vital to understand ubiquitin system for advancement of health care.

Keywords

Ubiquitin • Ubiquitin structure • Proteasomes • Ubiquitination • Deubiquitination

---

P. Yadav • A. Doshi • C. Ratna Prabha (✉)
Department of Biochemistry, Faculty of Science, The Maharaja Sayajirao University of Baroda, Vadodara 390002, India
e-mail: chivukula_r@yahoo.com

Y.J. Yoo
School of Life Sciences, Gwangju Institute of Science and Technology (GIST), Gwangju 500712, Republic of Korea

S. Chakraborti, N.S. Dhalla (eds.), *Proteases in Physiology and Pathology*,
DOI 10.1007/978-981-10-2513-6_26

26.1 Introduction

Ubiquitin is used by eukaryotic cells as a tag, and covalent attachment of ubiquitin to proteins marks them for different activities. Ubiquitin was first discovered in the mid-1970s [1], and for some time, it was seen as a molecule exclusively involved in proteasome-mediated protein degradation, whose use by proteasome enabled fast and extensive changes in the cell protein composition by selective protein degradation, enabling cell to adapt to changing environment [2–5]. In due course of time, however, many functions other than proteasomal degradation have also been associated with ubiquitin. Among them are DNA repair [6], transcription regulation [7, 8], translation regulation [9–11], cell signalling [12], autophagy [13] and endocytosis [14]. Hence, the emerging picture of ubiquitin system is a general post-translational modification system that greatly increases the diversity and functional ambit of proteome, analogous to other post-translational modifications such as phosphorylation, methylation and acetylation. Besides, its role as a tag meant to direct protein degradation has immense importance in regulating many of the functions listed above. Other components of ubiquitin pathway are the E1, E2 and E3 enzymes, involved in ubiquitination of target proteins, deubiquitinating enzymes and ubiquitin-binding partners. An extremely diverse group of E3 enzymes has evolved to accomplish the difficult task of selectively ubiquitinating a particular protein while leaving others unaffected, with each type of E3 specific for a particular protein or group of proteins. In fact E3 genes in humans outnumber the genes for even protein kinases and G-protein-coupled receptors. These E3s in turn are activated or deactivated depending on whether or not their substrate proteins are to be ubiquitinated. Such diverse and specific E3s also offer opportunities to be used as drug targets in numerous diseases ranging from cancer to neurological disorders [15–17]. Here we summarize the current understanding of ubiquitin system based on developments in the last few decades, with particular emphasis on its proteolytic role. We first describe the essential components of ubiquitin system in the context of their functions and the role their structures play in performing them and then move on to different functions of ubiquitin system as tagging and de-tagging machinery. Lastly, we highlight the roles that a defective ubiquitin system plays in different diseases, as well as use of its components directly as drug targets or indirectly in other modes of treatment.

26.2 Players in Ubiquitin System

26.2.1 Ubiquitin

26.2.1.1 Molecular Structure of Ubiquitin

Ubiquitin is a small, monomeric, single-domain protein of 76 residues, which makes it a good model system to study protein structure, folding and stability. Its X-ray crystallographic structure was first determined in the mid-1980s at 2.8 Å [18] and later at 1.8 Å [19]. The globular structure of ubiquitin is formed by a mixed β-sheet and α-helix held together in a β-grasp fold. The protein sports nine reverse turns and a 3_{10} helix. Additionally, ubiquitin has two β-bulges in its structure. β-bulges are regions in a β-sheet, where two residues in a β-strand are present opposite a single residue in the neighbouring strand. The space required by the extra residue causes this region to bulge out from the β-strand, hence the name. β-bulge affects β-sheet's structure firstly by accentuating the sheet's inherent twist and secondly by interfering with the alternating arrangement of side chains on the two sides of the backbone. The first β-bulge of ubiquitin is located in the N-terminal region between the two antiparallel β-strands forming a hairpin, while the second β-bulge is present in the β-sheet, adjacent to K63. The second β-bulge is a parallel G1β-bulge formed by residues E64 and S65 at positions 1 and 2 and Q2 at position X on the other strand.

Ubiquitin sequence shows a very high degree of conservation among all eukaryotes, with only three residues being replaced from yeast to humans.

In the 1990s, Makhatadze et al. studied the effects of different salts on ubiquitin's stability at pH 2.0, using differential scanning calorimetry, circular dichroism and fluorescence spectroscopy [20, 21]. They found that all salts tested increased the thermostability of ubiquitin through anion binding. They then studied the effect of surface charge on ubiquitin's stability using site-directed mutagenesis and specific chemical modifications. Robertson et al. too studied significance of charges and ion pairs on ubiquitin's surface using site-directed mutagenesis of specific surface residues and determining pKa of neighbouring charged residues by 2D NMR [22, 23]. Makhatadze et al. have produced an ubiquitin mutant more stable than its wild-type counterpart [24]. They first converted all arginine residues on ubiquitin surface to lysines and then carbamoylated their amino groups. They observed that ubiquitin was most stable when all these carboxyl groups were protonated and hence all surface charges were neutralized. Surface charges therefore do not appear to contribute to ubiquitin's stability. The outcome of this study helped in framing guidelines for engineering of surface charges to increase protein stability [25]. The importance of hydrophobic residues in ubiquitin's core was studied using site-directed mutagenesis and measurement of consequent heat capacity changes by differential scanning calorimetry [26, 27]. Replacing nonpolar residues with polar ones decreased the stability of ubiquitin, while replacing naturally occurring polar residues in core with nonpolar ones increased its stability. Replacing nonpolar residues with other nonpolar residues had no significant effect. There have been other studies too on the importance of hydrophobic residues which are present in the core of ubiquitin [28–30]. Significance of the interaction between I30 and I36 at the C-terminus of the α-helix has also been studied. Out of 16 variants produced in the study, none were found as stable as the wild type [31].

Ubiquitin residues essential for vegetative growth of yeast are clustered in three regions on ubiquitin surface. They are hydrophobic patches formed by L8, I44 and V70 and their surrounding residues, F4 and its surrounding residues and the C-terminal tail [32]. I44 patch is essential for proteasomal degradation and endocytosis [33], F4 patch is essential for endocytosis and for proteasomal degradation as well [32–35], and the C-terminal tail is essential for most ubiquitin functions, owing to its fundamental role in ubiquitination. The L8, I44 and V70 patch interacts with regulatory subunit(s) of proteasome, while this patch together with F4 patch may either form a common binding site for proteins involved in endocytosis, or both patches may bind separate sets of proteins is yet to be known.

26.2.1.2 Ubiquitination and Its Types

Ubiquitin, as the name suggests, is ubiquitously found across cell types and species, exhibiting sequence and structure conservation [36–41]. This indicates its vitality for cell survival. Cell uses ubiquitin as a post-translational modifier by forming an isopeptide bond between the ε amino group of a lysine on target protein and carboxyl group of C-terminal glycine of ubiquitin [42]. If only a single molecule of ubiquitin is attached to the target protein (which is often referred to as substrate protein), the process is called monoubiquitination. Alternatively, more than one ubiquitin can be attached to a substrate protein at different locations in multiubiquitination, or a linear chain of ubiquitins can be built on substrate-attached ubiquitin to produce chains of covalently linked ubiquitins, or polyubiquitins, attached to target protein in polyubiquitination [43]. Monoubiquitination is involved in numerous processes like DNA repair, transcriptional regulation, receptor transport, histone regulation, nuclear export, endosomal sorting and viral budding [44–48], while multiubiquitination is involved in endocytosis [14]. Polyubiquitination can be subclassified into different types based on the lysine residue ubiquitinated. Ubiquitin has seven lysine residues (K6, K11, K27, K29, K33, K48 and K63), all of which can be ubiquitinated [49, 50]. Hence, polyubiquitins may either have every ubiquitin in the chain ubiquitinated on the same lysine (homotypic chains), or different ubiquitin molecules in the same chain are ubiquitinated on different lysines (heterotypic chains). Moreover, the chain may sometimes contain not just ubiquitin but ubiquitin-like proteins such as SUMO [51] and Nedd8 [52] too (heterologous chains). The relative abundance of homotypic chains is in the order of K48 > K11 and K63 >> K6, K27, K29 and K33 [53]. Lys48 chains were the first to be discovered and are involved in proteasomal degradation [3–5]. Lys63 chains are involved in lysosomal degradation [13, 14], autophagy [13] and numerous nondegradative functions like DNA repair [6], regulation of ribosome activity [9, 10] and activation of protein kinases [12]. The functions of other polyubiquitins remain unclear. Mixed-linkage polyubiquitins are found in the lower proportion compared to homotypic polyubiquitins. Mixed-linkage polyubiquitins so far discovered are K6/11, K27/29, K29/48 and K29/33 [54]. Recently, 'linear ubiquitin chains' have also been documented, in which C-terminal glycine of one ubiquitin is covalently linked not to a lysine residue but to an N-terminal methionine residue of the preceding ubiquitin. These chains, assembled by linear ubiquitin chain assembly complex (LUBAC) [55, 56], are involved in NF-κB signalling [56–58].

We have focussed on the second G1 β-bulge of ubiquitin, as the residues present in the structure show low preference for these positions in G1 β-bulges in general. Moreover, G1 β-bulge itself is relatively rare. Yet the bulge and its residues are totally conserved in ubiquitin across all eukaryotes, highlighting their significance in ubiquitin biology. Due to its proximity to K63, residues of this G1 β-bulge may be necessary for the functions involving polyubiquitin chains with K63 linkage.

Functional studies with ubiquitin carrying E64G, S65D and Q2N substitutions and their combinations showed no effect on growth under normal conditions, survival under heat stress and adherence to N-end rule [59–62]. However, the mutations led to increased cycloheximide sensitivity. Absence of any significant structural changes in ubiquitin due to these substitutions establishes that increased cycloheximide sensitivity results from functional rather than structural defect. Ribosomal protein L28 is modified by K63-linked polyubiquitin chain, which is important for making the organism resistant to translational antibiotics. Here the mutations E64G, S65D and Q2N hamper the formation of K63 polyubiquitin chain, because of their spatial proximity to K63 [59–62]. PTEN-induced kinase 1 (PINK1) is a Ser/Thr kinase that phosphorylates S65 on ubiquitin. S65-phosphorylated ubiquitin in turn activates Parkin, which is an E3 implicated in Parkinson's disease, and ubiquitinates mitochondrial proteins. Phosphorylation of ubiquitin by PINK1 occurs in response to mitochondrial membrane depolarization. Parkin itself is also known to be phosphorylated by PINK1. Together, PINK1 and parkin are involved in mitochondrial membrane quality control [63]. Missense mutations in PINK1 gene are also implicated in autosomal recessive inherited Parkinson's disease.

Hence, the functional importance and in some cases the medical relevance of the conserved residues can be understood by generating mutations of ubiquitin and studying their differential effects. Ubiquitin gene was evolved in vitro, and a dosage-dependent lethal mutation UbEP42 was isolated in our laboratory [64]. UbEP42 carried amino acid substitutions in four positions, namely S20F, A46S, L50P and I61T. In spite of being incorporated into polyubiquitin chains, the mutant caused G1 phase arrest of the cell cycle by changing the Cdc28 protein kinase levels. Further, it displayed increased sensitivity towards heat stress and exposure to cycloheximide [65]. Ubiquitin's core is significantly hydrophobic, with 16 of its 21 leucine, isoleucine, methionine and valine residues buried in the interior. This may explain ubiquitin's high thermostability, which is necessary since the molecule's role in stress management often requires it to operate at high temperature.

26.2.2 Ubiquitin-Activating Enzyme or E1

E1 catalyses the first step in ubiquitination of a protein [35]. It binds an Mg ATP. ATP is hydrolysed into PP_i and AMP, and a high-energy mixed anhydride bond is formed between AMP and C-terminal glycine of ubiquitin [66, 67]. This step primes ubiquitin for subsequent attachment to substrate, as the energy released by ATP hydrolysis is used to form the high-energy mixed anhydride bond. Next, a high-energy thioester bond is formed between C-terminal glycine of the adenylated ubiquitin and a cysteine in E1, releasing AMP in the process [68–70]. This is a spontaneous step, as the thioester bond is at lower-energy level compared to mixed anhydride bond. From here, ubiquitin is transferred to an E2. The structure of E1 enzyme has been studied in great detail, and insights have been gained into its mechanism of action [71–73]. E1 can simultaneously bind to two ubiquitins, one adenylated and the other attached to cysteine. The C-termini of both ubiquitins are spatially close

[74], which probably allows easy transfer of ubiquitin from the first site to the second. Only a single E1 transfers primed ubiquitin to all the different E2 enzymes in most organisms. Catalysing the first step of the cascade, therefore, E1 is responsible for ensuring unlimited supply of primed ubiquitin to all the downstream conjugation reactions, and so it would be expected to be present in high concentrations. However, E1 concentration is less than total E2 concentration [4]. This requires E1 to be a highly efficient enzyme. Indeed, catalytic rate values for substrate ubiquitination are 10–100-fold slower than that of all E1-catalysed steps from ATP binding to thioester formation [75]. However, E1 affinity for ubiquitin increases tremendously after ATP binding [4, 66]. ATP binding causes a conformational change in E1, which makes the binding site more accessible to ubiquitin. The interactions between AMP-ubiquitin and E1 are extensive [76]. The only known residue in E1 active site is cysteine, but as arginine residues in ubiquitin are essential for its binding with E1, certain acidic residues may also be involved [77, 78]. E1s for ubiquitin and ubiquitin-like proteins (UBLs) are related. E1 for ubiquitin is a monomer of 110 K Da. Initially, the role of ubiquitin proteasome system in the degradation of a diverse array of short-lived proteins was demonstrated by using temperature-sensitive E1 mutants expressed in mammalian cell lines [79]. These mutants also help determine if a process is ubiquitin dependent [80, 81]. Hypomorphic allele of E1 was isolated, which produces mutant form of E1 that is less efficient than wild-type E1 in performing its function. It was demonstrated that the hypomorphic allele of E1 of yeast can successfully replace the temperature-sensitive variant in the experiments where ubiquitin dependence of degradation is investigated [35, 82].

26.2.3 Ubiquitin-Conjugating Enzymes or E2s

E2 accepts ubiquitin from E1. A thioester bond is formed between C-terminal glycine of the ubiquitin and active site cysteine in E2, as the Ub-E1 thioester bond breaks [4]. This step does not involve any change in energy level since energy stored in the thioester bond between ubiquitin and E1 cysteine is conserved in the thioester bond between ubiquitin and E2 cysteine. E2s are much more diverse than E1s. *S. cerevisiae* has 14 E2s, and higher organisms have even more. Some of these may be different isoforms of same E2 [83, 84], while others may have evolved independently [85–87]. This diversity confers first level of substrate specificity to ubiquitin system, the other being at the level of E3. Structural studies on E2 enzymes show that a core domain of 150 amino acids is conserved across all E2s and may account for their chief function of transferring ubiquitin to E3 enzymes. The core domain contains a four-stranded antiparallel β-sheet, four α-helices and a 3_{10} helix. Helix 2 and the β-sheet make up a central region bound by helices 3 and 4 on one side and helix 1 on the other. The loop connecting β-strand S4 to helix 2 contains the active site cysteine. It is located in a shallow groove formed by upstream residues of the same loop on one side and those of loop connecting helix 2–3 on the other. The region surrounding active site cysteine contains many of the most conserved residues [88–94], some of which interact with ubiquitin and others probably with E1.

The side opposite to the active site, on the other hand, contains most of the poorly conserved residues. Presently, it is not clear if this variation in sequence in different E2s is due to low selective pressure or because this region may be responsible for selecting specific E3s [95]. Some E2s also have N- or C-terminal extensions, which may be responsible for conferring either substrate or E3 specificity in these cases [96–98]. Recent structural studies on complexes of E3 with Ub-E2 conjugates are beginning to unveil the mechanism of ubiquitin transfer from E2 to E3 [99–104].

26.2.4 Ubiquitin Ligases or E3s

E3s catalyse the final step of ubiquitination, i.e. forming the isopeptide bond between the C-terminal glycine of ubiquitin accepted from an E2 and the lysine on the substrate protein. In some cases, substrate protein is ubiquitinated on the α-amino group of its N-terminal residue, which may not be lysine [105]. E3s are extremely diverse in their substrate specificity. This diversity enables selective ubiquitination of proteins and hence makes ubiquitination a powerful tool for regulating cellular activities. E3s are of two types, namely, RING E3s and HECT E3s. HECT E3s simply mimic the step of ubiquitin conjugation to E2, by breaking the thioester bond of Ub-E2 and forging another thioester bond between the C-terminus of ubiquitin and -SH group of the cysteine present in the active site of E3, leading to formation of Ub-E3. In this process, energy from E2-Ub thioester bond is conserved in E3-Ub thioester bond. Subsequently, Ub is transferred from E3 to a substrate protein, by forming an isopeptide bond between the carboxyl group of C-terminal glycine of ubiquitin and ε-amino group of lysine on the substrate protein. RING E3s, unlike HECT E3s, do not form a covalent bond with ubiquitin; instead, they act as adapters by binding to the substrate protein and E2 loaded with ubiquitin simultaneously. They facilitate transfer of ubiquitin from E2 to substrate protein directly.

Majority of E3s belong to RING domain type. RING domain was discovered and characterized in the early 1990s. The canonical sequence found in RING E3s is Cys-X2-Cys-X(9–39)-Cys-X(1–3)-His-X(2–3)-Cys-X2-Cys-X(4–48)-Cys-X2-Cys (where X is any amino acid). Cysteines 1, 2, 4 and 5 coordinate one Zn^{+2} ion, and cysteines 3, 6 and 7 along with the histidine coordinate a second Zn^{+2} ion. RING domain is thus structurally related to zinc finger domain. Unlike zinc finger, however, RING domain assumes a rigid and compact shape owing to the presence of the two zinc ions in the coordination sites. Sequence conservation in RING domain is not absolute. There are variants in which cysteines and histidines are swapped, as well as those in which another residue capable of coordinating zinc replaces a cysteine, e.g. Asp in Rbx1/Roc1. There is a third group of E3s known as U-box E3s, which also recruits E2s. They are often clubbed along with RING group of E3s, as the two are closely related. In U-box domain, which is structurally similar to RING domain, zinc ions are replaced by a network of hydrogen bonds [106].

HECT E3s have a modular architecture. HECT domain has a bilobal shape, with E2-binding site in the N-terminal lobe and the active site cysteine in the C-terminal lobe [86]. As the distance between the two lobes is large, both lobes must come

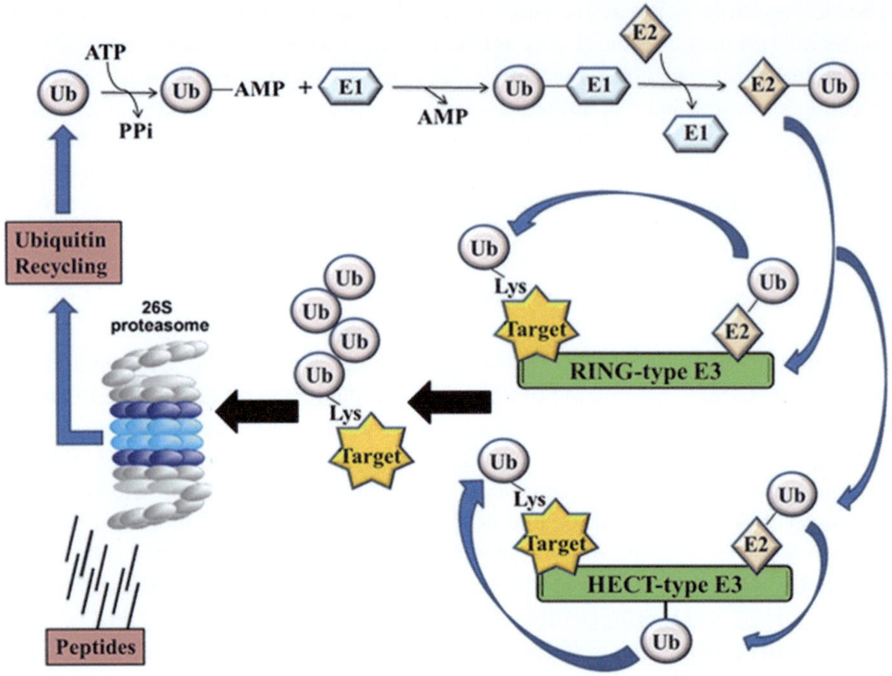

Fig. 26.1 The E1-E2-E3 pathway of ubiquitination of substrate proteins

close together for catalysis, perhaps covering a distance of around 50 Å [107]. Further, significant decrease in catalytic activity observed due to the mutations of the hinge between the lobes as a consequence of restricted movement supports the above model [108]. N-terminal region extending from N-terminal lobe of HECT domain differs from enzyme to enzyme and confers substrate specificity adding one more functional domain. Interestingly, substrate binding to E3 is not affected by the deletion of HECT domain [109–111]. Besides, isolated HECT domains retain binding to E2 enzymes. HECT domains do not show functional redundancy as they cannot substitute for each other [112], because they contribute to substrate specificity of their respective E3s [113] (Fig. 26.1).

26.2.5 Deubiquitinating Enzymes or DUBs

Deubiquitinating enzymes are proteases that remove substrate-conjugated ubiquitin from the target proteins. DUBs have several functions. Firstly, to maintain balance of any cellular process, negative regulation is as important as positive regulation. DUBs, by cleaving ubiquitin from target proteins, reverse the process of ubiquitination and hence act as negative regulators [114, 115]. Secondly, after the removal of polyubiquitin chains from target proteins either by other DUBs or by the proteasomes, DUBs replenish free ubiquitin pool by disassembling polyubiquitins into individual ubiquitins [116, 117]. They also cleave free polyubiquitin chains produced by ubiquitin conjugation machinery in the absence of any target protein. Thirdly, three of the four ubiquitin genes (*UBI1*, *UBI2* and *UBI3* in yeast) produce ubiquitin fused to ribosomal proteins, and the fourth (*UBI4*) produces a fusion of 4 or 5 ubiquitins in yeast [118–120]. Hence, DUBs are needed for producing free and functional ubiquitin from the fusions. Lastly, unwanted reactions involving ubiquitin in cell create species, which do not seem to serve any function. Such species include thiol esters between ubiquitin C-terminus and cellular thiols like glutathione, amide derivatives of ubiquitin formed with lysine and spermidine and free ubiquitin adenylate. DUBs release free ubiquitin from these species [121, 122].

There are five different families of DUBs in humans, which together constitute about 100 known DUBs. Four of these families are cysteine-like proteases. These are Josephin domain or MJD proteases, ubiquitin-specific proteases (USP/UBP), ubiquitin C-terminal hydrolases (UCH) and ovarian tumour (OTU) families. The fifth one is of JAB1/MPN/Mov34 metalloenzyme (JAMM) domain zinc-dependent metalloprotease family. The active site in DUBs of Cys protease families consists of the catalytic cysteine and a histidine. Histidine increases polarity of cysteine -SH group, making S more negative, enabling it to make a nucleophilic attack on the isopeptide bond linking ubiquitin to substrate protein. Cleavage of isopeptide bond is followed by formation of a transient acyl intermediate-linking carboxyl group of ubiquitin to catalytic Cys of DUB. This acyl intermediate is then cleaved by a water molecule, releasing ubiquitin. In many, though not all cases, a third residue, which usually is Asn or Asp, polarizes and aligns histidine. The catalytic Cys, His and Asn/Asp are called the catalytic triad. All the four Cys protease families have the same basic catalytic site structure and mechanism. What distinguishes them from one another is their tertiary structure.

Besides the DUBs, which cleave SUMO and Nedd8, ubiquitin-like proteases (ULPs) include Adenain family of cysteine proteases [123–127]. The members of Adenain family resemble adenovirus protease. The diversity seen with DUBs is mainly responsible for their immense substrate specificity, which enables selective deubiquitination of proteins. Similar to DUBs, the feature of 'structural diversity resulting in functional selectivity' is observed with E3 enzymes as well, while catalysing ubiquitination. This makes deubiquitination as useful as ubiquitination in the regulation of cellular activities such as regulation of DNA repair [128], gene expression [129], cell cycle regulation [130], kinase activation [131, 132], proteasome-mediated degradation [133, 134], lysosome-mediated degradation [134], microbial pathogenesis [135, 136] and myriad other activities. Many pathogenic bacteria [136, 137] and viruses [135, 138–145] have acquired DUB genes through parallel transfer from eukaryotic genomes. These microbes may use DUBs to shield their proteins against host's ubiquitination machinery. For example, DUBs in Adenain family have been acquired by bacteria and viruses to cleave ubiquitin and interferon-stimulated gene 15 (ISG15) conjugates [123–127]. Mutant DUBs have also been implicated in numerous diseases [128, 146–148]. To date, very few DUBs have been characterized in terms of their substrates and physiological roles, and much of this area remains to be explored.

In UCH domain family UCH catalytic core is formed by a domain of 230 amino acid residues. The salient feature of this family is that the active site is covered by a loop. The loop restricts access to ubiquitinated proteins and polyubiquitin chains as they are too big and ensures that ubiquitin attached through its C-terminal to small peptides, and small chemical groups only can access the active site of the DUB for processing. The space afforded by this arrangement is too small for large, folded ubiquitin conjugates or even polyubiquitin chains to fit in. In a study conducted on UCHL3, the loop was systematically extended until UCHL3 gained ability to hydrolyse polyubiquitin. The amount of extension needed was significant. Hence, UCHs act not on ubiquitin linkage with whole proteins but with small peptides produced as by-products of proteasomal or lysosomal degradation, molecules like aldehydes and C-terminal extensions of polymeric proubiquitin. However, unfolded whole proteins may also be targeted, if they can manage to thread through confined space provided by the loop [149]. In some UCHs, active site exists in an unproductive conformation in the absence of activation signals [150].

Members of USP domain family consist of three subdomains in their structure, namely, finger, palm and thumb [151]. CYLD is the only USP that lacks finger subdomain [152]. Though sequence similarity of USP domain across the family is low, it is structurally well conserved [151–155]. The active site lies in a cleft between palm and thumb, and the C-terminus of ubiquitin binds the active site, while the globular portion of ubiquitin binds the fingers [151, 153, 155]. Crystal structures of both ubiquitin-bound and ubiquitin-unbound forms of many USPs have been solved, and their comparison shows that some USPs exist in inactive conformation in the absence of ubiquitin, while others maintain active conformation irrespective of ubiquitin's presence. In the former case, inactive conformation of the USP may be a result of either improper positioning of catalytic triad or blocking of correctly positioned triad by ubiquitin-binding surface loops. In many cases, USP domain has insertions or terminal extensions capable of folding into independent domains with some functional relevance. In USP5, these domains have additional ubiquitin-binding sites, whereas in CYLD, they determine subcellular localization [152].

In the OTU domain family, the structure of Otu1 covalently bound to ubiquitin shows that most of its interactions with ubiquitin are mediated by a large surface loop. In the absence of ubiquitin, this surface loop has been found to be disordered in OTUB1 and the other members of the family OTUB2 and A20. Superposition of ubiquitin-bound Otu1 structure on the structure of A20 shows that a helical domain blocks binding site for ubiquitin moiety, suggesting architectural variation of this site in A20. The structure of OTUB1 apoprotein shows nonproductive alignment of His residue with catalytic Cys in the catalytic triad, suggesting that OTUs, like USPs and UCHs, may exist in catalytically inactive conformation in the absence of ubiquitin. In OTU core domain, five β-strands are sandwiched between helical domains. However, the lengths of the β-strands are variable within the OTU family [152].

MJD family has four members in humans, of which the best studied is Ataxin 3 [156–160]. Ataxin 3 probably acts specifically on K63 polyubiquitin. It has an extended helical arm, which may regulate access of polyubiquitin chain to the active site [157–159]. A second ubiquitin-binding site, which is distinct from the active site, lies at the back of this arm, suggesting that ataxin 3 may interact with two distal ubiquitins simultaneously in a polymer. Ataxin 3 also contains a polyglutamine stretch whose extension causes Machado-Joseph disease [161].

JAMM domain DUBs are commonly associated with large complexes [162–169]. Solving the crystal structure of one of the JAMM domain DUB, AMSH-like protease bound to a K63 diubiquitin has helped unravel the catalytic mechanism of JAMM domain DUBs [169]. JAMM domain coordinates two Zn ions, one of which primes an H_2O for hydrolysis of isopeptide bond, while the other is included in an AMSH-specific insert that forms a motif recognizing proximal ubiquitin of K63 diubiquitin. Another AMSH-specific insert recognizes distal ubiquitin. Hence, JAMM domain recognizes distal ubiquitin and sequence Gln62-Lys63-Glu64 in proximal ubiquitin, which makes AMSH-like protease K63 linkage specific. AMSH-specific inserts are absent in JAMM domain protcascs that are not specific for polyubiquitin.

26.3 Ubiquitin-Binding Domains

The binding partners of ubiquitin recognize and bind to substrate-conjugated ubiquitin using domains called ubiquitin-binding domains or UBDs through noncovalent interactions. Ubiquitin exerts its effects not by structural modification of ubiquitinated protein but by itself serving as an additional interacting surface, which makes ubiquitin-binding domains necessary. Structure and sequence information about these domains may help identify new ubiquitin-binding partners. More than 20 different UBDs are known at present, and more are expected to be discovered. But since a comprehensive discussion of this topic is beyond the scope of this chapter, only the earliest discovered UBDs are described and compared here, namely, ubiquitin-interacting motif (UIM), ubiquitin-associated domain (UBA); coupling ubiquitin to endoplasmic reticulum degradation (CUE); polyubiquitin-associated zinc finger (PAZ (ZnF-UBP)); Gga and Tom1 domain (GAT); Npl4 zinc finger motif (NZF); Vps27, HRS, STAM (VHS); GRAM-like ubiquitin binding in Eap45 (GLUE); and ubiquitin-conjugating enzyme variant (UEV). Structurally, these domains are quite unrelated. UIMs consist of a single helix [170]. NZFs have three residues on loops of four strands, which in turn are stabilized by a Zn ion [171]. UBA and CUE domains are so closely related that the structures of their complexes with ubiquitin are superimposable. Both CUE and UBA domains consist of three helix bundles, of which two helices recognize ubiquitin [172–174]. Even though GAT consists of three helix bundles with two helices recognizing ubiquitin, it is unrelated to UBA and CUE, as the ubiquitin-interacting helices in CUE and UBA are antiparallel; however, they are parallel in GAT [175]. UEV domain is made up of α-helices and a β-sheet, and ubiquitin is recognized by the loop between two helices and a part of β-sheet [176, 177]. This structural diversity suggests that different UBDs recognize different parts of ubiquitin, which in turn may explain the reason for high sequence and structure conservation observed with ubiquitin. Most

UBDs interact with the L8-I44-V70 patch of ubiquitin, making their footprints on ubiquitin overlapping, although some recognize other surfaces such as Asp58 and Gln62 [178] or the C-terminal residues [179]. Moreover, the footprints on I44 patch show marked variation in spite of some overlap. Although most UBDs characterized so far contact the I44 face of ubiquitin, high degree of conservation seen with other surfaces of ubiquitin suggests that many more UBDs remain to be discovered. Besides, many proteins involved in ubiquitin-dependent processes have been shown to have unidentified ubiquitin-binding sites.

Mechanisms of action of UBDs are unclear, but their properties give key insights into their functioning in vivo. Affinity of UBDs for ubiquitin is typically low with their K_d values in the range of 10–500 μM. This may make complexes based on UBD-Ub interactions capable of rapid assembly and disassembly, making them more dynamic [180]. Low affinity may also make regulation of these complexes easier, as disruption of even a single interaction may destabilize the complex. Besides, as ubiquitin concentration in cell is very high [181], low affinity may be a strategy to ensure availability of free UBDs. Proteins needed to bind strongly to ubiquitin may do so by having multiple UBDs. The overlapping footprints of different UBDs on ubiquitin may prevent simultaneous binding of more than one ubiquitin receptor to ubiquitinated protein. This may be desirable in pathways that need sequential handing over of ubiquitinated protein from one ubiquitin receptor to another. Different footprints may also help UBDs in distinguishing between monoubiquitination and polyubiquitination and also between different linkages of polyubiquitination [172, 182, 183]. Structural studies show that ubiquitin structure slightly changes on binding to different UBDs, increasing their affinity for ubiquitin [184]. Some UBDs appear to be linkage specific, while others are not [185, 186]. Linkage specificity can be conferred either by interaction of UBD with linker region of polyubiquitin [187] or by fixed length of the linker sequence between tandem UBDs that favourably positions them to recognize a particular polyubiquitin [158, 188, 189].

26.4 Proteasome-Mediated Degradation

Proteasomal degradation is the most well-studied consequence of ubiquitination. One of the most valuable attributes of a cell is its ability to quickly adapt to changes in its internal and external environments, which often requires fast and extensive changes in its proteome. The set of existing proteins, which are not useful in the changed environment, are quickly degraded by proteasome, and the amino acids are used to produce a new set of proteins to meet the challenge. Besides, proteasome also degrades misfolded, truncated and denatured proteins formed during normal function or under stress conditions like high temperature, UV exposure or starvation. Proteasome's function can therefore be described as constructive destruction. Most proteases nonspecifically and nonprocessively degrade proteins by recognizing intrinsic cleavage sites that cannot be modified and releasing the products after every cleavage. Unlike most proteases which are nonspecific, proteasome has been designed to specifically degrade certain proteins while leaving others unharmed. It is a hollow protein cylinder. The active sites are situated inside the cylinder, which ensures high processivity by preventing substrate proteins from diffusing away. The regulatory proteins associated with the two ends of the hollow cylinder, with few exceptions, recognize only ubiquitinated proteins. This confers specificity by ensuring that only ubiquitinated proteins are degraded.

The cylindrical protease also known as 20S proteasome is 15 nm in length and 11 nm in diameter and is made of four rings, two α-rings and two β-rings, stacked over one another in the order αββα. The α- and β-rings are made up of seven α-type and seven β-type subunits, respectively [190–196]. Amino acid sequences of α- and β-subunits share some homology. Proteasomes belong to a group of proteins called chambered proteases, all of which have same basic structure. In archaebacterial proteasome, all seven β-subunits are identical and have an active site for proteolysis on each of them [195, 197, 198]. In eukaryotic proteasome, however, not all β-subunits are identical, and only three of the seven β-subunits in each β-ring have proteolytic sites, presenting the proteasome with six of them. The three sites in each ring are located on β1, β2 and β5. The two β-rings therefore make a catalytic chamber lined by proteolytic sites, where proteins are degraded. β1, β2 and β5 sites have caspase-like, trypsin-like and chymotrypsin-like activities, respectively [196, 199, 200]. This diversity makes eukaryotic proteasome a highly efficient protease by increasing the probability of finding a cleavage site on substrate protein. Genetic studies have shown that the importance of these sites for cell growth decreases in the order β5 > β2 > β1. For example, β5β2 double mutants in yeast are lethal, while β5β1 and β2β1 are viable [199, 201, 202]. The α-rings form two additional chambers, called outer chambers or antechambers, which are present one on either side of the catalytic chamber. These outer chambers serve as antechambers for unfolded proteins before they can be inserted into catalytic chamber for degradation. The volume of outer chambers is 59 nm^3, while that of catalytic chamber is 84 nm^3 [197, 298]. Access to these chambers is controlled by pores, which may be either open or closed. When open, they are only 2 nm in diameter, which ensures that only unfolded proteins gain entry into chambers while leaving randomly colliding intact proteins unaffected. Besides, the N-terminal regions of α-subunits undergo conformational change to allow entry of the substrate protein into the proteolytic chamber [203, 204]. This conformational change is induced by the association of 19S complex with the 20S proteasome forming 26S proteasome. The 20S complex remains closed in the absence of 19S particle. PA28/11S, a complex involved in antigen processing by proteasomes, can also induce a similar conformational change. 28/11S helps proteasome generate immunopeptides from antigenic proteins that can be presented by MHC complexes for initiating an immune response [203, 205].

Attached on either end of the 20S complex is 19S complex. Composed of 19 subunits, it has two parts, the base and the lid [206, 207]. The base, composed of eight subunits, unfolds substrate proteins and inserts them into 20S complex. Since this process requires energy, six of the eight subunits in the base, namely, Rpt1–Rpt6, are ATPases [208]. Their C-terminal residues are also responsible for binding of 19S complex to 20S complex [209, 210]. These six subunits may assemble into a heterohexameric ring [211], though the order in which they are arranged in the ring is debated [212–214]. The two non-ATPase subunits are Rpn1 and Rpn2. The lid also has eight subunits, Rpn3 and from Rpn5 to Rpn11. Out of these, Rpn10 attaches lid to base. The 19S complex imparts substrate specificity to proteasome by favouring K48 polyubiquitinated proteins over others [215]. The K48 polyubiquitin must contain at least four to five ubiquitins for recognition by proteasome [215], which suggests that both ubiquitin structure and polyubiquitin topology determine recognition. Only a few proteins are known to be recognized by 19S in nonubiquitinated state [216–220]. Rpn10 [221] and Rpn13 [222] are the subunits, which are known to bind polyubiquitinated proteins. Rpn10 binds ubiquitin through its ubiquitin-interacting motif (UIM) consisting of three helices connected by flexible linkers [223]. The UIM does not display any definite tertiary structure due to the flexible linkers. Rpn13 binds ubiquitin through a pleckstrin-like receptor of ubiquitin (PRU) domain [224]. Ubiquitin binds to loops in PRU domain, unlike other UBDs in which it binds to secondary structures. Studies suggest that Rpn1, Rpn2 and Rpt5 are ubiquitin receptors as well [225–227]. Rpt5 is also known to bind ornithine decarboxylase, the nonubiquitinated substrate of proteasome [218]. Before the unfolded proteins are inserted into 20S complex, the polyubiquitin tail is cleaved off by Rpn11 using its zinc finger containing JAMM motif [162, 228]. Rpn11 functions only in association with the rest of the 19S, as its deubiquitinating activity is ATP dependent. The 19S particle also opens the gates formed by N-terminal regions of α-subunits at the mouth of 20S complex. The 20S and 19S complexes together make the 26S proteasome. Proteasomes can exist in the nucleus and cytoplasm and attached to the outer surface of the endoplasmic reticulum.

The tertiary structures of proteasomal subunits exhibit properties typical of and necessary for subunits of large protein assemblies. Proteasomal-activating nucleotidase (PAN), an orthologue of proteasomal AAA ATPases, has coiled coils protruding from an oligonucleotide binding (OB) fold called PAN-N and an AAA fold [213, 229]. The long and slender topology enables coiled coils of PAN to establish extensive contact with their binding proteins. Rpn1 and Rpn2 have proteasome cyclosome (PC) repeats [230], which may form two helix ARM/HEAT units, which in turn form α-solenoids [231]. The α-solenoids are superhelical quaternary structures having extensive surface area for establishing contact with binding partners. Rpn3, Rpn5, Rpn6, Rpn7, Rpn9, and Rpn12 have PCI domain [232], which consists of an N-terminal helical bundle fold and a C-terminal-winged helix fold. Repetitive bihelical blocks preceding PCI domain in these subunits may form α-solenoids together with helical bundle fold in PCI domain [232].

26.5 Role of Proteasomal Degradation in Cell Physiology, Disease and Therapy

26.5.1 Role in Cell Physiology

There are various aspects of cell physiology that are regulated by proteasomal degradation, including DNA repair, transcription, protein synthesis, cell signalling, cell cycle and autophagy. During the progression of cell cycle, ubiquitin system ensures timely degradation of cell cyclins, inhibitors of cyclin-dependent kinase and other important proteins. Often defective regulation of UPS in relation to degradation of these key proteins is a major cause of tumorigenesis. There are two main families of E3s involved in cell cycle: the SCF (Skp1/Cul1/F-box protein) complex and the anaphase promoting complex or cyclosome (APC/C). APC/C is responsible for exit from mitosis and establishing a stable G0/G1 phase, while SCF is involved in all stages of cell cycle. In transcription-coupled DNA repair, proteasomal degradation of the largest subunit of RNA polymerase II occurs after its ubiquitination by Rsp5, which is a HECT domain E3 [110]. Besides, transcriptional activators are also degraded by proteasomes [233, 234]. Transcriptional activation domains (TADs) are responsible for signalling proteasomal degradation. Transcriptional factors containing TADs may be both activated and degraded through the same pathway. Proteasomal ATPases unfold the yeast Gal4 activator irreversibly, leading to its proteasomal degradation [7]. Interestingly, monoubiquitination is sufficient to serve as a marker for degradation in the case of Gal4, instead of polyubiquitination with K48 linkage. Genome-wide ChIP-chip studies show that proteasomal ATPases are resident on hundreds of yeast genes, as several trans- and co-activators of transcription undergo monoubiquitination. Hence, ubiquitination is used by cells as a mechanism in transcriptional regulation. Proteasomal degradation is also involved in protein synthesis. In order to prevent formation of defective ribosomes, cell must produce all ribosomal proteins in equimolar amounts. This requires an unattainably high level of coordination between the large numbers of ribosomal protein genes. Cells solve this problem by producing excess of all ribosomal proteins and subjecting those subunits which fail to get incorporated into ribosomes to proteasome-mediated degradation [11]. Proteasomal degradation therefore sculpts the ribosomal protein set into equimolarity. Interestingly, ribosomal proteins might be protected from proteasomal degradation before being incorporated into ribosomes. For example, *Ubi1* and *Ubi2* genes are expressed as ubiquitin fusions with small ribosomal subunit protein S27a, while *Ubi3* gene is expressed as ubiquitin fused to large ribosomal subunit protein L40 [120, 121]. Ubiquitin fused to these proteins may protect them not only from proteasomal degradation but also from N-terminal-specific proteases, as the ubiquitin is fused to their N-terminal. Proteasomal degradation is also involved in regulating the action of two prosurvival switches in TNFR1-mediated cell signalling. NF-κB pathway is driven towards cell survival by proteasome-mediated degradation of Ik-Bα, the inhibitor of NF-κB. Degradation of Ik-Bα enables NF-κB to migrate to nucleus and upregulate its target genes [235]. The second switch in the same pathway is receptor-interacting serine/ threonine protein kinase 1 or RIP1. RIP1 serves as a dual switch. When RIP1 is conjugated with a K63-linked polyubiquitin chain, it functions as an inhibitor to proapoptotic pathway by NF-κB-independent mechanism initially and later by NF-κB-dependent mechanism [236]. Alternatively, a deubiquitinated RIP1 may also interact with FADD and caspase 8, causing apoptosis [237]. The enzyme A20 has both DUB and E3 activities and replaces the K63 chain on RIP1 with a K48 chain, causing its proteasomal degradation [238]. It is interesting to study the consequences when A20 fails to polyubiquitinate RIP1 with K48 chain, after removal of its K63 chain. Hence, the ubiquitination status of RIP1 acts as a checkpoint, as polyubiquitin chains with K48 linkage and K63 linkage serve as negative and positive regulators of NF-κB signalling, respectively, and in turn act as switches for apoptosis and cell survival. Ubiquitin system also plays a vital role in stress response. During stress, heat-shock proteins (HSPs) act as chaperones, folding unfolded proteins back to their native state. However, when the proteins are truncated or damaged beyond repair, these same HSPs associate with E3s like CHIP and Parkin and facilitate their degradation by proteasomes. If the rate of denaturation of proteins is so high that neither HSPs nor proteasomes can clear them quickly, the unfolded proteins form intracellular aggregates. Such potentially toxic aggregates may be responsible for neurodegenerative diseases like Parkinson's and Alzheimer's. Indeed, several protein aggregates have been shown to be ubiquitin positive. Interestingly, ubiquitin system plays a role not just in proteasomal degradation but also in lysosomal degradation [239–241]. Ubiquitin has been shown to be responsible for marking membrane proteins for selective degradation by lysosomes.

26.5.2 Role in Disease and Therapy

Owing to its complexity and involvement in multiple processes, ubiquitin system is associated with numerous diseases and also offers prime targets to cure them [15–17]. Mutation in E3s for specific target proteins can cause disease due to loss or gain of function associated with the target protein. Alzheimer's disease is characterized by extracellular amyloid plaques, containing misfolded β-peptides generated from the cleavage of amyloid precursor protein (APP) and intracellular neurofibrillary tangles containing Tau, a microtubule-associated protein. Defects in ubiquitin system may be involved in Alzheimer's disease in multiple ways. UBB^{+1} is a ubiquitin variant with a C-terminal extension of 19 amino acids. It cannot be conjugated to substrate lysines, although it may be incorporated into polyubiquitin chains. It was in the neurons of Alzheimer's disease that UBB^{+1} was first discovered, followed by other neurological disorders. Decreased proteasomal activity is observed in the severely affected regions of the brain of Alzheimer's patients. Besides, overexpression of UCH-L1, a DUB of ubiquitin hydrolase family, helped in reducing memory loss in mouse models of Alzheimer's disease, suggesting its role in the disease [15].

Parkinson's disease is a consequence of many unrelated causes, and studies implicate ubiquitin system as one of them. About 50% of juvenile patients of autosomal recessive Parkinson's disease have mutations in parkin gene, which encodes a RING finger E3. Parkin protein contains an N-terminal Ubl domain and two C-terminal RING finger domains flanking an in-between RING (IBR) domain. Parkin binds to proteasome through its S5a subunit, and this binding is reduced by mutations at R42 residue. Such mutations may compromise substrate degradation through parkin, and resultant parkin substrate accumulation may result in Parkinson's disease pathogenesis. Possible parkin substrates include Parkin-associated endothelin receptor-like receptor (PAEL-R), α-synuclein, synaptotagmin XI, cyclin E, tubulin, misfolded dopamine transporters and polyglutamine-repeat proteins. None of them, however, showed increased abundance in neurons of parkin-lacking mice [15].

Huntington's disease is a CAG repeat disorder. The protein Huntingtin is crucial to the function of neuronal cells. Polyglutamine (polyQ) tracks are expanded in Huntingtin protein giving rise to a mutant form. Role of ubiquitin system is suspected in Huntington's disease as the inclusion bodies formed contain ubiquitin, E2/E3s and proteasomal subunits. Moreover, Huntingtin undergoes ubiquitination, and overexpression of its mutant form inhibits proteasomal activity, leading to cell cycle arrest [15].

Defects in ubiquitin system have also been implicated in several types of cancers. Products of numerous oncogenes and suppressor genes undergo ubiquitination [81, 242, 243].

The protein p53 is a tumour suppressor. It is involved in numerous cell proliferation and apoptosis pathways, which makes it a good drug target. About 50% of all human tumours contain p53 mutations and many of those that do not have defects in other components of p53 network. MDM2 is a RING finger E3 that ubiquitinates p53 causing its proteasomal degradation. It also inhibits p53 activity by physically blocking its N-terminal transactivation domain and preventing its nuclear export. p53 transcriptionally induces MDM2, creating a negative feedback loop. Hence, p53 activity could be boosted in cancer cells by inhibiting MDM2, helping tumour suppression. Moreover, MDM2 may also be responsible for degradation of other antioncogenic proteins.

SCF E3s are multisubunit E3s composed of four components. The subunit of SCF, CUL1 serves as the scaffold for assembly of the rest of ubiquitin-conjugating machinery. RBX1 is a RING finger protein interacting with C-terminus of CUL1, while N-terminus of CUL1 binds to SKP1, and in turn SKP1 interacts with F-box proteins. At least 68 F-box proteins are found in human genome, each of which recognizes multiple substrates. SCF E3s interact with different F-box proteins to recognize different substrates. F-box proteins therefore modulate specificity of SCF E3s. Mutations in many F-box protein-substrate pairs are involved in cancers. SCF-FBW7 is an E3 promoting degradation of proteins involved in cell proliferation, like cyclin E, c-Myc, c-Jun, Notch and sterol regulatory element-binding proteins (SREBPs). Arginine residues in WD40 repeats of FBW7 interact with phosphodegrons in these substrate proteins. Mutations targeting these arginines cause many types of cancers. Mutations in substrate phosphodegron, preventing their recognition by FBW7, may also cause malignant transformations. SCF-SKP2 is involved in the degradation of several negative cell cycle regulators like p130 (a protein belonging to retinoblastoma family), FoxO (cell-cycle inhibitory transcription factor forkhead box protein O) and the CDK inhibitors p27, p21 and p57. Small CDK-interacting protein 1 (CKS1) is also an SCF-SKP2 subunit and is required for p27 degradation. It is probably needed for interaction of SKP2 with p27. SKP2 overexpression is involved in several human cancers and RNA interference. Intracellular injection of antiSKP2 antibodies can reduce cancer cell proliferation. IκB kinase phosphorylates IκB at S32 and S36, generating a phosphodegron, which is recognized by βTRCP, the E3 responsible for proteasomal degradation of IκB. βTRCP has numerous substrates, such as β-catenin (an oncogenic transcription factor), the cell-cycle regulatory proteins early mitotic inhibitor 1 (EMI1) and cell division cycle protein 25A (CDC25A) and progesterone receptor. Mutations in β-catenin phosphodegron are involved in the pathogenesis of several human cancers. VHL is an E3 that inhibits angiogenesis under normoxic conditions and is known for its antioncogenic role. Mutations in VHL are responsible for familial von Hippel-Lindau syndrome, a type of renal cancer, and somatic mutations of VHL genes are responsible for cancers like sporadic clear-cell renal carcinomas. E6-AP is a HECT domain E3 that degrades p53, proto-oncogene c-Myc and several other substrates. It is associated with E6 oncoprotein of human papillomavirus (HPV). E6-AP mutations are associated with Angelman's syndrome, and in certain sexually transmitted HPV serotypes like HPV-16 and HPV-18, P53 degradation by E6-AP causes transformation in uterine cervical epithelial cells, leading to cervical cancer. Mutations in components of ubiquitin system involved in DNA repair like E3 BRCA1 (discussed above) can also cause cancer [16].

Mutations and changes in the expression levels of various components of UPS are responsible for causing many diseases. Besides, ubiquitin system is also exploited by several pathogens for their benefit. Ubiquitin-mediated endocytosis of anthrax toxin-receptor complex leads to toxin activation [244]. *Yersinia pestis* produces a deubiquitinating enzyme YopJ, which prevents activation of NF-kB [245, 246]. Modulator of immune recognition 1 of Kaposi's sarcoma-associated herpes virus (KSHV) is an E3 enzyme, which ubiquitinates cysteines in MHC class I molecules, causing their endocytosis and degradation [247]. Certain tumour-causing viruses transform host cells by activating NF-κB pathway, using ubiquitin-dependent mechanisms. Tax protein of human T-cell leukaemia virus 1 (KSHV), an activator of IKK and NF-κB, is K63 polyubiquitinated by Ubc13 [248, 249]. However, no impairment is seen in Ubc13 knockdown cells [250]. STP-C of herpes virus *Saimiri* and LMP1 of Epstein-Barr virus also show ubiquitin-dependent activation of IKK [251, 252] by binding to TRAF6.

26.6 Conclusion

Ubiquitin-proteasome system is a very versatile protein-degrading machinery that selectively degrades polyubiquitinated proteins with a few exceptions. Unlike other proteases, this system gives cell the ability to make very specific changes in its proteome to adapt to changing environment. How much ubiquitin system increases cell's survivability is evident from lethality of ubiquitin mutations like EP42. And the high degree of ubiquitin's sequence and structure conservation across species underlines its functional relevance. Numerous proteins have evolved to interact with different surfaces on ubiquitin through various ubiquitin-binding domains. E1, E2 and E3 are the enzymes involved in ubiquitination, and high diversity of E3s is responsible for the specificity of ubiquitination. Deubiquitinating enzymes do not just recycle of ubiquitin but confer additional control and specificity to the system. The simple structure of ubiquitin makes it an ideal system to study protein folding, and many structural studies have been carried out on it. Proteasomal degradation is involved in numerous cellular processes including DNA repair, transcription, protein synthesis and cell signalling. Besides, ubiquitination is involved not just in proteasomal degradation but lysosomal degradation as well. Owing to its diverse functions, defects in ubiquitin system have been implicated in several diseases like Alzheimer's disease, Parkinson's disease, Huntington's disease, various cancers and microbial infections. Nevertheless ubiquitin proteasome system along with deubiquitinating enzymes offers several prime drug targets which could be exploited in future to treat these diseases. This has triggered intense research on ubiquitin system and makes it an important field of study in our quest to improve human health.

References

1. Goldstein G, Scheid M, Hammerling U et al (1975) Isolation of a polypeptide that has lymphocyte-differentiating properties and is probably represented universally in living cells. Proc Natl Acad Sci U S A 72:11–15
2. Ciechanover HA, Rose I (2006) The discovery of ubiquitin-mediated proteolysis. J Biol Chem 281(40):32
3. Ciechanover A, Elias S, Heller H et al (1980) Characterization of the heat-stable polypeptide of the ATP-dependent proteolytic system from reticulocytes. J Biol Chem 255:7525–7528
4. Hershko A, Heller H, Elias S et al (1983) Components of ubiquitin-protein ligase system. Resolution, affinity purification, and role in protein breakdown. J Biol Chem 258:8206–8214
5. Hershko A, Eytan E, Ciechanover A et al (1982) Immunochemical analysis of the turnover of ubiquitin-protein conjugates in intact cells. Relationship to the breakdown of abnormal proteins J Biol Chem 257:13964–13970
6. Ulrich HD, Walden H (2010) Ubiquitin signalling in DNA replication and repair. Nat Rev Mol Cell Biol 11:479–489
7. Kodadek T (2010) No splicing, no dicing: non-proteolytic roles of the ubiquitin-proteasome system in transcription. J Biol Chem 285:2221–2226
8. Zhang Y (2003) Transcriptional regulation by histone ubiquitination and deubiquitination. Genes Dev 17:2733–2740
9. Kwapisz M, Cholbinski P, Hopper AK et al (2005) Rsp5 ubiquitin ligase modulates translation accuracy in yeast Saccharomyces cerevisiae. RNA 11:1710–1718
10. Spence J, Gali RR, Dittmar G et al (2000) Cell cycle-regulated modification of the ribosome by a variant multiubiquitin chain. Cell 102:67–76
11. Shcherbik N, Pestov DG (2010) Ubiquitin and ubiquitin-like proteins in the nucleolus: multitasking tools for a ribosome factory. Genes Cancer 1:681–689
12. Chen ZJ, Sun LJ (2009) Nonproteolytic functions of ubiquitin in cell signaling. Mol Cell 33:275–286

13. Kirkin V, McEwan DG, Novak I et al (2009) A role for ubiquitin in selective autophagy. Mol Cell 34:259–269
14. Mukhopadhyay D, Riezman H (2007) Proteasome-independent functions of ubiquitin in endocytosis and signaling. Science 315(5809):201–205
15. Reinstein E, Ciechanover A (2006) Narrative review: protein degradation and human diseases: the ubiquitin connection. Ann Intern Med 145:676–684
16. Schwartz AL, Ciechanover A (2009) Targeting proteins for destruction by the ubiquitin system: implications for human pathobiology. Annu Rev Pharmacol Toxicol 49:73–96
17. Nalepa G, Rolfe M, Harper JW (2006) Drug discovery in the ubiquitin-proteasome system. Nat Rev Drug Discov 5:596–613
18. Vijay-Kumar S, Bugg CE, Wilkinson KD, Cook WJ (1985) Three-dimensional structure of ubiquitin at 2.8 A^0 resolution. Proc Natl Acad Sci U S A 82(11):3582–3585
19. Vijay-Kumar S, Bugg CE, Cook WJ (1987) Structure of ubiquitin refined at 1.8 Å resolution. J Mol Biol 194:531–544
20. Ibarra-Molero B, Loladze VV, Makhatadze GI et al (1999) Thermal versus guanidine-induced unfolding of ubiquitin. An analysis in terms of the contributions from charge-charge interactions to protein stability. Biochemistry 38:8138–8149
21. Sundd M, Iverson N, Ibarra-Molero B et al (2002) Electrostatic interactions in ubiquitin: stabilization of carboxylates by lysine amino groups. Biochemistry 41:7586–7596
22. Sundd M, Robertson AD (2003) Rearrangement of charge-charge interactions in variant ubiquitins as detected by double-mutant cycles and NMR. J Mol Biol 332:927–936
23. Loladze VV, Ibarra-Molero B, Sanchez-Ruiz JM et al (1999) Engineering a thermostable protein via optimization of charge-charge interactions on the protein surface. Biochemistry 38:16419–16423
24. Loladze VV, Makhatadze GI (2002) Removal of surface charge-charge interactions from ubiquitin leaves the protein folded and very stable. Protein Sci 11:174–177
25. Makhatadze GI, Loladze VV, Ermolenko DN et al (2003) Contribution of surface salt bridges to protein stability: guidelines for protein engineering. J Mol Biol 327:1135–1148
26. Loladze VV, Ermolenko DN, Makhatadze GI (2001) Heat capacity changes upon burial of polar and nonpolar groups in proteins. Protein Sci 10:1343–1352
27. Loladze VV, Ermolenko DN, Makhatadze GI (2002) Thermodynamic consequences of burial of polar and non-polar amino acid residues in the protein interior. J Mol Biol 320(2):343–357
28. Khorasanizadeh S, Peters ID, Roder H (1996) Evidence for a three-state model of protein folding from kinetic analysis of ubiquitin variants with altered core residues. Nat Struct Biol 3(2):193–205
29. Sosnick TR, Dothager RS, Krantz BA (2004) Differences in the folding transition state of ubiquitin indicated by phi and psi analyses. Proc Natl Acad Sci U S A 101:17377–17382
30. Went HM, Jackson SE (2005) Ubiquitin folds through a highly polarized transition state. Protein Eng Des Sel 18:229–237
31. Thomas ST, Makhatadze GI (2000) Contribution of the 30/36 hydrophobic contact at the C-terminus of the alpha-helix to the stability of the ubiquitin molecule. Biochemistry 39:10275–10283
32. Sloper-Mould KE, Jemc JC, Pickart CM et al (2001) Distinct functional surface regions on ubiquitin. J Biol Chem 276:30483–30489
33. Beal R, Deveraux Q, Xia G et al (1996) Surface hydrophobic residues of multiubiquitin chains essential for proteolytic targeting. Proc Natl Acad Sci U S A 93:861–866
34. Shih SC, Sloper-Mould KE, Hicke L (2000) Monoubiquitin carries a novel internalization signal that is appended to activated receptors. EMBO J 19:187–198
35. Schulman BA, Harper JW (2009) Ubiquitin-like protein activation by E1 enzymes: the apex for downstream signalling pathways. Nat Rev Mol Cell Biol 10:319–331
36. Gavilanes JG, Gonzalez de Buitrago G, Perez-Castells R, Rodriguez R (1982) Isolation, characterization, and amino acid sequence of a ubiquitin-like protein from insect eggs. J Biol Chem 257(17):10267–10270

37. Watson DC, Levy WB, Dixon GH (1978) Free ubiquitin is a non-histone protein of trout testis chromatin. Nature 276:196–198
38. Schlesinger DH, Goldstein G, Niall HD (1975) The complete amino acid sequence of ubiquitin, an adenylate cyclase stimulating polypeptide probably universal in living cells. Biochemistry 14:2214–2218
39. Schlesinger DH, Goldstein G (1975) Molecular conservation of 74 amino acid sequence of ubiquitin between cattle and man. Nature 255:423–424
40. Wilkinson KD, Cox MJ, O'Connor LB et al (1986) Structure and activities of a variant ubiquitin sequence from bakers' yeast. Biochemistry 25:4999–5004
41. Vierstra RD, Langan SM, Schaller GE (1986) Complete amino acid sequence of ubiquitin from the higher plant *Avena sativa*. Biochemistry 25(11):3105–3108
42. Nath D, Shadan S (2009) The ubiquitin system. Nature 458:421
43. Pickart CM, Fushman D (2004) Polyubiquitin chains: polymeric protein signals. Curr Opin Chem Biol 8:610–616
44. Hicke L (2001) Protein regulation by monoubiquitin. Nat Rev Mol Cell Biol 2:195–201
45. Di Fiore PP, Polo S, Hofmann K (2003) When ubiquitin meets ubiquitin receptors: a signalling connection. Nat Rev Mol Cell Biol 4:491–497
46. Haglund K, Sigismund S, Polo S et al (2003a) Multiple monoubiquitination of RTKs is sufficient for their endocytosis and degradation. Nat Cell Biol 5:461–466
47. Haglund K, Di Fiore PP, Dikic I (2003b) Distinct monoubiquitin signals in receptor endocytosis. Trends Biochem Sci 28:598–603
48. Hicke L, Dunn R (2003) Regulation of membrane protein transport by ubiquitin and ubiquitin-binding proteins. Annu Rev Cell Dev Biol 19:141–172
49. Johnson ES, Ma PC, Ota IM et al (1995) A proteolytic pathway that recognizes ubiquitin as a degradation signal. J Biol Chem 270:17442–17456
50. Peng J, Schwartz D, Elias JE et al (2003) A proteomics approach to understanding protein ubiquitination. Nat Biotechnol 21:921–926
51. Schimmel J, Larsen KM, Matic I et al (2008) The ubiquitin-proteasome system is a key component of the SUMO-2/3 cycle. Mol Cell Proteomics 7:2107–2122
52. Leidecker O, Matic I, Mahata B et al (2012) The ubiquitin E1 enzyme Ube1 mediates NEDD8 activation under diverse stress conditions. Cell Cycle 11:1142–1150
53. Ikeda F, Dikic I (2008) Atypical ubiquitin chains: new molecular signals. 'Protein modifications: beyond the usual Suspects' review series. EMBO Rep 9(6):536–542
54. Kim HT, Kim KP, Lledias F et al (2007) Certain pairs of ubiquitin-conjugating enzymes (E2s) and ubiquitin-protein ligases (E3s) synthesize nondegradable forked ubiquitin chains containing all possible isopeptide linkages. J Biol Chem 282:17375–17386
55. Kirisako T, Kamei K, Murata S et al (2006) A ubiquitin ligase complex assembles linear polyubiquitin chains. EMBO J 25:4877–4887
56. Iwai K, Tokunaga F (2009) Linear polyubiquitination: a new regulator of NF-kappaB activation. EMBO Rep 10:706–713
57. Tokunaga F, Sakata S, Saeki Y et al (2009) Involvement of linear polyubiquitylation of NEMO in NF-kappaB activation. Nat Cell Biol 11:123–132
58. Rahighi S, Ikeda F, Kawasaki M et al (2009) Specific recognition of linear ubiquitin chains by NEMO is important for NF-kappaB activation. Cell 136:1098–1109
59. Mishra P, Volety S, Rao Ch M et al (2009) Glutamate64 to glycine substitution in G1 beta-bulge of ubiquitin impairs function and stabilizes structure of the protein. J Biochem 146:563–569
60. Mishra P, Prabha CR, Rao Ch M et al (2011) Q2N and S65D substitutions of ubiquitin unravel functional significance of the invariant residues Gln2 and Ser65. Cell Biochem Biophys 61:619–628
61. Sharma M, Prabha CR (2015) Q2N and E64G double mutation of ubiquitin confers a stress sensitive phenotype on Saccharomyces cerevisiae. Indian J Exp Biol 53(9):617–620
62. Sharma M, Prabha CR (2011) Construction and functional characterization of double and triple mutants of parallel beta-bulge of ubiquitin. Indian J Exp Biol 49(12):919–924

63. Koyano F, Okatsu K, Kosako H et al (2014) Ubiquitin is phosphorylated by PINK1 to activate parkin. Nature 510(7503):162–166
64. Ratna Prabha C, Mishra P, Shahukar M (2010) Isolation of a dosage dependent lethal mutation in ubiquitin gene of Saccharomyces cereviae. Macromol Symp 287:89–94
65. Doshi A, Mishra P, Sharma M, Prabha CR (2014) Functional characterization of dosage-dependent lethal mutation of ubiquitin in Saccharomyces cerevisiae. FEMS Yeast Res 14(7):1080–1089
66. Haas AL, Rose IA (1982) The mechanism of ubiquitin activating enzyme. A kinetic and equilibrium analysis J Biol Chem 257:10329–10337
67. Haas AL, Warms JV, Rose IA (1983) Ubiquitin adenylate: structure and role in ubiquitin activation. Biochemistry 22(19):4388–4394
68. Haas AL, Warms JV, Hershko A et al (1982) Ubiquitin-activating enzyme. Mechanism and role in protein-ubiquitin conjugation J Biol Chem 257:2543–2548
69. Ciechanover A, Elias S, Heller H et al (1982) "Covalent affinity" purification of ubiquitin-activating enzyme. J Biol Chem 257:2537–2542
70. Ciechanover A, Heller H, Katz-Etzion R et al (1981) Activation of the heat-stable polypeptide of the ATP-dependent proteolytic system. Proc Natl Acad Sci USA 78:761–765
71. VanDemark AP, Hill CP (2003) Two-stepping with E1. Nat Struct Biol 10:244–246
72. Lee I, Schindelin H (2008) Structural insights into E1-catalyzed ubiquitin activation and transfer to conjugating enzymes. Cell 134:268–278
73. Hong SB, Kim BW, Lee KE et al (2011) Insights into noncanonical E1 enzyme activation from the structure of autophagic E1 Atg7 with Atg8. Nat Struct Mol Biol 18:1323–1330
74. Pickart CM, Kasperek EM, Beal R et al (1994) Substrate properties of site-specific mutant ubiquitin protein (G76A) reveal unexpected mechanistic features of ubiquitin-activating enzyme (E1). J Biol Chem 269:7115–7123
75. Mastrandrea LD, You J, Niles EG et al (1999) E2/E3-mediated assembly of lysine 29-linked polyubiquitin chains. J Biol Chem 274:27299–27306
76. Wilkinson KD, Smith SE, O'Connor L et al (1990) A specific inhibitor of the ubiquitin activating enzyme: synthesis and characterization of adenosyl-phospho-ubiquitinol, a nonhydrolyzable ubiquitin adenylate analogue. Biochemistry 29:7373–7380
77. Burch TJ, Haas AL (1994) Site-directed mutagenesis of ubiquitin. Differential roles for arginine in the interaction with ubiquitin-activating enzyme. Biochemistry 33:7300–7308
78. Whitby FG, Xia G, Pickart CM et al (1998) Crystal structure of the human ubiquitin-like protein NEDD8 and interactions with ubiquitin pathway enzymes. J Biol Chem 273:34983–34991
79. Ciechanover A, Finley D, Varshavsky A (1984) Ubiquitin dependence of selective protein degradation demonstrated in the mammalian cell cycle mutant ts85. Cell 37:57–66
80. Aviel S, Winberg G, Massucci M et al (2000) Degradation of the epstein-barr virus latent membrane protein 1 (LMP1) by the ubiquitin-proteasome pathway. Targeting via ubiquitination of the N-terminal residue. J Biol Chem 275:23491–23499
81. Ghaboosi N, Deshaies RJ (2007a) A conditional yeast E1 mutant blocks the ubiquitin-proteasome pathway and reveals a role for ubiquitin conjugates in targeting Rad23 to the proteasome. Mol Biol Cell 18:1953–1963
82. Swanson R, Hochstrasser M (2000) A viable ubiquitin-activating enzyme mutant for evaluating ubiquitin system function in Saccharomyces cerevisiae. FEBS Lett 477:193–198
83. Jensen JP, Bates PW, Yang M et al (1995) Identification of a family of closely related human ubiquitin conjugating enzymes. J Biol Chem 270:30408–30414
84. Rajapurohitam V, Morales CR, El-Alfy M et al (1999) Activation of a UBC4-dependent pathway of ubiquitin conjugation during postnatal development of the rat testis. Dev Biol 212:217–228
85. Wefes I, Mastrandrea LD, Haldeman M et al (1995) Induction of ubiquitin-conjugating enzymes during terminal erythroid differentiation. Proc Natl Acad Sci U S A 92:4982–4986
86. Liu Z, Haas AL, Diaz LA et al (1996) Characterization of a novel keratinocyte ubiquitin carrier protein. J Biol Chem 271:2817–2822

87. Hauser HP, Bardroff M, Pyrowolakis G et al (1998) A giant ubiquitin-conjugating enzyme related to IAP apoptosis inhibitors. J Cell Biol 141:1415–1422
88. Huang L, Kinnucan E, Wang G et al (1999) Structure of an E6AP-UbcH7 complex: insights into ubiquitination by the E2-E3 enzyme cascade. Science 286:1321–1326
89. Zheng N, Wang P, Jeffrey PD et al (2000) Structure of a c-Cbl-UbcH7 complex: RING domain function in ubiquitin-protein ligases. Cell 102:533–539
90. Cook WJ, Jeffrey LC, Xu Y et al (1993) Tertiary structures of class I ubiquitin-conjugating enzymes are highly conserved: crystal structure of yeast Ubc4. Biochemistry 32:13809–13817
91. Worthylake DK, Prakash S, Prakash L et al (1998) Crystal structure of the Saccharomyces cerevisiae ubiquitin-conjugating enzyme Rad6 at 2.6 a resolution. J Biol Chem 273:6271–6276
92. Jiang F, Basavappa R (1999) Crystal structure of the cyclin-specific ubiquitin-conjugating enzyme from clam, E2-C, at 2.0 A resolution. Biochemistry 38:6471–6478
93. Miura T, Klaus W, Gsell B et al (1999) Characterization of the binding interface between ubiquitin and class I human ubiquitin-conjugating enzyme 2b by multidimensional heteronuclear NMR spectroscopy in solution. J Mol Biol 290:213–228
94. Arai R, Yoshikawa S, Murayama K et al (2006) Structure of human ubiquitin-conjugating enzyme E2 G2 (UBE2G2/UBC7). Acta Crystallogr Sect F Struct Biol Cryst Commun 62:330–334
95. Yamanaka A, Hatakeyama S, Kominami K et al (2000) Cell cycle-dependent expression of mammalian E2-C regulated by the anaphase-promoting complex/cyclosome. Mol Biol Cell 11:2821–2831
96. Mathias N, Steussy CN, Goebl MG (1998) An essential domain within Cdc34p is required for binding to a complex containing Cdc4p and Cdc53p in Saccharomyces cerevisiae. J Biol Chem 273:4040–4045
97. Madura K, Dohmen RJ, Varshavsky A (1993) N-recognin/Ubc2 interactions in the N-end rule pathway. J Biol Chem 268:12046–12054
98. Xie Y, Varshavsky A (1999) The E2-E3 interaction in the N-end rule pathway: the RING-H2 finger of E3 is required for the synthesis of multiubiquitin chain. EMBO J 18:6832–6844
99. Yin Q, Lin SC, Lamothe B et al (2009) E2 interaction and dimerization in the crystal structure of TRAF6. Nat Struct Mol Biol 16:658–666
100. Wang J, Taherbhoy AM et al (2010) Crystal structure of UBA2(ufd)-Ubc9: insights into E1-E2 interactions in Sumo pathways. PLoS One 5(12):15805
101. Plechanovova A, Jaffray EG, Tatham MH et al (2012) Structure of a RING E3 ligase and ubiquitin-loaded E2 primed for catalysis. Nature 489:115–120
102. Pruneda JN, Littlefield PJ, Soss SE et al (2012) Structure of an E3:E2~Ub complex reveals an allosteric mechanism shared among RING/U-box ligases. Mol Cell 47:933–942
103. Dou H, Buetow L, Sibbet GJ et al (2012) BIRC7-E2 ubiquitin conjugate structure reveals the mechanism of ubiquitin transfer by a RING dimer. Nat Struct Mol Biol 19:876–883
104. Pickart CM (2001) Mechanisms underlying ubiquitination. Annu Rev Biochem 70:503–533
105. Ciechanover A, Ben-Saadon R (2004) N-terminal ubiquitination: more protein substrates join in. Trends Cell Biol 14:103–106
106. Deshaies RJ, Joazeiro CA (2009) RING domain E3 ubiquitin ligases. Annu Rev Biochem 78:399–434
107. Ogunjimi AA, Briant DJ, Pece-Barbara N et al (2005) Regulation of Smurf2 ubiquitin ligase activity by anchoring the E2 to the HECT domain. Mol Cell 19(3):297–308
108. Verdecia MA, Joazeiro CA, Wells NJ et al (2003) Conformational flexibility underlies ubiquitin ligation mediated by the WWP1 HECT domain E3 ligase. Mol Cell 11:249–259
109. Kumar S, Talis AL, Howley PM (1999) Identification of HHR23A as a substrate for E6-associated protein-mediated ubiquitination. J Biol Chem 274:18785–18792
110. Huibregtse JM, Yang JC, Beaudenon SL (1997a) The large subunit of RNA polymerase II is a substrate of the Rsp5 ubiquitin-protein ligase. Proc Natl Acad Sci U S A 94:3656–3661
111. Qiu L, Joazeiro C, Fang N et al (2000) Recognition and ubiquitination of notch by itch, a hect-type E3 ubiquitin ligase. J Biol Chem 275:35734–35737

112. Schwarz SE, Rosa JL, Scheffner M (1998) Characterization of human hect domain family members and their interaction with UbcH5 and UbcH7. J Biol Chem 273:12148–12154
113. Huibregtse JM, Scheffner M, Howley PM (1993) Localization of the E6-AP regions that direct human papillomavirus E6 binding, association with p53, and ubiquitination of associated proteins. Mol Cell Biol 13:4918–4927
114. Wilkinson KD (1997) Regulation of ubiquitin-dependent processes by deubiquitinating enzymes. FASEB J 11:1245–1256
115. Nijman SM, Luna-Vargas MP, Velds A et al (2005) A genomic and functional inventory of deubiquitinating enzymes. Cell 123:773–786
116. Wilkinson KD, Tashayev VL, O'Connor LB et al (1995) Metabolism of the polyubiquitin degradation signal: structure, mechanism, and role of isopeptidase T. Biochemistry 34:14535–14546
117. Piotrowski J, Beal R, Hoffman L et al (1997) Inhibition of the 26 S proteasome by polyubiquitin chains synthesized to have defined lengths. J Biol Chem 272:23712–23721
118. Baker RT, Board PG (1987) The human ubiquitin gene family: structure of a gene and pseudogenes from the Ub B subfamily. Nucleic Acids Res 15:443–463
119. Ozkaynak E, Finley D, Solomon MJ et al (1987) The yeast ubiquitin genes: a family of natural gene fusions. EMBO J 6:1429–1439
120. Wiborg O, Pedersen MS, Wind A et al (1985) The human ubiquitin multigene family: some genes contain multiple directly repeated ubiquitin coding sequences. EMBO J 4:755–759
121. Rose IA, Warms JV (1983) An enzyme with ubiquitin carboxy-terminal esterase activity from reticulocytes. Biochemistry 22:4234–4237
122. Pickart CM, Rose IA (1985) Ubiquitin carboxyl-terminal hydrolase acts on ubiquitin carboxyl-terminal amides. J Biol Chem 260:7903–7910
123. Amerik AY, Hochstrasser M (2004) Mechanism and function of deubiquitinating enzymes. Biochim Biophys Acta 1695:189–207
124. Love KR, Catic A, Schlieker C et al (2007) Mechanisms, biology and inhibitors of deubiquitinating enzymes. Nat Chem Biol 3:697–705
125. Gan-Erdene T, Nagamalleswari K, Yin L et al (2003) Identification and characterization of DEN1, a deneddylase of the ULP family. J Biol Chem 278:28892–28900
126. Li SJ, Hochstrasser M (1999) A new protease required for cell-cycle progression in yeast. Nature 398:246–251
127. Li SJ, Hochstrasser M (2000) The yeast ULP2 (SMT4) gene encodes a novel protease specific for the ubiquitin-like Smt3 protein. Mol Cell Biol 20:2367–2377
128. Kennedy RD, D'Andrea AD (2005) The Fanconi anemia/BRCA pathway: new faces in the crowd. Genes Dev 19:2925–2940
129. Daniel JA, Grant PA (2007) Multi-tasking on chromatin with the SAGA coactivator complexes. Mutat Res 618:135–148
130. Song L, Rape M (2008) Reverse the curse--the role of deubiquitination in cell cycle control. Curr Opin Cell Biol 20:156–163
131. Komada M (2008) Controlling receptor downregulation by ubiquitination and deubiquitination. Curr Drug Discov Technol 5:78–84
132. Adhikari A, Xu M, Chen ZJ (2007) Ubiquitin-mediated activation of TAK1 and IKK. Oncogene 26:3214–3226
133. Guterman A, Glickman MH (2004) Deubiquitinating enzymes are IN/(trinsic to proteasome function). Curr Protein Pept Sci 5:201–211
134. Schmidt M, Hanna J, Elsasser S et al (2005) Proteasome-associated proteins: regulation of a proteolytic machine. Biol Chem 386:725–737
135. Lindner HA (2007) Deubiquitination in virus infection. Virology 362:245–256
136. Rytkonen A, Holden DW (2007) Bacterial interference of ubiquitination and deubiquitination. Cell Host Microbe 1:13–22
137. Ye Z, Petrof EO, Boone D et al (2007) Salmonella effector AvrA regulation of colonic epithelial cell inflammation by deubiquitination. Am J Pathol 171:882–892

138. Arguello MD, Hiscott J (2007) Ub surprised: viral ovarian tumor domain proteases remove ubiquitin and ISG15 conjugates. Cell Host Microbe 2:367–369
139. Barretto N, Jukneliene D, Ratia K et al (2005) The papain-like protease of severe acute respiratory syndrome coronavirus has deubiquitinating activity. J Virol 79:15189–15198
140. Iyer LM, Koonin EV, Aravind L (2004) Novel predicted peptidases with a potential role in the ubiquitin signaling pathway. Cell Cycle 3:1440–1450
141. Kattenhorn LM, Korbel GA, Kessler BM et al (2005) A deubiquitinating enzyme encoded by HSV-1 belongs to a family of cysteine proteases that is conserved across the family Herpesviridae. Mol Cell 19:547–557
142. Ratia K, Pegan S, Takayama J et al (2008) A noncovalent class of papain-like protease/deubiquitinase inhibitors blocks SARS virus replication. Proc Natl Acad Sci U S A 105:16119–16124
143. Schlieker C, Korbel GA, Kattenhorn LM et al (2005) A deubiquitinating activity is conserved in the large tegument protein of the herpesviridae. J Virol 79:15582–15585
144. Sompallae R, Gastaldello S, Hildebrand S et al (2008) Epstein-barr virus encodes three bona fide ubiquitin-specific proteases. J Virol 82:10477–10486
145. Sulea T, Lindner HA, Purisima EO et al (2005) Deubiquitination, a new function of the severe acute respiratory syndrome coronavirus papain-like protease? J Virol 79:4550–4551
146. Fischer JA (2003) Deubiquitinating enzymes: their roles in development, differentiation, and disease. Int Rev Cytol 229:43–72
147. Jiang YH, Beaudet AL (2004) Human disorders of ubiquitination and proteasomal degradation. Curr Opin Pediatr 16:419–426
148. Shanmugham A, Ovaa H (2008) DUBs and disease: activity assays for inhibitor development. Curr Opin Drug Discov Devel 11:688–696
149. Komander D, Clague MJ, Urbé S (2009a) Breaking the chains: structure and function of the deubiquitinases. Nat Rev Mol Cell Biol 10(8):550–563
150. Maiti TK, Permaul M, Boudreaux DA et al (2011) Crystal structure of the catalytic domain of UCHL5, a proteasome-associated human deubiquitinating enzyme, reveals an unproductive form of the enzyme. FEBS J 278:4917–4926
151. Hu M, Li P, Li M et al (2002) Crystal structure of a UBP-family deubiquitinating enzyme in isolation and in complex with ubiquitin aldehyde. Cell 111:1041–1054
152. Komander D, Lord CJ, Scheel H et al (2008) The structure of the CYLD USP domain explains its specificity for Lys63-linked polyubiquitin and reveals a B box module. Mol Cell 29:451–464
153. Hu M, Li P, Song L et al (2005) Structure and mechanisms of the proteasome-associated deubiquitinating enzyme USP14. EMBO J 24:3747–3756
154. Avvakumov GV, Walker JR, Xue S et al (2006) Amino-terminal dimerization, NRDP1-rhodanese interaction, and inhibited catalytic domain conformation of the ubiquitin-specific protease 8 (USP8). J Biol Chem 281:38061–38070
155. Renatus M, Parrado SG, D'Arcy A et al (2006) Structural basis of ubiquitin recognition by the deubiquitinating protease USP2. Structure 14:1293–1302
156. Albrecht M, Golatta M, Wullner U et al (2004) Structural and functional analysis of ataxin-2 and ataxin-3. Eur J Biochem 271:3155–3170
157. Mao Y, Senic-Matuglia F, Di Fiore PP et al (2005) Deubiquitinating function of ataxin-3: insights from the solution structure of the Josephin domain. Proc Natl Acad Sci U S A 102:12700–12705
158. Chai Y, Berke SS, Cohen RE et al (2004a) Poly-ubiquitin binding by the polyglutamine disease protein ataxin-3 links its normal function to protein surveillance pathways. J Biol Chem 279:3605–3611
159. Nicastro G, Menon RP, Masino L et al (2005) The solution structure of the Josephin domain of ataxin-3: structural determinants for molecular recognition. Proc Natl Acad Sci U S A 102:10493–10498
160. Wang Q, Li L, Ye Y (2006) Regulation of retrotranslocation by p97-associated deubiquitinating enzyme ataxin-3. J Cell Biol 174:963–971

161. Riess O, Rub U, Pastore A et al (2008) SCA3: neurological features, pathogenesis and animal models. Cerebellum 7:125–137
162. Yao T, Cohen RE (2002a) A cryptic protease couples deubiquitination and degradation by the proteasome. Nature 419:403–407
163. McCullough J, Row PE, Lorenzo O et al (2006) Activation of the endosome-associated ubiquitin isopeptidase AMSH by STAM, a component of the multivesicular body-sorting machinery. Curr Biol 16:160–165
164. Cope GA, Suh GS, Aravind L et al (2002) Role of predicted metalloprotease motif of Jab1/Csn5 in cleavage of Nedd8 from Cul1. Science 298:608–611
165. Dong Y, Hakimi MA, Chen X et al (2003) Regulation of BRCC, a holoenzyme complex containing BRCA1 and BRCA2, by a signalosome-like subunit and its role in DNA repair. Mol Cell 12:1087–1099
166. Wang B, Elledge SJ (2007) Ubc13/Rnf8 ubiquitin ligases control foci formation of the Rap80/abraxas/Brca1/Brcc36 complex in response to DNA damage. Proc Natl Acad Sci U S A 104:20759–20763
167. Shao G, Lilli DR, Patterson-Fortin J et al (2009) The Rap80-BRCC36 de-ubiquitinating enzyme complex antagonizes RNF8-Ubc13-dependent ubiquitination events at DNA double strand breaks. Proc Natl Acad Sci U S A 106:3166–3171
168. Cooper EM, Cutcliffe C, Kristiansen TZ et al (2009) K63-specific deubiquitination by two JAMM/MPN+ complexes: BRISC-associated Brcc36 and proteasomal Poh1. EMBO J 28:621–631
169. Sato Y, Yoshikawa A, Yamagata A et al (2008) Structural basis for specific cleavage of Lys 63-linked polyubiquitin chains. Nature 455:358–362
170. Swanson KA, Kang RS, Stamenova SD, Hicke L, Radhakrishnan I (2003) Solution structure of Vps27 UIM ubiquitin complex important for endosomal sorting and receptor downregulation. EMBO J 22(18):4597–4606
171. Alam SL, Sun J, Payne M et al (2004) Ubiquitin interactions of NZF zinc fingers. EMBO J 23:1411–1421
172. Kang RS, Daniels CM, Francis SA et al (2003) Solution structure of a CUE-ubiquitin complex reveals a conserved mode of ubiquitin binding. Cell 113:621–630
173. Prag G, Misra S, Jones EA et al (2003) Mechanism of ubiquitin recognition by the CUE domain of Vps9p. Cell 113:609–620
174. Ohno A, Jee J, Fujiwara K et al (2005) Structure of the UBA domain of Dsk2p in complex with ubiquitin molecular determinants for ubiquitin recognition. Structure 13:521–532
175. Prag G, Lee S, Mattera R, Arighi CN et al (2005) Structural mechanism for ubiquitinated-cargo recognition by the Golgi-localized, gamma-ear-containing, ADP-ribosylation-factor-binding proteins. Proc Natl Acad Sci U S A 102(7):2334–2339
176. Sundquist WI, Schubert HL, Kelly BN et al (2004) Ubiquitin recognition by the human TSG101 protein. Mol Cell 13:783–789
177. Teo H, Veprintsev DB, Williams RL (2004) Structural insights into endosomal sorting complex required for transport (ESCRT-I) recognition of ubiquitinated proteins. J Biol Chem 279:28689–28696
178. Lee S, Tsai YC, Mattera R et al (2006) Structural basis for ubiquitin recognition and autoubiquitination by Rabex-5. Nat Struct Mol Biol 13:264–271
179. Reyes-Turcu FE, Horton JR, Mullally JE et al (2006) The ubiquitin binding domain ZnF UBP recognizes the C-terminal diglycine motif of unanchored ubiquitin. Cell 124:1197–1208
180. Praefcke GJ, Ford MG, Schmid EM et al (2004) Evolving nature of the AP2 alpha-appendage hub during clathrin-coated vesicle endocytosis. EMBO J 23(22):4371–4383
181. Haas AL, Bright PM (1985) The immunochemical detection and quantitation of intracellular ubiquitin-protein conjugates. J Biol Chem 260:12464–12473
182. Wilkinson CR, Seeger M, Hartmann-Petersen R et al (2001) Proteins containing the UBA domain are able to bind to multi-ubiquitin chains. Nat Cell Biol 3:939–943
183. Raasi S, Orlov I, Fleming KG et al (2004) Binding of polyubiquitin chains to ubiquitin-associated (UBA) domains of HHR23A. J Mol Biol 341:1367–1379

184. Lange OF, Lakomek NA, Fares C et al (2008) Recognition dynamics up to microseconds revealed from an RDC-derived ubiquitin ensemble in solution. Science 320:1471–1475
185. Raasi S, Varadan R, Fushman D et al (2005) Diverse polyubiquitin interaction properties of ubiquitin-associated domains. Nat Struct Mol Biol 12:708–714
186. Komander D, Reyes-Turcu F, Licchesi JD et al (2009b) Molecular discrimination of structurally equivalent Lys 63-linked and linear polyubiquitin chains. EMBO Rep 10:466–473
187. Varadan R, Assfalg M, Raasi S et al (2005) Structural determinants for selective recognition of a Lys48-linked polyubiquitin chain by a UBA domain. Mol Cell 18:687–698
188. Sobhian B, Shao G, Lilli DR et al (2007) RAP80 targets BRCA1 to specific ubiquitin structures at DNA damage sites. Science 316:1198–1202
189. Sims JJ, Cohen RE (2009) Linkage-specific avidity defines the lysine 63-linked polyubiquitin-binding preference of rap80. Mol Cell 33:775–783
190. Puhler G, Weinkauf S, Bachmann L et al (1992) Subunit stoichiometry and three-dimensional arrangement in proteasomes from Thermoplasma acidophilum. EMBO J 11:1607–1616
191. Grziwa A, Baumeister W, Dahlmann B et al (1991) Localization of subunits in proteasomes from Thermoplasma acidophilum by immunoelectron microscopy. FEBS Lett 290:186–190
192. Schauer TM, Nesper M, Kehl M et al (1993) Proteasomes from Dictyostelium discoideum: characterization of structure and function. J Struct Biol 111:135–147
193. Kopp F, Dahlmann B, Hendil KB (1993) Evidence indicating that the human proteasome is a complex dimer. J Mol Biol 229:14–19
194. Heinemeyer W, Trondle N, Albrecht G et al (1994) PRE5 and PRE6, the last missing genes encoding 20S proteasome subunits from yeast? Indication for a set of 14 different subunits in the eukaryotic proteasome core. Biochemistry 33:12229–12237
195. Lowe J, Stock D, Jap B et al (1995) Crystal structure of the 20S proteasome from the archaeon T. Acidophilum at 3.4 a resolution. Science 268:533–539
196. Groll M, Ditzel L, Lowe J et al (1997) Structure of 20S proteasome from yeast at 2.4 A^{0}resolution. Nature 386:463–471
197. Seemuller E, Lupas A, Stock D et al (1995) Proteasome from Thermoplasma acidophilum: a threonine protease. Science 268:579–582
198. Baumeister W, Walz J, Zuhl F et al (1998) The proteasome: paradigm of a self-compartmentalizing protease. Cell 92:367–380
199. Heinemeyer W, Fischer M, Krimmer T et al (1997) The active sites of the eukaryotic 20 S proteasome and their involvement in subunit precursor processing. J Biol Chem 272:25200–25209
200. Chen P, Hochstrasser M (1996) Autocatalytic subunit processing couples active site formation in the 20S proteasome to completion of assembly. Cell 86:961–972
201. Jager S, Groll M, Huber R et al (1999) Proteasome beta-type subunits: unequal roles of propeptides in core particle maturation and a hierarchy of active site function. J Mol Biol 291:997–1013
202. Heinemeyer W, Ramos PC, Dohmen RJ (2004) The ultimate nanoscale mincer: assembly, structure and active sites of the 20S proteasome core. Cell Mol Life Sci 61:1562–1578
203. Bajorek M, Glickman MH (2004) Keepers at the final gates: regulatory complexes and gating of the proteasome channel. Cell Mol Life Sci 61:1579–1588
204. Groll M, Bajorek M, Kohler A et al (2000) A gated channel into the proteasome core particle. Nat Struct Biol 7:1062–1067
205. Hill CP, Masters EI, Whitby FG (2002) The 11S regulators of 20S proteasome activity. Curr Top Microbiol Immunol 268:73–89
206. Glickman MH, Rubin DM, Coux O et al (1998) A subcomplex of the proteasome regulatory particle required for ubiquitin-conjugate degradation and related to the COP9-signalosome and eIF3. Cell 94:615–623
207. Finley D, Tanaka K, Mann C et al (1998) Unified nomenclature for subunits of the Saccharomyces cerevisiae proteasome regulatory particle. Trends Biochem Sci 23:244–245

208. Zhang F, Wu Z, Zhang P et al (2009a) Mechanism of substrate unfolding and translocation by the regulatory particle of the proteasome from Methanocaldococcus jannaschii. Mol Cell 34:485–496
209. Smith DM, Chang SC, Park S et al (2007) Docking of the proteasomal ATPases' carboxyl termini in the 20S proteasome's alpha ring opens the gate for substrate entry. Mol Cell 27:731–744
210. Gillette TG, Kumar B, Thompson D et al (2008) Differential roles of the COOH termini of AAA subunits of PA700 (19 S regulator) in asymmetric assembly and activation of the 26 S proteasome. J Biol Chem 283:31813–31822
211. Hanson PI, Whiteheart SW (2005) AAA+ proteins: have engine, will work. Nat Rev Mol Cell Biol 6(7):519–529
212. Hartmann-Petersen R, Tanaka K, Hendil KB (2001) Quaternary structure of the ATPase complex of human 26S proteasomes determined by chemical cross-linking. Arch Biochem Biophys 386:89–94
213. Djuranovic S, Hartmann MD, Habeck M et al (2009a) Structure and activity of the N-terminal substrate recognition domains in proteasomal ATPases. Mol Cell 34:580–590
214. Forster F, Lasker K, Beck F et al (2009) An atomic model AAA-ATPase/20S core particle sub-complex of the 26S proteasome. Biochem Biophys Res Commun 388:228–233
215. Thrower JS, Hoffman L, Rechsteiner M, Pickart CM (2000) Recognition of the polyubiquitin proteolytic signal. EMBO J 19(1):94–102
216. Rape M, Jentsch S (2002) Taking a bite: proteasomal protein processing. Nat Cell Biol 4:E113–E116
217. Asher G, Reuven N, Shaul Y (2006) 20S proteasomes and protein degradation "by default". BioEssays 28:844–849
218. Zhang M, Pickart CM, Coffino P (2003) Determinants of proteasome recognition of ornithine decarboxylase, a ubiquitin-independent substrate. EMBO J 22:1488–1496
219. Asher G, Shaul Y (2005) p53 proteasomal degradation: poly-ubiquitination is not the whole story. Cell Cycle 4:1015–1018
220. Shringarpure R, Grune T, Mehlhase J et al (2003) Ubiquitin conjugation is not required for the degradation of oxidized proteins by proteasome. J Biol Chem 278:311–318
221. Deveraux Q, van Nocker S, Mahaffey D et al (1995) Inhibition of ubiquitin-mediated proteolysis by the Arabidopsis 26 S protease subunit S5a. J Biol Chem 270:29660–29663
222. Husnjak K, Elsasser S, Zhang N et al (2008) Proteasome subunit Rpn13 is a novel ubiquitin receptor. Nature 453:481–488
223. Wang Q, Young P, Walters KJ (2005) Structure of S5a bound to monoubiquitin provides a model for polyubiquitin recognition. J Mol Biol 348:727–739
224. Schreiner P, Chen X, Husnjak K et al (2008) Ubiquitin docking at the proteasome through a novel pleckstrin-homology domain interaction. Nature 453:548–552
225. Tanaka K (2009) The proteasome: overview of structure and functions. Proc Jpn Acad Ser B Phys Biol Sci 85:12–36
226. Elsasser S, Gali RR, Schwickart M et al (2002) Proteasome subunit Rpn1 binds ubiquitin-like protein domains. Nat Cell Biol 4:725–730
227. Lam YA, Lawson TG, Velayutham M et al (2002) A proteasomal ATPase subunit recognizes the polyubiquitin degradation signal. Nature 416:763–767
228. Verma R, Aravind L, Oania R et al (2002) Role of Rpn11 metalloprotease in deubiquitination and degradation by the 26S proteasome. Science 298:611–615
229. Zhang F, Hu M, Tian G et al (2009b) Structural insights into the regulatory particle of the proteasome from Methanocaldococcus jannaschii. Mol Cell 34:473–484
230. Lupas A, Baumeister W, Hofmann K (1997) A repetitive sequence in subunits of the 26S proteasome and 20S cyclosome (anaphase-promoting complex). Trends Biochem Sciv 22:195–196
231. Kajava AV (2002) What curves alpha-solenoids? Evidence for an alpha-helical toroid structure of Rpn1 and Rpn2 proteins of the 26 S proteasome. J Biol Chem 277:49791–49798

232. Scheel H, Hofmann K (2005) Prediction of a common structural scaffold for proteasome lid, COP9-signalosome and eIF3 complexes. BMC Bioinformatics 6:71
233. Muratani M, Tansey WP (2003) How the ubiquitin-proteasome system controls transcription. Nat Rev Mol Cell Biol 4:192–201
234. Thomas D, Tyers M (2000) Transcriptional regulation: kamikaze activators. Curr Biol 10:R341–R343
235. Finley D, Bartel B, Varshavsky A (1989) The tails of ubiquitin precursors are ribosomal proteins whose fusion to ubiquitin facilitates ribosome biogenesis. Nature 338:394–401
236. Deng L, Wang C, Spencer E et al (2000) Activation of the IkappaB kinase complex by TRAF6 requires a dimeric ubiquitin-conjugating enzyme complex and a unique polyubiquitin chain. Cell 103:351–361
237. Newton K, Matsumoto ML, Wertz IE et al (2008) Ubiquitin chain editing revealed by polyubiquitin linkage-specific antibodies. Cell 134:668–678
238. Ting AT, Pimentel-Muinos FX, Seed B (1996) RIP mediates tumor necrosis factor receptor 1 activation of NF-kappaB but not Fas/APO-1-initiated apoptosis. EMBO J 15:6189–6196
239. Imai Y, Soda M, Hatakeyama S et al (2002) CHIP is associated with Parkin, a gene responsible for familial Parkinson's disease, and enhances its ubiquitin ligase activity. Mol Cell 10:55–67
240. Kopito RR (2000) Aggresomes, inclusion bodies and protein aggregation. Trends Cell Biol 10:524–530
241. Xie Z, Klionsky DJ (2007) Autophagosome formation: core machinery and adaptations. Nat Cell Biol 9:1102–1109
242. Imai Y, Soda M, Inoue H et al (2001) An unfolded putative transmembrane polypeptide, which can lead to endoplasmic reticulum stress, is a substrate of Parkin. Cell 105:891–902
243. Zuccato C, Tartari M, Crotti A et al (2003) Huntingtin interacts with REST/NRSF to modulate the transcription of NRSE-controlled neuronal genes. Nat Genet 35:76–83
244. Gewin L, Myers H, Kiyono T et al (2004) Identification of a novel telomerase repressor that interacts with the human papillomavirus type-16 E6/E6-AP complex. Genes Dev 18:2269–2282
245. Huibregtse JM, Scheffner M, Howley PM (1991) A cellular protein mediates association of p53 with the E6 oncoprotein of human papillomavirus types 16 or 18. EMBO J 10:4129–4135
246. Kishino T, Lalande M, Wagstaff J (1997) UBE3A/E6-AP mutations cause Angelman syndrome. Nat Genet 15:70–73
247. Matsuura T, Sutcliffe JS, Fang P et al (1997) De novo truncating mutations in E6-AP ubiquitin-protein ligase gene (UBE3A) in Angelman syndrome. Nat Genet 15:74–77
248. Duensing S, Munger K (2004) Mechanisms of genomic instability in human cancer: insights from studies with human papillomavirus oncoproteins. Int J Cancer 109:157–162
249. Ruffner H, Joazeiro CA, Hemmati D et al (2001) Cancer-predisposing mutations within the RING domain of BRCA1: loss of ubiquitin protein ligase activity and protection from radiation hypersensitivity. Proc Natl Acad Sci U S A 98:5134–5139
250. Abrami L, Leppla SH, van der Goot FG (2006) Receptor palmitoylation and ubiquitination regulate anthrax toxin endocytosis. J Cell Biol 172:309–320
251. Orth K, Xu Z, Mudgett MB et al (2000) Disruption of signaling by Yersinia effector YopJ, a ubiquitin-like protein protease. Science 290:1594–1597
252. Zhou H, Monack DM, Kayagaki N et al (2005) Yersinia virulence factor YopJ acts as a deubiquitinase to inhibit NF-kappa B activation. J Exp Med 202:1327–1332

A Brief Account of Structure-Function Relationship of the Traditional Cysteine Protease Inhibitor - Cystatin with a Special Focus on Human Family 1 and 2 Cystatins

27

Suman K. Nandy

Abstract

Cystatins are well-documented cysteine protease inhibitors with highly conserved structural folds, distributed in a variety of species. Involvement of cystatin in various biochemical pathways through regulation of protein degradation makes it an element of amazing therapeutic possibilities for treatment of a broad range of diseases. Cystatin superfamily has been divided into four groups: stefins or family 1; cystatins or family 2; kininogens or family 3; and family 4 cystatin. The cystatin superfamily shares a common cystatin fold constituting five antiparallel β-sheets enfolded around a five-turn α-helix forming a cuneus-shaped structure that blocks the access of the active site of papain-like cysteine proteases (CPs). Crystallographic and mutagenesis studies identify three conserved regions mainly involved in the interaction with papain (C1) family of CPs, namely, (a) N-terminal region, (b) L1 loop, and (c) L2 loop. Despite sharing the same structural fold and inhibiting through the same mechanism, cystatin demonstrates huge variation in inhibitory affinity toward C1 family of CPs. Relative contribution and sequential dissimilarity of three conserved sites controlled the diverse interaction patterns of cystatins, which in turn determined the wide-ranging affinity of cystatins toward papain family of CPs. Some of the members of family 2 cystatins show additional affinity toward legumain family of CPs through an alternate binding site compared to papains.

S.K. Nandy (✉)
Department of Biochemistry and Biophysics, University of Kalyani,
Kalyani, West Bengal 741235, India

Bioinformatics Infrastructure Facility (Bioinformatics Center), Department of Rural Development and Agricultural Production, North-Eastern Hill University, Tura Campus, Tura, Meghalaya 794002, India
e-mail: bhramoman@gmail.com

S. Chakraborti, N.S. Dhalla (eds.), *Proteases in Physiology and Pathology*,
DOI 10.1007/978-981-10-2513-6_27

Keywords
Cystatin • Cysteine protease inhibitor • Papain • C1 family • Cysteine protease

27.1 Introduction

Cystatins are a group of homologous proteins [1, 2] characterized by cystatin domains, largely known for their traditional inhibitory activity toward papain (C1) families of cysteine proteases (CPs) [3, 4]. These endogenous competitive protein inhibitors form an enzymatically inactive nonobligatory complex with CPs and share a strong transient association [5, 4]. Thus, cystatins play the role of typical emergency regulators of CPs, remain separated from the target enzyme, and mostly operate on escaped host proteases or foreign proteases of pathogens to control intra- and extracellular protein degradation [6, 7]. Through their competitive, reversible inhibitory activity, cystatins play crucial roles in numerous physiological and pathophysiological conditions ranging from cell survival and proliferation to differentiation, signaling, and immunomodulation [8]. In humans, decreased levels of cystatins and any kinds of imbalance to normal levels of CPs may result in pathological conditions including amyloidosis [9], tumorigenesis [10, 11], mammalian homeostasis, neurodegeneration, bone resorption, age-related macular degeneration, epilepsy, host-pathogen reactions, inflammation and immune responses, and destabilization of matrix metalloproteinases [12, 13] which in turn may well lead to cancer [8, 14, 15], arthritis, osteoporosis [16], diabetes, renal failure, neurological [17] or cardiovascular disorders [18], and periodontal diseases [19–21]. Thus, the cystatin superfamily has remarkable therapeutic potentials for our understanding of the biology of living organisms, particularly mammals.

Cystatins have surfaced in ancestors of eukaryotes and are widely expressed in a broad range of species, organisms, and tissues, from mammals, birds, fish, insects, and plants to bacteria, parasites, and viruses. In 1968 Fossum and Whitaker first isolated cystatin from chicken egg white and established its strong cysteine protease inhibitory activity against papain and ficin [22–24]. Chicken cystatin (cc) is also the first protein inhibitor with known sequence [25] and structure (PDB ID: 1CEW) [26]. The name "cystatin" was coined by Barrett et al. in 1981 [27] for cc, but afterward it became the collective name of all proteins in the superfamily. Cystatin C is the first human cystatin whose amino acid sequence is determined [28], and human stefin B (PDB ID: 1STF) in complex with papain [29] is the first known structure.

27.2 Classification of Cystatin Superfamily

Cystatin superfamily was first classified into three protein families based on evolutionary relationship, at least a 50% sequence identity [30] with cc [25] and presence or absence of disulfide bonds [1]. Later with the discovery of new members, the superfamily was classified into four types, on the basis of number of cystatin-like domains, disulfide bonds, and physiological localization [31]. Whereas the first three types, viz., stefins, cystatins, and kininogens, are cysteine protease inhibitors (CPIs), the fourth family includes members inactive to CP and with functions uncorrelated to CP inhibition [3, 4, 19, 31, 32]. Type 4 cystatins might serve as a classic example of loss of certain protein function due to sequential change, while the structural fold still remains conserved in the course of evolution, owing to much lower structural mutation sensitivity [33].

27.2.1 Type 1 Cystatins or Stefins

Stefins (also referred as family 1 or class 1 cystatins) are intracellular protease inhibitors, resided in cytoplasm of diverse cell types [32]. They are characteristically low molecular weight (10–11 kDa), about 100 amino acid long, single polypeptide chain protein, highly stable in wide pH range, and lacking disulfide bonds as well as carbohydrate moiety and even signal peptides [3]. Stefins also have been detected in extracellular medium (body fluids), where they possibly originated from dying cells releasing their content [34]. Stefins belong to most major eukaryotic supergroups [35]. Stefins A and B are the most well-studied members of this group, widely distributed in mammals [32]; bovine stefin C came up much later as the first Trp-containing stefin with an elongated N-terminus [36].

27.2.2 Type 2 Cystatins or Cystatins

Cystatins (also cited as family 2 or class 2 cystatins) are generally 120–125 amino acid protein (MW 13–14 kDa) with two intra-chain disulfide bonds [37]. In all mammalian type 2 cystatins, two disulfide bridges are located near C-terminal. Although cystatins are translated with a signal peptide and meant to found extracellularly [5, 38], they are also detected intracellularly [32, 34]. For example, significant uptake of extracellular cystatin C has been reported in human epithelial or neuroblastoma cells [39, 40]. Cystatins are broadly distributed compared to stefins and can be found in most of the body fluids [4, 35]. Presence of multigene families and the effect of polymorphism in coding sequences induce substantial diversity in sequence and function of the cystatins compared to stefins.

In addition, several posttranslational modifications are found in the members of this family. Asn-108 of cystatin M/E is glycosylated [41]; in cystatin S phosphorylation is detected at Ser1 and Ser3 [42, 43]. Cystatin S with variable states of phosphorylation is isolated from bronchoalveolar (BAL) and nasal fluids, but the implication of the variance is yet to be established [44].

27.2.3 Type 3 Cystatins or Kininogens

Kininogens (also known as family 3 or class 3 cystatins), the precursor protein of kinins, abundant in blood plasma and synovial fluid of mammals, are intravascularly expressed multifunctional, multidomain glycoproteins [34]. Kinin-kallikrein system and cathepsins are the two key routes of kinin release [45]. Kininogens are of three types: high-molecular-weight kininogen (HK), low-molecular-weight kininogen (LK), and an acute phase T-kininogen (TK) isolated only in rats, originated from TK gene [46, 47]. Both human HK and LK are released from the same gene through alternative mRNA splicing [48]. Kinin segment is released by kallikreins, and the rest of the protein has two parts: an N-terminal heavy chain and a C-terminal light chain linked through a disulfide bond [49]. HK (MW $\approx$ 90–120 kDa) and LK (MW $\approx$ 50–70 kDa) share an identical heavy chain and differ in light chain, which is much longer in HK [5, 32, 50, 51]. These proteins are equipped with three cystatin domains (D1, D2, and D3) [52] and eight disulfide bonds, six characteristic to cystatins and two additional between D2 and D3 in the heavy chains of HK and LK. The D2 and D3 domains possess the conserved QXVXG domain and in turn exhibit CP inhibitory activity [52]. Although kininogens are glycoproteins, no glycosylation sites are reported within cystatin domains. Both HK and LK bind two CPs per protein with high affinity including cathepsins and cruzipain [50, 51]. Kininogens apart from being CPI also act as substrates to a variety of serine proteinases, thus providing the nickname kininogenases.

27.2.4 Type 4 Cystatin

The type 4 cystatins (also quoted as family 4 or class 4 cystatins) represent cystatin homologs devoid of CP inhibitory activity, to wit fetuins [53], histidine-rich glycoprotein (HRG) [54], and alpha-2-Heremans-Schmid glycoprotein (α-2-HS glycoprotein), each with two cystatin-like domains [55]. Both glycosylation (N and O) and phosphorylation are observed in fetuins. Two cystatin-like domains built the N-terminal region linked through a disulfide bridge to the C-terminal region consisting of a histidine-rich domain surrounded by two proline-rich domains [44]. HRG exhibits substantial structural similarity with α-2-HS glycoprotein except the latter lacks the histidine-rich tandem repeat. In the course of evolution, these cystatins have lost their traditional inhibitory activity toward CPs due to mutations in the structurally conserved regions and gained new functions [35, 55], namely, inhibitory activity against subtilisin (S8) [56] and astacin/adamalysin (M12) [57] family of proteases.

27.3 MEROPS Classification

MEROPS database (http://merops.sanger.ac.uk/) is a manually organized web resource for proteases, its substrates, and inhibitors. Proteins and inhibitors are classified on the basis of their sequential and structural similarities. Here the comparisons are done only by considering the domain responsible for protease or inhibitory activity, referred as “peptidase unit” or “inhibitor unit” [58].

The classification is hierarchal – sequences are assembled in protein species, protein species are arranged into families, and families are gathered into clans. Members of each class in the hierarchy are assembled around a biochemically or structurally well-characterized representative, known as "holotype." A peptidase or inhibitor unit forms the starting point of the hierarchy. Identical proteins of different organisms (i.e., orthologues) are placed into a single protein species. Homologous protein species are grouped into a family, if a pair-wise alignment with the current type example is significant, i.e., expect value <0.001. Families clustered under a clan are required to have similar tertiary structure (z-score > 6.0 in DALI comparison w.r.t. the type example) as well as conservation of the order of catalytic or inhibitory site residues (when structure is not available) and are supposed to share the same ancestry [59, 60].

Currently inhibitors are grouped into 93 families assigned to 51 clans. In this scheme of classification, cystatins are positioned in clan IH, family I25 comprising three subfamilies: I25A, I25B, and I25C. Stefins belong to I25A, cystatins and kininogens are included in I25B, and metalloprotease inhibitors form I25C. Further, α-2-HS glycoprotein, HRG, carp fetuin, etc. remain unassigned to any subfamily. Human cystatin A (Uniprot ID: P01040), chicken egg-white cystatin (Uniprot ID: P01038), and snake venom metallopeptidase inhibitor (Uniprot ID: Q9DGI0) are chosen as holotype for subfamilies I25A, I25B, and I25C, respectively, whereas chicken egg-white cystatin (PDB ID: 1CEW) stands for IH clan type. Subfamilies I25A and I25B show characteristic CP inhibitory activities toward papain (C1) families of cysteine proteases, although some members of subfamily I25B inhibit legumain (C13) families of CP as well. On the contrary, the domains of subfamily I25C are typically non-inhibitors of CPs but have shown inhibitory activity against subtilisin or S8 family of serine protease [56] and snake venom metalloendo-protease of family M12 [57] as discussed before. Few transitive links among the subfamilies are also observed, such as the following: the sequence of soya phytocystatin (Uniprot ID: Q39842) relates subfamilies I25A and I25B; chicken egg-white cystatin (Uniprot ID: P01038) and the first cystatin-like unit of bovine H-kininogen (Uniprot ID: P01044, residues 27–131) document significant links between subfamilies I25B and I25C. Some examples of cystatins of different families are given in Table 27.1.

Table 27.1 Few members of cystatin superfamily

Cystatin family	Members	UniProt ID	MEROPS ID
Type 1 cystatins	Cystatin A/stefin A	P01040	I25.001
	Sarcocystatin	P31727	I25.013
	Stefin C	P35478	–
Type 2 cystatins	Cystatin C	P01034	I25.004
	Cystatin/ovocystatin	P01038	I25.011
	Cystatin 14/cystatin SC	Q8VIH8	I25.023
Type 3 cystatins	Kininogen-1/alpha-2-thiol proteinase inhibitor	P01042	I25.016
	Kininogen-1	P08934	I25.018
	T-kininogen 1/alpha-1-MAP	P01048	I25.019
Type 4 cystatins	Alpha-2-HS glycoprotein	P02765	I25.021
	Histidine-rich glycoprotein	P04196	I25.022
	Antihemorrhagic factor BJ46a	Q9DGI0	I25.026

27.4 Evolutionary Relationship

Phylogenetic analysis of the cystatin superfamily evidenced their existence even before the tripartite split of animals, plants, and fungi. Members of cystatin superfamily have been identified in *Eukaryota* and *Bacteria*; none reported in *Archaea*. The ancestor of this superfamily is supposed to be more like stefins, intracellular, devoid of disulfide bonds, and leader peptide, similar to Giardia cystatin. The two ancestral eukaryotic paralogs, cystatins and stefins, were produced in primeval gene duplication, while stefins stayed as a single gene or small multigene families all over the eukaryotes; cystatins have undergone active birth-death evolution [61] over the same species through numerous gene or domain duplications and loss. Only these two ancestral lineages are present throughout the eukaryotes; kininogens, fetuins, and HRGs emerged more recently and are restricted to the vertebrates. These multidomain cystatins are not monophyletic, rather originated several times in different eukaryotic ancestries independently through domain duplication, in the course of phylogenesis. In cystatin superfamily, 20 vertebrate-specific and 3 angiosperm-specific orthologous families are described – points out the occurrence of functional divergence within higher eukaryotes only. It is to be noted that functional diversification is mainly restricted in the divergent cystatin class, rather than the more conserved stefin lineage [35].

27.4.1 Functional Diversification

The differences in expression profiles and genetic organization of stefins and cystatins indicated early occurrence of neofunctionalization in the cystatin superfamily [35]. Considering the most of the eukaryotes are unicellular, the specialization of subcellular localization of cystatins and stefins might be of enormous importance for their hosts in the early days. It has been evidenced in several taxonomic groups that loss of any one of the ancestral lineages is compensated by the remaining one by gaining an additional function, i.e., inhibition of endogenous or exogenous CPs. For example, loss of stefins in plants is counterbalanced by cystatins, which additionally inhibits the endogenous CPs. Signal peptides have been found in many unicellular eukaryotic stefins (e.g., in *Hyperamoeba*, *Capsaspora*, and *Karlodinium*) which might be useful in novel host defense-related function. In major vertebrate orthologous families, cystatins are mostly subjected to loss of their traditional inhibitory activity and gaining new function in innate immunity. Stefins, kininogens, and cystatins C, M/E, and F conserved their inhibitory activity toward CPs, while some of them became more specialized in terms of tissue, cell type, and pathogen [3, 44, 62]. On the other hand, other members of the superfamily inhibit matrix metalloproteases (e.g. latexin) [63], gained new role in angiogenesis (e.g. HRG) [64], showed antimicrobial activity (e.g., HRG, latexin, and cathelicidin) [63–65], and participated in bone regulation and calcification [63, 66, 67]. The number of modified functions of cystatins in the vertebrate families points toward the very diverse protein-protein interaction module of cystatin fold and its ability to interact with novel targets which might also be useful in protein designing [3, 44, 62].

27.5 General Account of Human Family 1 and 2 Cystatins

27.5.1 Cystatin A

Common name: Stefin A
Class: Family I/I25A
Gene name: CSTA
Sequence length: 98
Chromosomal localization: Human chromosome 3
Subcellular location: Cytoplasm [68]
Tissue specificity: Expressed in the skin throughout the epidermis [68], spleen, and liver.
Function:

- Intracellular cysteine proteinase inhibitor (CPI).
- Plays an essential role in desmosome-mediated cell-cell adhesion [68].
- Stefin A has been reported as prognostic and diagnostic marker for cancer [69].

Disease association:

- Ichthyosis, autosomal recessive, exfoliative, ichthyosis bullosa of Siemens-like (AREI) [MIM: 607936] [68]
- Association with psoriasis [70] and atopic dermatitis [71]

27.5.2 Cystatin B

Common name: Stefin B
Class: Family I/I25A
Gene name: CSTB
Sequence length: 98
Chromosomal localization: Human chromosome 21
Subcellular location: Cytoplasm, nucleus [72], and lysosome [73]. Localization of stefin B depends on the differentiation status of the cell; on myotube differentiation stefin B resided only in cytoplasm [73].
Tissue specificity: Broadly distributed, typically intracellular, but also found extracellularly [34]
Function:

- Intracellular CPI
- Potent inhibitor of papain and cathepsins B, H, and L [3, 5]
- Acts against proteases liberated from lysosomes [73]
- Forms inactive dimer stabilized by nonbonded interactions [74]

Disease association: Epilepsy, progressive myoclonic 1 (EPM1) [MIM: 254800] induced by a mutation in the gene which causes loss of association with lysosome and sets up the molecular pathogenesis of EPM1 [73, 75]

27.5.3 Cystatin C

Common name: Gamma-trace
Class: Family II/I25B
Gene name: CST3
Sequence length: 146
Chromosomal localization: Human *chromosome* 20
Subcellular location: Secreted (located outside cell membrane) [43]
Tissue specificity:

- The most abundant extracellular CPI found in high concentrations in a variety of body fluids, for instance, plasma and cerebrospinal fluid, and expressed in virtually all organs of the body [28, 76].
- High expression is observed in the brain, epididymis, ovary, vas deferens, and thymus, while the submandibular gland records the lowest [43, 77].

Posttranslational modification: The Thr-25 variant is O-glycosylated with a core 1 or possibly core 8 glycan. The signal peptide of the O-glycosylated Thr-25 variant is cleaved between Ala-20 and Val-21 [78].
Function:

- Inhibits both papain (C1) [3, 5] and legumain (C13) [79] family of cystatin proteases.
- In diagnosis of Creutzfeldt-Jakob disease, cystatin C is used as prospective cerebrospinal fluid marker [77].

Disease association:

- Cerebral amyloid angiopathy [MIM: 105,150] – genetic disorder due to L68Q mutation in CST3 gene [80]
- Age-related macular degeneration 11 (ARMD11) [MIM:611,953] [81]

27.5.4 Cystatin D

Common name: Cystatin 5
Class: Family II/I25B
Gene name: CST5
Sequence length: 142
Chromosomal localization: Human chromosome 20
Subcellular location: Secreted (located outside cell membrane) [43]
Tissue specificity:

- Found in submandibular, sublingual saliva, parotid gland, and tears [43, 82]

Function:

- Acts as CPI; mainly its activity remains limited to oral cavity.
- Exhibits potential inhibition against cathepsins B, L, H, and S in ascending order [83].

27.5.5 Cystatin F

Cystatin F is the exception in its class, synthesized and released as an inactive dimeric precursor linked by interchain disulfide bridge [84] and subsequently reduced to regain its monomeric active form [85].

Common name: Cystatin 7, leukocystatin
Class: Family II/I25B
Gene name: CST7
Sequence length: 145
Chromosomal localization: Human chromosome 20
Subcellular location: Secreted (located outside cell membrane), cytoplasm [86]
Tissue specificity:

- Cystatin F expression is primarily confined to T cells, natural killer cells, and dendritic cells, and selective expression is also noted in hematopoietic cells [87–90].
- Cystatin F is scarcely present in the blood but secreted in significant amounts from several T and myeloid cell lines [88].

Posttranslational modification: Cystatin F modified by N-linked glycosylation on Asn62 and Asn115 [91]
Function:

- Illustrates inhibitory activity against papain and cathepsins F, K, L, V, S, and H but in lesser extent compared to other cystatins [85, 88].
- The glycosylated cystatin F acts as immune regulator by targeting the hematopoietic system [90].

27.5.6 Cystatin M/E

Common name: Cystatin M, cystatin E, cystatin E/M, cystatin 6
Class: Family II/I25B
Gene name: CST6
Sequence length: 149
Chromosomal localization: Human chromosome 11
Subcellular location: Secreted (located outside cell membrane)
Tissue specificity:

- Found in the stratum granulosum of normal skin, stratum spinosum of psoriatic skin, and the secretory coils of eccrine sweat glands [92].
- Low expression levels are found in the nasal cavity [93].

Posttranslational modification: Cystatin M/E has a functional N-glycosylation site (N108) within the conserved L2 loop and thus acts as a substrate for transglutaminases and in turn gets acetylated [41, 94].

*Function: M*oderately inhibits cathepsin B and legumain but shows no significant activity against cathepsin C [3, 95]

Disease association: A correlation between disease progression and downregulation of cystatin M/E is observed in breast cancer as loss of expression is witnessed in metastatic breast tumor cells in comparison to the primary one [94].

27.5.7 Cystatin S

Common name: Cystatin 4, salivary acidic protein 1
Class: Family II/I25B
Gene name: CST4
Sequence length: 141
Chromosomal localization: Human chromosome 20
Subcellular location: Secreted (located outside cell membrane) [43]
Tissue specificity:

- Cystatin S is expressed highly in the submandibular gland; fairly in the lacrimal gland, parotid gland, and gallbladder; and less than moderately in the kidney and seminal vesicles; and yet the prostate records the lowest expression [96].

Posttranslational modification: Phosphorylation detected in both terminal regions [43]

Function:

- Cystatin S reported high inhibitory activity toward papain (noncompetitively) and ficin, moderate to stem bromelain and bovine cathepsin C, and almost none for porcine cathepsin B or clostripain [97].
- Binds with hydroxylapatite and possibly has a specialized role in mineralization [96].

27.5.8 Cystatin SA

Common name: Cystatin 2, cystatin S5
Class: Family II/I25B
Gene name: CST2
Sequence length: 141
Chromosomal localization: Human chromosome 20
Subcellular location: Secreted (located outside cell membrane) [43]
Tissue specificity:

- Expressed in the submandibular gland and parotid gland [43, 96]
- Found at high levels in saliva, tears, and seminal plasma [42].

Function:

- Cystatin SA acts against the harmful dietary CPs and protects salivary proteins from degradation [44].
- Cystatin SA is a physiologic inhibitor of acid ceramidase that contributes to regulation of cellular ceramide content [98].

27.5.9 Cystatin SN

Common name: Cystatin 1, cystatin SA-I, salivary cystatin SA-I
Class: Family II/I25B
Gene name: CST1
Sequence length: 141
Chromosomal localization: Human chromosome 20
Subcellular location: Secreted (located outside cell membrane) [43]
Tissue specificity:

- Expression level is highest in the submandibular gland; lower in the parotid gland, lacrimal gland, and gallbladder; and very low in the trachea [96].
- Found in saliva, tears, urine, and seminal fluid [42, 43].

Function:

- Cystatin SN binds more tightly to papain and dipeptidyl peptidase I than other S-type cystatins but binds equally to ficin [99].
- Plays important role as emergency inhibitor and regulator against both self and foreign CPs in oral cavity [44].
- Might serve as potential biomarker for early detection of pancreatic cancer [100].

27.6 The Cystatin Fold

Despite considerable differences in amino acid sequences, all members of cystatin superfamily share conserved regions including a cystatin fold of five-stranded antiparallel β-sheets enfolded around a core of a five-turn α-helix at the center lying almost perpendicular to the sheets [26, 29, 35]. A conserved connectivity of secondary structures is also observed, (N)-β1-α1-β2-L1-β3-(AS)-α2-β4-L2-β5-(C), where AS denotes the wide "appending structure" positioned at the opposite end of the β-sheet in comparison to the N-terminus and β-hairpin loops L1 and L2 (Fig. 27.1). Generally, in between AS loop and β4, a second short α-helix (Fig. 27.1) is also observed. Two disulfides bonds, one between β3 and AS loop and the other between β4 and β5, both formed near C-terminal end around AS loop and in turn stabilize it (Fig. 27.1 shown in stick but not marked). Even the plant inhibitor oryzacystatin-I demonstrates the same cystatin fold as the animal cystatins [101].

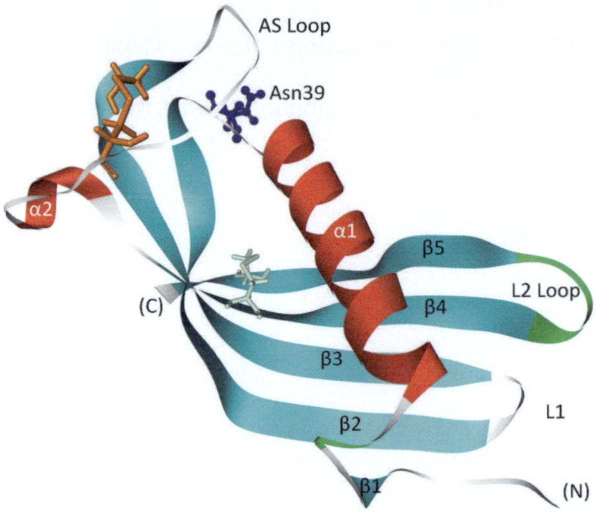

Fig. 27.1 Cystatin fold of human cystatin C [155]. Papain (N-terminal, L1 and L2 loop) and legumain (Asn-39 containing post-α1-helix region and part of AS loop) binding sites are labeled, and disulfide bonds are shown in stick

27.7 Mechanism of Inhibition

Mutational and X-ray crystallographic studies of three CPIs, cc [26, 102, 103], stefin B in complex with papain [29], and stefin A [104], marked three conserved regions of cystatins distributed at the edges of the cuneal-shaped structure that prevent the access of the active site of C1 family of CPs [32]. The exposed L1 loop consists of a highly conserved Glu_Val_Gly (QXVXG) region flanked between the protruding N-terminal, and a second C-terminal β-hairpin loop (L2) with highly conserved Pro-Trp (PW) residues (Figs. 27.1 and 27.2) constitutes the tripartite, hydrophobic, cuneal-shaped structure – highly complementary to the active site of C1 CPs. It should, however, be noted that in human stefins, the PW motif is replaced by PG in stefin A and PH in stefin B [26, 29] (Fig. 27.2a). The cystatin-CP complexes are largely stabilized by the hydrophobic side-chain interactions. The interactions in the S2 subsite, second binding site at the catalytic site of papain toward N-terminal starting from scissile bond of the substrate [105], further fortify the complexes significantly [106]. Cystatins are exosite inhibitors, bind around the catalytic cleft without interacting directly with the active site residues, and block the entry of substrate [107]. This marks a fundamental difference in mechanism of inhibition of cystatins in contrast to serine proteases and their inhibitors [20].

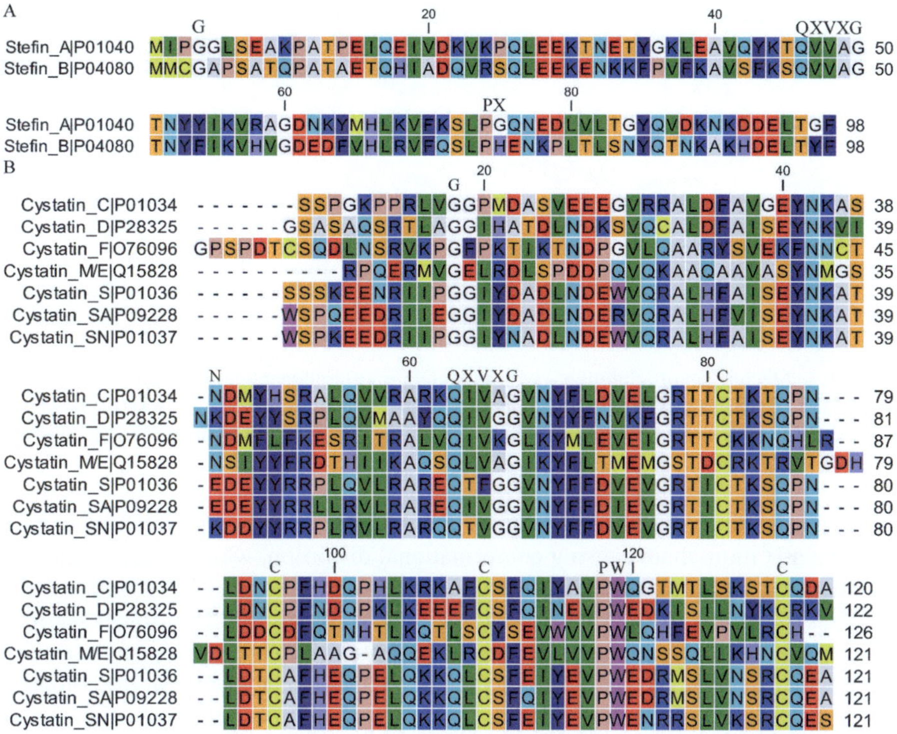

Fig. 27.2 Alignment of human family 1 and 2 cystatins. Disulfide bonds, conserved regions, and reactive sites for papain (G, QXVXG, and PX/PW) and legumain (N) binding are marked

27.7.1 Endopeptidases vs. Exopeptidases: A Two-Step Binding

Despite being the exosite inhibitors, cystatins are able to distinguish between endopeptidases and exopeptidases, by recognizing the differences in the active site scenario of these enzymes. The easily accessible catalytic site clefts of endopeptidases, like cathepsins K, L, and S and papain, readily accommodate cystatins. The more rigid β-hairpin loops (L1 and L2) participate in initial binding, and the flexible N-terminal region stabilizes the cystatin-endopeptidase complex in the second step by reducing the dissociation rate (Fig. 27.3a) [108]. This two-step binding is further demonstrated in stefin A-papain complex, where N-terminal interaction is preceded by the two β-hairpin loops binding in the catalytic core of papain [109].

In contrast, exopeptidase cathepsin B also illustrates a two-step binding, but in reverse order, initial weak interactions through N-terminal segment of cystatin A or C trigger a conformational change by displacing the occluding loop of cathepsin B and in turn disrupting a major salt bridge between His110 and Asp22 to facilitate β-hairpin loops binding in the catalytic site of the endopeptidase [108, 110–112]. Cystatin C displaces the loop more efficiently than stefin A [111]. Similar conformational change is also observed in cathepsin H-stefin A complex, where the

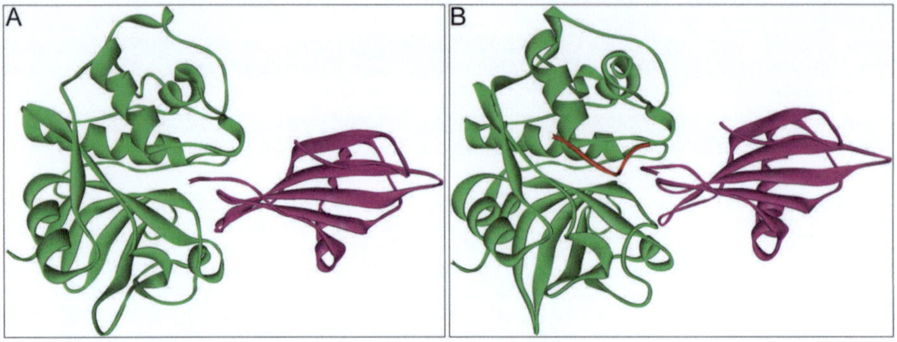

Fig. 27.3 Mechanism of inhibition – two-step binding. (**a**) Endopeptidase (papain-stefin B complex, PDB ID-1STF) and (**b**) exopeptidase (cathepsin H-stefin A complex, PDB ID-1NB3) complex. Papain and cathepsin H are shown in green, stefins A and B are shown in purple, and cathepsin H mini-chain is *red* in color

N-terminal residue of stefin A adopted a hooklike form and marginally displaced the cathepsin H mini-chain causing conformational distortion, which becomes characteristic for exopeptidases (Fig. 27.3b) [113].

Thus the comparison of the kinetics of binding shows that N-terminal segment interacts differently with endo- and exopeptidases. For endopeptidases the N-terminal stabilizes the complex by reducing the dissociation rate of the complex. In exopeptidases, N-terminal region of cystatins contributes to the association rate of the inhibitors, whereas the conformational change determines the rate-limiting step (Fig. 27.3). Although the rate of interaction of β-hairpin loops of cystatins remains almost the same with exo- and endopeptidases, the steric hindrance caused by occluding loops in carboxypeptidases (e.g., cathepsins B, X) and propeptide regions in aminopeptidases (e.g., cathepsins H, C) decreases the overall affinity of cystatins toward exopeptidases [20].

27.8 Relative Importance of Conserved Regions

The inhibitory interactions of cystatins of the first three families toward few selected members of C1 family of CPs are summarized in Table 27.2. Human cystatin C (hcc) and cc are the most potent inhibitors, whereas SD-type cystatins are the back benchers. Exopeptidases report lower binding affinity for cystatins in comparison to endopeptidases [116] (Table 27.2). Moreover, the affinities of the cystatins differ remarkably, from μM (10^{-6}) to pM (10^{-12}) range, even toward a particular CP despite sharing the same cystatin fold and mechanism of inhibition (Table 27.2). These huge affinity differences between CPs cannot only be attributed to the differences in the two-step binding between endo- and exopeptidases [7, 117, 118]. Further, S-type cystatins illustrate ≈90% sequence similarity, and still cystatin S is recognized as considerably poorer inhibitor than the other two (Table 27.2). It may be viewed as a gradual specialization of SD-type cystatins in order to protect mucosa from foreign

Table 27.2 Binding affinity of cystatins toward papain (C1) family of cysteine proteases

	K_i or $K_{i, app*}$ (nM)			
Inhibitor	Papain	Cathepsin B	Cathepsin H	Cathepsin L
Stefin A	0.019	8.2	0.31	1.3
Stefin B	0.012	73	0.58	0.23
Cystatin C	0.00001	0.27	0.28	<0.005
Cystatin D	1.2	>1000	7.5	18
Cystatin E/M	0.39	32	n.d.	1.78#
Cystatin F	1.1	>1000	n.d.	0.31
Cystatin S	108	>6000*	>6000*	>6000*
Cystatin SA	0.32	>6000*	>6000*	7300*
Cystatin SN	0.016	19	900*	400*
Chicken cystatin	0.005	1.7	0.06	0.019
L-kininogen	0.015	600	0.72	0.017

Inhibitor constants [K_i] for human cystatins [3, 5], chicken cystatins [5], *apparent inhibitor constant ($K_{i, app.*}$) [114]; # [95]; *n.d.* not determined
Note: Determination of the equilibrium dissociation constant provides a measure of the affinity of an inhibitor for an enzyme. In most cases, cystatins bind so tightly that K_i cannot be measured directly, so it is determined as the ratio of the separately measured rate constants or as the inhibition constant K_i, obtained by measurement of the substrate-dependent apparent inhibition constant $K_{i, app}$, followed by a mathematical correction for the presence of the inhibitor [115]

CPs without interfering with the endogenous CPs in a great extent. Collectively, the diverse role of the three conserved regions to the particular CP-cystatin binding and sequential variation in the binding interface may be the major determinant of the dissimilarities in inhibition constant (K_i) values, i.e., for the free energy change in complex formation between different cystatins and CPs [119].

27.8.1 N-Terminal Segment

Molecular docking studies of cc and papain point toward the conserved Gly9 residue that might provide the additional flexibility to the N-terminal region to maximize the binding contribution [26]. N-terminally truncated forms of cc showed 5000- to 10,000-fold lower affinities for papain [120] and further confirm the vital role of the region in binding [121].

Consistent with this model, mutation in conserved Gly11 residue of cystatin C (Fig. 27.2b) decreases binding affinity, and the size of the side chain of the substituted amino acid determines the magnitude of the effect [122]. Truncated forms of these mutated cystatin C show slightly lower affinities, in contrast to truncated wild-type cystatin C forms where binding affinity decreases significantly and further affirms the importance of conserved Gly residue. Inconsistent reduction of binding affinity of modified cystatin C is observed toward cathepsins, dubious reports are obtained about the extent of decrease [123, 124]. Apart from the conserved Gly11, Val10 is found to contribute largely to the binding of cathepsins B, H, L, and S; Arg8 and Leu9 are also known to do minor, enzyme-specific contributions to binding affinity (Fig. 27.2b). Effect of some mutations are found to be highly enzyme dependent, for example, Val10Trp substitution causes about twofold decrease in cathepsin S affinity but triggers a tenfold increase in cathepsin L affinity [124, 125]. Thus, N-terminal Arg8, Leu9, and Val10 determine the specificity of cystatin C binding, and by substituting these residues suitably, cathepsin-specific cystatins might be designed. In fact, N-terminal cystatin C decapeptide substitution in place of the first three residues of stefin A results in about 15-fold increase in binding affinity in stefin A-cathepsin B complex. The higher binding affinity is predominantly due to an increase in overall association rate constant resulting from better displacement of cathepsin B occluding loop by the elongated N-terminal. This substitution neither affect the dissociation rate in cathepsin B-stefin A interaction nor the modified stefin A's binding capability toward endopeptidases [126].

Truncated forms of human salivary cystatins show the N-terminal region is not of that importance for inhibition [127–130]. Three rat cystatin S forms of varying N-terminal length differ only moderately (<50-fold) in their inhibitory activity toward papain and ficin [131]. Two conserved consecutive Gly to Ala mutation (Fig. 27.2b) in human cystatin SN also illustrate only minute change in binding affinity toward papain [132]. In contrast a six-residue truncated form of cystatin SA shows 1000-fold poorer inhibitory activity to cathepsin L [127]. A cystatin D form lacking all residues including conserved Gly reported to be virtually inactive against cathepsins H, L, or S ($K_i > 1$ mM) pointed two to four orders of contribution of N-terminal in binding affinity in an enzyme-specific manner. Exchange of N-terminal regions between S-type cystatins and human cystatin C/D only moderately alters binding affinities for cathepsins [133]. Thus we can say N-terminal region of SD-type cystatin is more important for cathepsin binding than papain.

The requirement of N-terminal region of human stefin A, not only for specificity of binding toward CPs but also for maintenance of conformational integrity of tripartite reactive wedge, is well documented [134]. Even the smallest replacement of evolutionarily conserved Gly-4 residue, by Ala (Fig. 27.2a), causes $\approx 10^3$-fold decrease in binding affinity for papain and cathepsin B, while cathepsin L stays least affected [135]. N-terminally truncated variants, sequential deletion mutant of the first three amino acids, show progressive decrease of cystatin A affinity and estimate about 40% contribution in total binding free energy of cystatin A to papain and cathepsin B. Pro3 and Ile2 are the most prominent contributor in the interaction. For stefin B, in contrast to the previous observations [120], Pol et al. have described that N-terminal region is accountable for 12–40% of the total binding affinity in stefin B-papain and stefin B-cathepsin B and L complexes; cathepsin H remains least affected [136]. Cys3 of stefin B contributes most in papain, cathepsin L and H inhibition; the contribution of the rest of the residues of N-terminal varies with the concerned CP [136, 137].

An earlier study of oryzacystatin-papain interaction indicates that N-terminal or C-terminal truncation of oryzacystatin fails to affect its inhibitory activity [138]. In the contrary to this report, the latter studies notify the necessity of N-terminal region, in particular the highly conserved Gly10 residue of oryzacystatin in papain inhibition [139]. A mature form of sunflower cystatin (SCA), lacking 15 amino acids at the N-terminus, illustrates a higher dissociation rate constant and lower affinity toward papain in comparison to the full-length recombinant SCA (rSCA) due to the lack of additional stabilization of the complex by the elongated N-terminal region [140]. Another sunflower cystatin SCB also evidences the instrumental role of the first four N-terminal amino acids (IPGG) manifesting the inhibitory activity toward papain [141].

27.8.2 First β-Hairpin Loop (QXVXG Region)

Substitution in conserved QXVXG motif of cc causes an increase in inhibition constant principally due to enhanced dissociation rate [142]. Yet different mutations show incoherent effects with CPs: for example, mutation of the QXVXG loop exhibits minor effect on cathepsin L inhibition and inhibits papain temporarily but triggers a more than three-order increase in inhibition constant for cathepsin B [119].

N-terminal truncation and L2 loop mutation specify that L1 loop is responsible for 40–60% of the total binding free energy of hcc-actinidin; hcc-cathepsins B, H, and S; and hcc-papain complexes [124, 143]. Modification of the QVVAG loop of rat cystatin S and mutation in QTVGG loop of human cystatin SN cause radical increase in inhibition constant [129, 132, 144, 145]. Appreciable difference in inhibitory activities toward plant CPs including papain is also observed between allelic variants of cystatin SA; difference by single amino acid (QIVGG and QIVDG) in conserved L1 loop region [130] also portrays the importance of the motif.

Modification in the QXVXG region of stefin A fails to affect its inhibitory affinity for cathepsins B, H, and L and papain [146]. On the other hand, conserved pentapeptide of oryzacystatin is the primary region of interaction in papain inhibition, and Gln appears to be the key residue [138].

27.8.3 Second β-Hairpin Loop (PW Motif)

The highly conserved Pro-Trp motif of cystatin L2 loop contributes to the free energy of binding by lowering the dissociation rate in an enzyme-specific manner. For cc-cathepsin B and cc-papain complexes, L2 loop contributes moderately but rather lowly in case of cathepsin L, where the N-terminal and L2 loop are the major determinant of the interaction [119]. Modification in conserved Trp residue of cc lowers its affinity toward papain by 10^5-fold [147].

Substitution of Trp106 with Gly in hcc results decrease in binding affinity for actinidin, papain, and cathepsins B and H by 300- to 900-fold; a two-order increase in inhibition constant is also reported in case of cathepsin B inhibition for Trp106Ser mutation though cathepsins H and L remain less affected [143, 148, 149]. These studies estimate L2 loop is responsible for 20–30% of the total binding affinity [143, 148]. PW motif substitutions by Gly in cystatin SN show a 100-fold decrease in binding free energy in case of cathepsin C inhibition but of negligible interest for papain binding [132].

Mutation of highly conserved family 1 cystatin residue Leu73, the residue just preceding the conserved PX motif (Fig. 27.2a) of stefin A known for establishing hydrophobic interaction with CPs, causes decrease in binding affinity by ≈4000-fold, >10-fold, and ≈300-fold for cathepsin B, L and papain inhibition, respectively. Alteration of Pro74 results in ≈10-fold increase in K_i for cathepsin B but shows negligible effect to cathepsin L and papain inhibition [148]. Trp/His substitutions in place of Gly75 to make the L2 loop more like cystatin C/B, respectively, give mixed results – a ≈10-fold increase in affinity is observed for papain but nothing notable for cathepsin B. Causing local conformational changes, Gly75His alteration slows down the dissociation rate of papain complexes, while Gly75Trp substitution boosts the rate of association [126]. Similar substitution in stefin B, Leu73Gly, and His75Gly (Fig. 27.2a) results significant decrease in cathepsin B, H and papain binding affinity; only cathepsin L emerges as an exception. The contribution of the second β-hairpin loop varies with CPs, it is largest (≈45% of total energy of binding) for stefin A-cathepsin B interaction and only around 20% for papain inhibition, for stefin B the contribution varies between 20 and 30% for all CPs, and irrespective of CPs Leu73 plays a vital role in every stefin inhibition [148, 150].

27.8.4 Additional Domains

Cystatins can also act as inhibitor of mammalian legumain – a family C13 CP with dual protease-ligase activity, evolutionarily unrelated to C1 family of CP and equipped with distinctly different catalytic site and structural fold [79]. Dall et al. [151] recently crystallographically evidenced the much speculated inhibition of legumain by cystatin M/E through dual interaction of legumain reactive center loop (RCL – residues 38–43 in post-α1-helix region, hcc numbering) and legumain exosite loop (LEL – residues 74–94 in AS loop region, hcc numbering) (Figs. 27.2 and 27.4). RCL exhibits a canonical, substrate-like binding mode, typical within the type 2 cystatin family, while overall hydrophobic LEL serves as exosite binder, and Asn39 residue (hcc numbering) plays a very important role in ligase activity of legumain as hypothesized previously [151, 152]. Cystatin M/E emerges as the most potent inhibitor of legumain (K_i = 0.0016 nM), and cystatins C and F come next (K_i = 0.2 nM and 10 nM with pig legumain, respectively) [152]. Cystatin D does not show any inhibitory activity toward legumain despite the presence of Asn around this region, possibly because of the positively charged Lys residue in the very next position, instead of the highly conserved negatively charged residue in cystatins (Fig. 27.2b). The S-type cystatins are devoid of Asn in the post-helix region and appear to be non-inhibitor of legumain (Fig. 27.2b).

Cystatin SN is reported to bind stably in a second site of papain, other than its active site, without affecting the proteolytic activity of the enzyme. Even when the catalytic site is preoccupied with an irreversible inhibitor like E-64, binding of cystatin SN at the second site is not hampered. Unfortunately the details of these interactions are still unknown [153].

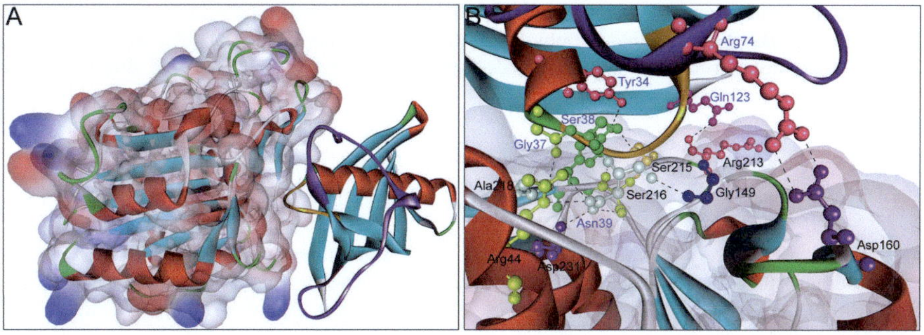

Fig. 27.4 Binding conformation of legumain-cystatin M/E complex (PDB ID: 4N6N). (**a**) Legumain (*left*)-cystatin M/E (*right*) interaction is mediated by the reactive center loop (RCL in yellow) containing Asn39 and the legumain exosite loop (LEL in *purple*). (**b**) Binding site scenario. Amino acid residues of legumain and cystatin M/E are labeled in *black* and *blue*, respectively

The C-terminal residue Tyr-97 of stefin B is reported to contribute in CP binding by providing 6–12% of binding energy. Tyr97Ala substitution causes decrease in binding affinity of cathepsins H and L and papain; however, cathepsin B remains least affected [29, 150].

27.9 Conclusion

To summarize, the mutation, substitution, deletion, and grafting studies point toward the differential inhibitory approaches of cystatins toward CPs. The N-terminal segment and QXVXG loop make the most significant contributions in cc-papain (≈87% of binding energy) and cc-cathepsin B complexes; in case of cathepsin L complexes, the N-terminal holds the key [119, 120]. Two β-hairpin loops are jointly recognized by Björk et al. [143] for the major part (60–90%) of human cystatin C affinity toward endopeptidase papain and exopeptidase cathepsin B; in contrast to Hall et al.'s report [124], the side chains of N-terminal segment and the Trp residue of L2 loop are mainly responsible for cystatin C affinity toward endopeptidase cathepsin L and exopeptidase cathepsins B and H. Arg8, Leu9, and Val10 of hcc are recognized for their contribution in selectivity and affinity [124, 125]. In case of SD-type cystatins, the first β-hairpin loop plays the major role in papain inhibition, whereas all three conserved regions chip in for cathepsin CPs [44, 2002]. In human family 1 cystatins, the flexible N-terminal and the PW motif of L2 loop together contribute most of the binding energy to papain and cathepsin B. The highly conserved Leu73 in type 1 cystatins, Cys3, Tyr97 for stefin B and Ile2, and Pro3 for stefin A are the key residues for inhibition [29, 108, 135–137, 148, 150]. All three conserved regions of plant cystatins participate in papain inhibition, although N-terminal and first hairpin loop plays the central role [154].

Acknowledgment I would like to thank Dr. Alpana Seal, Retired Professor in Biophysics, Department of Biochemistry and Biophysics, University of Kalyani, for her help and encouragement. Thanks are also due to Bioinformatics Centre, University of Kalyani and Bioinformatics Centre, NEHU, Tura campus.

References

1. Barrett AJ, Fritz H, Grubb A et al (1986) Nomenclature and classification of the proteins homologous with the cysteine-proteinase inhibitor chicken cystatin. Biochem J 236:312
2. Müller-Esterl W, Fritz H, Kellermann J et al (1985) Genealogy of mammalian cysteine proteinase inhibitors. FEBS Lett 191:221–226
3. Abrahamson M, Alvarez-Fernandez M, Nathanson CM (2003) Cystatins. In Biochemical Society Symposia Portland Press Limited 70:179–199
4. Turk V, Stoka V, Turk D (2008) Cystatins: biochemical and structural properties, and medical relevance. Front Biosci 13:5406–5420
5. Barrett AJ, Rawlings ND, Davies ME et al (1986) Cysteine proteinase inhibitors of cystatin superfamily. In: Barrett AJ, Rawlings ND (eds) Proteinase inhibitors Elsevier, Amsterdam, pp 519–569
6. Klotz C, Ziegler T, Daniłowicz-Luebert E, Hartmann S (2011) Cystatins of parasitic organisms. In: Cysteine proteases of pathogenic organisms. Springer US, New York, pp 208–221
7. Turk B, Turk D, Salvesen GS (2002) Regulating cysteine protease activity: essential role of protease inhibitors as guardians and regulators. Curr Pharm Design 8:1623–1637
8. Keppler D (2006) Towards novel anti-cancer strategies based on cystatin function. Cancer Lett 235:159–176
9. Revesz T, Ghiso J, Lashley T et al (2003) Cerebral amyloid angiopathies: a pathologic, biochemical, and genetic view. J Neuropathol Exp Neurol 62:885–898
10. Joyce JA, Baruch A, Chehade K et al (2004) Cathepsin cysteine proteases are effectors of invasive growth and angiogenesis during multistage tumorigenesis. Cancer Cell 5:443–453
11. Mignatti P, Robbins E, Rifkin DB (1986) Tumor invasion through the human amniotic membrane: requirement for a proteinase cascade. Cell 47:487–498
12. Coussens LM, Werb Z (1996) Matrix metal loproteinases and the development of cancer. Chem Biol 3:895–904
13. Dollery CM, McEwan JR, Henney AM (1995) Matrix metalloproteinases and cardiovascular disease. Circ Res 77:863–868
14. Gocheva V, Joyce JA (2007) Cysteine cathepsins and the cutting edge of cancer invasion. Cell 6:60–64
15. Turk V, Kos J, Turk B (2004) Cysteine cathepsins (proteases)—on the main stage of cancer? Cancer Cell 5:409–410
16. Yasuda Y, Kaleta J, Brömme D (2005) The role of cathepsins in osteoporosis and arthritis: rationale for the design of new therapeutics. Adv Drug Deliv Rev 57:973–993
17. Nakanishi H (2003) Neuronal and microglial cathepsins in aging and age-related diseases. Ageing Res Rev 2:367–381
18. Lutgens SP, Cleutjens KB, Daemen MJ, Heeneman S (2007) Cathepsin cysteine proteases in cardiovascular disease. FASEB J 21:3029–3041
19. Ochieng J, Chaudhuri G (2010) Cystatin superfamily. J Health Care Poor Underserved 21:51
20. Turk V, Turk B (2008) Lysosomal cysteine proteases and their protein inhibitors: recent developments. Acta Chim Slov 55:727–738
21. Vasiljeva O, Reinheckel T, Peters C et al (2007) Emerging roles of cysteine cathepsins in disease and their potential as drug targets. Curr Pharm Design 13:387–403
22. Fossum K, Whitaker JR (1968) Ficin and papain inhibitor from chicken egg white. Arch Biochem Biophys 125:367–375

23. Keilova H, Tomášek V (1974) Effect of papain inhibitor from chicken egg white on cathepsin B1. Biochim Biophys Acta 334:179–186
24. Sen LC, Whitaker JR (1973) Some properties of a ficin-papain inhibitor from avion egg white. Arch Biochem Biophys 158:623–632
25. Turk V, Brzin J, Longer M et al (1983) Protein inhibitors of cysteine proteinases. III. Amino-acid sequence of cystatin from chicken egg white. H-S Z Physiol Chem 364:1487–1496
26. Bode W, Engh R, Musil DJ et al (1988) The 2.0 A X-ray crystal structure of chicken egg white cystatin and its possible mode of interaction with cysteine proteinases. EMBO J 7:2593
27. Barrett AJ (1981) Cystatin, the egg white inhibitor of cysteine proteinases. Methods Enzymol 80:771–778
28. Grubb A, Löfberg H (1982) Human gamma-trace, a basic microprotein: amino acid sequence and presence in the adenohypophysis. Proc Natl Acad Sci U S A 79:3024–3027
29. Stubbs MT, Laber B, Bode W et al (1990) The refined 2.4 A X-ray crystal structure of recombinant human stefin B in complex with the cysteine proteinase papain: a novel type of proteinase inhibitor interaction. EMBO J 9:1939
30. Dayhoff MO, Barker WC, Hunt LT (1983) Establishing homologies in protein sequences. Methods Enzymol 91:524–545
31. Rawlings ND, Barrett AJ (1990) Evolution of proteins of the cystatin superfamily. J Mol Evol 30:60–71
32. Turk V, Bode W (1991) The cystatins: protein inhibitors of cysteine proteinases. FEBS Lett 285:213–219
33. Wood TC, Pearson WR (1999) Evolution of protein sequences and structures. J Mol Biol 291:977–995
34. Abrahamson M, Barrett A, Salvesen G, Grubb A (1986) Isolation of six cysteine proteinase inhibitors from human urine. Their physicochemical and enzyme kinetic properties and concentrations in biological fluids. J Biol Chem 261:11282–11289
35. Kordiš D, Turk V (2009) Phylogenomic analysis of the cystatin superfamily in eukaryotes and prokaryotes. BMC Evol Biol 9:1
36. Turk B, Krizaj I, Kralj B (1993) Bovine stefin C, a new member of the stefin family. J Biol Chem 268:7323–7329
37. Bjoerk I, Ylinenjaervi K (1992) Different roles of the two disulfide bonds of the cysteine proteinase inhibitor, chicken cystatin, for the conformation of the active protein. Biochemistry 31:8597–8602
38. Paraoan L, Grierson I, Maden BE (2003) Fate of cystatin C lacking the leader sequence in RPE cells. Exp Eye Res 76:753–756
39. Ekström U, Wallin H, Lorenzo J (2008) Internalization of cystatin C in human cell lines. FEBS J 275:4571–4582
40. Wallin H, Bjarnadottir M, Vogel LK et al (2010) Cystatins–extra-and intracellular cysteine protease inhibitors: high-level secretion and uptake of cystatin C in human neuroblastoma cells. Biochimie 92:1625–1634
41. Ni J, Abrahamson M, Zhang M et al (1997) Cystatin E is a novel human cysteine proteinase inhibitor with structural resemblance to family 2 cystatins. J Biol Chem 272:10853–10858
42. Isemura S, Saitoh E, Sanada K, Minakata K (1991) Identification of full-sized forms of salivary (type) cystatins (cystatin SN, cystatin SA, cystatin S, and two phosphorylated forms of cystatin S) in human whole saliva and determination of phosphorylation sites of cystatin S. J Biochem 110:648–654
43. Ryan CM, Souda P, Halgand F et al (2010) Confident assignment of intact mass tags to human salivary cystatins using top-down Fourier-transform ion cyclotron resonance mass spectrometry. J Am Soc Mass Spectrom 21:908–917
44. Dickinson DP (2002) Salivary (SD-type) cystatins: over one billion years in the making—but to what purpose? Crit Rev Oral Biol Med 13:485–508
45. Veillard F, Lecaille F, Lalmanach G (2008) Lung cysteine cathepsins: intruders or unorthodox contributors to the kallikrein–kinin system? Int J Biochem Cell Biol 40:1079–1094

46. Cadena RA, Colman RW (1991) Structure and functions of human kininogens. Trends Pharmacol Sci 12:272–275
47. Kakizuka A, Kitamura N, Nakanishi S (1988) Localization of DNA sequences governing alternative mRNA production of rat kininogen genes. J Biol Chem 263:3884–3892
48. Kitamura N, Kitagawa H, Fukushima D et al (1985) Structural organization of the human kininogen gene and a model for its evolution. J Biol Chem 260:8610–8617
49. Müller-Esterl W, Iwanaga S, Nakanishi S (1986) Kininogens revisited. Trends Biochem Sci 11:336–339
50. Turk B, Stoká V, Björk I et al (1995) High-affinity binding of two molecules of cysteine proteinases to low-molecular-weight kininogen. Protein Sci 4:1874–1880
51. Turk B, Stoka V, Turk V et al (1996) High-molecular-weight kininogen binds two molecules of cysteine proteinases with different rate constants. FEBS Lett 391:109–112
52. Salvesen G, Parkes C, Abrahamson M et al (1986) Human low-Mr kininogen contains three copies of a cystatin sequence that are divergent in structure and in inhibitory activity for cysteine proteinases. Biochem J 234:429–434
53. Reynolds JL, Skepper JN, McNair R et al (2005) Multifunctional roles for serum protein fetuin-a in inhibition of human vascular smooth muscle cell calcification. J Am Soc Nephrol 16:2920–2930
54. Jones AL, Hulett MD, Parish CR (2005) Histidine-rich glycoprotein: a novel adaptor protein in plasma that modulates the immune, vascular and coagulation systems. Immunol Cell Biol 83:106–118
55. Brown WM, Dziegielewska KM (1997) Friends and relations of the cystatin superfamily—new members and their evolution. Protein Sci 6:5–12
56. Cornwall GA, Cameron A, Lindberg I et al (2003) The cystatin-related epididymal spermatogenic protein inhibits the serine protease prohormone convertase 2. Endocrinology 144:901–908
57. Valente RH, Dragulev B, Perales J et al (2001) BJ46a, a snake venom metalloproteinase inhibitor. Eur J Biochem 268:3042–3052
58. Rawlings ND, Tolle DP, Barrett AJ (2004) Evolutionary families of peptidase inhibitors. Biochem J 378:705–716
59. Rawlings ND, Waller M, Barrett AJ, Bateman A (2014) MEROPS: the database of proteolytic enzymes, their substrates and inhibitors. Nucleic Acids Res 42:D503–D509
60. Rawlings ND, Barrett AJ, Bateman A (2012) MEROPS: the database of proteolytic enzymes, their substrates and inhibitors. Nucleic Acids Res 40:D343–D350
61. Nei M, Rooney AP (2005) Concerted and birth-and-death evolution of multigene families. Annu Rev Genet 39:121
62. Cornwall GA, Hsia N (2003) A new subgroup of the family 2 cystatins. Mol Cell Endocrinol 200:1–8
63. Aagaard A, Listwan P, Cowieson N et al (2005) An inflammatory role for the mammalian carboxypeptidase inhibitor latexin: relationship to cystatins and the tumor suppressor TIG1. Structure 13:309–317
64. Dixelius J, Olsson AK, Thulin Å et al (2006) Minimal active domain and mechanism of action of the angiogenesis inhibitor histidine-rich glycoprotein. Cancer Res 66:2089–2097
65. Zhu S (2008) Did cathelicidins, a family of multifunctional host-defense peptides, arise from a cysteine protease inhibitor? Trends Microbiol 16:353–360
66. Bennett CS, Khorshid HR, Kitchen JA et al (2004) Characterization of the human secreted phosphoprotein 24 gene (SPP2) and comparison of the protein sequence in nine species. Matrix Biol 22:641–651
67. Toroian D, Price PA (2008) The essential role of fetuin in the serum-induced calcification of collagen. Calcif Tissue Int 82:116–126
68. Blaydon DC, Nitoiu D, Eckl KM et al (2011) Mutations in CSTA, encoding Cystatin A, underlie exfoliative ichthyosis and reveal a role for this protease inhibitor in cell-cell adhesion. Am J Hum Genet 89:564–571

69. Parker BS, Ciocca DR, Bidwell BN et al (2008) Primary tumour expression of the cysteine cathepsin inhibitor Stefin A inhibits distant metastasis in breast cancer. J Pathol 214:337–346
70. Vasilopoulos Y, Walters K, Cork MJ et al (2008) Association analysis of the skin barrier gene cystatin A at the PSORS5 locus in psoriatic patients: evidence for interaction between PSORS1 and PSORS5. Eur J Hum Genet 16:1002–1009
71. Vasilopoulos Y, Cork MJ, Teare D et al (2007) A nonsynonymous substitution of cystatin A, a cysteine protease inhibitor of house dust mite protease, leads to decreased mRNA stability and shows a significant association with atopic dermatitis. Allergy 62:514–519
72. Riccio M, Di Giaimo R, Pianetti S et al (2001) Nuclear localization of cystatin B, the cathepsin inhibitor implicated in myoclonus epilepsy (EPM1). Exp Cell Biol 262:84–94
73. Alakurtti K, Weber E, Rinne R et al (2005) Loss of lysosomal association of cystatin B proteins representing progressive myoclonus epilepsy, EPM1, mutations. Eur J Hum Genet 13:208–215
74. Žerovnik E, Pompe-Novak M, Škarabot M et al (2002) Human stefin B readily forms amyloid fibrils in vitro. Biochim Biophys Acta 1594:1–5
75. Lalioti MD, Mirotsou M, Buresi C et al (1997) Identification of mutations in cystatin B, the gene responsible for the Unverricht-Lundborg type of progressive myoclonus epilepsy (EPM1). Am J Hum Genet 60:342
76. Abrahamson M (1988) Human cysteine proteinase inhibitors: isolation, physiological importance, inhibitory mechanism, gene structure and relation to hereditary cerebral hemorrhage. Scand J Clin Lab Invest 48:21–31
77. Sanchez JC, Guillaume E, Lescuyer P et al (2004) Cystatin C as a potential cerebrospinal fluid marker for the diagnosis of Creutzfeldt-Jakob disease. Proteomics 4:2229–2233
78. Nilsson J, Rüetschi U, Halim A et al (2009) Enrichment of glycopeptides for glycan structure and attachment site identification. Nat Methods 6:809–811
79. Chen JM, Dando PM, Rawlings ND et al (1997) Cloning, isolation, and characterization of mammalian legumain, an asparaginyl endopeptidase. J Biol Chem 272:8090–8098
80. Snorradottir AO (2006) Hereditary cystatin C amyloid angiopathy: genetic, clinical, and pathological aspects. Brain Pathol 16:55–59
81. Zurdel J, Finckh U, Menzer G et al (2002) CST3 genotype associated with exudative age related macular degeneration. Br J Ophthalmol 86:214–219
82. Freije JP, Abrahamson M, Olafsson I et al (1991) Structure and expression of the gene encoding cystatin D, a novel human cysteine proteinase inhibitor. J Biol Chem 266:20538–20543
83. Balbin M, Hall A, Grubb A et al (1994) Structural and functional characterization of two allelic variants of human cystatin D sharing a characteristic inhibition spectrum against mammalian cysteine proteinases. J Biol Chem 269:23156–23162
84. Cappello F, Gatti E, Camossetto V et al (2004) Cystatin F is secreted, but artificial modification of its C-terminus can induce its endocytic targeting. Exp Cell Biol 297:607–618
85. Langerholc T, Zavašnik-Bergant V, Turk B et al (2005) Inhibitory properties of cystatin F and its localization in U937 promonocyte cells. FEBS J 272:1535–1545
86. Nathanson CM, Wasselius J, Wallin H, Abrahamson M (2002) Regulated expression and intracellular localization of cystatin F in human U937 cells. Eur J Biochem 269:5502–5511
87. Hashimoto SI, Suzuki T, Nagai S et al (2000) Identification of genes specifically expressed in human activated and mature dendritic cells through serial analysis of gene expression. Blood 96:2206–2214
88. Ni J, Fernandez MA, Danielsson L et al (1998) Cystatin F is a glycosylated human low molecular weight cysteine proteinase inhibitor. J Biol Chem 273:24797–24804
89. Obata-Onai A, Hashimoto SI, Onai N et al (2002) Comprehensive gene expression analysis of human NK cells and CD8+ T lymphocytes. Int Immunol 14:1085–1098
90. Halfon S, Ford J, Foster J et al (1998) Leukocystatin, a new class II cystatin expressed selectively by hematopoietic cells. J Biol Chem 273:16400–16408
91. Schüttelkopf AW, Hamilton G, Watts C, van Aalten DM (2006) Structural basis of reduction-dependent activation of human cystatin F. J Biol Chem 281:16570–16575

92. Zeeuwen PL, Vlijmen-Willems V, Egami H, Schalkwijk J (2002) Cystatin M/E expression in inflammatory and neoplastic skin disorders. Br J Dermatol 147:87–94
93. Zeeuwen PL, van Vlijmen-Willems IM, Jansen BJ et al (2001) Cystatin M/E expression is restricted to differentiated epidermal keratinocytes and sweat glands: a new skin-specific proteinase inhibitor that is a target for cross-linking by transglutaminase. J Invest Dermatol 116:693–701
94. Sotiropoulou G, Anisowicz A, Sager R (1997) Identification, cloning, and characterization of cystatin M, a novel cysteine proteinase inhibitor, down-regulated in breast cancer. J Biol Chem 272:903–910
95. Cheng T, Hitomi K, van Vlijmen-Willems IM et al (2006) Cystatin M/E is a high affinity inhibitor of cathepsin v and cathepsin l by a reactive site that is distinct from the legumain-binding site a novel clue for the role of cystatin m/e in epidermal cornification. J Biol Chem 281:15893–15899
96. Dickinson DP, Thiesse M, Hicks MJ (2002) Expression of type 2 cystatin genes CST1-CST5 in adult human tissues and the developing submandibular gland. DNA Cell Biol 21:47–65
97. Isemura S, Saitoh E, Ito S et al (1984) Cystatin S: a cysteine proteinase inhibitor of human saliva. J Biochem 96:1311–1314
98. Eliyahu E, Shtraizent N, He X et al (2011) Identification of cystatin SA as a novel inhibitor of acid ceramidase. J Biol Chem 286:35624–35633
99. Isemura S, Saitoh E, Sanada K (1986) Characterization of a new cysteine proteinase inhibitor of human saliva, cystatin SN, which is immunologically related to cystatin S. FEBS Lett 198:145–149
100. Jiang J, Liu HL, Liu ZH et al (2015) Identification of cystatin SN as a novel biomarker for pancreatic cancer. Tumor Biol 36:3903–3910
101. Nagata K, Kudo N, Abe K et al (2000) Three-dimensional solution structure of oryzacystatin-I, a cysteine proteinase inhibitor of the rice, *Oryza sativa* L japonica. Biochemistry 39:14753–14760
102. Dieckmann T, Mitschang L, Hofmann M et al (1993) The structures of native phosphorylated chicken cystatin and of a recombinant unphosphorylated variant in solution. J Mol Biol 234:1048–1059
103. Engh RA, Dieckmann T, Bode W (1993) Conformational variability of chicken cystatin: comparison of structures determined by X-ray diffraction and NMR spectroscopy. J Mol Biol 234:1060–1069
104. Martin JR, Craven JC, Jerala R (1995) The three-dimensional solution structure of human stefin A. J Mol Biol 246:331–343
105. Schechter I, Berger A (1967) On the size of the active site in proteases. I. Papain. Biochem Biophys Res Commun 27:157–162
106. Berger A, Schechter I (1970) Mapping the active site of papain with the aid of peptide substrates and inhibitors. Philos Trans R Soc Lond Ser B Biol Sci 257:249–264
107. Rzychon M, Chmiel D, Stec-Niemczyk J (2004) Modes of inhibition of cysteine proteases. Acta Biochim 51:861–873
108. Estrada S, Pavlova A, Björk I (1999) The contribution of N-terminal region residues of cystatin a (stefin a) to the affinity and kinetics of inhibition of papain, cathepsin B, and cathepsin L. Biochemistry 38:7339–7345
109. Estrada S, Raub-Segall E, Björk I, Olson ST (2000) The N-terminal region of cystatin A (stefin A) binds to papain subsequent to the two hairpin loops of the inhibitor. Demonstration of two-step binding by rapid-kinetic studies of cystatin A labeled at the N-terminus with a fluorescent reporter group. Protein Sci 9:2218–2224
110. Nycander M, Estrada S, Mort JS et al (1998) Two-step mechanism of inhibition of cathepsin B by cystatin C due to displacement of the proteinase occluding loop. FEBS Lett 422:61–64
111. Pavlova A, Krupa JC, Mort JS et al (2000) Cystatin inhibition of cathepsin B requires dislocation of the proteinase occluding loop. Demonstration by release of loop anchoring through mutation of His110. FEBS Lett 487:156–160

112. Renko M, Požgan U, Majera D, Turk D (2010) Stefin A displaces the occluding loop of cathepsin B only by as much as required to bind to the active site cleft. FEBS J 277:4338–4345
113. Jenko S, Dolenc I, Gunčar G et al (2003) Crystal structure of Stefin a in complex with cathepsin H: N-terminal residues of inhibitors can adapt to the active sites of endo-and exopeptidases. J Mol Biol 326:875–885
114. Baron AC, DeCarlo AA, Featherstone JD (1999) Functional aspects of the human salivary cystatins in the oral environment. Oral Dis 5:234–240
115. Bieth JG (1995) Theoretical and practical aspects of proteinase inhibition kinetics. Methods Enzymol 248:59
116. Mihelič M, Teuscher C, Turk V, Turk D (2006) Mouse stefins A1 and A2 (Stfa1 and Stfa2) differentiate between papain-like endo-and exopeptidases. FEBS Lett 580:4195–4199
117. Turk D, Gunčar G, Podobnik M, Turk B (1998) Revised definition of substrate binding sites of papain-like cysteine proteases. Biol Chem 379:137–148
118. Turk V, Stoka V, Vasiljeva O et al (2012) Cysteine cathepsins: from structure, function and regulation to new frontiers. Biochim Biophys 1824:68–88
119. Auerswald EA, Nägler DK, Assfalg-Machleidt I et al (1995) Hairpin loop mutations of chicken cystatin have different effects on the inhibition of cathepsin B, cathepsin L and papain. FEBS Lett 361:179–184
120. Machleidt W, Thiele U, Assfalg-Machleidt I et al (1990) Molecular mechanism of inhibition of cysteine proteinases by their protein inhibitors: kinetic studies with natural and recombinant variants of cystatins and stefins. Biomed Biochim Acta 50:613–620
121. Machleidt W, Thiele U, Laber B et al (1989) Mechanism of inhibition of papain by chicken egg white cystatin. FEBS Lett 243:234–238
122. Hall A, Dalbøge H, Grubb A, Abrahamson M (1993) Importance of the evolutionarily conserved glycine residue in the N-terminal region of human cystatin C (Gly-11) for cysteine endopeptidase inhibition. Biochem J 291:123–129
123. Abrahamson M, Mason RW, Hansson H et al (1991) Human cystatin C. Role of the N-terminal segment in the inhibition of human cysteine proteinases and in its inactivation by leucocyte elastase. Biochem J 273:621–626
124. Hall A, Håkansson K, Mason RW et al (1995) Structural basis for the biological specificity of cystatin C identification of leucine 9 in the n-terminal binding region as a selectivity-conferring residue in the inhibition of mammalian cysteine peptidases. J Biol Chem 270:5115–5121
125. Mason WR, Katia SC, Abrahamson M (1998) Amino acid substitutions in the N-terminal segment of cystatin C create selective protein inhibitors of lysosomal cysteine proteinases. Biochem J 330:833–838
126. Pavlova A, Björk I (2003) Grafting of features of cystatins C or B into the N-terminal region or second binding loop of cystatin A (stefin A) substantially enhances inhibition of cysteine proteinases. Biochemistry 42:11326–11333
127. Baron AC, Gansky SA, Ryder MI, Featherstone JD (1999) Cysteine protease inhibitory activity and levels of salivary cystatins in whole saliva of periodontally diseased patients. J Periodontal Res 34:437–444
128. Blankenvoorde MF, Henskens YM, Van't Hof W et al (1996) Inhibition of the growth and cysteine proteinase activity of Porphyromonas gingivalis by human salivary cystatin S and chicken cystatin. Biol Chem Hoppe Seyler 377:847–850
129. Bobek LA, Ramasubbu N, Wang X et al (1994) Biological activities and secondary structures of variant forms of human salivary cystatin SN produced in Escherichia coli. Gene 151:303–308
130. Saitoh E, Minaguchi K, Ishibashi O et al (1998) Production and characterization of two variants of human cystatin SA encoded by two alleles at theCST2Locus of the type 2 cystatin Gene family. Arch Biochem Biophys 352:199–206
131. Nishiura T, Ishibashi K, Abe K (1991) Isolation of three forms of cystatin from submandibular saliva of isoproterenol-treated rats, its properties and kinetic data. Biochim Biophys Acta 1077:346–354

132. Tseng CC, Tseng CP, Levine MJ, Bobek LA (2000) Differential effect toward inhibition of papain and cathepsin C by recombinant human salivary cystatin SN and its variants produced by a baculovirus system. Arch Biochem Biophys 380:133–140
133. Hall A, Ekiel I, Mason RW et al (1998) Structural basis for different inhibitory specificities of human cystatins C and D. Biochemistry 37:4071–4079
134. Shibuya K, Kaji H, Itoh T et al (1995) Human cystatin A is inactivated by engineered truncation. The NH2-terminal region of the cysteine proteinase inhibitor is essential for expression of its inhibitory activity. Biochemistry 34:12185–12192
135. Estrada S, Nycander M, Hill NJ et al (1998) The role of Gly-4 of human cystatin A (stefin A) in the binding of target proteinases. Characterization by kinetic and equilibrium methods of the interactions of cystatin A Gly-4 mutants with papain, cathepsin B, and cathepsin L. Biochemistry 37:7551–7560
136. Pol E, Björk I (2003) Contributions of individual residues in the N-terminal region of cystatin B (stefin B) to inhibition of cysteine proteinases. Biochim Biophys 1645:105–112
137. Pol E, Björk I (2001) Role of the single cysteine residue, Cys 3, of human and bovine cystatin B (stefin B) in the inhibition of cysteine proteinases. Protein Sci 10:1729–1738
138. Arai S, Watanabe H, Kondo H et al (1991) Papain-inhibitory activity of oryzacystatin, a rice seed cysteine proteinase inhibitor, depends on the central Gln-Val-Val-Ala-Gly region conserved among cystatin superfamily members. J Biochem 109:294–298
139. Urwin PE, Atkinson HJ, McPherson MJ (1995) Involvement of the NH2-terminal region of oryzacystatin-I in cysteine proteinase inhibition. Protein Eng 8:1303–1307
140. Kouzuma Y, Tsukigata K, Inanaga H et al (2001) Molecular cloning and functional expression of cDNA encoding the cysteine proteinase inhibitor Sca from sunflower seeds. Biosci Biotechnol Biochem 65:969–972
141. Doi-Kawano K, Kouzuma Y, Yamasaki N, Kimura M (1998) Molecular cloning, functional expression, and mutagenesis of cDNA encoding a cysteine proteinase inhibitor from sunflower seeds. J Biochem 124:911–916
142. Auerswald EA, Genenger G, Assfalg-Machleidt I et al (1992) Recombinant chicken egg white cystatin variants of the QLVSG region. European J Biochem 209:837–845
143. Björk I, Brieditis I, Raub-Segall E et al (1996) The importance of the second hairpin loop of cystatin C for proteinase binding. Characterization of the interaction of Trp-106 variants of the inhibitor with cysteine proteinases. Biochemistry 35:10720–10726
144. Hiltkel TR, Lee TC, Bobekl LA (1999) Structure/function analysis of human cystatin SN and comparison of the cysteine proteinase inhibitory profiles of human cystatins C and SN. J Dent Res 78:1401–1409
145. Bedi GS, Zhou T, Bedi SK (1998) Production of rat salivary cystatin S variant polypeptides in Escherichia coli. Arch Oral Biol 43:173–182
146. Nikawa T, Towatari T, Ike Y, Katunuma N (1989) Studies on the reactive site of the cystatin superfamily using recombinant cystatin a mutants. FEBS Lett 255:309–314
147. Nycander M, Björk I (1990) Evidence by chemical modification that tryptophan-104 of the cysteine-proteinase inhibitor chicken cystatin is located in or near the proteinase-binding site. Biochem J 271:281–284
148. Pavlova A, Estrada S, Björk I (2002) The role of the second binding loop of the cysteine protease inhibitor, cystatin a (stefin a), in stabilizing complexes with target proteases is exerted predominantly by Leu73. Eur J Biochem 269:5649–5658
149. Cimerman N, Prebanda MT, Turk B et al (1999) Interaction of cystatin C variants with papain and human cathepsins B, H and L. J Enzym Inhib 14:167–174
150. Pol E, Björk I (1999) Importance of the second binding loop and the C-terminal end of cystatin B (stefin B) for inhibition of cysteine proteinases. Biochemistry 38:10519–10526
151. Dall E, Fegg JC, Briza P, Brandstetter H (2015) Structure and mechanism of an Aspartimide-dependent peptide ligase in human Legumain. Angew Chem Int Edit 54:2917–2921
152. Alvarez-Fernandez M, Barrett AJ, Gerhartz B et al (1999) Inhibition of mammalian legumain by some cystatins is due to a novel second reactive site. J Biol Chem 274:19195–19203

153. Baron AC, Barrett-Vespone NA, Featherstone JD (1999) Purification of large quantities of human salivary cystatins S, SA and SN: their interactions with the model cysteine protease papain in a non-inhibitory mode. Oral Dis 5:344–353
154. Benchabane M, Schlüter U, Vorster J et al (2010) Plant cystatins. Biochimie 92:1657–1666
155. Nandy SK, Bhuyan R, Seal A (2013) Modelling family 2 cystatins and their interaction with papain. J Biomol Struct Dyn 31:649–664

Solid Support Synthesis of a Dnp-Labeled Peptide for Assay of Matrix Metalloproteinase-2

28

Amritlal Mandal, Atanu Maiti, Tapati Chakraborti, and Sajal Chakraborti

Abstract

Herein, synthesis of a Dnp-labeled peptide, Dnp-Pro-Gln-Gly-Ile-Ala-Gly-Gln-D-Arg-$CONH_2$, is described by the Fmoc solid-phase method. Post-synthesis of the peptide was purified by reversed-phase HPLC. The purity of the peptide was determined by nuclear magnetic resonance total correlation spectroscopy (NMR TOCSY), and the validity of the peptide as a specific synthetic substrate for matrix metalloproteinase-2 (MMP-2) was also assessed by measuring the specific activities of the MMP-2 using the peptide as a substrate. It was found to be a suitable substrate with respect to MMP-2 and correlated well with the [^{14}C]-gelatin degradation assay of MMP-2. Pretreatment of the pure MMP-2 with tissue inhibitor of metalloproteinase-2 (TIMP-2), a specific endogenous inhibitor of MMP-2, prevented both the Dnp-labeled peptide substrate degradation and the [^{14}C]-gelatin degradation. Thus, the Dnp-labeled peptide can be used as a synthetic substrate for in vitro assay of MMP-2.

A. Mandal
Department of Biochemistry and Biophysics,
University of Kalyani, Kalyani 741235, West Bengal, India

Department of Physiology, University of Arizona,
1501 N Campbell Ave, Tucson AZ85724, USA

A. Maiti
Department of Biophysics, Bose Institute, P-1/12 C.I.T. Scheme VIIM, Kolkata 700054, India

Department of Biochemistry and Molecular Biology and Greenebaum Cancer Center,
University of Maryland School of Medicine, Baltimore, MD 21201, USA

T. Chakraborti • S. Chakraborti (✉)
Department of Biochemistry and Biophysics,
University of Kalyani, Kalyani 741235, West Bengal, India
e-mail: saj_chakra@rediffmail.com

S. Chakraborti, N.S. Dhalla (eds.), *Proteases in Physiology and Pathology*,
DOI 10.1007/978-981-10-2513-6_28

Keywords
Dnp-labeled peptide • Solid-phase peptide synthesis • MMP-2 • TIMP-2 • HPLC • TOCSY

Abbreviations

APMA	Aminophenylmercuric acetate
DIPEA	Diisopropylethylamine
DMF	Dimethylformamide
Dnp	2,4-Dinitorophenyl
Fmoc	Fluorenylmethyloxycarbonyl
HOBt	1-Hydroxybenzotriazole
HPLC	High-performance liquid chromatography
MBHA resin	4-Methylbenzhydrylamine resin
MMP-2	Matrix metalloproteinase-2
NMR	Nuclear magnetic resonance
OPfp ester	Pentaflurophenyl ester
PyBOP	(Benzotriazol-1-yloxy)tris(pyrrolidino)phosphoniumhexafluoro phosphate
TFA	Trifluoroacetic acid
TIMP-2	Tissue inhibitor of metalloproteinase-2
TNBS	Trinitrobenzenesulfonic acid
TOCSY	Total correlation spectroscopy

28.1 Introduction

Studies on the substrate specificity of collagenases, for example, matrix metalloproteinases using peptides having the same or closely similar sequences to that around the cleavage site of collagen molecule, suggested that the enzymes preferentially hydrolyze the peptide Dnp-Pro-Gln-Gly-Ile-Ala-Gly-Gln-D-Arg-OH, at the Gly-Ile bond [1, 2]. Therefore, several attempts have been made to synthesize a peptide substrate for the quantitative assay of matrix metalloproteinases by different investigators. Based on the accumulated knowledge of peptide chemistry, a new strategy for peptide synthesis, the solid-phase method, was developed by Merrifield in 1963 [3]. The solid-phase method permitted the simple and rapid preparation of peptides and made an enormous contribution to the development of synthetic substrate for proteases [4, 5]. Along with the development of related technologies such as reversed-phase high-performance liquid chromatography (RP-HPLC) and mass spectrometry, the solid-phase method actually became a major technique in peptide synthesis. As an advancement over N-terminal protection and de-protection of the constituent amino acids using t-butoxycarbonyl (t-BOC) amino acid derivatives, a new solid-phase method that uses *Nα*-9-fluorenylmethoxycarbonyl (Fmoc) amino acid derivatives [6] has been developed [7, 8] which provides better protection and easy de-protection employing mild basic condition.

Since the pioneering work of Merrifield in solid-phase peptide synthesis [3], several different methods for synthetic peptide substrates, suitable for protease(s) assay have been developed [9]. The cleavage of the peptide from the resin is one of the key steps in solid-phase peptide synthesis. Solid-phase chemistry is currently a developing area particularly with regard to synthesis of small organic molecules, and it has some advantages over solution-phase strategy. First, easy removal of the large excess of reagents is achieved by filtration and washing. Second, because of easy separation of reagents and products, solid-phase chemistry can be automated more conveniently than solution phase.

Separation of compounds bound to the solid support from those in solution is accomplished by simple filtration. The solid support refers to the insoluble material reversibly bound to the starting reactants. Solid-phase reactions can occur on the surface or inside of the solid particles. Several types of materials are being used as solid supports that allow reactions only on the surface, for example, beads made from glass and cellulose fibers [10].

MMPs are a family of zinc containing calcium-dependent endopeptidases secreted by both normal and transformed cells and are capable of degrading the components of extracellular matrix (ECM) [11–15]. Several members of the MMP family have been identified, and that includes stromelysins, collagenases, and gelatinases [13]. They differ markedly from substrate specificities and varieties of available inhibitors. For example, MMP-2 (~72 kDa) is produced as a zymogen; is activated in vitro by organomercurials, for example, 4-aminophenylmercuric acetate (APMA), resulting in the autolytic cleavage of 6–10 kDa amino terminal fragments [16]; and is inhibited by TIMP-2 [11–15]. TIMP-2 is a non-glycosylated protein with a molecular mass of ~21 kDa. Upon activation with APMA, MMP-2 shows a marked increase in gelatinolytic activity and that can be inhibited by TIMP-2 [17–19].

The use of Dnp-labeled peptide prepared by solution-phase peptide synthesis for the assay of collagenase activity in synovial fluids from patients with osteoarthritis and rheumatoid arthritis was first reported by Masui et al. [1]. The results indicated that the peptide could be used as a substrate for the assay of collagenases. Herein, an improved method for the synthesis of the peptide has been reported for the assay of MMP-2. Furthermore, the peptide was used and compared with the conventional [^{14}C]-gelatin degradation assay of MMP-2.

28.2 Materials and Methods

28.2.1 General

MBHA resin, HOBt, PyBOP, and Fmoc-protected amino acids except Fmoc-protected proline were obtained from Novabiochem (Darmstadt, Germany);

Dnp-labeled Fmoc-protected proline was obtained from Sigma Chemical Co. (St. Louis, MO, USA); and DIPEA, TFA (spectrofluorometric grade), water (HPLC grade), acetonitrile, DMF, piperidine, TNBS, ethane dithiol, anisole, phenol, diethyl ether, *t*-amyl alcohol, and glacial acetic acid were obtained from E. Merck (Darmstadt, Germany). Molecular sieve 4a was obtained from Delta Adsorbents (Roselle, IL, USA). Pure MMP-2 and TIMP-2 were obtained from Chemicon International Inc. (Temecula, CA, USA). All other chemicals used were of analytical grade and purchased from E. Merck (Darmstadt, Germany).

28.2.2 Solid-Phase Peptide Synthesis Using Fmoc Chemistry

The solid-phase peptide synthesis was done by following the method described by Merrifield et al.[3] with some modifications [6]. The Fmoc chemistry for the solid-phase peptide synthesis was adopted from Sheppard et al. [20–22]. For the synthesis of the peptide of interest, amide MBHA resin was chosen. To achieve the targeted quantity of the final synthesis, each constituent amino acids used were in excess (3 times of the final concentration). To synthesize 0.05 mmole peptide, 0.15mmole of individual amino acids were used.

28.2.3 Swelling of the Resin

Amide MBHA (4-methylbenzhydrylamine hydrochloride salt) resin (90 mg) was loaded in a clean and dry peptide synthesizer column, and to it 1.5 ml DMF was added very slowly. After that the column was closed and rolled horizontally to swell the resin. Vigorous agitation was avoided to minimize the possibility of air bubble entrapment. The column containing the resin was kept overnight at 4 °C for complete swelling of the resin. Due to the use of amide MBHA resin, the product contained –$CONH_2$ group at the C-terminus, unless hydrolyzed.

28.2.4 Priming the Column

After equilibration at 4 °C, the column was allowed to attain room temperature (23–25 °C) to match with the operational temperature of the peptide synthesizer (Biolynx 4175 (LKB Biochrom, Cambridge, UK). Unless stated otherwise, the entire procedure of the solid support peptide synthesis was performed in an inert nitrogen atmosphere. Near-zero moisture content of the solvents used in the synthesis process has been ascertained by passing the solvents through molecular sieve 4a. The column was washed with DMF for 10 min (3.5mlmin^{-1}) to prime the entire system. Following this priming step, 20% piperidine in DMF was run through the column with the same flow rate for 5 min, and finally the column was again washed with DMF for 15 min.

Since the peptide is synthesized from C-terminus to N-terminus, the first amino acid to be coupled was D-Arg-OH following the target sequence. The complete availability of the N-terminus of the resin was confirmed by trinitrobenzenesulfonic acid (TNBS) test before loading the column with the first Fmoc-D-Arg-OH.

28.2.5 TNBS Test

Freshly prepared TNBS reagent was used. DMF (450 μl) was taken in a microcentrifuge tube protected from light, and to it DIPEA (50 μl) was added. In the resulting solution, TNBS (50 μl) was added. The resulting reddish orange solution was used as TNBS reagent to check the availability of N-terminus of the resin to couple with the first amino acid. A small quantity of the pre-swollen and primed resin from the peptide synthesizer column was carefully withdrawn and placed in a small wide-mouth glass vial. The resin was washed with a sequential addition (2 ml each) of DMF, *t*-amyl alcohol, glacial acetic acid, *t*-amyl alcohol, DMF, and diethyl ether. Freshly prepared TNBS reagent (500 μl) was added to the resin and allowed to react for 1 min at room temperature. The resin was washed repetitively with DMF and finally washed twice with diethyl ether. A bright orange color of the bead indicates the presence of free amino group available for coupling of the first Fmoc-D-Arg-OH.

28.2.6 Loading the Column with First Fmoc-Protected Amino Acid (Fmoc-D-Arg-OH)

Fmoc-D-Arg-OH (98 mg, MW 648.8), PyBOP (80 mg, MW 520.3), and HOBt (20 mg, MW 135.13) were taken in a clean and dry glass vial. To it DMF (500 μl) was added and the entire material was carefully dissolved. The final volume of this amino acid reagent was made up to 2 ml with the addition of requisite amount of DMF. Before loading the column with Fmoc-amino acid reagent, the eluate of the column was carefully monitored by a Nova spectrophotometer (Biochrom, Holliston, MA, USA) at 360 nm. Before loading the column, the absorbance at 360 nm read negative. The Fmoc-D-Arg-OH reagent was then loaded into the column slowly to minimize the possibility to trap any air bubble. DIPEA (50 μl) was directly applied over the column. The loading flow rate of the column was 2 ml/min. The absorbance at 360 nm increased gradually from negative to positive values, indicating amino acid coupling reaction with the resin. After the completion of the loading, the column was recirculated (3.5 ml/ min) for another 5 h for complete coupling reaction to maximize the yield. Next, TNBS test was performed to ascertain completion of the coupling as indicated by a negative result. The column was again washed (3.5 ml/ min) with DMF (10 min), 20% piperidine (10 min), and finally DMF (15 min). TNBS test was again performed as described. A positive test indicated the presence of free N-terminus of the first coupled amino acid, an indication for the readiness of the resin for next amino acid coupling.

28.2.7 Loading the Column with Second and Subsequent Fmoc-Protected Amino Acid

The second and subsequent amino acids (0.05mmole each) were loaded following the method described for the first amino acid. In case of glycine, Pentaflurophenyl ester (OPfp) ester form was used keeping other conditions unaltered. After completion of the synthesis of the peptide, it was cleaved from the solid resin by the following method.

28.2.8 Peptide-Resin Cleavage

The synthesized peptide from the solid resin support was cleaved following the method of Guy et al. [23]. Briefly, the resin-bound peptide was withdrawn and taken in a sintered glass funnel connected to a low vacuum pump via Buchner flask. The resin was washed in the sequence as follows, DMF, *t*-amyl alcohol, acetic acid, *t*-amyl alcohol, DMF, and diethyl ether followed by drying under vacuum and subsequently taken for the peptide cleavage. The peptide cleavage reagent consists of TFA, ethane dithiol, anisole, and phenol (94:2:2:2 v/v) and is prepared fresh. The peptide cleavage reagent was added slowly to the dried peptide attached to resin over the Buchner funnel, and the filtration was continued. This step was repeated three times to ascertain complete recovery of the cleaved peptide. The pooled combined filtrate was transferred to a round-bottomed flask and evaporated to dryness using a refrigerated rotary evaporator (Savant, Holbrook, NY, USA). The peptide was purified further by adding diethyl ether which helps removal of organic contaminants. Diethyl ether was aspirated carefully to get rid of contaminants. The residual diethyl ether was evaporated with rotary evaporator. The entire procedure for contaminant removal using diethyl ether was repeated 7–8 times, and finally the dried peptide was stored at 70 °C for purification.

28.2.9 Purification of the Peptide

The peptide was purified by HPLC following the method of Miranda et al. [24] with some modifications. Preparative reversed-phase HPLC was performed with Waters 510 solvent delivery system (Waters Corporation, Milford, MA, USA) using Waters Baseline 810 software at an operating pressure of 4 × 1000PSI (sample injection volume, 200 μl). The peptide solution was thoroughly filtered through Whatman filter paper (0.45 μm) by pressure filtration prior to loading to avoid particulate contamination. The chromatographic separation was monitored using a Waters 484 Lambda Max HPLC UV-Visible Detector (Waters Corporation, Milford, MA, USA) at 210 nm. Chromatographic separations were achieved with a 1% min^{-1} linear gradient of buffer B in A (A = 0.1% TFA in HPLC grade water; B = 60% CH_3CN in A) over 55 min at a flow rate of 8 ml min^{-1} using Waters C18 preparative reversed-phase HPLC (RP-HPLC) column (10 μm, 2.2 cm × 25 cm) (Waters Corporation, Milford, MA, USA).

The 210 nm peak fraction (resolved at 35.35 min) was collected in a thoroughly cleaned and dried round-bottomed flask, evaporated to dryness by refrigerated rotary evaporator (Savant, Holbrook, NY, USA) for 3 h, solubilized with HPLC grade water, and lyophilized for future use.

28.2.10 Assessment of Peptide Purity

28.2.10.1 NMR Study

All ^{1}H NMR experiments were performed on a Bruker DRX 500 MHz spectrometer (Bruker Daltonics Inc., Billerica, MA, USA). NMR spectra of the Dnp-labeled peptide were obtained with 2 mM peptide in 50 mM potassium phosphate buffer (pH 7.5) and 5% D_2O at 27 °C. Water suppression was achieved by using WATERGATE pulse sequence [25] for all experiments. TOCSY [26] experiment was performed using standard protocols [27]. The number of data points used in TOCSY was F2xF1 = 2048x512.

28.2.10.2 Activation of MMP-2

Activation of MMP-2 was carried out as described by Murphy et al. [28]. MMP-2 (0.1 μg) was incubated with APMA (2 mM) in Tris-HCl (50 mM, pH 7.5) buffer containing NaCl (200 mM) and $CaCl_2$(10 mM).

28.2.10.3 Peptide Degradation Study

The purified peptide was analyzed by using the peptide as a substrate for MMP-2. Briefly, the peptide was dissolved in Tris-NaCl-$CaCl_2$ (50 mM Tris-HCl, pH 7.5) buffer containing NaCl (200 mM), $CaCl_2$(10 mM), and BSA (0.02%, w/v). The peptide (100 μl) was mixed with an equal volume of pure MMP-2 (0.1 μg) (Chemicon International Inc., Temecula, CA, USA) and incubated for 1 h at 37 °C. The enzymatic reaction was stopped by adding 0.5 ml HCl (1 M); the Dnp-labeled peptide fragments released was extracted by vigorous shaking with ethyl acetate (1 ml) followed by centrifugation at 5000 × g at room temperature for 10 min to separate into two layers. The degree of hydrolysis was determined by measuring the absorbance of the organic layer at 365 nm.

To study the effect of TIMP-2 on MMP-2, the enzyme was preincubated with TIMP-2 (5 μg/ml) for 1 h at 37 °C after APMA activation prior to synthetic substrate degradation.

28.2.10.4 [^{14}C]-Gelatin Degradation

Assay of protease activity by [^{14}C]-gelatin degradation of MMP-2 was determined as follows. The radio-labeled gelatin substrate was prepared by diluting 20 μl (1.2 mCi) of [^{14}C]-labeled gelatin with 480 μl of cold gelatin (1 mg/ ml). The substrate mixture was then heated at 55 °C for 25 min and allowed to cool down slowly to room temperature. The final assay reaction contained 40 μl of [^{14}C]-gelatin substrate, $CaCl_2$ (10 mM), and pure MMP-2 (0.1 μg). Samples were incubated for 1 h at 37 °C, and the reaction was stopped by adding 250 mM EGTA (20 μl). Undigested gelatin was precipitated by the addition of 10% TCA (60 μl). After cooling down on ice for 10 min, samples were centrifuged at 10,000 × g for 10 min, and the radioactivity in the supernatant was determined [29].

To study the effect of TIMP-2 on MMP-2, the enzyme was preincubated with TIMP-2 (5 μg/ml) for 1 h at 37 °C after APMA activation prior to [^{14}C]-gelatin degradation.

28.2.11 Statistical Analysis

Results are expressed as mean ± SE of data from a specified number of independent experiments. Statistical comparison was made by two-sample student "t"-test and by one-way analysis of variance followed by the Bonferroni post hoc multiple comparison test. A probability (P) value of <0.05 was considered significant.

28.3 Results and Discussion

The general scheme of the solid support peptide synthesis was shown in Fig. 28.1. The Dnp-labeled peptide was purified by RP-HPLC. The peak fractions of the HPLC run were collected by measuring the absorbance at 210 nm (Fig. 28.2). The HPLC purified and lyophilized peptide was analyzed by 1D ^{1}H NMR and TOCSY (Fig. 28.3). The outcome of the ^{1}H 1D NMR and the TOCSY was shown in Fig. 28.3. The presence of the purified peptide in the 35.35 min peak (Fig. 28.2) was also confirmed by the TOCSY spectra (Fig. 28.3). The 1D spectra were assigned using the backbone amide ^{1}H chemical shift (ppm) for the random coil [30]. The 1D ^{1}H NMR spectra confirm the presence of the following amino acids Pro, Gln, Gly, Ile, Ala, Gln, and D-Arg (Fig. 28.3). The peak at the 8.73 ppm corresponds to Dnp. From the data it was found that all amino acids were incorporated in the synthesized peptide and since the peptide was synthesized in the solid support method so the chance of misincorporation of any amino acid is ruled out and hence all of the amino acids were added in the sequence Dnp-Pro-Gln-Gly-Ile-Ala-Gly-Gln-D-Arg-$CONH_2$ (Fig. 28.3). The presence of proline was not visible due to the absence of any αH atom (Fig. 28.3). The 35.35 min peak was intense yellow colored and confirmed as purified peptide (Figs. 28.2 and 28.3) and validated as an MMP-2 substrate by degradation of the peptide by MMP-2 which was also compared with [^{14}C]-gelatin degradation studies (Fig. 28.4a, b). Pretreatment of the pure MMP-2 with TIMP-2 prevents both the Dnp-labeled peptide substrate degradation and [^{14}C]-gelatin degradation (Fig. 28.4c, d). Thus, it can be concluded that the Dnp-labeled peptide synthesized in the solid-phase peptide synthesis method can be effectively used as a synthetic substrate for MMP-2 (Fig. 28.4).

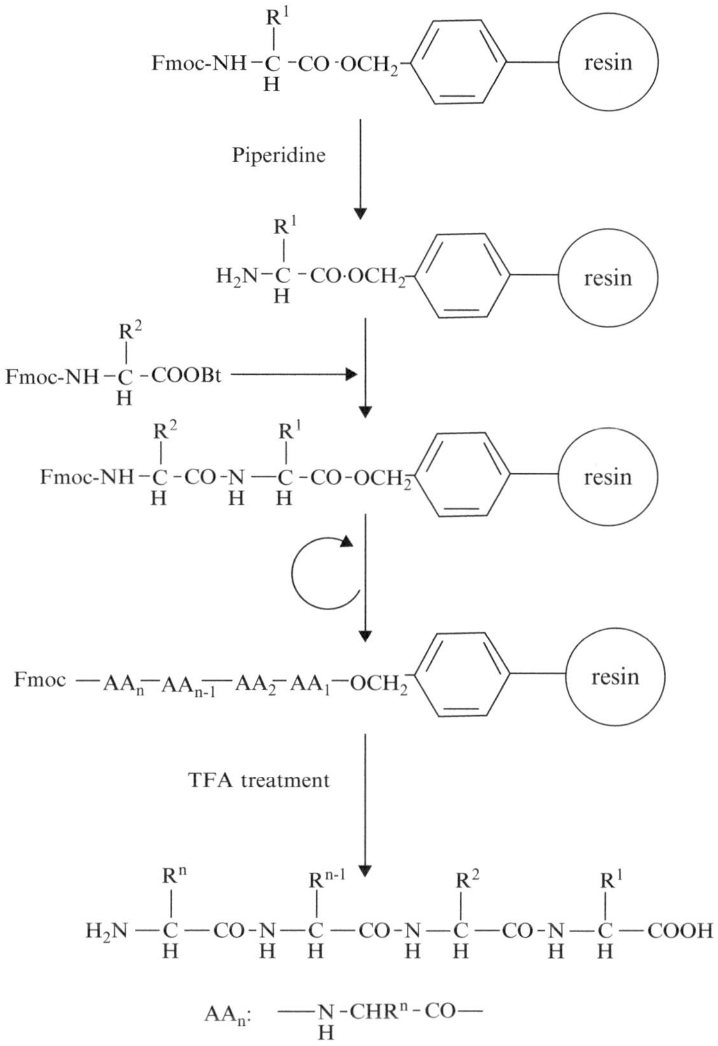

Fig. 28.1 Schematic representation of the individual reaction steps associated with solid-phase peptide synthesis employing the use of Fmoc-protected amino acids

Herein, a comprehensive description of the synthesis, purification, and validity as a preferred substrate for in vitro MMP-2 assay was provided. The results obtained were also analyzed and compared with reference to the conventional [^{14}C]-gelatin degradation studies for the protease assay. The metalloproteinase assay employing the synthesized peptide correlated well with the data obtained with [^{14}C]-gelatin degradation (Fig. 28.4). This overall strategy typically allowed spectrophotometric assay of MMP-2 using Dnp-labeled synthetic peptide substrate as a safe and reliable alternative to the radioactive assay of MMP-2.

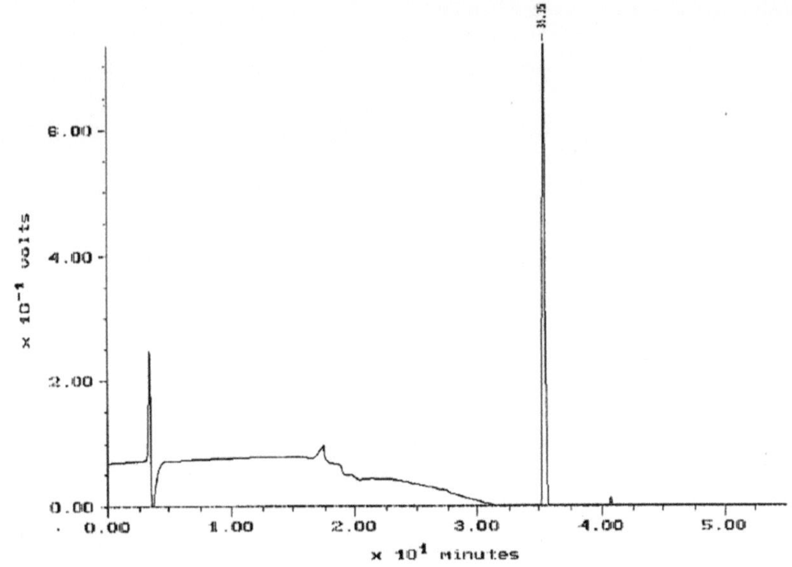

Fig. 28.2 HPLC purification chromatogram of the synthesized peptide. Purification was performed in 0–60% CH_3CN (in H_2O containing 0.1% TFA) and monitored at 210 nm. Chromatographic separations were performed with a 1% min^{-1} linear gradient of buffer B in A (A = 0.1% TFA in HPLC grade water; B = 60% CH_3CN in A) over 55 min at a flow rate of 8 ml/min using Waters C18 preparative column (10 μm, 2.2 cm × 25 cm). The chromatogram shows the purified peak fraction resolved at 35.35 min

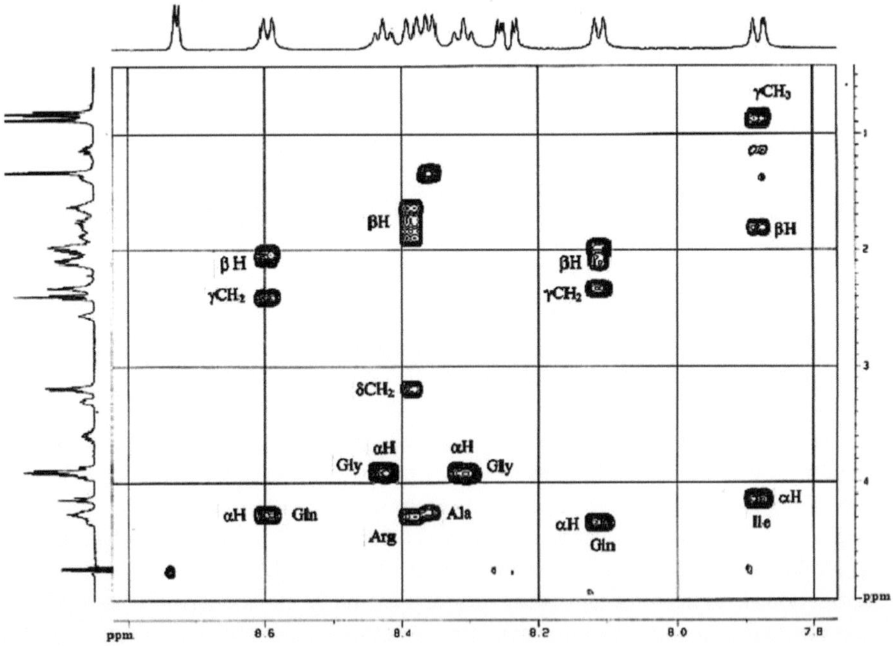

Fig. 28.3 The TOCSY spectra of the HPLC-purified peptide at 27 °C. The 1D ^{1}H NMR spectra confirm the presence of the amino acids Pro, Gln, Gly, Ile, Ala, Gln, and D-Arg. The peak at the 8.73 ppm corresponds to Dnp. Proline was not detected in this spectra due to the absence of αH atom

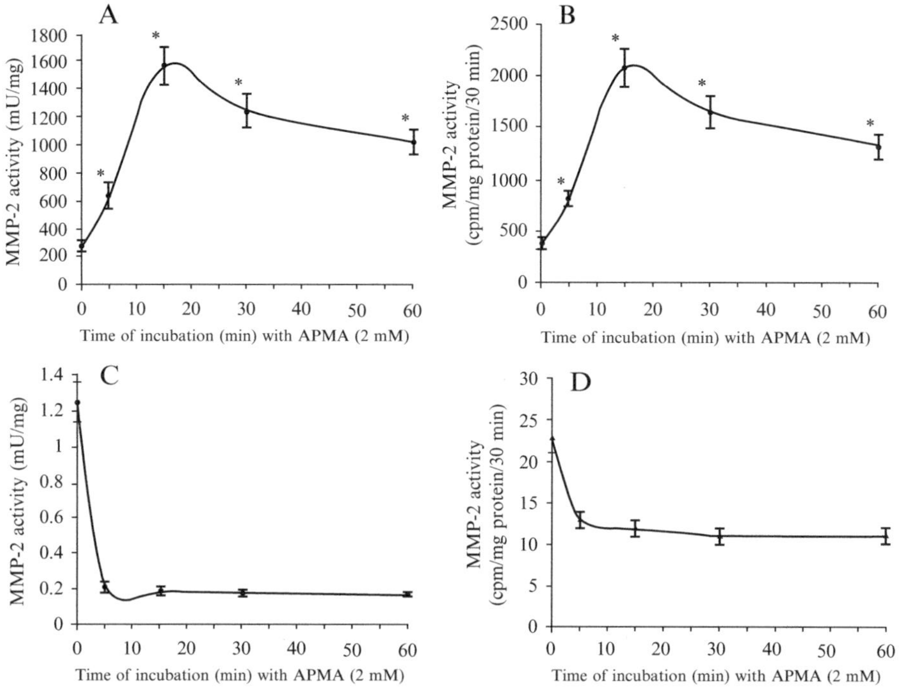

Fig. 28.4 Relative degradation study of Dnp-labeled peptide with [^{14}C]-gelatin by APMA (2 mM)-activated pure MMP-2 (0.1 μg) at different times in the absence and presence of pure TIMP-2 (5 μg/ml). Pure MMP-2 was activated with APMA at 37 °C for indicated times. Then, the APMA-activated MMP-2 was (**a**) incubated with the Dnp-labeled synthetic peptide (—●—) for 1 h at 37 °C, (**b**) incubated with the [^{14}C]-gelatin (—■—) for 1 h at 37 °C, (**c**) preincubated with TIMP-2 (5 μg/ml) for 1 h at 37 °C prior to synthetic substrate degradation (—◆—), and (**d**) preincubated with TIMP-2 (5 μg/ml) for 1 h at 37 °C prior to [^{14}C]-gelatin degradation (—▲—). The degree of hydrolysis of the synthetic peptide substrate was determined by measuring the absorbance of the organic layer at 365 nm. 1 U gelatinase catalyzes the hydrolysis of 1 μmol Dnp-labeled peptide/30 min at 37 °C. The [^{14}C]-gelatinolytic activity was determined by taking the radioactive count and is expressed as cpm/mg protein/30 min. Results are mean ± SE (n = 4). *$p < 0.001$ compared with basal value

Acknowledgments Financial assistance from the Department of Biotechnology (DBT) (Government of India); the Council of Scientific and Industrial Research (CSIR), New Delhi; and the Indian Council of Medical Research (ICMR), New Delhi, is gratefully acknowledged. Thanks are due to Dr. Sudip Das (Department of Biochemistry and Biophysics, University of Kalyani, Kalyani, West Bengal, India) and Mr. Smriti Ranjan Maji (Bose Institute, Kolkata, West Bengal, India) for their expert technical help in this work.

References

1. Masui Y, Takemoto T, Sakakibara Set al (1977) Synthetic substrates for vertebrate collagenase. Biochem med 17: 215-221
2. Nagai Y, Masui Y, Sakakibara S (1976) Substrate specificity of vertebrate collagenase. Biochim Biophys Acta 445:521–524

3. Merrifield RB (1963) Solid phase peptide synthesis. I. The synthesis of a tetrapeptide. J Am Chem Soc 85: 2149–2154
4. Doscher MS (1977) Solid-phase peptide synthesis. Academic Press, Methods Enzymol, pp 578–617
5. Kent SBH (1988) Chemical synthesis of peptides and proteins. Annu Rev Biochem 57:957–989
6. Carpino LA, Han GY (1970) 9-Fluorenylmethoxycarbonyl function, a new base-sensitive amino-protecting group. J Am Chem Soc 92:5748–5749
7. Atherton E, Meienhofer J (1972) Synthesis of the four isomers of α,β-diaminobutyric acid. J Antibiot (Tokyo) 25: 539–540
8. Chang C-D, Meienhofer J (1978) Solid-phase peptide synthesis using mild base cleavage of Nα-fluorenylmethyloxycarbonylamino acids, exemplified by a synthesis of dihydrosomatostatin Int J Pept Protein Res 11:246–249
9. Aimoto S (2001) Contemporary methods for peptide and protein synthesis. Curr Org Chem 5:45–87
10. Frank R, Döring R (1988) Simultaneous multiple peptide synthesis under continuous flow conditions on cellulose paper discs as segmental solid supports. Tetrahedron 44:6031–6040
11. Woessner JF (1991) Matrix metalloproteinases and their inhibitors in connective tissue remodeling the. FASEB J 5:2145–2154
12. Matrisian LM (1990) Metalloproteinases and their inhibitors in matrix remodeling. Trends in Genet 6:121–125
13. Sternlicht MD, Werb Z (2001) How matrix metalloproteinases regulate cell behavior. Annu Rev Cell Devl Biol 17:463–516
14. Murphy G, Docherty AJP (1992) The matrix metalloproteinases and their inhibitors. Am J Resp Cell Mol Biol 7:120–125
15. Ray JM, Stetler-Stevenson WG (1994) The role of matrix metalloproteinases and their inhibitors in tumour invasion, metastasis and angiogenesis. Eur Resp J 7:2062–2072
16. Stetler-Stevenson WG, Krutzsch HC, Wacher MP, Margulies IM, Liotta LA (1989) The activation of human type IV collagenase proenzyme. Sequence identification of the major conversion productfollowing organomercurial activation J Biol Chem 264:1353–1356
17. Murphy G, Willenbrock F (1995) Tissue inhibitors of matrix metalloendopeptidases. Academic Press, Methods Enzymol, pp 496–510
18. Kolkenbrock H, Orgel D, Hecker-Kia A et al (1991) The complex between a tissue inhibitor of metalloproteinases (TIMP-2) and 72-kDa progelatinase is a metalloproteinase inhibitor. Eur J Biochem 198:775–781
19. Howard EW, Bullen EC, Banda MJ (1991) Regulation of the autoactivation of human 72-kDa progelatinase by tissue inhibitor of metalloproteinases-2. J Biol Chem 266:13064–13069
20. Atherton E, Clive DLJ, Sheppard RC (1975) Polyamide supports for polypeptide synthesis. J Am Chem Soc 97:6584–6585
21. Sheppard RC (1980) Peptides and proteins. Biochem Soc Trans 8:744–747
22. Gait MJ, Sheppard RC (1979) Rapid synthesis of oligodeoxyribonucleotides. III. Effect of added carboxylate ion on the efficiency of internucleotide bond formation. Solid-phase synthesis of the dodecanucleotide, d(pT-A-A-C-T-G-C-T-C-A-C-T). Nucleic Acids Res 6:1259–1268
23. Guy CA, Fields GB (1997) Trifluoroacetic acid cleavage and deprotection of resin-bound peptides following synthesis by Fmoc chemistry. Academic Press, Methods Enzymol, pp 67–83
24. Miranda LP, Alewood PF (1999) Accelerated chemical synthesis of peptides and small proteins. Proc Natl Acad Sci 96:1181–1186
25. Piotto M, Saudek V, Sklenář V (1992) Gradient-tailored excitation for single-quantum NMR spectroscopy of aqueous solutions. J Biomol NMR 2:661–665
26. Bax A (1969) Davis DG (1985) MLEV-17-based two-dimensional homonuclear magnetization transfer spectroscopy. J MagnReson 65:355–360
27. Feeney J, Birdsall B (1993) NMR studies of protein-ligand interactions. In: Roberts GCK (ed) Oxford University Press, New York, pp 183–215

28. Murphy G, Cockett MI, Stephens PE et al (1987) Stromelysin is an activator of procollagenase. A study with natural and recombinant enzymes. Biochem J 248:265–268
29. Mandal A, Chakraborti T, Das S et al (2004) Matrix metalloproteinase-2-mediated inhibition of Na^+-dependent Ca^{2+} uptake by superoxide radicals ($O_2^{\bullet -}$) in microsomes of pulmonary smooth muscle. IUBMB Life 56:267–276
30. Wagner G, Neuhaus D, Wörgötter E et al (1986) Sequence-specific 1H-NMR assignments in rabbit-liver metallothionein-2. Eur J Biochem 157:275–289

MIX
Papier aus verantwortungsvollen Quellen
Paper from responsible sources
FSC® C105338

If you have any concerns about our products,
you can contact us on
ProductSafety@springernature.com

In case Publisher is established outside the EU,
the EU authorized representative is:
Springer Nature Customer Service Center GmbH
Europaplatz 3, 69115 Heidelberg, Germany

Printed by Libri Plureos GmbH
in Hamburg, Germany